IMMUNOGLOBULIN GENE EXPRESSION IN DEVELOPMENT AND DISEASE

[ANNALS OF THE NEW YORK ACADEMY OF SCIENCES]

Volume 764

IMMUNOGLOBULIN GENE EXPRESSION IN DEVELOPMENT AND DISEASE

Edited by Paolo Casali and Leslie E. Silberstein

The New York Academy of Sciences
New York, New York
1995

∞ *The paper used in this publication meets the minimum requirements of American National Standard for Information Sciences—Permanence of Paper for Printed Library Materials, ANSI Z39.48-1984.*

Front cover depicts the complex between the Fv fragment of monoclonal antibody D1.3 and its specific antigen, hen egg-white lysozyme. Atoms are represented in the "space filling" convention. Lysozyme is shown in red, the V_H segment in green, and the V_L segment in yellow. Blue spheres represent the water molecules that are most tightly bound to the complex. Note the location of many water molecules around the antigen–antibody interface.

By the courtesy of Dr. Roberto J. Poljak, Center for Advanced Research in Biotechnology, University of Maryland Biotechnology Institute and National Institute of Standards and Technology, Rockville, Maryland 20850.

Library of Congress Cataloging-in-Publication Data

Immunoglobulin gene expression in development and disease / edited by Paolo Casali and Leslie E. Silberstein.
p. cm. — (Annals of the New York Academy of Sciences ; v. 764)
Includes bibliographical references and index.
ISBN 0-89766-933-9 (cloth : alk. paper). — ISBN 0-89766-934-7 (pbk. : alk. paper)
1. Immunoglobulin genes—Congresses. 2. Immunoglobulins—Congresses. 3. Gene expression—Congresses. I. Casali, Paolo. II. Silberstein, Leslie E. III. Series.
Q11.N5 vol. 764
[QR186.7]
500 s—dc20
[616.07'93]
95-37032
CIP

BC/PCP
Printed in the United States of America
ISBN 0-89766-933-9 (cloth)
ISBN 0-89766-934-7 (paper)
ISSN 0077-8923

ANNALS OF THE NEW YORK ACADEMY OF SCIENCES

Volume 764
September 29, 1995

IMMUNOGLOBULIN GENE EXPRESSION IN DEVELOPMENT AND DISEASE[a]

Editors and Conference Organizers
PAOLO CASALI AND LESLIE E. SILBERSTEIN

CONTENTS

[a] This volume is the result of a conference entitled **Immunoglobulin Gene Expression in Development and Disease**, which was sponsored by the New York Academy of Sciences and held on July 13 through 17, 1994 in Montreal, Quebec, Canada.

Part III. V(D)J Gene Recombination and Ig Class Switch

Part VI. Structural Antibody Correlates for Antigen Binding

Part VII. Ig V Genes and Cellular Origin of Autoantibodies

Part VIII. Ig Gene Expression in B-Cell Neoplasias

Part IX. Approaches to Recombinant Human Ig Gene Expression

Financial assistance was received from:

Supporter

- KAPLAN COMPREHENSIVE CANCER CENTER, NEW YORK UNIVERSITY SCHOOL OF MEDICINE

Contributors

- BASF BIORESEARCH CORPORATION
- BRISTOL-MYERS SQUIBB PHARMACEUTICAL RESEARCH INSTITUTE
- GENPHARM INTERNATIONAL
- GIBCO-BRL LIFE TECHNOLOGIES, INC.
- IMRE CORPORATION
- NATIONAL INSTITUTE OF ALLERGY AND INFECTIOUS DISEASES, NIH
- PFIZER CENTRAL RESEARCH
- PHARMINGEN
- R. W. JOHNSON
- SANDOZ RESEARCH INSTITUTE

Others

- PHARMACIA BIOTECH
- IMMUCOR
- CALBIOCHEM CORP.

Preface

PAOLO CASALI[a] AND LESLIE E. SILBERSTEIN[b]

[a]*Division of Molecular Immunology*
Department of Pathology
Cornell University Medical College
New York, New York 10021

[b]*Department of Pathology and Laboratory Medicine*
University of Pennsylvania Hospital
Philadelphia, Pennsylvania 19104

Until the mid-nineteen eighties, very little structural information existed on human antibodies. This scarcity of information was largely due to the difficulties encountered in amino acid sequencing and the limited availability of homogeneous antibody populations, that is, myeloma proteins. The advent in the last decade of effective B-cell immortalization techniques as well as the polymerase chain reaction has led to the rapid generation of a tremendous amount of structural information on human Ig variable heavy (V_H) and light (V_L) chain genes.

In the mouse, knowledge of the Ig V_H and V_L genes and their genomic organization has been crucial in analyzing the mechanisms underlying the composition and maturation of the immune responses to self and foreign antigens. In addition, Ig V gene chromosomal organization has been essential in appreciating the contribution of discrete Ig V_H and/or V_L genes in shaping the B-cell repertoire at different ontogenic stages. Selected J_H-proximal V_H genes are preferentially expressed at early stages of fetal and neonatal life, and they are utilized in the adult B-cell repertoire to encode anti-self antigen reactivities. The anti-self reactivity of the V regions encoded in these J_H-proximal genes has raised the issue as to whether this very same self-reactivity plays a major role in the high frequency of expression of these genes in the early and, possibly, late B-cell repertoires. Finally, in addition to chromosomal location and reactivity of the gene product, other factors, including recombinase accessibility and regulatory DNA elements affecting transcription, may influence the expression of certain Ig V_H and, perhaps, V_L genes.

In the human, most studies of expressed Ig V genes have focused on autoimmune and B-cell neoplastic conditions. In these conditions, the repertoire of expressed Ig V genes overlaps significantly with that expressed at early stages of life. The desire to understand the significance of these studies to the normal human B-cell repertoire prompted us to organize a conference entitled Immunoglobulin Gene Expression in Development and Disease. Our primary focus was to address the structural and functional features of Ig genes in the human, but we included a number of studies in the mouse wherever needed to complement and/or extend a particular topic. A major intended endeavor of the conference was that of defining the overall Ig gene expression in normal human development and relating it to disease. The evidence presented at the conference on the expression, organization, and polymorphism of the human Ig V gene loci indicated that the skewed Ig V gene usage observed in early human development and in autoimmune and neoplastic diseases likely reflects the skewed Ig gene expression inherent to the normal human B-cell repertoire and is not dependent on V segment chromosomal position. This information derived in particular from studies related to the V_H4.21 (V_H4-34) and V_H26 (V_H3-23) genes.

Other aspects important to the understanding of Ig gene expression were discussed at the conference. Surface phenotypic analysis has helped better define distinct stages of early B-cell development. These distinct developmental stages are underpinned by the on-off expression of discrete genes, including RAG-1, RAG-2, and λ5. The role of some of these developmentally regulated genes is underscored by the abnormal expression of the genes in immunodeficiency states. Discussion of the mechanisms of V(D)J recombination focused on a family of genes that are ubiquitously expressed and possibly used for other functions in DNA damage repair. The molecular mechanisms of Ig switch recombination were discussed with respect to the DNA initiation and regulatory elements, and with respect to single and multistep processes involved in isotype switching. Sequence analyses offered insight into biases of the somatic mutation processes; these biases are related to inherent structural features of certain V_H but not V_L genes. With respect to the structural correlates to antigen binding, strong evidence was provided for a major contribution by somatically generated structures, such as the Ig H chain CDR3. Somatically generated structures are less likely to play a role in unconventional or "superantigen" binding by B cells. The optimization of antigen–antibody surface contact appears to be influenced to a large extent by the presence of water molecules. Structural analysis of autoantibodies characteristic of different autoimmune conditions suggested that these autoantibodies utilized the same restricted pool of V genes expressed in the neonatal and adult B-cell repertoires. The same pool of Ig V genes is also utilized in certain human leukemias and lymphomas. In both autoimmunity and B-cell neoplasia, Ig V gene DNA templates can be expressed in unmutated configuration, or they can accumulate somatic point-mutations, the nature and distribution of which are in some cases suggestive of antigen selection of the gene products. In general, the mechanisms by which autoantibodies are generated point to the levels of B-cell development at which regulation (tolerance) must act. Finally, emerging recombinant and transgenic technologies were discussed with respect to production of antigen-specific human antibodies in mice, as well as to the *in vitro* generation of human Ig gene expression libraries. These approaches may be applied to the development of new human antibody reagents, as well as to dissection of the human antibody response to a variety of antigens.

The New York Academy of Sciences has a long tradition of bringing together investigators from all over the world to exchange knowledge and define new directions in specialized fields. The Academy was very responsive and supportive when we proposed an international conference devoted to the discussion of immunoglobulin gene expression in development and disease. We would like to thank the Academy staff for their enthusiastic and effective help, in particular Linda Hotchkiss Mehta, our devoted associate editor. We appreciate the contribution of all attendees in making this conference successful. Most of all, we hope that the information provided and discussed at the conference will help investigators to better define current and future directions in the field. Finally, on a personal note, we would like to offer the following dedications:

To my daughter, Nathalie Fiammetta, inexhaustible source of love and joy (P.C.)

To my family and friends for their continued love and support and to Robert Schwartz, whose example and guidance nurtured my interest in experimental immunohematology (L.E.S.)

Human B-Cell Development

LINDA G. BILLIPS, KAISS LASSOUED, CESAR NUÑEZ, JIYANG WANG, HIROMI KUBAGAWA, G. LARRY GARTLAND, PETER D. BURROWS, AND MAX D. COOPER

Division of Developmental and Clinical Immunology
University of Alabama at Birmingham, and
Howard Hughes Medical Institute
Birmingham, Alabama

The development of B lymphocytes has been the focus of our research since the discovery in birds of the B and T lymphocyte lineages. The primary sites of B-cell development in mammals are the fetal liver and adult bone marrow. Here, stem cell precursors of all hemopoietic lineages receive developmental cues from the "inductive microenvironment" that guides their growth and differentiation.[1,2] This allows for continual renewal of the B-cell repertoire throughout life. Interruptions in key events occurring in this differentiative pathway may result in antibody deficiencies, and cumulative mutations of genes affecting growth and differentiation often lead to B-cell malignancies.

The immunoglobulin (Ig) gene rearrangements required for production of functional heavy (H) and light (L) chain genes have been thought to be tightly regulated, in order to guarantee allelic exclusion, and highly ordered, beginning at the heavy chain locus, before proceeding to the light chain locus, first κ and then, if necessary, λ. Experimental support for the first of these concepts is fairly convincing, especially for the heavy chain locus. For example, introduction of a functionally rearranged IgH chain gene into the germline of transgenic mice inhibits rearrangements of the endogenous H chain locus. A similar, but less efficient shutdown occurs at the light chain locus in κ transgenic mice.[3] A reasonable mechanism to explain how production of a μ chain protein could provide a signal that would prevent further rearrangements of the heavy chain locus was not immediately forthcoming, especially since free μ chains appeared to be retained in the endoplasmic reticulum and degraded.[4,5] The discovery of the ψ-light chain complex, light chain–like proteins encoded by monomorphic pre-B-cell-specific genes that do not require rearrangement for expression,[6–11] seemed to provide the missing piece of this puzzle. Several groups proposed that the μ chains produced following a productive VDJ_H rearrangement would associate with the ψ light chain (ψLC) proteins and be transported onto the cell surface where the μ/ψLC receptor would interact with a ligand on stromal cells. The signal provided by this interaction was envisioned to have two consequences: (i) it would terminate IgH gene rearrangement, and (ii) it would protect the B-cell precursor from programmed cell death, which is apparently the fate of cells that fail to make a productive heavy chain rearrangement on either chromosome. It is clear that the ψLC is important since B-cell development is severely impaired when the λ5 gene is disrupted.[12] However, the mechanism for the ψLC effect is a subject of great controversy, with the number of models seeming to exceed the number of laboratories engaging this issue (FIG. 1).

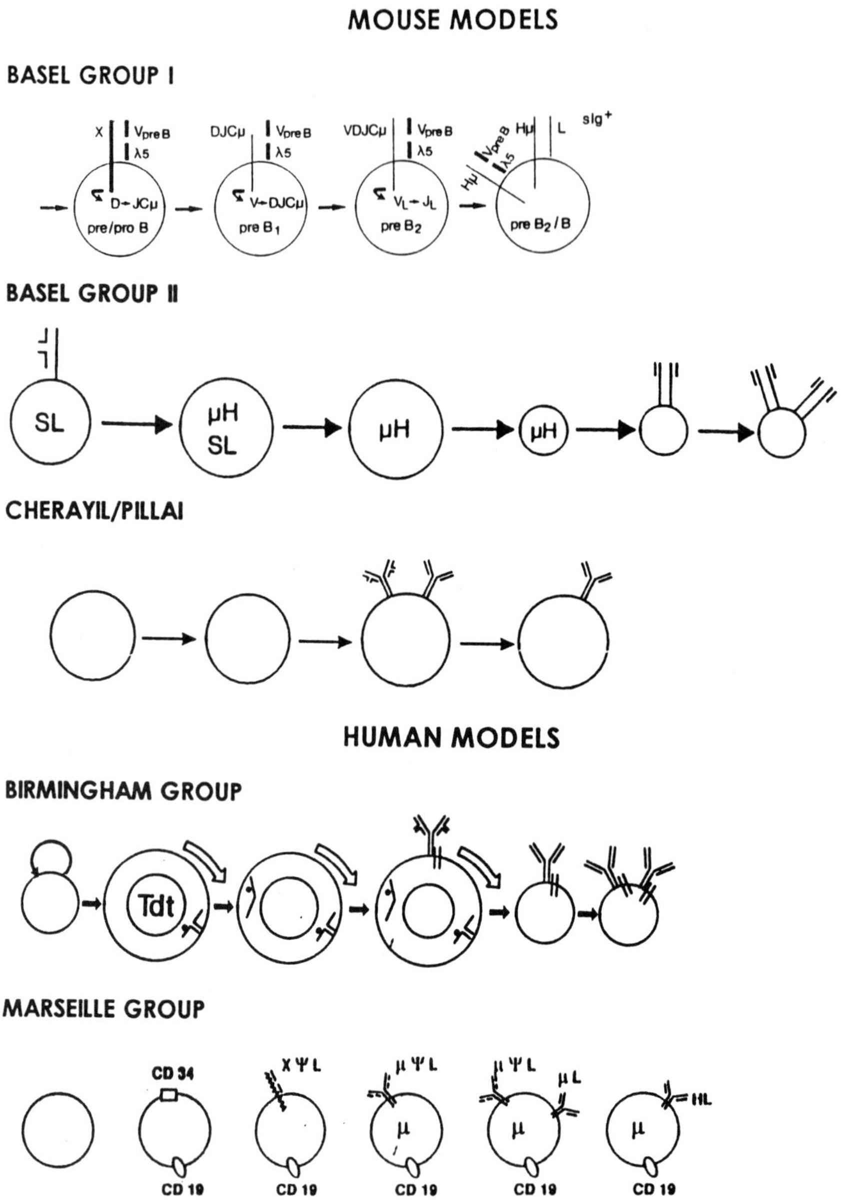

FIGURE 1. Models for the role of the ψLC in B-cell development.

Our studies were undertaken when we discovered the human version of the ψLC proteins paired with μ heavy chain (μHC) on acute lymphocytic leukemia cell lines,[11] and were able to show their expression of a subpopulation of pre-B cells.[13] Later, we isolated enough of the protein to make mAbs with which we could detect ψLC/μ chain receptors at relatively low levels on the surface of normal human B-cell precursors during a limited window of their differentiation, predominantly the late pre-B-cell stage.[13,14]

Other models (FIG. 1) proposed in mice and humans range from (i) those in which the ψLC is expressed on the cell surface of all developing precursors, first in association with a hypothetical ψ-heavy chain, and then with either DJ-μ or full length μ chain (see FIG. 1) to (ii) those in which the ψLC is never expressed on the cell surface, but functions intracellularly. A critical difference among the various models is when the μ/ψLC receptor might be expressed on the surface, potentially to subject the B-lineage cell to positive or negative selection by antigen. This would have an important effect in molding the repertoire of antigen specificities, and hence subsequent immune responses. The issue of receptor assembly and the identification of essential components necessary for this process in pro- and pre-B cells thus remains an important goal.

To address the controversial issue of the existence of a pro-B-cell receptor in which ψLC proteins couple with a complex of ψHC proteins to reach the cell surface, we have performed an analysis of the ψLC proteins in pro-B-cell lines. Pilot pulse-chase experiments, in which cells of the RS4;11 pro-B-cell line were labeled for brief periods (5–15 min), have yielded interesting leads. They suggest that the ψLC protein complex (14.1/λ5 22-kDa protein and the 16- and 18-kDa Vpre-B proteins) has a relatively short half-life (<30 min), does not associate with Igα and Igβ chains, and fails to reach the cell surface (unpublished data). The ψLC proteins transiently associate with grp78/Bip and are also associated with two unknown proteins of M_r 40 kDa and 98 kDa. By contrast the calnexin and grp94 chaperonins do not appear to associate with the ψLC in either pro-B or pre-B-cell lines.

It has been possible to detect cell surface expression of the pre-B-cell receptor (pre-BCR, ψLC/μHC/Igαβ) only by a subpopulation of the normal pre-B cells in bone marrow.[13,14] Although this issue continues to be controversial, this finding in humans has recently been confirmed in mouse bone marrow pre-B cells following their incubation at 37°C for one hour (F. Melchers personal communication). Our previous studies indicate that pre-BCR expression occurs relatively late in pre-B-cell differentiation.[13,14] Our results suggest that (i) the μHC and all three LC proteins (16- and 18-kDa Vpre-B proteins, and the 22-kDa 14.1/λ5 product) are rapidly synthesized and have relatively long half-lives (>4 hr); (ii) a limited portion of the μHC and ψLC pools rapidly associate with Igα and Igβ to form oligomeric units; and (iii) calnexin, grp78/Bip, and grp94 are transiently associated with the newly synthesized μHC, whereas grp78/Bip is the only identifiable chaperone that associates with the newly synthesized ψLC proteins (unpublished data); (iv) a novel 17.5-kDa protein associated with μ chain can be detected after a brief pulse labeling when anti-μ or anti-Bip, but not anti-ψLC, antibodies are used for immunoprecipitation (unpublished data). This may be another chaperonin, a ψLC gene product in a conformation not recognizedd by our mAb, or a new member of the ψLC gene family. After 30 min of chase, this protein is no longer detected, and only conventional ψLC proteins are seen.

The assumption that rearrangements must proceed in an ordered and self-regulating manner, from H to L chain genes, can no longer be accepted as entirely true. Although this is the major pathway of B-cell development, precursor cells exist that rearrange IgL genes before IgH genes. This was first shown in humans, using the Epstein-Barr virus (EBV) to transform B-cell precursors *in vitro,* where cells were identified that had rearranged a κ-chain gene and produced κ light chains, yet had not performed any rearrangements at the IgH locus.[15] Subsequent studies using retrovirus-transformed B-cell precursors confirmed that this order of Ig gene rearrangement also occurs in the mouse.[16] In fact the B cells that are generated in the ψLC gene-targeted mice are postulated to have differentiated along this alternate pathway,[12] since cells that rearrange the IgH locus first could not express the μ/ψLC complex and so may be doomed to programmed cell death. Cells that express a κ chain before IgH chain gene rearrangement might be expected to have a mechanism for regulating gene rearrangement in an analogous fashion to the μ/ψLC receptor of the major pathway. Indeed, one of our EBV-transformed clones expresses κ light chains on the cell surface,[15] and we have identified a pre-B leukemia cell line with the same phenotype. These cells express none of the known heavy chain isotypes but yet express κ light chains in association with Igα/β. Immunoprecipitation of labeled cell surface proteins was used to identify associated molecules, and a glycoprotein of approximately 60,000 daltons was observed. We hypothesize that this is the "surrogate heavy chain" that links the Igα/β and LC into an alternate pathway pre-BCR.

We have found that generation of B-lineage cells is a multifocal process in the human fetus. Although most pro- and pre-B cells are generated in the fetal liver and bone marrow, cells at all stages in this developmental pathway are also present in the intestinal mesentery, omentum, lungs, and kidneys during fetal life, albeit in relatively small numbers.[17] Nevertheless, the bone marrow is the major generation site for B-lineage cells from around mid-gestation onward. Using a combination of cell surface and intracellular markers to examine progression of bone marrow cells along the B lineage pathway as a function of age, we have observed that B cells of immature phenotype are continuously produced throughout life.

B-lineage precursors were isolated from fetal and adult bone marrow samples as early progenitors, pro-B cells, early and late pre-B cells, and B lymphocytes based on their expression of CD34, CD19, CD20, and sIgM: (i) early progenitors as $CD34^+CD19^-$ lymphoid cells; (ii) pro-B cells as $CD34^+CD19^+$ lymphocytes; (iii) early pre-B cells as $CD20^+$ pre-BCR (μ, ψLC, Igα/β) negative lymphocytes; (iv) late pre-B cells as $CD19^{hi}$, pre-BCR^{lo} lymphocytes; and (v) immature B cells as IgM^+ $CD19^{hi}$ cells (FIG. 2). Equal amounts of RNA from each population were converted to cDNA using oligo dT and reverse transcriptase, and then analyzed by PCR. For these studies we designed primers to detect expression of (i) the Ig family of genes; μHC, κ, λ, and ψLC (14.1/λ5); (ii) the Ig recombinase genes, RAG-1 and RAG-2; (iii) the Ig somatic diversification gene, terminal deoxynucleotidyl transferase (TdT;) (iv) the genes encoding the surface Ig-associated molecules Igα (mb-1) and Igβ (B29); and (v) the IL-7 receptor gene, because this cytokine plays an important but ill-defined role in early human B-lineage development. The PCR products were analyzed on ethidium bromide-stained agarose gels, and actin primers were used as a control for the relative amount of cDNA present in each sample. To estimate gene expression levels, the gels were blotted to nylon membranes that were hybridized with ^{32}P-labeled oligonucleotides whose sequence lay between the amplification primers. A phosphorimager was used to quantify the hybridization signals, which were then normalized to the signal obtained with actin.

The results for the analyzed genes were often as expected, but there were several interesting new findings (FIG. 2): (i) RAG-1/2 and TdT showed near coordinate expression that was highest in pro-B or early pre-B cells, declining quickly thereafter. (ii) RAG-2 message was detectable later in development than RAG-1, a some-

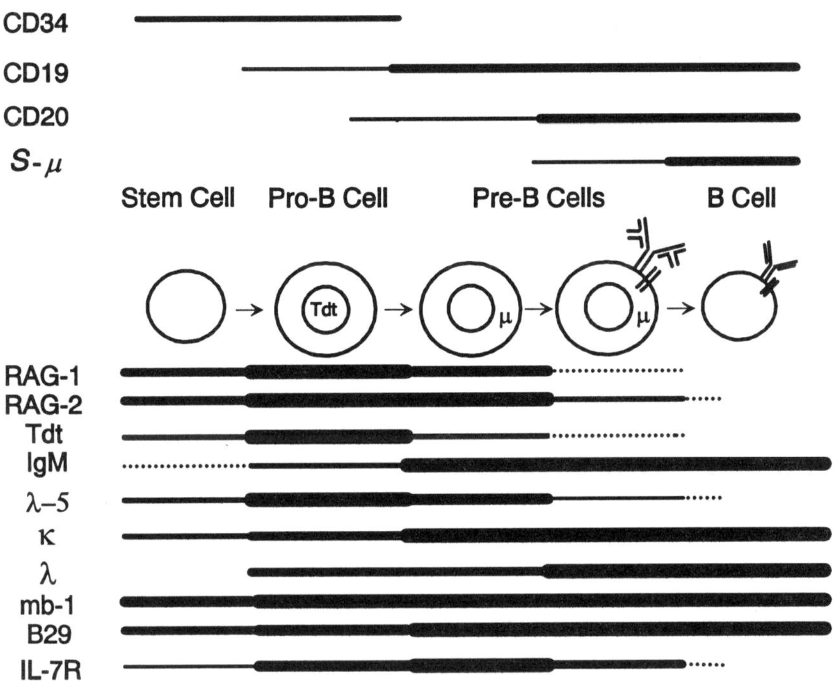

FIGURE 2. Expression of developmentally regulated genes in the human B lineage. The top of the figure illustrates the pattern of CD34, CD19, CD20, and surface μ chain expression. RT-PCR analysis of FACS-sorted fetal BM cells was performed on oligo-dT-primed cDNA using actin expression as a control. After hybridization with a ^{32}P-labeled internal oligo, signal intensity was quantified with a phosphorimager, and expression levels normalized to the actin signal. The thickness of the lines is proportional to the expression level.

what unexpected finding, because expression of the two genes has been thought to be coregulated. However, other exceptions have been observed including in the avian bursa[18] and murine central nervous system[19] and in a subset of thymocytes that express RAG-1 but not RAG-2.[20] There could be several explanations for the apparent extended expression of RAG-2 in the precursor B cells: (a) since both RAG gene products are required for Ig gene rearrangement,[21,22] only one gene may require stringent regulation. (b) The activity of RAG-2 protein is regulated post-translationally by phosphorylation,[23] and there may be differences in mRNA stability; thus, message levels may not accurately reflect levels of functional protein. (iii) Message for μ heavy chain could be detected as early as the CD34$^+$ CD19$^-$ precursor cell stage and reached maximum levels in B cells. The μ chain primers used in these initial experiments did not distinguish full-length, functional message from any sterile intermediates, which may be the major species in the pro-B-cell samples. (iv) Perhaps the most intriguing finding from this analysis was that κ mRNA could be detected in the earliest precursors examined, the CD34$^+$ CD19$^-$ population, which likely contains the earliest committed progenitors for all blood lineages as well as pluripotent stem cells.[24] It is unlikely that the detection of κ mRNA is due to contamination of the samples with more mature cells, inasmuch as the sorted cells were ~99% pure by FACScan analysis. Moreover, we detected only very low levels of μ message, which is mostly sterile because no productive VDJ transcripts and no conventional λ message were observed. The κ primers used in these studies were derived from the Cκ exon sequence and would thus not discriminate between sterile and functional messages. Primers able to provide this discrimination will be used in future studies. (v) The genes encoding the sIg-associated molecules mb-1(Igα) and B29 (Igβ) were also expressed at relatively high levels in very early precursors. Early expression of mb-1 has also been observed in mice, both in freshly isolated pre-pro-B cells (fraction A, B220$^+$/HSA$^-$,BP-1$^-$) by Hardy and colleagues[25] and in pro-B-cell clones grown by Palacios and Samsridis.[26] The expression of the sIg-associated proteins begins well before productive rearrangement of conventional heavy or light chain genes has occurred, thus they are ready to associate with the surrogate immunoglobulin chains to form pre-B-cell receptors.

Most of this data has come from the analysis of fetal bone marrow samples, where B lymphoid and other hemopoietic cell precursors predominate. Analysis of pediatric and adult bone marrow samples from ribs resected during thoracic surgery suggest that this differentiative pattern exists throughout life, although we have observed a decline in the relative numbers of B-cell progenitors with advancing age.

The steady-state production of B cells appears to be dependent on a cascade of regulatory cells and cytokines. We have examined the effects of the stromal cell-derived cytokine, interleukin-7 (IL-7), on human B-cell development. When fetal bone marrow mononuclear cells (FBM) were incubated with IL-7 and examined for their expression of B-lineage differentiation markers (CD34, CD19, CD10, CD20, CD24, CD45, and IgM), the most striking effect was an upregulation of CD19 expression, as shown earlier by Wolf *et al.*[27] Particularly affected was the CD34$^+$/CD19$^+$ subpopulation, in which CD19 levels increased three- to fourfold. Other tested cytokines, IL-3, TGF-β, and stem cell factor, had no effect on CD19 expression. Because CD19 is an important signal transduction molecule on mature B cells and can transduce a mitogenic signal on leukemic pro- and pre-B cells,[28,29] we hypothesized that its upregulation on pro-B cells by IL-7 might be associated with an alteration in expression levels of other genes.

Using the RT-PCR assay, we found that genes involved in Ig gene rearrangement, RAG-1, RAG-2, and TdT, were dramatically downregulated by IL-7 treatment.[30] However, the kinetics of these two effects were different. TdT message levels declined within 24 hours of IL-7 treatment, while RAG message expression was not affected until day 3.[30] This suggests that IL-7R-mediated signaling may have a direct effect on TdT transcription, whereas RAG-1/2 expression is regulated by a more complex, indirect pathway.

Further evidence that different mechanisms are involved in downregulating RAG and TdT gene expression was obtained when we examined the role of CD19 in this process. Anti-CD19 antibodies alone had no effect, but they could block the IL-7-induced suppression of RAG-1/2 gene expression, while having no effect on the IL-7-induced suppression of TdT.[30]

Other studies on the growth and development of murine B-cell precursors *in vitro* may provide clues for the interpretation of these observations. The developing precursors undergo three progressive differentiation stages in culture. The earliest cells are stromal cell contact dependent, later becoming dependent on both IL-7 and stromal cell contact, and finally IL-7 dependent, stromal cell-independent.[2] TdT is expressed only in the earliest stages of the B lineage, thus accounting for the absence of N sequences in the variable regions of light chains in cells that rearrange HC before LC genes. Another pertinent finding was made recently by Doug Fearon who showed that a bone marrow stromal cell line, which supports lymphopoiesis and produces IL-7, can bind purified CD19 and thus apparently expresses a CD19 ligand (personal communication). Finally, the overexpression of human CD19 in a transgenic mouse resulted in a partial block in development at the pre-B cell stage.[31] Taking these observations together, we propose a hypothetical model in which $CD34^+/CD19^{lo}$ precursors bind a CD19 ligand on stromal cells to begin expression of RAG1/2 and TdT and the onset of Ig gene rearrangement by an unknown signaling mechanism. Subsequent production of IL-7 by the stromal cell would downregulate TdT, but the RAG genes would continue to be expressed, even in the presence of IL-7, because of the continued ligation of CD19. As the cells mature to the contact-independent stage, IL-7 would still be produced in the local environment because of its other required functions at this differentiation stage, but in the absence of CD19 ligation this cytokine would downregulate the RAG genes, and Ig gene rearrangement would cease.

REFERENCES

1. KINCADE, P. W. 1994. B lymphopoiesis: Global factors, local control. Proc. Natl. Acad. Sci. USA **91:** 2888.
2. NISHIKAWA, S. I., T. ERA, M. OGAWA, S. NISHIKAWA, N. OHNO, S. I. HAYASHI & T. KUNISADA. 1992. Control of intramarrow B-cell genesis by stromal cell-derived molecules. Cur. Topics Microbiol. Immunol. **182:** 27.
3. STORB, U. 1987. Transgenic mice with immunoglobulin genes. Annu. Rev. Immunol. **5:** 151.
4. HAAS, I. G. & M. WABL. 1983. Immunoglobulin heavy chain binding protein. Nature **306:** 387.
5. HENDERSHOT, L., D. BOLE, G. KÖHLER & J. F. KEARNEY. 1987. Assembly and secretion of heavy chains that do not associate post-translationally with immunoglobulin heavy chain-binding protein. J. Cell Biol. **104:** 761.
6. SAKAGUCHI, N. & F. MELCHERS. 1986. λ5, a new light-chain-related locus selectively expressed in pre-B lymphocytes. Nature **329:** 579.
7. PILLAI, S. & D. BALTIMORE. 1987. Formation of disulfide-linked $\mu2$ $\omega2$ tetramers in pre-B cells by the 18K ω-immunoglobulin light chain. Nature **329:** 172.
8. KUDO, A., N. SAKAGUCHI & F. MELCHERS. 1987. Organization of the murine Ig related λ5 gene transcribed selectively in pre-B lymphocytes. EMBO J. **6:** 103.
9. SCHIFF, C., M. BENSMANA, P. GUGLIELMI, M. MILLI, M.-P. LEFRANC & M. FOUGERAU. 1990. The immunoglobulin λ-like gene cluster (14.1, 16.1, and $F_\lambda 1$) contains gene(s) selectively expressed in pre-B cells and is the human counterpart of the mouse λ5 gene. Int. Immunol. **2:** 201.
10. HOLLIS, G. F., R. J. EVANS, J. M. STAFFORD-HOLLIS, S. J. KORSMEYER & J. P. MCKEARN. 1989. Immunoglobulin λ light-chain-related genes 114.1 and 16.1 are expressed in pre-B cells and may encode the human immunoglobulin ω light chain protein. Proc. Natl. Acad. Sci. USA **86:** 5552.
11. KERR, W. G., M. D. COOPER, L. FENG, P. D. BURROWS & L. M. HENDERSHOT. 1989. Mu heavy chains can associate with a pseudo-light chain complex (ψLC) in human pre-B cell lines. Int. Immunol. **1:** 355.
12. KITAMURA, D., A. KUDO, S. SCHAAL, W. MULLER, F. MELCHERS & K. RAJEWSKY. 1992. A critical role of λ5 protein in B cell development. Cell **69:** 823.
13. NISHIMOTO, N., H. KUBAGAWA, T. OHNO, G. L. GARTLAND, A. K. STANKOVIC & M. D. COOPER. 1991. Normal pre-B cells express a receptor complex of μ heavy chains and surrogate light-chain protein. Proc. Natl. Acad. Sci. USA **88:** 6284.
14. LASSOUED, K., C. A. NUÑEZ, L. BILLIPS, H. KUBAGAWA, R. C. MONTEIRO, T. W.

LeBien & M. D. Cooper. 1993. Expression of surrogate light chain receptors is restricted to a late stage in pre-B cell differentiation. Cell **73:** 73.

15. Kubagawa, H., M. D. Cooper, A. J. Carroll & P. D. Burrows. 1989. Light chain gene expression before heavy chain gene rearrangement in pre-B cells transformed by Epstein-Barr virus. Proc. Natl. Acad. Sci. USA **86:** 2356.
16. Schlissel, M. S. & D. Baltimore. 1989. Activation of immunoglobulin kappa gene rearrangement correlates with induction of germline kappa gene transcription. Cell **58:** 1001.
17. Nuñez, C. A., N. Nishimoto, G. L. Gartland, H. Kubagawa & M. D. Cooper. 1995. The generation of human B-lineage cells as a function of age. Submitted for publication.
18. Carlson, L. M., M. A. Oettinger, D. G. Schatz, E. L. Masteller, E. A. Hurley, W. T. McCormack, D. Baltimore & C. B. Thompson. 1991. Selective expression of RAG-2 in chicken B cells undergoing immunoglobulin gene conversion. Cell **64:** 201.
19. Chun, J. J., D. G. Schatz, M. A. Oettinger, R. Jaenisch & D. Baltimore. 1991. The recombination activating gene-1 (RAG-1) transcript is present in the murine central nervous system. Cell **64:** 189.
20. Campbell, J. J. & Y. Hashimoto. 1993. Recombinase activating gene expression in thymic subpopulations. J. Immunol. **150:** 1307.
21. Shinkai, Y., G. Rathbun, K-P. Lam, E. M. Oltz, V. Stewart, M. Mendelsohn, J. Charron, M. Datta, F. Young & F. W. Alt. 1992. RAG-2 deficient mice lack mature lymphocytes owing to inability to initiate V(D)J rearrangement. Cell **68:** 855.
22. Mombaerts, P., J. Iacomini, R. S. Johnson, K. Herrup, S. Tonegawa & V. E. Papaioannou. 1992. RAG-1-deficient mice have no mature B and T lymphocytes. Cell **68:** 869.
23. Lin, W-C. & S. Desiderio. 1993. Regulation of V(D)J recombination activator protein RAG-2 by phosphorylation. Science **260:** 953.
24. Greaves, M. F., J. Brown, H. V. Molgaard, J. K. Spurr, D. Robertson, D. Delia, & D. R. Sutherland. 1992. Molecular features of CD34: A hemopoietic progenitor cell-associated molecule. Leukemia **6**(Suppl. 1): 31.
25. Li, Y-S., K. Hayakawa & R. R. Hardy. 1993. The regulated expression of B lineage associated genes during B cell differentiation in bone marrow and fetal liver. J. Exp. Med. **178:** 951.
26. Palacios, R. & J. Samaridis. 1992. Fetal liver pro-B and pre-B lymphocyte clones: Expression of lymphoid-specific genes, surface markers, growth requirements, colonization of the bone marrow, and generation of B lymphocytes *in vivo* and *in vitro*. Mol. Cell. Biol. **12:** 518.
27. Wolf, M. L., W-K. Weng, K. T. Stieglbauer, N. Shah & T. W. LeBien. 1993. Functional effect of IL-7-enhanced CD19 expression on human B cell precursors. J. Immunol. **151:** 138.
28. Chalupny, N. J., S. B. Kanner, G. L. Schieven, S. Wee, L. K. Gilliland, A. Aruffo & J. A. Ledbetter. 1993. Tyrosine phosphorylation of CD19 in pre-B and mature B cells. EMBO J. **12:** 2691.
29. Uckun, F. M., A. L. Burkhardt, L. Jarvis, X. Jun, B. Stealey, I. Dibirdik, D. E. Myers, L. Tuel-Ahlgren & J. B. Bolen. 1993. Signal transduction through the CD19 receptor during discrete development of human B-cell ontogeny. J. Biol. Chem. **268:** 21172.
30. Billips, L. G., C. A. Nuñez, F. E. Bertrand III, A. K. Stankovic, G. L. Gartland, P. D. Burrows & M. D. Cooper. 1995. Immunoglobulin recombinase gene activity is modulated reciprocally by interleukin-7 and CD19 in B cell progenitors. Submitted for publication.
31. Zhou, L-J., H. M. Smith, T. J. Waldschmidt, R. Schwarting, J. Daley & T. F. Tedder. 1994. Tissue-specific expression of the human CD19 gene in transgenic mice inhibits antigen-independent B-lymphocyte development. Mol. Cell. Biol. **14:** 3884.

In Vitro Studies of Human B Lymphopoiesis[a]

JULIE R. PRIBYL, BONNIE N. DITTEL,[b]
AND TUCKER W. LEBIEN[c]

Department of Laboratory Medicine and Pathology
University of Minnesota Medical School
Minneapolis, Minnesota 55455

HUMAN LYMPHOHEMATOPOIESIS

Human bone marrow (BM) is a complex mixture of lymphohematopoietic cells at multiple stages of development, in close proximity to a microenvironment consisting of stromal cells (a generic term encompassing fibroblasts, endothelial cells, adipocytes, osteoclasts, and macrophages) and components of the extracellular matrix (i.e., fibronectin, laminin, and various collagens). A vast body of information has accrued characterizing the commitment of pluripotent stem cells to the granulocyte, monocyte, erythrocyte, and platelet lineages, as well as the colony-stimulating factors and cytokines that regulate self-renewal and differentiation.[1] In contrast, considerably less is known about the developmental biology of pluripotent stem cell commitment to the B lineage, and regulation of B lymphopoiesis once commitment has occurred. In this contribution we will briefly review the more salient aspects of human B-cell ontogeny, with an emphasis on work in our laboratory that has focused on development of *in vitro* model systems for identifying and characterizing how human B-cell ontogeny is regulated.

HUMAN B-CELL ONTOGENY: GENERAL PATTERNS OF GENE EXPRESSION

Pre-B cells (i.e., cells expressing cytoplasmic μ heavy [H] chains but no cell surface μ/κ or μ/λ immunoglobulin (Ig) receptors[2,3]) can be initially detected in fetal liver and omentum at ~8 weeks gestation.[4,5] From the second trimester through neonatal and adult life, the BM is the primary, if not exclusive, site of B-cell ontogeny, encompassing all the elements of commitment to the B lineage and Ig gene rearrangement/expression.[6] Ig gene rearrangement during human (and murine) B-cell ontogeny begins at the H chain locus (D to J and then V to DJ), and continues through κ and λ light (L) chain loci.[7] The H to L chain rearrangement

[a] Supported by Grants R01 CA-31685 and P01 CA-21737 from the National Cancer Institute/National Institutes of Health, the Leukemia Task Force, and the Minnesota Medical Foundation. B.N.D. was the recipient of a predoctoral fellowship from NIH Immunology Training Grant T32 AI-07313.

[b] Current address: Section of Immunobiology, Yale University School of Medicine, New Haven, CT 06510.

[c] Address for correspondence: Department of Laboratory Medicine and Pathology, University of Minnesota Medical School, Box 609 UMHC, Minneapolis, MN 55455.

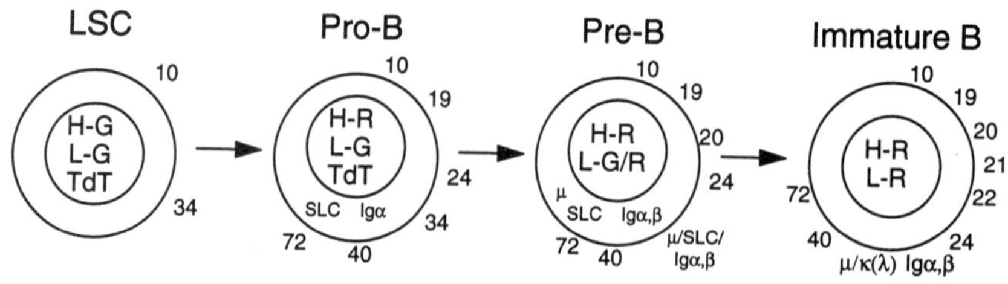

FIGURE 1. Minimalistic model of B-cell ontogeny in human fetal BM. The numbers represent CD designations. LSC = lymphoid stem cells, H = heavy chain locus, L = light chain locus, G = germline, R = rearranged. The μ/SLC/Igα,β complex is only expressed on ~10% of the pre-B cells.

hierarchy is probably the predominant sequence that occurs in the majority of developing B-cell precursors *in vivo,* although alternative pathways have been identified in man[8,9] and mouse.[10] FIGURE 1 defines several lymphoid populations present in human fetal BM and their most probable patterns of Ig gene rearrangement and expression. Some of these rearrangements are inferred from studies of human B-cell precursor acute lymphoblastic leukemias,[7] the neoplastic counterparts of normal B-cell precursors. The developmental stage designated LSC (lymphoid stem cell) is a heterogeneous population that may contain precursors of B, T, and natural killer cells.[11] Although we designate these cells as having germline H chain genes, they could be in the early stages of D to J rearrangement, since they express the enzyme terminal deoxynucleotidyl transferase (TdT).[11] Interestingly, as reported elsewhere in this volume (Bertrand *et al.*), germline DQH52 transcripts and DQH52 rearrangements can be detected in CD34$^+$/CD19$^-$ fetal BM cells, which may share some developmental overlap with LSC. Commitment of a LSC to the B lineage results in a developmentally committed cell designated the pro-B cell. Pro-B cells undergo functional (or nonfunctional) VDJ rearrangements, but their L chain genes remain in germline configuration. Pro-B cells that make a functional VDJ rearrangement on one allele differentiate into pre-B cells, whereas pro-B cells that fail to make a functional rearrangement on either allele cannot develop any further and probably die by apoptosis. The hallmark phenotypic attribute of pre-B cells is expression of cytoplasmic μ,[2,3] and L chain rearrangement (likely somehow regulated by the μ H chain protein) begins at this stage. A functional κ or λ rearrangement would yield an immature B cell expressing either L chain associated with μ H chain, although rare (1–2%) fetal BM and splenic B cells exist that can express *both* κ and λ chains at the level of a single B cell.[9] FIGURE 1 also portrays the expression of Igα and Igβ, signal transduction mediators associated with cell surface Ig, and the so-called surrogate light chains (SLC). The function of SLC is a matter of continuing debate, but recent studies have revealed that SLC surface expression is confined to a minor percentage of normal human pre-B cells expressing low levels of μ H chains[3] and surface μ^- murine pro-B cells.[12] Thus, if SLC associate with μ H chains (in humans) and non-μ Ig proteins (in mice), their putative receptor function is restricted to a very small number of early B-lineage cells.

A simplified antigenic map of human fetal BM B-cell ontogeny is also shown in FIGURE 1. These same phenoptypes can be detected in adult BM, although their frequencies are 5- to 10-fold less on a per nucleated cell basis. This model reflects data generated from several laboratories, as we have discussed elsewhere.[13] Briefly, since a large percentage (>50%) of putative LSC are CD10$^+$/CD19$^-$,[11] the expression of CD19 may herald commitment to the B lineage, thereby distinguishing LSC from pro-B cells. No experimental evidence exists, however, to support the hypothesis that CD19 is some type of master switch whose function (still unknown) is essential for commitment. Pro-B cells can be distinguished from pre-B cells by the expression of CD34 and TdT in the former and the expression of cytoplasmic μ and the absence of CD34 and TdT in the latter.[14,15] These are not absolute, distinguishing phenotypes since CD34$^+$ pro-B cells are >80% TdT$^+$ and 5 to 10% cytoplasmic μ^+, whereas CD34$^-$ pre-B cells are >95% cytoplasmic μ^+ and 1 to 2% TdT$^+$ (Ref. 15 and our unpublished observations). Differentiation of pre-B cells into immature B cells is defined by the acquisition of μ/κ or μ/λ Ig receptors, as well as the appearance of cell surface CD21 and CD22.

IN VITRO STUDIES OF HUMAN B-CELL ONTOGENY

Mammalian B-cell ontogeny in fetal liver and BM is regulated by a microenvironment consisting of stromal cells, secreted and membrane-bound cytokines, membrane-bound adhesion molecules, and extracellular matrix components. Lifelong production of B cells is absolutely dependent on a stromal cell microenvironment, as numerous *in vitro* and *in vivo* studies in the mouse have demonstrated.[16] The development of the Whitlock-Witte culture was an important technical advancement that provided the means to study the *in vitro* dependence of murine B-cell precursor growth on a functional BM stromal cell microenvironment.[17] As discussed earlier,[18] technical problems impeded the development of comparable human BM cultures with the capacity to support the growth and/or differentiation of human B-cell precursors *in vitro*. Nevertheless, the last several years have witnessed important progress in resolving these technical problems. Our published contributions to this area[9,15,19–23] and preliminary new data are presented below. Other laboratories have made important contributions,[24–33] but we shall largely restrict our discussions from here on to our own work.

Differentiation of Pre-B Cells

For the last five years, our studies of human B-cell ontogeny have used highly enriched B-cell precursors isolated from fetal BM. A strategy using magnetic bead depletion to eliminate the vast majority of myeloid/monocytoid/erythroid cells, T cells, and surface L chain$^+$ B cells was developed that consistently yields a population that is >90% enriched for B-cell precursors.[15] Flow cytometric analysis of a representative enrichment is shown in FIGURE 2. Several points are worth noting. First, ≤10% of these cells express cell surface μ and SLC, consistent with the recently published study by Lassoued *et al.*[3] As shown in the two-color dot plot (*middle panel*) <10% of the enriched population expresses very low levels of κ and λ L chains, which represents residual immature B cells that are not removed by magnetic bead depletion. As reported previously,[15] these enriched B-cell precursors contain 60 to 70% cytoplasmic μ^+ pre-B cells and 30 to 40% cytoplasmic μ^- pro-B cells. Several other selected combinations of antigens are also shown in FIGURE 2. The hematopoietic stem cell antigen CD34 readily distinguishes CD19$^+$/CD34$^+$ pro-B cells from CD19$^+$/CD34$^-$ pre-B cells, and TdT ex-

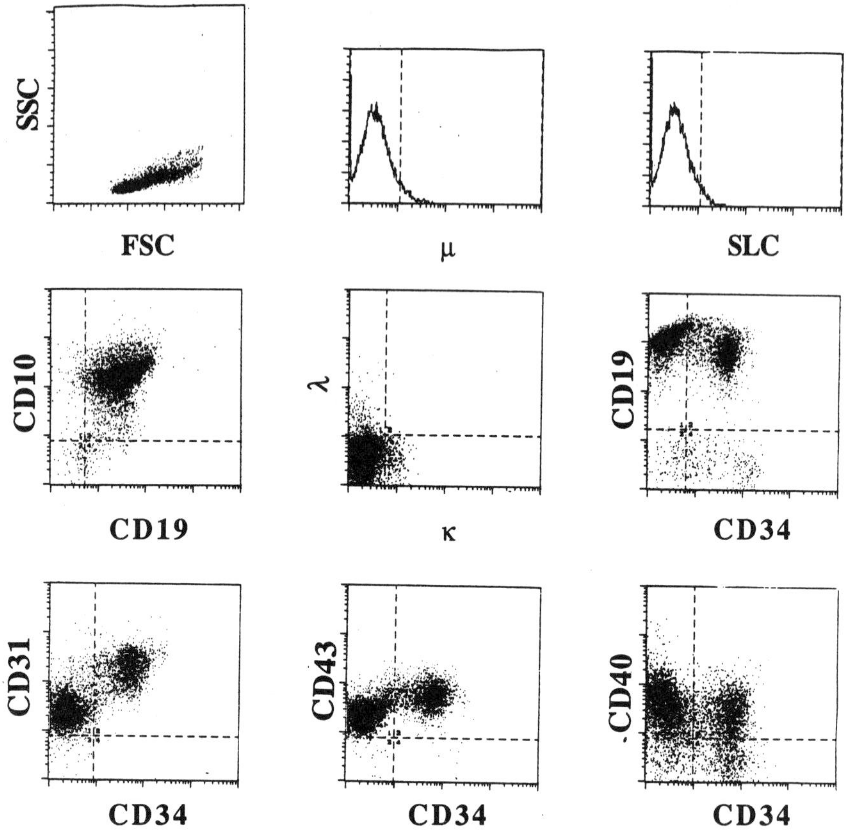

FIGURE 2. Antigen expression on B-cell precursors. FACS analysis of CD10$^+$/CD19$^+$/L chain$^-$ B-cell precursors isolated by magnetic bead depletion from human fetal BM. The upper left dot plot shows the light scatter profile (SSC = side scatter, FSC = forward scatter) of the small lymphoid cell/lymphoblast population. The two histograms in the top row represent single-color analysis of μ H chain and SLC expression. The bottom two rows represent two-color dot plots of various antigen combinations. The dashed lines in all analyses represent the gates set to define background staining.

pression is restricted to the CD19$^+$/CD34$^+$ pro-B-cell population.[14] Interestingly, CD34$^+$ pro-B cells express higher levels of CD31 and CD43 compared to CD34$^-$ pre-B cells. The higher level of CD43 is consistent with the presence of CD43/S7 on so-called fractions A through C in early murine B-cell ontogeny.[34] CD40 is expressed on CD34$^+$ pro-B and CD34$^-$ pre-B cells,[22] and engagement of CD40 by mAb exerts biological effects on B-cell precursors (Larson and LeBien[22] and see below). It is important to emphasize that this magnetic bead depletion-derived population of B-cell precursors does contain small numbers (1–10% depending on the experiment) of non-B-lineage cells. Indeed, this population contains varying numbers of CD34$^+$ hematopoietic stem cells that can be readily detected by their robust response to IL-3 (see below).

Judy Villablanca was the first to develop an assay that supports the differentiation of normal human surface L chain$^-$ pre-B cells into surface L chain$^+$ immature B cells.[15] CD19$^+$/surface L chain$^-$ pre-B cells cultured for two days differentiate into surface L chain$^+$ cells in the absence of BM stromal cells and exogenous cytokines. An unexpected characteristic of pre-B-cell differentiation in this assay was a skewing toward λ L chain expression, resulting in a κ : λ ratio of from 1 : 1 to 1 : 5, in contrast to the 2 : 1 ratio extant in peripheral blood. We therefore determined whether specific factors could influence the κ : λ ratio. We found that the multifunctional polypeptide-transforming growth factor-β (TGF-β) exerted a selective inhibitory effect on the acquisition of cell surface λ L chains, giving rise to an ~1 : 1 ratio of κ^+/to λ^+ cells that approximates the fetal BM ratio *in vivo*.[21] A variety of neuropeptides, interleukins, and mAbs to B-cell surface molecules had no effect on L chain expression, suggesting a potentially selective regulatory role for TGF-β on pre-B-cell differentiation.

In order to determine whether the predicted $\kappa \rightarrow \lambda$ hierarchy was operative in our *in vitro* model, Mary Pauza generated 143 EBV-transformed cell lines from cultures undergoing pre-B-cell differentiation.[9] One highly unusual transformant, designated 30.30, was found to coexpress cell surface κ and λ L chains; 30.30 cells had a normal diploid karyotype, both κ alleles rearranged, and a single λ rearrangement. Immunoprecipitation/SDS-PAGE confirmed that 30.30 cells synthesized and expressed κ and λ L chains. Multiparameter flow cytometry and immunofluorescence microscopy were used to discover the existence of κ^+/λ^+ cells in fetal BM and spleen at frequencies of 2 to 3% of the total surface Ig$^+$ population.[9] The stability of κ/λ coexpression on these immature B cells is unknown. However, these κ^+/λ^+ cells may be undergoing L chain switching, and the coexpression of κ and λ may be a very transient event. Relevant to this possibility, Rolink and colleagues have demonstrated that some κ^+ immature B cells isolated from bcl-2 transgenic mice can become λ^+ when cultured for several days *in vitro*.[35] If a κ^+/λ^+ cell were in the process of switching from, for example, κ expression to λ expression, one could hypothesize a change in the specificity of the Ig receptor on that B cell. The receptor-editing model of Nemazee and colleagues,[36] in which autoantigen-reactive immature B cells are proposed to undergo further L chain rearrangement following strong cross-linking by membrane-bound autoantigen, could explain the existence of κ^+/λ^+ immature B cells.

B-Cell Precursor Growth

Martin Wolf successfully overcame technical problems that had hindered attempts to grow normal human B-cell precursors *in vitro*.[19] A crucial variable in the development of an IL-7/BM stromal cell-dependent culture was the use of serum-free medium.[19] Fetal BM CD10$^+$/L chain$^-$ B-cell precursors isolated by magnetic bead depletion will grow for three to four weeks on BM stromal cells supplemented with exogenous IL-7 at 10 ng/ml. Although IL-7 supplementation is necessary to achieve optimal growth when B-cell precursors are plated at high density (i.e., 5×10^4 cells/well), plating lower numbers of B-cell precursors may at least partially alleviate the exogenous IL-7 requirement (see below). Wolf's original report documented the growth of CD10$^+$/CD19$^+$/HLA-DR$^+$/CD7$^-$ B-cell precursors.[19] Wolf's observations have now been extended to definitively show that CD34$^+$ pro-B cells proliferate in the IL-7/BM stromal cell-dependent culture, whereas CD34$^-$ pre-B cells are completely nonresponsive.[23] Pro-B-cell growth was BM stromal cell contact dependent. IL-7/BM stromal cell–stimulated pro-B

TABLE 1. Survival and Growth of Fetal BM CD19$^+$ Cells in a Limiting Dilution Assay[a]

Cells/Well (Plated)	Cells/Well (Recovered)	
	+IL-7	−IL-7
25000	45275 ± 3620	11980 ± 1381
6250	18937 ± 4364	5285 ± 734
1600	10258 ± 2433	1827 ± 285
400	3847 ± 3188	1049 ± 892
100	1154 ± 866	429 ± 217
25	22 ± 3	87 ± 11
Stroma alone	ND ND	1 ± 1

[a] CD10$^+$/CD19$^+$/surface L chain$^-$ cells (>90% B lineage) were isolated by magnetic bead depletion and cultured at the indicated concentrations on BM stroma in 96-well flat-bottom plates, in the absence or presence of 10 ng/ml IL-7. On day 20, cells were harvested from each concentration and CD19 expression was quantified using the microsphere flow cytometry assay.[22] Each value represents the mean ± SD of triplicate analyses.

cells partially differentiated into a pre-B-cell population, based on loss of CD34 and TdT expression and the acquisition of cytoplasmic μ H chains. Surprisingly, pro-B and pre-B cells expressed comparable levels of IL-7 receptors, as measured in flow cytometry using mAb to IL-7 receptors or biotinylated IL-7.[23] Recently published data has demonstrated that the IL-2 receptor γ chain is a physical and functional component of the IL-7 receptor.[37,38] Therefore, an important question is whether the difference between pro-B and pre-B cell responses to IL-7 can be attributed to differences in the expression and/or signaling function of IL-2 receptor γ chain.

The inability of pre-B cells to respond to IL-7 raised the obvious question of whether other ligands promote survival or growth of these cells. To this end, Andy Larson investigated the potential role of CD40 in the growth of pre-B cells using the L cell/CD40 system developed by Banchereau and his colleagues.[39] Larson found that stimulation of human B cells in the L cell/CD40 system (IL-4 + anti-CD40) yielded a hierarchy of responsiveness: high-density tonsillar B cells > fetal splenic B cells > fetal BM surface Ig$^+$ immature B cells > fetal BM surface Ig$^-$ B-cell precursors.[22] However, whereas FACS-purified pre-B cells responded poorly or not at all to IL-4 and anti-CD40, they did respond to IL-3 and anti-CD40. Pre-B cells did not respond to IL-3 alone or anti-CD40 alone. In six separate experiments there was an average 96-fold difference in the day 12 pre-B-cell numbers present in IL-3 and anti-CD40-stimulated cultures, compared to medium alone. Therefore, coengagement of CD40 and IL-3 receptors delivers a synergistic survival and/or growth stimulus to normal human pre-B cells. We are currently determining whether soluble CD40 ligand synergizes with IL-3 to promote the survival and/or growth of normal human pre-B cells.

As part of an effort to improve our capacity to evaluate the growth of small numbers of B-cell precursors isolated from fetal and adult BM, we are coupling limiting dilution analysis with the microsphere assay. An example of our progress is shown in TABLE 1. CD10$^+$/L chain$^-$ cells (>90% B lineage) were isolated by magnetic bead depletion and cultured on BM stromal cells in the absence or presence of 10 ng/ml IL-7. On day 20 the number of CD19$^+$ B-lineage cells was quantified. Two important messages emerged from this experiment. First, addition of exogenous IL-7 led to an increase in the number of CD19$^+$ B-lineage cells at every concentration plated except 25 cells/well. Interestingly, the stimulation index (cells recovered/cells plated) increased as the number of cells plated decreased. The cloning efficiency of IL-7 stimulated cells was ~1 cell per 160 cells using frequency estimates based on the number of plated cells required to give growth in 63% of the wells (data not shown). Second, and of potentially greater importance, cells cultured in the absence of IL-7 showed evidence of growth at plated concentrations of 1600 cells/well and less. Again, the stimulation index increased as the plated cell number decreased. This indicates that BM stromal cells maintained in serum-free medium may harbor the capacity to support the growth of B-cell precursors in the absence of exogenously added IL-7, albeit less effectively. We have screened for IL-7 in a variety of BM stromal cell supernatants using an IL-7 ELISA, but found no detectable IL-7 at a cutoff sensitivity of 50 pg/ml. It is conceivable that BM stromal cells secrete a small amount of IL-7 that associates with extracellular matrix components and stimulates B-cell precursor growth through this mechanism. If this occurs, anti-IL-7 antibodies may inhibit the growth of B-cell precursors under conditions of limiting dilution, and experiments to test this possibility are in progress.

Because our serum-free culture supports the growth of $CD34^+$ pro-B cells in the presence of IL-7, we were interested in determining how a mixed population of $CD34^+$ pro-B cells and hematopoietic stem cells would behave in this culture. The approach we used was to initially enrich for B-cell precursors by magnetic bead depletion, and then positively select the $CD34^+$ cells by FACS. The initial magnetic bead depletion step employs a mAb to CD33, thereby eliminating the bulk of the committed myeloid cells. The final $CD34^+$-sorted population would therefore contain >90% $CD19^+$ cells, as well as <10% $CD19^-$ cells that would include $CD34^+/CD33^-$ hematopoietic stem cell populations. TABLE 2 shows the results of plating these FACS-purified $CD34^+$ cells onto BM stromal cells supplemented with varying combinations of IL-7, IL-3, and insulin-like growth factor-

TABLE 2. Survival and Growth of $CD34^+$ Progenitors on Human Bone Marrow Stromal Cells[a]

Culture Condition	Cells/Well	
	$CD19^+$	$CD33^+$
Medium alone	14546 ± 289	663 ± 189
IL-3	28315 ± 8070	44821 ± 13148
IL-7	37948 ± 11555	3935 ± 549
IL-3 + IL-7	54207 ± 8989	117740 ± 33203
IGF-1 (10 ng)	19430	1194
IGF-1 (50 ng)	17463 ± 4303	1154 ± 135
IGF-1 (100 ng)	18449 ± 5280	1315 ± 138
IL-3 + IGF-1 (10 ng)	24414 ± 3550	44289 ± 16874
IL-7 + IGF-1 (10 ng)	38571 ± 822	6000 ± 2167
IL-3 + IL-7 + IGF-1 (10 ng)	47513 ± 16475	104757 ± 10502

[a] FACS-purified $CD34^+$ progenitors containing hematopoietic stem cells (~10%) and pro-B cells (~90%) were plated on BM stromal cells at 2.5×10^4 cells/well in the absence or presence of various cytokines. Cells were harvested on day 14, and $CD19^+$ or $CD33^+$ cells were quantified using the microsphere flow cytometric assay. IL-3 and IL-7 were used at 10 ng/ml. Values represent the mean ± SD of triplicate values. Only a single well was available for analysis of cells cultured in 10 ng/ml of IGF-1. Hence, there is no SD for this condition.

1 (IGF-1). The latter cytokine was chosen because it potentiates the murine pro-B-cell proliferative response to IL-7.[40] Analysis of day 14 cell growth using the microsphere assay yielded several interesting results. As expected, IL-7 promoted the expansion of CD19$^+$ cells, as did, to a lesser extent, IL-3. The combination of IL-3 and IL-7 had a nearly additive effect on the growth of CD19$^+$ cells, consistent with recently published studies.[22] Interestingly, multiple concentrations of IGF-1 alone had no effect on the growth of CD19$^+$ cells, and IGF-1 did not enhance the response to IL-3, IL-7, or IL-3 + IL-7. Similar results with IGF-1 were obtained in two separate experiments. Thus, there is no evidence that IGF-1 potentiates the IL-3/IL-7-induced growth of human pro-B cells, in contrast to its potentiating effect on IL-7 stimulation of mouse pro-B cells. Analysis of CD33$^+$ cell growth gave different results. As expected, IL-3 induced a veritable explosion of CD33$^+$ cells. Somewhat surprisingly, IL-7 exerted a small but significant effect ($p = 0.004$) on the growth of CD33$^+$ cells compared to medium alone, but the combination of IL-3 + IL-7 exerted a potentially synergistic effect. As was the case with CD19$^+$ cell growth, IGF-1 had no effect on the CD33$^+$ myeloid cell response to IL-3 and/or IL-7. These results argue that CD19$^+$ and CD33$^+$ B lymphoid and myeloid cells can be co-expanded by a combination of cytokine and BM stromal cell signals. This has important implications for future studies that will attempt to elucidate the differences and similarities in myeloid and B lymphoid commitment from pluripotent stem cell compartments.

The long-term goal of our laboratory is to completely recapitulate all stages of normal human B-cell ontogeny *in vitro*. We propose to accomplish this goal by identifying and characterizing the requisite signals for committing the development of B-lineage cells from pluripotent stem cells and continuing through successive stages of B-cell development culminating in Ig secretion. The development of such an *in vitro* model (i.e., pluripotent stem cells → Ig secretion) will facilitate in-depth analyses of the mechanisms underlying certain types of immunodeficiency diseases and acute lymphoblastic leukemia that are characterized by perturbations in B-cell ontogeny.

REFERENCES

1. METCALF, D. 1993. Hematopoietic regulators: Redundancy or subtlety? Blood **82:** 3515–3523.
2. RAFF, M. C., M. MEGSON, J. J. T. OWEN & M. D. COOPER. 1976. Early production of intracellular IgM by B-lymphocyte precursors in mouse. Nature **259:** 224–226.
3. LASSOUED, K., C. A. NUNEZ, L. BILLIPS, H. KUBAGAWA, R. C. MONTEIRO, T. W. LEBIEN & M. D. COOPER. 1993. Expression of surrogate light chain receptors is restricted to a late stage in pre-B cell differentiation. Cell **73:** 73–86.
4. GATHINGS, W. E., A. R. LAWTON & M. D. COOPER. 1977. Immunofluorescent studies of the development of pre-B cells, B lymphocytes, and immunoglobulin isotype diversity in humans. Eur. J. Immunol. **7:** 804–810.
5. SOLVASON, N. & J. F. KEARNEY. 1992. The human fetal omentum: A site of B cell generation. J. Exp. Med. **175:** 397–403.
6. COOPER, M. D. 1987. B lymphocytes—normal development and function. N. Engl. J. Med. **317:** 1452–1456.
7. KORSMEYER, S. J., P. A. HIETER, J. V. RAVETCH, D. G. POPLACK, T. A. WALDMANN & P. LEDER. 1981. Developmental hierarchy of immunological gene rearrangements in human leukemic pre-B cells. Proc. Natl. Acad. Sci. USA **78:** 7096–7100.
8. KUBAGAWA, H., M. D. COOPER, A. J. CARROLL & P. D. BURROWS. 1989. Light-chain gene expression before heavy-chain gene rearrangement in pre-B cells transformed by Epstein-Barr virus. Proc. Natl. Acad. Sci. USA **86:** 2356–2360.

9. PAUZA, M. E., J. A. REHMANN & T. W. LEBIEN. 1993. Unusual patterns of immunoglobulin gene rearrangement and expression during human B cell ontogeny: Human B cells can simultaneously express cell surface κ and λ light chains. J. Exp. Med. **178:** 139–149.
10. EHLICH, A., S. SCHAAL, H. GU, D. KITAMURA, W. MULLER & K. RAJEWSKY. 1993. Immunoglobulin heavy and light chain genes rearrange independently at early stages of B cell development. Cell **72:** 695–704.
11. GORE, S. D., M. B. KASTAN & C. I. CIVIN. 1991. Normal human bone marrow precursors that express terminal deoxynucleotidyl transferase include T-cell precursors and possible lymphoid stem cells. Blood **77:** 1681–1690.
12. KARASUYAMA, H., A. ROLINK, Y. SHIMIZU, F. YOUNG, F. W. ALT & F. MELCHERS. 1994. The expression of Vpre-B/λ5 surrogate light chain in early bone marrow precursor B cells of normal and B cell-deficient mutant mice. Cell **77:** 133–143.
13. LEBIEN, T. W. 1995. Lymphopoiesis. *In* Leukemia. E. S. Henderson, T. A. Lister & M. F. Greaves, Eds. W. B. Saunders. New York. In press.
14. LEBIEN, T. W., B. WÖRMANN, J. G. VILLABLANCA, C.-L. LAW, L. M. STEINBERG, V. O. SHAH & M. R. LOKEN. 1990. Multiparameter flow cytometric analysis of human fetal bone marrow B cells. Leukemia **4:** 354–358.
15. VILLABLANCA, J. G., J. M. ANDERSON, M. MOSELEY, C.-L. LAW, R. L. ELSTROM & T. W. LEBIEN. 1990. Differentiation of normal human pre-B cells in vitro. J. Exp. Med. **172:** 325–334.
16. KINCADE, P. W. 1990. The lymphopoietic microenvironment in bone marrow. Adv. Cancer Res. **54:** 235–272.
17. WHITLOCK, C. A. & O. N. WITTE. 1982. Long-term culture of B lymphocytes and their precursors from murine bone marrow. Proc. Natl. Acad. Sci. USA **79:** 3608–3612.
18. LEBIEN, T. W. 1989. Growing human B-cell precursors in vitro: the continuing challenge. Immunol. Today **10:** 296–298.
19. WOLF, M. L., J. A. BUCKLEY, A. GOLDFARB, C.-L. LAW & T. W. LEBIEN. 1991. Development of a bone marrow culture for maintenance and growth of normal human B cell precursors. J. Immunol. **147:** 3324–3330.
20. DITTEL, B. N., J. B. MCCARTHY, E. A. WAYNER & T. W. LEBIEN. 1993. Regulation of human B-cell precursor adhesion to bone marrow stromal cells by cytokines that exert opposing effects on the expression of vascular cell adhesion molecule-1 (VCAM-1). Blood **81:** 2272–2282.
21. REHMANN, J. A. & T. W. LEBIEN. 1994. Transforming growth factor beta regulates human pre-B cell differentiation. Intl. Immunol. **6:** 315–322.
22. LARSON, A. W. & T. W. LEBIEN. 1994. Cross-linking CD40 on human B cell precursors inhibits or enhances growth depending on the stage of development and the interleukin costimulus. J. Immunol. **153:** 584–594.
23. DITTEL, B. N. & T. W. LEBIEN. 1995. The growth response to IL-7 during normal human B cell ontogeny is restricted to B-lineage cells expressing CD34. J. Immunol. **154:** 58–67.
24. UCKUN, F. M. & J. A. LEDBETTER. 1988. Immunobiologic differences between normal and leukemic human B-cell precursors. Proc. Natl. Acad. Sci. USA **85:** 8603–8607.
25. MCGINNES, K., E. KEYSTONE, E. BOGOCH, D. HASTINGS, H. A. MESSNER, N. JAMAL & C. J. PAIGE. 1990. Growth and detection of human bone marrow B-lineage colonies. Blood **76:** 896–905.
26. MCGINNES, K., V. QUESNIAUX, J. HITZLER & C. PAIGE. 1991. Human B-lymphopoiesis is supported by bone marrow-derived stromal cells. Exp. Hematol. **19:** 294–303.
27. RYAN, D. H., B. L. NUCCIE, C. N. ABBOUD & J. M. WINSLOW. 1991. Vascular cell adhesion molecule-1 and the integrin VLA-4 mediate adhesion of human B cell precursors to cultured bone marrow adherent cells. J. Clin. Invest. **88:** 995–1004.
28. RYAN, D. H., B. L. NUCCIE & C. N. ABBOUD. 1992. Inhibition of human bone marrow lymphoid progenitor colonies by antibodies to VLA integrins. J. Immunol. **149:** 3759–3764.
29. PANDRAU, D., S. SAELAND, V. DUVERT, I. DURAND, A. M. MANEL, M. T. ZABOT, N. PHILLIPE & J. BANCHEREAU. 1992. Interleukin-4 inhibits in vitro proliferation of leukemic and normal human B cell precursors. J. Clin. Invest. **90:** 1697–1706.

30. Saeland, S., V. Duvert, I. Moreau & J. Banchereau. 1993. Human B cell precursors proliferate and express CD23 following CD40 ligation. J. Exp. Med. **178:** 113–120.
31. Moreau, I., V. Duvert, J. Banchereau & S. Saeland. 1993. Culture of human fetal B-cell precursors on bone marrow stroma maintains highly proliferative $CD20^{dim}$ cells. Blood **81:** 1170–1178.
32. Punnonen, J., G. Aversa & J. E. deVries. 1993. Human pre-B cells differentiate into immunoglobulin-secreting plasma cells in the presence of interleukin-4 and activated $CD4^+$ T cells or their membranes. Blood **82:** 2781–2789.
33. Ryan, D. H., B. L. Nuccie, I. Ritterman, J. L. Liesveld & C. N. Abboud. 1994. Cytokine regulation of early human lymphopoiesis. J. Immunol. **152:** 5250–5258.
34. Hardy, R. R., C. E. Carmack, S. A. Shinton, J. D. Kemp & K. Hayakawa. 1991. Resolution and characterization of pro-B and pre-pro-B stages in normal mouse bone marrow. J. Exp. Med. **173:** 1213–1225.
35. Rolink, A., U. Grawunder, D. Hassner, A. Strasser & F. Melchers. 1993. Immature surface Ig^+ B cells can continue to rearrange κ and λ chain gene loci. J. Exp. Med. **178:** 1263–1270.
36. Tiegs, S. L., D. M. Russell & D. Nemazee. 1993. Receptor editing in self-reactive bone marrow B cells. J. Exp. Med. **177:** 1009–1020.
37. Noguchi, M., Y. Nakamura, S. M. Russell, S. F. Ziegler, M. Tsang, X. Cao & W. J. Leonard. 1993. Interleukin-2 receptor γ chain: A functional component of the interleukin-7 receptor. Science **262:** 1877–1879.
38. Kondo, M., T. Takeshita, M. Higuchi, M. Nakamura, T. Sudo, S.-I. Nishikawa & K. Sugamura. 1994. Functional participation of the IL-2 receptor γ chain in IL-7 receptor complexes. Science **263:** 1453–1455.
39. Banchereau, J., P. de Paoli, A. Valle, E. Garcia & F. Rousset. 1991. Long-term human B cell lines dependent on interleukin 4 and antibody to CD40. Science **251:** 70–72.
40. Landreth, K. S., R. Narayanan & K. Dorshkind. 1992. Insulin-like growth factor-1 regulates pro-B cell differentiation. Blood **80:** 1207–1212.

B-Lineage Differentiation Stages Resolved by Multiparameter Flow Cytometry

RICHARD R. HARDY AND KYOKO HAYAKAWA

Institute for Cancer Research
Fox Chase Cancer Center
7701 Burholme Avenue
Philadelphia, Pennsylvania 19111

B lymphopoiesis begins in the embryonic yolk sac, shifts to the liver by day 14 of gestation, and becomes established in the bone marrow within a few weeks after birth, where it continues throughout life. The differentiation of B lymphocytes from hematopoietic stem cells is a complex process involving several intermediate cell stages characterized by distinctive molecular and functional features. B-lineage cells in mouse are readily identified by expression of the high-molecular-weight form of the common leukocyte antigen, CD45, known as B220.[1] Of the several monoclonal antibodies that recognize B220, the clone RA3-6B2 is particularly B-lineage specific, with expression initiating very early in the B-cell pathway.[2,3] Staining of mouse bone marrow with antibodies to B220 and IgM clearly delineates two fractions of $B220^+$ cells, one IgM^+ (immature and mature B cells), the other IgM^- (pre-B cells). Expression of the cell surface molecule known variously as leukosialin, sialophorin, or CD43, recognized by the S7 monoclonal antibody,[4] divides the IgM^- pre-B cells into two fractions, one $CD43^+$, the other $CD43^-$.[5] Most mature B cells express low to undetectable levels of CD43(S7), and this is also true of the immature B cells in bone marrow. Thus, expression of this determinant is lost during B-cell generation, at the pre-B-cell stage. Analysis of the expression of several other molecules on $B220^+CD43(S7)^+$ cells was consistent with their assignment to the B lineage, but also demonstrated considerable heterogeneity within this population. For example, expression of the heat-stable antigen (HSA), used for defining stages of T-cell development,[6] is found at several distinctive levels. FIGURE 1 summarizes the phenotypes of seven $B220^+$ cell subsets in mouse bone marrow and presents the expression of several other molecules that we have found to be differentially expressed on these fractions.

In order to determine the precursor/progeny relationships of the seven bone marrow subsets, each fraction was isolated by cell sorting and deposited onto a stromal cell line that had previously been shown to promote the proliferation and differentiation of early B-lineage cells.[7] This analysis revealed a progression within the $CD43^+$ fractions from HSA^- to HSA^+ and from $BP\text{-}1^-$ and $BP\text{-}1^+$. Thereafter, CD43 was downregulated. Analysis of the $B220^+CD43^-$ cells revealed three readily discernible fractions with a known order: IgM^-IgD^- cells are small pre-B cells, IgM^+IgD^- cells are newly generated B cells, and IgM^+IgD^+ cells are mature B cells, possibly recirculating back through the bone marrow. For ease of discussion, we have labeled these fractions alphabetically from Fr. A (the HSA^- stage) through Fr. F (the mature B-cell stage). These are diagrammed in FIGURE 2.

Work with Abelson virus transformed B lineage major milestones in the development of B cells: completion of heavy chain D-J rearrangement, productive V-DJ

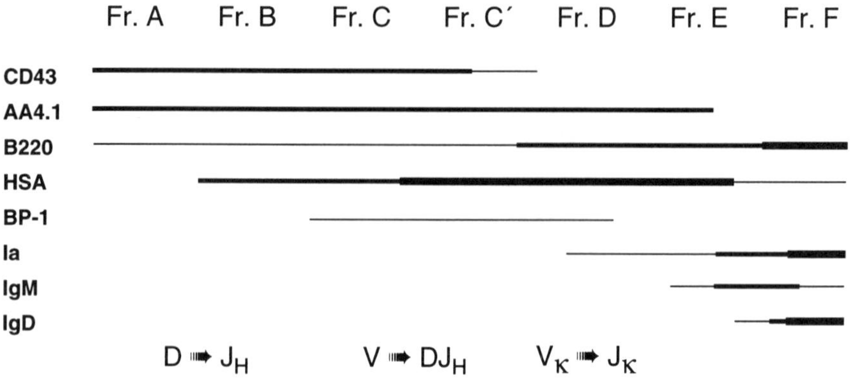

FIGURE 1. Expression of cell surface molecules during B lymphopoiesis in mouse bone marrow. Line thickness corresponds approximately to relative expression level. Predominant Ig rearrangements are also indicated.

rearrangement, and productive light chain rearrangement (mostly at the kappa locus in mouse). We have used two approaches, both involving polymerase chain reaction (PCR) amplification of DNA to determine rearrangement status in these fractions. One uses primers that amplify DNA that is deleted upon rearrangement, either between the D and J loci or between the V and D loci.[10] Because Ig heavy chain rearrangement occurs by deleting intervening DNA, then rearrangement of any D to any J or of any V to any D will result in a loss of the corresponding PCR band. A limitation of this technique is that the DNA must be lost from the cell (either by degradation or dilution upon cell division) so excised circles retained in the cell[11] may mask some rearrangement. An alternative approach is to amplify generated rearrangements using primers with some homology to many V segments or many D segments.[12,13] Before rearrangement, the distance between the V locus or D locus and the J locus is too great for PCR amplification, so signal derives only from rearranged genes. A limitation of this technique is the need for primers

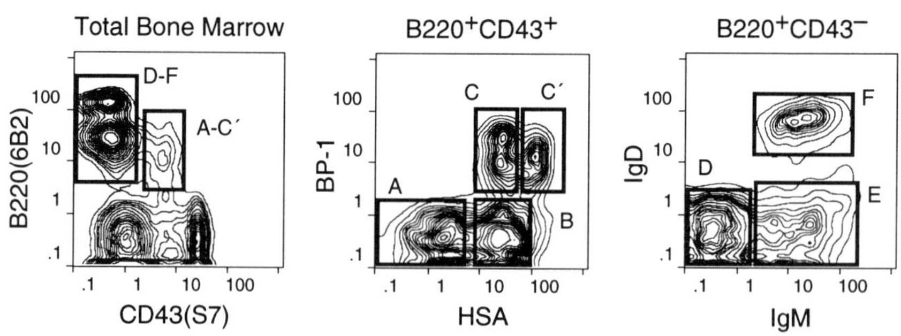

FIGURE 2. Flow cytometry analysis of mouse B-lineage cells in bone marrow. Two combinations of four-color staining reagents (B220/CD43 with either HSA/BP-1 or with IgM/IgD) serve to resolve the seven fractions described in FIGURE 1.

with good homology to numerous V or D segments that possess considerable sequence diversity, so sampling may not be uniform. Nevertheless, both methods show that there is relatively little rearrangement in Fr. A (undetectable by the first, a few percent by the second). Both approaches also demonstrate very extensive D-J rearrangement in Fr. B, but little V-DJ rearrangement (undetectable by the first, about 10% by the second). V-DJ rearrangement becomes apparent by both techniques in Fr. C (5–20%) and plateaus in Fr. D, as expected. Thus, the $B220^+CD43^+HSA^+$ fractions (B and C) consist largely of D-J rearranged cells in the process of V-DJ rearrangement, and we have termed this stage pro-B (as distinct from the stage where heavy chain rearrangement is complete, pre-B).

A number of gene products important in B-lineage differentiation are not detectable on the cell surface by fluorescence staining, but expression of these can be investigated at the message level using the technique of reverse transcription of total RNA followed by PCR (RT-PCR). Analysis of certain genes can serve to further test the order of B-lineage stages proposed above, whereas analysis of other genes may provide useful new information. MB-1, the gene encoding Ig-α (one of the CD3-like Ig accessory molecules), is found at high levels from Fr. B and serves to confirm the B-lineage nature of most of the fractions. The distinctive timing of the recombinase activating genes (Rag-1 and Rag-2)[14,15] and terminal deoxynucleotidyl transferase (TdT) serve to explain a novel difference between the joints of heavy and light chain genes. That is, TdT is responsible for the nongermline nucleotides (N-sequence) found at the heavy chain D-J and V-D junctions,[16] and the relative rarity of such N-sequence in light chain junctions is simply explained by the absence of TdT expression after Fr. C, since most light chain rearrangement occurs in Fr. D.

Many years ago we noticed that $CD5^+$ B cells were poorly generated by transfer of total bone marrow into lethally irradiated allotype congenic recipients, even though $CD5^-$ B cells and T cells were restored to normal levels.[17] Curiously, similar transfers employing liver from newborn mice did reconstitute this subset. One can take advantage of the ability to resolve subsets of B-lineage precursors to repeat this experiment with a more defined cell population. Thus, pro-B cells are isolated from both fetal liver and bone marrow of adult BALB/c mice based on the $B220^+CD43^+HSA^+$ phenotype (Fr. B, C) and then transferred into lightly irradiated (300R) SCID recipients. This method allows careful comparison of the progression from pro-B to B cell in a relatively normal microenvironment.[18] The spleens of such transferred mice are analyzed by flow cytometry 3 to 8 weeks (or longer) after injection for extent of lymphoid engraftment and for cell surface phenotype of the lymphoid cells generated.[19] The results, shown in FIGURE 3, are quite striking: whereas bone marrow pro-B cells generate B cells ($B220^+IgM^+$) in adult SCID spleen three weeks after transfer that are predominantly IgD^+ and $CD5^-$, transfers from fetal liver yield comparable numbers of B cells that show lower levels of IgD, and half are $CD5^+$. These results have led us to suggest that there is a developmental switch in B lymphopoiesis, similar to that proposed for the T lineage.[20]

One can test this model by asking whether there are differences discernible between pro-B cells isolated from fetal liver and adult bone marrow. RT-PCR analysis of the expression of a number of genes that are important to B-lineage differentiation reveals similarities and differences. Rag-1, Rag-2, λ5, and VpreB are all expressed at similar levels. In contrast, TdT and a recently described gene known as the precursor lymphocyte regulated myosin-like light chain (PLRLC)[21] are both highly expressed during the pro-B stage in bone marrow, but absent from fetal liver pro-B cells. Although the TdT result was anticipated, based on the low

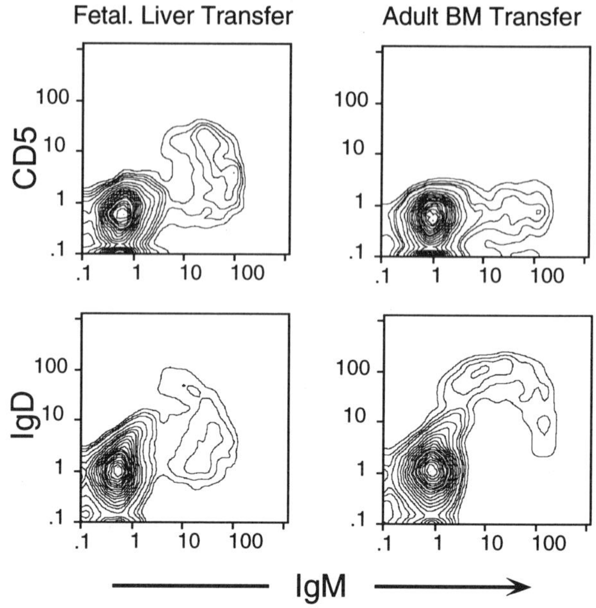

FIGURE 3. Distinct B-cell phenotypes generated in immunodeficient SCID mouse spleen after transfer of 10^5 pro-B cells (Fr. B/C) from either fetal liver or bone marrow of adult mice. Repopulation analyzed three weeks after transfer.

level of N-regions found in sequences of Ig heavy chains from B cells in neonatal mice,[22,23] the all-or-none nature of the expression was quite striking, suggesting that all lymphopoiesis in fetal liver at day 16 is distinctive. The significance of the absence of PLRLC is unknown at present, although the molecule has been suggested to play a structural role, perhaps in mediating the extent of capping of the pre-B receptor complex upon crosslinking. This is an important issue to pursue, because $CD5^+$ B cells appear to have undergone striking selection for certain kinds of self-reactivities.[24,25]

Analysis of the pre-B and B cells generated *in vitro* during stromal cell culture has demonstrated differences in the timing of expression of MHC class II between fetal and adult B lymphopoiesis.[26] During normal B-cell differentiation in bone marrow, class I becomes detectable at the pre-B-cell stage. A curious observation, made a number of years ago, is that most B cells in the spleen of neonatal mice have low to undetectable levels of class II. This apparent paradox was explained by a simple culture experiment with fetal liver pro-B cells. The pre-B and immature B cells generated in this culture lack class II as detected by either cell surface immunofluorescence with monoclonal antibodies specific for class II or by RT-PCR analysis of the individual class II α and β chains.[26] This contrasts with readily detectable levels on pre-B and B cells generated on similar culture of bone marrow pro-B. The progeny of fetal pro-B cells eventually do express class II as assessed by SCID transfer experiments, so it is the timing of expression of class II, not the presence or absence of class II, that distinguished fetal from adult B lymphopoiesis.

In summary, we have devised a scheme for resolving stages of B-lineage cells in the mouse and used several approaches for characterizing these fractions. The phenotypically distinct cell subsets show functional distinctions, and analysis of Ig rearrangement provides evidence of the ordered nature of this process in normal cells. Finally, comparison of corresponding fractions from adult bone marrow with those from fetal liver reveals many similarities, but also several differences. Considering the association of the progeny of this fetal wave, $CD5^+$ B cells, with autoreactivity and neoplasia, we feel that further investigation of these differences is merited.

REFERENCES

1. Coffman, R. L. & I. L. Weissman. 1981. B220: A B-cell-specific member of the T200 glycoprotein family. Nature **289:** 681–683.
2. Coffman, R. L. 1982. Surface antigen expression and immunoglobulin gene rearrangement during mouse pre-B cell development. Immunol. Rev. **69:** 5–23.
3. Coffman, R. L. & I. L. Weissman. 1983. Immunoglobulin gene rearrangement during pre-B cell differentiation. J. Mol. Cell. Immunol. **1:** 31–41.
4. Gulley, M. L., L. C. Ogata, J. A. Thorson, M. O. Dailey & J. D. Kemp. 1988. Identification of a murine pan-T cell antigen which is also expressed during the terminal phases of B cell differentiation. J. Immunol. **140:** 3751–3757.
5. Hardy, R. R., J. D. Kemp & K. Hayakawa. 1989. Analysis of lymphoid population in scid mice: Detection of a potential B lymphocyte progenitor population present at normal levels in scid mice by three-color flow cytometry with B220 and S7. Curr. Top. Microbiol. Immunol. **152:** 19–25.
6. Crispe, I. N., M. W. Moore, L. A. Husmann, L. Smith, M. J. Bevan & R. P. Shimonkevitz. 1987. Differentiation potential of subsets of CD4-8-thymocytes. Nature **329:** 336–339.
7. Hardy, R. R., T. Kishimoto & K. Hayakawa. 1987. Differentiation of B cell progenitors in vitro: Generation of surface IgM^+ B cells, including Ly-1 B cells, from Thy-1-asialo$GM1^+$ cells in newborn liver. Eur. J. Immunol. **17:** 1769–1774.
8. Alt, F., N. Rosenberg, S. Lewis, E. Thomas & D. Baltimore. 1981. Organization and reorganization of immunoglobulin genes in A-MULV-transformed cells: Rearrangement of heavy but not light chain genes. Cell **27:** 381.
9. Alt, F. W., G. D. Yancopoulos, T. K. Blackwell, C. Wood, E. Thomas, M. Boss, R. Coffman, N. Rosenberg, S. Tonegawa & D. Baltimore. 1984. Ordered rearrangement of immunoglobulin heavy chain variable region segments. Embo J. **3:** 1209–1219.
10. Hardy, R. R., C. E. Carmack, S. A. Shinton, J. D. Kemp & K. Hayakawa. 1991. Resolution and characterization of pro-B and pre-pro-B cell stages in normal mouse bone marrow. J. Exp. Med. **173:** 1213–1225.
11. Toda, M., T. Hirama, S. Takeshita & H. Yamagishi. 1989. Excision products of immunoglobulin gene rearrangements. Immunol. Lett. **21:** 311–316.
12. Ehlich, A., S. Schaal, H. Gu, D. Kitamura, W. Muller & K. Rajewsky. 1993. Immunoglobulin heavy and light chain genes rearrange independently at early stages of B cell development. Cell **72:** 695–704.
13. Li, Y. S., K. Hayakawa & R. R. Hardy. 1993. The regulated expression of B lineage associated genes during B cell differentiation in bone marrow and fetal liver. J. Exp. Med. **178:** 951–960.
14. Schatz, D. G., M. A. Oettinger & D. Baltimore. 1989. The V(D)J recombination activating gene, RAG-1. Cell **59:** 1035–1048.
15. Oettinger, M. A., D. G. Schatz, C. Gorka & D. Baltimore. 1990. RAG-1 and RAG-2, adjacent genes that synergistically activate V(D)J recombination. Science **248:** 1517–1523.
16. Desiderio, S. V., G. D. Yancopoulos, M. Paskind, E. Thomas, M. A. Boss,

N. LANDAU, F. W. ALT & D. BALTIMORE. 1984. Insertion of N regions into heavy-chain genes is correlated with expression of terminal deoxytransferase in B cells. Nature **311:** 752–755.
17. HAYAKAWA, K., R. R. HARDY, L. A. HERZENBERG & L. A. HERZENBERG. 1985. Progenitors for Ly-1 B cells are distinct from progenitors for other B cells. J. Exp. Med. **161:** 1554–1568.
18. DORSHKIND, K., K. A. DENIS & O. N. WITTE. 1986. Lymphoid bone marrow cultures can reconstitute heterogeneous B and T cell-dependent responses in severe combined immunodeficient mice. J. Immunol. **137:** 3457–3463.
19. HARDY, R. R. & K. HAYAKAWA. 1991. A developmental switch in B lymphopoiesis. Proc. Natl. Acad. Sci. U.S.A. **88:** 11550–11554.
20. IKUTA, K., T. KINA, I. MACNEIL, N. UCHIDA, B. PEAULT, Y. H. CHIEN & I. L. WEISSMAN. 1990. A developmental switch in thymic lymphocyte maturation potential occurs at the level of hematopoietic stem cells. Cell **62:** 863–874.
21. OLTZ, E. M., G. D. YANCOPOULOS, M. A. MORROW, A. ROLINK, G. LEE, F. WONG, K. KAPLAN, S. GILLIS, F. MELCHERS & F. W. ALT. 1992. A novel regulatory myosin light chain gene distinguishes pre-B cell subsets and is IL-7 inducible. Embo J. **11:** 2759–2767.
22. FEENEY, A. J. 1990. Lack of N regions in fetal and neonatal mouse immunoglobulin V-D-J junctional sequences. J. Exp. Med. **172:** 1377–1390.
23. GU, H., I. FORSTER & K. RAJEWSKY. 1990. Sequence homologies, N sequence insertion and JH gene utilization in VHDJH joining: implications for the joining mechanism and the ontogenetic timing of Ly1 B cell and B-CLL progenitor generation. Embo J. **9:** 2133–2140.
24. PENNELL, C. A., T. J. MERCOLINO, T. A. GRDINA, L. W. ARNOLD, G. HAUGHTON & S. H. CLARKE. 1989. Biased immunoglobulin variable region gene expression by Ly-1 B cells due to clonal selection. Eur. J. Immunol. **19:** 1289–1295.
25. CARMACK, C. E., S. A. SHINTON, K. HAYAKAWA & R. R. HARDY. 1990. Rearrangement and selection of VH11 in the Ly-1 B cell lineage. J. Exp. Med. **172:** 371–374.
26. HAYAKAWA, K., D. TARLINTON & R. R. HARDY. 1994. Absence of MHC class II expression distinguishes fetal from adult B lymphopoiesis in mice. J. Immunol. **152:** 4801–4807.

Identification of Genes Specifically Expressed at Discrete Stages of B-Cell Development

YU ZHANG,[a] IAN McDERMOTT, MICHAEL ATKINSON, AND CHRISTOPHER PAIGE

Wellesley Hospital Research Institute and Department of Medical Biophysics University of Toronto Toronto, Ontario Canada M4Y 1J3

The progression from the pluripotent hemopoietic stem cell to the antigen-responsive mature B cell is the consequence of coordinate activation and suppression of many genes. Our understanding of B-cell differentiation has been greatly increased in recent years due to the isolation and characterization of lineage- and/or stage-specific genes, such as λ5/VpreB, RAG1/RAG2, MB-1/B29, and others (reviewed in Cumano *et al.*[1]). The genetic basis of B-cell development on the whole, however, is not understood. Identification of additional genes specifically expressed at discrete developmental stages should allow better understanding of the pathway.

We are particularly interested in the genes differentially expressed around the transition from the recombination-active to recombination-silent stage. IIB4 and 70Z/3 have been selected as the representative cell lines in our study. IIB4 is virally transformed, and 70Z/3 is chemically induced. These two lines are derived from the same genetic background, (C57BL6XDBA/2)F1. They have a similar expression pattern in terms of many known surface markers and genes. Both cell lines are AA4.1+, B220+, κ- and λ-, and express λ5 and μ mRNA. On the other hand, they differ significantly in the status of their Ig gene rearrangements. While IIB4 is actively undergoing V(D)J recombination, 70Z/3 has ceased further rearrangements of its Ig genes. Accordingly, RAG1/RAG2 were found to be expressed in IIB4 cells, but not in 70Z/3 cells. To isolate the genes that are differentially expressed between IIB4 and 70Z/3, the mRNA differential display technique[2] was applied. With the 80 combinations of primers tested to date, nine genes were identified as differentially expressed between IIB4 and 70Z/3. Among them, three are IIB4-specific, the other six are 70Z/3-specific. Based on the DNA sequence data available, eight of the nine genes are novel.

The expression pattern of these candidate genes was investigated in various cell lines and primary tissues. Northern blot analysis revealed the stage and/or lineage specificity of the candidate genes. For example, IIG15 is present in all Abelson pre-B-cell lines, but generally absent from cell lines representative of later stages of B cell development except to the terminally differentiated J558 cell. 7G9, a 70Z/3-specific gene, appears to be both lineage and stage specific as suggested by the restricted expression in pre-B cells and absence in mature B

[a] Address for correspondence: Department of Medical Biophysics, University of Toronto, 160 Wellesley St., Toronto, ON M4Y 1J3, Canada.

cells as well as other lineages of hemopoietic cells and primary tissues. 7G10 has two transcripts of different size. The larger one is only detected in 70Z/3 cells, while the smaller is present both in IIB4 and 70Z/3. 7G3 is another 70Z/3-specific gene. Although not lineage specific, it is stage specific in B-cell development. It is not expressed either in earlier or later stages of B-cell development.

The expression study was extended to individual primary cells. They were obtained by FACS sorting from spleen (B220+ and B220−), 14-day fetal liver (B220+), or 12-day fetal liver (AA4.1+, Mac1−, B220−, Scal+ and AA4.1+, Mac1−, B220−, Scal−). mRNA in each cell was amplified using poly A+ PCR.[3] The amplified samples were then probed for the presence of our candidate genes as well as known genes. The stage specificity of 7G3 was confirmed in this study. Eight out of 11 d14 B220+ fetal liver cells were 7G3 positive, whereas the pro-B or the mature B cells were negative. This method allows us to correlate the expression pattern of novel genes with that of known ones.

REFERENCES

1. Cumano, A., B. Kee, D. Ramsden, A. Marshall, C. Paige & G. Wu. 1994. Immunol. Rev. **137:** 5–33.
2. Liang, P. & A. Parder. 1992. Science **257:** 969–971.
3. Brady, G., M. Barbara & N. Iscove. 1990. Meth. Mol. Cell. Biol. **2:** 17–25.

The Role of Bruton's Tyrosine Kinase in B-Cell Development and Function in Mice and Man[a]

WASIF N. KHAN,[b,c] PASCHALIS SIDERAS,[d]
FRED S. ROSEN,[c] AND FREDERICK W. ALT[b,c]

[b]*Howard Hughes Medical Institute*
The Children's Hospital
300 Longwood Avenue
Boston, Massachusetts 02115

[c]*Center for Blood Research*
200 Longwood Avenue
Boston, Massachusetts 02115

[d]*Unit of Applied Cell and Molecular Biology*
Umeå University
S-901 87 Umeå, Sweden

X-LINKED AGAMMAGLOBULINEMIA AND X-LINKED IMMUNODEFICIENCY

X-linked agammaglobulinemia (*XLA*)[1] is an inherited humoral immunodeficiency disease caused by an intrinsic B-cell defect as demonstrated by the unilateral inactivation of the X chromosome in the female carriers. The affected males suffer from recurrent bacterial infections.[2] Typical *XLA* is characterized by a severe deficiency of mature B cells (less than 1%: 5–15% in normal) and very low levels of serum immunoglobulins of all isotypes (about 10% of normal). T cell dependent immune responses are undetectable, and germinal follicles in the lymph nodes are not found (reviewed in Conley[3] and Rosen *et al.*[4]). Numbers of pre-B cells in the bone marrow are normal or decreased. The pro-B to pre-B cell ratio in the bone marrow is inverted as reflected by increased numbers of $TdT^+/CD34^+$ B cells[5,6] and severe deficiency of surface IgM^+ B cells. The *XLA* defect may reflect an inefficient differentiation to the pre-B stage or inefficient expansion of pre B and to mature B cells (FIG. 1).

X-linked immunodeficient (*xid*) mice (FIG. 1, Refs. 7, 11) share features with *XLA*; however, the defect is milder. In *xid* mice, numbers of B cells are only slightly reduced, and they show an immature phenotype (IgM^{lo}/IgD^{hi} population is reduced or absent.[7,8] The $CD5^+$ cells are absent, and there is a selective Ig deficiency; that is, levels of IgM and IgG3 are very low.[9] *Xid* mice do not respond to TI-II antigens[10,11] and show reduced response to some TD antigens.[12,13] *In vitro* studies showed that *xid* B cells do not proliferate in response to antigen receptor ligation[14] and showed a defective response to mitogenic activity of LPS.[11]

[a] This work was supported by the Howard Hughes Medical Institute and by National Institutes of Health Grants AI20047 and PO1AI35714. W. N. Khan is a fellow of the Swedish Natural Science Research Council.

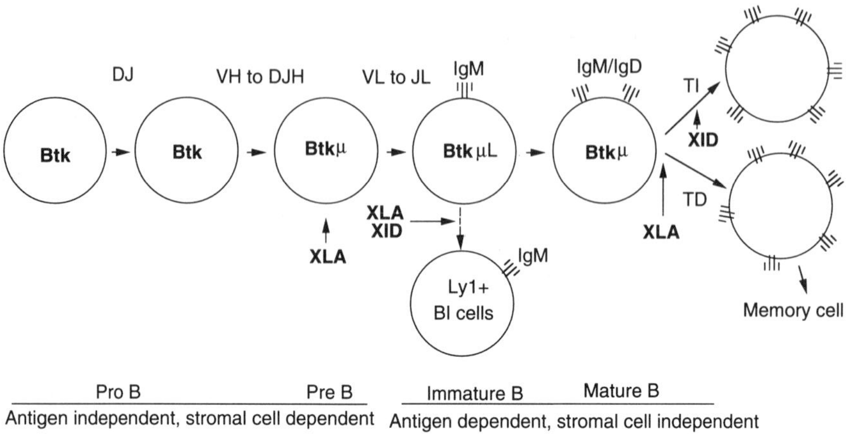

FIGURE 1. *Btk* expression in B-cell ontogeny and *XLA* and *xid* defects in B-cell development. The developmental blocks at pre-B and the absence of humoral immune responses in *XLA* and *xid* are shown by arrows. B1 lineage is absent in *xid* and *XLA*; the point of separation of the B1 and B2 lineages is hypothetical.

THE BRUTON'S TYROSINE KINASE

The genetic defect in *XLA* in man and *xid* in mice was found to be mutations in the Bruton's tyrosine kinase (*Btk*) gene[15–18] (FIG. 2). *Btk* belongs to a subfamily comprising *Btk*, *Tec*, and *Itk* that is related to but distinct from the *Src* family of cytoplasmic tyrosine kinases. In addition to the *Src* homology (SH) domain 1 (kinase domain), SH2, and SH3, the members of this novel subfamily have a unique NH_2-terminal region that contains a pleckstrin homology (PH) domain of 117 amino acids.[19–21] The PH domain has been found in many proteins involved in intracellular signaling pathways. Adjacent to the PH domain at the COOH-terminal end of the unique NH_2-terminal region, another protein domain, called the *Tec* homology (TH) domain[22] containing a proline-rich region, has been found that may serve as a target for SH3 domains of other tyrosine kinases. The function of this domain, however, is yet to be determined. *Btk* lacks a myristylation signal, which is used for membrane anchorage in other cytoplasmic tyrosine kinases, at its NH_2 terminus and a negative regulatory tyrosine at its COOH terminus (FIG. 2).

RESULTS AND DISCUSSION

Expression Pattern and Regulation of Btk

Recent studies have demonstrated the presence of *btk* transcripts of approximately 2.6 kb in cell lines that represent the pre-B and B cell stage; these transcripts were also found in spleen and bone marrow.[15,16,23,24] *Btk* transcripts also have been found in lung, pancreas, and a monocytic and myelomonocytic cell line; however, no specific defect has been described in these tissues in *XLA* patients.

TABLE 1. Summary of Phenotypes in Bruton's Tyrosine Kinase (*Btk*) PH Domain and Kinase Domain Knock Outs Compared with *xid* and *XLA* Phenotypes

	PH KO	Kinase KO	*xid*	*XLA*
Pre B cells in the bone marrow	Relatively normal	Relatively normal	Relatively normal	Relatively normal
Peripheral B cells	Reduced	Reduced	Reduced	Drastically reduced
LY1$^+$ B cells (B1) in the peritonium	Absent	Absent	Absent	Absent
IgM^{low}/IgD^{high} B cells in the spleen	Absent	Absent	Absent	
T cells in thymus and peripheri	Normal	Normal	Normal	Normal
Immunoglobulins				
IgM	Very low	Very low	Very low	Very low
IgG3	Very low	Very low	Very low	Very low
IgG1	Low	Low	Normal	Very low
IgG2b	Low	Low	Normal	Very low
IgG2a	Low	Low	Normal	Very low
IgA	Low	Low	Normal	Very low
B-cell proliferation in response to anti-IgM	Undetectable	Undetectable	Undetectable	
B-cell proliferation in response to LPS	Defective	Defective	Defective	
Humoral response to thymus independent Type-II antigens	Undetectable	Undetectable	Undetectable	No detectable humoral response
Humoral response to TD antigens	Normal	Normal	Normal	No detectable humoral response

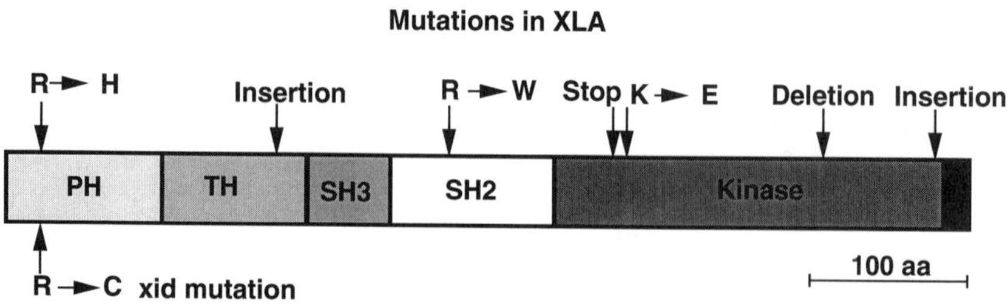

FIGURE 2. Structure of *Btk* and some mutations identified in *XLA* patients and the *xid* mutation are shown. PH, pleckstrin homology domain; TH, Tec homology domain; SH, Src homology domain and catalytic kinase domain are illustrated as boxes. Scale in amino acids is shown.

In those studies, *Btk* transcripts were not detected in T-cell lines, thymus, peripheral T cells, or other tissues.

We have also performed an extensive analysis of *btk* expression in cell lines, normal tissues, and tissues from mice with other types of immunodeficiencies; in general our results confirmed those recently published, but we also made some novel observations. We found expression of *btk* in newborn liver, fetal liver day 17 and 21, pre-B, transitional pre-B to B, and surface IgM$^+$ B-cell lines but not in plasmacytoma lines or T cell lines. We also found expression in multiple A-MμLV pre-B lines from RAG-1$^{-/-}$, RAG-2$^{-/-}$ or JH$^{-/-}$ mice, demonstrating *btk* expression before the onset of Ig gene rearrangement (FIG. 3). This pattern of expression is consistent with the notion that the *btk* gene product is involved in early B-cell developmental stages in the bone marrow, but could also play an important role at later stages of differentiation as suggested from the analyses of the *xid* mutation. Although the lack of expression of *btk* function in plasmacytomas suggests that this protein may not be employed in the terminal stages of B-cell differentiation, we found no significant diminution of *btk* expression upon cross-linking of surface IgM on splenocytes or upon activating these cells with LPS or LPS plus IL-4 (FIG. 3). Therefore, potential roles of *btk* in terminal B-cell differentiation bears further examination. Finally, in preliminary experiments, we have observed very low level expression of *btk* transcripts in the DN thymocytes of RAG-2-deficient mice; however, no defect in T cell development was obvious in *Btk* knockout mice (see below).

Putative transcription start sites and potential promoters of *Btk* have been characterized.[25] No obvious TATAA-like or CAAAT elements were found in the upstream sequences of the transcription start site. Nevertheless, the promoter activity could be localized within 300 bp of the transcription initiation site in chloramphenicol acetyl transferase (CAT) assays. These findings will facilitate further characterization of regulation of *Btk*.

Mutations in Btk *and Function of Different Domains*

Mutations in the *Btk* gene of *XLA* patients resulting in deletions, insertions, and point mutations leading to amino acid substitution or premature stop codon in virtually all the domains have been identified (FIG. 2). Most patients show a typical *XLA* phenotype with less than 1% B cells and undetectable Igs. Within families, the phenotype is highly variable, including cases that would have not been considered *XLA* if they did not belong to the families with affected males showing a typical *XLA* phenotype.

Atypical *XLA* patients with a less severe phenotype have also been reported.[26] After the identification of *Btk* as the gene for *XLA*, some of the patients classified under common variable immune deficiency (CVID) were found to have mutations in their *btk* genes. It is interesting to note that patients carrying the same mutation can display considerable variation in the severity of phenotype. Efforts are made to correlate different domains to a function on the basis of mutations identified and the phenotype observed. However, a clear picture is yet to emerge.

Determination of Btk *Function through Gene-Targeting Experiments*

To determine the role of *Btk* in B-cell maturation and function, we generated two mutations in the *Btk* molecule. We deleted most of the PH domain in embryonic

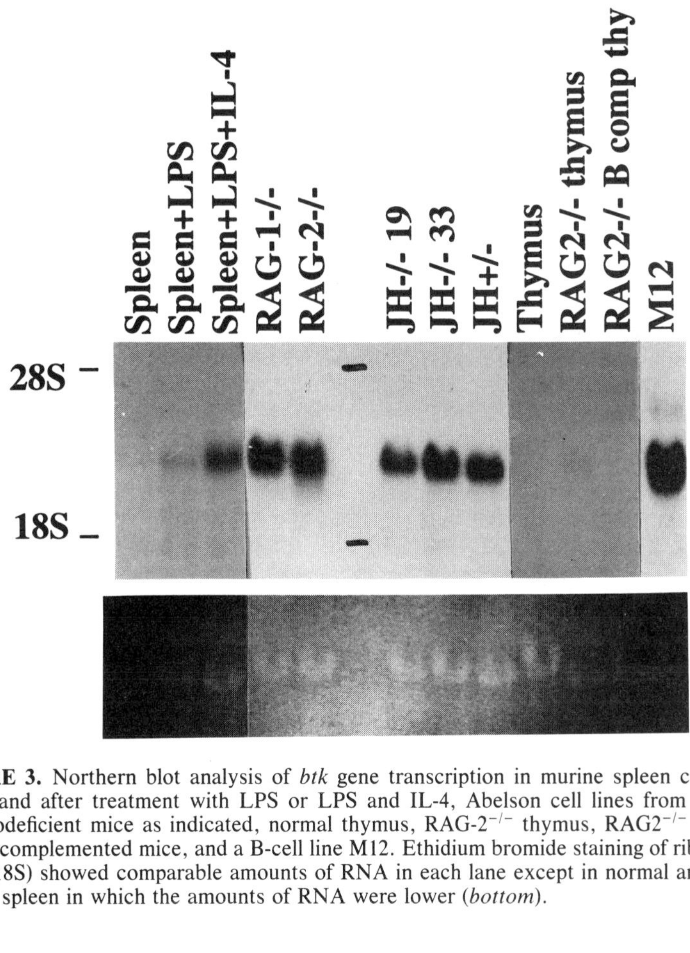

FIGURE 3. Northern blot analysis of *btk* gene transcription in murine spleen cells and before and after treatment with LPS or LPS and IL-4, Abelson cell lines from various immunodeficient mice as indicated, normal thymus, RAG-$2^{-/-}$ thymus, RAG$2^{-/-}$ thymus from β-complemented mice, and a B-cell line M12. Ethidium bromide staining of ribosomal RNA (18S) showed comparable amounts of RNA in each lane except in normal and LPS-treated spleen in which the amounts of RNA were lower (*bottom*).

stem cells (PH knockout [KO]), and we also deleted the entire kinase domain in embryonic stem cells (kinase knockout) by homologous recombination. Our preliminary results indicate that both of these mutations of the *Btk* gene result in the development of the *xid*, as opposed to the *XLA*, phenotype.

Generation of the Btk *Plekstrin Homology Domain and Kinase Domain Mutant Embryonic Stem Cells and Chimeric Mice by the RAG2-Deficient Blastocyst Complementation System*

The *Btk* genomic locus is encoded by 19 exons (FIG. 4**A**). The first three coding exons encode the PH domain.[25] There are four potential translation initiation codons in exons 2, 4, 5, and 6, which upon *in vitro* translation give proteins of 77, 66, 58, and 50 kDa. Nevertheless, only the 77-kDa protein starting at the first ATG is detectable *in vivo*.[16] To delete most of the PH domain of the *Btk* gene, the first two coding exons encoding for 80 of the 117 amino acids were deleted and replaced with neomycin gene (FIG. 4**B**). This mutation deletes the first translation initiation codon in exon 2 and should therefore either permit generation of a truncated *Btk* protein lacking a PH domain or will completely ablate *Btk* protein production. PH mutant embryonic stem (ES) cells were generated through homologous recombination (HR) with the frequency of HR being 1 in 250. These *Btk* PH mutant ES cells were used to generate chimeras by injecting into RAG2-deficient blastocysts and implanting them into foster mothers. RAG-2-deficient mice lack mature lymphocytes because of their inability to initiate VDJ recombination[27]; therefore, all the mature lymphocytes in chimeric mice are mutant ES cell derived.[28]

We also deleted exons 13 and 14, which code for the last 20 amino acids of the SH2 domain, and the first two exons of the kinase domain were deleted and replaced with neomycin (FIG. 4**C**). This deletion also disrupts the open reading frame and introduces a stop codon, thereby inactivating the entire kinase domain. ES cells were generated with HR carrying the kinase mutation with an HR frequency of 1 in 50 and were used to generate somatic chimeras in the RAG2-deficient blastocyst complementation system.[28]

Generation of Btk *Kinase Domain-Deficient Mice*

ES cells carrying the kinase mutation were also used to generate chimeric mice that transmitted the *Btk*-mutated X allele to their female offspring. Males carrying the mutated allele were obtained by interbreeding of the heterozygous females to wild-type litter mates. In preliminary analyses (Khan *et al.*, in preparation), mutant males and homozygous females showed indistinguishable phenotypes that are summarized below.

Lack of Btk *Expression in PH and Kinase Mutant B Cells*

Btk transcripts of apparently similar size compared to normal could be detected in Northern blot analyses in B cells from both KOs. In PH KO, using reverse transcriptase PCR assay, we could also show that a bona fide transcript as a result of splicing of exon 1 to exon 4, deleting the neomycin gene, was present. Conclusive results could not be obtained by similar analyses on kinase KO B cells. To investigate whether a truncated protein is made in these KOs, Western blotting and

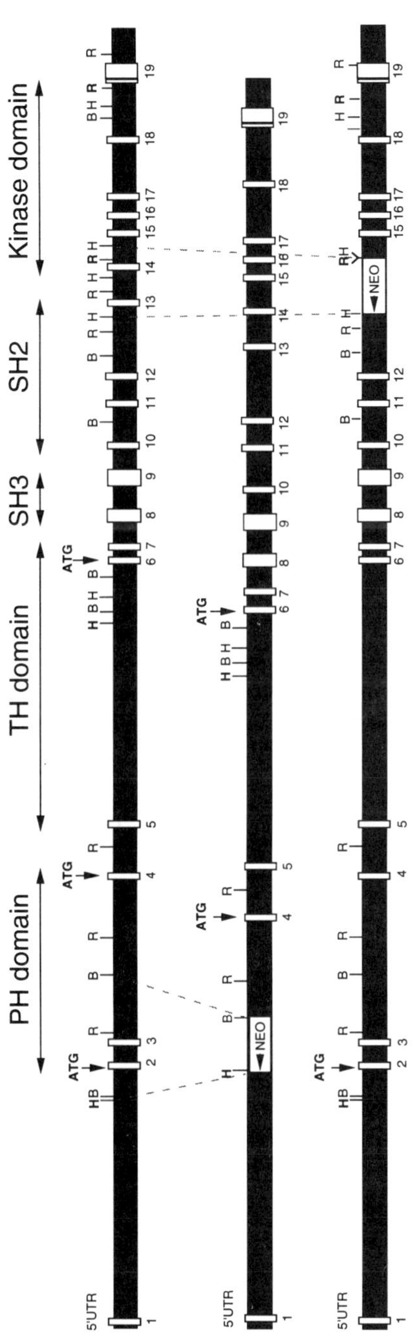

FIGURE 4. Physical map of the *Btk* locus **(A)**. In PH-targeted locus, exons 2 and 3 are replaced by neomycin resistance (NEO) gene **(B)**, and in kinase-targeted locus, exons 13 and 14 are replaced by neomycin resistance (NEO) gene **(C)**. Single-letter codes used for restriction endonuclease are: B, *Bam*HI; H, *Hin*dIII; R, *Eco*RI.

immunoprecipitation experiments were performed. *In vivo Btk* is a 77-kDa protein. To examine whether a truncated protein, using a second (a protein of 66 kDa) or third translation initiation codon (a protein of 58 kDa), is synthesized, Abelson leukemia virus-transformed pre-B-cell lines were generated from PH-mutant chimeric mice. Immunoprecipitation and Western blotting experiments were performed on these mutant Abelson cell lines using affinity-purified antibodies raised against the COOH-terminal region of *Btk* as previously described.[16] Using anti-NH_2-terminus antibodies, similar experiments were performed to investigate the presence of a truncated protein in kinase KO Abelson lines. A 77-kDa *btk* protein was present in RAG-$2^{-/-}$ A-MμLV cell line used as positive control, but no protein was detectable in either of the two knockouts.

B-Cell Development in the Btk *Mutant Mice*

Subpopulations representing different stages of B-cell development in the bone marrow can be recognized using antibodies against defined surface markers. The normal bone marrow contains Pro-B ($B220^{lo}/CD43^+$), Pre-B ($B220^{lo}/CD43^-$), newly generated B ($B220^{lo}/IgM^+$), and recirculating mature B cells ($B220^{hi}/IgM^+$).[29] B-cell development in the bone marrow of RAG$2^{-/-}$ mice is blocked at the $B220^+/CD43^+$ pro-B-cell stage as a result of an inability to rearrange immunoglobulin genes.[27] Populations corresponding to the Pro, Pre, surface Ig^+, and circulating B-cell subsets could be recognized in the bone marrow of chimeric mice derived from the *Btk*-PH^- and Kin^- ES cells. However, the numbers of cells in the compartments past Pro-B cell stage ($B220^+/CD43^+$ to $B220^+/CD43^-$) appear to be reduced. Consequently, the frequency of surface immunoglobulin expressing mature B cells ($B220^+/IgM^+$) is also reduced. The accumulation of pro-B cells in the bone marrow of PH^-/RAG-$2^{-/-}$ and Kin^-/RAG-$2^{-/-}$ chimeric mice may be difficult to interpret because of the accumulation of this population in RAG$2^{-/-}$ BM. Bone marrow analyses of kinase mutant mice carrying the mutation in their germline confirmed the results obtained in the chimeras generated by the RAG2-deficient blastocyst complementation system. This minimal impairment from pro- to pre-B-cell development was also seen in *Xid* (CBA/N) mice, demonstrating that, unlike man, normal function of *Btk* is required but not essential for early B-cell development in mice.

Analyses of the peripheral B-cell compartment showed that mature IgM^+ B cells are present in the spleen of *Btk*-PH^- chimeric and kinase mutant mice, albeit in slightly lower numbers (30 to 40% in contrast to 50 to 60% found in normal 129sV mice). In normal mice three populations of IgM^+/IgD^+ are found: IgM^{lo}/IgD^{hi}, IgM^{hi}/IgD^{hi}, and IgM^{hi}/IgD^{lo}. Like *Xid* splenocytes,[7] a significant decrease in the frequency of B cells exhibiting the IgM^{lo}/IgD^{hi} profile was observed in the spleens of PH^- and $kinase^-$ mice. Mature IgM^+ B cells were found in the lymph nodes as well as the peritoneal cavity. However, again similar to *xid*, the $CD5^+$ subset of B cells (B1) was not detected in the peritoneal cavity of the *Btk*-PH^- and $kinase^-$ animals.

BtK *PH^- and $Kinase^-$ B Cells Show Functional Defects*

Defective *Btk* function has been correlated with abnormalities in circulating antibody levels. In *XLA* the levels of all isotypes are dramatically reduced, whereas in *Xid,* IgM and IgG3 levels are drastically reduced, and the rest of the isotypes are relatively normal. In both of the mutant mice, the serum levels of all isotypes were found to be similar to *xid*.

In order to assess the capacity of the Btk-PH⁻ or kinase⁻ B cells to respond to mitogenic signals, PH mutant, kinase mutant, *xid*, and normal (129sV) purified B cells were subjected to stimulation with soluble anti-IgM and different doses of lipopolysaccharide (LPS). *Xid* mutant B cells have previously been shown not to respond to anti-IgM treatment.[14] Proliferation of anti-IgM-treated *Btk*-PH and kinase mutant B cells, as measured by [^{3}H]thymidine incorporation, was comparable to that of *xid* B cells but was an order of magnitude lower than normal B cells. Similarly, the proliferative response of *btk* PH or kinase-deficient B cells to LPS was comparable to that of *xid* and was significantly lower than normal B cells.

In normal mice TI-II antigens elicit immune responses largely restricted to IgM and IgG3 isotypes. *Xid* mice do not respond to TI-II antigens (TNP-Ficol), whereas their response to TI-I antigens (TNP-LPS) is not defective, and thymus-dependent (TD) responses are selectively diminished depending on the immunogen. Typical *XLA* patients show no immune responses; however, atypical *XLA* patients have been shown to mount specific immune responses to TD antigens.[30] To investigate the humoral responses of the *Btk* mutant mice of *Btk* PH⁻ (chimera) and kinase mutant mice were immunized with TI-II and TD antigens together with *xid* and normal controls. Both TD and TI-II responses of *Btk* mutant mice were comparable to *xid*, demonstrating that absence and dysfunction of *Btk* have similar consequences on the ability of mice to mount a humoral immune response (Khan *et al.*, in preparation).

Requirement of Btk *in B-Cell Development and Function Is Different in Mice and Man*

Preliminary characterizations have indicated that both our *Btk* PH mutant chimeras and our *Btk* kinase domain mutant mice appeared to lack expression of the *btk* protein. Notably, the maturation and function of B cells are defective in both types of mutant mice, and this defect is similar to that described for *xid* unequivocally proving that defective *Btk* function results in a *xid* disease; *xid* mice carry an amino acid substitution mutation ($Arg_{28}Cys$) in the P_H domain of their *Btk* molecule.[17,18] Initially, we speculated that the milder phenotype of the *xid* as compared to the *XLA* mutations might reflect multiple roles for the *btk* protein in B-cell differentiation—with the PH mutation only affecting an activity required at later stages in the context of some forms of B cell activation. In this context, the *XLA* phenotype was postulated to result from more severe inactivation of the *btk* protein, which blocked a required function in early B-cell development.[17,18] This speculation became less likely, however, when the same amino acid mutated in murine *xid* was found mutated in a man ($Arg_{28}His$)[31] in association with the typical *XLA* phenotype, suggesting a less stringent requirement for *btk* for early murine B-cell development. Our current findings clearly support this alternative possibility and suggest a compensatory mechanism in mice during early B-cell development that can rescue B-cell maturation and proliferation from the early pre-B stage to the sIg$^+$ stage.

Various scenarios can be envisioned: first and simplest could be the presence of a tyrosine kinase in mice that substitutes for the activity of *Btk*. This tyrosine kinase does not have to be unique for murine B cells, but the temporal difference in expression could be sufficient to drive early B-cell transitions. Similarly, spatiotemporal expression of a cytokine or its receptor that can mediate the transition of precursors to mature B cells in mice may explain the difference in phenotype. An interesting possibility is the presence of a factor produced by stromal cells or T cells in mice that is not available to support B cell maturation in man. As we know, early B cell development in *Xid* is minimally impaired. However, *Xid*, when expressed together with the mouse nude mutation, severely affects both B and T cell development.[32,33] In *xid*/nude mice, the B cell development is blocked at pro-B stage,[34] which is similar to the block in human *XLA*. The nude defect does not affect a stromal cells' capability to support the differentiation of pro-B cells into mature, $B220^+/IgM^+$ cells *in vitro*.[35] However, *xid* bone marrow cells do not repopulate the periphery of thymectomized mice unless T cells are provided.[36] The T-cell factor required for differentiation of *xid* B cells may be the CD40 ligand; hopefully, breeding of *Btk* mutant mice with tyrosine kinases expressed in early B cells or with ligands of receptors expressed on B cells will help elucidate the mechanism of *Btk* function.

REFERENCES

1. Bruton, O. C. 1952. Agammaglobulinemia. Pediatrics **9:** 722–727.
2. Lederman, H. M. & J. A. Winkelstein. 1985. X-linked agammaglobulinaemia: An analysis of 96 patients. Medicine (Baltimore) **64:** 145–156.
3. Conley, M. E. 1985. B cells in patients with X-linked agammaglobulinemia. J. Immunol. **134:** 3070–3074.
4. Rosen, F. S., M. D. Cooper & R. J. P. Wedgwood. 1984. The primary immunodeficiencies (first of two papers). N. Engl. J. Med. **311:** 235–242.
5. Milili, M., F. Le Deist, G. de Saint-Basile, A. Fischer, M. Fougereau & C. Schiff. 1993. Bone marrow cells in X-linked agammaglobulinaemia express pre-B-specific genes (lambda-like and V pre-B) and present immunoglobulin V-D-J gene usage strongly biased to a fetal-like repertoire. J. Clin. Invest. **91:** 1616–1629.
6. Pearl, E. R., L. B. Vagler, A. J. Okos, A. R. Lawton & M. D. Cooper. 1978. B lymphocyte precursors in bone marrow: An analysis of normal individuals and patients with antibody deficiency states. J. Immunol. **120:** 1169.
7. Scher, I. 1982. The CBA/N mouse strain: An experimental model illustrating the influence of X chromosome on immunity. Adv. Immunol. **47:** 241–244.
8. Hardy, R. R., K. Hayakawa, D. R. Parks, & L. A. Herzenberg. 1983. Demonstration of B-cell maturation in X-linked immunodeficient mice by simultaneous three colour immunofluorescence. Nature **306:** 270–272.
9. Perlmutter, R. M., M. Nahm, K. E. Stein, J. Slack, I. Zitron, W. E. Paul, & J. M. Davie. 1979. Immunoglobulin subclass-specific immunodeficiency in mice with an X-linked B-lymphocyte defect. J. Exp. Med. **149:** 993–998.
10. Scher, I., A. D. Steinberg, A. K. Berning & W. E. Paul. 1975. X-linked B lymphocyte immune defect in CBA/N mice. II. Studies of the mechanisms underlying the immune defect. J. Exp. Med. **142:** 637.
11. Amsbaugh, D. F., C. T. Hansen, B. Prescot, P. W. Stashak, D. R. Barthold & P. J. Baker. 1972. Genetic control of the antibody response to type III pneumonococcal polysaccharides in mice. I. Evidence that an X-linked gene plays a decisive role in determining responsiveness. J. Exp. Med. **136:** 931–936.
12. Press, J. L. 1981. The CBA/N defect defines two classes of T cell dependent antigens. J. Immunol. **126:** 1234–1240.
13. Press, J. L. & C. A. Giorgetti. 1986. Clonal analysis of the primary and secondary B cell responses of neonatal, adult and xid mice to (T,G)-A--L. J. Immunol. **137:** 784–790.
14. Mond, J. J., M. Schaefer, J. Smith & F. D. Finkleman. 1983. Lyb-5- B cells of CBA/N mice can be induced to synthesize DNA by culture with insolubilized but not soluble anti-Ig. I. Immunol. **131:** 2107.

15. VETRIE, D., I. VORECHOVSKY, P. SIDERAS, J. HOLLAND, A. DAVIES, F. FLINTER, L. HAMMARSTRÖM, K. KINNON, R. LEVINSKY, M. BOBROW, C. I. E. SMITH & D. R. BENTLEY. 1993. The gene involved in X-linked agammaglobulinaemia (XLA) is a member of the *src* family of protein-tyrosine kinases. Nature **361:** 226.
16. TSUKADA, S., D. C. SAFFRAN, D. J. RAWLINGS, O. PAROLINI, R. C. ALLEN, I. KLISAK, R. S. SPARKES, H. KUBAGAWA, T. MOHANDAS, S. QUAN, J. W. BELMONT, M. D. COOPER, M. E. CONLEY & O. N. WITTE. 1993. Deficient expression of a B cell cytoplasmic tyrosine kinase in human X-linked agammaglobulinemia. Cell **72:** 279.
17. THOMAS, J., P. SIDERAS, I. VORECHOVSKY, C. I. E. SMITH, V. CHAPMAN & W. E. PAUL. 1993. A missence mutation in the X-linked agammaglobulinemia gene colocalizes with the mouse X-linked immunodeficiency gene. Science **261:** 355.
18. RAWLINGS, D. J., D. C. SAFFRAN, S. TSUKADA, D. A. LARGAESPADA, J. C. GRIMALDI, L. COHEN, R. N. MOHR, J. F. BAZAN, M. HOWARD, N. G. COPELAND, N. A. JENKINS & O. N. WITTE. 1993. Mutations of unique region of Bruton's tyrosine kinase in immunodeficient XID mice. Science **261:** 358.
19. MAYERS, B., R. REN, K. CLARK & D. BALTIMORE. 1993. A putative modular domain present in diverse signaling proteins. Cell **73:** 629.
20. MUSACCHIO, A., T. GIBSON, P. RICE, J. THOMPSON & M. SARASTE. 1993. The PH domain: A common piece in the structural patchwork of signalling proteins. TIBS **18:** 343.
21. SHAW, G. 1993. Identification of novel pleckstrin homology (P_H) domains provides a hypothesis for P_H domain function. Biochem. Biophys. Res. Commun. **195:** 1145.
22. VIHINEN, M., L. NILSSON & C. I. E. SMITH. 1994. Tec homology domain (TH) adjacent to PH domain. FEBS Lett. **350:** 263–265.
23. SMITH, C. I. E., B. BASKIN, P. HUMIRE-GREIFF, J. ZHOU, P. G. OLSSON, H. S. MANIAR, P. KJELLEN, J. D. LAMBRIS, B. CHRISTIANSEN, L. HAMMARSTROM, D. BENTLEY, D. VETRIE, K. B. ISLAM, I. VORECHOVSKY & P. SIDERAS. 1994. Expression of Bruton's tyrosine kinase gene, BTK, is selectively down-regulated in T lymphocytes and plasma cells. J. Immunol. **152:** 557–565.
24. DE WEERS, M., M. C. M. VERSCHUREN, M. E. M. KRAAKMAN, R. G. J. MENSIK, R. K. B. SCHUURMAN, J. J. M. VAN DONGEN & R. W. HENDRIKS. 1993. The Bruton's tyrosine kinase gene is expressed throughout B cell differentiation, from early precursor B cell stages preceding immunoglobin gene rearrangement up to mature B cell stages. Eur. J. Immunol. **23:** 3109–3114.
25. SIDERAS, P., S. MULLER, H. SHIELDS, H. JIN, W. N. KHAN, L. NILSSON, E. PARKINSON, J. D. THOMAS, L. BRANDEN, I. LARSSON, W. E. PAUL, F. S. ROSEN, F. W. ALT, D. VETRI, C. I. E. SMITH & K. G. XANTHOPOLOUS. 1995. Genomic organization of mouse and human Bruton's agammaglobulinemia tyrosine kinase (*Btk*). J. Immunol. **153:** 5607–5617.
26. SAFFRAN, D. C., O. PAROLINI, M. E. FITCH-HILGENBERG, D. J. RAWLINGS, D. E. H. AFAR, O. N. WITTE & M. E. CONLEY. 1994. Brief report: A point mutation in the S_H2 domain of Bruton's tyrosine kinase in atypical X-linked agammaglobulinemia. N. Engl. J. Med. **330:** 1488–1491.
27. SHINKAI, Y., G. RATHBUN, K.-P. LAM, E. M. OLTZ, V. STEWART, M. MENDELSOHN, J. CHARRON, M. DATTA, F. YOUNG, A. M. STAHL & F. W. ALT. 1992. RAG-2-deficient mice lack mature lymphocytes owing to inability to initiate V(D)J rearrangement. Cell **68:** 855–867.
28. CHEN, J., R. LANSFORD, V. STEWART, F. YOUNG & F. W. ALT. 1993. RAG-2 deficient blastocyst complementation: An assay of gene function in lymphocyte development. Proc. Natl. Acad. Sci. USA **90:** 4528–4532.
29. HARDY, R. R., C. E. CARMACK, S. A. SHINTON, D. J. KEMP & K. HAYAKAWA. 1991. Resolution and characterization of pro-B and pre-pro-B cell stages in normal mouse bone marrow. J. Exp. Med. **173:** 1213.
30. ZHU, Q., M. ZHANG, D. J. RAWLINGS, M. VIHINEN, T. HAGEMANN, D. C. SAFFRAN, S.-P. KWAN, L. NILSSON, C. I. E. SMITH, O. N. WITTE, S.-H. CHEN & H. D. OCHS. 1994. Deletion within the Src homology domain 3 of Bruton's tyrosine kinase resulting in X-linked agammaglobulinemia (XLA). J. Exp. Med. **180:** 461–470.

31. DE WEERS, M., R. G. J. MENSINK, M. E. M. KRAAKMAN, R. K. B. SCHUURMANN & R. W. HENDRIKS. 1994. Mutation analysis of Bruton's tyrosine kinase gene in X-linked agammaglobulinemia: Identification of a mutation which affects the same codon as is altered in immunodeficient *xid* mice. Hum. Mol. Genet. **3:** 161–166.
32. WORTIS, H. H., L. BURKLY, D. HUGHES, S. ROSCHELLE & G. WANECK. 1982. Lack of mature B cells in nude mice with X-linked immune deficiency. J. Exp. Med. **155:** 903–913.
33. MOND, J. J. 1982. Use of the T lymphocyte-regulated type II antigens for the analysis of responsiveness of $Lyb5^{+}$ and $Lyb5^{-}$ B lymphocytes to T lymphocyte derived factors. Immunol. Rev. **64:** 99.
34. KARAGOGEOS, D., N. ROSENBERG & H. H. WORTIS. 1986. Early arrest in B cell development in nude, X-linked immunodeficient mice. Eur. J. Immunol. **16:** 1125–1130.
35. DONG, Z. & H. H. WORTIS. 1994. Function of bone marrow stromal cell lines derived from nude mice. J. Immunol. **153:** 1441–1448.
36. SPRENT, J. & J. BRUCE. 1984. Physiology of the B cells in mice with X-linked immunodeficiency (Xid). III. Disappearance of xid B cells in double bone marrow chimeras. J. Exp. Med. **160:** 711–720.

Ig μ Heavy Chains with V_H81X Variable Regions Do Not Associate with $\lambda5$[a]

ULRIKE KEYNA,[b] STEVEN E. APPLEQUIST,[b]
JAN JONGSTRA,[c] GABRIELE B. BECK-ENGESER,[b]
AND HANS-MARTIN JÄCK[b]

[b]*Department of Microbiology and Immunology*
Loyola University Chicago
Stritch School of Medicine
Maywood, Illinois 60153

[c]*Toronto Western Hospital*
Toronto, Ontario, Canada M5T 2S8

As a long-term goal we seek to understand how pre-B cells are activated to differentiate into mature B cells. Pre-B cells express immunoglobulin (Ig) μ heavy chains and surrogate light (SL) chains but not conventional Ig light chains. μ and SL chains, which are composed of the two noncovalently associated polypeptides VpreB and $\lambda5$, form a disulfide-linked heterodimeric complex that can be detected on the surface of pre-B cells.[1] Rajewski and coworkers have demonstrated by targeted deletion of gene segments that the membrane form of μ (μ_m) chains, as well as $\lambda5$,[2,3] are required for normal B-cell development. For example, mice that carry homozygous deletions of the μ membrane exons do not possess mature B cells because of a block in the transition from large $CD43^+$ pro-B/pre-B cells (Hardy's fraction A–C[4]) to small $CD43^-$ pre-B cells (Hardy's fraction D[4]). On the basis of these observations, the same authors suggested that μ_m and $\lambda5$ are required during B-cell development, because they form a μ/SL complex that signals a pre-B cell to develop into a mature B cell. If this is true, then μ chains that are unable to form a complex with SL chains should not be detected in mature B cells, because pre-B cells that express these μ chains are not positively selected.

One example for a H chain that does not combine with SL chains might be a μ chain with the V_H81X region, because V_H81X/μ chains, although they are frequently expressed in pre-B cells,[5] are rarely detected in mature B cells.[6] To test this hypothesis, we studied the V_H81X/μ-positive pre-B cell line F, which is a subclone of the Abelson virus-transformed 18-81 cell line (TABLE 1). If our idea is correct, we predict that μ chains do not associate in F cells with SL chains and are, therefore, not expressed on the surface of F cells. Indeed, when we stained F cells with FITC-labeled anti-μ antibodies and analyzed them by flow cytometry, we found that μ chains are not expressed on the surface (TABLE 1). To exclude the possibility that F cells have lost the ability to transport μ chains to the surface

[a] This work was supported by grants from the American Cancer Society of Illinois, the National Cancer Institute, and the Schweppe Foundation of Chicago to H.-M.J.. U.K. is supported by a fellowship from the Deutsche Forschungsgemeinschaft, and S.E.A. by a Schmitt Fellowship.

TABLE 1. Analysis of μ and Surrogate Light Chain Expression in Pre-B and B Cells

			μ Expression		Expression of[c]		Association of μ[c]	
Cell Line	Cell Type	V_H Region	Cytoplasm[a]	Surface[b]	λ5	VpreB	with λ5	with VpreB
NYC 31[d]	B	V_HNYC	+	+	–	–	–	–
107.2[e]	Pre-B	?	+	+	+	+	+	+
2G4	Pre-B	V_H17.2.25[f]	+	+	+	+	+	+
Bine 4.8	Pre-B	none	–	–	+	+	–	–
F[g]	Pre-B	V_H18X[h]	+	–	+	+	–	+
23.4[i]	Pre-B	V_H81X[h]	+	–	+	+	–	+
9.3[i]	Pre-B	V_H81X	+	–	+	+	–	+

[a] Determined by Western blot analysis.
[b] Determined by flow cytometry.
[c] Determined by SDS/PAGE of anti-μ and anti-λ5 precipitated proteins.
[d] Jäck *et al.* 1992. Proc. Natl. Acad. Sci. USA **89:** 8482–8486.
[e] Misener *et al.* 1991. Int. Immunol. **3**(11): 129–136.
[f] Groscchedl *et al.* 1984. Cell **38:** 647–658.
[g] Wabl *et al.* 1984. Proc. Natl. Acad. Sci. USA **81:** 867–870.
[h] V_H regions were amplified from mRNA by PCR and sequenced.
[i] Provided by Patricia Gearhart (Lawler *et al.* 1987. Proc. Natl. Acad. Sci. USA **84:** 2454–2458).

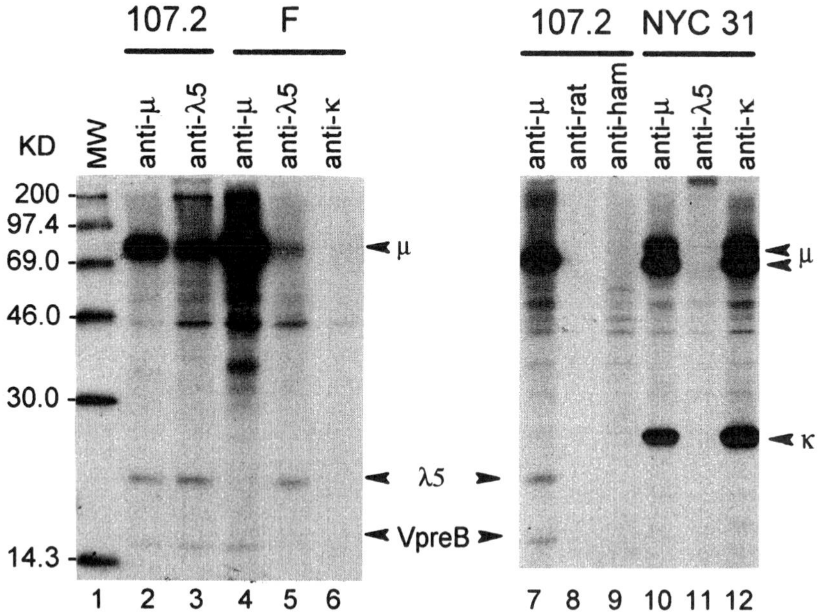

FIGURE 1. Analysis of μ and surrogate light chain expression in various B-cell lines. In panel **a**, equal aliquots of [^{35}S]methionine-labeled cell extracts were first incubated with the rat monoclonal anti-mouse (Cμ2) antibody, b7-6 (anti-μ), the rat monoclonal anti-mouse (κ) antibody, 187.1, and the hamster monoclonal anti-mouse (λ5) antibody, FS-1 (anti-λ5), followed by appropriate secondary antibodies (goat anti-hamster [IgG] = anti-ham and goat anti-rat [IgG] = anti-rat) and *Staphylococcus aureus*. Solubilized proteins were resolved by SDS/PAGE on a 12.5% gel, and labeled proteins were detected by fluorography. In panel **b**, the specificity of all antibodies used in this study was tested as described in **a**. The protein with a molecular weight (MW) of 16K was identified as VpreB, because it has the predicted MW of VpreB and coprecipitates with μ (*lanes 2, 3, 4, 5,* and *7*), as well as with λ5 chains (*lane 5*).

regardless of their V_H region, we isolated a μ-negative variant of F cells (Bine 4.8), transfected those cells with a μ gene that contains the V_H 17.2.25 region, which is a member of the V_HJ558 family and is different from V_H81X, and analyzed the transfectants by flow cytometry. We found that the transfectants (e.g., 2G4) express V_H17.2.25/μ chains on the cell surface (TABLE 1). We therefore conclude that F cells can transport V_H17.2.25/μ chains, but not V_H81X/μ chains, to the surface. One explanation for this might be that the latter μ chains do not associate with SL chains and are, therefore, trapped in the endoplasmic reticulum via the heavy-chain binding protein, BiP.[7] To determine whether V_H81X/μ chains associate with SL, we metabolically labeled proteins in F cells with [^{35}S]methionine, prepared cellular extracts, and immunoprecipitated labeled proteins from cell lysates with monoclonal anti-μ and anti-λ5 antibodies. When we analyzed the solubilized proteins of F cells by SDS/PAGE under reducing conditions, we found that anti-μ antibodies precipitated only μ chains and VpreB but not λ5 (FIG. 1, *lane 4*), whereas anti-λ5 antibodies precipitated VpreB, λ5, and a few μ chains

(FIG. 1, *lane 5*), which might be due to a weak, noncovalent interaction of μ/V_H81X chains with VpreB/λ5 complexes. We obtained the same result when we analyzed two other pre-B cell lines, BFL 23.4 and BFL 9.3, that express μ/V_H81X chains (TABLE 1). In contrast, when we analyzed the surface μ-positive pre-B cell line 107.2, we found that anti-μ (FIG. 1, *lane 2*) and anti-λ5 antibodies (FIG. 1, *lane 3*) precipitated μ chains, λ5, and VpreB from cellular lysates. From these results, we conclude that V_H81X/μ chains are not expressed on the surface of pre-B cells, because V_H81X regions prevent the association of μ chains with λ5, which traps the μ chain within the cell. We speculate further that pre-B cells that express these μ chains do not receive a positive signal and, therefore, do not differentiate into mature B cells.

REFERENCES

1. MELCHERS, F., *et al.* 1993. Immunol. Today **14:** 60–69.
2. KITAMURA, D. & K. RAJEWSKI. 1991. Nature **350:** 423–426.
3. KITAMURA, D., *et al.* 1992. Cell **69:** 823–831.
4. LI *et al.* 1993. J. Exp. Med. **178:** 951–960.
5. WU *et al.* 1990. Semin. Immunol. **2:** 207–216.
6. DECKER, D. J., N. E. BOYLE & N. R. KLINMAN. 1991. J. Immunol. **147:** 1406–1411.
7. HAAS, I. & M. WABL. 1983. Nature **306:** 387–389.

A Complete Map of the Human Immunoglobulin V_H Locus

IAN M. TOMLINSON,[a,b] GRAHAM P. COOK,[a,b]
GERALD WALTER,[a,b] NIGEL P. CARTER,[c]
HAROLD RIETHMAN,[d] LAKJAYA BULUWELA,[b]
TERENCE H. RABBITTS,[b] AND GREG WINTER[a,b]

[a]MRC Centre for Protein Engineering
[b]MRC Laboratory of Molecular Biology
[c]Cambridge University Department of Pathology
Cambridge, United Kingdom
and
[d]The Wistar Institute
Philadelphia, Pennsylvania

The structural diversity required for antigen recognition is generated during B-cell development by the combinatorial rearrangement of a relatively small number of basic building blocks: V_H (variable), D (diversity), and J_H (joining) segments for the heavy chain and V_L and J_L segments for the light chain.[1] The rearranged genes are further diversified during an immune response by somatic mutation.[2]

Using the polymerase chain reaction, we have amplified, cloned, and sequenced 88 V_H segments from the genomic DNA of a single person[3] (I.M.T., unpublished results). Twenty of these segments are located on chromosomes 15q11.2 and 16p11.2, outside the functional locus on chromosome 14.[4] A further 54 segments could be located on recently published maps of the J_H-proximal region of the V_H locus on chromosome 14q32.3.[5–7] The remaining segments, several of which are known to be functional, are presumably located at the telomeric end of the V_H locus.

Analysis of the telomeric region of chromosome 14q has enabled us to complete a map of the human immunoglobulin V_H locus that locates almost all V_H segments known to rearrange in B-lymphocytes[8] (FIG. 1). The human germline V_H repertoire consists of approximately 50 functional V_H segments, the exact number depending on the haplotype, spanning 1100 kb upstream of the J_H segments. Functional and nonfunctional members of each of the seven V_H families are interspersed throughout the locus. The limited structural diversity encoded by the functional V_H segments[3,9] demonstrates the importance of combinatorial diversity produced by VDJ joining, by association with different light chains, and by somatic mutation. The map also provides a framework to which specific haplotypes can be assigned.

In contrast to the limited sequence polymorphism within the human V_H locus (I.M.T., unpublished results), several functional segments have been shown to be associated with insertion/deletion polymorphisms,[5,7,8,10] suggesting that certain persons will have "holes" in their germline repertoire, while others may encode an abundance of cross-reactive or anti-self antibodies. The fact that some of the alternative haplotypes are rather rare (for example, the large insertion towards the telomeric end of the locus is only present in 28% of Caucasian haplotypes[8,10]) suggests a possible association with autoimmune disease susceptibility.

We have compiled a comprehensive directory of germline heavy and light

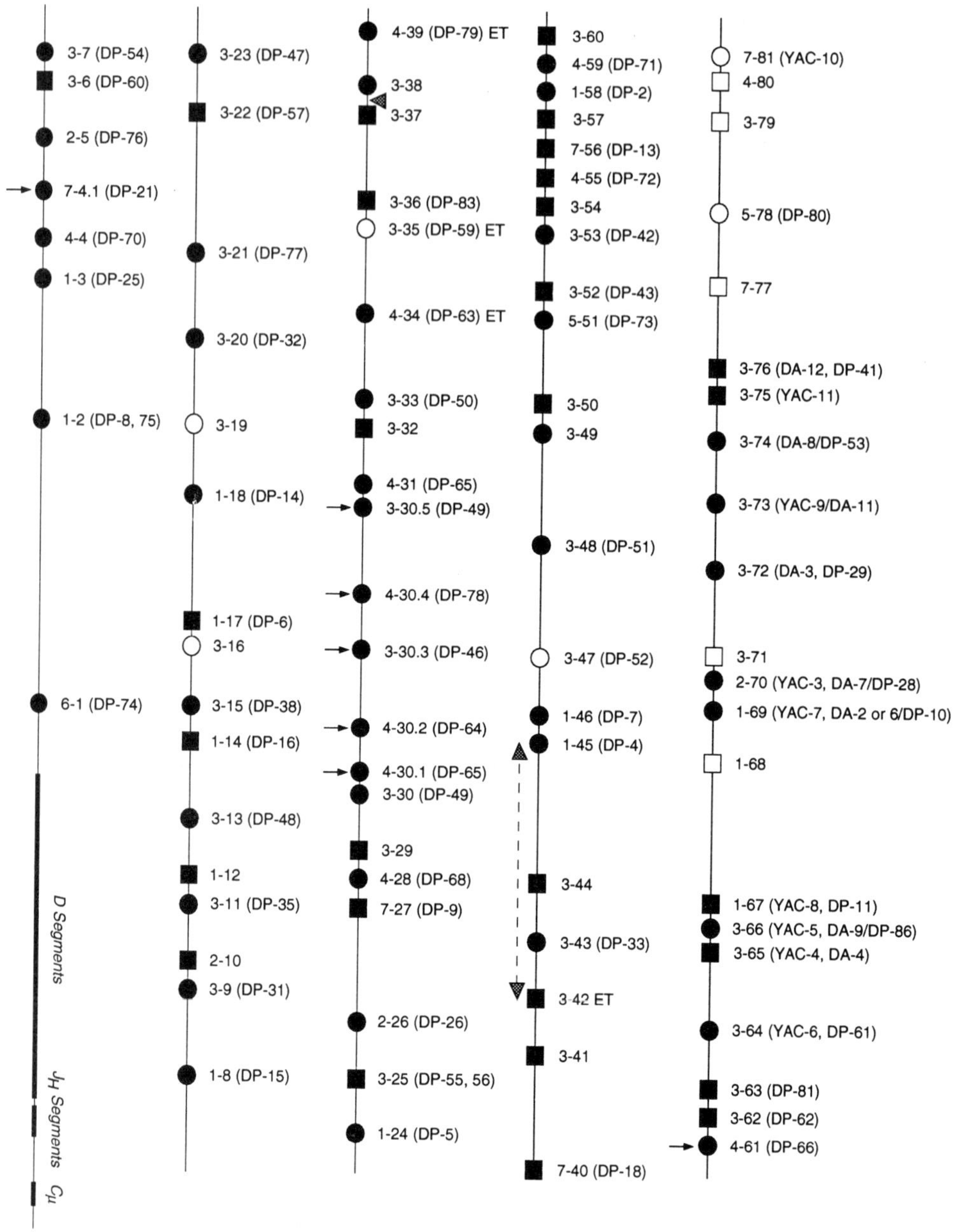

3-7 (DP-54)
3-6 (DP-60)
2-5 (DP-76)
7-4.1 (DP-21)
4-4 (DP-70)
1-3 (DP-25)
1-2 (DP-8, 75)
6-1 (DP-74)
D Segments
J_H Segments
C_μ
3-23 (DP-47)
3-22 (DP-57)
3-21 (DP-77)
3-20 (DP-32)
3-19
1-18 (DP-14)
1-17 (DP-6)
3-16
3-15 (DP-38)
1-14 (DP-16)
3-13 (DP-48)
1-12
3-11 (DP-35)
2-10
3-9 (DP-31)
1-8 (DP-15)
4-39 (DP-79) ET
3-38
3-37
3-36 (DP-83)
3-35 (DP-59) ET
4-34 (DP-63) ET
3-33 (DP-50)
3-32
4-31 (DP-65)
3-30.5 (DP-49)
4-30.4 (DP-78)
3-30.3 (DP-46)
4-30.2 (DP-64)
4-30.1 (DP-65)
3-30 (DP-49)
3-29
4-28 (DP-68)
7-27 (DP-9)
2-26 (DP-26)
3-25 (DP-55, 56)
1-24 (DP-5)
3-60
4-59 (DP-71)
1-58 (DP-2)
3-57
7-56 (DP-13)
4-55 (DP-72)
3-54
3-53 (DP-42)
3-52 (DP-43)
5-51 (DP-73)
3-50
3-49
3-48 (DP-51)
3-47 (DP-52)
1-46 (DP-7)
1-45 (DP-4)
3-44
3-43 (DP-33)
3 42 ET
3-41
7-40 (DP-18)
7-81 (YAC-10)
4-80
3-79
5-78 (DP-80)
7-77
3-76 (DA-12, DP-41)
3-75 (YAC-11)
3-74 (DA-8/DP-53)
3-73 (YAC-9/DA-11)
3-72 (DA-3, DP-29)
3-71
2-70 (YAC-3, DA-7/DP-28)
1-69 (YAC-7, DA-2 or 6/DP-10)
1-68
1-67 (YAC-8, DP-11)
3-66 (YAC-5, DA-9/DP-86)
3-65 (YAC-4, DA-4)
3-64 (YAC-6, DP-61)
3-63 (DP-81)
3-62 (DP-62)
4-61 (DP-66)

FIGURE 1. A map of the human V_H locus on chromosome 14. V_H segments are numbered according to the nomenclature of Matsuda *et al.*[6] Data from Shin *et al.*,[5] Matsuda *et al.*,[6] Walter *et al.*,[7] and Cook *et al.*[8] Forty-seven V_H segments are classified as functional because they contain an open reading frame and have been observed in VDJ rearrangements (*filled circles*); six V_H segments have open reading frames but have not been seen in VDJ rearrangements (*open circles*); twenty-nine segments are pseudogenes (*filled squares*); and five segments are unsequenced (*open squares*). Note that the segment 1–58 (DP-2) has now been seen rearranged in the donor DP and is therefore considered functional. Mapped insertion polymorphisms are shown with arrows (Refs. 5, 7 and I.M.T., unpublished results). Each line corresponds to approximately 210 kb; the total distance between 7–81 and J_H is approximately 1000 kb. The exons, ET_1 and ET_2, which splice onto V_H transcripts in T cells (replacing the V_H leader exon), were positioned using cosmid clones from genomic DNA (ET_1, triangle between 3–37 and 3–38; ET_2, between arrows, from 3–42 to 1–45). Segments whose transcripts have been observed spliced to ET in T cells are indicated (ET). Insertion/deletion polymorphisms contain at least six additional segments (four of which are functional) not shown on this map. These include four segments from the derivative 8 translocation product in the "Daudi" cell line, which contains the telomeric end of the V_H locus[8] and also the V_H segments VH32 and V_H^{SP}/DP-67 (see Cook *et al.*[8]). (Reproduced from Cook *et al.*[8]; used with permission from *Nature Genetics*.)

chain V gene sequences (V BASE), which includes our data[3,4,7,8,11,12] and data gathered from the literature, so that rearranged genes can be assigned germline counterparts, thereby enabling V gene usage and somatic mutation to be assessed. In addition, by helping to identify the position of the V-(D)-J join and the extent of N- and P- segment addition, it should facilitate investigation of the mechanisms involved in recombination. These factors may be particularly important when constructing antibody repertoires for use in phage display libraries or in transgenic animals.

REFERENCES

1. Tonegawa, S. 1983. Somatic generation of antibody diversity. Nature **302:** 575–581.
2. Berek, C. & C. Milstein. 1988. The dynamic nature of the antibody repertoire. Immunol. Rev. **105:** 5–26.
3. Tomlinson, I. M., G. Walter, J. D. Marks, M. B. Llewelyn & G. Winter. 1992. The repertoire of human germline V_H segments reveals about fifty groups of V_H segments with different hypervariable loops. J. Mol. Biol. **227:** 776–798.
4. Tomlinson, I. M., G. P. Cook, N. P. Carter, R. Elaswarapui, S. Smith, G. Walter, L. Buluwela, T. H. Rabbitts & G. Winter. 1994. Human immunoglobulin V_H and D segments on chromosomes 15q11.2 and 16p11.2. Hum. Mol. Genet. **3:** 853–860.
5. Shin, E. K., F. Matsuda, H. Nagaoka, Y. Fukita, T. Imai, K. Yokoyama, E. Soeda & T. Honjo. 1991. Physical map of the 3′ region of the human immunoglobulin heavy chain locus: Clustering of autoantibody-related variable segments in one haplotype. EMBO J. **10:** 3641–3645.
6. Matsuda, F., E. K. Shin, H. Nagaoka, R. Matsumara, M. Haino, Y. Fukita, S. Taka-Ishi, T. Imai, J. H. Riley, R. Anand, E. Soeda & T. Honjo. 1993. Structure and physical map of 64 variable segments in the 3′ 0.8-megabase region of the human immunoglobulin heavy-chain locus. Nature Genet. **3:** 88–94.
7. Walter, G., I. M. Tomlinson, G. P. Cook, G. Winter, T. H. Rabbitts & P. H. Dear. 1993. HAPPY mapping of a YAC reveals alternative haplotypes in the human immunoglobulin V_H locus. Nucleic Acids Res. **21:** 4524–4529.

8. COOK, G. P., I. M. TOMLINSON, G. WALTER, H. RIETHMAN, N. P. CARTER, L. BULUWELA, G. WINTER & T. H. RABBITTS. 1994. A map of the human immunoglobulin V_H locus completed by analysis of the telomeric region of chromosome 14q. Nature Genet. **7:** 162–168.
9. CHOTHIA, C., A. M. LESK, E. GHERARDI, I. M. TOMLINSON, G. WALTER, J. D. MARKS, M. B. LLEWELYN & G. WINTER. 1992. The structural repertoire of the human V_H segments. J. Mol. Biol. **227:** 799–817.
10. WALTER, M. A., U. SURTI, M. H. HOFKER & D. W. COX. 1990. The physical organisation of the human immunoglobulin heavy chain gene complex. EMBO J. **9:** 3303–3313.
11. WILLIAMS, S. C. & G. WINTER. 1993. Cloning and sequencing of human immunoglobulin V_λ gene segments. Eur. J. Immunol. **23:** 1456–1461.
12. COX, J. P. L., I. M. TOMLINSON & G. WINTER. 1994. A directory of human germline V_κ segments reveals a strong bias in their usage. Eur. J. Immunol. **24:** 827–836.

LIGM-DB/IMGT: An Integrated Database of Ig and TcR, Part of the Immunogenetics Database[a]

M.-P. LEFRANC,[b] V. GIUDICELLI,[b] C. BUSIN,[b]
A. MALIK,[c] I. MOUGENOT,[d] P. DÉHAIS,[d] D. CHAUME[e]

[b]*Laboratoire d'ImmunoGénétique Moléculaire*
LIGM, UMR 9942, IGMM
34033 Montpellier Cedex 1, France

[c]*EMBL Outstation*
The EBI
Cambridge, United Kingdom

[d]*Laboratoire d'Informatique, de Robotique et de Microélectronique*
LIRMM, UMR 9928
Montpellier, France

[e]*CNUSC*
Montpellier, France

An integrated immunogenetics database (LIGM-DB/IMGT) specializing in immunoglobulins (Ig) and T-cell receptors (TcR) is under development through collaboration between the EMBL Data Library, EBI (Cambridge, UK) and LIGM (Montpellier, France). LIGM-DB will combine Ig and TcR nucleotide and protein sequences, detailed annotation of these sequences, mapping data, and the results of comparative sequence analyses. It will provide the tools for automation of annotations, sequence classification, and modeling. This project has been started by the Laboratoire d'ImmunoGénétique Moléculaire (LIGM), in collaboration with the Laboratoire d'Informatique, de Robotique et de Microélectronique de Montpellier (LIRMM), which insures the informatic realization, and the Centre National Universitaire Sud de Calcul (CNUSC), which takes charge of the exploitation. Collaboration with ICRF in London (S. Marsh, J. Bodmer) will allow integration of HLA proteins and genes, and with IFG in Köln (W. Müller), integration of murine Ig alignments in the IMGT (ImmunoGenetics DataBase).

The main objectives of IMGT are the following:

- to provide immunologists with an integrated tool that will allow them to manage and analyze the complex information of interest to them;
- to provide a consistent view on all data presently available concerning immunoglobulins and T-cell receptors (LIGM-DB) and HLA (HLA-DB);
- to propose a basis for international discussion to establish standardization of data description in this field; and
- to provide mapping and genetics data useful in the search of candidate genes in diseases.

A user-friendly interface for LIGM-DB/IMGT has been developed according to biologists' requirements and ACeDB interface.

[a] IMGT is funded by the European Economic Community's BIOMED1 program and the Groupement de Recherches et d'Etudes sur les Génomes.

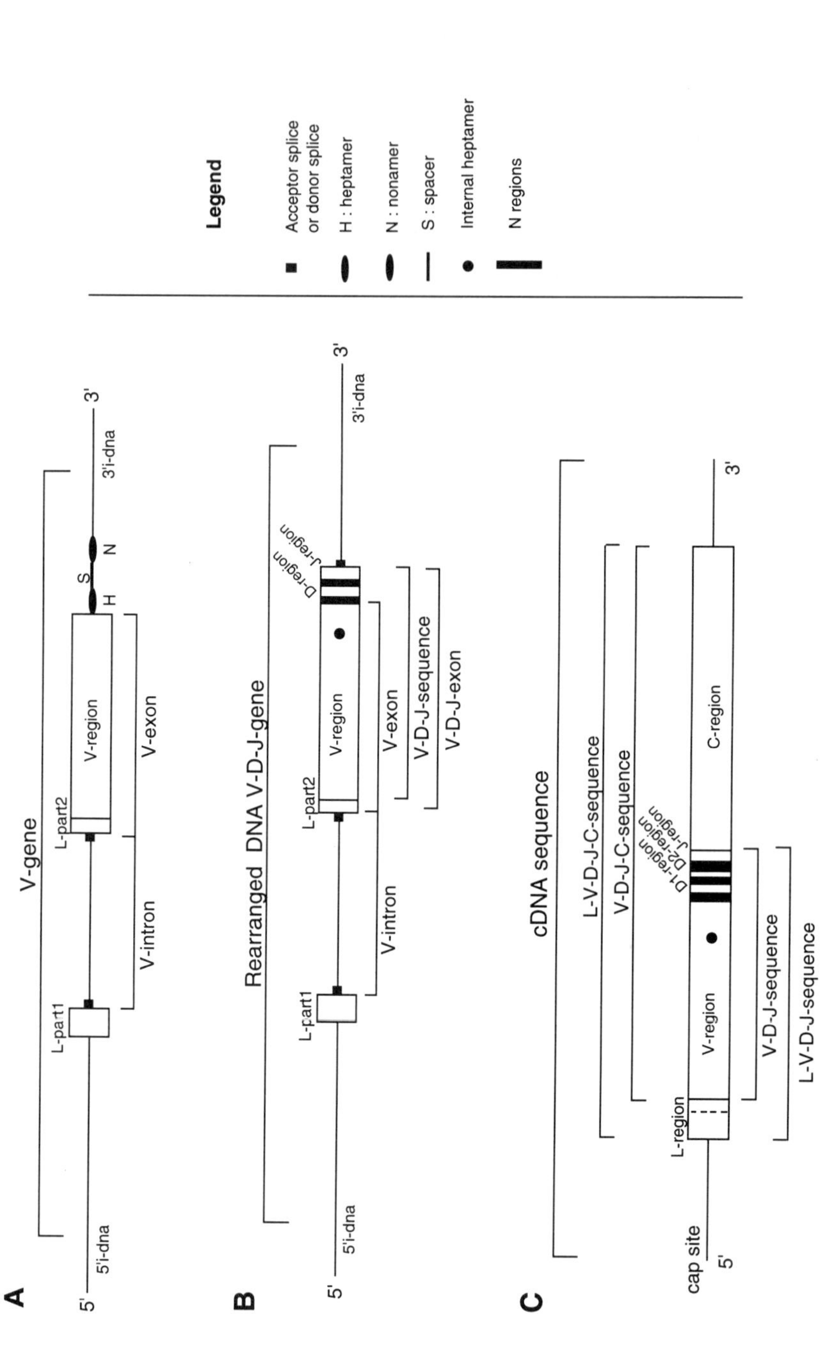

FIGURE 1. Schematic representation of the variable genes of the Ig and TcR with LIGM-DB labels: **(A)** germline DNA, **(B)** rearranged DNA, **(C)** complementary DNA.

TABLE 1. LIGM-DB Standardized Key Word List

General	Antigen receptor	Immunoglobulin superfamily	
Receptor type	**T cell receptor**	**Immunoglobulin**	**Chimeric**
Receptor class	TCR alpha-beta TCR gamma-delta	IgM, IgD, IgG, IgG1, IgG2, IgG2a, IgG2b, IgG3, IgG4, IgA, IgA1, IgA2, IgE	
Chain type	T cell alpha chain T cell delta chain T cell gamma chain T cell beta chain	immunoglobulin heavy chain immunoglobulin light chain immunoglobulin kappa chain immunoglobulin lambda chain immunoglobulin mu chain immunoglobulin delta chain immunoglobulin alpha chain immunoglobulin gamma chain immunoglobulin epsilon chain immunoglobulin surrogate chain immunoglobulin X chain immunoglobulin Y chain	
Region type	constant region alpha constant region beta constant region beta 1 constant region beta 2 constant region gamma constant region gamma 1 constant region gamma 2 constant region delta constant region	constant region alpha constant region delta constant region gamma constant region mu constant region epsilon constant region gamma 1 constant region gamma 2 constant region gamma 2a constant region gamma 2b constant region gamma 3 constant region gamma 4 constant region alpha 1 constant region alpha 2 constant region lambda-like constant region V pre B region	
	variable region diversity region joining region		

Particular key words *(not exhaustive)*

orphon;	processed gene;	pseudogene;	enhancer	silencer
repeat unit;	rheumatoid factor;	secreted immunoglobulin;	systemic lupus erythematosus;	
translocation;	sequence tagged site;	autoantibody;	cell surface glycoprotein;	glycoprotein;
leukemia;	heavy chain disease;	lymphoma;	aberrant;	tumor;
hybridom;	delta rec; . . .			

The procedure for entering new sequences into LIGM-DB/IMGT is the following: sequences are first sent to EMBL, following EMBL standard submission and using LIGM-DB key words (TABLE 1). If authors agree that their data can be made available in the database before they appear in print, EMBL will integrate the sequences to LIGM-DB, where they will be annotated (FIG. 1). If authors prefer to wait for publication, sequences will be integrated into LIGM-DB only when EMBL is authorized by them to release the sequences. Requirements for access to LIGM-DB are an Internet network connection and a X-terminal or emulation.

Polymorphism and Utilization of Human V_H Genes[a]

ERIC C. B. MILNER, WENDY O. HUFNAGLE, ANNUSKA M. GLAS, IVY SUZUKI, AND CATHERINE ALEXANDER

Virginia Mason Research Center and Department of Immunology University of Washington Seattle, Washington

INTRODUCTION

The completion of the physical map of the V_H locus[1,2] has answered many questions and has raised a set of new ones. To a large extent, the issue now is to delineate allelic variation among V_H genes and to determine the degree to which variation (or polymorphism) of V_H genes contributes to differences in V gene utilization or to disease susceptibility or resistance. Polymorphic variation within the V_H locus takes two fundamental forms. The first is conventional allelic variation in coding region sequences. Putative alleles of a number of genes have been identified by sequence analysis of the same gene isolated from different individuals. The second form of polymorphism and the more problematic to analyze is the result of numerous duplications, insertions, and deletions. This phenomenon leads to the unusual situation in which clearly distinct genes derived by duplication may be absolutely identical but may differ from their respective alleles by one or more nucleotide substitutions.

It is of considerable interest and importance to delineate the alleles of each V_H gene as well as the haplotype structure including the insertion deletion polymorphisms. In the context of understanding the expressed antibody repertoire, it will be important to know and understand the germline repertoire. To address these issues, we have focused on the identification of specific genes and the discrimination between closely related genes and alleles. We have used synthetic oligonucleotides corresponding to human V_H sequences to analyze sequence variation in human genomic DNA. By employing conditions under which single nucleotide mismatches abrogate hybridization, it has been possible to unambiguously identify individual germline V_H genes and to identify variation in closely related germline genes. We have demonstrated that analysis of the human V_H locus with sequence-specific oligonucleotide probes reveals hybridization patterns of limited complexity and striking heterogeneity.[3,4] Within regions (FR1, CDR1, CDR2) different oligonucleotide probes are generally "mutually exclusive" by virtue of the nucleotide substitutions that distinguish the probes. Thus, a sequence detected by a given probe will not be detected by any mutually exclusive probe directed to the homolo-

[a] This research was supported in part by National Institutes of Health Grants AR39918, AI31241, and American Cancer Society Grant IM485.

[b] Address for correspondence: Eric C. B. Milner, Virginia Mason Research Center, 1000 Seneca St., Seattle, WA 98101.

gous region of a different V_H gene segment. We have found that the detection of specific sequence motifs by hybridization of restriction enzyme digested DNA in a modified Southern analysis using short oligonucleotide probes allows the specific identification of germline genes.[3–7] Based on data obtained from genomic hybridization using such probes, it is possible to assign individual genes to specific restriction fragments. By this approach, germline sequences of multiple individuals can be compared with a very high degree of precision.

POLYMORPHISM OF THE V_H4 GENE FAMILY

A large number of germline sequences of V_H4 genes have now been determined.[6,8–11] When the set of sequences is viewed as a whole, one striking feature of the heterogeneity that emerges is that most differences between any pair of sequences can be best described as the sharing or nonsharing of common conserved sequence motifs.[6] The V_H4 gene family could be resolved into 12 loci, 11 of which were located on chromosome 14 within the V_H locus, and one located on chromosome 15.[2,6,12,13] Hybridization with a full-length V_H4 probe revealed 14 *Bgl*II restriction fragments[12] (FIG. 1). Individual elements could be assigned to the *Bgl*II restriction fragments by sequence-specific oligonucleotide probe hybridization[6] (FIG. 2). The various V_H4 loci and the corresponding *Bgl*II bands are cata-

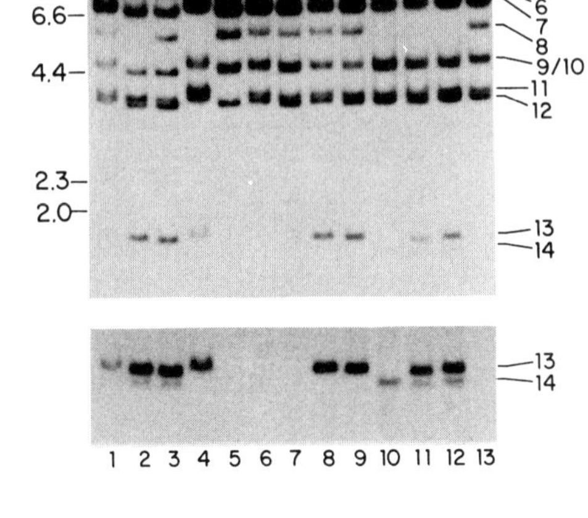

FIGURE 1. Detection of V_H4 RFLP. DNA extracted from PBL derived from members of three families was digested with *Bgl*II, run on 1% agarose, and transferred to nylon membrane. Probing, washing, and exposure were performed as described in methods using the full-length V_H4 probe. Bands are numbered from higher to lower MW. A poorly resolved triplet at band 9 is assigned both 9 and 10. Two rare fragments, designated 1L and 6H, migrating just below and just above bands 1 and 6, are not present in the persons in this figure. Note that migration of DNA in lanes 1 and 4 is slightly retarded relative to the rest of the gel. Slow migration is characteristic of these DNA preparations and can be seen in lanes 1 and 4, FIGURE 3, as well. The *upper panel* is an 18-hr exposure; the *lower panel* is a three-day exposure included to enhance visibility of bands 13 and 14. Band 1, lane 4 is poorly visible in this figure but was easily visible on the original film. (Reprinted from Willems van Dijk *et al.*[12] with permission from *The Journal of Immunology*, © 1991, *The Journal of Immunology*.)

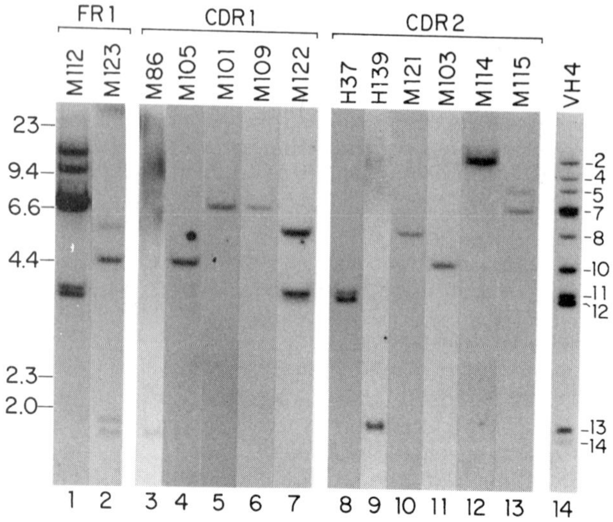

FIGURE 2. Sequence-specific hybridization of V_H4-specific oligonucleotide probes. Genomic DNA derived from donor 3116 was digested with *Bgl*II and hybridized to the indicated oligonucleotide probes (*lanes 1–13*) or a cloned V_H4 probe (*lane 14*). Compensation for differences in migration distance on the different gels used to generate this figure was accomplished by photographic enlargement or reduction to align the molecular weight markers. All hybridizations were performed at least twice, and comigration of elements was assessed by sequential hybridization of single gels. (Reprinted from van der Maarel *et al.*[6] with permission from *The Journal of Immunology*, © 1993, *The Journal of Immunology*.)

logued in TABLE 1. In the person studied, 9 of 11 loci were diploid, and two were single-copy insertion polymorphisms.[6,12,13] Alleles of the diploid locus, V4-61, could be discriminated by an RFLP as *Bgl* bands 8 and 11.[6] The remaining diploid loci comigrated. By sequence analysis and by sequence-specific hybridization, alleles of loci V4-28, carried on *Bgl* band 13, and V4-31, carried on *Bgl* band 10, could be discriminated. *Bgl* band 10 in this person carried two loci, V4-30.4 and V4-31. *Bgl* band 7 carried two chromosome 14 loci, V4-34 and V4-55, as well as the chromosome 15 locus. The alleles of V4-28 (band 13) differed by four nucleotides, resulting in two amino acid interchanges, a glutamine for an aspartic acid in FR1, and an isoleucine for a threonine in CDR2. The alleles of V4-31 differed by a single nucleotide in the first codon of framework 3, resulting in a leucine substitution for the conserved arginine at the beginning of framework 3. The apparent alleles of V4-61 (band 11 and band 8) differed profoundly. The band 8 allele differed from the band 11 allele by 11 nucleotides resulting in five amino acid substitutions, a glutamine for a glutamic acid in FR1, an isoleucine for a valine in framework 1, alanine for a proline in FR2, and an arginine for a tyrosine and a threonine for a tyrosine in CDR2. It is possible that band 11 and band 8 are not true alleles, but rather represent insertion deletion polymorphisms that map to the same location.

It has been reported on the basis of familial segregation that *Bgl* bands 1 and 6 were allelic,[14] although a population analysis did not suggest this.[12] Neither of these bands were present in the donor analyzed in the current studies. By sequence-specific hybridization analysis of a population of 75 unrelated subjects, band 6

TABLE 1. Prevalence of V_H4 Genes

Band[a]	kb	Gene[b]		Oligonucleotide Probes[c] FR1	CDR1	CDR2	FR3	Prevalence[d]	Allele Freq.[e]
1	22	V4-4?[f]		M112?	H125, E7?	M121		61	38
2	14	V4-39	4d154	M112	M99	M114		96	80
3	11.5	nd		H125?				23	12
4	9.4	V4-80		M112	M69	H139		100	
5	8.2	V4-4b	4d68	E6, M110	E1	M115		85	61
6	7.5	V_H^{SP}				M113		51	30
7	6.7	V4-34	4d76	M112	M109	M108		100	
7		V4-55	4d64	M112	M101	M102		100	
7		CH15	4d255	M112	M69	M115		100	
8	5.6	V4-61b	3d279d	M123, M67	M122	M121		36	20
9	4.5	V4-28c		M111	M71	H139		79	54
10	4.5	V4-31	3d75d	M123	M98	M103	E23	70	45
10		V4-31b	3d277d	M123	M98	M103	E24		
10		V4-30.4	3d230d	M123	M105	M103		70	45
11	3.7	V4-61	nc11	M112, M67	E13, M122	M120		83	59
12	3.6	V4-59	3d197d	M112, M67	E7, H125	M120		100	
13	1.8	V4-28	3d28d	M111	M100	H139		37	21
13		V4-28b	3d24d	M123	M100	H139		35	19
14	1.7	V4-30.2	3d216d	M123	M86	M87		69	44

[a] Numbering of bands corresponding to restriction fragments is according to FIGURE 1.

[b] Nomenclature from Refs. 1, 2, and 6.

[c] Probes are listed in 5′ to 3′ orientation. Some probes, indicated by (AS), are oriented in the antisense direction with respect to the coding sequence of the V_H gene segment. E1, CTGGCGGACCCAACTCCACCA (AS); E6, AAGCCTCCGGGGACCCTGTCC; E7, TCCATCAGTAGTTACTACTGG; E8, TCCATCAGCAGTGGTAGTTAC; E13, TCCGTCAGCAGTGGTAGTTAC; E23, AAGAGTCGAGTTACCATATCA; E24, AAGAGTCTAGTTACCATATCA; H37, TATATCTATTACAGTGGGAGCACCAACTAC; H125, AGTAGTTACTACTGGAGCTGG; H139, TACATCTATTATAGTGGGAGC; M69, TGGTGGAGCTGGGTCCGCCAG; M71, TGGTGGGGCTGGATCCGGCAG; M86, GGTGGTTACTCCTGGAGCTGG; M87, GCTCCCACTATGATAGATGTA (AS); M98, AGTGGTGGTTACTACTGGAGC; M99, AGTAGTAGTTACTACTGGGGC; M100, AGCAGTAGTAACTGGTGGGGC; M101, AGCAGTGGTAACTGGTGAATC; M102, GCTCCCACTATGATGGATTTC (AS); M103, GCTCCCACTGTAATAGATGTA (AS); M105, AGTGGTGATTACTACTGGAGT; M108, GCTTCCACTATGATTGATTTC (AS); M109, AGTGGTTACTACTGGAGCTGG; M110, AAGCCTTCGGGGACCCTGTCC; M111, AAGCCTTCGGACACCCTGTCC; M112, AAGCCTTCGGAGACCCTGTCC; M113, GCTCCCACTATGATAGATACT (AS); M114, GCTCCCACTATAATAGATACT (AS); M115, GCTCCCACTATGATAGATTTC (AS); M120, GCTCCCACTGTAATAGATATA (AS); M121, GCTCCCACTGGTATAGATACG (AS); M122, GGTAGTTACTACTGGAGCTGG; M123, GGACAGGGTCTGTGAAGGCTT (AS).

[d] The presence of the indicated bands in DNA samples from 75 unrelated persons was determined by hybridization to the cloned V_H4 probe 58p2. Band 9 was identified by sequence-specific probe H139 and represents a subset of band 9/10. Prevalences of individual genes detected by sequence-specific probes was, in some cases, assessed on less than the full panel. The six parents of the families shown in FIGURE 1 were included in the panel.

[e] Allele frequencies were calculated from prevalences and assume all alleles are in Hardy-Weinberg equilibrium.

[f] ? indicates weak hybridization that may indicate a nucleotide substitution from the sequence of the indicated probe.

was found to carry the sequence of the $V_H{}^{SP}$ germline gene. In rare individuals a duplication of the $V_H{}^{SP}$ gene appears to have occurred, and, when present, the duplicated $V_H{}^{SP}$ gene is carried on band 7. The $V_H{}^{SP}$ gene has not been formally assigned to one of the known loci but appears to map in the vicinity of V4-39.[9,15] By sequence-specific hybridization, band 1 was not found to carry the sequence of $V_H{}^{SP}$ in any person tested. By hybridization analysis the sequence carried in band 1 corresponds to the V4-4 sequence. The V4-4 sequence is one codon shorter in CDR1 than V4-4b, and differs from V4-4b by 21 nucleotides. These nucleotide substitutions result in 12 amino acid interchanges.

No sequence has yet been assigned to band 3. In hybridization analysis band 3 does not hybridize to the commonly used oligonucleotide probes. The prevalence in the population of the *Bgl*II bands carrying V_H4 genes is shown in TABLE 1. Although the genes carried in band 6 and band 3 have not been mapped, they were found to be in significant linkage disequilibrium, suggesting that they are close together.[12]

POLYMORPHISM OF V_H3 GENES

At least 48 V_H3 genes have been mapped to the V_H locus.[1,2,14] Slightly more than half of these have been found to be pseudogenes.[1,2] Variation among the V_H3 genes is similar to that seen among the V_H4 genes but is somewhat more difficult to delineate because of the larger number of elements. By assessing the occurrence of specific sequence motifs using oligonucleotide probes, however, it has been possible to identify many of the V_H3 genes and to detect variation in their occurrence. We have analyzed the occurrence of several elements mapping in the vicinity of, or allelic to, V3-30 (designated 1.9III, 56p1, hv3005).[4,5,7] Hybridization analysis of these elements is shown in FIGURE 3. The results shown in FIGURE 3 in conjunction with a population analysis[4] and deletion mapping studies[5] indicate that all three elements cannot be alleles of the same locus. Calculations based on the prevalence of the individual elements 1.9III, 56p1, and hv3005, and considering the known haplotype structures, indicate that 1.9III (V3-30) and hv3005 (V3-30b) are appropriately considered to be alleles and 56p1 (V3-30.3) an insertion/deletion element. Allele frequencies for hv3005 and 1.9III are, respectively, 0.19 and 0.72. Because these allele frequencies do not add up to one, it seems likely that an additional haplotype or haplotypes with an allele frequency of 0.08 that lacks both 1.9III and hv3005 occurs (i.e., a deletion of V3-30). No additional allelic sequence has been identified, however, suggesting that this putative allele is a blank haplotype. The putative third allele is definitely not 56p1, which in our analysis always occurs on a haplotype carrying one or two copies of 1.9III.

We found that a 50-kb insertion/deletion polymorphism profoundly affected the haplotype structure and allelic relationships in this region.[5] The 50-kb insertion that carries 56p1 and the second copy of 1.9III has been localized to the interval between V3-30 and V4-31. There is some evidence for heterogeneity in the exact break points among the population. Using a probe adjacent to one of the *Sfi*I sites that border the insertion, we have found that approximately 93% of the population carry the insertion, an allele frequency of 0.73. At least six haplotypes can be identified in this region, four of which include the 50-kb insertion (FIG. 4). The occurrence of additional haplotypes is likely.

A second duplication and insertion/deletion involves the element V3-23 (also known as 30p1[16] or V_H26[17]). All persons thus far tested carry a copy of V3-23 on a 5-kb *Taq*I fragment. Approximately 40% of the population carries a duplicated

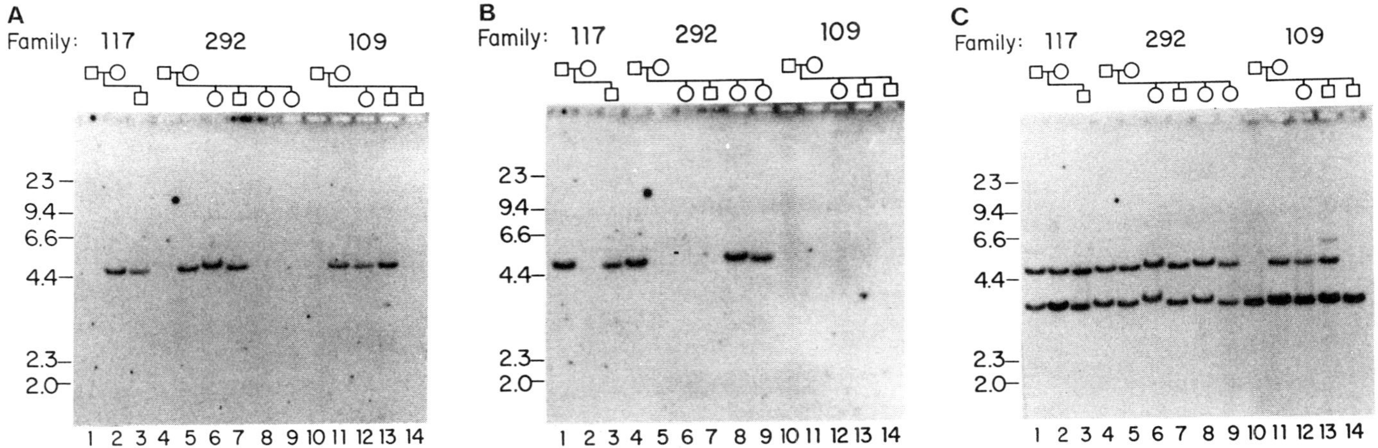

FIGURE 3. Hybridization analysis of V3-30.3 (56p1), V3-30b (hv3005), and V3-30 (1.9III). Samples of genomic DNA from Virginia Mason Research Center families 117, 292, and 109 were digested with *Taq*I and separated on an agarose gel. The gel was then dried and sequentially hybridized with the V3-30.3-specific probe M19 **(A)**; the V3-30b-specific probe M41 **(B)**; and a probe, H61, that detects V3-30, V3-30b, and V3-30.3 **(C)**. (Modified from Sasso *et al.*[7] with permission from *The Journal of Immunology*, © 1990, *The Journal of Immunology*.)

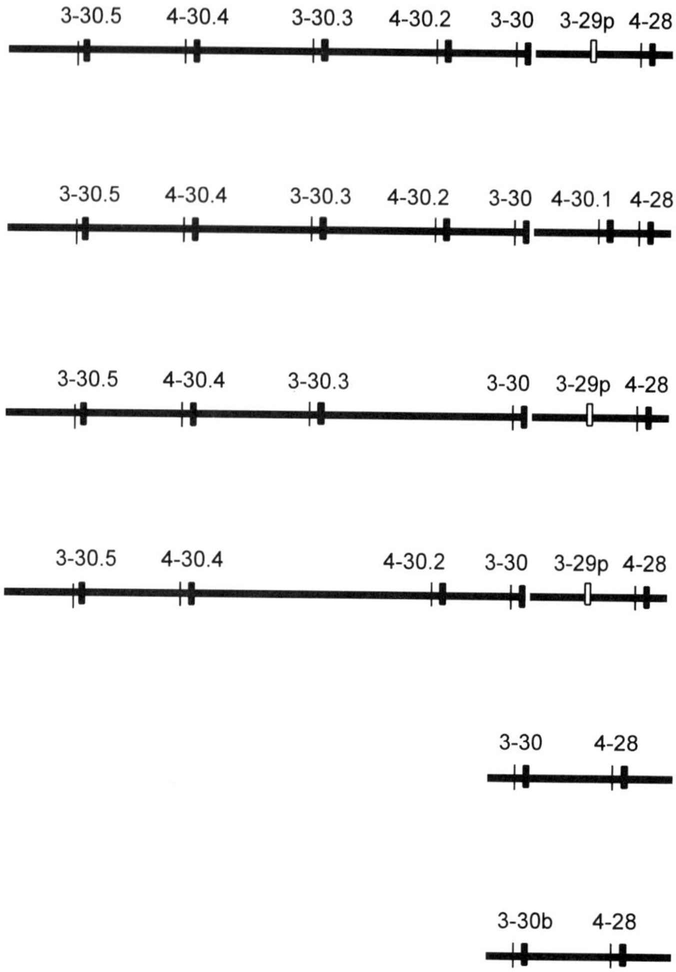

FIGURE 4. Haplotype structures in the region between V4-28 and V3-31. Putative haplotypes were deduced from mapping studies[1,2,5,7,13] and oligonucleotide probe analysis.[4,5,7,13]

element on a 7-kb *Taq*I fragment. Phosphorimage analysis of band intensities indicate that the 5-kb band may carry alleles of both loci, only one of which is allelic to the 7-kb band. By deletional analysis in B-cell lines, both of the V3-23 loci map to the same interval, suggesting that they may be tandemly arrayed.

Yet another V_H3 duplication involves V3-53, (also known as 60p2[16]). The V3-53 locus occurs on two sets of polymorphic *Taq*I fragments. The more 3′ locus, which we have mapped by deletional analysis to the same interval as V3-53, is carried on *Taq*I fragments of approximately 3 kb and 4 kb. The second locus that maps 5′ to the known V3-53 locus is carried on *Taq*I fragments of approximately 12 kb and 15 kb. No nucleotide or amino acid substitutions have as yet been associated with these RFLPs.

V_H1 POLYMORPHISM

The most complex allelic relationships that we have identified belong to the V1-69 locus. On the basis of sequence-specific hybridization, RFLP, or both, 13 variants that map to V1-69 are identified[18] (FIG. 5). These variants encode seven distinct amino acid sequences. Only five codons are involved in the amino acid substitutions, one in CDR1, three in CDR2, and one in FR3 (FIG. 5). At each codon, a single amino acid interchange may be observed. These are a threonine for an alanine at position 33 in CDR1; an arginine for a glycine at position 50, a leucine for a phenylalanine at position 55, and an isoleucine for a threonine at position 57, all in CDR2; and a lysine for a glutamic acid at position 73 in FR3.[18] Three haplotypes have been identified that carry duplications of the V1-69 gene. The most common duplication pairs variants 1 and 7 on a haplotype that occurs with a prevalence of 0.46 and a calculated allele frequency of 0.265 (FIG. 5). In the second duplication, variants 1 and 9 are paired on a haplotype that occurs with a prevalence of 0.04 and a calculated allele frequency of 0.0202. The third duplication pairs variants 2 and 6 on a haplotype that occurs with a prevalence of 0.02 and a calculated allele frequency of 0.0101. The sum of the allele frequencies for all variants, including duplications as single alleles, is 1.037, which is slightly greater than 1.0 and suggests that an additional duplication occurs that was undetected. The most likely duplication would be a haplotype pairing variants 5 and 11. Variant 11 occurred only in persons who carried variant 5, but in the population studied none of the three persons who carried variant 11 carried a third variant, as would be expected in the case of a duplication. However, these persons may have been homozygous for variant 5, because variant 5 occurs with a prevalence of 58% and variant 11 with a prevalence of just 6%. Taking into account this fourth putative duplication, the total allele frequencies equal 1.006, which is close to the expected value of 1.0.

HAPLOTYPE STRUCTURE

The existence of numerous insertion/deletion polymorphisms raises the possibility that different haplotypes of the V_H locus may carry substantially different coding information. In addition, we and others have found little evidence for linkage disequilibrium between polymorphic elements in the V_H locus. In several instances where linkage disequilibrium is evident, the elements involved have been physically closely linked, in several instances elements in linkage disequilibrium were carried on the same λ-phage clone placing them within about 15 kb of one another. In an analysis of 11 families in which 42 (of 44) haplotypes could be identified with respect to five loci, one of which is a VNTR located 5′ to the J_H region, 36 clearly distinct haplotypes could be identified. A group of three haplotypes were identical, and four haplotypes were ambiguous because they were uninformative at one or more loci (E. C. B. Milner and C. M. Alexander, unpublished observations). It seems likely that by analyzing additional markers, all 42 haplotypes would be found to be distinct. We have argued on the basis of similar observations that the haplotype structure in the population is extraordinarily complex.[4,6,12]

V_H EXPRESSION AND POLYMORPHISM

Given the degree of polymorphism and the complexity of the V_H locus, it becomes of considerable interest to know what influence these factors have on

Haplotypes

Variant	CDR1	CDR2	FR3	Variant	CDR1	CDR2	FR3
	A	G F T	E		A	G F T	E
1	-	- - -	-	7	-	- - -	K
1	-	- - -	-	9	-	R L I	K
2	-	- - -	-	6	-	R L -	-
5	-	- - -	-	11	-	- - -	-
				5	-	- - -	-
				7	-	- - -	K
				3	-	R L I	K
				4	-	R L I	K
				8	T	R L -	K
				10	T	R L I	K
				13	T	R L I	K
				12	-	- L I	K

Allele Frequency

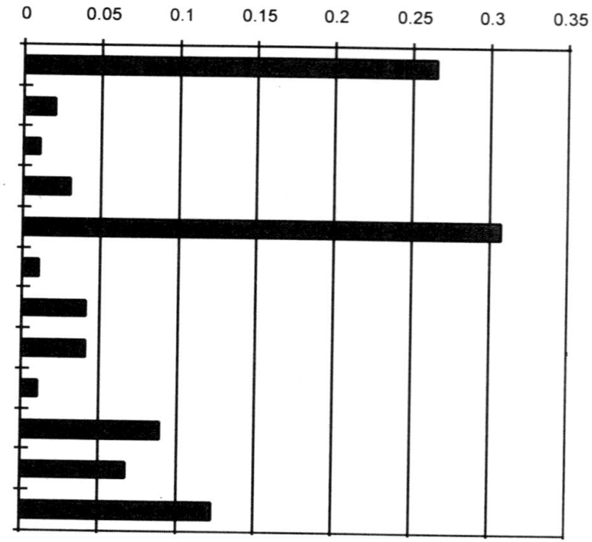

FIGURE 5. Allelic variants of V1-69. Amino acid sequences at variable sites in the V1-69 allelic set. The reference sequence is that of V1-69 (51p1),[16,18] and only differences are shown.

the composition of the antibody repertoire. It has been argued that the adult repertoire exhibits biased utilization of individual V_H gene segments. Notably, the V3-23 gene segment (also called 30p1, V_H26, or 18/2) and the V4-34 gene segment (also designated 4.21) are overexpressed relative to other genes.[19–22] We have found that thc rcpertoire in adult peripheral blood B cells exhibits intrinsic preferences for individual gene segments[23] (FIG. 6). The repertoires of different persons exhibit preferences that presumably reflect their V_H germline and genetic background, but which usually include V3-23 and V4-34 as predominant elements. These preferences could be the result of positive or negative selective processes occurring during B-cell differentiation, or could be the result of increased efficiency of rearrangement. In both V_H3 and V_H4 compartments, the genes represented most frequently among rearrangements, V3-23 and V4-34, are highly conserved elements that have not been found to be deleted in any subjects studied. The possibility that this reflects some evolutionary selection is uncertain. The nonfunctional gene, V4-55, has not been found to be deleted in any persons studied, and some elements that are highly expressed in persons possessing them have been found to be absent from the genome of other persons. An example of the latter is V4-31, and to a lesser extent V4-30.4, both of which are deleted in some persons, but which contribute significantly to rearrangements in persons possessing them.

One obvious insight made possible by the completion of the physical map of the V_H locus is that the total number of functional V_H genes is smaller than previously thought. Analyses of the expressed repertoire indicate that relatively few members of this, perhaps surprisingly, limited germline repertoire actually contribute to the expressed repertoire. In fact, we have found that fewer than eight V_H3 genes make major contributions to the repertoire (each gene accounting for 8 to 20% of rearrangements), whereas the remainder of functional V_H3 genes contribute at substantially lower levels (each gene accounting for 0 to 3% of

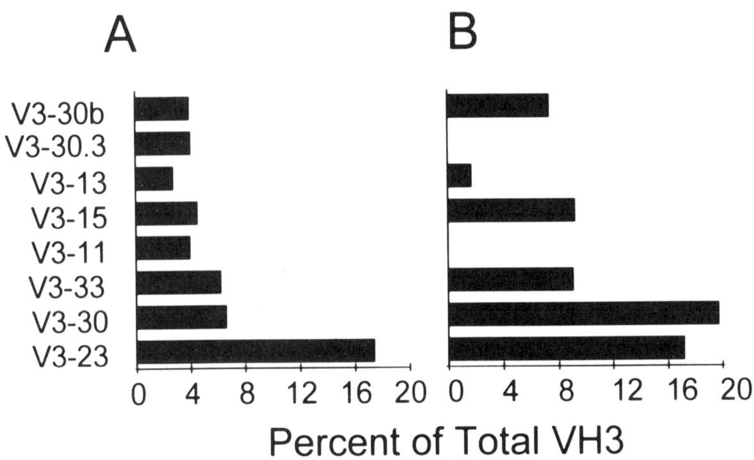

FIGURE 6. Utilization of V_H3 repertoire in B cells from two subjects. The incidence of representation of eight specific V_H3 gene segments in phagemid clones of quantitatively amplified rearranged V_H3 genes from peripheral blood B cells (Subject 3116, *panel A*; Subject 4882, *panel B*). In *panel A*, 662 clones were analyzed, and in *panel B*, 742 clones were analyzed.

rearrangements) (FIG. 6 and W. O. Hufnagle and E. C. B. Milner, unpublished observations). In a preliminary experiment in which both the germline copy number and the occurrence of rearrangement of individual V_H genes in peripheral blood B cells were determined, it was found that the number of rearrangements of the preferentially expressed genes was roughly proportional to the copy number of those genes. Knowing the influence of the V_H germline on the repertoire, it will now be of interest to determine the extent to which variation in the V_H germline contributes to variation in immune responsiveness.

SUMMARY

The human V_H germline repertoire comprises approximately 100 elements, which can be grouped into seven families based on nucleotide sequence similarity. Members of different families are interspersed throughout the complex, with limited sets of alleles identified for most loci. Linkage disequilibrium between most elements is weak. Variation within the population can be attributed to differences in nucleotide sequence between allelic genes as well as to differences in the number of genes present. Gene number per haplotype varies as a result of the common occurrence of insertion/deletion polymorphisms, which may be small, involving a single element, or may be extensive, involving four or five elements. In some cases, such polymorphisms may involve duplication of a functional V_H gene segment on some haplotypes and deletion of the gene on others. The resulting variation in germline composition of the V_H locus may have profound effects on V_H gene utilization.

ACKNOWLEDGMENTS

We thank former members of the laboratory, especially Drs. Ko Willems van Dijk and Eric H. Sasso, who have contributed immensely to these studies.

REFERENCES

1. MATSUDA, F., E. K. SHIN, H. NAGAOKA, R. MATSUMURA, M. HAINO, Y. FUKITA, S. TAKA-ISHI, T. IMAI, J. H. RILEY, R. ANAND, E. SOEDA & T. HONJO. 1993. Structure and physical map of 64 variable segments in the 3′ 0.8-megabase region of the human immunoglobulin heavy-chain locus. Nature Genet. **3:** 88–94.
2. COOK, G. P., I. M. TOMLINSON, G. WALTER, H. RIETHMAN, N. P. CARTER, L. BULUWELA, G. WINTER & T. H. RABBITTS. 1994. A map of the human immunoglobulin V_H locus completed by analysis of the telomeric region of chromosome 14q. Nature Genet. **7:** 162–168.
3. WILLEMS VAN DIJK, K., H. W. SCHROEDER, JR., R. M. PERLMUTTER & E. C. B. MILNER. 1989. Heterogeneity in the human immunoglobulin VH locus. J. Immunol. **142:** 2547–2554.
4. SASSO, E. H., K. WILLEMS VAN DIJK & E. C. B. MILNER. 1990. Prevalence and polymorphism of human V_H3 genes. J. Immunol. **145:** 2751–2757.
5. WILLEMS VAN DIJK, K., L. A. MILNER, E. H. SASSO & E. C. B. MILNER. 1992. Chromosomal organization of the heavy chain variable region gene segments comprising the human fetal antibody repertoire. Proc. Natl. Acad. Sci. USA **89:** 10430–10434.
6. VAN DER MAAREL, S. M., K. WILLEMS VAN DIJK, C. M. ALEXANDER, E. H. SASSO,

A. B. BULL & E. C. B. MILNER. 1993. Chromosomal organization of the human V_H4 gene family: Location of individual gene segments. J. Immunol. **150:** 2858–2868.

7. SASSO, E. H., K. WILLEMS VAN DIJK, A BULL, S. M. VAN DER MAAREL & E. C. B. MILNER. 1992. V_H genes in tandem array comprise a repeated germline motif. J. Immunol. **149:** 1230–1236.
8. SANZ, I., P. KELLY, C. WILLIAMS, S. SCHOLL, P. TUCKER & J. D. CAPRA. 1989. The smaller human V_H gene families display remarkably little polymorphism. EMBO J. **8:** 3741–3748.
9. BAER, R., K.-C. CHEN, S. D. SMITH & T. H. RABBITTS. 1985 Fusion of an immunoglobulin variable gene and a T cell receptor constant gene in the chromosome 14 inversion associated with T cell tumors. Cell **43:** 705–713.
10. WENG, N.-P., J. G. SNYDER, L.-Y. YU-LEE & D. M. MARCUS. 1992. Polymorphism of human immunoglobulin V_H4 germline genes. Eur. J. Immunol. **22:** 1075–1082.
11. VAN ES, J. H., M. HEUTINK, H. AANSTOOT & T. LOGTENBERG. 1992. Sequence analysis of members of the human Ig V_H4 gene family derived from a single V_H locus. Identification of novel germ-line members. J. Immunol. **149:** 492–497.
12. WILLEMS VAN DIJK, K., E. H. SASSO & E. C. B. MILNER. 1991. Polymorphism of the human immunoglobulin V_H4 gene family. J. Immunol. **146:** 3646–3651.
13. WILLEMS VAN DIJK, K., L. A. MILNER & E. C. B. MILNER. 1992. Mapping of human H chain V region genes (V_H4) using deletional analysis and pulsed field gel electrophoresis. J. Immunol. **148:** 2923–2931.
14. WALTER, M. A., U. SURTI, M. H. HOFKER & D. W. COX. 1990. The physical organization of the human immunoglobulin heavy chain gene complex. EMBO J. **9:** 3303–3313.
15. BAER, R., A. FORSTER & T. H. RABBITTS. 1987. The mechanism of chromosome 14 inversion in a human T cell lymphoma. Cell **50:** 97–105.
16. SCHROEDER, H. W., JR., J. L. HILLSON & R. M. PERLMUTTER. 1987. Early restriction of the human antibody repertoire. Science **238:** 791–793.
17. MATTHYSSENS, G. & T. H. RABBITTS. 1980. Structure and multiplicity of genes for the human immunoglobulin heavy chain variable region. Proc. Natl. Acad. Sci. USA **77:** 6561–6565.
18. SASSO, E. H., K. WILLEMS VAN DIJK, A. BULL & E. C. B. MILNER. 1993. A fetally expressed V_H1 gene belongs to a complex set of alleles. J. Clin. Invest. **91:** 2358–2367.
19. HUANG, C. & B. D. STOLLAR. 1993. A majority of Ig H chain cDNA of normal human adult blood lymphocytes resembles cDNA for fetal Ig and natural autoantibodies. J. Immunol. **151:** 5290–5300.
20. STEWART, A. K., C. HUANG, B. D. STOLLAR & R. S. SCHWARTZ. 1993. High-frequency representation of a single V_H gene in the expressed human B cell repertoire. J. Exp. Med. **177:** 409–418.
21. HUANG, C., A. K. STEWART, R. S. SCHWARTZ & B. D. STOLLAR. 1992. Immunoglobulin heavy chain gene expression in peripheral blood lymphocytes. J. Clin. Invest. **89:** 1331–1343.
22. PASCUAL, V., K. VICTOR, D. LELSZ, M. B. SPELLERBERG, T. J. HAMBLIN, K. M. THOMPSON, I. RANDEN, J. NATVIG, J. D. CAPRA & F. K. STEVENSON. 1991. Nucleotide sequence analysis of the V regions of two IgM cold agglutinins. Evidence that the VH4-21 gene segment is responsible for the major cross-reactive idiotype. J. Immunol. **146:** 4385–4391.
23. SUZUKI, I., L. PFISTER, A. GLAS, C. NOTTENBURG & E. C. B. MILNER. 1995. Representation of rearranged V_H gene segments in the human adult antibody repertoire. J. Immunol. In press.

Characterization and Genomic Mapping of a Novel Leader Peptide Associated with the Human V_H4–21 (V_H4–34) Gene Segment[a,b]

GEORGE F. WIDHOPF II,[c,d] VIRGINIA PASCUAL,[c]
RICHARD BAER,[c] AND J. DONALD CAPRA[c,e]

[c]Department of Microbiology and
[d]Graduate Program in Immunology
University of Texas Southwestern Medical Center
Dallas, Texas, 75235-9048

INTRODUCTION

The immunoglobulin molecule, the antigen-specific cell surface receptor of the B lymphocyte, comprises two pairs of identical polypeptide chains, the heavy and light, each of which contain both a highly variable and a constant region. Antigenic specificity of the heavy chain begins with the combinatorial rearrangement of three distinct gene segments, variable (V_H), diversity (D), and joining (J_H),[1] which are located in three distinct subloci on human chromosome 14q32.[2–4] The V_H locus contains >100 gene segments extending over a distance of approximately 1 mb.[5–7] The V_H gene segments have been classified into six to seven families based on the homology of their nucleotide coding regions (review in Pascual and Capra[8] and Van Dijk *et al.*[9]). Certain of these V_H segments are of particular interest as a result of their frequent and sometimes varied patterns of expression. One such gene segment is V_H4–21 (V_H4–34), one of 10 to 15 members of the V_H4 family.[10] V_H4–21 (V_H4–34) has been seen expressed in a variety of circumstances including early fetal development,[11] the autoantibody repertoire,[12–16] and in a highly restricted manner in antibodies that recognize alloantigens on the surface of red blood cells.[17] Most interesting, however, is the expression of this gene segment in T cells as a semi-germline transcript in association with a nontraditional V_H leader peptide. This nonhydrophobic leader sequence, termed "Et" for exon in T cells, was first described as a 5.2-kb transcript found in a T-cell leukemic cell line.[18] Subsequently, the same transcript was detected in several other transformed T and one B-cell line. This exon encodes a four-amino-acid open reading frame that, via a 3′ RNA splice site and the acceptor site upstream of the unrearranged

[a] J.D.C. holds the Edwin L. Cox Distinguished Chair in Immunology and Genetics at the University of Texas Southwestern Medical Center and is supported in part by grants from the National Institutes of Health (AI-12127 and GM-31689). G.F.W. is supported in part by the Texas Department Ladies Auxiliary, Veterans of Foreign Wars.

[b] V_H4–34 is the new designation assigned to V_H4–21 by Matsuda *et al.*[6] based on its location within the human V_H locus.

[e] Address for correspondence: J. Donald Capra, University of Texas Southwestern Medical Center, Department of Microbiology, 5323 Harry Hines Blvd., Dallas, TX 75235-9048.

V_H coding region, replaces the normal V_H leader sequence.[18] Upon screening a variety of normal cells, as well as several cell lines, we identified the same Et/V_H4–21 (V_H4–34) transcript in T-cell populations, but not in B cells (unpublished observations). Although others have described sterile transcripts from unrearranged V_H gene segments in B cells, they have not been associated with Et. The role that the Et/V_H4–21 (V_H4–34) transcript may play in T cells is unknown at this time, although one possibility is that it could be involved in suppression of V_H gene segment rearrangement in T cells. Here, we describe further characterization of Et at the genomic level. We found it located near the V_H4–21 (V_H4–34) gene segment. Additionally, two distinct Et genes were found within 100 kb of each other. This report concerns the characterization of these genomic clones.

MATERIALS AND METHODS

Oligonucleotides

The oligonucleotide primers used for PCR and sequencing were synthesized on a DNA synthesizer (Applied Biosystems, Foster City, CA). The following primer pairs were used for both the library screenings and screening of individual clones: (1) V_H4–21 (V_H4–34) 5′-TACAGCAGTGGGGCGCAGGACTGT-3′ and 5′-CCCGGGCTCACACTCACCTCCCCT-3′, (2) V_H4 family 5′-CCCCTAGACCCAGGGCTCACTGTGGG and 5′-CCCGGGCTCACACTCACCTCCCCT-3′, and (3) Et 5′-GCCGAT CCCAGGGCTGCAGCT-3′ and 5′-ACCTC CTGTCACCATTTCAGTG-3′.

Polymerase Chain Reaction

Polymerase chain reactions (PCRs) for screening of clones were performed essentially via the method recommended by Perkin-Elmer Cetus. Nanogram quantities of DNA were added to a 200 μM solution of each dATP, dCTP, dGTP, and dTTP, with 600 ng of each primer and 2.25U of *Taq* DNA polymerase (Promega Corp., Madison, WI). PCR cycles consisted of the following conditions: one cycle of denaturation at 94°C for 4 min, annealing at appropriate temperature for 1 min, extension at 72°C for 2 min and 29 cycles of denaturation at 94°C for 1 min, annealing for 1 min, and extension at 72°C for 2 min. Annealing temperatures for V_H4, V_H4–21 (V_H4–34), and Et are 72°C, 72°C, and 56°C, respectively. Amplifications were carried out in a programmable thermal controller (MJ Research Inc., Watertown, MA).

YAC Library Screening

Yeast artificial chromosome (YAC) clones were obtained from The Center for Genetics in Medicine of the Washington University School of Medicine and its associates. A human YAC library[19] was screened using both PCR oligonucleotides[20] specific for the V_H4–21 (V_H4–34) gene segment and conventional colony hybridization.[19] Five positive clones were screened by PCR for the presence of V_H4–21 (V_H4–34) and the Et exon. Two clones, yWPR344 and yWPR345, were positive for both.

P1 Library Screening

P1 clones were obtained from Genome Systems (St. Louis, MO) by screening their human genomic library constructed in the bacteriophage P1[21] via PCR with oligonucleotides specific for the V_H4 family. Two of six clones (Dupont Merck Pharmaceutical Co., human foreskin fibroblast [DMPC-HFF B] #s 92 and 113, here designated P183 and P185, respectively) also were positive for the Et exon when screened via PCR.

Southern Analysis

Genomic DNA was isolated from YAC clones and human liver via standard methods. P1 plasmid DNA was isolated by alkaline lysis. Five to ten micrograms of DNA was digested overnight with 10 units of *Eco*RI or *Hin*dIII (Boehringer Mannheim, Indianapolis, IN) and electrophoresed in 0.6% agarose for 6 hr using a Probe Tech 2 (Oncor, Gaithersburg, MD). DNA was transferred via the manufacturer's instructions to Sureblot membrane (Oncor, Gaithersburg, MD). After a 30-min prehybridization, ^{32}P-labeled probes were hybridized overnight at 42°C. Membranes were rinsed in 2× SSC for 10 min and then washed in 0.1× SSC + 0.1% SDS at 65°C for 30 min. Films were exposed for either 4 hr or overnight.

Cloning and Sequencing

P183 and P185 were digested to completion with *Eco*RI. After phenol extraction and precipitation, resuspended DNA was ligated into the *Eco*RI site of Bluescript phagemid vector (Stratagene, La Jolla, CA), transformed into $CaCl_2$-competent XL-1 blue bacteria (Stratagene, La Jolla, CA), and screened with the Et-5′ oligo. DNA was prepared from positive colonies using the Wizard Miniprep system (Promega Corp., Madison, WI). Sequencing was performed in duplicate for all three subclones on an Applied Biosystems automated sequencer, model 373A (Applied Biosystems, Foster City, CA) using oligonucleotides designed to previously generated sequence. Completed sequences were analyzed using DNASTAR (DNASTAR Inc., Madison, WI).

RESULTS

Two Et Genes Lie within 100 kb of the V_H4–21 (V_H4–34) Gene Segment

In order to better study the relationship between V_H4–21 (V_H4–34) and Et, we screened both a YAC library and a P1 library using primers to specifically identify clones that contain V_H4–21 (V_H4–34) and quite possibly Et. From the YAC library, five clones were obtained that contained V_H4–21 (V_H4–34) and of those, two were also positive for Et on screening by PCR. These clones are approximately 250 kb in length, and based on the sequences of the V_H gene segments PCR amplified from them, correspond to the 5′ end of the map created by Matsuda *et al.*[6] Screening the P1 library yielded six clones of ~100 kb, each of which contain members of the V_H4 family. Of the six, one contains the V_H4–21 (V_H4–34) gene segment and two of them [including the one containing V_H4–21

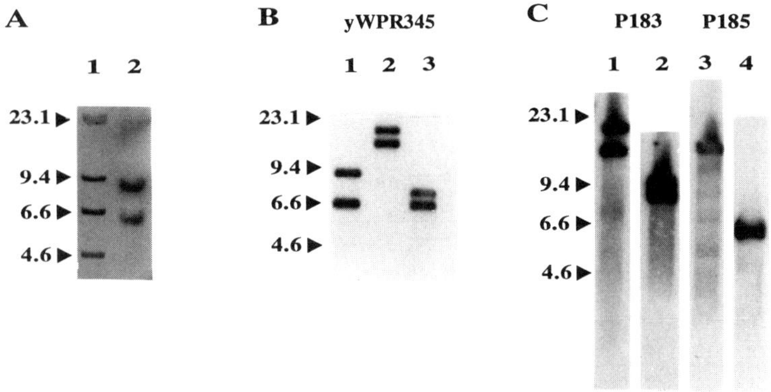

FIGURE 1. Southern filter hybridization of human liver **(A)**, YAC DNA **(B)**, and P1 DNA **(C)** with a probe specific for the Et exon. **(A)** Hybridization of an Et specific probe to human liver DNA digested with *Eco*RI (*lane 2*). The molecular size of the two bands is determined by comparison to ^{35}S-labeled molecular weight marker (*lane 1*). **(B)** Hybridization of the same probe to DNA from YAC clone yWPR345 digested with *Eco*RI (*lane 1*), *Hin*dIII (*lane 2*), or both (*lane 3*). The same two bands seen in liver are present both on this clone and the second YAC clone, yWPR344 (data not shown). **(C)** DNA from P183 (*lanes 1 and 2*) and P185 (*lanes 3 and 4*) digested with *Eco*RI (*lanes 2 and 4*) or *Hin*dIII (*lanes 1 and 3*) and probed with the same Et fragment.

(V_H4–34)] harbor an Et exon. Sequencing of the PCR-amplified product from each set of clones identified a single product. Because it was previously shown that two copies of the Et exon are located within the human genome, this result indicated that either each clone contains only a single copy of Et, thus placing them >250 kb apart, or more likely that they are almost identical in the region sequenced.

In order to further clarify whether just a single exon was present, DNA from human liver and each set of clones was digested with *Eco*RI and *Hin*dIII and subjected to Southern analysis. Probing with a PCR-amplified Et product revealed two bands of ~9 kb and ~6 kb in human liver (FIG. 1), identical to those previously reported.[18] Similarly, two bands of the same size were seen in both YAC clones, indicating that the two Et exons were reasonably close to one another in the genome. Of the two P1 clones, P185 contains only the smaller 6-kb *Eco*RI fragment, indicating that at least one copy of the Et exon must be within at least 100 kb of V_H4–21 (V_H4–34). The other, P183, contains the larger of the two Et *Eco*RI fragments (9 kb), but also has a second band that is ~8 kb, slightly larger than the expected size.

Further characterization of both the precise location and similarity of the two Et exons was investigated via restriction mapping of the two P1 clones described above. Each P1 clone was digested to completion with *Eco*RI and subcloned into Bluescript phagemid vector. Screening of transformants with oligonucleotides specific for Et identified all three of the bands seen on Southern filter hybridization. Restriction mapping of each subclone revealed a striking homology between all three that spans >4 kb (FIG. 2). The restriction pattern of the 6-kb clone is identical to that previously described.[18] The 9-kb clone from P183 overlaps the smaller over a distance of ~5 kb, which begins roughly 3 kb upstream of Et. The second Et subclone from P183 also has an identical restriction map in the region upstream of Et, but diverges from both of the others at the opposite end. Here the restriction pattern resembles that of the region adjacent to the cloning site of the P1 vector. Sequencing of this end of the 8-kb clone confirmed this suspicion, indicating that one of the Et exons is located within 3 kb of the end of P183.

Sequencing Reveals >97% Homology between Each Subclone

Sequencing of each of these three clones confirmed the homology indicated by the restriction mapping. As indicated earlier by the sequence generated from the PCR products, the Et exons themselves are almost identical in the region initially defined by sequenced mRNAs. The Et exon from P185 is identical in sequence over the span of the 315 bp that was previously published.[18] The other exon that is located on the 9-kb *Eco*RI fragment of P183 differs by four nucleotides. In addition to the two mismatches identified by PCR analysis, two additional differences are located further upstream of the 5′ PCR oligonucleotide. The second Et exon from P183, that contained on the 8-kb *Eco*RI subclone, is identical to that found on P185 (data not shown). Sequence generated from both subclones over a region of ~3 kb (FIG. 3) yields only minor differences, which is somewhat unexpected in an area that is thought to contain primarily intron. Flanking DNA both upstream and downstream over a span of >2.8 kb contains approximately 80 nucleotide substitutions. Four small gaps of 6 bp, 2 bp, 3 bp, and 3 bp also occur in the 3′ region, but overall the two clones are >97% homologous. About

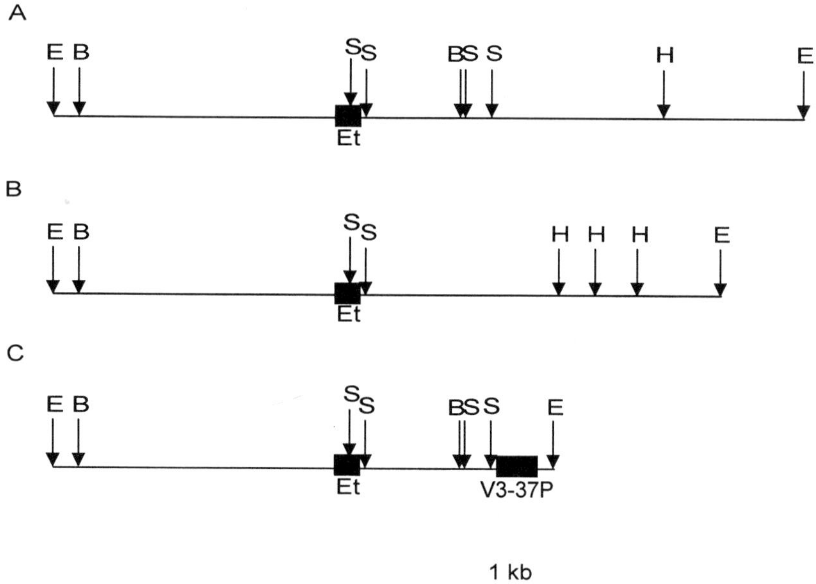

FIGURE 2. Restriction maps of the three *Eco*RI subclones that contain the Et exon. **A** and **B** are from P183, whereas **C** was subcloned from P185. Arrows indicate the position of restriction sites of *Eco*RI (E), *Bam*HI (B), *Hin*dIII (H), and *Sac*I (S). The positions of Et and V3–37P are indicated by solid boxes.

1.5 kb downstream of Et on the P185 subclone is a V_H gene segment initially referred to as ΨV_H4.[18] It is a V_H3 pseudogene that has >95% homology to the V3–37p gene segment described by Matsuda *et al.*[6] This gene segment is not present on the 9-kb Et subclone, implying that one of the breakpoints involved in this probable duplication event lies between the end of our present sequence and this particular gene segment.

DISCUSSION

By analysis of P1 and YAC genomic DNA clones, we have been able to better characterize the Et exon, a short nonhydrophobic peptide that replaces the normal leader peptide in semi-germline V_H transcripts found in various T-cell populations. Two copies of this exon are located in the 5′ half of the human V_H locus; one located on a 6-kb *Eco*RI fragment and the other on a 9-kb *Eco*RI fragment. Restriction mapping and sequencing of the P185 genomic clone indicates that the Et exon located on the 6-kb *Eco*RI restriction fragment is located a short distance upstream of V_H3–37p. Thus, according to the map of the V_H locus created by Matsuda *et al.*,[6] it is approximately 40 kb upstream of V_H4–21 (V_H4–34), the V_H gene segment with which it has predominantly been associated. Restriction mapping of P183, which contains both copies of the Et exon, indicates that the same Et exon from P185 is located on the 8-kb *Eco*RI restriction fragment, and thus at the extreme 3′ end of this clone. Therefore, the second exon that is contained on the 9-kb subclone is located further 5′ in the locus, but at a distance no greater than 100 kb. The mechanism by which Et and V_H4–21 (V_H4–34) become associated is thought to occur via RNA splicing involving the donor splice site 3′ of Et and the acceptor site of the V_H gene segment normally used by the V_H leader exon. Sequencing of each of the Et exons revealed four differences in the region that is 5′ of Et and contained on the mature mRNA transcripts. None of these changes occur in what are likely the coding and functional regions of Et and therefore are unlikely to impair either's ability to function, since both have identical 3′ donor splice sites and initiation codons. In order to become associated with V_H4–21 (V_H4–34), it appears that this exon must "jump" over at least three additional gene segments, two pseudogenes (V_H3–36P and V_H3–37P) and one functional member of the V_H3 family (V_H3–35).[6] All three appear to have functional acceptor splice sites, but only V_H3–35 has been seen associated with Et, and at a much lower frequency than V_H4–21 (V_H4–34)[18] (unpublished observations).

Finally, sequence analysis of the flanking regions of both subclones indicates a high degree of conservation, with >97% homology in the region sequenced. As described above, most of the differences are minor, and do not appear to be in areas that are thought to encode the potential functional properties of the Et exon. Further sequencing at the 5′ end of each of the three subclones indicates homology similar to that shown here, and indicates that the region of overlapping sequence for the two exons extends further upstream of the *Eco*RI site used here for subcloning. A search of this region for open reading frames yielded no other potential exons and confirmed that both Et exons have stop codons in all reading frames except that of the four amino acids described previously. The role of these transcripts in T cells at this time is unclear. One possibility is that the protein product of Et/V_H4–21 (V_H4–34) may play a role in preventing certain rearrangements of the V_H locus in T cells. Further analysis and characterization of this exon both at the genomic and protein level, particularly in various T-cell subsets,

```
P185                       AAGACATTTTCATATCGTTGTCATTGTTTTCTTGTGTGC   -784
P183                       ................A......................

P185  ATTTTATTCCTCGCTATTGTATACATCATTGGAAATTCTAAGTATTCTTTTGAAATATCT   -724
P183  G...........................................................

P185  AGTCTTTCTAGATGTTCTGAAGTGCCTGATGTATGTTAAAATTACAGGTGGTAAAATAAT   -664
P183  ............................................................

P185  AAATTTTGTAAATGTCTTTTTGTTAAAATTCATATGCAGTGTTTTATTTTATTTTGATGT   -614
P183  .............A..............................................

P185  TGGTGGGGGTGGGGAGGATGGCCAAATCCCTGCTTGATCAACACACACTGCGTTTGTGC    -554
P183  ................................................T...........

P185  TCTGGTTCAGGGGAGGAGAGAGGAGGAGAAAGTGCAGACTTCCAGGCCTCTGTGCGCACC   -494
P183  ............................................................

P185  GGGAGGAGAGATTAATGATCATCTCTTCTGTCTGTGTGTTTGTTTTATTTATCTATGGAT   -434
P183  ............................................................

P185  ATCCTGTGTATAAAGGATGAACAAGTCCTATTTATAACATCTAATCTTTTCAGGTGTTAA   -374
P183  ............................................................

P185  GTTGCCAGCGTATGACGGAAGTAGCATTATTAAACGAATGCAGCTTGTACATATTGTGTT   -314
P183  ........T................G............C.....................

P185  AAAATTCATAAAAAGCCAGTCTTCTGTAAAGAACCCTTGGGCCCTCCCTCCGCAGCCTCC   -254
P183  G.........G...A.............................................

P185  GGCGCCAGCTCCAGGAAGACCTGTCCAGGGGAAGGCGGGCGACAGAAACTCCTCCTGTCT   -194
P183  ..........G.......................T.........C........G......

P185  GTGGGGAAGGCGGTGGAGACCAGGCCTCCACAGGCTCCCCACAGGCTCCCCACAGGCTCG   -134
P183  ..................C.........................................

P185  CGCGGCTCTGGTTGCTGGGCTCCGCACGGCCGCGTGGGGGCGCCTCCTGGGAGCGAGGA   -74
P183  ............C...............................................

P185  GGCGCCATCCTGGCTCGCCTGCTCCAGGAGGACGCTCTGGGCCTGCGCAGATACAGTTCT   -14
P183  ............................................................
                  +1
P185  CCAGGATGGCGCGGGCGCTGGGGGCGGCGCTCCTTGGGCTGCTGCGCCTCCACTGAGCC   +47
P183  ..........................C.................................

P185  GCCTGGGTGTGAGGACCTCGCGCCGGCGCCTCCGGGAAGGACGCGGAGCCGCCGCGGCCG   +107
P183  ............................................................

P185  ATCCCAGGGCTGCAGCTGCTGCCCGACGCCCGCGGGAAGGCCCTTCCGCCACCGCCCGGG   +167
P183  .G................................A.......................C.

P185  GGTTCGAACTTGGAGCAGCTGCTGCCAGCTGAGGCGCCAGCGGCGTCGGGAGCTCCTCAG   +227
P183  ............................................................

P185  AGTCAGTGTTTTGTGCTGGATTCGGCATGTCCTCTGAAAACTCGGGAATTTGTCACTGAA   +287
P183  ............................................................
       M  V  T  G
P185  ATGGTGACAGGAGGTGAGAAGTTTTTTTAGACGTTGCCTTCGCTGGCCTGGCGCGGGGCT   +347
P183  .............................................A..G.....C.....

P185  CAGGCCTGGAATCCCAGCACTTTGGGAGGCTGAGGCGGGCGGATCGCTTGAGCTCCGGAG   +407
P183  ..............................C.............................

P185  GTGGAGAAGCTGCCTTCGTGGTAACTGTGGTCTTAAGTTCAGCTGAGAACGATAAATGGC   +467
P183  ............................................................

P185  TTTTCCTTGAATTGGCTGC------TTTTGTGATTTCTCTCAGGCTCATCTTTTGTCAGT   +527
P183  ..................TGAATGG............................C......
```

FIGURE 3. Sequence comparison of the 6-kb (P185) and 9-kb (P183) subclones of the Et exon. Similarity between the two is designated by dashes, whereas differences are indicated in the lower sequence. The sequence of the Et exon previously published is underlined. The position marked +1 denotes the nucleotide that corresponds to the 5′ end of the Et/V_H4–21 (V_H4–34) mRNA. The transcriptional start site and coding region are marked.

```
P185  TAAAAATCCAGTGGTAGGTGTAGCTTAGAGACGGGAAATTTTTGGTTTTGGTTTTGGCTA  +587
P183  ......................................................G.....

P185  CACGTAAGTGTTGGAAATTATTTTCTTTTACGTTCCAGTGTGAGCAAATCCAGAAGGATG  +647
P183  ...A.....C.................,.........A......................

P185  TCAGATTAAACACCGAATTTAACAAATTCAGCCGGGCACGGTGGCTCACGCCTGTAATCC  +707
P183  ......................................................G.....

P185  CAGCACTTTGAGAGGCCGAGGCGGGTGGATGATGAGGTCAGGAGTTTGAGAACAGCCTGA  +767
P183  ............................................................

P185  CCAACATAGTGAAACTCCGTCTCTACTAAAAATACAAAAATTAGCCGGGCGTGGTTGTGC  +827
P183  ............................................................

P185  GCGCCTGTAATCCCAGCTACTCAGGAGGCTTGAGGCAGGAGAATCGCCGAGATTGCGCCA  +887
P183  ............................................................

P185  CTGCACTCAGCCTGGGCGGACAGAGCGAGACTTTCGTCTCAAAAAAACAAAAAAAAA--T  +947
P183  .........................................................AA.

P185  CAATCATTGGAATACTGTTGTTCATTACAATTAATGAACGTTTGATAGACGCATAAACGA  +1007
P183  ...............................................C.........T.C

P185  ACTAAATTCACAAGTACATATAAGTAAAATAAGCCAAAAAAAAAAAACTACATCCCGTAT  +1067
P183  .......................................................T....

P185  GATGACACTTTTATACATTCTACAGAATAGAAACTAAAGTGATATGAAGGTTGGGAGTTG  +1127
P183  ...................................G......C........C........

P185  AATGGAGAAAAGGTGAGCGAAAAGAGGGAGGACGGAGGAGTTGCAAGAGAACAAGAGAAA  +1187
P183  .................A..............---...C............T........

P185  ATGTTGAGGGTAATTGATCTGTTTTCTATCTTAATAATGGTTATGGTTACATCAATAATT  +1247
P183  ..................T......................G..................

P185  TCTAAATTGTACATTTTGTGTGTTAATTATAATTTTATATTATTCAAGCTATTACAAATA  +1307
P183  C.C..........................G..............................

P185  AATACCTGTATAAATTTCTTAGGGACGGTTTAAGGGAGATGGATAAATCACATAAAAAGT  +1367
P183  .............................G..............................

P185  CAGATATTTCTGAATGTACAATGAATAACCCTTAGCTCTATCTTTATTTTTTGGTAATTC  +1427
P183  ...........................................C................

P185  TAAAATGCAGCAGATTTTTTTAATGTTTTTATTACAAGAAGGGTTTCTTCAAACCACACC  +1487
P183  .............T..............................................

P185  AGGTGTGTGTAACTCTGGGCTAAAGTTGGTTAAGGGAGCAGTGGGATCCGGTGGAGAGGA  +1547
P183  ....A..C...........A.........C.........T....................

P185  GCACAGGCCTCGATCCTGGAGCCCATTCTTGTCACATGTGAGCTCCTGGACACATCTCAT  +1607
P183  .......T.C...............................C..................

P185  AGCTCCCTCATGCTTCTGCGTTTACATTCTCATCTGTAAATGGAGGACCAATGGGCATCT  +1667
P183  ..................G..........A..............................

P185  GCCTCTAACAATGTATTTTCATGTGTTAAGATGATGTATGTTAAGTGTTTACTATAGCAC  +1727
P183  ....................G...........A.......C...A.......C.C.....

P185  AATCTTTTAATAAGAAATTACATTTTTTTCCAAATACTATCATTGTCTTCAAGCCCCACC  +1787
P183  .........C..................-.......T.......................

P185  ACCCACCAGAAAACTTAATATGCTCTGTGCCCTGCCCTGTCCTCAGACATCCAGCCCAAG  +1847
P183  ......T.........C..C...........T....................T.......

P185  AGACTTTTATATAGTAGGAGATATGCAAATAGTTCCCTCCCTCTGCTGATGATGAAAACC  +1907
P183  ......C..............C..............................---.....

P185  AGCCCAGCCGTGACCGTGCAGCTCTGGGAGACGAGCTCAAGTGCCAGGATTCCCAAGTGT  +1967
P183  .........T.....C...............G.......................G....

P185  TTTCACTTGGTGATCAGAACTTAACACAGAGGAATCACCATGTTG
P183  .................................C........G..
```

FIGURE 3 (*Continued*).

is in progress and should yield insight into both the function and mechanism that this exon mediates.

SUMMARY

The human IgV_H locus is located on chromosome 14, band q32 and spans approximately 1 mb. Within this locus are approximately 120 V_H gene segments that are subdivided into six to seven families based on sequence homology of their coding regions. V_H4–21 (V_H4–34) is a member of the V_H4 family, a family that contains 10 to 15 members. It is expressed in a variety of circumstances including early fetal development, the autoantibody repertoire, and in a highly restricted manner in antibodies that recognize alloantigens on the surface of human red blood cells. Most interesting, however, is the expression of this gene segment in T cells as a semi-germline transcript in conjunction with a nontraditional V_H leader peptide. This nonhydrophobic leader sequence, termed "Et" for exon in T cells, has previously been shown to reside within the V_H locus. Using YAC and P1 clones, we have identified two copies of this exon, both of which are located in the region of the locus that contains V_H4–21 (V_H4–34). Characterization of the two exons suggests that they arose by duplication, as flanking DNA is almost identical over a distance of >5 kb. Preliminary data suggests they are both located >20 kb upstream of V_H4–21 (V_H4–34).

ACKNOWLEDGMENTS

The authors thank Shirley Hall for her technical expertise in the operation of the ABI-automated sequencer.

REFERENCES

1. Tonegawa, S. 1983. Nature **302:** 575–581.
2. Kirsch, I. L., C. C. Morton, K. Nakahara & P. Leder. 1982. Science **216:** 301–303.
3. Cox, D. W., V. D. Markovic & I. E. Teshima. 1982. Nature **297:** 428–430.
4. McBride, O. W., J. Battey, G. F. Hollis, D. C. Swan, U. Siebenlist & P. Leder. 1982. Nucl. Acids Res. **10:** 8155–8169.
5. Walter, M. A., U. Surti, M. Hofker & D. W. Cox. 1990. EMBO J. **10:** 3303–3313.
6. Matsuda, F., E. K. Shin, H. Nagoaka, R. Matsumura, M. Haino, Y. Fukita, S. Taka-ishi, T. Imai, J. H. Riley, R. Anand, E. Soeda & T. Honjo. 1993. Nature Genet. **3:** 88–94.
7. Cook, G. D., I. Tomlinson, G. Walter, H. Riethman, N. Carter, L. Buluwela, G. Winter & T. Rabbitts. 1994. Nature Genet. **7:** 162–168.
8. Pascual, V. & J. D. Capra. 1991. Adv. Immunol. **49:** 1–74.
9. Van Dijk, K. W., F. Mortari, P. M. Kirkham, H. W. Schroeder, Jr & E. C. B. Milner. 1993. Eur. J. Immunol. **23:** 832–839.
10. Sanz, I., P. Kelly, C. Williams, S. Scholl, P. Tucker & J. D. Capra. 1989. EMBO J. **8:** 3741–3748.
11. Pascual, V. & J. D. Capra. 1992. Arthritis Rheum. **35:** 11–18.
12. Pascual, V., K. Victor, D. Lelsz, M. B. Spellerberg, T. J. Hamblin, K. M. Thompson, I. Randen, J. B. Natvig, J. D. Capra & F. K. Stevenson. 1991. J. Immunol. **146:** 4385–4391.

13. PASCUAL, V., K. VICTOR, M. SPELLERBERG, T. J. HAMBLIN, F. K. STEVENSON & J. D. CAPRA. 1992. J. Immunol. **149:** 2337–2344.
14. VAN ES, J., F. H. J. G. MEYELING, W. VAN DE AKKER, H. AANSTOOT, R. H. W. M. DERSKIN & T. LOGTENBERG. 1991. J. Exp. Med. **173:** 461–470.
15. SILBERSTEIN, L. E., L. C. JEFFERIES, J. GOLMAN, D. FRIEDMAN, J. S. MOORE, P. C. NOWELL, D. ROELCKE, W. PRUZANSKI, J. ROUDIER & G. J. SILVERMAN. 1991. Blood **78:** 2372–2386.
16. PASCUAL, V., K. VICTOR, I. RANDEN, K. THOMPSON, M. STEINITZ, Ø. FØRRE, S.-M. FU, J. B. NATVIG & J. D. CAPRA. 1992. Scand. J. Immunol. **36:** 349–362.
17. K. M. THOMPSON, J. SUTHERLAND, G. BARDEN, M. D. MELAMED, I. RANDEN, J. B. NATVIG, V. PASCUAL, J. D. CAPRA & F. K. STEVENSON. 1991. Scand. J. Immunol. **34:** 509–518.
18. BAER, R., A. FORSTER, I. LAVENIR & T. H. RABBITTS. 1988. J. Exp. Med. **167:** 2011–2016.
19. BROWNSTEIN, B. H., G. A. SILVERMAN, R. D. LITTLE, D. T. BURKE, S. J. KORSMEYER, D. SCHLESSINGER & M. V. OLSON. 1989. Science **244:** 1348–1351.
20. GREEN, E. D. & M. V. OLSON. 1990. Proc. Natl. Acad. Sci. USA **87:** 1213–1217.
21. PIERCE, J. C. & N. L. STERNBERG. Methods Enzymol. **216:** 549–574.

Ethnic Differences in V_H Gene Polymorphism

E. H. SASSO, J. H. BUCKNER, AND L. A. SUZUKI

Harborview Medical Center
University of Washington
Seattle, Washington 98125

The Ig V_H3 gene V_H26 (also known an 30p1 or 3-23) is duplicated and occupies distinct A and B loci.[1] Polymorphism at each locus has not been fully described. In this report, distinct patterns of polymorphism for each V_H26 locus are described in three ethnic groups. Sequence-specific RFLP analysis[2] of genomic DNA from 72 Caucasians, 52 Asians (mostly Chinese, Korean, Japanese), and 35 African Americans was performed with a 21-mer oligonucleotide probe targeting CDR1 of V_H26. Four different *Taq*I restriction fragments (RF) were detected. Phenotype hybridization results appear in TABLE 1.

Concordant hybridization results with a V_H26 CDR2 probe showed that 7.0-kb blanks signify homozygosity for locus A null alleles. The 6.8- and 3.7-kb bands, seen predominantly in Asians, were allelic to the locus B 5.0-kb band, based on the distribution of these bands in the population and their segregation in families. Locus B null alleles and duplicated 5.0-kb or 7.0-kb alleles were seen infrequently. In Asians only, the 3.7-kb band always had a high intensity consistent with three or four gene copies, suggesting it arose by multiple duplicaiton. The gene in the 3.7-kb band was cloned. Its sequence differed from V_H26 by eight nucleotides in CDR2 and one in FR3, and it appears to have risen by gene conversion from another V_H3 gene, 8-1B. The presence of this novel sequence in the germline was confirmed by hybridization with CDR2 probes (S1 and S2) specific for the cloned sequence. In all 15 subjects possessing a 3.7-kb band, probe S2 detected only the 3.7-kb band. Probe S1 detected the 3.7-kb bands plus additional non-V_H26 bands, which might contain 8-1B. A V_H26 gene was also cloned from a subject having only a 5.0-kb band and had an exact V_H26 (i.e., V_H26c^3) sequence. These data indicate that the predominant V_H26 alleles, locus A at 7.0 kb and locus B at 5.0 kb, and the rarer 6.8-kb locus B allele, probably all have V_H26 sequences. In contrast, the 3.7-kb allele is a sequence variant that belongs to a different canonical structural group[4] and, in Asians, exists as a multigene cluster. The alleles of V_H26 are shown in TABLE 2.

TABLE 1. Phenotype Hybridization

	TaqI RF	Caucasian	Asian	Black	χ^2	p
Locus A	7.0 kb	89%	75%	**26%**	56	0.001
Locus B	6.8 kb	0	**10%**	3%	7.4	0.02
	5.0 kb	100%	96%	97%		
	3.7 kb	0.7%	**21%**	9%	7.7	0.001

TABLE 2. Alleles of V_H26

Locus A	Locus B
7.0 kb	6.8 kb
7.0 kb duplicated	5.0 kb
null	5.0 kb duplicated
	3.7 kb
	3.7 kb duplicated
	3.7 kb multi-duplicated
	null

We conclude that both V_H26 loci are polymorphic and that ethnic differences exist at each locus. Locus A polymorphism occurred mostly as null alleles, which were most prevalent in African Americans. The locus B variation occurred mostly as RFLP, one of which is also a sequence variant. African Americans tended to have fewer copies of V_H26, and Asians were the most likely to have novel locus B alleles.

REFERENCES

1. Willems van Dijk, K., L. Milner, E. Sasso & E. Milner. 1992. Chromosomal organization of the heavy chain variable region gene segments comprising the human fetal antibody repertoire. Proc. Natl. Acad. Sci. USA **89:** 10430–10434.
2. Sasso, E., K. Willems van Dijk & E. Milner. 1990. Prevalence and polymorphism of human V_H3 genes. J. Immunol. **145:** 2751–2757.
3. Chen, P. P., M.-F. Liu, S. Sinha & D. A. Carson. 1988. Arthritis Rheum. **31:** 1429–1431.
4. Chothia, C., A. Lest, E. Gherardi, I. Tomlinson, G. Walter, J. Marks, M. Llewelyn & G. Winter. 1992. Structural repertoire of the human V_H segments. J. Mol. Biol. **227:** 799–817.

Expression and Hypermutation of Human Immunoglobulin κ Genes[a]

RALPH KLEIN AND HANS G. ZACHAU[b]

Institut für Physiologische Chemie
Universität München
D-80336 München, Germany

INTRODUCTION

The immunoglobulin κ locus on chromosome 2p11-12 comprises a C_κ proximal (p) copy consisting of the single constant (C_κ), five joining (J_κ), and 40 variable (V_κ) gene segments, and a distal (d) copy consisting of 36 V_κ gene segments. The d copy arose by duplication of a large part of the p copy; and, consequently, there are in addition to 10 solitary V_κ genes 33 gene pairs whose sequences are 95 to 100% identical in the coding regions. According to their sequences, 32 of the 76 V_κ genes are potentially functional (pf), 16 have minor defects (md), 25 are pseudogenes (ps), and for three genes both pf and md alleles have been found. The minor defects are defined as one or two 1-bp alterations in a gene, for example, a stop codon in the coding part and/or a deviation from the canonical sequences of regulatory elements, splice or recombination sites. In addition to the V_κ genes of the κ locus, 24 so-called V_κ orphons have been cloned and sequenced, that is, V_κ genes on chromosome 2 outside the locus or on other chromosomes. The existence of up to 12 further V_κ orphons was made likely.

cDNAs derived from 22 V_κ genes and five pairs of identical duplicated V_κ genes have been found up to now. The corresponding numbers for rearranged genomic V_κ genes and κ proteins are 17 plus 4 and 7 plus 7, respectively. No pseudogenes or V_κ orphons but some md genes have been found to be expressed.

Some features of the κ locus and the expression products of its genes are summarized in FIGURE 1, which constitutes the basis of the present contribution. Structural and functional aspects of the κ genes were summarized in a lecture review,[1] a somewhat more detailed review,[2] two special reviews of the germline sequences,[3,4] and a publication on the expressed κ genes.[5] The purpose of the present contribution is to discuss, with as little overlap as possible with previous publications, the expression and hypermutation patterns of the κ genes. Some compilations of data are taken from a doctoral thesis.[6]

THE USE OF DIFFERENT J_κ AND V_κ GENE SEGMENTS IN EXPRESSION

The five J_κ genes are used with quite different frequencies, as can be seen in TABLE 1. The relatively rare usage of $J_\kappa3$ may be related to the presence of a Pro

[a] This work was supported by Bundesministerium für Forschung and Technologie, Center Grant 0316200A, and by Fonds der Chemischen Industrie.

[b] Address for correspondence: Professor H. G. Zachau, Institut für Physiologische Chemie, Universität München, Schillerstrasse 44, D-80336 München, Germany.

TABLE 1. J_κ Usage in Different V_κ Subgroups[a]

	V_κI (49)	V_κII (32)	V_κIII (55)	V_κIV (12)
J_κ1 (52)	20	10	15	7
J_κ2 (37)	7	10	18	2
J_κ3 (11)	4	1	5	1
J_κ4 (39)	17	7	13	2
J_κ5 (9)	1	4	4	0

[a] The analysis is based on 67 cDNAs sequenced in our laboratory and on 81 genomic V_κ–J_κ joints compiled from the literature.[5,6] The total numbers of V_κI to V_κIV sequences and J_κ1 to J_κ5 sequences, on which the analysis is based, are given in parentheses.

in position 100 (instead of Gln or Gly) and an Asp in position 105 (instead of Glu) and/or a 22-bp instead of a 23-bp spacer between the hepta- and nonanucleotide recognition sequences. J_κ5 has an Arg in position 103 instead of a Lys present in the other J_κ genes. The five J_κ genes are also used to different extents by the V_κ genes of the four major subgroups. This can probably be explained only on the basis of three-dimensional structures of the resulting κ proteins.

It can be seen in FIGURE 1 that up to now not all pf genes were found to be expressed. In this context it should be kept in mind that our definition of potential functionality is based on the gene sequences and does not include the presence or absence of codons for so-called invariant amino acids[7] in certain positions of the κ proteins. On the other hand, expression products of some md genes, possibly alleles of the sequenced genes, were observed as was described before.[5]

As to the numbers of expressed κ genes a recent survey of the literature by Cox *et al.*[8] comes to conclusions similar to those of our survey (FIG. 1 and Klein *et al.*[5] and Klein[6]): 24 germline V_κ genes are listed as expressed[8] as compared to 21 genes plus five gene pairs[5]; only 11 genes are considered to be expressed frequently,[8] whereas we counted 16 genes and gene pairs with five or more expression products; five or, in our counting, six genes of the d copy of the locus are expressed. The only addition provided by the survey of Cox *et al.*[8] is the recording of an expression product of the md gene L9, which increases the number of expressed genes to 22 plus five pairs of identical duplicated genes. We would expect that expression products of one or the other further germline genes will still be found, but we believe that FIGURE 1 is complete with respect to listing all frequently expressed genes.

SEQUENCE VARIATION AT AND NEAR V_κ–J_κ JUNCTIONS

The sequence variation at and near V_κ–J_κ junctions is a major factor in the generation of antibody diversity. An analysis of the situation is presented in TABLES

FIGURE 1. The human immunoglobulin κ locus. The structure of the locus and the different types of genes (pf, ps, md) are described in the text. *Open boxes,* pf genes; *filled boxes,* ps genes; *boxes with crossed lines,* md genes; *boxes with one diagonal line,* genes for which pf and md alleles are known; numbers of cDNAs and full-length κ proteins are boxed and encircled, respectively. Seventy-one cDNAs sequenced recently in our laboratory and 51 V_κ–J_κ joints, 171 cDNAs and 74 κ proteins from the literature are included in the survey. The figure is from Klein *et al.*[5] and was updated by the addition of a cDNA derived from L9 (X13 in Ref. 16). (*Figure on following page.*)

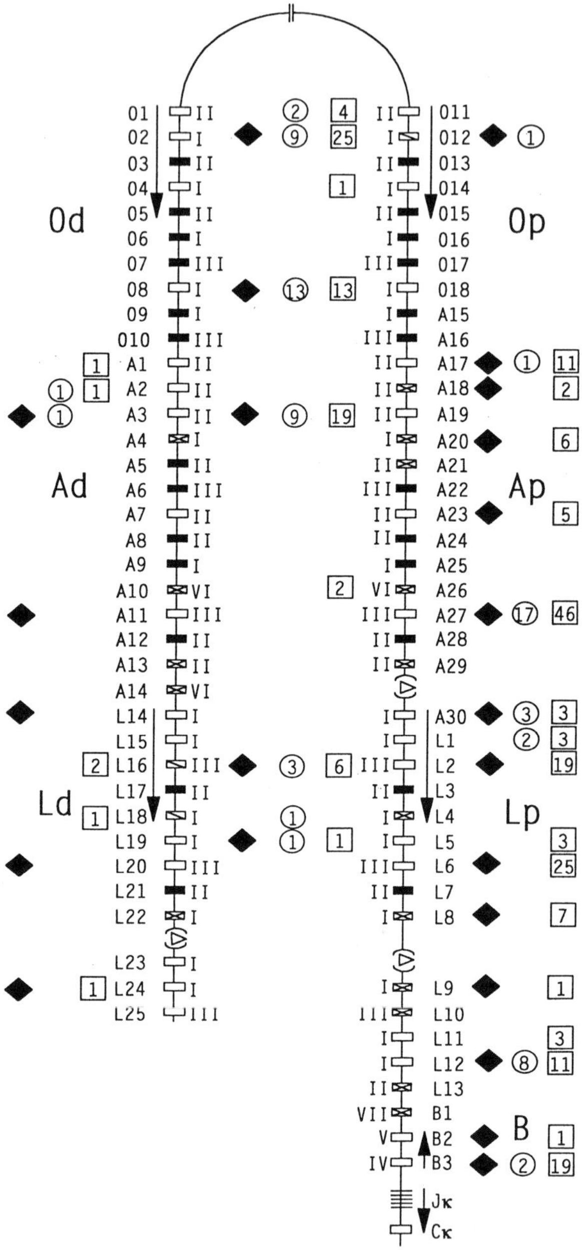
Od
Op
Ad
Ap
Ld
Lp
B
Jκ
Cκ

TABLE 2. Sequence Variations at and near the V_κ–J_κ Recombination Site

Type of Variation	Number of Clones[a]	
Codon 95 and 96 directly joined	99	
V_κ and/or J_κ shortened	16	
Additional nucleotides between Codon 95 and 96	33	
+1:		2 (2 | 0)
+2:		1 (1 | 0)
+3:		24 (14 | 10)
+5:		2 (0 | 2)
+6:		3 (0 | 3)
+8:		1 (0 | 1)

[a] The additional nucleotides that can be attributed fully or partially to germline sequences are given in parentheses (in front or behind the vertical bars).

2 and 3. In the majority of sequences a complementarity-determining region CDR3 consisting of seven codons was found (112/148), but in some sequences eight codons were observed (26/148). In rare cases 2, 5, 6, and 9 codons were seen (FIG. 2). In 24 sequences three nucleotides had been added to form a codon 95a (TABLE 2), which can be fully attributed to germline sequences in 14 clones and partially in 10 clones. Some of the additional nucleotides could have been added by the action of terminal deoxy transferase (TdT) or by other processes, but a germline derivation is the simplest explanation. Also the high proportion of G formation in codon 95 (TABLE 3) may be discussed in terms of a TdT involvement. Alternatively, κ chains with amino acids derived from G-containing codons 95 are selected during maturation. It is remarkable that 49 of the 58 changes to G (TABLE 3) are T → G changes, 35 of them being CCT→CCG changes. In the other codons near the V_κ–J_κ junction, the nucleotide changes are similar to those found in L′-FR3 (leader, framework; TABLE 3 compared to Table A3 in Klein *et al.*[5]).

EXTENT OF SOMATIC HYPERMUTATION IN DIFFERENT GERMLINE GENES AND IN DIFFERENT INDIVIDUALS

Many of the sequence changes near the V_κ–J_κ junctions are probably also caused by somatic hypermutation, but they have been described separately in the preceding section because here hypermutation is superimposed on other processes as truncation and repair of the gene segments.

In FIGURES 3 and 4, the extent and distribution of mutational changes is

TABLE 3. Nucleotide Exchanges near the V_κ–J_κ Recombination Site[a]

	Codons						
	92	93	94	95	96	97	98
G,C,T→A	16	10	10	13	13	2	–
A,C,T→G	9	21	4	58	31	6	–
A,G,T→C	9	21	4	18	33	7	2
A,C,G→T	10	15	16	7	12	5	1

[a] 67 cDNA clones of our laboratory and 81 clones from the literature were evaluated.

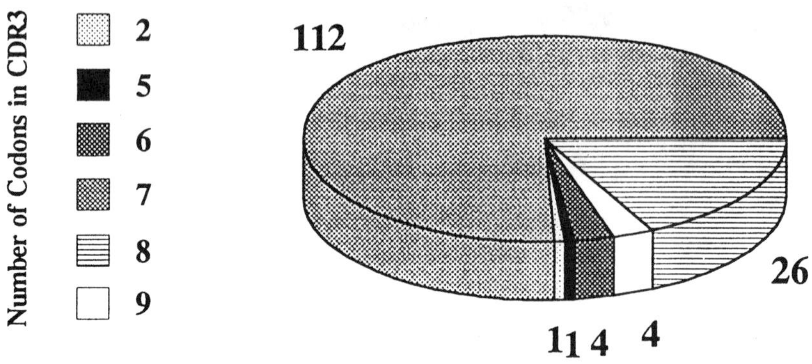

FIGURE 2. Occurrence of κ genes with 2 to 9 codons in CDR3. One hundred forty-eight cDNA and genomic V_κ–J_κ clones were used in the compilation. Numbers refer to the different types of CDR3; for example, there are 26 clones with 8 codons in CDR3.

described. It can be seen in FIGURE 3 that the derivatives of some germline genes (e.g. O12/O2) are, on an average, more highly mutated than the derivatives of other genes (e.g. A19/A3). The distinction between the cDNA clones of a library sequenced in our laboratory and sequences from the literature is made in order to stress a possibly interesting aspect of the hypermutation process. Our cDNA library was prepared from spleen tissue of an 86-year-old person (individual ML),[5] whereas the clones and sequences of the literature are probably derived from persons with a random age distribution. It is apparent from FIGURE 4 that the proportion of highly mutated sequences is considerably higher in the clones from the cDNA library of individual ML than in the other clones. The highly mutated clones may have accumulated during the lifetime of the 86-year-old person. They have, on an average, 10.3 mutations per clone compared to 6.6 in sequences from the literature. Sequences from patients with autoimmune diseases, which are abundant in the literature, are indicated separately in FIGURE 4.

TYPES OF MUTATIONAL CHANGES

It is well known that the ratio of replacement over silent mutations (R/S) is higher in the CDRs than in the FRs. In addition, there are codons with a particular mutation behavior, that is, codons with no mutations at all and codons with low as well as very high R/S values (TABLE 4). Codon 31 had been classified as an intrinsic mutational hot spot in mouse κ chains.[9] The high R/S ratios in codons 93 and 94 are remarkable. The Pro codon 95 was most frequently changed from CCT to CCG.

The relatively high abundance of mutations in adjacent positions called block mutations has been discussed before as an interesting feature of the hypermutation and κ chain maturation processes.[5] Now, the incidence of isolated 1-bp mutations was compared to the one of block mutations (TABLE 5): 21% and 25%, respectively, of the mutated nucleotides were found in blocks. Block mutations comprising two nucleotides are found about twice as often as expected statistically.[6] They were most frequently observed in codons 29 through 32.[6] If one assumes hot spots

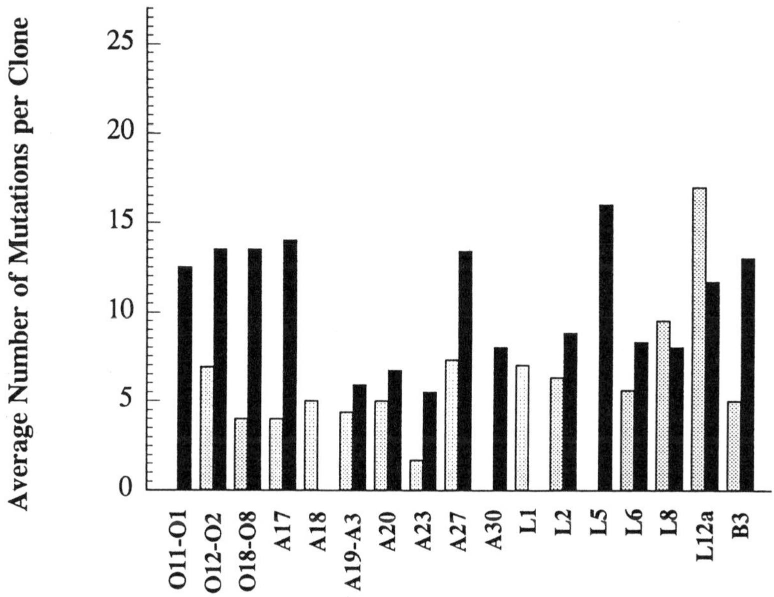

FIGURE 3. Average number of mutations per clone. Only those germline genes are shown for which more than one derivative is known; 69 cDNA clones of individual ML (*black bars*) and 123 cDNAs taken from the literature (*dotted bars*) are included. Block mutations are counted as single events.

of mutation in this region, the possibility of independent mutations in adjacent nucleotides may be seriously considered. This would be a mechanistically interesting feature as would be the alternative, that is, the occurrence of block mutations as distinct single steps. There seems to be a preference of A → G and G → A transitions in the two-nucleotide blocks,[6] but it cannot be decided whether this is caused by a property of the hypermutation machinery or by a selection, for instance, against Ser encoded by AGC or AGT or against another amino acid with A- or G-containing codons. Hypothetical mechanistic explanations of block mutations and the possible role of selection have been discussed before.[5] Both selection for functionality of the κ chains and selection by antigen can be envisaged.

The types of mutation were previously analyzed jointly for our own cDNA clones and the clones from the literature (Table A3 in Klein *et al.*[5]), and the numbers of transversions and transitions were not specified. This is now done in TABLE 6. It can be seen that the ratio of transversions to transitions is >1 only in CDR3 and that the overall ratio is approximately the same for both groups of clones.

Deletions and insertions seem to be relatively rare events compared to nucleotide exchanges. Only six deletions and one insertion were found in 70 of our own cDNA clones and 150 clones from the literature (TABLE 7). V41 is a genomic V_κ–J_κ clone, and the others are cDNAs. The reading frame is preserved in all clones, which is probably due to the fact that the others were selected against. Complex mutations including deletions and insertions may be the result of transient misalignments during replication.[10] Extensive quasipalindromic sequences, as were observed in some systems,[11] have not been found in the V_κ genes, but the participation of small stem-loop structures in the mutational process cannot be excluded. It has not been possible to attribute the block mutations in a convincing way to transient misalignments or stem-loop structures.

THE EXTENT OF ALLELIC VARIATIONS OF κ GENES

The analysis of somatically mutated κ genes would be extremely difficult if the extent of allelic variation in these genes were high. But this seems not to be the case, at least not for the three V_κ genes of the B region and for C_κ (reviewed in Zachau[2]). The alleles of the other 73 germline V_κ genes known up to 1993 have been compiled in a list.[3] Twenty-two alleles of 19 genes were sequenced from the DNA of different individuals. The gene regions were identical in 9 alleles and different in 13 alleles (all together 3 and approximately 30 differences in exons I and II, respectively; numbers based on comparison with both the p and d copies of the κ locus). In only three V_κ genes, five or more differences were found between alleles. For three V_κ genes, defective and potentially functional alleles are known (FIG. 1).

The relatively low extent of allelic variation was confirmed in a recent study by Cox *et al.*,[8] in which 37 V_κ genes were amplified from genomic DNA by the polymerase chain reaction (PCR), and large parts of the second exons were

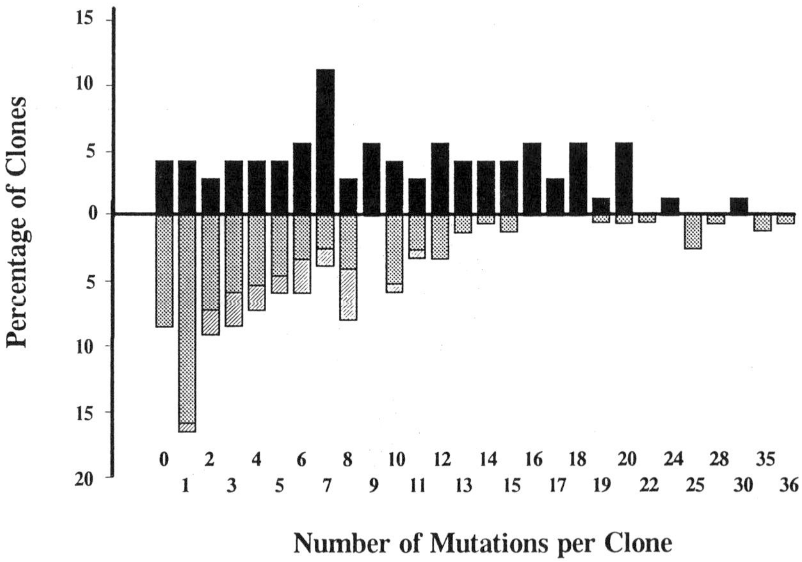

FIGURE 4. "Mutation profiles" seen in the 70 cDNA clones of individual ML (*black bars*) and in 150 cDNA and genomic V_κ–J_κ clones of the literature (*dotted bars*). Derivatives of the genes A27, L2, and L6 are shown by diagonal shading, if the respective clones were from patients with autoimmune diseases (19 clones). Block mutations are counted as single events.

TABLE 4. Replacement (R) and Silent (S) Mutations in V_κ Gene Regions and Particular Codons[a]

	Position	R	S	R/S
L'		**13**	**7**	**1,9**
	−1	0	0	0
FR1		**52**	**33**	**1,6**
	6	0	0	0
	10	0	0	0
	15	5	5	1
CDR1		**118**	**34**	**3,5**
	28	11	2	5,5
	30	22	4	5,5
	31	7	0	7/0
FR2		**58**	**47**	**1,2**
	35	0	0	0
	40	5	8	0,6
	44	0	3	0
	45	12	5	2,5
	47	1	5	0,2
CDR2		**54**	**24**	**2,3**
	50	16	4	4
	51	4	9	0,4
	53	16	4	4
FR3		**85**	**52**	**1,6**
	80	0	0	0
	81	5	5	1
	85	13	2	6,5
CDR3		**74**	**53**	**1,4**
	89	2	6	0,3
	91	12	5	2,4
	93	25	1	25
	94	13	1	13
	95	8	28	0,3

[a] 704 codon changes in our cDNA clones[5] are listed. Some block mutations affect more than one codon.

sequenced. Most of the sequences were found to be identical to the published ones (reviews in Refs. 3, 4) including DPK2 and DPK23 (identical to our sequences of L14 and L25, respectively). Twenty-one nucleotide substitutions were attributed to errors in the PCR reactions. Those four sequences, which were considered different or new by the authors because they were confirmed with clones from independent PCR amplifications, turned out in computer-aided comparisons to be orphons or closely related alleles of known V_κ genes (K. F. Schäble and R. Klein, unpublished):

- LFVK431 is identical to L1, except for an A instead of a G in position 182, which results in a Lys to Arg change.
- LFVK5 is identical to the V_κ orphon V118,[12] except for a C instead of a G in codon 87, converting a stop to a Tyr codon.
- DPK14 is an allele of the pseudogene pair A21/A5 with the characteristic

TABLE 5. Single Nucleotide and Block Mutations[a]

Clones Derived from	Single Nucleotide Mutations	Block Mutations[b]							
		2	3	4	5	6	8	Σ_{Blocks}	$\Sigma_{\text{Mutations in Blocks}}$
Ind. ML	644	62	6	4	1	1	–	74	169
Others	807	95	16	2	3	–	1	117	269

[a] Mutations in L′-FR3 were compiled for 70 cDNAs from individual ML and 129 cDNA and 21 genomic V_κ–J_κ clones from the literature.
[b] Numbers of blocks comprising two to eight nucleotides are listed.

insertion of an A in codon 39, which is not found in any potentially functional V_κII gene. The insertion is compensated for by a deletion in codon 45 in DPK14 and A5 but not in A21. In addition, DPK14 differs from both duplicates in codon 49 and from one of the duplicates in codons 21 and 83.

- DPK37 is said to be most similar to the V_κ orphon Z3,[8] but for orphons it can be decided only by extensive genomic mapping whether slightly differing sequences are alleles of known or of still-unknown structures.[13] This reservations holds in principle also for the above-mentioned LFVK5 sequence.

Several pieces of evidence indicate that we have found all V_κ genes of the locus,[1–5] but the evidence remains circumstantial until the approximately 2 Mb of the locus and its immediate surroundings have been sequenced. It was, therefore, gratifying to see that with a different approach, that is, by PCR amplification of 37 V_κ genes from genomic DNA,[8] no additional genes were found, and our conclusion concerning the low extent of allelic variation was confirmed.

In analyzing hypermutation, it has to be kept in mind that, in principle, the unrearranged maternal and paternal alleles of a germline V_κ gene of the same person should be known from whom the mutated V_κ–J_κ gene or cDNA is obtained. This demand has to be taken seriously if one is interested in the mutation behavior of specific single V_κ genes. It is less important if one considers the average extent and type of mutations of large numbers of rearranged genes or cDNAs, as it was done previously[5] and in the present contribution.

CONCLUDING REMARKS

The two aspects of this contribution, expression and hypermutation of κ genes, are closely related. The analysis of our own cDNA sequences and of the rearranged and expressed sequences of the literature was extensive and provides a basis for

TABLE 6. Transversions and Transitions[a]

	L	L′-FR3	CDR3	J	Σ
Transversions	10 | 8	298 | 229	80 | 111	22 | 23	410 | 379
Transitions	5 | 2	338 | 331	73 | 69	27 | 35	443 | 437
Transversions/Transitions	2 | 4	0,9 | 0,7	1,1 | 1,6	0,8 | 0,7	0,9 | 0,9

[a] The numbers in front and behind the vertical bars refer to mutations in cDNAs from individual ML and to clones from the literature, respectively.

TABLE 7. Deletions and Insertions

Clones[a]	Germline Gene	Deletions	Insertion	Codon
II.2	A19-A3	3 bp	—	94 (CDR1)
III.14	A27	—	3 bp	29/30 (CDR1)
V41	A27	3 bp	—	30 (CDR1)
III.7	L6	3 bp	—	90 (CDR3)
LS	L6	6 bp	—	27 + 28 (CDR1)
C4S5	L6	3 bp	—	30 (CDR1)
I.27	L12	15 bp	—	90-95 (CDR3)

[a] Clone designations are as in Klein *et al.*[5]

further studies. Why certain germline V_κ gene segments are used more frequently than others largely depends on the demands to the immune systems. Promoter strength as well as the three-dimensional structures of the resulting κ proteins are also important factors in determining the abundance, or, for a few potentially functional V_κ genes, also the absence of expression products.

After the types of sequence changes by hypermutation have been described in much detail, further progress must come from other directions. The sequence elements important for directing the hypermutation process, which seem to be located downstream of the mutated V_κ genes,[14,15] have to be analyzed further. It is a most interesting and challenging task to elucidate the mechanisms of the V_κ–J_κ rearrangement and of the subsequent hypermutation. Ultimately, the enzymes and accessory proteins involved in the processes should become known.

REFERENCES

1. ZACHAU, H. G. 1993. Gene **135:** 167–173.
2. ZACHAU, H. G. 1995. *In* Immunoglobulin Genes. 2nd Edit. T. Honjo & F. W. Alt, Eds. Academic Press. London. In press.
3. SCHÄBLE, K. F. & H. G. ZACHAU. 1993. Biol. Chem. Hoppe-Seyler **374:** 1001–1022.
4. SCHÄBLE, K. F., R. THIEBE, A. FLÜGEL, A. MEINDL & H. G. ZACHAU. 1994. Biol. Chem. Hoppe-Seyler **375:** 189–199.
5. KLEIN, R., R. JAENICHEN & H. G. ZACHAU. 1993. Eur. J. Immunol. **23:** 3248–3262; Appendix. R. Klein & H. G. Zachau. 1993. Eur. J. Immunol. **23:** 3263–3271.
6. KLEIN, R. 1994. Doctoral thesis. Fakultät für Biologie, Universität München, Germany.
7. KABAT, E. A., T. T. WU, H. M. PERRY, K. S. GOTTESMAN & C. FOELLER. 1991. NIH Publication Vol. II, No. 91-3242.
8. COX, J. P. L., I. M. TOMLINSON & G. WINTER. 1994. Eur. J. Immunol. **24:** 827–836.
9. BETZ, A. G., M. S. NEUBERGER & C. MILSTEIN. 1993. Immunol. Today **14:** 405–412.
10. KUNKEL, T. A. & A. SONI. 1988. J. Biol. Chem. **263:** 14784–14789.
11. RIPLEY, L. S. 1982. Proc. Natl. Acad. Sci. USA **79:** 4128–4132.
12. LÖTSCHER, E., F.-J. ZIMMER, T. KLOPSTOCK, K.-H. GRZESCHIK, R. JAENICHEN, B. STRAUBINGER & H. G. ZACHAU. 1988. Gene **69:** 215–223.
13. RÖSCHENTHALER, F., K. F. SCHÄBLE, R. THIEBE & H. G. ZACHAU. 1992. Biol. Chem. Hoppe-Seyler **373:** 177–186.
14. KLOBECK, H.-G., G. COMBRIATO & H. G. ZACHAU. 1987. Nucleic Acids Res. **15:** 4877–4888.
15. BETZ, A. G., C. MILSTEIN, A. GONZÁLEZ-FERNÁNDEZ, R. PANNELL, T. LARSON & M. S. NEUBERGER. 1994. Cell **77:** 239–248.
16. TIMMERS, E., M. M. HERMANS, M. E. M. KRAAKMAN, R. W. HENDRIKS & R. K. B. SCHUURMAN. 1993. Eur. J. Immunol. **26:** 619–624.

Regulation of Human λ Light Chain Gene Expression[a]

BONNIE B. BLOMBERG,[b] MICHELE A. GLOZAK, AND MARY E. DONOHOE

Department of Microbiology and Immunology
University of Miami School of Medicine
Miami, Florida 33101

ORGANIZATION AND EXPRESSION OF THE HUMAN λ LOCUS

Human lambda (λ) light chains, found on ~40% of immunoglobulin (Ig) molecules, are the products of gene segments that encode variable (V), joining (J), and constant (C) regions. The Vλ gene complex has been originally estimated, by Southern blotting,[1,2] to consist of ~70 members, of which about two-thirds are potentially functional.[3,4] The Vλ proteins encoded by the Vλ gene segments were initially divided by Kabat *et al.*[5,6] into six subgroups (I–VI) based on their amino acid homology (>80%) within the first framework region (FR1); and, more recently, studies at the molecular level have revealed the presence of other Vλ subgroups—for example, VλVII,[4,7–9] VλVIII[10] (also L. Y. Ch'ang *et al.*[10a]), VλIX,[4] and VλX.[11] The multigene nature of certain of the human Vλ subgroups or gene families has been recognized. The VλI subgroup is composed of three sub-subgroups (a, b, c), and germline genes for each of these have been identified.[3,4,8,12] At least 12 potentially functional VλII genes have been recognized.[4,13,14] On the basis of serologic and amino acid sequence data, the VλIII subgroup has been divided into sub-subgroups (a, b, c), and germline genes for each subgroup identified.[14–16] Germline genes encoding VλIV,[15] VλVI,[17] VλVII,[4,7–9] VλVIII,[10,10a] VλIX,[4] and VλX[11] have also been identified. Combriato and Klobeck physically linked several Vλs to Jλ1, only one of which was potentially functional (gene 3.1). They also determined that these Vλs are in the same orientation as the JCs and thereby must rearrange by deletion of the intervening sequence.[14] The organization of the human lambda locus to date is shown (FIG. 1). Evidence for both clustering[3,7] and nonclustering[3,8,14,15] of the Vλ gene families has been found. From the above data, the estimated number of members of each of the Vλ gene families is given (TABLE 1). Until all the Vλ genes are cloned and sequenced, some necessary fluctuation of the subgroup names and gene numbers will occur.

Studies from many laboratories have indicated that particular Ig V gene families are preferentially expressed during fetal development[24] and in certain disease states—for example, B-cell-related malignant or reactive immunoproliferative disorders.[25–27] In the latter, the nonstochastic representation of Ig V genes (reviewed in Refs. 28 and 29) has been taken as evidence that antigenic (external or

[a] This work was supported by National Institutes of Health Grant AI21870 to B.B.B.

[b] To whom correspondence should be sent at Department of Microbiology and Immunology, P.O. Box 016960 (R138), Miami, FL 33101. Internet address: bblomber@-molbio.med.miami.edu.

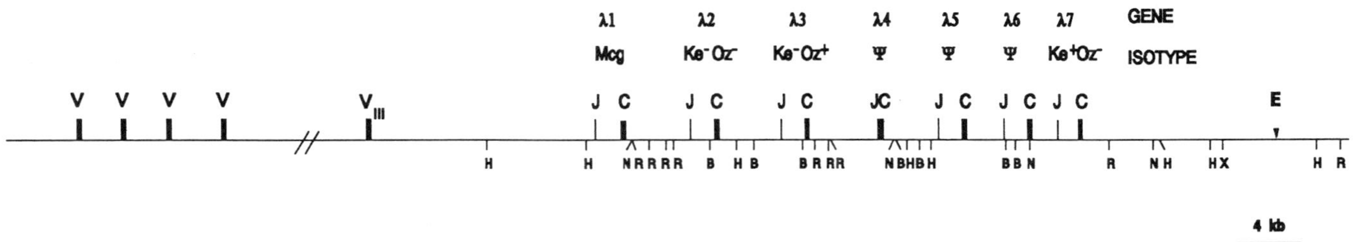

FIGURE 1. Organization of the human λ light chain gene complex. V, J, and C exons are denoted by black boxes. Distance within the V region is not to scale. The 3′ λ enhancer is indicated as "E" and is 11.7 kb 3′ of Cλ7. (Based on data from Refs. 14, 18, 19, 20, and 21.)

TABLE 1. Genes in Vλ Families

Vλ Family	Approximate Number of Genes
VλI (a, b, c)	12–19
VλII	10–17
VλIII (a, b, c)	10–14
VλIV	2
VλVI	2–6
VλVII	3–10
VλVIII	4
VλIX	1
VλX	2
VλT1[a]	1

NOTE: Data for VλI, VλII, VλIII, VλIV, and VλVI are as defined by Kabat *et al.*[5,6] Data for VλVII, VλVIII, VλX, and VλT1 from Refs. (4, 7–9), 10, 11, and 22, respectively. Within the first framework region, VλII and VλV are homologous,[14,23] and these sequences have been placed into subgroup II.[9]

[a] VλT1 was placed into a new VλV subgroup.[9]

autoantigen) stimulation is involved in the disease process. Alternatively, the disease may arise at a particular stage of B-cell development or involve a specific stage of B-cell activation, as is marked by the expression of the CD5 cell surface marker. An additional alternative may be that particular V genes are activated coordinately with other genes (oncogenes or transcription factors) responsible for the disease, either as a cause or as an effect. For example, particular transcription factors involved in tumorigenesis (e.g., c-myc) may preferentially cause transcriptional activation and rearrangement of particular Vs and Cs. Whatever the mechanism, determining the expression of Ig genes in normal and disease states, including autoimmune disorders,[30–32] may provide new information regarding disease pathogenesis and novel therapeutic strategies.

The availability of anti-human Vλ subgroup-specific polyclonal and monoclonal antibodies has made it possible to determine Vλ subgroup expression serologically in normal and disease samples[33,34] (TABLE 2). Of note was the finding that certain Vλ subgroups, particularly VλII and VλVI, are nonrandomly expressed in λ-type multiple myeloma (MM), light chain-associated (AL) amyloidosis, and Waldenström's macroglobulinemia (WM). Whereas the concentration of IgVλII molecules

TABLE 2. Vλ Subgroup Expression

Subgroup[a]	Normal	AL	MM	WM
VλI	40	10	27	25
VλII	3	20	28	63
VλIII	43	29	39	6
VλIV	5	0	5	6
VλVI	5	41	0	0
VλVIII	3	0	1	0

NOTE: Vλ-subgroup distribution (%) in sera from normal persons, patients with amyloidosis (AL), multiple myeloma (MM), and Waldenström's macroglobulinemia (WM). Data from A. Solomon laboratory, Ozaki *et al.*[34]

[a] VλI, VλII, VλIII, VλIV, and VλVI designation as per Kabat *et al.*[5,6]; VλVIII designation as per Deftos *et al.*[10]

in normal sera is only 3%, this subgroup represented 28, 20, and 63% of monoclonal Igλ proteins from patients with MM, AL amyloidosis, and WM, respectively. The VλVI subgroup is also overrepresented in AL amyloidosis (41% vs. 5% in serum Igλ of normal persons) and, to date, VλVI monoclonal serum or urinary proteins have only been found in patients with AL amyloidosis. These data suggest that the nonstochastic expression of Vλ genes may result from differential clonal selection due to antigenic stimulation by particular pathogens or, in the case of amyloidosis, unique properties of certain λ chains to form pathologic deposits. Alternatively, the (apparently) biased Vλ expression in these diseases may reflect a developmental state at which the Vλs are overly represented (i.e., immature, IgM^+ B cells).

Our laboratory has been studying molecular mechanisms for Cλ gene expression in normal and disease states. In particular, we wish to determine the expression of the λ locus so as to better understand and potentially regulate abnormal λ expression. Our approach has been first to determine the representation of Cλ genes in normal adults. To do this, we have developed a PCR gene-specific probe strategy to discriminate the products of this very homologous family of genes. FIGURE 2 shows the sequences of the functional Cλ1, Cλ2, Cλ3, and Cλ7 genes, three of the primers used for PCR of the reverse transcriptase (RT) products, and the 15-mer oligonucleotide probes used for hybridization. As seen in FIGURE 3, Cλ gene products were readily identified with specific nucleotide probes. Notably, Cλ2 and Cλ3, which differ by only one nucleotide, were readily distinguished using 15-bp oligonucleotide probes differing by one base pair at position 7. We have performed Cλ expression studies on human hybridomas (HuHy) derived from tonsils of normal children and on PBL from normal adults, by hybridization of the isotype-specific probes to the RT-PCR products (FIG. 3). In the hybridomas and PBL, Cλ2 is the predominant gene expressed; the amount of Cλ3 is approximately one-half the level (or one-half the number of cells) of Cλ2. Among 28 hybridomas and cell lines we have tested, 19 were Cλ2, eight were Cλ3, and only one was Cλ7. Based on agarose gel electrophoresis and the indicative probes, we found that RT-PCR products of PBL also showed the level of Cλ2 to be about twice (66%) that of Cλ3 (33%), whereas Cλ1 and Cλ7 were undetectable by the methods employed. From our previous studies,[19] we showed the amount of Cλ7 in a PBL sample to be 3 to 6% of the amount of Cλ2, Cλ3, and Cλ7 combined. We do not believe that the polymorphism of the Cλ2-Cλ3 region, duplicating probably the Cλ2 gene, with an allele frequency of 0.05 for two copies, 0.196 for three copies, and 0.01 for four copies[37] could explain the overexpression of Cλ2 in myeloma patients, normal tonsil hybridomas, peripheral blood, and bone marrow, but we are currently testing this hypothesis by correlating Cλ2 expression in nonduplicated homozygotes versus heterozygotes with duplicated Cλ2 genes.

We have extended these studies to include a fetal liver library. A 15-gestation week (105-day) human fetal liver cDNA library (furnished by Dr. H. Schroeder, U. Alabama, Birmingham) was screened by hybridization with a Cλ3 probe (0.8-kb *Bgl*II-*Eco*RI, Ref. 18). Out of 4×10^5 plaques, 51 positive clones were identified, five of which have been sequenced. Four of the clones were Cλ1, and one expressed the λ-like gene, 14.1. In view of the fact that the B-lineage cell population at 13 to 23 weeks has been shown to contain pro/pre-B cells and B cells in the ratio 1.2 to 1,[38] these results indicate that the fetal Cλ gene expression may be dramatically different from that in adult tissue where Cλ2 and Cλ3 predominate.

To study further whether the skewed expression of Cλ2 in adult tissue could be due to preferential transcription and rearrangement of Cλ2, we sorted adult bone marrow cells into $CD19^+/\mu^-$ (pre-B) and $CD19^+/\mu^+$ (B) cell populations

```
                   c2-7                                                    cLam
          -------------------->                                   -------------------->
                    Mcg   Mcg     *       AciI          *          *          *       80
Cλ1  GGTCAGCCCAAGGCCAACCCCACTGTCACTCTGTTCCCGCCCTCCTCTGAGGAGCTCCAAGCCAACAAGGCCACACTAGT
Cλ2  --------------TGC----T-G--------------------------------T--------------------G--
Cλ3  --------------TGC----T-G--------------A-----------------T--------------------G--
Cλ7  --------------TGC---AT-G--------------A-----------------T--------------------G--

                               Cλ1
                       ---------------
              *         *  AvaI   *         *         *    Kern*         *AciI   160
Cλ1  GTGTCTGATCAGTGACTTCTACCCGGGAGCTGTGACAGTGGCCTGGAAGGCAGATGGCAGCCCCGTCAAGGCGGGAGTGG
Cλ2  ------C--A--------------------C------------------------A------------------------
Cλ3  ------C--A--------------------C------------------------A------------------------
Cλ7  ------CG-A--------------------C----------------------------------------T--------

                                                          Cλ2
                                                   --------------
              Mcg       *         *      AciI      *         *      AvaI         240
Cλ1  AGACCACCAAACCCTCCAAACAGAGCAACAACAAGTACGCGGCCAGCAGCTACCTGAGCCTGACGCCCGAGCAGTGGAAG
Cλ2  ---------C------------A-----------------------------T--------------T------------
Cλ3  ---------C------------A--------------------------------------------T------------
Cλ7  ----------------------A--------------T---------------------------------------G--

          Cλ3
     --------------      Cλ7
           Oz *     ---------------        *         *         *         *     318
Cλ1  TCCCACAGAAGCTACAGCTGCCAGGTCACGCATGAAGGGAGCACCGTGGAGAAGACAGTGGCCCCTACAGAATGTTCA
Cλ2  ------------------------------------------------------------------------------
Cλ3  -------A----------------------------------------------------------------------
Cλ7  ----------------------G-------------------------------------------G-------C--T

                                                                      c3pall
     Term    FokI   *         *         *                     <-------------------393
Cλ1  TAGGTTCCCAACTCTAACCCCACCCACGGGAGCCTGGAGCTGCAGGATCCCAGGGGAGGGGTCTCTCTCCCCATC
Cλ2  -------T----C--C-----CA---------A--A-------------------------------CT----C-
Cλ3  -------T--T-C--C-----CA---------A--A-------------------------------CT----C-
Cλ7  ----CC---G--C--C-----------A--G---------------T--------------------GC-------
```

FIGURE 2. Nucleotide sequence of the functional human Cλ1, Cλ2, Cλ3, and Cλ7 genes. RT-PCR products were produced with either the C2-7 and c3pall or cLam and c3pall primers, as indicated, to give products of 379 bp or 327 bp, respectively. The λ-like gene 14.1 may be amplified with the cLam and c3pall primers, but it is easily distinguished by restriction enzyme sites and a distinct oligomer probe and is not considered in this report. Probe locations are denoted by Cλ1, Cλ2, Cλ3, and Cλ7. Certain restriction enzyme sites that distinguish the Cλ gene products (i.e. *Aci*I, *Ava*I, *Fok*I) and the positions of the isotypic markers (Mcg, Ke, and Oz) are shown, with the codon underlined. The termination codon is marked "Term" and underlined, at position 319–321.

(FIG. 4) and performed RT-PCR and restriction enzyme analysis (FIG. 5). Both the pre-B- and B-cell populations showed mostly Cλ2 and, to a much lesser extent, Cλ3 expression (the lack of a 141-bp fragment with the *Aci*I digest [see legend to FIG. 3] indicated that Cλ7 is below our level of detection in these experiments). The *Ava*I digest showed only the 283-bp fragment, confirming the presence of Cλ2 and Cλ3, but not Cλ1 or Cλ7 gene products. The *Fok*I digest gave predominately the uncut 327-bp fragment, consistent with genes Cλ1, Cλ2, and Cλ7; the presence of Cλ3 can be inferred by the 253-bp fragment consistent with Cλ3. From these results, we conclude that Cλ2 transcripts are more prevalent in both adult B and pre-B cells and that in pre-B cells, they likely represent sterile transcripts. Studies are in process to confirm this hypothesis.

Whether or not there is preferential expression of Cλ genes in monoclonal

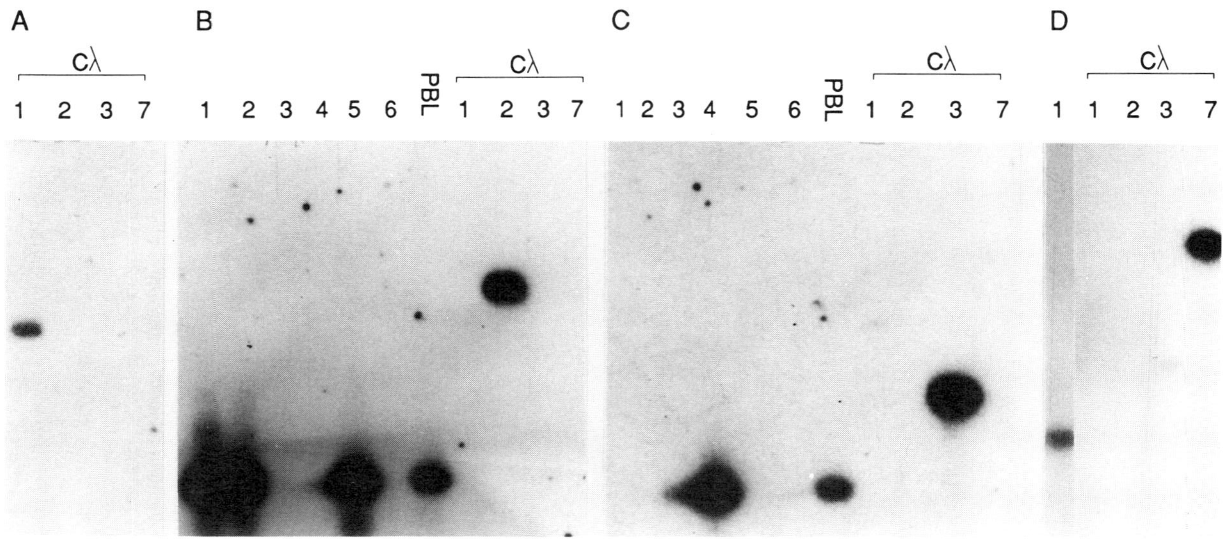

FIGURE 3. Southern blotting analysis of human Cλ gene expression. The RT-PCR from HuHy RNA hybridized to Cλ isotype-specific oligonucleotide probes is shown. RNA from different hybridomas (HuHy) from fusion of a derivative of WI-L2-729[35,36] was reverse transcribed (RT) and submitted to the polymerase chain reaction (PCR) with the C2-7 and c3pall primers (see FIG. 2). Different human hybridomas, *lanes 1–6* in panels **B, C** and *lane 1* in panel **D,** and PBL RT-PCR products were separated by electrophoresis on a 1.5% agarose gel, blotted to nitrocellulose, and hybridized with oligonucleotide probes specific for Cλ1 **(panel A),** Cλ2 **(panel B),** Cλ3 **(panel C)** or Cλ7 **(panel D)** labeled with λ-^{32}P-ATP. Sequence and location of probes is shown in FIGURE 2. Positive controls (Cλ1, 2, 3, 7 plasmid DNAs), indicated as the Cλ bracket over the numbers, 1, 2, 3, and 7, are included for each probe to show probe specificity. (Cλ1 was not detected in any cell line.)

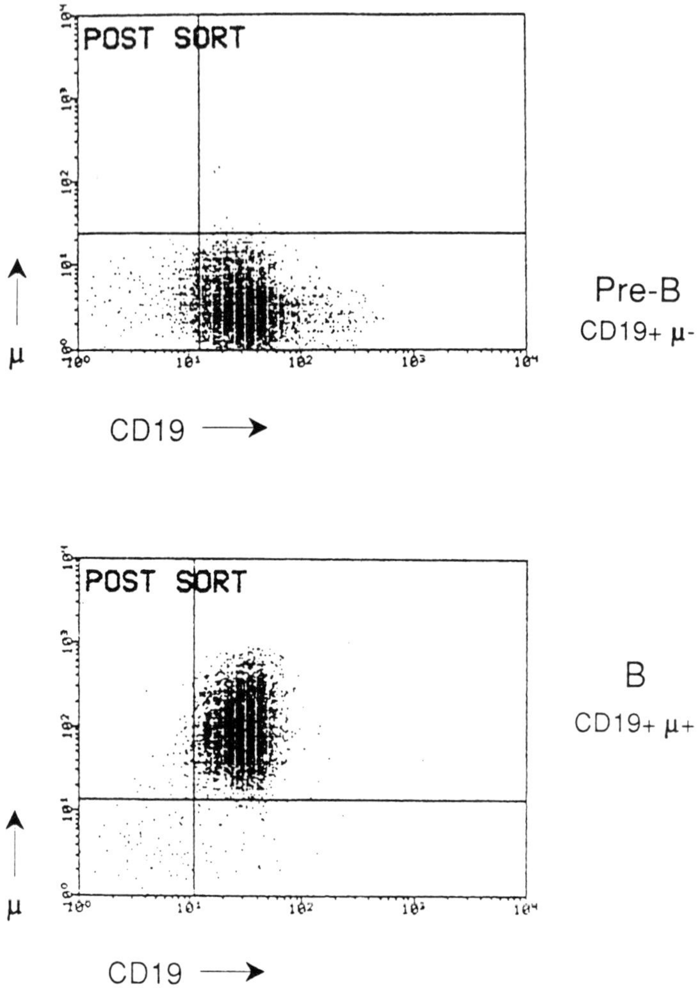

FIGURE 4. Adult bone marrow pre-B/B separation. Adult bone marrow was frozen in 10% DMSO, 10% FCS in RPMI medium, thawed and separated on Ficoll-hypaque to obtain leukocytes, and stained with anti-CD19, which marks all pro-B, pre-B and B cells, and anti-μ marking B cells. Cells were sorted into the CD19$^+$/μ^- (pro-B/pre-B) (*upper panel*) and CD19$^+$/μ^+ (B) (*lower panel*) cell populations. The post-sorted profiles are shown. The pre-B-cell population had 0.22% CD19$^+$/μ^+ cells, and the B-cell population had 1.1% CD19$^+$/μ^- cells. Sorted cells were counted and resuspended in the extraction buffer for QuickPrep micro-mRNA purification kit from Pharmacia.

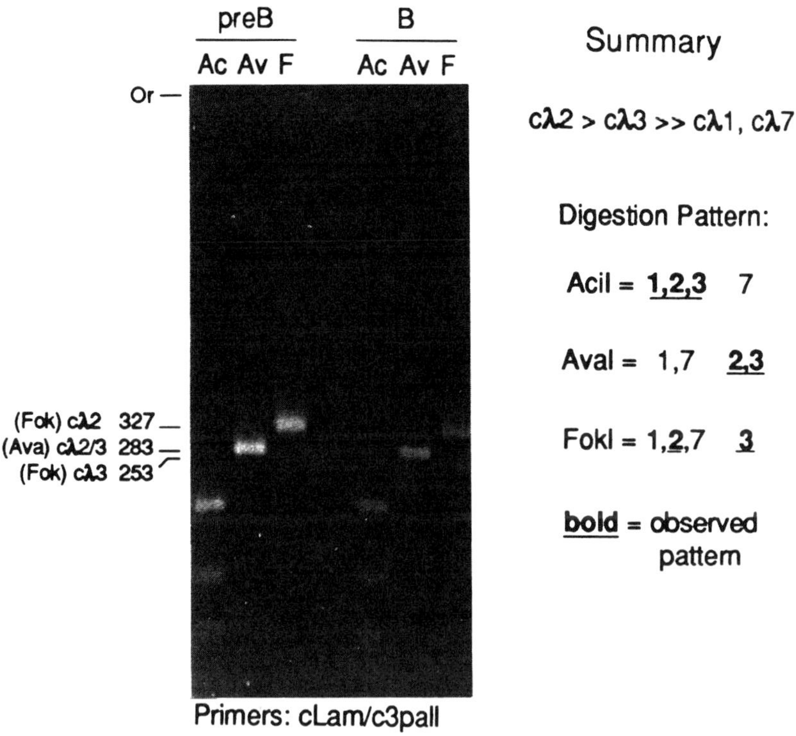

FIGURE 5. Cλ gene expression in adult bone marrow pre-B and B cells. RNA from cells separated as in FIGURE 4 was reverse transcribed and amplified by PCR according to the procedure from Perkin-Elmer. The PCR product was digested with the enzymes indicated, electrophoresed through a 1.5% agarose and stained with EtBr. The *Aci*I digest produces a pattern of 186 bp, 93 bp, and 48 bp for genes Cλ1, Cλ2, Cλ3, and 186 bp and 141 bp for gene Cλ7. The *Ava*I digest produces a pattern of 159 bp, 124 bp, and 44 bp for genes Cλ1 and Cλ7 and 283 bp and 44 bp for Cλ2 and Cλ3. The *Fok*I digest produces an uncut fragment of 327 bp for genes Cλ1, Cλ2, and Cλ7 and fragments of 253 bp and 74 bp for gene Cλ3.

malignant or benign immunoproliferative disorders has not been established. In the case of MM, 15, 61, and 18% of myeloma proteins or Bence Jones proteins have been classified chemically or serologically as Cλ1 (Mcg), Cλ2 (Ke^-Oz^-), and Cλ3 (Ke^+Oz^-).[6,18,39,40] The remaining 6% have been designated Ke^+Oz^-. However, although the Cλ7 gene does encode Ke^+Oz^- determinants (FIGS. 1, 2), it is unlikely that the Cλ7 gene encodes the Ke^+Oz^- myeloma proteins because 5-bp changes in Cλ7 would be needed. We have proposed an alternative, that the Ke^+Oz^- protein is encoded by a 1-bp mutation from the Cλ2 Ke^-Oz^- gene (see FIG. 2).[19] To date, there has been no conclusive demonstration that an authentic Cλ7 gene product is expressed by malignant monoclonal plasma cell populations as evidenced from sequence analyses of myeloma proteins or Bence Jones proteins.[6,19] Our data, in comparison with that of multiple myeloma, suggest a slight elevation of Cλ1, from less than 5% to 15%, and a slight decrease of Cλ7 from 3 to 6% to less than 1%, for normal versus multiple myeloma, respectively. If this differential expression bears out, it may be another indicator of an ontogenetically immature B cell induced to multiple myeloma (see also L. Kunkel, these proceedings).

REGULATORY MECHANISMS OF THE λ LOCUS

We are especially interested in the transcriptional regulation of the entire Cλ locus, as it is most likely important in accessibility and VJ recombination, as well as increased transcriptional expression after VJ rearrangement, and, thus, important for expression of the complete λ light chain gene.

We and others have contributed to the knowledge of the organization and expression of the human Cλ locus. Hieter *et al.*[20] initially showed that the human λ gene complex consisted of six clustered C-region gene segments, three of which they determined encoded the nonallelic isotypes Mcg (Cλ1), Ke^-Oz^- (Cλ2), and Ke^+Oz^- (Cλ3) (see FIG. 6). Remarkably, the organization of the Jλ and Cλ gene segments differ from the corresponding elements in the kappa (κ) locus, where the J_κ segments are clustered together and precede the single C_κ gene. We showed

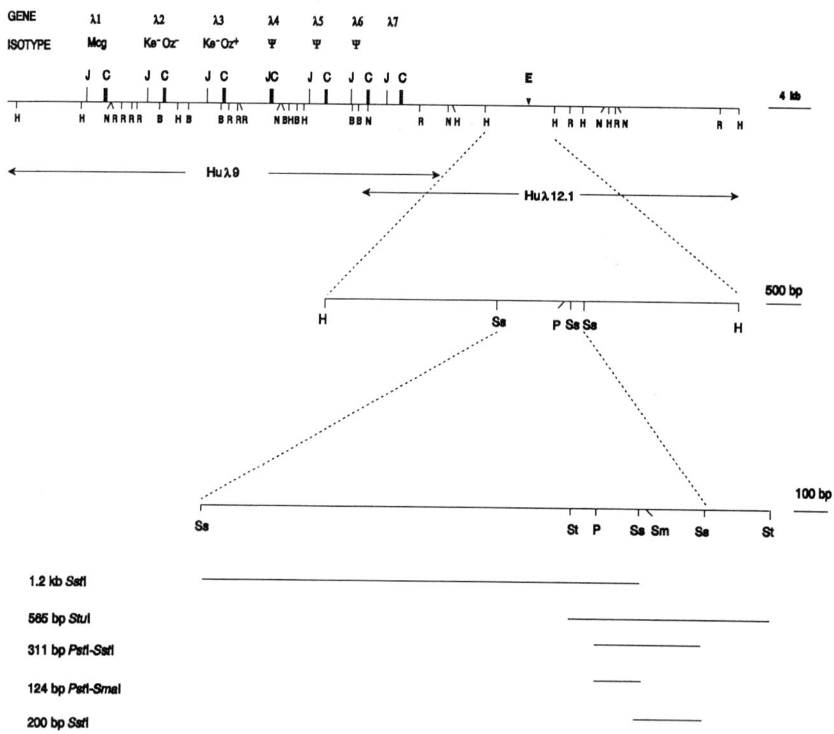

FIGURE 6. Map of Cλ constant region showing 3′ enhancer and constructs. Restriction enzymes: B = *Bam*HI, H = *Hin*dIII, N = *Nhe*I, P = *Pst*I, R = *Eco*RI, Sm = *Sma*I, Ss = *Sst*I, and St = *Stu*I. Lines at the bottom of figure denote fragments tested for enhancer activity in the reporter vector OPλCAT (see TABLES 3 and 4).

TABLE 3. Transfection of the Human B-Cell Line Namalawa Using the Described Fragments in the Reporter Vector OPλCAT[a]

	Relative CAT Activity
1.2-kb *Sst*I	10X
565-bp *Stu*I	20X
311-bp *Pst*I-*Sst*I	20X
124-bp *Pst*I-*Sma*I	10X
200-bp *Sst*I	1X
OPλCAT	1X

[a] All data are the result of two independent transfections. Transfection efficiency was normalized by cotransfection with pSVβGAL and subsequent β-Gal assay. Extracts containing equivalent amounts of β-Gal activity were then assayed for CAT activity. CAT conversion was determined by scintillation counting of the TLC plate. Relative CAT activity is expressed as fold increase over the vector alone. Fragment locations shown in FIGURE 6.

that the three Cλ regions were each preceded by a J segment (i.e., Jλ1-Cλ1, Jλ2-Cλ2, Jλ3-Cλ3)[18,41] similar to that found for the mouse Jλ and Cλ gene segments.[42–46] Additionally, the presence of other Jλ gene segments (Jλ4, Jλ5, Jλ6, and Jλ7) was recognized.[14,19,21,47] The JCλ4, JCλ5, and JCλ6 genes are pseudogenes; however, we[19] and others[14,21] have demonstrated that JCλ7 is potentially functional as evidenced by an open reading frame. Additionally, we have found it expressed in mRNA obtained from PBL, its association with Vλ in PBL cDNA, and its expression in a human B-cell lymphoma[19] that contained a secondary Vλ rearrangement to JCλ7 (see Berinstein *et al.*[22]).[c] Based on Southern and cosmid analyses, we conclude the four functional JCλ units (JCλ1, JCλ2, JCλ3, and JCλ7), the three pseudogenes (JCλ4, JCλ5, JCλ6), and ΨCλ1 (not on chromosome 22[48]) represent all of the human Cλ genes. These Cλ genes, as well as the λ-like genes (14.1, 16.1, and 16.2), have been mapped and physically linked on chromosome 22.[49,50]

Although transcriptional regulation of the human heavy chain and κ light chain loci has been well studied,[51,52] relatively little is known about regulation of the λ locus. Our laboratory has identified the first transcriptional enhancer for the human λ light chain locus. The human λ enhancer (HuEλ) is located 11.7 kb downstream of Cλ7, the most 3′ λ constant region gene (see FIG. 6), similar in position to the murine λ enhancers that are located downstream of each JC2-JC4 and JC3-JC1 gene cluster.[54] Enhancer activity was originally described as residing on a 1.2-kb *Sst*I fragment, with partial activity retained in the most 3′ 111 bp. This 3′ portion shares 72 to 74% identity with the murine λ enhancers and is called the enhancer core. Transient transfection and CAT assays showed that HuEλ is tissue specific, as it is functional in B, but not in T, cell lines. This enhancer is also independent of NF-κB, as it is functional in the mouse myeloma S107, which lacks NF-κB. As with all known enhancers, HuEλ is position and orientation independent.

More recently, we have determined that the complete human λ enhancer resides on a 311-bp fragment, which includes 200 bp downstream of our originally de-

[c] The JCλ6 gene complex, thought initially to be functional,[47] was shown to be a pseudogene[14,19,21] and not simply the result of a polymorphism in the human population. This was demonstrated by analysis of DNA obtained from more than 30 individuals and the determination that Cλ6 had a restriction enzyme pattern of the pseudogene, with insertion of 4 bp creating a termination codon, in all tested samples.[12]

TABLE 4. Transfection of the Human Pre-B-Cell Line 697 Using the Described Fragments in the Reporter Vector OPλCAT[a]

	Relative CAT Activity
124-bp *Pst*I-*Sma*I	6X
311-bp *Pst*I-*Sst*I	17X
565-bp *Stu*I	2X
OPλCAT	1X

[a] All data are the result of two independent transfections. Transfections were normalized for protein content. CAT conversion was determined by scintillation counting of the TLC plate as in legend to TABLE 3.

scribed core enhancer (TABLE 3). The 111-bp core enhancer, as well as a slightly larger 124-bp core fragment, retains one-third to one-half the activity of the complete enhancer. The remaining 200-bp fragment has no activity when assayed alone, but is necessary for complete enhancer activity. This suggests that motifs important for the upregulation of enhancer activity reside within the 200-bp fragment. Methylation interference analysis revealed a protein interaction site downstream of the core enhancer that we have designated the HELP (human enhancer lambda protein) motif. The addition of this motif has been shown to augment the activity of the core element. Protein binding to the HELP motif appears to be restricted to cells of hematopoietic lineage.[55]

We have also found the 311-bp complete enhancer to be active in pre-B-cell lines. Interestingly, in the presence of 5′ and 3′ flanking sequences, its activity in these cells (but not mature B cells) was greatly reduced (TABLE 4). This suggests the presence of developmentally regulated negative elements flanking the human λ enhancer that prevent its activity at the pre-B-cell stage. By methylation interference analysis, we have identified protein-binding sites within both the 5′ and 3′ flanking sequences that may participate in the downregulation. The two sites are 67% identical and are able to compete with each other for binding to the same factor(s). The presence of such negative elements may prevent the activity of HuEλ at a developmentally incorrect time, that is, when heavy chain genes are rearranging but light chain genes are recombinationally inactive.

The major role of the HuEλ is likely to upregulate transcription at the B-cell and plasma cell stages when large quantities of Ig are needed, as has been found for the 3′ Ig enhancers for κ[56] and heavy[57,58] chains. The 3′ λ enhancer, HuEλ, may also be involved in early B-cell development, but other intronic enhancers will also likely be implicated.

CONCLUSIONS AND FUTURE DIRECTIONS

The lambda locus consists of multiple V region gene segments upstream of four functional JCλ units, JCλ1, JCλ2, JCλ3, and JCλ7. The Cλ2 gene is preferentially expressed in normal PBL and adult B-lineage cells. Transcriptional regulation of the human λ locus involves a strong 3′ enhancer, 11.7 kb 3′ of Cλ7. Complete enhancer activity resides on a 311-bp fragment. A larger fragment containing the 311-bp enhancer as well as 5′ and 3′ flanking sequences is not active in pre-B cells, but is active in B cells. We expect that additional positive regulatory regions will be important for the increased (sterile) transcription of Cλ2 in pre-B lymphocytes, and studies to test this hypothesis are currently being performed.

ACKNOWLEDGMENTS

We are indebted to Dr. Alan Solomon for many thoughtful discussions and for critically reading the manuscript. We wish to thank Laura Clapp Eiras and Melvin Denis for help with experiments, figures, and references. We appreciate Drs. Alan Solomon and Neil Berinstein for sharing their unpublished work.

REFERENCES

1. Tsujimoto, Y. & C. M. Croce. 1984. Molecular cloning of human immunoglobulin λ chain variable sequence. Nuc. Acids. Res. **12:** 8407–8414.
2. Anderson, M. L. M., L. Brown, E. McKenzie, J. E. Kellow & B. D. Young. 1985. Cloning and sequence analysis of an Igλ light chain mRNA expressed in the Burkitt's lymphoma cell line EB4. Nuc. Acids Res. **13:** 2931–2941.
3. Daley, M. D., T. Olee, H. Deng, R. W. Soto-Gil, P. P. Chen & K. A. Siminovitch. 1992. Molecular characterization of the human immunoglobulin VλI germline gene repertoire. Mol. Immunol. **29:** 1031–1042.
4. Williams, S. C. & G. Winter. 1993. Cloning and sequencing of human immunoglobulin Vλ gene segments. Eur. J. Immunol. **23:** 1456–1461.
5. Kabat, E. A., T. T. Wu & H. Bilofsky. 1979. Sequence of Immunoglobulin Chains. Tabulation and Analysis of Amino Acid Sequences of Precursors, V-regions, C-regions, J-chain, and β_2-Microglobulins. U.S. Department of Health, Education and Welfare, Washington, DC.
6. Kabat, E. A., T. T. Wu, H. M. Perry, K. S. Gottesman & C. Foeller. 1991. *In* Sequences of Proteins of Immunological Interest. Fifth edit. U.S. Department of Health and Human Services. Bethesda, MD.
7. Anderson, M. L. M., M. F. Szajnet, J. C. Kaplan, L. McColl & B. D. Young. 1984. The isolation of a human Ig Vλ gene from a recombinant library of chromosome 22 and estimation of its copy number. Nuc. Acids. Res. **12:** 6647–6661.
8. Alexandre, D., P. Chuchana, F. Brockly, A. Blancher, G. Lefranc & M.-P. Lefranc. 1989. First genomic sequence of a human Ig variable λ gene belonging to subgroup I. Functional genes, pseudogenes & vestigial sequences are interspersed in the IGLV locus. Nuc. Acids. Res. **17:** 3975.
9. Chuchana, P., A. Blancher, F. Brockly, D. Alexandre, G. Lefranc & M.-P. Lefranc. 1990. Definition of the human immunoglobulin variable lambda (IGLV) gene subgroups. Eur. J. Immunol. **20:** 1317–1325.
10. Deftos, M., R. Soto-Gill, M. Quan, T. Olee & P. P. Chen. 1994. Utilization of a potentially universal downstream primer in the rapid identification and characterization of Vλ genes from two new human Vλ gene family. Scand. J. Immunol. **39:** 95–103.
10a. Ch'ang, L. Y., M. Schell, C. Ringelberg, D. T. Weiss & A. Solomon. 1995. Molecular characterization of a human VλVIII germline gene. Mol. Immunol., in press.
11. Stiernholm, B. J. N., N. Kuzniar & N. L. Berinstein. 1994. Identification of a new human Vλ gene family—VλX. J. Immunol. **152:** 4969–4975.
12. Bernard, F., P. Chuchana, J. P. Frippiat, L. Bulewala & M.-P. Lefranc. 1990. Genomic sequence of IGLV1S2 a human Ig variable lambda gene belonging to subgroup I. Nuc. Acids Res. **18:** 7139.
13. Brockly, F., D. Alexandre, P. Chuchana, S. Huck, G. Lefranc & M.-P. Lefranc. 1989. First nucleotide sequence of a human Ig variable λ gene belonging to subgroup II. Nuc. Acids. Res. **17:** 3976.
14. Combriato, G. & H.-G. Klobeck. 1991. Vλ and Jλ-Cλ gene segments of the human

immunoglobulin λ light chain locus are separated by 14 kb and rearrange by a deletion mechanism. Eur. J. Immunol. **21:** 1513–1522.
15. Daley, M. D., V. Misener, X-Y. Liu, P. P. Chen & K. A. Siminovitch. 1992. Molecular analysis of human immunoglobulin Vλ germline genes: Subgroups VλIII and VλIV. Mol. Immunol. **29:** 1515–1518.
16. Fang, Q., C. C. Kannapell, F. Gaskin, A. Solomon, W. J. Koopman & S. M. Fu. 1994. Human rheumatoid factors with restrictive specificity for rabbit immunoglobulin G: Auto- and multi-reactivity, diverse V_H gene segment usage and preferential usage of V_λ IIIb. J. Exp. Med. **179:** 1445–1456.
17. Ch'ang, L. Y., Y. Chin-Ping, L. Best, M. Schell & A. Solomon. 1994. Identification and characterization of a functional human Ig VλVI germline gene. Mol. Immunol. **31:** 531–536.
18. Udey, J. A. & B. Blomberg. 1987. Human λ light chain locus: Organization and DNA sequences of three genomic J regions. Immunogenetics **25:** 63–70.
19. Bauer, T. R. Jr. & B. Blomberg. 1991. The human lambda light chain immunoglobulin locus: Recharacterization of JCλ6 and identification of a functional JCλ7. J. Immunol. **146:** 2813–2820.
20. Hieter, P. A., G. F. Hollis, S. J. Korsmeyer, T. A. Waldmann & P. Leder. 1981. Clustered arrangement of immunoglobulin λ constant region genes in man. Nature **294:** 536–540.
21. Vasicek, T. & P. Leder. 1990. Structure and expression of the human immunoglobulin λ genes. J. Exp. Med. **172:** 609–620.
22. Berinstein, N., S. Levy & R. Levy. 1989. Activation of an excluded immunoglobulin allele in a human B lymphoma cell line. Science **244:** 337–339.
23. Solomon, A. & D. T. Weiss. 1987. Serologically defined V region subgroups of human λ light chains. J. Immunol. **139:** 824–830.
24. Schroeder, Jr., H. W., J. L. Hillson & R. M. Perlmutter. 1987. Early restriction of the human antibody repertoire. Science **238:** 791–793.
25. Humphries, C. G., A. Shen, W. A. Kuziel, J. D. Capra, F. R. Blatner & P. W. Tucker. 1988. A new human immunoglobulin V_H family preferentially rearranged in immature B-cell tumors. Nature **331:** 446–449.
26. Kipps, T. J., E. Tomhave, L. F. Pratt, S. Duffy, P. P. Chen & D. A. Carson. 1989. Developmentally restricted immunoglobulin heavy chain variable region gene expressed at high frequency in chronic lymphocytic leukemia. Proc. Natl. Acad. Sci. USA **86:** 5913–5917.
27. Logtenberg, T., M. E. M. Schutte, G. Inghirami, J. Berman, F. H. J. Gmelig-Meyling, R. A. Insel, D. M. Knowles & F. W. Alt. 1989. Immunoglobulin V_H gene expression in human B cell lines and tumors: Biased V_H gene expression in chronic lymphocytic leukemia. Int. Immunol. **1:** 361–365.
28. Kipps, T. J., L. Z. Rassenti, S. Duffy, T. Johnson, R. Kobayshi & D. A. Carson. 1992. Immunoglobulin V gene expression in CD5 B-cell malignancies. Ann. N.Y. Acad. Sci. **651:** 373–383.
29. Schroeder, Jr., H. W. & G. Dighiero. 1994. The pathogenesis of chronic lymphocytic leukemia: Analysis of the antibody repertoire. Immunol. Today **15:** 288–294.
30. Casali, P., B. S. Parbhakar & A. L. Notkins. 1988. Characterization of multireactive autoantibodies and identification of Leu-1$^+$ B lymphocytes as cells making antibodies binding multiple self and exogenous molecules. Int. Rev. Immunol. **3:** 17–45.
31. Pascual, V. & J. D. Capra. 1991. Human immunoglobulin heavy-chain variable region genes: Organization, polymorphism and expression. Adv. Immunol **49:** 1–74.
32. Logtenberg, T., M. E. M. Scutte, F. H. J. Gmelig-Meyling & J. H. van Es. 1992. Molecular approaches to the study of human B-cell and (auto) antibody repertoire generation and selection. Immunol. Rev. **128:** 23–47.
33. Abe, M., M. S. Ozaki, D. Wolfenbarger, M. dePrana-Hart, D. T. Neiss & A. Solomon. 1994. Variable-region subgroup distribution among λ-type immunoglobulins in normal human serum. J. Clin. Lab. Anal. **8:** 4–9.
34. Ozaki, S., M. Abe, D. Wolfenbarger, D. T. Weiss & A. Solomon. 1994. Preferential expression of human λ-light chain variable-region subgroups in multiple myeloma,

AL amyloidosis and Waldenström's macroglobulinemia. Clin. Immunol. Immunopath. **71:** 183–189.
35. HEITZMANN, J. G. & M. COHN. 1993. The WI-L2-729-HF_2 human hybridoma system: Stable hybrids at high frequency. Mol. Biol. Med. **1:** 236–243.
36. HEITZMANN, J. G., B. BLOMBERG & M. COHN. 1986. The WI-L2-729-HF_2 human hybridoma system: Arrangement of lambda genes in monoclonal hybrids. Mol. Biol. Med. **3:** 339–350.
37. TAUB, R. A., G. F. HOLLIS, P. A. HIETER, S. KORSMEYER, T. A. WALDMANN & P. LEDER. 1983. Variable amplification of immunoglobulin λ light-chain genes in human populations. Nature **304:** 172–174.
38. SOLVASON, N. & J. F. KEARNEY. 1992. The human fetal omentum: A site of B cell generation. J. Exp. Med. **175:** 397–404.
39. FRANGIONE, B., T. MOLOSHOK, F. PRELLI & A. SOLOMON. 1985. Human λ light chain constant region gene Cλ *Mor*: The primary structure of λVI Bence Jones protein *Mor*. Proc. Natl. Acad. Sci. USA **82:** 3415–3419.
40. GHISO, J., A. SOLOMON & B. FRANGIONE. 1986. Association of human λ light chain V/J/C segments: Serologic analysis and primary structure of the λVI Bence Jones protein *Tho*. J. Immunol. **136:** 716–719.
41. UDEY, J. A. & B. BLOMBERG. 1988. Intergenic exchange maintains identity between two human lambda light chain immunoglobulin gene intron sequences. Nuc. Acids. Res. **16:** 2959–2969.
42. BLOMBERG, B., A. TRAUNECKER, H. N. EISEN & S. TONEGAWA. 1981. Organization of four mouse λ light chain immunoglobulin genes. Proc. Natl. Acad. Sci. USA **78:** 3765–3769.
43. BLOMBERG, B. & S. TONEGAWA. 1982. DNA sequences of the joining regions of mouse λ light chain immunoglobulin genes. Proc. Natl. Acad. Sci. USA **79:** 530–533.
44. MILLER, J., A. BOTHWELL & U. STORB. 1981. Physical linkage of the constant region genes for immunoglobulins λI and λIII. Proc. Natl. Acad. Sci. USA **78:** 3829–3833.
45. MILLER, J., E. SELSING & U. STORB. 1982. Structural alterations of J regions of mouse immunoglobulin λ genes are associated with differential gene expression. Nature **295:** 428–430.
46. SELSING, E., J. MILLER, R. WILSON & U. STORB. 1982. Evolution of mouse immunoglobulin λ genes. Proc. Natl. Acad. Sci. USA **79:** 4681–4685.
47. DARIAVACH, P., G. LEFRANC & M.-P. LEFRANC. 1987. Human immunoglobulin Cλ6 gene encodes the $Kern^+Oz^-$ λ chain and Cλ4 and Cλ5 are pseudogenes. Proc. Natl. Acad. Sci. USA **84:** 9074–9078.
48. HOLLIS, G. F., P. A. HIETER, O. W. MCBRIDE, D. SWAN & P. LEDER. 1982. Processed genes: A dispersed human immunoglobulin gene bearing evidence of RNA-type processing. Nature **296:** 321–325.
49. BAUER, T. R. JR., H. E. MCDERMID, M. L. BUDARF, M. L. VAN KEUREN & B. B. BLOMBERG. 1993. Physical location of the human immunoglobulin lambda-like genes, 14.1, 16.1 and 16.2. Immunogenetics **38:** 387–399.
50. BOSSY, D., M. MILILI, J. ZUCMAN, G. THOMAS, M. FOUGEREAU & C. SCHIFF. 1991. Organization and expression of the λ-like genes that contribute to the μ-ψ light chain complex in human pre-B cells. Int. Immunol. **3:** 1081–1090.
51. STAUDT, L. M. & M. J. LENARDO. 1991. Immunoglobulin gene transcription. Ann. Rev. Immunol. **9:** 373–398.
52. NELSEN, B. & R. SEN. 1992. Regulation of immunoglobulin gene transcription. Internat. Rev. Cytol. **133:** 121–149.
53. BLOMBERG, B. B., C. M. RUDIN & U. STORB. 1991. Identification and localization of an enhancer for the human lambda light chain immunoglobulin gene complex. J. Immunol. **147:** 2354–2358.
54. HAGMAN, J., C. M. RUDIN, D. HAASCH, D. CHAPLIN & U. STORB. 1990. A novel enhancer in the immunoglobulin λ locus is duplicated and functionally independent of NFκB. Genes Dev. **4:** 978–992.
55. GLOZAK, M. A. & B. B. BLOMBERG. 1995. Characterization of the human λ enhancer: Identification of a novel site which augments enhancer activity. Manuscript submitted.

56. MEYER, K. B. & M. S. NEUBERGER. 1989. The immunoglobulin κ locus contains a second, stronger B-cell-specific enhancer which is located downstream of the constant region. EMBO J. **8:** 1959–1964.
57. DARIAVACH, P., G. T. WILLIAMS, K. CAMPBELL, S. PETTERSON & M. S. NEUBERGER. 1991. The mouse IgH 3′-enhancer. Eur. J. Immunol. **21:** 1499–1504.
58. LIEBERSON, R., S. L. GIANNINI, B. K. BIRSHTEIN & L. A. ECKHARDT. 1991. An enhancer at the 3′ end of the mouse immunoglobulin heavy chain locus. Nuc. Acids Res. **19:** 933–937.

V(D)J Recombination: Double-Strand Break Repair Gene Products Used in the Joining Mechanism

DAVID WEAVER,[a] NIKOLAI BOUBNOV,
ZACHARY WILLS, KATHRYN HALL,
AND JANE STAUNTON

Division of Tumor Immunology
Dana-Farber Cancer Institute
Boston, Massachusetts 02115
and
Department of Microbiology and Molecular Genetics
Harvard Medical School
Boston, Massachusetts 02115

THE V(D)J RECOMBINATION PATHWAY: AN UPDATE

The V(D)J recombination pathway is the most intensely studied mechanism of recombination in mammals. The functioning of this important pathway allows the somatic construction of immunoglobulin (Ig) and T-cell receptor (TCR) genes by joining of V, D, and J gene segments. By the combinatorial association of these gene segments from extended gene families, new antigen recognition structures are formed. By the heterogeneity created in the V(D)J recombination coding sequence junctions, a repertoire of subtly different epitopes is created for individualizing antigen recognition.

A great deal of interest has focused on the V(D)J recombination mechanism *per se*. Several gene products have been shown to be particularly important for initiating the pathway in lymphoid cells. These include the lymphoid-restricted RAG1 and RAG2 genes,[1-3] which will not be discussed extensively here. Knockout mutations of RAG1 and RAG2 eliminate all V(D)J recombination *in vivo*[4,5]; however, the molecular function of these interesting genes has not yet been described.

Two cassettes of recombination signal sequences (RSSs) are required for V(D)J recombination (FIG. 1). Each Ig or TCR gene segment that has been observed to undergo rearrangement *in vivo* contains one RSS at its 3′ or 5′ border. Each RSS consists of a highly conserved heptamer and a less-conserved nonamer where the consensus heptamer (CACAGTG) is palindromic and separated from the nonamer by a spacer region in one of two ways. RSS-12 contains a 12-bp spacer, and RSS-23 contains a 23-bp spacer. One RSS-12 and one RSS-23 are needed to signal V(D)J rearrangement, and these are the only essential elements for the reaction.[6,7] The features of the recombination products help to define the reaction steps. The two DNA products are structurally different, and therefore the joining pathway is thought to diverge following initiating cleavages. Coding segment joining is

[a] Address for correspondence: David Weaver, Division of Tumor Immunology, Dana-Farber Cancer Institute, 44 Binney St., Boston, MA 02115.

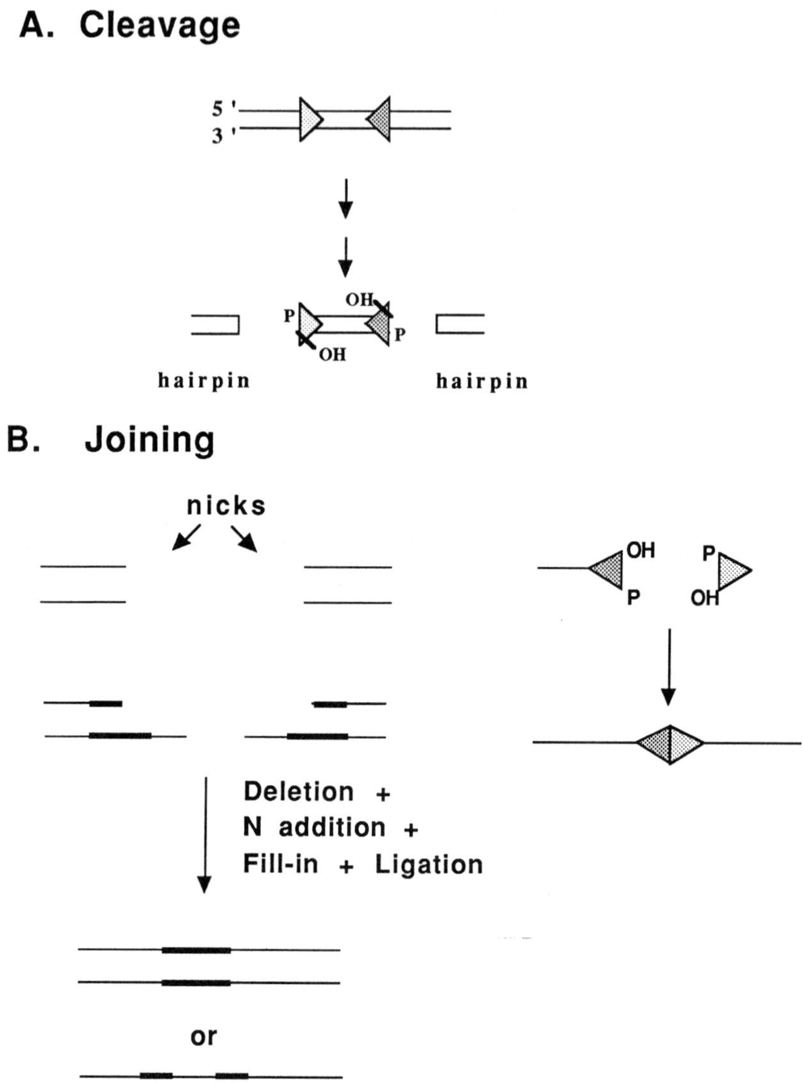

FIGURE 1. The mechanism of V(D)J recombination. (A) Cleavage at the border of each RSS (*triangles*) is shown. RSS ends are 5′-phosphorylated and blunt ended. **(B)** Joining steps. Coding junction formation is indicated starting from hairpin-ended coding DNA. Nicking of the hairpins leads to a variety of end structures. DNA homology between the ends (*bold line*) may be used in the joining events some of the time. RSS joints are normally precise.

discussed below. Chromosomal signal joints are formed by the precise fusion of the two RSSs. Usually these junctions are totally dispensable.

The cut sites for V(D)J recombination are likely to occur at the borders of RSS heptamers where coding sequence from V, D, or J elements is found (FIG. 1). New findings suggest that both RSS ends and coding ends may pass through double-strand break intermediates in the pathway. RSS ends have been detected in association with V(D)J recombination of several of the rearranging gene families.[8,9] These studies have shown a uniformity in end structures for the RSS: blunt ended and 5′ phosphorylated. Similarly, several studies argue that coding DNA is uniquely configured following V(D)J recombination cleavages. The origin of junctional P nucleotides is hypothesized to be synthesized from the resolution of hairpins at coding DNA ends during the reaction. In fact, hairpin ends of coding DNA were observed for *scid* rearrangements.[10] With hairpin intermediates, coding ends with P nucleotides could be formed by the asymmetric cleavage of the hairpin as shown (FIG. 1B). This process would generate a heterogeneous population of DNA ends for joining. Coding joints include P nucleotides only a fraction of the time, suggesting that terminal nibbling usually removes these residues.

Limited DNA homology between the coding ends may be used to direct joining. The variety of coding junction products formed *in vivo* and with recombination substrates in cell culture indicates that the joining process can occur between coding DNA that has little or no homology.[11–14] The restriction on the extent of deletion is probably limited by the ability to hold the two coding ends in a synaptic complex by protein associations, although there is no proof yet. N region addition is very likely to occur after terminal deletion. Terminal deoxynucleotidyl transferase (TdT) knockout mice produce Ig and TCR junctions using microhomology between coding ends.[15,16]

The V(D)J recombination pathway of vertebrates requires the functioning of a family of genes, corresponding to the complexity of the overall reaction. Many of these genes are now appreciated to be ubiquitously expressed and possibly used for other functions in DNA damage repair. Double-strand breaks (DSBs) are deleterious to cells unless repaired. DNA damage occurs spontaneously as errors in DNA synthesis and recombination, or may be induced experimentally by ionizing radiation (IR). DSB repair of IR breaks occurs by an end-joining process that resembles V(D)J joining.

THE *scid* GENE IS REQUIRED FOR DSB REPAIR AND V(D)J RECOMBINATION AND LOCATED ON HUMAN CHROMOSOME 8

Our lab and others have previously shown that the *scid* mutation portrays how the two processes of DSB repair and V(D)J recombination intersect. The mouse *scid* mutation, leading to a profound immunodeficiency, is caused by a defect in V(D)J recombination coding junction formation (reviewed in Gellert[6]), and possibly hairpin resolution.[10] *Scid* cells are severely IR-sensitive, specifically containing an IR-induced DSB repair defect.[26–28] Considering the V(D)J recombination and IR-sensitivity defects together, our data about the *scid* mutation strongly suggests that a common repair pathway is used for these two processes. The *scid* defect serves as a paradigm for this mechanism.

We first tested whether human cells would be capable of complementing the *scid* mutation by using somatic cell fusion analysis. We made pools of somatic cell radiation hybrids between *scid* fibroblasts (*SCGR11*) and human cells (HeLa) and tested their level of IR resistance. Exponentially growing HeLa cells were lethally irradiated at 8000 rads and then fused to an equal number of unirradiated *SCGR11* cells. The fused cells were irradiated and plated to determine cell survival. *SCGR11*/HeLa fusions produced 20-fold greater cell survival than *SCGR11*/ *SCGR11* controls. Thus, human cells, and presumably the human *scid* gene, complements the X-ray sensitivity of *scid* cells.

Microcell-mediated chromosome transfer was conducted to determine the chromosome able to complement the mouse *scid* mutation. These experiments were done collaboratively with S. Banga and R. Athwal.[29] We found that human chromosome 8 complemented the IR sensitivity of *scSV3* cells. Loss of human chromosome 8 in this cell line restored the *scid* sensitivity of the cell line. Similarly, V(D)J recombination coding junction formation was observed to be specifically restored when human chromosome 8 was added. DNA sequencing of recovered recombinant junctions proved that the coding junctions were normal. Therefore, both molecular defects of the *scid* mutation were complemented in our assays, and the human copy of the *scid* gene is very likely to reside on chromosome 8. This information will be useful for future mapping strategies for *scid*.

NEW DSB REPAIR MUTANTS OF V79-4 ARE DEFICIENT IN JOINING OF BOTH PRODUCTS OF V(D)J RECOMBINATION

Several mutations of V(D)J joining have been isolated in cell culture. The *xrs*, *XR-1*, and *sxi* mutants were generated in screens for IR sensitivity.[17–20] The *xrs* mutation has been shown to be deficient in a DNA end-binding function,[21] where DSBs may be recognized by the *xrs* gene product. *Xrs* cells containing a transferred human chromosome 2 complement both phenotypes: IR resistance and normal V(D)J recombination.[23,24] The radiosensitivity of *XR-1* is complemented by transfer of human chromosome 5.[22] *XR-1*: Ch5 cells also revert the V(D)J recombination deficiency for both RSS and coding junction formation.[23] These previously characterized IR-sensitive mutants were also shown to possess V(D)J recombination defects by activating gene rearrangement in cell lines. The *xrs* and *XR-1* CHO cell mutants were deficient in both coding and RSS joining of V(D)J recombination.[23,25] Because of the DSB repair defects of *xrs* and *XR-1*, these genes may normally be used for joining rather than initiation steps of V(D)J recombination.

Two mutants in DSB repair and V(D)J recombination were recently isolated in a screen for IR-sensitive mutations from the V79-4 cell line (*sxi-3* and *sxi-2*).[20] We examined the V(D)J recombination properties of these two cell lines. Specific phenotypes of V(D)J joining mutants had previously been informative about the reaction. To determine if RSS or coding junction formation, or both, were impacted, we cotransfected the RAG1 and RAG2 genes with V(D)J recombination templates. This method has previously been shown to promote V(D)J recombination in cell lines that otherwise have no ability to undergo V(D)J joining.[3] Efficient V(D)J recombination coding junction synthesis was found for parental cells, but not *sxi-2* and *sxi-3* (25-fold reduced) (FIG. 2). The infrequent coding junctions of *sxi-2* and *sxi-3* showed that the recombination products recovered were structurally normal junctions. RSS junction formation was also deficient (25-fold reduced) in *sxi-2* and *sxi-3* in similar assays. These events were usually aberrant because the signal junctions were not precisely joined. Therefore, the two mutants of DSB

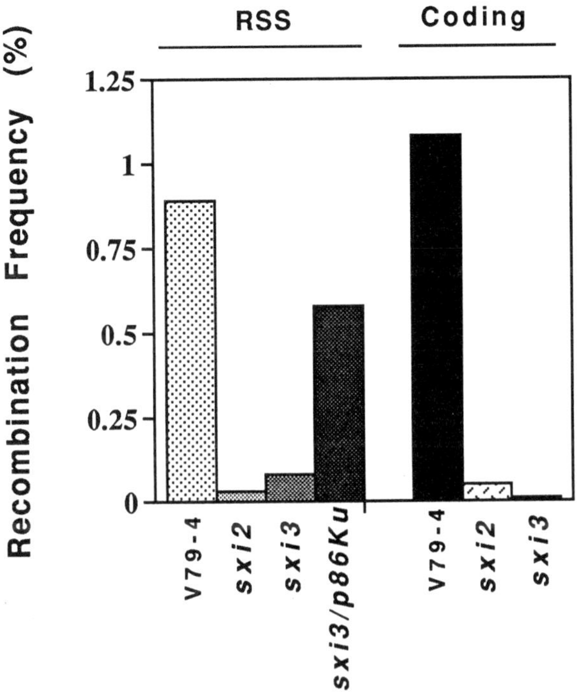

FIGURE 2. V(D)J recombination assays of *sxi-2* and *sxi-3* mutants. V(D)J recombination assays were performed according to published procedures. RSS junction frequency was examined for V79-4, *sxi-2*, and *sxi-3* cells, as well as *sxi-3*/p86Ku cells. Coding junction frequency was measured for V79-4, *sxi-2*, and *sxi-3* cells.

repair, *sxi-2* and *sxi-3*, are deficient in both joining events of V(D)J recombination. The V(D)J recombination defect of these mutants was very similar to *xrs* and *XR-1* mutants, which also affected both joining products.[23]

sxi-3 AND *sxi-2* ARE MEMBERS OF THE *XRCC5* COMPLEMENTATION GROUP

Mutations in DSB repair have been ordered into complementation groups by examining whether the fusion of two different cells lines is able to complement mutant phenotypes of the cell lines. This analysis is able to give a first-order approximation of whether cell lines are in the same complementation group. Previous analysis had demonstrated that *scid*, *xrs*, and *XR-1* were in separable complementation groups, and it was confirmed by the findings above and elsewhere that the human chromosome that complements each mutation was different. *Sxi-2* and *sxi-3* cells were fused to other mutants: *scSV3*, *xrs-6*, *XR-1*, and to each other. Fusion of *sxi-3* or *sxi-2* to V79-4 parental cells gave IR^r cells. Fusion of *sxi-3* to

TABLE 1. Complementation of IR Sensitivity of *sxi-2* and *sxi-3* by Cell Fusion

Cell Line	*xrs-6*	*scid*	*XR-1*	V79-4	*sxi-2*	*sxi-3*
sxi-2	–	+	+	+	–	–
sxi-3	–	+	+	+	–	–

sxi-2 only yielded IR^s cells like the mutant parents (TABLE 1). Similarly, IR^r clones were shown in fusions of *sxi-3* with *XR-1* and *scSV3*, but not with *xrs-6* (TABLE 1). Likewise, *sxi-2* fusions with *xrs-6* produced all IR^s clones, whereas each of the other fusions yielded IR^r clones. Therefore, *sxi-3*, *sxi-2*, and *xrs-6* most likely are members of the same DSB repair complementation group, *XRCC5*.

DNA END-BINDING ACTIVITY IS DEFICIENT IN SOME DSB REPAIR MUTANTS BUT NOT OTHERS

It was previously observed that eukaryotic cells contain abundant nuclear proteins that bind to ends of linearized double-stranded DNA.[21,32] This DNA end-binding activity binds to DNA ends without any preferences for one type of end structure over another. End-binding proteins would be valuable for repair mechanisms where any type of DSB induced by IR would need to be repaired. Likewise, V(D)J recombination coding junctions, which may have variable sequence composition and structure, may also use this function. We tested several of the cell culture mutants for DNA end-binding using mobility shift gel electrophoresis (TABLE 2). P^{32}-labeled restriction fragments (25–300 bp) bind to nuclear protein(s) in the presence of 1000-fold excess circular DNA. These DNA–protein complexes are resolved by fractionating on gel electrophoresis because the complexes may possess a distinct mobility. Several of the mutants have normal DNA end-binding activity, and others are deficient in this activity (TABLE 2). *Scid*, *XR-1*, and *V-3* cells, although defective in V(D)J joining, are normal for DNA end

TABLE 2. Mutants of V(D)J Recombination and DSB Repair

Mutation Group	IR Survival	IR DSB Repair	DNA End Binding	V(D)J Recombination Signal	V(D)J Recombination Coding	Gene Identity	Hu Chrm Location
scid							
scid	Low	–	+	+/–	–		8q11-p12
V-3	Low	–	+	+/–	–		
XRCC5							
xrs-5.xrs-6	Low	–	–	–	–	p82^{KU}	2q34-36
XR-V79B	Low	–	–				
XR-V15B	Low		–				
sxi-2	Low	–	–	–	–		
sxi-3	Low	–	–	–	–		
sxi-1, sxi-4							
sxi-1	Low	–	–	–	–	p70^{KU}?	
sxi-4	Low	–	–	–	–		
XR-1	Low	–	+	–	+/–		5

binding relative to their control cell lines. Also, all mammalian cells and cell types that we have examined are positive in this assay. The mutant cell lines *xrs-6*, *sxi-2*, *sxi-3*, *sxi-1*, and *sxi-4* were all missing any DNA end-binding activity (TABLE 2). Hybrid cell lines resulting from the fusion of *sxi-3* and *xrs-6* were also defective for DNA end binding, in addition to retaining IR^s phenotypes (data not shown).

COMPLEMENTATION OF IR SENSITIVITY DEFECT OF *sxi-3* BY p86KU

The molecular mechanism of V(D)J recombination and DSB repair can best be advanced by identifying the gene products that are required for completion of the reaction pathways. DNA end binding has previously been attributed to an abundant nuclear protein complex called Ku. The Ku protein complex is immunoprecipitated by autoantibodies produced by a subset of schleroderma patients.[32,33] Ku forms a stable heterodimer (p70 and p86 kDa subunits) that binds to DNA ends and alterations in DNA structure with high affinity.[34] Because Ku may well be required for DNA repair, we examined whether the *sxi* mutant phenotypes could be complemented by introduction of Ku genes. *Sxi-3* cells were stably transfected with the p70 and p86Ku subunit genes in expression constructs.

We found that transfection of p86Ku rescued the IR^s phenotype of *sxi-3* cells (FIG. 3). *Sxi-3* cells were also partially complemented by cotransfections of

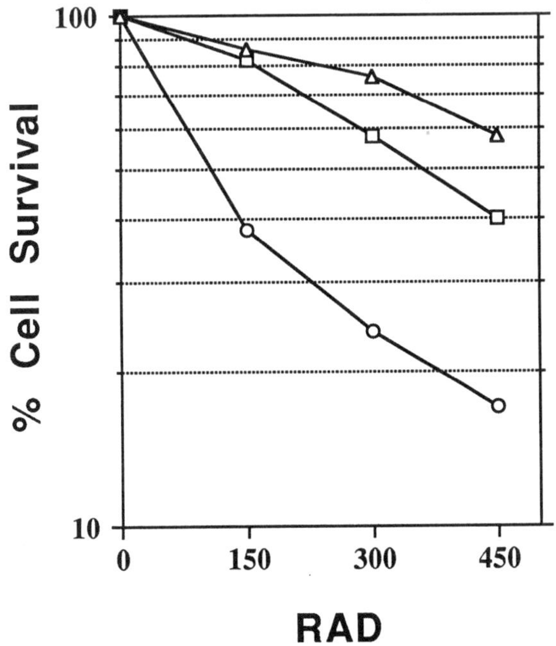

FIGURE 3. The p86Ku gene complements the IR^s defect of *sxi-3*. Cell survival profiles are displayed as surviving cell colonies following exposure to IR. V79-4, *sxi-3*, and *sxi-3*/p86Ku cells were examined. V79-4 (Δ), *sxi-3* (○), and *sxi-3*/p86Ku (□).

p70 + p86Ku and p70Ku alone, but not to as great an extent as p86Ku (data not shown). Therefore, Ku subunits can reconstitute one defect of DSB repair mutants.

RESTORATION OF DNA END-BINDING ACTIVITY IN *sxi-3* BY p86Ku

The IR-resistance of *sxi-3*/p86Ku should correlate with an increased DNA end-binding capacity; *sxi-3*/p86Ku cells reconstituted DNA end-binding activity to the level shown with the parental cells to the mutation (V79-4). In contrast, another cell line containing an introduced p70Ku, complements DNA end binding poorly. The low level of DNA end-binding activity was not increased in radioresistant cells derived following IR treatment. The monoclonal antibody, 111, that was raised against the mouse p86Ku, strongly cross-reacts with human p86Ku.[35] We found that the Ku complex that was reconstituted in *sxi-3*/p86Ku was supershifted in the presence of the antibody, strongly suggesting that the complex contained the human p86Ku subunit in these cells. We also found that extended electrophoresis distinguished between a "hamster" and "human" form of Ku. The complex with *sxi-3*/p86Ku yielded the same mobility as the human extracts showing that *sxi-3*/p86 cells contain a functionally active p86Ku subunit expressed from the human gene. Also, methionine-labeled lysates from *sxi-3*/p86Ku cells immunoprecipitated the human p86Ku, and identified a coprecipitating p70 protein that is presumably the hamster p70Ku subunit. *Sxi-3* control cells do not have any immunoprecipitating material in the p86 and p70 regions. These data indicate that human p86Ku can stably associate with hamster p70Ku.

V(D)J RECOMBINATION PRODUCT FORMATION IS ALSO RESTORED IN *sxi-3*/p86Ku

An important additional criterion for complementation of the *sxi-3* mutation is the ability to restore V(D)J recombination capacity. Transient transfection of human p86 and/or p70Ku expression plasmids in conjunction with RAG1/RAG2 and a plasmid substrate for V(D)J recombination (pJH200) into *sxi-3* cells did not yield any significant complementation that may have been due to an inability to get enough plasmid DNA into the cells by transfection. Transient transfection of p86Ku into *xrs-6* cells, another *XRCC5* complementation group member, did restore V(D)J recombination potential significantly (data not shown). Therefore, p86Ku complements the V(D)J recombination deficiency of *xrs-6* cells. RSS joining using pJH200, however, was also examined in cells complemented by p86Ku (*sxi-3*/p86Ku) We observed an increased recombination frequency (0.58%) and an increased proficiency of precise RSS junctions (100%, FIG. 2). The cell lines containing the other Ku subunit (*sxi-3*/p70Ku) showed no complementation of V(D)J recombination RSS junction formation (FIG. 2). Thus, p86Ku complements the V(D)J recombination deficiency as effectively as the IR-sensitivity and DNA end binding properties of the *sxi-3* mutant.

Our preliminary data indicates that the molecular defects of *sxi-3* cells may be equivalent to a null mutation. RNA analysis of this mutant using Ku hybridization probes indicates that the hamster p86Ku mRNA is absent in *sxi-3* cells. Also, it appears that *xrs-6* and *sxi-2* cells have wild-type levels of p86Ku mRNA and are likely to be point mutations (data not shown).

No information is yet available on the human chromosome able to complement the *sxi-1,4* IR-sensitivity and V(D)J recombination defects (TABLE 2). A likely candidate is human chromosome 22, where p70Ku is located.[36] *Sxi-1* and *sxi-4* may be mutant in the gene encoding p70Ku because both cell lines are defective in the DNA end-binding complex associated with Ku heterodimers, and because *sxi-1,4* do complement *XRCC5* group cells in cell fusion tests (Boubnov *et al.*, in preparation).

Ku FUNCTION IN THE V(D)J RECOMBINATION PATHWAY

The DNA end-binding activity of Ku is likely to be highly significant for V(D)J recombination. A parallel function may be expected for damaged DSBs in DNA in DSB repair. In V(D)J joining, Ku monomers could associate with coding and/or RSS ends (FIG. 4). Because of the extremely high affinity of Ku for DNA ends, the binding function of the heterodimer is likely to be essential for function. This property of Ku may be a means to synapse DNA ends before ligation by holding the DNA strands together via protein–protein interactions. Ku also binds to transitions between double-stranded and single-stranded DNA as efficiently as to DSBs.[37] Hairpin-ended DNA is an efficient substrate and competitor for Ku binding.[38,37] Thus, a logical mechanism would be that Ku complexes to coding end hairpins once they are formed. Binding may be sufficiently internal to the coding end termini that hairpin resolution and processing by other factors is not interfered with in the presence of Ku binding. Secondly, Ku-Ku associations may be formed between the two coding ends, allowing synapsis between the ends that are to be joined, and promoting the completion of joining. Ku may also associate with RSS ends because *XRCC5* group mutants and *sxi-1,4* mutants also affect RSS junction formation. Ku may have a similar ability to recognize and bind to fragmented DNA from IR damage. Ku-mediated repair of chromosome damage may yield recombinant joining products by nonhomologous recombination pathways that have small deletions and/or filling-in.

Ku is able to translocate along DNA.[38–41] This property may have ramifications for V(D)J recombination and DSB repair where Ku can transit to the sites of recombination synapsis or DNA damage (FIG. 4). This property could explain how Ku may only be involved in late or joining stages of V(D)J recombination.

FUTURE DIRECTIONS

Ku forms a stable complex with a 350-kDa serine/threonine protein kinase (DNA-PK[41]; reviewed in Anderson[34]). A novel feature of this protein kinase is that it is activated by binding to DNA, particularly DNA ends.[42] DNA-PK is nuclear, but it is uncertain whether its cellular localization is entirely dictated by Ku. The features of DNA-PK thus far observed are consistent with a dual function in cell metabolism: transcription regulation and DNA repair. DNA-PK activates transcription factors by phosphorylation.[34] DNA conformational changes may be recognized by Ku during transcription assembly. The activation of transcription initiation would then be facilitated by localized phosphorylation of transcription factor substrates by DNA-PK. In fact, DNA-PK/Ku complexes have been isolated based on the property of stimulation of basal transcription by several investigators.[37,43,44]

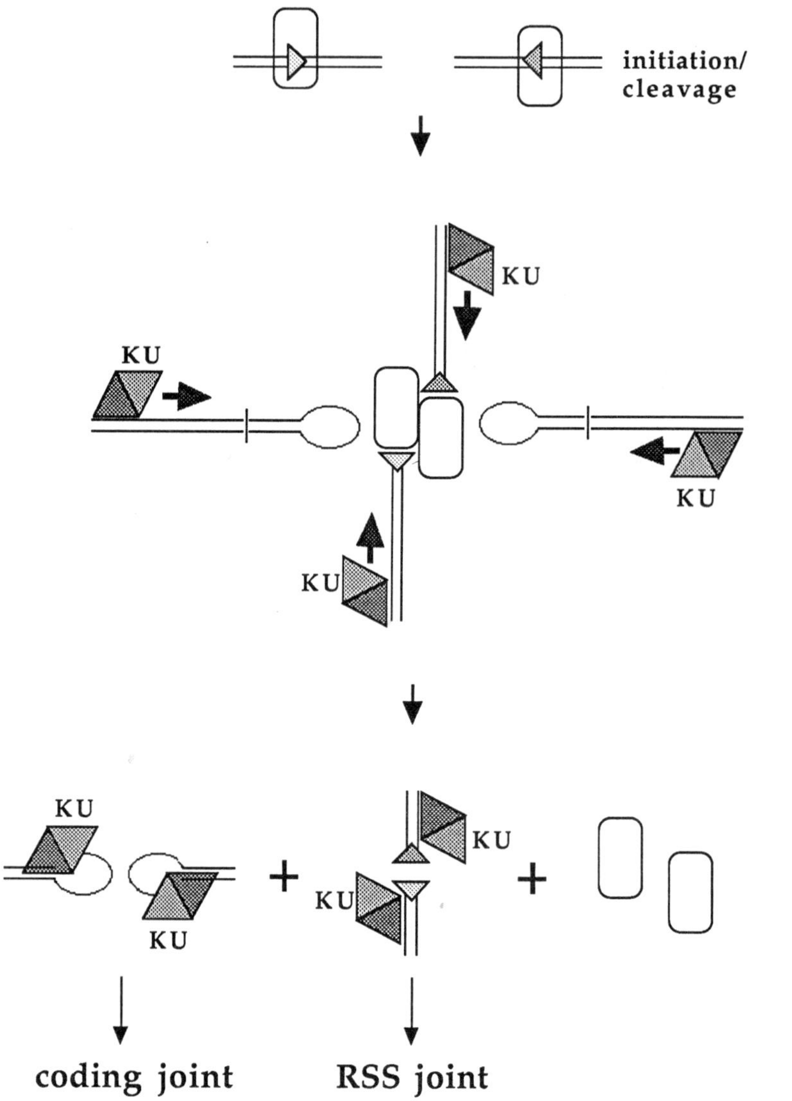

FIGURE 4. Ku complexes may control V(D)J joining steps. Following initiation and cleavages of V(D)J recombination, an intermediate complex consisting of hairpin-ended coding DNA and proteins are hypothesized. Movement of Ku complexes along DNA may bind to these terminal sites and effect joining. Ku may well associate with both coding DNA and RSS DNA ends.

The anchoring of a DNA-dependent protein kinase via Ku heterodimers very likely will have exciting ramifications for V(D)J recombination and DNA repair. The tumor suppressor protein, p53, that is part of the G1-S phase DNA damage-induced cell cycle checkpoint is phosphorylated by DNA-PK.[45,46] DNA-PK could act as an immediate mediator of DNA damage responses by binding to DNA ends and signaling cell cycle arrest by a p53-dependent pathway.[34] Thus, because of the DNA end-targeting properties of Ku, DNA-PK is likely to be locally available to DNA damage sites.

Ku/DNA-PK complexes may regulate the completion of product formation and end joining of V(D)J recombination. Also, an activated DNA-PK may phosphorylate other proteins relevant to V(D)J recombination to direct the reaction towards completion of products or to maximize reaction efficiency. Candidate substrates could be the RAG1 and RAG2 proteins. RAG2 is phosphorylated at several serine and threonine residues in a cell-cycle-dependent manner.[47,48] If there is a separation between initiation and cleavage steps of the pathway, then Ku binding to hairpin ends may bring DNA-PK into the vicinity of RAG proteins to be inactivated. Additional DNA-PK substrates may include other proteins in the V(D)J complexes where phosphorylation may regulate the efficiency of recombinational synapses or the completion of products. For example, Ku is phosphorylated by DNA-PK,[45] which may be significant for its V(D)J recombination activity. Finally, DNA-PK may be another V(D)J recombination/DSB repair complementation group. Mutations in DNA-PK may be expected to be defective in gene rearrangement much like Ku mutations, if the DNA-PK : Ku complexes are relevant to the pathway.

REFERENCES

1. Schatz, D. & D. Baltimore. 1988. Stable expression of immunoglobulin gene V(D)J recombinase activity by gene transfer into 3T3 fibroblasts. Cell **53:** 107–115.
2. Schatz, D., M. A. Oettinger & D. Baltimore. 1989. The V(D)J recombination activating gene RAG-1. Cell **59:** 1035–1048.
3. Oettinger, M. A., D. G. Schatz, C. Gorka & D. Baltimore. 1990. RAG-1 and RAG-2, adjacent genes that synergistically activate V(D)J recombination. Science **248:** 1517–1523.
4. Shinkai, Y., G. Rathbun, K. Lam, E. Oltz, V. Stewart, M. Mendelsohn, J. Charon, M. Datta, F. Young, A. Stall & F. Alt. 1992. RAG-2 deficient mice lack mature lymphocytes due to an inability to initiate V(D)J recombination. Cell **68:** 855–867.
5. Mombaerts, P., J. Iacomini, R. S. Johnson, K. Herrup & S. Tonegawa. 1992. RAG-1-deficient mice have no mature B and T lymphocytes. Cell **68:** 869–877.
6. Gellert, M. 1992. Molecular analysis of V(D)J recombination. Annu. Rev. Genet. **22:** 425–446.
7. Akira, S., K. Okazaki & H. Sakano. 1987. Two pairs of recombination signals are sufficient to cause immunoglobulin V-(D)-J joining. Science **238:** 1134–1138.
8. Roth, D. B., C. Zhu & M. Gellert. 1993. Characterization of broken DNA molecules associated with V(D)J recombination. Proc. Natl. Acad. Sci. USA **90:** 10788–10792.
9. Schlissel, M., A. Constantinescu, T. Morrow, M. Baxter & A. Peng. 1993. Double-strand signal sequence breaks in V(D)J recombination are blunt, 5′-phosphorylated, RAG-dependent, and cell cycle regulated. Genes Dev. **7:** 2520–2532.
10. Roth, D. B., J. P. Menetski, P. B. Nakajima, M. J. Bosma & M. Gellert. 1992. V(D)J recombination: broken DNA molecules with covalently sealed (hairpin) coding ends in scid mouse thymocytes. Cell **70:** 983–991.
11. Gu, H., I. Foster & K. Rajewsky. 1990. Sequence homologies, N sequence insertion and J_H gene utilization in V_HDJ_H joining: Implications for the joining mechanism and

the ontogenetic timing of Ly1 B cell and B-CLL progenitor generation. EMBO J. **9:** 2133–2140.

12. FEENEY, A. 1990. Lack of N regions in fetal and neonatal mouse immunoglobulin V-D-J junctional sequences. J. Exp. Med. **172:** 1377–1390.
13. GERSTEIN, R. M. & M. R. LIEBER. 1993. Extent to which homology can constrain coding exon junctional diversity in V(D)J recombination. Nature **363:** 625–627.
14. BOUBNOV, N. V., Z. P. WILLS & D. T. WEAVER. 1995. Coding sequence composition flanking either signal element alters V(D)J recombination efficiency. Nucl. Acids Res. In press.
15. KOMORI, T., A. OKADA, V. STEWART & F. W. ALT. 1993. Lack of N regions in antigen receptor variable region genes of TdT-deficient lymphocytes. Science **261:** 1171–1175.
16. GILFILLAN, S., A. DIERICH, M. LEMEUR, C. BENOIST & D. MATHIS. 1993. Mice lacking TdT: Mature animals with an immature lymphocyte repertoire. Science **261:** 1175–1178.
17. JEGGO, P. A. & L. M. KEMP. 1983. X-ray-sensitive mutants of Chinese hamster ovary cell line. Isolation and cross-sensitivity to other DNA-damaging agents. Mutation Res. **112:** 313–327.
18. JEGGO, P. 1990. Studies on mammalian mutants defective in rejoining double-strand breaks in DNA. Mutation Res. **239:** 1–16.
19. STAMATO, T. D., R. WEINSTEIN, A. GIACCIA & L. MACKENZIE. 1983. Isolation of cell cycle-dependent gamma-ray-sensitive Chinese hamster ovary cell. Somat. Cell Genet. **9:** 165–173.
20. LEE, S. E., C. R. PULASKI, M. J. VOSS, J. UM, D. M. HE, E. R. KARPINSKI, E. Y. M. RHEW & E. A. HENDRICKSON. 1994. Isolation and characterization of four mammalian cell mutants generated by insertional mutagenesis that are x-ray sensitive and defective in DNA double-strand break repair. Submitted.
21. RATHMELL, W. K. & G. CHU. 1994. A DNA end-binding factor involved in double-strand break repair and V(D)J recombination. Mol. Cell. Biol. **14:** 4741–4748.
22. GIACCIA, A. J., N. DENKO, R. MACLAREN, D. MIRMAN, C. WALDREN, I. HART & T. D. STAMATO. 1990. Human chromosome 5 complements the DNA double-strand break-repair deficiency and gamma-ray sensitivity of the XR-1 hamster variant. Am. J. Hum. Genet. **47:** 459–469.
23. TACCIOLI, G. E., G. RATHBUN, E. OLTZ, T. STAMATO, P. A. JEGGO & F. W. ALT. 1993. Impairment of V(D)J recombination in double-strand break repair mutants. Science **260:** 207–210.
24. JEGGO, P., M. HAFEZPARAST, A. F. THOMPSON, B. C. BROUGHTON, G. P. KAUR, M. Z. ZDZIENICKA & R. S. ATHWAL. 1992. Localization of a DNA repair gene (XRCC5) involved in double-strand-break rejoining to human chromosome 2. Proc. Natl. Acad. Sci. USA **89:** 6423–6427.
25. PERGOLA, F., M. Z. ZDZIENICKA & M. R. LIEBER. 1993. V(D)J recombination in mammalian cell mutants defective in DNA double-strand break repair. Mol. Cell. Biol. **13:** 3464–3471.
26. FULOP, G. M. & R. A. PHILLIPS. 1990. The *scid* mutation in mice causes a general defect in DNA repair. Nature **347:** 479–482.
27. BIEDERMANN, K. A., J. SUN, A. J. GIACCIA, L. M. TOSTO & J. M. BROWN. 1991. *scid* mutation in mice confers hypersensitivity to ionizing radiation and a deficiency in DNA double-strand break repair. Proc. Natl. Acad. Sci. USA **88:** 1394–1397.
28. HENDRICKSON, E. A., X.-Q. QIN, E. A. BUMP, D. G. SCHATZ, M. OETTINGER & D. T. WEAVER. 1991. A link between double-strand break-related repair and V(D)J recombination: The scid mutation. Proc. Natl. Acad. Sci. USA **88:** 4061–4065.
29. BANGA, S., K. HALL, A. SANDHU, D. WEAVER & R. ATHWAL. 1994. Complementation of V(D)J recombination defect and x-ray sensitivity of scid mouse cells by human chromosome 8. Mutat. Res. **315:** 239–247.
30. TACCIOLI, G., G. RATHBURN, Y. SHINKAI, E. OLTZ, H. CHENG, G. WHITMORE, T. STAMATO, P. JEGGO & F. W. ALT. 1992. Activities involved in V(D)J recombination. Curr. Top. Microbiol. Immunol. **182:** 107–114.

31. WEAVER, D. 1994. V(D)J recombination and double-strand break repair. Adv. Immunol. **58:** 29–85.
32. MIMORI, T. & J. A. HARDIN. 1986. Mechanism of interaction between Ku protein and DNA. J. Biol. Chem. **261:** 10375–10379.
33. MIMORI, T., J. A. HARDIN & J. A. STEITZ. 1986. Characterization of the DNA-binding protein antigen Ku recognized by autoantibodies from patients with rheumatic disorders. J. Biol. Chem. **261:** 2264–2278.
34. ANDERSON, C. W. 1993. DNA damage and the DNA-activated protein kinase. TIBS **18:** 433–437.
35. REEVES, W. 1985. Use of monoclonal antibodies for the characterization of novel DNA-binding proteins recognized by human autoimmune sera. J. Exp. Med. **161:** 18–39.
36. CAI, Q. Q., A. PLET, J. IMBERT, M. LAFAGEPOCHITALOFF, C. CERDAN & J. M. BLANCHARD. 1994. Chromosomal location and expression of the genes coding for Ku p70 and p80 in human cell lines and normal tissues. Cytogen. Cell Genet. **65:** 221–227.
37. FALZON, M., J. W. FEWELL & E. L. KUFF. 1993. EBP-80, a transcription factor closely resembling the human autoantigen Ku, recognizes single- to double-strand transitions in DNA. J. Biol. Chem. **268:** 10546–10552.
38. PAILLARD, S. & F. STRAUSS. 1991. Analysis of the mechanism of interaction of simian Ku protein with DNA. Nucleic Acids Res. **19:** 5619–5624.
39. DEVRIES, E., W. VAN DRIEL, W. G. BERGSMA, A. C. ARNBERG & P. C. VAN DER VLIET. 1989. HeLa nuclear protein recognizing DNA termini and translocating on DNA forming a regular DNA–multimeric protein complex. J. Mol. Biol. **208:** 65–78.
40. ZHANG, W.-W. & M. YANEVA. 1992. On the mechanisms of Ku protein binding to DNA. Biochem. Biophys. Res. Commun. **186:** 574–579.
41. GOTTLIEB, T. M. & S. P. JACKSON. 1993. The DNA-dependent protein kinase: Requirement for DNA ends and association with Ku antigen. Cell **72:** 131–142.
42. WALKER, A. I., T. HUNT, R. J. JACKSON & C. W. ANDERSON. 1985. Double-stranded DNA induces the phosphorylation of several proteins including the 90,000 M_r heat-shock protein in animal cell extracts. EMBO J. **4:** 139–145.
43. JACKSON, S. P., J. J. MACDONALD, S. LEES-MILLER & R. TJIAN. 1990. GC box binding induces phosphorylation of Sp1 by a DNA-dependent protein kinase. Cell **63:** 155–165.
44. DVIR, A., L. Y. STEIN, B. L. CALORE & W. S. DYNAN. 1993. Purification and characterization of a template-associated protein kinase that phosphorylates RNA polymerase II. J. Biol. Chem. **268:** 10440–10447.
45. LEES-MILLER, S. P., Y. CHEN & C. W. ANDERSON. 1990. Human cells contain a DNA-activated protein kinase that phosphorylates simian virus 40 T antigen, mouse p53, and the human Ku autoantigen. Mol. Cell. Biol. **10:** 6472–6481.
46. LEES-MILLER, S. P., K. SAKAGUCHI, S. J. ULLRICH, E. APPELLA & C. W. ANDERSON. 1992. Human DNA-activated protein kinase phosphorylates serines 15 and 37 in the amino-terminal transactivation domain of human p53. Mol. Cell. Biol. **12:** 5041–5049.
47. LIN, W.-C. & S. DESIDERIO. 1993. Regulation of V(D)J recombination activator protein RAG-2 by phosphorylation. Science **260:** 953–959.
48. LIN, W.-C. & S. DESIDERIO. 1994. Cell cycle regulation of V(D)J recombination activating protein RAG-2. Proc. Natl. Acad. Sci. USA **91:** 2733–2737.

Regulation of N-Region Diversity in Ig Heavy Chain Transcripts Expressed by Murine Fetal Liver Cells Cultured with IL-7[a]

PIERETTE M. APPASAMY,[b] ALBERT B. DELEO, YEHUA WENG, THOMAS W. KENNISTON, JR., LIN TANG, AND MEGHAN A. BURKE

Pittsburgh Cancer Institute and Department of Pathology University of Pittsburgh Pittsburgh, Pennsylvania 15213

T-cell receptor (TCR) and immunoglobulin (Ig) N-region diversity is primarily a consequence of terminal deoxynucleotidyl transferase (TdT) activity and results in a larger repertoire of antigenic specificities.[1] TdT activity and N-region diversity are minimal in the fetus and neonate, but are upregulated postnatally.[2,3] The mechanisms by which N-region diversity is developmentally regulated are not well understood. The regulation of diversity cannot be explained solely on the existence of different fetal and adult lymphoid precursors, since fetal thymocytes can give rise to cells with N-region diversity when they develop in an adult thymic microenvironment.[4] Rather, changes in TdT activity during development may be due to environmental cues.[4] We found that N-region diversity and TdT expression can be induced *in vitro* in fetal liver (FL) cells cultured with IL-7.

METHODS

FL cultures were established and reverse transcription-polymerase chain reaction (RT-PCR) was performed as described previously.[5] Human rIL-7 was a gift of Sterling Research Group, Malvern, Pennsylvania. For PCR of V_H7183 to J_H4 (Ig), the sense primer was V_H7183 5′-CGGGATCCCGTGGAGTCTGGGGGAGGCTTA-3′ and the antisense primer was J_H4 5′-GCGCTCGAGAATTCGGTGACTGAGGT-3′. Restriction enzyme sites used in cloning are underlined. For TdT, primers used were 5′: 5′-CTCAGGGTCAGACGTCCTGGAGT-3′ and 3′: 5′-TTCTGGAAATGGTCGAGAGCATC-3′. PCR products were cloned into pBluescript II and sequenced on both strands with Sequenase™ V.2.0 (U.S. Biochemical Corp., Cleveland, OH).

[a] Supported by National Institutes of Health Grant AI31515 to P.M.A.

[b] Address for correspondence: Dr. P. Appasamy, Pittsburgh Cancer Institute, W1055 Biomedical Science Tower, DeSoto and O'Hara Streets, Pittsburgh, PA 15213.

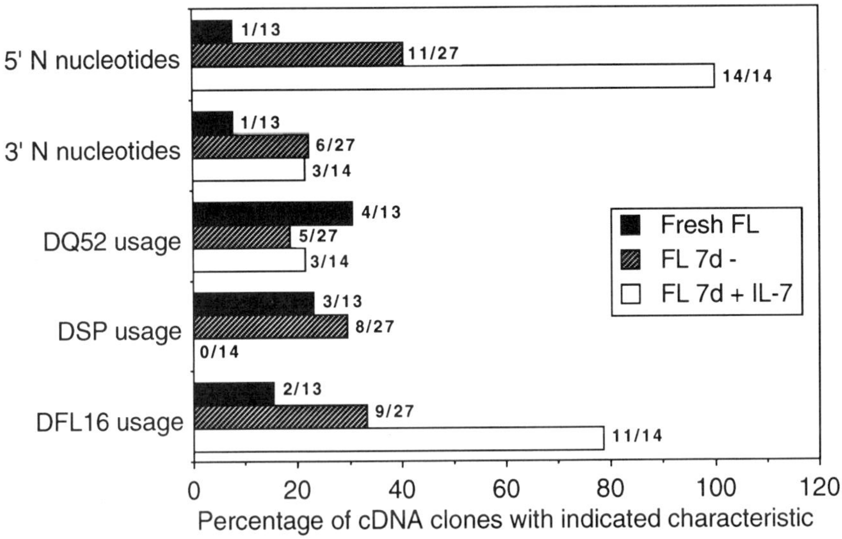

FIGURE 1. N region diversity and D region usage by FL cells. Transcripts were identified by RT-PCR from the indicated FL cells, and cDNA products were cloned and sequenced. Assignments of N or D region nucleotides was based on published sequences of V, D, and J region sequences. The 5′ N region is that region proximal to the V region; the 3′ N region is that region proximal to the J region. The presence of nucleotides in the 5′ N region was significantly different between each group: FL cells cultured with IL-7 versus fresh FL cells, $p < 0.001$ by chi square; FL cells cultured with IL-7 versus FL cells cultured without IL-7, $p < 0.001$ by chi square; FL cells cultured for 7 days without IL-7 versus fresh FL cells, $p < 0.05$ by Fisher's exact test.

RESULTS AND DISCUSSION

We previously showed that culture of FL cells for 7 days with IL-7 induced expression of TCR γ variable (V) region genes Vγ4 and Vγ6, but not Vγ5 or Vγ7.[5] We cloned and sequenced the RT-PCR products we detected in response to IL-7 and observed that more than 50% had N-region nucleotides (data not shown). The presence of N-region diversity in TCR RT-PCR products from FL was striking, since fetal and neonatal TCR transcripts generally lack N-region nucleotides.[3] We then reverse-transcribed RNA from cultured FL cells and amplified, by PCR, the V_H7183 to J_H4 Ig products. Eleven of 27 clones from FL cells cultured in medium alone had N-region nucleotides proximal to the V_H7183 region, and 6 of 27 had nucleotides in the N region proximal to J_H4 (FIG. 1). In contrast, all of the 14 clones isolated from FL cells cultured with IL-7 had nucleotides in the N region proximal to V_H7183, and three of the 14 clones had N-region insertions between the D and J_H4 regions. Thus, N-region insertions were more frequent in Ig transcripts from FL cells cultured with IL-7, as we observed for TCR Vγ4 and Vγ6. The DFL16.1 region was preferentially used for recombination in the cells cultured with IL-7, but not without IL-7. When we cloned and sequenced 14 different PCR products from freshly isolated day 14 FL, we found only one clone of 14 had

N-region nucleotides in either N region. There was no restricted use of D regions by the fresh FL cells.

Because template-independent insertion of nucleotides between the V-(D)-J regions is catalyzed by TdT, these data suggested that TdT levels might be higher in FL cells cultured with IL-7, compared to FL cells cultured in medium alone. To test this hypothesis, we isolated RNA from FL cells cultured with or without IL-7 and performed semiquantitative RT-PCR analyses of TdT mRNA. TdT mRNA levels were relatively higher in FL cells cultured with IL-7, than without (FIG. 2). As predicted from the RT-PCR experiments, a relatively high level of TdT protein was detected in FL cells cultured with IL-7, whereas no TdT was detected in fresh FL cells or FL cells cultured with IL-7 (data not shown). We conclude from these data that IL-7 can regulate N-region diversity of Ig and TCR genes by stimulating TdT expression. These observations provide further evidence for environmental influences on diversity. Moreover, this provides a novel system with which to study the developmental regulation of N-region diversity.

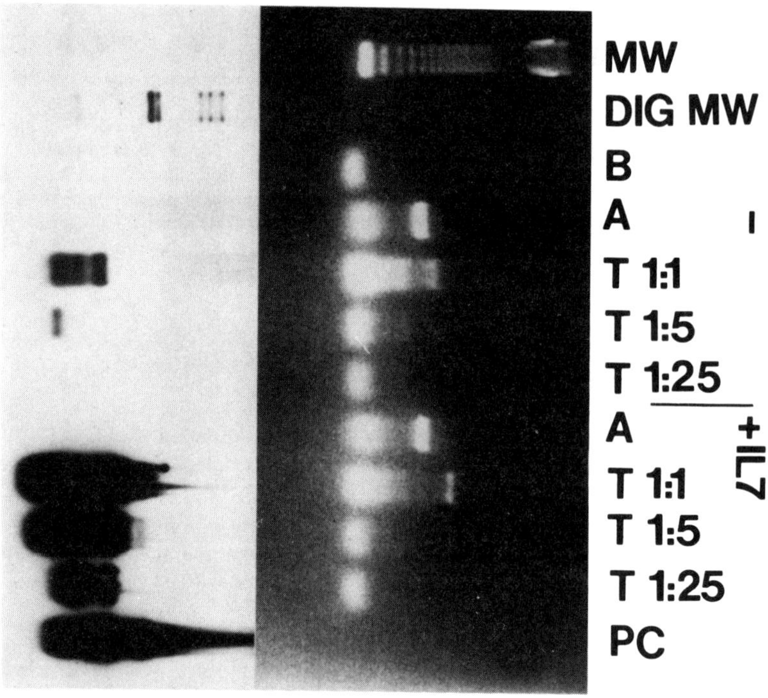

FIGURE 2. Expression of TdT mRNA by FL cells in response to culture with IL-7. RT-PCR analysis of actin and TdT expression using RNA from FL cells or adult thymus. Ten micrograms of total cellular RNA from FL cells cultured 7 days without IL-7 (−) or with 250 U/ml IL-7 (+IL-7), was reverse transcribed to make cDNA. PCR was then performed on undiluted cDNA (1 : 1), or cDNA product diluted 1 : 5 or 1 : 25. *Top panel:* ethidium bromide-stained agarose gels; *bottom panel:* chemiluminescent Southern blot analysis using TdT cDNA probe. MW = 123 bp molecular weight markers; DIG MW = digoxigenin-labeled MW markers; B = blank (no RNA); A = RT-PCR using actin primers; PC = positive control TdT cDNA for Southern blot.

REFERENCES

1. KOMORI, T., A. OKADA, V. STEWART & F. W. ALT. 1993. Science **261:** 1171–1175.
2. FEENEY, A. 1990. J. Exp. Med. **172:** 1377–1390.
3. LAFAILLE, J. J., A. DECLOUX, M. BONNEVILLE, Y. TAKAGAKI & S. TONEGAWA. 1989. Cell **59:** 859–870.
4. BOGUE, M., H. MOSSMANN, U. STAUFFER, C. BENOIST & D. MATHIS. 1993. Eur. J. Immunol. **23:** 1185–1188.
5. APPASAMY, P. M., T. W. KENNISTON, JR., Y. WENG, E. C. HOLT, J. KOST & W. H. CHAMBERS. 1993. J. Exp. Med. **178:** 2201–2206.

Mutations in Immunoglobulin V Gene Promoters May Cause Reduced Germline Transcription and Diminished Recombination Frequencies

B. J. NICLAS STIERNHOLM AND NEIL L. BERINSTEIN

Departments of Immunology and Medicine
University of Toronto
Toronto, Ontario M5S 1A8, Canada

The process by which discontinuous gene segments are sequentially joined to form the intact genes that code for the surface immunoglobulin (sIg) receptors on B lymphocytes, often referred to as V(D)J recombination, is central to the humoral immune system. From relatively small pools of germline gene segments, variable (V), diversity (D), and joining (J) gene segments, intact genes coding for receptors with a seemingly infinite number of specificities are generated. In contrast to what one may suspect, it is becoming increasingly evident that the V gene segments used in the expressed repertoire do not truly represent the available repertoire (reviewed in Krawinkel *et al.*[1] and Schwartz and Stollar[2]). That is, of all the germline V gene segments at the various human Ig loci, some appear more often than others in the expressed V gene repertoire. This phenomenon has been observed in fetal liver B cells[3,4] and in B-cell malignancies,[5] as well as in normal and autoreactive adult B cells (reviewed in Refs. 6–8). It also appears that genetic factors play a role in determining the expressed Ig repertoire, since the Ig repertoire differs between various mouse strains.[9,10] Several explanations for skewed V gene usage in the expressed repertoire have been proposed and include high copy numbers of individual V gene segments in the germline as well as cellular selection (reviewed in Refs. 7, 8, and 11). A third explanation is that certain V gene segments are rearranged at higher frequencies than others.[12,13]

A number of factors have been shown to influence rearrangement frequency *in vitro*. Chromosomal proximity of the recombining elements is one of them, in that many overrepresented V_H gene segments have been shown to be located in the 3′ end of the locus, proximal to the J_H locus (reviewed in Pascual and Capra[6]). Although it is evident that proximity of the recombining gene segments may play an important role, it has been shown that this cannot be the only factor determining the frequency of rearrangement.[8,14,15] Other factors that have been shown to affect the frequency of V(D)J recombination include the qualities of the RSSs,[16,17] short regions of homology between the two recombining coding ends,[18,19] as well as the actual nucleotide sequences of the coding ends.[20] Accessibility of the DNA to the recombinase is yet another factor thought to influence rearrangement frequency. There are several reports correlating V(D)J recombination with germline transcription, which is thought to "open" the chromatin structure and render the DNA more accessible to the recombinase.[21–27] In addition, CpG methylation has been implicated in the control of V(D)J recombination. An inverse association between CpG methylation and V(D)J recombination has been documented,[28–30] suggesting that hypomethylation also creates a more accessible target for the recombinase.

TABLE 1. Comparison of Relevant Features between the Frequently Rearranged VλVI-3.6 Gene and the Infrequently Rearranged VλX-4.4 Gene

<table>
<tr><th>Feature</th><th>VλVI-3.6</th><th>VλX-4.4</th></tr>
<tr><td>Rearrangement frequency
(p = 0.009)[a]</td><td>22/27</td><td>5/27</td></tr>
<tr><td>Relative chromosomal position</td><td>5′</td><td>3′</td></tr>
<tr><td>Recombination signal sequences

Consensus</td><td>CACAGTG | 23 | ACAGAAACT
| | | | | | | | | | | | | |
CACAGTG | 23 | ACAAAAACC</td><td>CACAGTG | 23 | ATAAAAACT
| | | | | | | | | | | | | |
CACAGTG | 23 | ACAAAAACC</td></tr>
<tr><td>Coding end sequences[b]</td><td>CAATCA</td><td>TGCTCA</td></tr>
<tr><td>Regions of homology</td><td>none</td><td>TGCTCA . . . . . TGCTGT
Vλ   Jλ</td></tr>
<tr><td>CpG methylation</td><td>hypo</td><td>hypo</td></tr>
<tr><td>Germline transcription</td><td>++++</td><td>+</td></tr>
<tr><td>Promoter

Consensus</td><td>AnATTTGCAT . . . . . GATAA
| | | | | | | | | | | | |
AnATTTGCAT . . . . . TATAA</td><td>TnATTTAAAT . . . . . TATAA
| | | | | | | | | | | |
AnATTTGCAT . . . . . TATAA</td></tr>
</table>

[a] Statistical analysis showed that the probability of having this distribution randomly is less than 1%.

[b] A correlation between a redundancy of "Ts," and to some extent "As," in the coding end and a reduced efficiency of rearrangement has been demonstrated.[20]

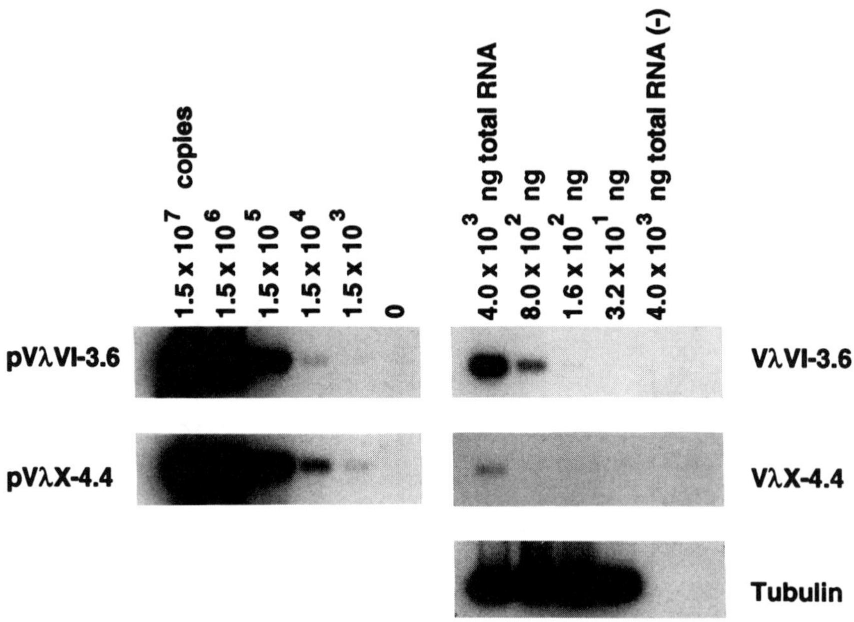

FIGURE 1. Reverse-transcriptase PCR assessing the levels of germline transcription of the VλVI-3.6 and the VλX-4.4 gene before rearrangement. Reverse transcriptase was omitted as a negative control, and tubulin was amplified as a positive control. Cloned genomic Vλ germline fragments were amplified as standards. One representative of three independent experiments is shown.

To assess the extent to which secondary Ig light chain gene rearrangements can contribute to the diversification of the expressed Ig repertoire, we studied a large number of secondary Igλ gene rearrangements in identical subclones of a human B-cell line. This analysis demonstrated that only two different Vλ gene segments were used in these rearrangements and that there was a bias against rearrangement of one of them. To uncover the mechanism behind this bias, we systematically investigated the parameters that have previously been shown to affect the efficiency of V(D)J recombination (TABLE 1). Through deletional mapping, using Vλ-specific probes on Southern blots, we established that the infrequently rearranged Vλ gene is located 3′ of a frequently rearranged Vλ gene segment in the germline. Analysis of the germline nucleotide sequences of the two Vλ gene segments revealed that there was no reason to believe that the nucleotide sequences of the RSSs or the coding ends contributed to this bias. Similarly, analysis of the CpG methylation status indicated that both Vλ gene segments are hypomethylated in the germline. When we assessed the levels of germline transcripts of the two Vλ genes using RT-PCR, we found that the infrequently rearranged Vλ gene is transcribed at 50 to 100 times lower levels compared to the frequently rearranged Vλ gene (FIG. 1). Characterization of the promoters of the two gene segments showed that the former has three mutations in the octamer motif, which likely explains the reduction in germline transcription.

We thus provide an example where a low rearrangement frequency of an individual V gene can be directly correlated with a reduced level of germline transcription before rearrangement. We also demonstrate that this reduction of germline transcription can be linked to a mutated promoter motif. The latter represents a genetic element that may be, at least to some extent, responsible for biased V gene usage in the expressed Ig repertoire.

REFERENCES

1. KRAWINKEL, U., G. ZOEBELEIN, M. BRUGGEMAN, A. RADBRUCH & K. RAJEWSKY. 1983. Recombination between antibody heavy chain variable-region genes: Evidence for gene conversion. Proc. Natl. Acad. Sci. USA **78:** 2072–2076.
2. SCHWARTZ, R. S. & B. D. STOLLAR. 1994. Heavy-chain directed B-cell maturation: Continuous clonal selection beginning at the pre-B cell stage. Immunol. Today **15:** 27–32.
3. SCHROEDER, H. W., J. L. HILLSON & R. M. PERLMUTTER. 1987. Early restriction of the human antibody repertoire. Science **238:** 791–793.
4. SCHROEDER, H. W. & J. Y. WANG. 1990. Preferential utilization of conserved immunoglobulin heavy chain variable gene segments during human fetal life. Proc. Natl. Acad. Sci. USA **87:** 6146–6150.
5. LOGTENBERG, T., M. E. M. SCUTTE, G. INGHIRAMI, J. E. BERMAN, F. H. G. GMELIG-MEYLING, R. A. INSEL, D. M. KNOWLES & F. W. ALT. 1989. Immunoglobulin VH gene expression in human B cell lines and tumours: Biased VH gene expression in chronic lymphocytic leukemia. Int. Immunol. **1:** 362–366.
6. PASCUAL, V. & J. D. CAPRA. 1991. Human immunoglobulin heavy-chain variable region genes: Organization, polymorphism and expression. Adv. Immunol. **49:** 1–74.
7. STEWART, A. K., C. HUANG, A. A. LONG, B. D. STOLLAR & R. S. SCHWARTZ. 1992. VH-gene representation in autoantibodies reflects the normal human B-cell repertoire. Immunol. Rev. **128:** 101–122.
8. LOGTENBERG, T., M. E. M. SCUTTE, S. B. EBELING, F. H. J. GMELIG-MEYLING & J. H. VAN ES. 1992. Molecular approaches to the study of human B-cell and (auto)antibody repertoire generation and selection. Immunol. Rev. **128:** 23–47.
9. WU, G. E. & C. J. PAIGE. 1986. VH gene family utilization in colonies derived from B and pre-B cells detected by the RNA colony blot assay. EMBO J. **5:** 3475–3481.
10. YANCOPOULOS, G. D., B. A. MALYNN & F. W. ALT. 1988. Developmentally regulated and strain-specific expression of murine VH gene families. J. Exp. Med. **168:** 417–435.
11. MARION, T. N., D. M. TILLMAN, N.-T. JOU & R. J. HILL. 1992. Selection of immunoglobulin variable regions in autoimmunity to DNA. Immunol. Rev. **128:** 123–149.
12. CHEN, P. P., R. W. SOTO-GIL & D. A. CARSON. 1990. The early expression of some human autoantibody associated heavy chain variable region genes is controlled by specific regulatory elements. Scand. J. Immunol. **31:** 673–678.
13. RASSENTI, L. Z., L. F. PRATT, P. P. CHEN, D. A. CARSON & T. J. KIPPS. 1991. Autoantibody-encoding κ L chain genes frequently rearranged in λ L chain-expressing chronic lymphocytic leukemia. J. Immunol. **147:** 1060–1066.
14. STEWART, A. K., C. HUANG, B. D. STOLLAR & R. S. SCHWARTZ. 1993. High-frequency representation of a single VH gene in the expressed human B cell repertoire. J. Exp. Med. **177:** 409–418.
15. CUISINIER, A. M., L. GAUTHIER & L. BOUBLI. 1993. Mechanisms that generate human immunoglobulin diversity operate from the 8th week of gestation in fetal liver. Eur. J. Immunol. **23:** 110–118.
16. HESSE, J. E., M. R. LIEBER, K. MIZUUCHI & M. GELLERT. 1989. V(D)J recombination: A functional definition of the joining signals. Genes Dev. **3:** 1053–1061.
17. RAMSDEN, D. A. & G. E. WU. 1991. Mouse κ light-chain recombination signal sequences mediate recombination more frequently than do those of λ light chain. Proc. Natl. Acad. Sci. USA **88:** 10721–10725.
18. GU, H., I. FORSTER & K. RAJEWSKY. 1990. Sequence homologies, N sequence insertions and JH gene utilization in VHDJH joining: Implications for the joining mecha-

nism and the ontogenic timing of Ly1 B cell and B-CLL progenitor generation. EMBO J. **9:** 2133–2140.

19. ICHIHARA, Y., H. HAYASHIDA, S. MYAZAWA & Y. KUROSAWA. 1989. Only DFL16, DSP2, and DQ52 gene families exist in the mouse immunoglobulin heavy chain diversity gene loci, of which DFL16 and DSP2 originate from the same primordial DH gene. Eur. J. Immunol. **19:** 1849–1854.
20. GERSTEIN, R. M. & M. R. LIEBER. 1993. Coding end sequences can markedly affect the initiation of the V(D)J recombination. Genes Dev. **7:** 1459–1469.
21. FONDELL, J. D. & K. B. MARCU. 1992. Transcription of germline Vα segments correlates with ongoing T-cell receptor α-chain rearrangmeent. Mol. Cell. Biol. **12:** 1480–1489.
22. YANCOPOULOS, G. D. & F. W. ALT.1985. Developmentally controlled and tissue-specific expression of unrearranged VH gene segments. Cell **40:** 271–281.
23. LENNON, G. G. & R. P. PERRY. 1990. The temporal order of appearance of transcripts from unrearranged and rearranged Ig genes in murine fetal liver. J. Immunol. **144:** 1983–1987.
24. ALT, F. W., T. K. BLACKWELL, R. A. DEPINHO, M. G. RETH & G. D. YANCOPOULOS. 1986. Regulation of genome rearrangement events during lymphocyte differentiation. Immunol. Rev. **89:** 5–30.
25. SCHLISSEL, M. S. & D. BALTIMORE. 1989. Activation of immunoglobulin kappa gene rearrangement correlates with induction of germline kappa gene transcription. Cell **58:** 1001–1007.
26. MARTIN, D., R. HUANG, T. LEBIEN & B. VAN NESS. 1991. Induced rearrangement of κ genes in the BLIN-1 human pre-B cell line correlates with germline J-C_κ and V_κ transcription. J. Exp. Med. **173:** 639–645.
27. GOLDMAN, J. P., D. M. SPENCER & D. H. RAULET. 1993. Ordered rearrangement of variable region genes of the T cell receptor γ locus correlates with transcription of the unrearranged genes. J. Exp. Med. **177:** 729–739.
28. ENGLER, P., P. ROTH, J. Y. KIM & U. STORB. 1991. Factors affecting the rearrangement efficiency of an Ig test gene. J. Immunol. **146:** 2826–2835.
29. HSIEH, C. L. & M. R. LIEBER. 1992. CpG methylated mini-chromosomes become inaccessible for V(D)J recombination after undergoing replication. EMBO J. **11:** 315–325.
30. GOODHARDT, M., P. CAVELIER, N. DOYEN, S. KALLENBACH, C. BABINET & F. ROUGEON. 1993. Methylation status of immunoglobulin κ gene segments correlates with their recombination potential. Eur. J. Immunol. **23:** 1789–1795.

High Frequency of Normal IgH Rearrangement in *scid* Mice

JACQUELINE PENNYCOOK,[a] AARON MARSHALL,[a]
YENHUI CHANG,[a] ROBERT A. PHILLIPS,[b]
AND GILLIAN E. WU[a,c]

[a]*Department of Immunology*
University of Toronto at the
Wellesley Hospital Research Institute and

[b]*Hospital for Sick Children*
Toronto, Ontario, Canada M4Y 1J3

The *scid* mouse has a mutation that affects the V(D)J recombinase system leading to a general absence of mature B and T cells.[1,2] Most models of B-cell development propose that maturation is dependent on the immunoglobulin (Ig) products created during the developmental process; thus the *scid* mouse provides an ideal system to determine to what degree this is true. Preliminary studies of *scid* Abelson murine leukemia virus (AMuLV) transformed lines and long-term bone marrow cultures demonstrated that this defect specifically affected coding joint formation, while signal joint formation appears relatively unaffected.[1,3,4] The *scid* defect is leaky. Approximately 15% of all *scid* mice develop detectable levels of serum Ig and some T-cell activity.[5–8]

In order to understand the nature of the *scid* defect and its effect on V(D)J recombination, we analyzed Ig heavy-chain gene rearrangement in *scid* mice. The extent of endogenous Ig gene rearrangement was determined using two quantitative PCR assays. DJH rearrangement in nonleaky *scid* whole bone marrow was found to be 1 to 10% of the normal number in control C.B-17 mice, with some unequal usage of J_H elements varying from mouse to mouse.[9] In contrast, VDJH rearrangement was found to be 1 to 100% of the normal number in control C.B-17 mice with even more pronounced unequal usage of the four J_H segments.

Characterization of the *scid* DJH structures showed that individually they were identical to those generated in normal mice. As a population, however, many of the DJH structures were the same, suggesting a restricted repertoire. Moreover, the *scid* structures exhibited a profound lack of nucleotide addition[9] in spite of normal TdT expression.[10] Characterization of the *scid* VDJH structures is continuing. A single VDJH structure has been isolated repeatedly from separate amplifications of a single mouse bone marrow sample. This structure is functionally rearranged, exhibiting normal numbers of nucleotide deletions and N insertions. Hence, there is extensive normal-type gene rearrangement in *scid* mice bone marrow, leaving in question the absence of detectable B and T cells in these mice.

[c] Address for correspondence: Wellesley Hospital Research Institute, 160 Wellesley Street East, Toronto, Ontario, Canada M4Y 1J3.

REFERENCES

1. Malynn, B. A., *et al.* 1988. Cell **54:** 453–460.
2. Shinkai, Y., *et al.* 1992. Cell **68:** 855–867.
3. Kim, M. G., W. Schuler, M. J. Bosma & K. B. Marcu. 1988. J. Immunol. **141:** 1341–1347.
4. Okazaki, K., S. I. Nishikawa & H. Sakano. 1988. J. Immunol. **141:** 1348–1352.
5. Bosma, G. C., *et al.* 1988. J. Exp. Med. **167:** 1016–1033.
6. Gibson, D. M., G. C. Bosma & M. J. Bosma. 1989. Curr. Top. Microbiol. Immunol. **152:** 125–136.
7. Kearney, J. F., *et al.* 1989. Curr. Top. Microbiol. Immunol. **152:** 137–146.
8. Carroll, A. M., R. R. Hardy, J. Petrini & M. J. Bosma. 1989. Curr. Top. Microbiol. Immunol. **152:** 117–123.
9. Pennycook, J. L., Y. Chang, J. Celler, R. A. Phillips & G. E. Wu. 1993. J. Exp. Med. **178:** 1007–1016.
10. Osmond, D. G., *et al.* 1992. Blood **79:** 1695–1703.

Regulation of Antibody Class Switching to IgE: Characterization of an IL-4-Responsive Region in the Immunoglobulin Heavy-Chain Germline ε Promoter[a]

SANDRA DELPHIN AND JANET STAVNEZER[b]

Department of Molecular Genetics and Microbiology
University of Massachusetts Medical School
Worcester, Massachusetts 01655-0122

INTRODUCTION

Antibody class switching is effected by a DNA recombination event that results in the substitution of one Ig heavy-chain constant region (C_H) gene for another, thus changing the C_H region and thereby the effector function of the antibody, but keeping the antigen binding specificity unchanged. The specificity of class switching is regulated by cytokines that direct switching by inducing or repressing transcription from unrearranged, or germline, C_H genes before switch recombination to the same C_H gene.[1–7] For example, a large body of data indicate that IL-4 induces transcription of the germline Cε gene in B cells activated by polyclonal activators, for example, lipopolysaccharide (LPS), Epstein-Barr virus (EBV) infection, or anti-CD40 antibody and that subsequently cells in this population will undergo switch recombination to the Cε gene, thereby allowing expression of IgE.[7–11] Furthermore, it has recently been shown that if a DNA segment that contains the promoter and first exon (I, or germline exon) of the germline transcript is deleted, switching does not occur on that chromosome.[5,6,12] Although the sequences necessary and sufficient for switching have not been defined, these gene knock-out experiments indicate that transcription of germline transcripts is probably necessary for switching. Therefore, it is important to understand how transcription of unrearranged C_H genes is regulated.

A map of the mouse germline ε DNA segment and transcript is shown in FIGURE 1. The DNA sequences that regulate transcription of the mouse and human germline ε RNAs are currently being studied, and initial characterizations of DNA sequences that regulate induction by IL-4 have been published.[13–15] In this report we characterize an additional IL-4-responsive region (IL-4RR) in the mouse germline ε promoter that overlaps with the previously characterized IL-4-responsive

[a] This research was supported by National Institutes of Health Grant R01-AI23283 and by Grant IM-679 from the American Cancer Society.

[b] Address for correspondence: Department of Molecular Genetics and Microbiology, University of Massachusetts Medical Center, 55 Lake Avenue North, Worcester, MA 01655.

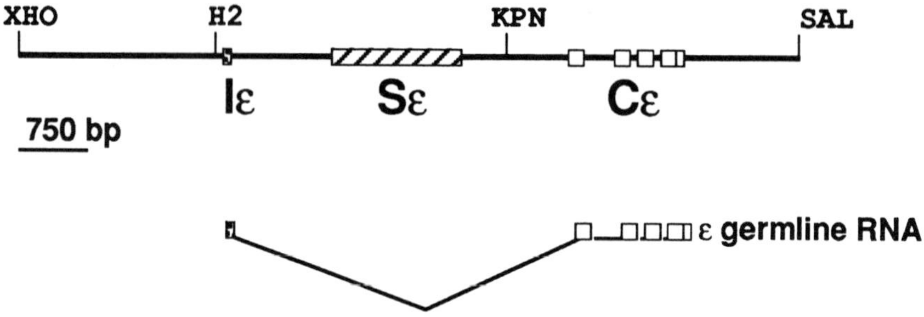

FIGURE 1. Germline Ig epsilon DNA and transcript. Structure of the mouse immunoglobulin heavy-chain germline ε gene and ε transcript. Restriction map of unrearranged Cε gene and germline ε transcript with location of exons and switch region (Sε) marked.

region of the human germline ε promoter.[14] This ε IL-4RR resides within a DNA sequence highly conserved between human and mouse and also has elements similar to those within an IL-4-responsive region within the promoter for mouse germline $\gamma 1$ transcripts.[16,17]

RESULTS

Induction of Germline ε Transcripts in Two Mouse B-Cell Lines

For these studies, we used two B lymphoma cell lines that inducibly express germline ε RNA. The I.29μ mouse B lymphoma expresses surface IgM and can be induced to undergo class switching to IgA by treatment with LPS. Switching can be increased by treatment with TGFβ, which increases transcription from the unrearranged Cα genes. I.29μ cells will switch to IgE expression if treated with IL-4 + LPS, although switching to IgE is about 100 times less frequent than optimal switching to IgA. I.29μ cells synthesize germline ε transcripts constitutively, and this can be further increased 5- to 20-fold by treatment with IL-4[2] (and data not shown). Addition of LPS does not further increase the level of transcripts. The other B-cell lymphoma used for these studies is M12.4.1, a class II$^+$, sIg-negative variant that was derived from the M12 B lymphoma.[18] M12.4.1 cells do not express germline ε transcripts constitutively, but germline ε transcripts can be induced by treatment with IL-4 + LPS.[13] Unlike I.29μ, IL-4 alone does not induce germline ε transcripts in M12.4.1 cells.

Transient Transfection of Luciferase Reporter Plasmids Containing Various Lengths of 5′ Flanking Region from the Germline ε Segment

To identify DNA segments necessary for regulation of transcription of germline ε RNA, a series of luciferase reporter plasmids containing various lengths of the 5′ flank of the Iε exon were prepared, varying from having 1.8 kb to only 15 bp of 5′ flanking sequences. The plasmids were transiently transfected into I.29μ or M12.4.1 cells. Optimum luciferase activity was obtained when cells were harvested 12 hr after transfection. An internal control CAT plasmid driven by a SV40 promoter/enhancer (pSV2CAT) was used to correct for transfection efficiency when comparing the basal activity of different plasmids. Maximal basal expression is not affected much by deletion of the 5′ flanking segment to −127 relative to the RNA initiation site, but further deletion reduces basal expression in both I.29μ and M12.4.1 cell lines.[19]

To test whether expression of these promoter-luciferase plasmids could be induced by IL-4, cells were divided after transfection into two or more aliquots and treated with recombinant mouse IL-4 for various times. Optimal responses were obtained with a dose of 240 ng/ml, which is equivalent to 2500 units IL-4 per ml, and we routinely use 1000 units per milliliter. LPS treatment does not affect the induction by IL-4 of any of the promoter constructs, nor does LPS alone induce luciferase activity in either I.29μ or M12.4.1 cells (data not shown). The IL-4 inducibility of plasmids with the longest 5′ flanks, those having 1.8 and 1.2 kb of ε promoter segment, is two to four times less than plasmids having from 640 to 162 bp of 5′ flank, which are approximately equivalently inducible by IL-4: eightfold in I.29μ and threefold in M12.4.1 cells. Deletion past −162 reduces inducibility of the luciferase plasmids in both cell lines, and expression of a construct with 115 nucleotides of the ε promoter segment cannot be induced by IL-4 (data not shown). Therefore, the DNA segment residing between −162 and the start site of transcription has one or more DNA elements required for IL-4 inducibility of the ε promoter luciferase plasmids. The nucleotide sequence of this region is shown in FIGURE 2A.

Effects of a Series of Linker Scanning Mutations on Expression of the Germline ε Promoter

Linker scanning mutations were introduced into the −162/+53 ε promoter segment contained within the −162 Luc reporter plasmid in order to identify DNA sequences required for IL-4 inducibility and for basal activity of the germline ε promoter. The nucleotides affected by the mutations, either due to substitution or deletion, are shown in FIGURE 2B, and their positions are indicated to the left of each sequence.

Plasmids bearing the linker scanning mutations were transiently transfected into I.29μ and M12.4.1 cells in order to test for their ability to drive luciferase expression. Each construct was tested at least three times in both cell lines. Most of the linker scanning mutations reduce basal level expression of the promoter relative to that obtained with the wild-type −162Luc plasmid when transfected into I.29μ cells, although for most mutations the effect is not dramatic.[19] The effects of the mutations differ in M12.4.1 cells, in that nearly all mutations located 3′ to −87 abolish basal expression, although mutations located 5′ to −92 do not. These results suggest there are several DNA elements residing between −136 and the start site of transcription that are necessary for normal levels of basal expression of the germline ε promoter in both cell lines, as assayed in the luciferase plasmid. Consistent with this is the fact that the nucleotide sequence between −144 and −14 is highly conserved, as it is almost identical to the sequence of the human germline ε promoter sequence.[14,20]

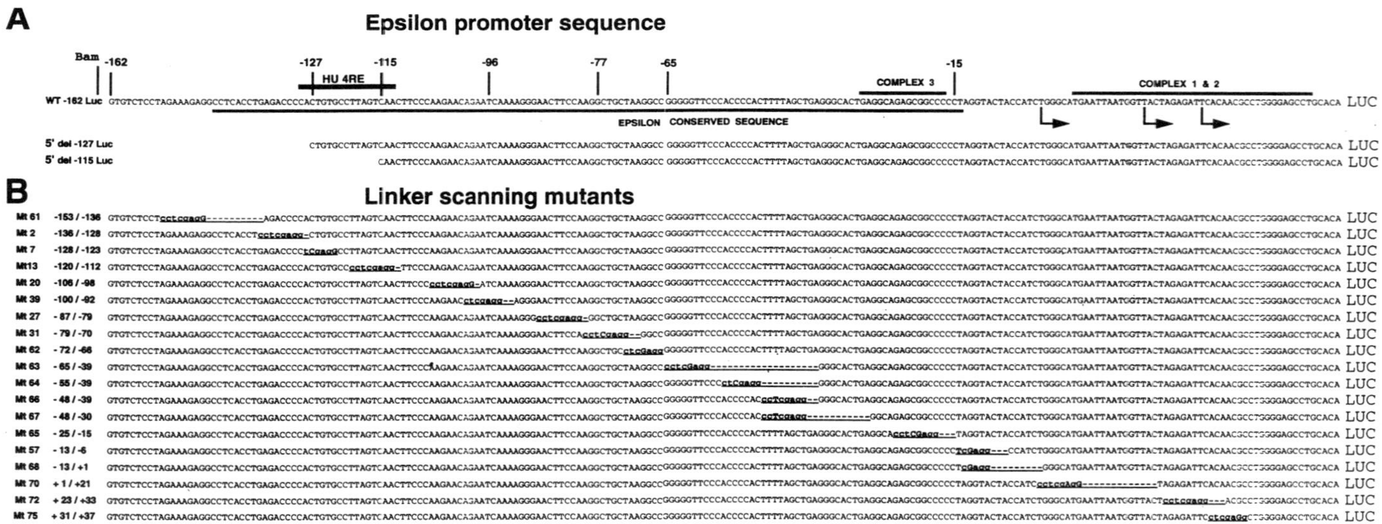

FIGURE 2. **(A)** Sequence of 5′ flank of germline ε RNA and first 53 nucleotides of the Iε exon, present in the ε promoter −162Luc plasmid. The plasmid contains all three transcription start sites detected in I.29μ cells. The first 53 bp of the Iε exon does not have a translation initiation ATG codon. Regions indicated above the sequence are *Hu 4RE*, a region necessary for induction of the human germline ε promoter by IL-4[14]; *complexes 1 and 2*, a region that binds nuclear factors in B cells/B cell lines stimulated by IL-4[13]; *complex 3*, the binding site for BSAP[15,32] and necessary for expression of the germline ε promoter-CAT constructs after LPS + IL-4 induction.[13,15] *Epsilon conserved sequence:* the segment highly homologous (>80%) between human and mouse sequences. This sequence differs at two sites from that previously reported[33]: at +40/+43 we find GGGG instead of GGG and at −15/−19 CCCCC instead of CCCC. **(B)** Sequences of the −162Luc plasmids with linker scanning mutations. Names are given on the left, followed by nucleotides that are either substituted or deleted in the various mutants.

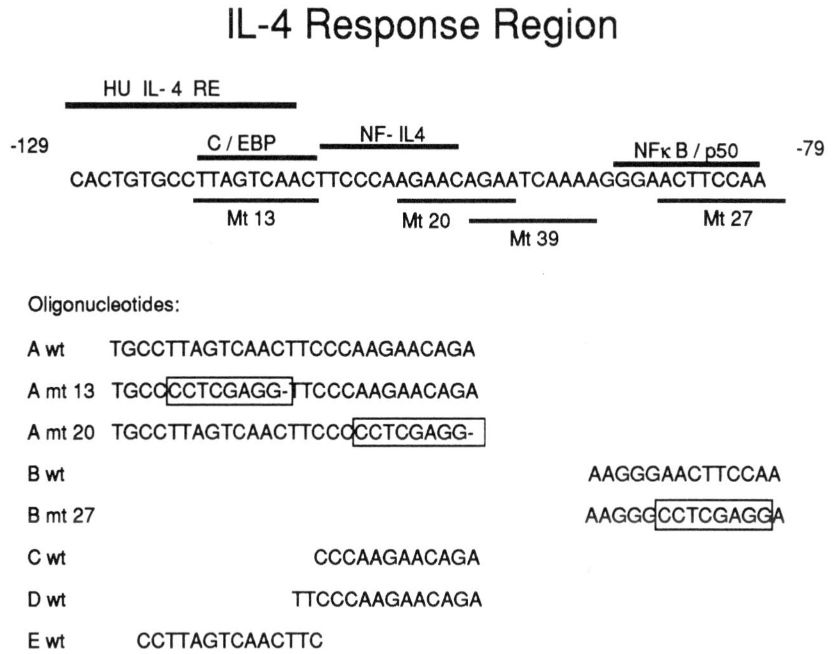

FIGURE 3. Nucleotide sequence of the IL-4RR in the mouse germline ε promoter. Three transcription factor binding sites are detected: A C/EBP consensus element (−120/−112), the sequence defined as an NF-IL-4 site (−111/−102), and a NF-κB/p50 site (−90/−81). Indicated below the DNA sequence are the positions of the linker scanning mutations tested. *Hu IL-4 RE* is the segment that when mutated disrupts IL-4 responsiveness of a human germline ε promoter-CAT constructs.[14] Beneath the sequence are the wild-type and mutated double-stranded oligonucleotides that were tested for the ability to transfer IL-4 inducibility to a minimal c-*fos* promoter and/or to bind to nuclear proteins in an EMSA. The mutations of the oligos are the same as those present in the linker scanning mutations.

Localization of an IL-4-Responsive Region (IL-4RR) Using Linker Scanning Mutations

The sequences in the germline ε promoter that control induction by IL-4 appear to be much more localized than those required for basal expression, since only four of the linker scanning mutations abolish or markedly reduce IL-4 inducibility in both I.29μ and M12.4.1 cells, mt 13: at −120/−112, mt 20: at −106/−98, mt 27: at −87/−79, and mt 63: at −65/−39.[19]

The sequence of the −126/−79 segment that contains three of the linker scanning mutations that abolish or reduce IL-4 inducibility (mts: 13, 20, and 27) is shown in FIGURE 3. Both mts 13 and 20 completely abolish the IL-4 response of the plasmid in both cell lines. Mt 27 reduces the response by 70%.[19] Interestingly, these three mutations alter potential binding sites for three transcription factors. Mt 13 alters a consensus binding site for the C/EBP family of transcription factors.[21–24] Mt 20 alters a consensus element for a newly described IL-4-inducible transcription factor, NF-IL4 or IL-4NAF.[25,26] Mt 27 alters a consensus binding site for NF-κB.[27] This third mutation also abolishes constitutive expression. The fourth linker-scanning mutation mt 63 (−65/−38) reduces IL-4 inducibility in I.29μ cells by 60% and basal expression by 90%, although it completely abolishes inducibility and basal expression in M12.4.1 cells. This mutation should abolish the binding site for the B-cell-specific DNA-binding protein BSAP, which binds to the germline ε promoter at −42/−15 and is required for IL-4 + LPS inducibility of a CAT reporter gene driven by the mouse germline ε promoter in M12.4.1 cells.[13,15] Nevertheless, four other mutations covering the region from −55 to −15 also mutate the BSAP-binding site, but have little or no effect on induction by IL-4.

A DNA Segment Containing Binding Sites for C/EBP and NF-IL4 is Sufficient to Confer IL-4 Inducibility

We wished to determine whether the DNA sequences shown by the linker scanning mutations are necessary for IL-4 induction of the germline ε promoter are also sufficient to confer IL-4 inducibility on a heterologous promoter. To test this, we inserted a series of double-stranded oligonucleotides containing wild-type or mutated sequences of the IL-4RR shown in FIGURE 3 into a luciferase reporter plasmid driven by a minimal c-*fos* promoter. Some of the plasmids have multiple copies of the oligonucleotides in sense and anti-sense orientations. The plasmids were transiently transfected into I.29μ cells, and the cells were stimulated with IL-4. Only the wild-type oligonucleotide A, which contains both the consensus binding sites for C/EBP and NF-IL4, is able to confer IL-4 inducibility upon the c-*fos* promoter.[19] The activity of a plasmid with one copy of the A oligo is induced threefold by IL-4, and plasmids with two or three copies are induced about 20-fold. If either the C/EBP or the NF-IL4 site is mutated, the plasmid is not inducible by IL-4. Therefore, both sequence elements are required for IL-4 inducibility, consistent with the linker scanning analysis, in which it was found that mutation of either of these sequence elements eliminates inducibility. The wild-type B oligo, which has the consensus κB site, does not transfer IL-4 inducibility to the c-*fos* promoter. Results are similar when the plasmids are transfected into M12.4.1 cells, although the induction by IL-4 is not as great as in I.29μ cells.

Electrophoretic Mobility Shift Assays (EMSAs) Demonstrate That the IL-4RR Does Indeed Contain Binding Sites for NF-IL4, Ig/EBP −1, and NF-κB/p50

To determine if these consensus binding sites bind the predicted proteins, a series of EMSAs were performed using the double-stranded oligonucleotides shown in FIGURE 3 as probes and/or as competitors. When wild-type oligonucleotide A is incubated with nuclear extracts from unstimulated or IL-4-treated I.29μ cells, a low mobility complex is induced by IL-4 treatment.[19] Competition experiments demonstrate that this complex binds to a site that matches the consensus sequence for a complex induced by IL-4 (NF-IL4/IL-4 NAF) in human monocyte and B-cell lines.[25,26] This complex is competed by wild-type oligo A, by A mt 13, and by oligo D, but not with mt 20, which has nucleotide substitutions in the putative NF-IL4 binding site, nor with oligo E. Oligo C does not compete, apparently because it lacks the first T of the binding site, which has been shown to be important for binding of NF-IL4[25] (data not shown). A kinetic experiment showed that the IL-4-inducible complex is detected after 30 min of IL-4 treatment, is maximally induced by 4 hr, and is maintained for 24 hr. No later time points were examined.[19] The kinetics of induction are slower and more sustained than those reported for the binding activity in monocytes.[26]

The IL-4-inducible complex is also detected in splenic B cells treated with IL-4 alone for 12 hr, and LPS has no additional effect.[19] Thus, although LPS is required for induction of germline ε transcripts in splenic B cells, NF-IL4 can be induced by IL-4 alone, indicating that induction of NF-IL4 is not sufficient to induce transcription of the endogenous germline ε RNA.

Mt 13 disrupts a consensus sequence for the C/EBP family of transcription factors. This family of proteins possesses a basic region and leucine zipper, and all members bind the same DNA sequence, although not all cells have all family members. B cells express two members of the family: Ig/EBP-1 (also called C/EBPγ) which is most abundant in pre-B cells, and C/EBPβ (also called NF-IL6, LAP, AGP/EBP, CRP-2, or IL6DBP) which is expressed in mature B-cell lines and in splenic B cells induced with LPS.[24] To determine if C/EBP family members bind the consensus C/EBP site, we tested binding of recombinant mouse Ig/EBP-1.[28] Both wild-type oligonucleotide A and mt 20 bind Ig/EBP-1, whereas mt 13 does not.[19] Thus, oligonucleotides with a wild-type, but not a mutated, consensus site for C/EBP bind Ig/EBP. We have not yet determined if I.29μ cells contain C/EBPβ and/or Ig-EBP.

We demonstrated that the putative κB site binds a member of the NF-κB/rel family of transcription factors by performing EMSAs using oligonucleotide B as a probe with nuclear extracts from untreated or IL-4-treated I.29μ cells, competing with a DNA fragment containing the NF-κB site from the Ig κ enhancer.[19] Specificity of binding was also demonstrated by competition with wild-type B or with mt 27, which contains the mutated κB site. Three complexes form with oligo B that can all be competed with the wild-type B oligo or with a κB-binding site, but not with mt 27. In addition, when mt 27 is labeled and used as a probe, it fails to bind any complex (data not shown). The binding to oligo B is not induced by IL-4. This is consistent with the finding that oligo B does not transfer IL-4 inducibility to the c-*fos* promoter, although the κB site contributes to the IL-4 response, since the linker-scanning mutation at this site reduces the response to IL-4 by threefold in both I.29μ and M12.4.1 cells.

Different members of the NF-κB/rel family differ somewhat in their DNA sequence requirements for binding.[27] In order to determine which member of this family binds to the putative binding site at −90/−81, we tested a series of NF-κB antisera for their effect in EMSAs using oligo B as the probe. An antiserum to the p50 homodimer supershifts the complex, whereas antisera to p65, or p50B do not.[19] Linker scanning mutation 39, which mutates −100/−91 has no effect on expression of the promoter in these assays, suggesting that there is no additional transcription factor binding site between the sites for NF-IL4 and for NF-κB/p50.

DISCUSSION

The Results Reported Here Extend but Also Contrast with Previous Studies of the Regulation of the Ig Germline ε Promoter in Mouse and Human

Rothman *et al.*[13] analyzed regulation of the mouse germline ε promoter using a chloramphenicol acetyl transferase (CAT) reporter plasmid transiently transfected into M12.4.1 cells. This group found that the germline ε promoter is induced by IL-4 + LPS, but not by IL-4 nor LPS alone, as is also true for the endogenously expressed germline ε transcripts. We also find that induction of endogenous germline ε transcripts at the steady-state level requires both IL-4 and LPS in M12.4.1 cells, but that our luciferase reporter construct is inducible by IL-4 alone although LPS further increases expression. By contrast, in I.29μ cells the endogenous transcripts are inducible at the steady-state level by IL-4 alone, and LPS does not further increase their expression. Another difference from Rothman *et al.* is that a 5′ deletion that deleted the C/EBP site did not reduce IL-4 inducibility of their reporter gene, whereas we find that mutations of the C/EBP site abolish inducibility. Furthermore, we find that four out of five mutations that affect the BSAP binding site in the promoter do not reduce IL-4 inducibility in either cell line, whereas they find the BSAP binding site is essential for IL-4 plus LPS inducibility in M12.4.1 cells.[13,15] Our findings agree with preliminary results on the promoter for the human germline ε transcripts.[14]

Characterization of an IL-4RR in the Germline ε Promoter

We have identified three transcription factors that appear to regulate basal and IL-4 induction of transcription of the mouse germline ε promoter: one or more members of the C/EBP family of transcription factors, the p50 subunit of NF-κB, and a newly described IL-4-inducible transcription factor, termed NF-IL4 or IL-4 NAF.[25,26] We demonstrate that mutation of any of the sequence elements that bind these factors, within the context of the germline ε promoter, abolishes or reduces induction by IL-4. A fragment containing binding sites for C/EBP and NF-IL4 is sufficient to transfer IL-4 inducibility to a minimal c-*fos* promoter, but a fragment containing only the C/EBP consensus element or only the NF-IL4 element is not sufficient. The binding sites for these three factors and their positions relative to each other are conserved within the human germline ε promoter. FIGURE 4 shows an alignment of the C/EBP and NF-IL4 consensus elements in the mouse and human germline ε promoters. Furthermore, substitutions in the C/EBP consensus element of the human germline ε promoter eliminates IL-4 responsiveness.[14] Effects of mutations in the NF-IL4 or NF-κB consensus elements of the human promoter have not been tested.

Consensus binding sites for the IL-4-inducible complex NF-IL4/IL-4 NAF have been identified in various IL-4-inducible promoters.[25,26] Although Kohler and Rieber[25] demonstrated that mutation of the binding site reduced the IL-4 inducibility of a CD23 promoter transfected into a human B-cell line, no evidence has been presented to indicate that the binding site for this factor by itself can transfer IL-4 responsiveness to another promoter.

Interestingly, binding sites for C/EBP and NF-IL4 are also found in the promoter for mouse germline γ1 RNA at about the same distance from the first RNA initiation site (−122/−108) as in the ε promoter. The segment containing these binding sites is necessary for induction of the promoter-luciferase reporter plasmids by phorbol ester and IL-4.[16] Furthermore, both C/EBPβ and C/EBPγ have been shown to bind the C/EBP site in the germline γ1 promoter.[17] Unlike the ε promoter, the germline γ1 promoter is not induced by IL-4 alone, although IL-4 synergizes with phorbol ester to activate this promoter. FIGURE 4 shows that in the γ1 promoter the C/EBP and NF-IL4 elements overlap and have a different position relative to each other than in the ε promoter. Switching to IgG4 in humans is inducible by IL-4, but the germline γ4 promoter and transcription start site have not been defined. These same two consensus elements are present within an evolutionarily conserved region 5′ to the Sγ4 tandem repeats, which has been

```
                                            C/EBP      NF-IL-4
A.  Mouse GL Epsilon    -124    TGCCTTAGTCAACTTCCCAAGAACAG   -99
    Human GL Epsilon    -168   CGCTGTTGCTCAATCGACTTCCCAAGAACA -141

                                       NF-IL-4   C/EBP
B.  Mouse GL Gamma1     -125   ACATTCACATGAAGTAATCTAAGTC   -101

    Human GL Gamma4     1538   GATTTCCTAGGAAGACAATAGCGGC   1665


C.  Mouse GL Epsilon      +2    GGGCATGAATTAAT       +16
    Mouse GL Gamma1       -8    TGTTCTTAGTCAAT       +6
```

FIGURE 4. Comparison of DNA sequences 5′ to the I exons of IL-4 responsive mouse and human germline C_H genes. **(A)** Alignment of the mouse and human germline ε promoters to show homology between the C/EBP and NF-IL4 sites. **(B)** Alignment of NF-IL4 and C/EBP sites in the mouse germline γ1 promoter with sequences upstream of the human γ4 switch region. This C/EBP site in the γ1 promoter has been demonstrated to bind C/EBPβ and possibly C/EBPγ in nuclear extracts.[17] The positions of the nucleotides from the γ4 gene are according to the published sequence.[20] **(C)** Alignment of the C/EBP consensus sequences located at the start sites of transcription of the mouse germline ε and γ1 RNAs. The γ1 sequence has been shown to bind C/EBPγ in nuclear extracts.[17]

postulated to encode the Iγ4 exon and its promoter[20] (FIG. 4). Thus, it appears possible that the proteins that bind the C/EBP and NF-IL4 sites interact and function together to effect IL-4 inducibility in all four of these promoters. Factors induced by LPS + IL-4 in splenic B cells that bind the C/EBP and NF-IL-4 sites of the germline γ1 promoter appear to interact in EMSAs.[17]

An additional C/EBP consensus element is present at the start site of transcription in both the mouse ε and γ1 germline promoters[17] (see FIG. 4). Lundgren *et al.*[17] have shown that C/EBP proteins bind at this second C/EBP site in the γ1 promoter. The second C/EBP site in the ε promoter is located where an IL-4 inducible complex, termed STF-IL4, has been shown to bind by footprinting and by EMSA competition analyses.[13,29] Unlike the C/EBP sites located upstream in each of these promoters, no obvious consensus binding sites for NF-IL4, as previously defined,[25,26] are nearby. This is consistent with our finding that linker-scanning mutations in this region do not reduce IL-4 inducibility, although they do reduce or abolish basal activity, in I.29μ or M12.4.1 cells, respectively.

C/EBPβ/NF-IL6 has been demonstrated to interact (via its leucine zipper) with RelA (NF-κB/p65) and to synergistically activate transcription when binding the IL-8 promoter. By contrast, NF-κB/p50 when bound to the same κB site had little or no ability to synergize with C/EBPβ.[30,31] We have been unable to detect binding of p65 to the κB site in the ε promoter, although binding of p50 is readily detected. An attractive hypothesis is that induction of NF-IL-4 by IL-4 aids the interaction between C/EBPβ and p50, thus creating an effective transcriptional activator involving C/EBPβ, NF-κB/p50, and NF-IL4 (FIG. 5). Our data suggest that p50 may interact with a C/EBP protein and/or with NF-IL4, since we have shown that mutation of the κB/p50 site reduces IL-4 responsiveness by threefold,

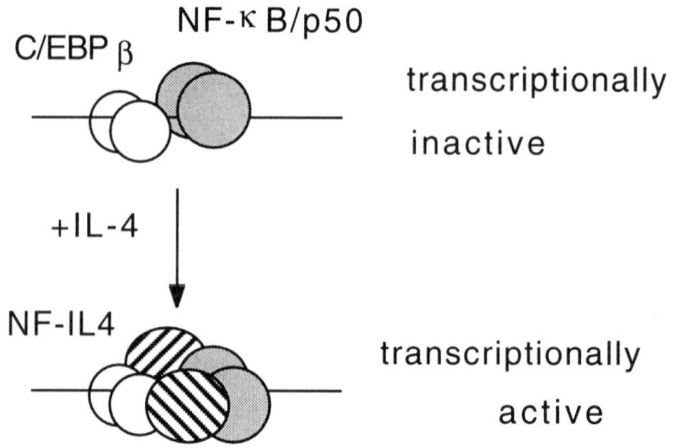

FIGURE 5. Speculative model for how NF-IL4 activates transcription by binding with C/EBPβ and NF-κB/p50 bound to DNA. It is known that C/EBPβ and p50 bind each other[34] and that they both bind the IL-6 and IL-8 promoters. However, the combination does not activate transcription better than C/EBPβ alone, although NF-κB/p65 will synergize with C/EBPβ to activate these promoters. Furthermore, p50 can synergize with C/EBPβ to activate transcription from a multimerized oligonucleotide containing both binding sites.[30,31] Thus, we speculate that binding of NF-IL4 to the ε promoter enables p50 to synergistically activate transcription with C/EBPβ.

although fragment B, which contains the κB site but not the C/EBP or NF-IL4 sites, is not sufficient to transfer IL-4 inducibility to a minimal c-*fos* promoter.

In conclusion, we have defined and characterized an IL-4RR in the promoter for germline ε RNA that is necessary for induction of the promoter by IL-4 when assayed in a luciferase reporter plasmid. This IL-4RR is sufficient to transfer IL-4 inducibility to another promoter as long as binding sites for NF-IL4 and C/EBP transcription factors are both intact. Binding sites for these same two transcription factors are also closely spaced in promoters of other germline transcripts inducible by IL-4. Future studies will be directed towards determining if these factors form a complex that effects IL-4 induction.

SUMMARY

A large body of data indicate that antibody class switching is directed by cytokines by inducing or repressing transcription from unrearranged, or germline, C_H genes. IL-4 induces transcription of the germline Cε genes in activated B cells, and subsequently cells in this population will undergo switch recombination to IgE. Furthermore, the data suggest that transcription of germline Cε genes is required for class switching. In this paper we define DNA elements required for induction of transcription of the germline Cε genes by IL-4. To do this, segments of DNA from the 5′ flank of the initiation sites for germline ε RNA were ligated to a luciferase reporter gene and transfected into two mouse B-cell lines, one of which can be induced to switch to IgE. By analysis of a series of 5′ deletion constructs and linker-scanning mutations, we demonstrate that a 46-bp segment (residing at −126/−79 relative to the first RNA initiation site) contains an IL-4 responsive region. This segment binds three transcription factors: the recently described NF-IL4, one or more members of the C/EBP family of transcription factors, and NF-κB/p50. Mutation of any of the binding sites for these three factors abolishes or reduces IL-4 inducibility of the ε promoter. A 27-bp segment within this IL-4 response region containing binding sites for NF-IL4 and a C/EBP factor is sufficient to transfer IL-4 inducibility to a minimal c-*fos* promoter.

ACKNOWLEDGMENTS

We thank present and past members of our laboratory, especially Shih-Chang Lin for performing the EMSA with the splenic B-cell extracts. We thank Drs. Yi-Chaung A. Lin, Minzhen Xu, and Penny Shockett for advice and for sharing of results and reagents. We thank Dr. Steve Gillis of Immunex Corp. and also the Sterling Research Group for recombinant mouse IL-4. We thank Drs. Ranjan Sen (Brandeis University), Ulrich Siebenlist (NIH), and Nancy Rice (Frederick Cancer Center) for antibodies to NF-κB. We thank Dr. Paul Rothman, Columbia University, for the M12.4.1 line. We thank Steven Baker for help with statistical analysis. We thank Drs. Jeffrey Purkerson and Meng Jiao Shi for providing helpful criticisms of the manuscript.

REFERENCES

1. LUTZKER, S., P. ROTHMAN, R. POLLOCK, R. COFFMAN & F. W. ALT. 1988. Mitogen- and IL-4-regulated expression of germline Igγ2b transcripts: Evidence for directed heavy chain class switching. Cell **53:** 177.
2. STAVNEZER, J., G. RADCLIFFE, Y.-C. LIN, J. NIEUTUPSKI, L. BERGGREN, R. SITIA & E. SEVERINSON. 1988. Immunoglobulin heavy-chain switching may be directed by prior induction of transcripts from constant-region genes. Proc. Natl. Acad. Sci. USA **85:** 7704.
3. BERTON, M. T., J. W. UHR & E. VITETTA. 1989. Synthesis of germline γ1 immunoglobulin heavy-chain transcripts in resting B cells: Induction by interleukin 4 and inhibition by interferon γ. Proc. Natl. Acad. Sci. USA **86:** 2829.
4. SEVERINSON, E., C. FERNANDEZ & J. STAVNEZER. 1990. Induction of germ-line immunoglobulin heavy chain transcripts by mitogens and interleukins prior to switch recombination. Eur. J. Immunol. **20:** 1079.
5. JUNG, S., K. RAJEWSKY & A. RADBRUCH. 1993. Shutdown of class switch recombination by deletion of a switch region control element. Science **259:** 984.
6. ZHANG, J., A. BOTTARO, S. LI, V. STEWART & F. W. ALT. 1993. A selective defect in IgG_{2b} switching as a result of targeted mutation of the $I_{\gamma 2b}$ promoter and exon. EMBO J. **12:** 3529.
7. COFFMAN, R. L., D. A. LEBMAN & P. ROTHMAN. 1993. The mechanism and regulation of immunoglobulin isotype switching. Adv. Immunol. **54:** 229.
8. ROTHMAN, P., S. LUTZKER, W. COOK, R. COFFMAN & F. W. ALT. 1988. Mitogen plus interleukin 4 induction of Cε transcripts in B lymphoid cells. J. Exp. Med. **168:** 2385.
9. GAUCHAT, J.-F., J. G. G. AVERSIO & J. E. DE VRIES. 1992. Modulation of IL-4 induced germline ε RNA synthesis in human B cells by tumor necrosis factor-α, anti-CD40 monoclonal antibodies or transforming growth factor-β correlates with levels of IgE production. Int. Immunol. **4:** 397.
10. SHAPIRA, S. K., D. VERCELLI, H. H. JABARA, S. M. FU & R. S. GEHA. 1992. Molecular analysis of the induction of immunoglobulin E synthesis in human B cells by interleukin 4 and engagement of CD40 antigen. J. Exp. Med. **175:** 289.

11. JABARA, H. H., L. C. SCHNEIDER, S. K. SHAPIRA, C. ALFIERI, C. T. MOODY, E. KIEFF, R. S. GEHA & D. VERCELLI. 1990. Induction of germ-line and mature Cε transcripts in human B cell stimulated with rIL-4 and EBV. J. Immunol. **145:** 3468.
12. BOTTARO, A., R. LANSFORD, L. XU, J. ZHANG, P. ROTHMAN & F. ALT. 1994. I region transcription (per se) promotes basal IgE class switch recombination but additional factors regulate the efficiency of the process. EMBO J. **13:** 665.
13. ROTHMAN, P., S. C. LI, B. GORHAM, L. GLIMCHER, F. ALT & M. BOOTHBY. 1991. Identification of a conserved LPS/IL-4 responsive element located at the promoter of germline ε transcripts. Mol. Cell. Biol. **11:** 5551.
14. ICHIKI, T., W. TAKAHASHI & T. WATANABE. 1993. Regulation of the expression of human Cε germline transcript. J. Immunol. **150:** 5408.
15. LIAO, F., B. BIRSHTEIN, M. BUSSLINGER & P. ROTHMAN. 1994. The transcription factor BSAP (NF-HB) is essential for immunoglobulin germ-line ε transcription. J. Immunol. **152:** 2904.
16. XU, M. & J. STAVNEZER. 1992. Regulation of transcription of immunoglobulin germ-line γ1 RNA: Analysis of the promoter/enhancer. EMBO J. **11:** 145.
17. LUNDGREN, M., C. LARRSON, A. FEMINO, M. XU, J. STAVNEZER & E. SEVERINSON. 1994. Activation of the immunoglobulin germ line γ1 promoter: Involvement of C/EBP transcription factors and their possible interaction with a NF-IL4 site. J. Immunol. **153:** 2983.
18. GLIMCHER, L. H., T. HAMANO, R. ASOFSKY, E. HEBER-KATZ, R. H. SCHWARTZ & W. E. PAUL. 1982. I region-restricted antigen presentation by B cell–B lymphoma hybridomas. Nature **298:** 283.
19. DELPHIN, S. & J. STAVNEZER. 1994. Characterization of an IL-4 responsive region in the immunoglobulin heavy chain ε promoter: Regulation by NF-IL4, a C/EBP family member and NF-κB/p50. J. Exp. Med. **181:** 181.
20. MILLS, F. C., J. S. BROOKER & R. D. CAMERINI-OTERO. 1990. Sequences of human immunoglobulin switch regions: Implications for recombination and transcription. Nucleic Acid Res. **18:** 7305.
21. POLI, P., F. P. MANCINI & R. CORTESE. 1990. IL-6DBP, a nuclear protein involved in interleukin-6 signal transduction defines a new family of leucine zipper proteins related to C/EBP. Cell **63:** 643.
22. WILLIAMS, S. C., C. A. CANTWELL & P. F. JOHNSON. 1991. A family of C/EBP related proteins capable of forming covalently linked leucine dimers in vitro. Genes Dev. **5:** 1553.
23. AKIRA, S., H. ISSHIKI, T. SUGITA, O. TANABE, S. KINOSHITA, Y. NISHIO, T. NAKAJIMA, T. HIRANO & T. KISHIMOTO. 1990. A nuclear factor for IL-6 expression (NF-IL6) is a member of a C/EBP family. EMBO J. **9:** 1897.
24. COOPER, C., D. JOHNSON, C. ROMAN, N. AVITAHL, P. TUCKER & K. CALAME. 1992. The C/EBP family of transcriptional activators is functionally important for Ig V_H promoter activity in vivo and in vitro. J. Immunol. **149:** 3225.
25. KOHLER, I. & E. P. RIEBER. 1993. Allergy-associated Iε and Fcε receptor II (CD23b) genes activated via binding of an interleukin-4-induced transcription factor to a novel responsive element. Eur. J. Immunol. **23:** 3066.
26. KOTANIDES, H. & N. C. REICH. 1993. Requirement of tyrosine phosphorylation for rapid activation of a DNA binding factor by IL-4. Science **262:** 1265.
27. KUNSCH, C., S. M. RUBEN & C. A. ROSEN. 1992. Selection of optimal κB/Rel DNA-binding motifs: Interaction of both subunits of NF-κB with DNA is required for transcriptional activation. Mol. Cell. Biol. **12:** 4412.
28. ROMAN, C., J. S. PLATERO, J. SHUMAN & K. CALAME. 1990. Ig/EBP-1: A ubiquitously expressed immunoglobulin enhancer binding protein that is similar to C/EBP and heterodimerizes with C/EBP. Genes Dev. **4:** 1404.
29. SCHINDLER, C., H. KASHLEVA, A. PERNIS, R. PINE & P. ROTHMAN. 1994. STF-IL-4: A novel IL-4-induced signal transducing factor. EMBO J. **13:** 1350.
30. STEIN, B., P. C. COGSWELL & J. A. S. BALDWIN. 1993. Functional and physical associations between NF-κB and C/EBP family members: A Rel domain-bZIP interaction. Mol. Cell. Biol. **13:** 3964.
31. MATSUSAKA, T., K. FUJIKAWA, Y. NISHIO, N. MUKAIDA, K. M. T. KISHIMOTO &

S. AKIRA. 1993. Transcription factors NF-IL6 and NF-κB synergistically activate transcription of the inflammatory cytokines, IL-6 and IL-8. Proc. Natl. Acad. Sci. USA **90:** 10193.

32. ADAMS, B., P. DORFLER, A. AGUZZI, Z. KOZMIK, P. URBANEK, I. MAURER-FOGY & M. BUSSLINGER. 1992. Pax-5 encodes the transcription factor BASP and is expressed in B lymphocytes, the developing CNS, and adult testis. Genes Dev. **6:** 1589.

33. ROTHMAN, P., Y.-Y. CHEN, S. LUTZKER, S. C. LI, V. STEWART, R. COFFMAN & F. W. ALT. 1990. Structure and expression of germline immunoglobulin heavy chain ε transcripts: Interleukin-4-plus lipopolysaccharide-directed switching to Cε. Mol. Cell. Biol. **10:** 1672.

34. LECLAIR, K. P., M. A. BLANAR & P. S. SHARP. 1992. The p50 subunit of NF-κB associated with NF-IL6 transcription factor. Proc. Natl. Acad. Sci. USA **89:** 8145.

Human Immunoglobulin Isotype Switching

Studies at the DNA Level

E. E. MAX, F. C. MILLS, AND N. HARINDRANATH

Center for Biologics Evaluation and Research
Food and Drug Administration
Bethesda, Maryland 20892

INTRODUCTION

The immunoglobulin isotype switch occurs as part of a developmental program by which a B cell expressing IgM advances to express immunoglobulin of a different isotype but identical antigen binding specificity. The expression of a new isotype is associated with a rearrangement of the lymphocyte DNA; in this rearrangement the gene encoding the constant region of the μ heavy chain (Cμ) is replaced by the gene encoding the newly expressed constant region, and most of the DNA lying between the V(D)J region and the new CH region is deleted. The order of the human constant regions in germline DNA is μ-δ-γ3-γ1-$\Psi\varepsilon$-α1-$\Psi\gamma$-γ2-γ4-ε-α2. The breakpoints of the deletion usually fall in or near a segment of DNA known as the "switch region." Switch regions lie upstream of all heavy chain C region genes (except Cδ) and are characterized by internally repeated motifs. Before the deletional recombination to a new CH gene occurs, the target CH gene is transcribed into an mRNA that includes the CH gene flanked by a noncoding 5′ exon known as the I exon; this exon is transcribed from a segment of DNA that lies upstream of the switch region and that is deleted during the recombination event.

We have been interested in the switch from IgM to IgE in human B cells, a switch that underlies all allergic phenomena and is dependent on the lymphokine IL-4. To study this at the DNA level, we have developed an assay for the switch recombination to Cε. The assay takes advantage of the fact that during this recombination, the 5′ region of Sμ becomes joined to the 3′ region of Sε, forming a composite Sμ-Sε junction. We can detect the presence of such junctions by using the polymerase chain reaction (PCR) to amplify across Sμ-Sε from a primer sequence 5′ of Sμ to a downstream primer 3′ of Sε.[1] Before the μ-ε isotype switch recombination occurs, these two primer sequences are separated by over 200 kb of genomic DNA, a distance much too great to allow amplification by PCR. As switching occurs in a population of cells, composite Sμ-Sε junctions of different sizes will be created, with the size of each composite junction dependent on the exact position within Sμ and Sε where recombination occurs.

By amplifying, cloning, and analyzing the nucleotide sequence of composite Sμ-Sε junctions from human B-cell lines, we previously identified a class of junctions in which a segment of Sγ sequence was interposed between the Sμ and Sε sequences.[1] This Sμ-Sγ-Sε structure suggests that one pathway of switching to ε involves an initial switch from μ to a γ gene, followed by a subsequent switch to ε. Such a pathway has been reported to be the predominant, if not exclusive, route to ε expression in the mouse[2–4]; in that species γ1 is the γ subtype involved.

In our studies of human Sμ-Sγ-Sε switching, the Sγ sequence was recognized by its similarity to Sγ4, the only human Sγ region for which the sequence was available; but we expected that the switch regions of the other three γ subtypes were probably quite similar, so it was impossible to conclude which of the Sγ regions had contributed a particular interposed Sγ segment. In order to define this pathway in more detail, we have now determined the complete sequence of the other three subtype Sγ regions and analyzed the sequence of nine additional Sμ-Sγ-Sε regions; comparisons between these sequences allow us to identify the subtype contributing to each Sμ-Sγ-Sε composite switch region. Furthermore, examination of the Sγ subtype sequences has revealed unexpected features in the comparisons between these loci.

Because some workers have suggested that in the mouse the Sμ sequence does not routinely rearrange directly with Sε (i.e., that switching to Sγ1 is an obligatory intermediate),[2] we were interested to determine whether direct Sμ-Sε occurs in human B cells. We approached this question by adapting our PCR strategy to amplify the reciprocal circular DNA that is formed—in a least some cells—by the circularization of the segment of DNA deleted by the switch recombination event. Because a deletion circle containing a reciprocal Sε-Sμ junction could be formed only by a direct recombination between Sμ and Sε and not by sequential events involving other isotypes, we asked whether we could detect any reciprocal junctions of this type as evidence for the direct μ-ε switch.

MATERIALS AND METHODS

Cells

Human B lymphocytes were isolated from lymphapheresis samples as previously described[1] and were cultured with 20 U/ml of recombinant human IL-4 for varying periods to induce switching to ε. After culture, the B cells were isolated by positive selection for CD19 expression using antibody-coated beads (Dynal).

PCR Amplification

DNA was isolated from the purified B cells by standard methods. Samples of 0.5 μg of DNA were amplified in 25 μl for 32 cycles using the 5′ Sμ primer and 3′ Sε primer previously described[1] and illustrated in FIGURE 1A. For preparative amplification of Sμ-Sγ-Sε junctions, 2 μg of DNA was amplified in 100 μl for 40 cycles. Reciprocal switch junctions were amplified for 30 cycles from 0.5 μg of DNA using the strategy shown in FIGURE 1B.

Cloning of Sμ-Sγ-Sε Junctions

Preparative amplification samples were digested with *Xba*I and *Eco*RI to expose sticky ends from the corresponding sites in the amplification primers. The digested amplification mix was electrophoresed on 0.7% agarose, and DNA was prepared from a slice corresponding to 0.4 kb to 8 kb; this material was ligated into *Xba*I/*Eco*RI-digested Bluescript vector and transformed into HB101. Clones positive

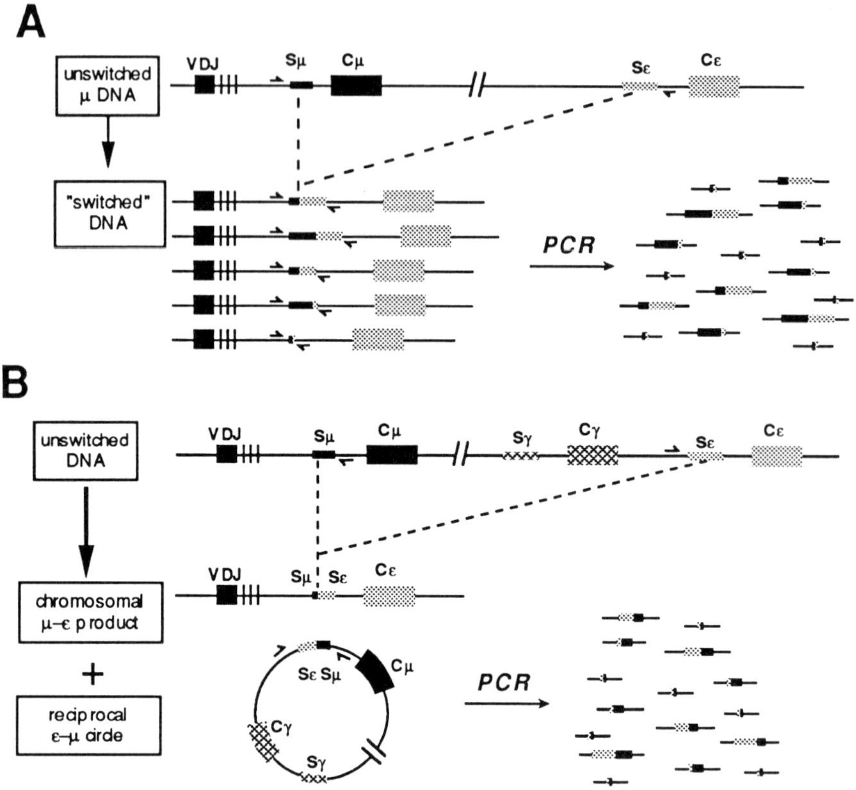

FIGURE 1. Amplification of switch junctions. (A) The strategy for amplifying the Sμ-Sε composite switch region from chromosomal DNA is shown by the maps indicating Sμ and Sε, the switch regions associated with the corresponding CH genes. Oligonucleotides representing the sequence of the sense strand upstream of Sμ (5′-GGCCTCTAGACAAGGG-GACCTGCTCATTTTTATC) and the antisense strand downstream of Sε (5′-GGCCGAATT-CTAGTGCGGTCTGTACAGCGTGGC) were chosen to prime amplification by the polymerase chain reaction. **(B)** The reciprocal recombination product on DNA circles contains an Sε-Sμ junction. This was amplified by oligonucleotides representing the sense sequence upstream of Sε (5′-GGGCTCTAGAGCTGATCTTGGCAAGTCCGAGCTGGGCGACTG) and the antisense sequence downstream of Sμ (5′-GGCCGAATTCCCTGAAGTCTCCC-CAGGCAGACAACCCAGG).

for hybridization to an Sγ4 probe (designated MGE clones) were prepared for restriction and sequence analysis.

Sequence Analysis

Using primers from known Sμ or Sε sequence, switch junctions of nine different MGE clones were sequenced in order to identify the Sγ subtype and to define the Sμ-Sγ and the Sγ-Sε borders. For sequence comparison, analysis of the

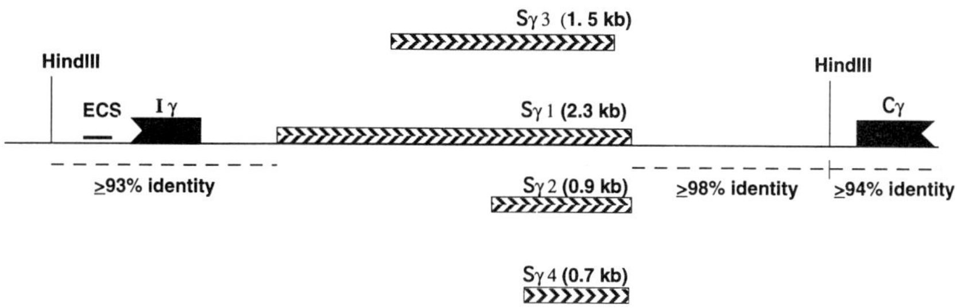

FIGURE 2. Summary of sequence similarities between different segments surrounding the four human Sγ regions. The four Sγ regions are shown in their genomic order (γ3, γ1, γ2, and γ4). The degree of sequence identity in homologous segments bounded by the shared *Hin*dIII sites is indicated. The evolutionary conserved sequences (ECS) represent segments of DNA that lie upstream of the Iγ exon and that show relatively high sequence conservation between human and mouse; they are presumed to participate in promoter function.

unrearranged Sγ1, Sγ2 and Sγ3 regions was accomplished by subcloning the corresponding *Hin*dIII fragments from genomic clones in phage λ kindly provided by Dr. I. Kirsch. Regions containing the Sγ repeat units were difficult to sequence by standard methods, presumably because of G + C richness or secondary structure, and required subcloning into M13 and high-temperature protocols using Taq polymerase for reliable sequence determination. For the longest and most highly repetitive Sγ1 region, nested deletion M13 subclones were obtained using an Exonuclease III + S1 nuclease protocol (Erase-A-Base, Promega).

RESULTS AND DISCUSSION

Sequence Analysis of Sγ Regions

As a basis for examining the role of μ to γ switching in cells that ultimately express the ε isotype (and for studying the μ to γ switch itself), we have completed the sequence analysis of all four human Sγ regions, extending the analysis to the shared *Hin*dIII sites 1.6 kb upstream and 1.3 kb downstream of the repetitive Sγ regions. These sequences include the four Iγ regions[5] and some of their 5′ flanking regions, undoubtedly including parts of the Iγ promoters. One striking finding from these analyses is that the sequences of switch regions themselves are considerably less conserved than either their 5′ or 3′ flanking regions. As diagrammed schematically in FIGURE 2, the sequence of the flanking regions are at least 93% identical 5′ of the switch and 98% identical 3′. The contrast between this high conservation and that observed in the switch regions themselves is indicated by the dot matrix analysis of FIGURE 3, in which the Sγ1 region is compared with itself and the other three Sγ regions. In each comparison between different Sγ regions, the strong diagonal representing high sequence similarity is absent over a region in which internal repeats can be visualized. This suggests that the rate of sequence change in the Sγ regions may be higher than that in the flanking regions, perhaps reflecting mutations induced by the repeat structure or selection for sequence divergence of the Sγ subtypes. We have also examined the comparisons between each flanking sequence and the other homologous sequences in the hope of deducing the evolutionary order of duplications; that is, did a γ-γ-ε-α segment duplicate, or did a γ-ε-α segment duplicate, followed by two individual duplications (γ3/γ1 and γ2/γ4)? We find that the comparisons show nearly identical degrees of sequence divergence, prohibiting a deduction about the sequence of duplications. This result is consistent with a scenario in which the two types of duplication occurred within a rather short interval of evolutionary time.

The Sγ1 region is the largest of the four Sγ regions and contains the most numerous and most highly conserved internal repeats. These characteristics may contribute to the high frequency of switching to this subtype, reflected in the high level of IgG1 in normal human serum. Similar numerous internally conserved repeats are found in the Sγ region of the murine subtype most abundantly expressed. The structure of the Iγ1 promoter may also play an important role in the predominant expression of IgG1 in both species. In the human we find that a cluster of sequence discrepancies between the subtype sequences occurs in an evolutionarily conserved sequence (ECS) that has been proposed to contribute to promoter function. These discrepancies could allow discrimination between subtype-specific promoter binding factors, although this possibility remains to be explored.

The 79-bp unit motif of the Sγ1 switch region is repeated most precisely at the 3′ end of this segment and becomes less precise as one moves 5′ (see FIG. 3). Interestingly, as shown below, the 3′ region of precise repeats is the locus of almost all the recombination breakpoints we have detected, supporting a functional role for the repeats in targeting recombination. The 79-bp consensus Sγ1 repeat bears some similarity to the 49-bp repeat prominent in murine Sγ regions (FIG. 4), but the similarity is not as extensive as one might find in cross-species comparisons of motifs binding homologous nuclear regulatory proteins. Of two motifs identified in the murine Sγ regions as possible binding motifs for nuclear proteins,[6] the SNAP site is much better conserved in the human Sγ1 repeat than is the SNIP site (thought to be recognized by NF-κB or a related protein); but we have not been able to document *in vitro* binding of nuclear proteins to either human sequence. The 79-bp human repeat motif is extremely G-rich (42/79 bp) and T-poor (4/79). The sequence CAGG occurs eight times in the repeat unit, involving almost half of the 17 A residues. The significance of these sequence features is not clear. The sequence CTGG and its complement CCAG each appear once in the canonical repeat; these sequences were reported[7] to occur with unexpectedly high frequency near switch recombination breakpoints.

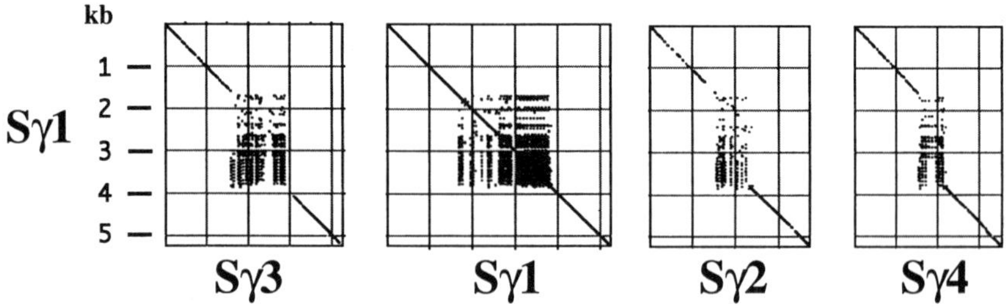

FIGURE 3. Dot matrix sequence comparison between Sγ1 and the other human Sγ regions and flanking segments. The diagonal line indicates the high degree of inter-subtype sequence similarity in the flanking segments, which is interrupted in the repetitive switch regions.

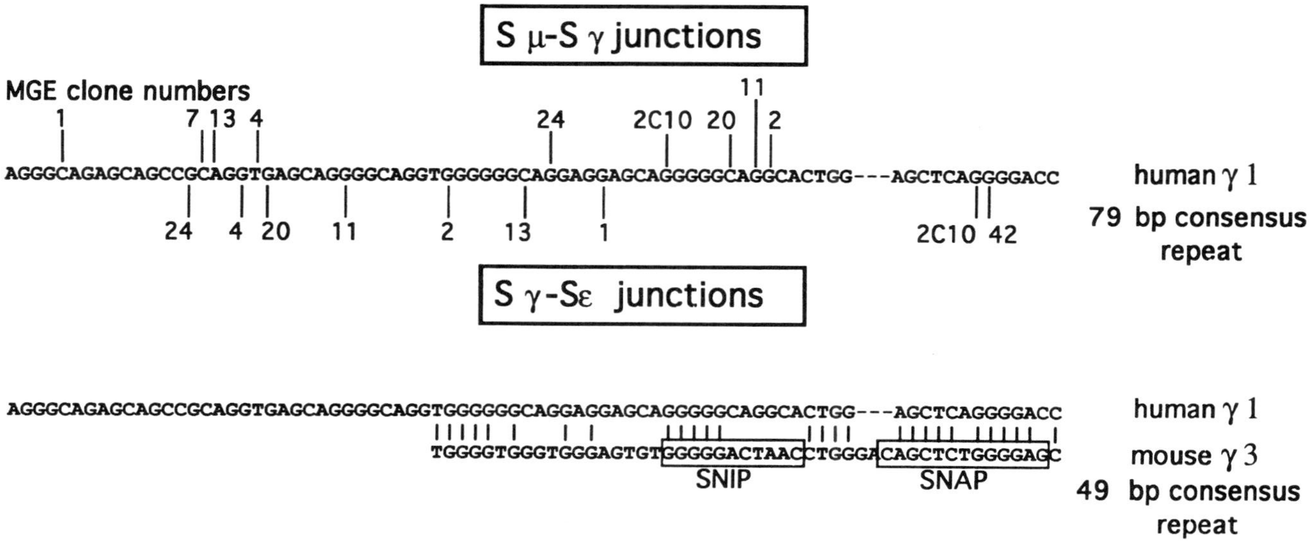

FIGURE 4. Sequence comparison between the consensus 79-bp repeat of the human Sγ1 region and the consensus 49-bp repeat of the murine Sγ3 region. Vertical tick marks highlight identical nucleotides. Vertical lines show the position of recombination breakpoints in our MGE clones (specified by the indicated MGE number) between Sμ and Sγ and between Sγ and Sε. The boxed segments of the murine sequence identify putative binding sites for nuclear proteins.[6]

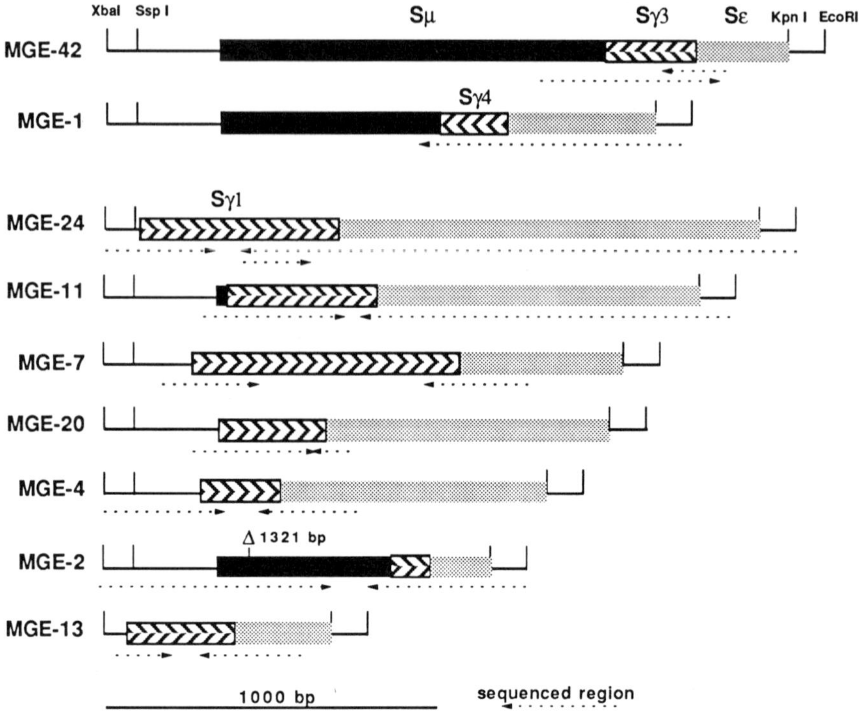

FIGURE 5. Structures of μ-γ-ε junctions in the MGE clones. The contribution of Sμ (*black rectangle*), Sγ (*herringbone pattern*) and Sε (*grey*) is shown for each clone. The orientation of the Sγ regions is indicated by the direction of the herringbone. Arrows under the clone maps indicate the region over which the sequence is known.

Sequence Analysis of MGE Clones

We cultured PBLs from a normal donor with IL-4 to induce switching to ε, then isolated B cells and prepared DNA from them. From the resulting DNA samples the μ-ε switch junctions were amplified using primers 5′ to Sμ and 3′ to Sε; and a subset of the resulting clones was selected based on hybridization to a human Sγ probe. The structures of nine hybridizing clones are shown diagrammatically in FIGURE 5, as determined by restriction analysis and sequencing of the μ-γ and γ-ε junctions. The sequence of each interposed Sγ segment was compared with the four germline Sγ sequences in order to identify its origin; this comparison allowed unambiguous identification of seven of the interposed Sγ sequences as deriving from Sγ1. The Sγ sequence of clone MGE42 apparently derived from Sγ3 (its sequence was 98% identical to Sγ3 and only 90% identical to Sγ1). The Sγ sequence of MGE1 was 89% identical to Sγ4, 88% to Sγ1 and only 85% to Sγ2 or Sγ3; we tentatively identify this interposed sequence as deriving from Sγ4. The fact that some of the interposed Sγ sequences did not exactly match any of our germline sequences may be due to germline polymorphisms in the Sγ regions or to somatic changes, possibly influenced by the switch recombination event.[8,9]

The predominance of Sγ1 inserts in the MGE clones raises an interesting question: Are cells that have switched to γ1—in comparison to the other γ subtypes—favored in some way for subsequent switches to ε? In mouse, that seems to be the case in that the same culture stimuli that activate the γ1 switch also upregulate the switch to ε.[10] It is important to note, however, that the human γ1 gene is no more homologous to the murine γ1 than is any other human γ gene; it is possible that the last common ancestor of these two species possessed only a single γ gene and that the four subtypes of both species independently diverged by duplication and mutation after the mouse–human split. Therefore, the fact that in both species the γ subtype most frequently involved as an intermediate in ε expression is designated γ1 may be fortuitous. In humans the γ subtype that is regulated most similarly to ε is γ4; the expression of both isotypes is activated by IL-4 plus T cell-derived signals.[11–13] In fact, some investigators have suggested that μ-γ-ε junctions reflect sequential switching involving γ4,[14] an interpretation that is not supported by our results. Although a sequential switch pathway favoring γ1 may exist in humans, an alternative possibility is that the γ1 predominance in our μ-γ-ε junctions simply reflects the higher frequency of switching from μ to γ1 as opposed to the other γ subtypes; if μ-γ recombined chromosomes subsequently switch to ε with equal frequency for all four subtypes, the larger pool of μ-γ1 rearrangements would favor the overrepresentation of this subtype in our μ-γ-ε junctions. The four murine γ subtypes may also be equally disposed to sequential switch to ε.

The first μ-γ-ε junction described in a human B cell was derived from the EBV immortalized lymphoblastoid line 2C10.[1] In this junction the interposed Sγ sequence, which we now identify as deriving from Sγ1, had a peculiar palindromic structure: the 5′ portion of the Sγ sequence was in inverted orientation, while the 3′ portion was in the "normal" orientation found in the unrearranged DNA. Although a complex speculative mechanism for generation of such a structure was formulated, it was not clear how frequently such structures occurred; and it was a goal of the present study to approach this question. In fact, none of the μ-γ-ε junctions obtained in the present study showed a palindromic structure. Two of the interposed Sγ inserts were in reverse orientation, presumably reflecting inversional recombination events. (These inverted segments appeared in the only MGE clones with Sγ inserts derived from a subtype other than γ1, but our sample is too small to know whether this reflects a subtype distinction in inversion frequency.) Similar inverted structures have been reported previously,[15] and our finding of two Sγ inversions out of a sample of nine reinforces the notion that inversions are not rare. The palindromic structure of 2C10 may, however, reflect a very unusual mechanism.

The locations of the μ-γ and γ-ε breakpoints in the MGE clones are illustrated in FIGURES 5 and 6. As shown in FIGURE 5, the length of Sμ retained in the composite μ-γ-ε junction varies considerably: MGE42 includes about 1.2 kb of Sμ sequence, whereas in five clones (MGE24, MGE7, MGE20, MGE4, and MGE13) the μ-γ breakpoint actually falls upstream of the internal repeats defining the Sμ sequence. The longest Sμ segments are in the only two clones (MGE42 and MGE1) that are derived from Sγ regions other than Sγ1. This may be fortuitous, but if the spectrum of Sμ breakpoints reproducibly differs for different isotypes this would provide a mechanistic clue suggesting that the DNA cleavage in Sμ may occur after the downstream CH gene recombination target has been selected and be influenced by that target, or alternatively that the position of the Sμ breakpoint can influence which Sγ subtype is selected for rearrangement. FIGURE 5 demonstrates that the γ-ε junctions also occur over a fairly wide area in the 3′ portion

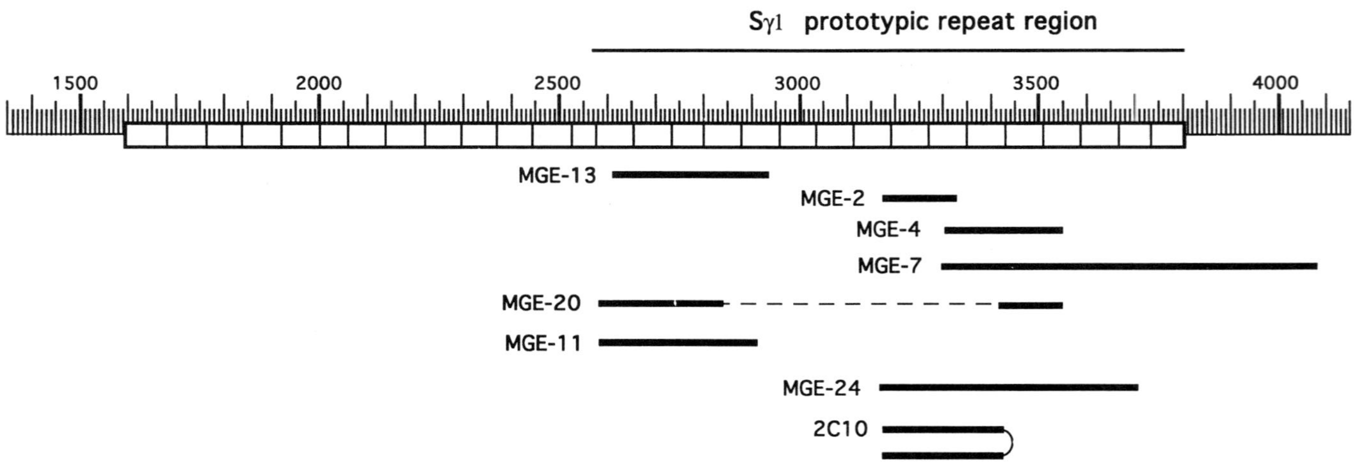

FIGURE 6. Positions of MGE Sγ1 sequences. Origin of the Sγ1 inserts in the MGE clones. The Sγ1 region is illustrated by the hatched rectangle. *Horizontal bars* indicate the region of Sγ1 sequence that best matches the Sγ inserts of the indicated MGE clones. The *dashed line* in the diagram of MGE20 represents an internal deletion of Sγ1 sequence.

of the Sε region. The entire Sε region is about 2 kb long; it is possible that some μ-γ-ε recombinations involving the 5′ portion of the Sε region occur, but such junctions would tend to be longer and perhaps beyond our current size limit for PCR amplification (about 4 kb).

FIGURE 6 indicates what segment of the Sγ1 sequence contributed the Sγ1 insert in each MGE alone. Strikingly, all the inserts derive from the dense highly conserved repeats at the 3′ end of the Sγ1 sequence. This observation reinforces the importance of these repeats in targeting the recombination site.

Because of the suggestion[6] that certain positions within the murine Sγ repeats represent hotspots for recombination (putatively reflecting the sites of DNA binding proteins), we have examined the distribution of recombination breakpoints within the canonical Sγ1 repeat. As shown in FIGURE 4, the breakpoints do not show any obvious clustering.

Analysis of Reciprocal Switch Junctions

When clones of μ-ε amplification products were hybridized to an Sγ probe, approximately 2% were positive. This represents a minimum estimate for the fraction of ε-expressing cells that underwent a sequential switch involving a γ gene, since the final recombination of an Sμ-Sγ junction with Sε might remove Sγ sequence that was present. However, in view of the suggestion that the predominant or exclusive pathway for ε expression in the mouse involves a sequential switch,[2–4] we were interested to know whether *any* direct μ-ε switching occurred in human B cells. To address this question, we used the strategy shown in FIGURE 1B to amplify the reciprocal Sε-Sμ recombination product that could only be formed if the deleted DNA from a direct μ-ε switch recircularized. Such reciprocal circles would not be expected to contain a replication origin or centromere, and so would be expected to be lost from a culture as each circular DNA molecule was diluted out by proliferation of the cells. As shown in FIGURE 7, this strategy successfully detected amplification products in cultures of B cells switching under the influence of IL-4, but not in control B cells. This indicates that direct μ-ε switching occurs in human B cells; the relative frequency of direct versus sequential switching remains to be determined.

CONCLUSION

We have demonstrated that one pathway for switching of human B lymphocytes to IgE expression involves a sequential switch via a γ gene. Although the γ4 subtype resembles ε in being subject to IL-4-regulated expression and this subtype has been proposed as an intermediate in the switch to ε, our evidence suggests that of the four γ subtypes, γ1 is the one most frequently involved in μ-γ-ε switching. The Sγ1 region is also the longest of the four Sγ regions and the one with the most numerous and highly conserved internal repeats; such features may contribute to the predominant expression of this subtype in humans, and perhaps also in mouse. Our evidence confirms that, in switch junctions that are not selected for maintenance of immunoglobulin expression, inversional recombination is not rare. Finally, we have shown that ε expression in the human does not involve an obligatory switch to a γ isotype; that is, direct μ-ε switching can occur.

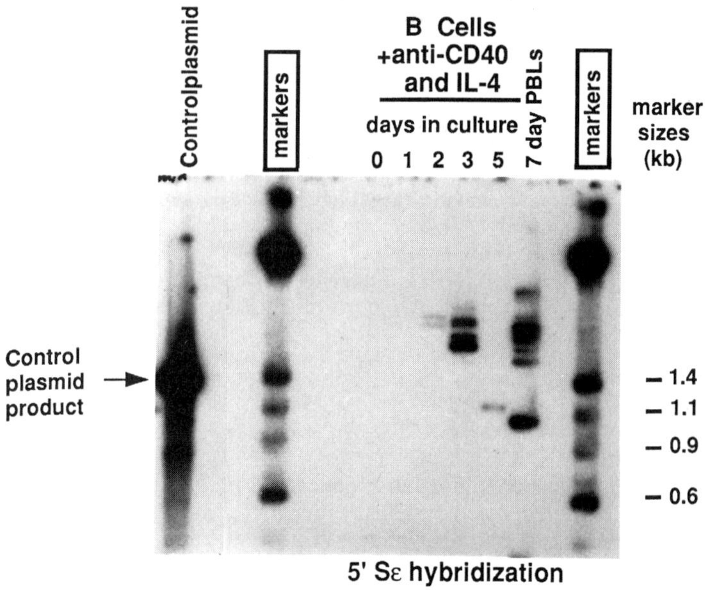

FIGURE 7. Amplification of Sε-Sμ reciprocal recombination products. The strategy shown in FIGURE 1B was used to amplify reciprocal products that are formed as transient circular DNAs in B cells undergoing a direct μ-ε switch. The products were electrophoresed, blotted, and probed with a radioactively labeled fragment from just upstream of the Sμ region.

REFERENCES

1. MILLS, F. C., G. THYPHRONITIS, F. D. FINKELMAN & E. E. MAX. 1992. Ig mu-epsilon isotype switch in IL-4-treated human B lymphoblastoid cells. Evidence for a sequential switch. J. Immunol. **149(1075):** 1075–1085.
2. YOSHIDA, K., M. MATSUOKA, S. USUDA, A. MORI, K. ISHIZAKA & H. SAKANO. 1990. Immunoglobulin switch circular DNA in the mouse infected with *Nippostrongylus brasiliensis*: Evidence for successive class switching from mu to epsilon via gamma 1. Proc. Natl. Acad. Sci. USA **87(20):** 7829–7833.
3. MANDLER, R., F. D. FINKELMAN, A. D. LEVINE & C. M. SNAPPER. 1993. IL-4 induction of IgE class switching by lipopolysaccharide-activated murine B cells occurs predominantly through sequential switching. J. Immunol. **150(2):** 407–418.
4. SIEBENKOTTEN, G., C. ESSER, M. WABL & A. RADBRUCH. 1992. The murine IgG1/IgE class switch program. Eur. J. Immunol. **22(7):** 1827–1834.
5. KITANI, A. & W. STROBER. 1993. Regulation of C gamma subclass germ-line transcripts in human peripheral blood B cells. J. Immunol. **151(7):** 3478–3488.
6. WUERFFEL, R., C. E. JAMIESON, L. MORGAN, G. V. MERKULOV, R. SEN & A. L. KENTER. 1992. Switch recombination breakpoints are strictly correlated with DNA recognition motifs for immunoglobulin S gamma 3 DNA-binding proteins J. Exp. Med. **176(339):** 339–349.
7. CHOU, C. L. & S. L. MORRISON. 1993. A common sequence motif near nonhomologous recombination breakpoints involving Ig sequences. J. Immunol. **150(12):** 5350–5360.
8. DUNNICK W., M. WILSON & J. STAVNEZER. 1989. Mutations, duplication, and deletion of recombined switch regions suggest a role for DNA replication in the immunoglobulin heavy-chain switch. Mol. Cell. Biol. **9(1850):** 1850–1856.

9. DUNNICK, W. & J. STAVNEZER. 1990. Copy choice mechanism of immnoglobulin heavy-chain switch recombination. Mol. Cell. Biol. **10(397):** 397–400.
10. SNAPPER, C. M., F. D. FINKELMAN & W. E. PAUL. 1988. Differential regulation of IgG1 and IgE synthesis by interleukin 4. J. Exp. Med. **167(183):** 183–196.
11. ISHIZAKA, A., Y. SAKIYAMA, M. NAKANISHI, K. TOMIZAWA, E. OSHIKA, K. KOJIMA Y. TAGUCHI, E. KANDIL & S. MATSUMOTO. 1990. The inductive effect of interleukin-4 on IgG4 and IgE synthesis in human peripheral blood lymphocytes. Clin Exp Immunol. **79(3):** 392–396.
12. ISHIZAKA, A., *et al.* 1990. Regulation of IgE and IgG4 synthesis in patients with hyper IgE syndrome. Immunology **70(3):** 414–416.
13. NUSSLEIN, H. G. & H. L. SPIEGELBERG. 1990. Interleukin-4 induces both IgG4 and IgE secretion by peripheral blood B cells. J. Clin. Lab. Anal. **4(6):** 414–419.
14. JABARA, H. H., R. LOH, N. RAMESH, D. VERCELLI & R. S. GEHA. 1993. Sequential switching from mu to epsilon via gamma 4 in human B cells stimulated with IL-4 and hydrocortisone. J. Immunol. **151(9):** 4528–4533.
15. LAFFAN, M. & L. LUZZATTO. 1992. Anomalous rearrangements of the immunoglobulin heavy chain genes in human leukemias support the loop-out mechanism of class switch. J. Clin. Invest. **90(6):** 2299–2303.

Role of Oct-2 in Immunoglobulin Gene Silencing[a]

HANNA S. RADOMSKA,[b,c] MARIA D. HINES,[b,c] AND LAUREL A. ECKHARDT[b,d]

[b]*Department of Biological Sciences Hunter College of The City University of New York New York, New York 10021*

[c]*Graduate School of Arts and Sciences Columbia University New York, New York 10027*

When B lineage cells are fused to non-B cells (fibroblasts or T lymphoma), immunoglobulin (Ig) genes contributed by the B cells are silenced at the transcriptional level,[1] as are other B-cell-specific genes (for example, J chain and B29).[2] The *cis* elements marking Ig genes for extinction have been identified as the Ig (heavy and light chain) gene promoters and the IgH intron enhancer. Sequence analyses revealed that a common motif shared by those elements, as well as by J-chain and B29 promoters, is an eight-nucleotide sequence (ATGCAAAT) called the octamer.[1,2] In B cells, the octamer serves as a binding site for two known transcription factors: ubiquitous Oct-1 and mainly B cell-limited Oct-2. It has been shown that the latter factor is required for full activity of the Ig gene promoters and IgH intron enhancer.[1]

In the present study, we found that Oct-2 was no longer expressed in hybrids resulting from fusions of the murine myeloma MPC11 to the murine T lymphoma BW5147. We designed experiments to determine whether loss of this transactivator was responsible for the lack of expression of other B-cell-specific genes. We found that an exogenously provided human Oct-2 (hOct-2) gene, under the control of a viral (CMV) promoter, was incapable of reactivating silenced Ig, J-chain, or endogenous mOct-2 genes when introduced into previously established MPC11 × BW5147 hybrids (data not shown). However, when the same hOct-2 gene was provided to either of the parental lines before fusion, there was a pronounced effect on myeloma gene expression. hOct-2 expressed in the T lymphoma led to a high percentage of hybrids that continued expressing all tested B-cell-specific genes. Supplying hOct-2 to the myeloma parent before fusion resulted in 100% hybrids with a myeloma-like phenotype (expressing Ig, J chain, and mOct-2). In summary, hOct-2 could prevent the silencing of B-cell-specific genes in myeloma × T cell hybrids, but it could not reactivate previously silenced loci. Furthermore, hOct-2's effect on gene silencing was more pronounced when introduced into the myeloma, rather than into the T lymphoma, before cell fusion.

One model that we are considering to explain these findings is outlined in FIGURE 1. We postulate that Oct-2 serves an essential role in myeloma Ig (and other) gene expression but that it requires a B-cell-restricted accessory factor/

[a] This work was funded by grants from the American Cancer Society and from The City University of New York PSC-CUNY Research Award Program (to L.A.E.).

[d] Author to whom correspondence should be addressed.

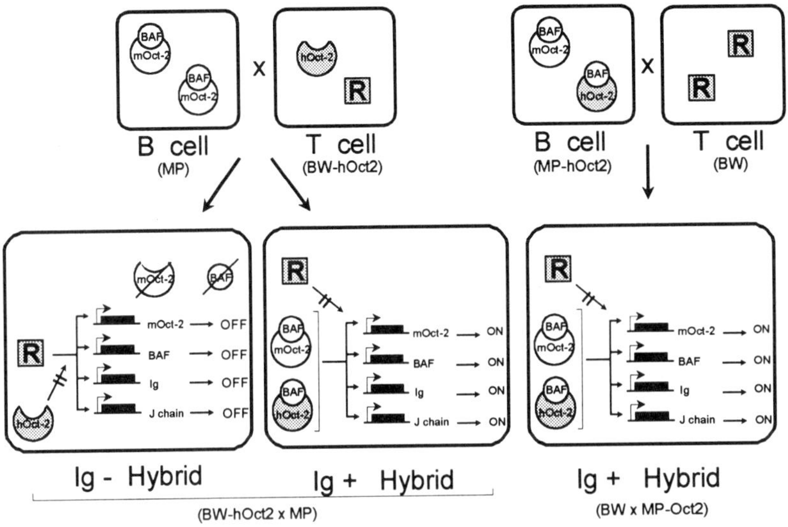

FIGURE 1. Model of hOct-2 function in preserving myeloma-specific gene expression. Abbreviations: hOct-2, transfected human Oct-2; mOct-2, endogenous mouse Oct-2; BAF, B cell accessory factor/modifier; R, T cell-derived repressor; Ig, immunoglobulin; BW-hOct2, hOct-2-expressing BW5147 T lymphoma; MP-hOct2, hOct-2 expressing MPC11 myeloma. See text for full description of model.

modifier(s) (BAF) for this function. The T cell produces an antagonistic factor(s) that can repress Ig genes (R). When hOct-2 is expressed in the T cell, it is expressed in the absence of BAF. When the hOct-2^+ T cell is fused to the myeloma, two outcomes are possible. If the T cell's hOct-2 can interact with BAF from the myeloma before the T-cell-encoded repressor (R) takes effect, the myeloma-specific genes are rescued from silencing. If, conversely, the hOct-2/BAF interaction lags behind the activity of the repressor, the myeloma's tissue-specific genes (including endogenous mOct-2 and presumably BAF) are silenced. Providing hOct-2 to the myeloma before fusion insures that hOct-2 and BAF will interact before introduction of the T-cell-provided repressor. Interestingly, the endogenous mOct-2 gene is among the loci affected by constitutive expression of hOct-2, suggesting that it is autoregulatory.

Regardless of the precise mechanism involved, our results provide fundamental insights into the function of Oct-2 in B cells. Considered together with the data from Oct-2 gene "knockout" mice,[3] these studies show that although Oct-2 is not sufficient to activate Ig genes, it is essential for maintaining their activity (as well as other B-cell-specific genes) in the late stage, Ig-secreting plasmacyte, and therefore likely plays a central role in the preservation of this cell's genetic program.

ACKNOWLEDGMENTS

We would like to thank Ryszard Stawowy for excellent technical assistance and Jacqueline Kim for help in hybrid analyses.

REFERENCES

1. Eckhardt, L. A. 1992. Immunoglobulin gene expression only in the right cells at the right time. FASEB J. **6:** 2553–2560.
2. Lieberman, S. A., M. D. Hines, P. Bergsagel, W. Kuehl & L. A. Eckhardt. 1993. Coordinate silencing of myeloma-specific genes in myeloma × T lymphoma hybrids. J. Immunol. **151:** 2588–2600.
3. Corcoran, L., M. Karvelas, G. Nossal, Z. Ye, T. Jacks & D. Baltimore. 1993. Oct-2, although not required for early B-cell development, is critical for later maturation and for postnatal survival. Genes Dev. **7:** 570–582.

Molecular Evidence That *in Vivo* Isotype Switching Occurs within the Germinal Centers

HÉCTOR MARTINEZ-VALDEZ, FLORENCE MALISAN, ODETTE DE BOUTEILLER, CHRISTIANE GURET, JACQUES BANCHEREAU, AND YONG JUN LIU

Laboratory for Immunological Research
Schering-Plough
27 Chemin des Peupliers
69571 Dardilly Cedex, France

As a result of isotype switching, B lymphocytes express antibodies with identical antigen specificity but with a different isotype, which determines their biological function. This recombination process moves the variable heavy chain (V_H) segment so that it associates with a different set of constant (C_H) regions. This is mediated by the junction of switch regions (S-regions), which are arrays of short tandem repeats, located 5′ of each constant region (except Cδ). Early work by Kraal, Weissman, and Butcher[1] suggested that germinal centers (GC) might be the place where isotype switching occurs. Despite the large body of correlative data supporting this hypothesis, no molecular evidence had been directly shown.

Using specific phenotypic markers, which included surface immunoglobulins (sIg) CD10, CD23, CD38, CD44, CD77, Bcl2, and Ki67, we have recently been able to identify five distinct mature tonsillar B-cell (Bm) subpopulations corresponding to defined stages of differentiation, that is, naive follicular mantle, germinal center, and memory cells.[2,3] Thus, and as shown in FIGURE 1, naive B lymphocytes are $sIgM^+IgD^+CD38^-$ cells. These cells are further divided into Bm1 and Bm2, which are $CD23^-$ and $CD23^+$, respectively. GC $CD38^+$ cells have lost IgD and can be further subdivided into $CD77^+$ centroblasts (Bm3) and $CD77^-$ centrocytes (Bm4). Finally, memory cells (Bm5) are identified as double negative cells, lacking IgD and CD38.

As the transcription of germline Ig has been demonstrated to be a prerequisite during isotype switching,[4,5] we devised a polymerase chain reaction (PCR) strategy to determine isotype switching activity *in vivo*. Primary germline Ig transcripts, as schematically shown in FIGURE 2, are initiated at a segment denominated by the I exon, which is located upstream of corresponding S region. Following transcription and primary transcript processing, the I exon is juxtaposed to downstream C_H domains. We combined the distinctive tonsillar B-cell subset purification with our sterile RT-PCR, to investigate whether the initiation of isotype switching, determined by specific sterile transcript expression, correlates with the acquisition of distinctive phenotype. We therefore carried out amplifications targeting the I segment in combination with either Cγ or Cα primers. As also shown in FIGURE 2, germline Iγ as well as Iα transcripts were specifically detected within the Bm4 subset, providing direct molecular evidence that isotype switching does occur within germinal center centrocytes.

As a result of antigen stimulation, secondary lymphoid organs display B lymphocytes at distinct stages of differentiation that distinguishes naive cells from plasma and memory cells. The landmark of secondary lymphoid organ physiology is the antigen-dependent education of B lymphocytes which warrants the generation of a functional antibody repertoire. To do this, B lymphocytes undergo stages of proliferation and affinity maturation by somatic hypermutation on the V_H genes. They are then subjected to a stringent selection that rescues useful and eliminates the useless and potentially harmful cells. This is followed by the acquisition of a distinct Ig isotype that confers novel functions, while keeping the antigen specificity for which it was selected. Interestingly, this distinctive purification of tonsillar B-cell subpopulations has also permitted us to demonstrate that transcripts for Bm3 centroblasts display somatic mutations,[3] which is in agreement with studies performed on isolated germinal center B cells.[6–8]

Our findings that isotype switching activity can be mainly detected within the centrocyte (Bm4) subpopulation is consistent with the current model of germinal center reaction.[9] Somatic mutation occurs in proliferating centroblasts within the dark zone. They give rise to nonproliferating centrocytes, which then migrate to the light zone. These cells undergo selection, based on the affinity of their receptors

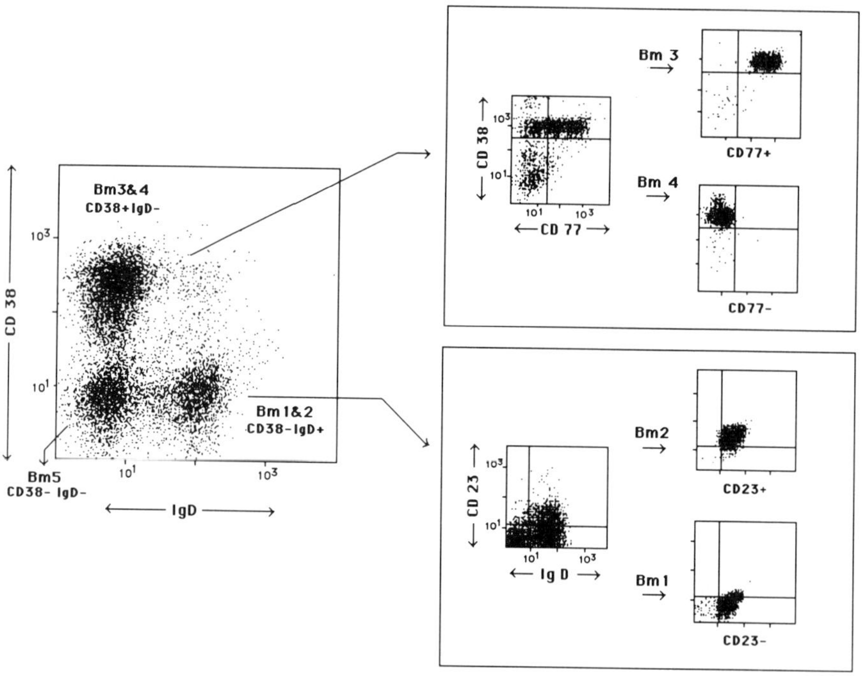

FIGURE 1. FACS sorting of IgD^+CD38^- follicular mantle B cells into $CD23^-$ (Bm1) and $CD23^+$ (Bm2) subsets, and IgD^-CD38^+ germinal center B cells into $CD77^+$ (Bm3) and $CD77^-$ (Bm4) subsets. High-density cells were double stained with anti IgD and anti-CD23 and sorted into Bm1 and Bm2 populations. Low-density cells were double stained with anti-CD38 and anti-CD77 and sorted into Bm3 and Bm4 cells. Bm5 memory cells were purified as an IgD/CD38 double negative subset.

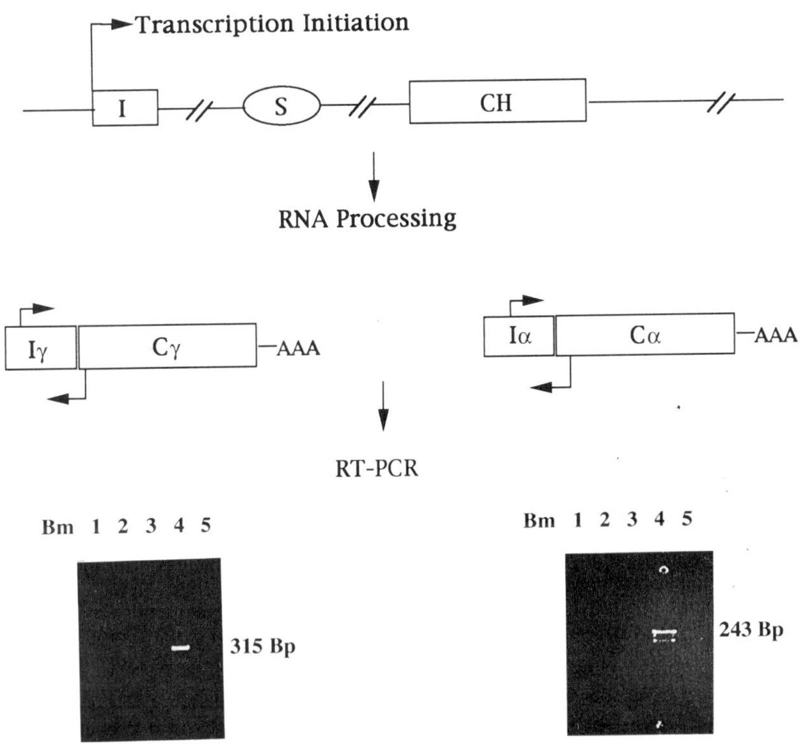

FIGURE 2. Sterile transcript expression by mature B-cell (Bm) subpopulations. Structural organization of a primary germline transcript, depicting the transcription initiation site at the I exon, located upstream of the S-region. Following processing of primary transcript, the I exon is spliced to corresponding isotype. Shown here are the location of amplifying primers, spanning the I exon and the first exon of the C_H domain. Following oligo-d(T)-based reverse transcription of RNA from 500 cells/subset (indicated herein as Bm1 to 5), Iγ and Iα expression was determined by 35 cycles (1 min at 94°C for denaturation, 2 min at 60° for primer annealing, and 2 min at 72°C for extension) of PCR amplification. Predicted sterile transcripts were identified on 1.6% agarose gels stained with ethidium bromide.

for antigen (Ag), trapped by follicular dendritic cells (FDC). Positively selected high-affinity centrocytes pick up Ag and present it to T cells in the light zone. Accordingly, germinal center T cells are known to express the ligand for CD40 as well as IL4, which are key elements for isotype switching.[10,11] Therefore, the T cell/centrocyte/FDC interaction results in the expansion of high-affinity B-cell clones and their isotype switch in a CD40/cytokine-dependent fashion. The role of IL4 at that level is further demonstrated by the presence of CD23β mRNA in the Bm4 centrocytes (Martinez-Valdez *et al.*, unpublished results), which is a signature of cytokine signaling.[12] Taken together, these data stress the crucial role of cognate B/T cell interactions during isotype switching within the light zone of germinal centers.

REFERENCES

1. KRAAL, G., I. L. WEISSMAN & E. C. BUTCHER. 1982. Nature **298:** 377–379.
2. LIU, Y-J., *et al.* 1994. Adv. Exp. Med. Biol. In press.
3. PASCUAL, V., *et al.* 1994. J. Exp. Med. **180:** 329–339.
4. ZHANG, J., *et al.* 1993. EMBO J. **12:** 3529–3537.
5. LIXING, X., *et al.* 1993. Proc. Natl. Acad. Sci. USA **90:** 3705–3709.
6. BEREK, C., A. BERGER & M. APEL. 1991. Cell **67:** 1121–1129.
7. JACOB, J., *et al.* 1991. Nature **354:** 389–392.
8. KÜPPERS, R., *et al.* 1993. EMBO J. **12:** 4955–4967.
9. LEDERMAN, S., *et al.* 1992. J. Immunol. **149:** 3817–3826.
10. FULER, K. A., O. KANAGAWA & M. H. NAHM. 1993. J. Immunol. **151:** 4505–4512.
11. LIU, Y-J., *et al.* 1992. Immunol. Today **13:** 17–21.
12. YOKOTA, A., *et al.* 1988. Cell **55:** 611–618.

Diversification of Bovine λ-Light Chain Genes

CHUEN-LEI PARNG,[a] SUSAN HANSAL,[a]
RICHARD A. GOLDSBY,[a,b]
AND BARBARA A. OSBORNE[a,b,c]

[a]*Program in Molecular and Cellular Biology*
[b]*Department of Veterinary and Animal Sciences*
University of Massachusetts at Amherst
Amherst, Massachusetts 01003

Here we report that the gene conversion, followed by somatic hypermutation, diversifies the bovine λ-light chain repertoire. The diversification mechanisms of immunoglobulin (Ig) molecules varies among the species studied thus far. Somatic recombination of different germline V(D)J genes in human and mouse, gene conversion in chicken and rabbit, and somatic hypermutation in sheep have been shown to generate a diverse Ig repertoire in these species. Studies of Ig diversification in different species has provided us insight into the potential mechanisms used by vertebrates to create a functional Ig repertoire.

We used cattle as an animal model to study λ-light chain diversification mechanisms.[1] cDNA sequences of Vλ and germline Vλ sequences were obtained, respectively, from the ileal Peyer's patches cDNA libraries of animals at different ages and from a bovine liver genomic library. A variable region (V) probe, a joining region (J) oligo probe, a constant region (C) probe, and a JC-intron probe were used to detect the distribution of germline V, J, C genes and the frequency of the gene rearrangement. The results obtained are as follows: (A) Extraordinarily high homology is observed between genomic Vλ genes as well as between the Vλ cDNAs. Differences are clustered in the three complementary determining regions (CDRs). (B) A gene family we designated B4 is preferentially used in the cDNAs.[2] Only one rearranged band is detected in genomic Southern blots. As well, only one J region and one C region are found in the cDNA sequences. These results indicate that restricted gene rearrangement and restricted RNA splicing occur in cattle. Recombinational diversification using multiple gene segments does not appear to play a major role in the generation of bovine Ig repertoire. (C) A number of pseudogenes are found in the germline that possess V fused to J; other pseudogenes have defects in either the 5′-transcriptional initiation site or 3′-recombinational signals. They are potential donors for gene conversion. These pseudogenes cannot undergo gene rearrangement; however, they can act as donor sequences for gene conversion via short flanking homologous sequences. Additionally, those pseudogenes with a V(J) fusion may act to provide perfect homology for gene conversion to generate a diverse CDR3. Gene conversion is able to introduce multiple gene fragments from different germline Vλ genes into a single rearranged functional VJ gene in cattle (FIG. 1). (D) Single base mutations are

[c] Address for correspondence: Department of Veterinary and Animal Sciences, Paige Laboratory, Box 36410, University of Massachusetts, Amherst, MA 01003-6410.

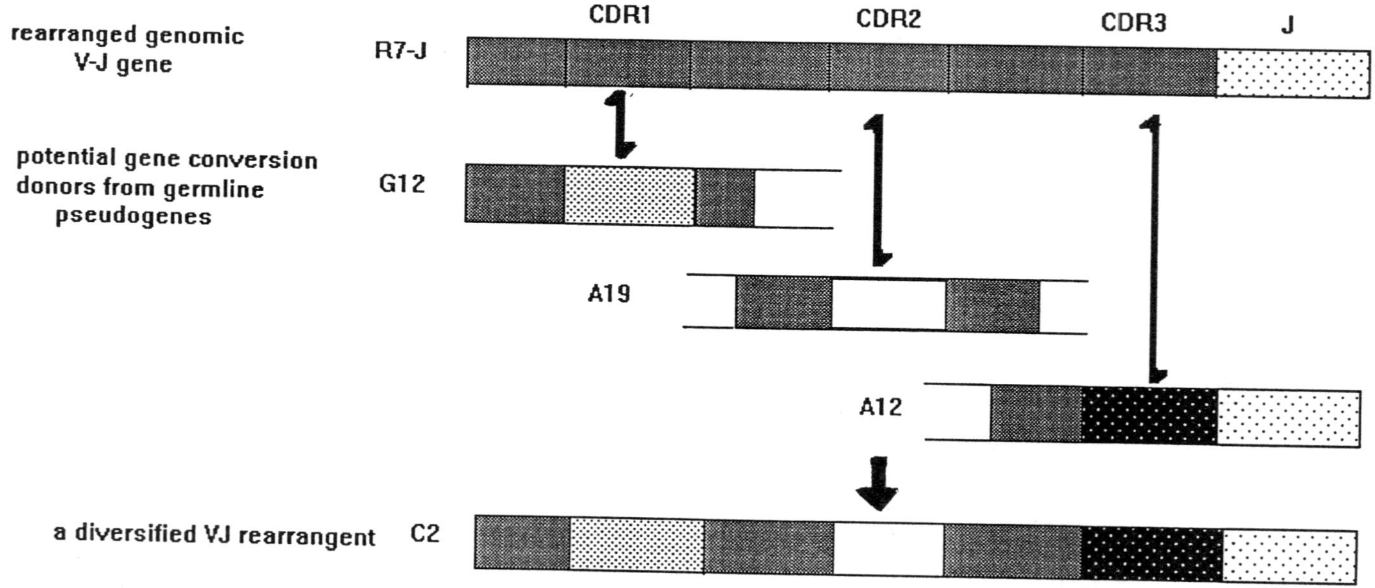

FIGURE 1. Example of multiple gene conversion events occurring in cattle. Gene conversion can take place between sequences containing short flanking homologous regions. Potential donors for gene conversion are found in cattle that possess high homology in the framework regions and clustered differences in the CDRs. Introduction of CDRs from germline pseudogenes G12, A19, and A12 into the rearranged R7-J complex results in the diversified C2-cDNA sequence. The different CDR sequences are shown in different patterns in this figure, and the flanking sequences around these CDRs are homologous to the germline R7 gene.

observed among the potential diversified cDNA sequences generated by gene conversion. The frequency of single-base mutation events increases as the animal increases in age. Somatic hypermutation may create further diversity of Ig genes following exposure to antigens, thus further diversifying the repertoire.

REFERENCES

1. PARNG, C-L., S. HANSAL, R. A. GOLDSBY & B. A. OSBORNE. 1994. Gene conversion generates immunoglobulin diversity in cattle. Submitted.
2. PARNG, C-L., R. A. GOLDSBY & B. A. OSBORNE. 1994. Diversification of the bovine immunoglobulin repertoire during development. In preparation.

Somatic Mutation of Human Immunoglobulin V Genes: Bias, Rate, and Regulation

RICHARD A. INSEL,[a,b] WILLIAM S. VARADE,[b]
YU-WAYE CHU,[b] ELIDES MARIN,[b]
RAMSAY FULEIHAN,[c] AND RAIF S. GEHA[c]

[b]*Department of Pediatrics*
University of Rochester School of Medicine and Dentistry
Rochester, New York 14642

[c]*Harvard Medical School*
Department of Pediatrics
Boston, Massachusetts 02115

Immunoglobulin variable regions in B lymphocytes are diversified by a process of somatic hypermutation following antigenic stimulation.[1,2] Somatic mutation is the primary mechanism by which diversity is introduced into antibody following immunization and accounts for the affinity maturation of antibodies associated with immunity. This mutation process is activated in the first week after antigenic stimulation in peripheral lymphoid organs and persists for only about two weeks after activation.[3–5] Mutation is most prominent in germinal centers of lymphoid tissue, but it is unclear whether the mutation process initiates upon entry into the germinal center of B cells activated in extrafollicular sites or whether the process initiates in these activated B cells before germinal center entry.[6,7] In the germinal center, B cells with mutated immunoglobulin receptors that give rise to higher affinity antibodies escape apoptosis by binding of native antigen on the surface of follicular dendritic cells and by costimulation from the follicular dendritic cells and antigen-specific T cells. The surviving B cells then give rise to memory B cells and antibody-secreting cells. Following reimmunization, the mutation process appears to be reactivated in B cells, although this has been a controversial concept.[8]

Somatic hypermutation is highly regulated at both the cellular as well as at the molecular level. The mutation events are usually point mutations that occur at an estimated rate of approximately 10^{-3} to 10^{-4} per base pair per generation.[1,2,9,10] Mutations are targeted to rearranged heavy and light chain immunoglobulin variable regions and occur at both the productive and nonproductive immunoglobulin gene alleles in the B cell.[11,12] The cellular processes responsible for this high mutation frequency and the mechanisms of its regulation, initiation, and cessation are unknown, but several models have been proposed.[1,2,10,13–16] The minimal genetic elements required for a high rate of mutation include the immunoglobulin gene intron enhancer, the Vκ 3′ enhancer, and the homologous or a heterologous promoter.[17,18]

We have used several model systems to characterize the nature of somatic

[a] Address for correspondence: 601 Elmwood Avenue, Box 777, Rochester, New York 14642.

mutation of human immunoglobulin genes. The intrinsic properties of translationally silent mutations of a nonpolymorphic human V gene—V_{H6}—have been characterized to gain an understanding of intrinsic properties of the mutation process. To determine the rate of mutation of human V genes, we have developed a system to induce, recover, and characterize mutations that occur in a germline human V gene. To better understand regulation of and initiation of the mutation process, we have analyzed the status of mutations in B cells from patients with a primary human immunodeficiency disease associated with absence of germinal centers. The results of these studies have provided new insights into the mutation process.

INTRINSIC PROPERTIES OF THE MUTATION PROCESS

V_{H6} is a single-member human V_H family that is preferentially expressed in the fetus as well as in B-cell chronic lymphocytic leukemia.[19] We have recently shown that V_{H6} is also expressed in post-fetal life with a high rate of somatic mutation and with a distribution and characteristics of mutation suggesting an antigen-driven process.[20–23] Mutations were observed in human V_{H6} gene rearrangements expressed with IgM, IgG, or IgA; and the same mutations were commonly shared among different rearrangements. The high mutation frequency and the absence of polymorphism of the germline human V_{H6} gene makes V_{H6} an excellent substrate for study of the mechanism of somatic hypermutation.

A total of 63 independent and unique V_{H6}DJ recombinant clones were derived by RT-PCR and cloning using RNA from four human spleens surgically removed for hematologic indications.[23] The isotype distribution of these clones was IgM-25, IgG-26, and IgA-12; and they arose from 38 independent V_{H6}DJ rearrangements. Multiple unique clones arose from each of four rearrangements with evidence of intraclonal diversification, consistent with their derivation from germinal centers.[20–23] These multimember clonal rearrangements included eight IgA, two IgM, and two IgG clones; 11 IgG and 1 IgA clone; three IgM clones; and two IgM clones, respectively.

All 63 clones showed greater than 80% sequence identity with germline V_{H6}, and all but one had an open reading frame. Overall, the mutation frequency in the V_{H6} region was 4.5% (767 of 17,068 bases sequenced). Infidelity of reverse transcription or Taq polymerase-mediated DNA amplification accounted for only 1/15 to 1/30 of this frequency, which is similar in magnitude to that described by others.[7] The distribution of mutations across the V_{H6} gene showed that the majority were concentrated in CDR1 (9.2%) and CDR2 (6.6%) versus the FR (3.4%). Approximately two-thirds of the mutations were replacement and one-third were translationally silent mutations (R/S = 2.1). The replacement to silent mutation ratio (R/S) across the V_{H6} gene showed the highest ratio in CDR1 (4.6) and CDR2 (2.7) as compared to the FR (1.6). The high frequency and R/S ratio in the CDRs versus the FR (3.2 vs. 1.6), were consistent with antigen selection of these rearrangements.

The mutation frequency was higher for clones that express IgG (5.4%) and IgA (5.2%) than those that express IgM (3.2%). The R/S ratio for the total gene and for the CDR of IgM clones (3.0, 5.1) was higher than IgG (1.8, 2.3) or IgA clones (2.0, 3.8). The higher frequency of mutations and the lower R/S ratios in the IgG and IgA clones suggest that repetitive antigenic stimulation may have induced new rounds of somatic mutation and resulted in isotype switching. Excess replacement mutations, which may be deleterious for preservation of immunoglob-

TABLE 1. Bias in Silent Mutations in V_{H6} CDR versus FR and A versus T Nucleotides

A	FR1	CDR1	FR2	CDR2	FR3	Total CDR	Total FR
Observed	26	14	22	40	65	54	113
(%)	(16)	(8)	(13)	(24)	(39)	(32)	(68)
Expected	47.4	8.5	25.4	26.2	59.3	34.7	132.3
(%)	(28)	(5)	(15)	(16)	(36)	(21)	(79)

	Germline Nucleotide			
B	A	T	C	G
Observed	55	21	55	36
(%)	(33)	(13)	(33)	(22)
Expected	39.1	28.7	58.4	40.7
(%)	(23)	(17)	(35)	(24)

ulin function,[24] may have been selected against with silent mutations accumulating, leading to lower R/S ratios. These highly mutated IgM clones, which were probably derived from IgD-negative memory splenic B cells, have the potential to develop more appropriate effector function after another antigenic challenge that elicits isotype switching.

To better understand intrinsic properties of somatic hypermutation, silent mutations, which are not antigen selected, were analyzed.[23] Unique mutations for each rearrangement event were counted only once in the analysis of bias to avoid the skewing that would arise from shared mutations of multiple clones arising from the same rearrangement. An expected frequency of replacement and silent mutations, transitions, and transversions for each base position for the germline V_{H6} gene was calculated based on the germline V_{H6} sequence assuming randomness without negative or positive selection except if a stop codon was generated, which was discounted as an event.

There were 167 unique silent mutations in these 38 V_{H6} rearrangements. Bias existed in the distribution of these silent mutations across the V_{H6} gene. There were 54 silent mutations in the CDR compared to the 34.7 expected, and, similarly, 113 FR mutations occurred compared to the 132.3 expected ($p < 0.001$) (TABLE 1A). The bias was primarily due to an increased frequency of silent mutations in both CDR1 and CDR2 and a decreased frequency of silent mutations in FR1. The excess silent mutations in the CDR could be explained either by coselection of silent mutations with antigenic selection of replacement mutations in the CDR or, alternatively, by the CDR of this V gene being more mutagenic. Nonexpressed immunoglobulin murine V_L transgenes were recently shown to display an excess frequency of CDR1 mutations.[25,26] Here, we have observed bias in targeting of mutations to both CDR1 and CDR2. The targeting of CDR mutations in V_{H6} could represent an evolutionary selection of mutable CDRs for this gene whose overall structure has been highly conserved in evolution.[27]

There was an increase of silent mutations involving adenine residues and a decrease of mutations involving thymine residues on the coding strand compared to the expected frequency (TABLE 1B). Over one-half of the silent mutations involved an adenine compared to the expected 39.1%, while only 21% of thymine residues were mutated compared to the 28.7% expected ($p < 0.05$). A bias of adenine over thymine replacement mutations was also observed. Only 16% of

TABLE 2. Silent Transitions (TS) and Transversions (TV) in (A) V_{H6}DJ Clones and at (B) Codons with Third Position Threefold Wobble

A

From	To: TS	To: TV	Total
A	29	26	55
T	18	3	21
C	43	12	55
G	25	11	36
Total	115	52	167
Observed (%)	69	31	
Expected (%) ($p < 0.001$)	46	54	

B

From	To: TS	To: TV	Total
A	20	25	45
T	9	3	12
C	5	11	16
G	10	11	21
Total	44	50	94
Observed	47	53	
Expected (%) ($p < 0.05$)	33	66	

replacements involved a thymine, compared to an expected thymine replacement at twice this frequency, while 43% of replacements involved an adenine, compared to the 27% expected. This bias against mutation of thymine is similar to what has been recently described for a murine passenger V_L transgene and in murine germinal centers.[26,28] The bias in mutations targeting adenines and disfavoring thymine nucleotide substitutions suggests that there is strand polarity in the mutation process. Replication of the leading and lagging strands of double-stranded DNA is an asymmetric process, and the data suggests there is preferential targeting of the mutation process to one of the two strands.

Bias for silent transitions versus transversions was observed. The expected transition to transversion ratio for all potential silent mutations in germline V_{H6} was 0.84, and, yet, there were 115 transitions and 52 transversions, giving a ratio of 2.2 (TABLE 2), which was significantly different from the expected ($p < 0.001$). An excess of silent transitions was observed for each of the bases, representing 53%, 69%, 78%, and 86% of silent mutations of A, G, C, and T, respectively. There were 94 unique silent mutations at the third position in codons in which mutation at the third position will yield a silent mutation regardless of the substitution, or threefold wobble. Compared to the expected one-third transition and two-thirds transversion frequency that would occur randomly at this position, there was almost an equal number of transitions and transversions (44 transitions, 50 transversions) ($p < 0.05$). This finding of an excess of transitions to transversions is similar to that observed in mutations of mouse immunoglobulin gene sequences selected for antigenic specificity and in murine passenger transgenes.[26,28,29]

TABLE 3. Bias in V_{H6} Transversions in Mutated Nucleotides C, G, and A

C	G	A
AA$\underline{C}$(N)→ AA$\underline{G}$(K)-8	CA$\underline{G}$(Q) → CA$\underline{C}$(H)-8	
AA$\underline{A}$(K)-1	→ CA$\underline{T}$(H)-1	
AG$\underline{C}$(S) → AG$\underline{G}$(R)-3	AA$\underline{G}$(K) → AA$\underline{C}$(N)-4	
TT$\underline{C}$(F) or TT$\underline{G}$(L)	→ AA$\underline{T}$(N)-2	
→ AG$\underline{A}$(R)-0		
or TT$\underline{A}$(L)		
Silent $\underline{C}$ → $\underline{G}$ -7	Silent $\underline{G}$ → $\underline{C}$ - 9	Silent $\underline{A}$ → $\underline{C}$ - 16
$\underline{A}$ -4	$\underline{T}$ - 2	$\underline{T}$ - 9
Total $\underline{C}$ → $\underline{G}$ - 18	Total $\underline{G}$ → $\underline{C}$ - 21	Total $\underline{A}$ → $\underline{C}$ - 16
→ A - 5	→ $\underline{T}$ - 5	→ $\underline{T}$ - 9

Bias for specific types of transversions was also observed based on analysis of the third position of codons that had an equal chance of giving rise to one of two replacement mutations and of silent transversions. As shown in TABLE 3, there was bias for C → G over C → A transversions, for G → C over G → T transversions, and for A → C over A → T transversions. Thus, 18 of 23 transversions used a C → G instead of a C → A replacement, 21 of 26 transversions used a G → C instead of a G → T transversion, and 16 of 25 transversions used an A → C instead of an A → T transversion. This bias for C → G, G → C, and A → C transversions was not explained by codon usage in the human genome.

Some nucleotide positions were heavily targeted for mutation. In 15 unrelated rearrangements, silent mutations were observed at the third base position of the valine codon GTA at amino acid position 61 in CDR2 (TABLE 4). This GTA codon also had 20 replacement mutations for a total of 35 mutations, which was far in excess of any codon or any three bases in any reading frame. Inverted repeats, or palindromes, and direct repeats have been suggested as sites susceptible to mutation by introduction of base errors or error-prone repair.[29,30] There was potential stem-loop formation at codon 60–62 CAG > TAT < CTG, with the tip (TAT) encoding 15 silent and 12 replacement mutations and the bordering G and C

TABLE 4. Hot Spots of Somatic Mutation

R/S	CDR2-60-62 Stem Loop	R/S
0/15	A	
11/0	61 T $\overline{T}$	1/0
9/0	$\underline{G}$------C 62	3/0
0/4	$\underline{A}$------$\overline{T}$	0/1
3/0	60 C------$\overline{G}$	2/0
R/S	**FR3-80-83 Stem Loop**	**R/S**
6/5	G------C	1/4
0/0	81 A------T 82	0/0
0/0	$\underline{C}$------$\underline{G}$	0/0
0/1	$\overline{G}$ A	4/0
0/0	80 T------A 83	7/0

another 12 replacement mutations (TABLE 4). Codons 80–83 in FR3 TGCAG-CTGAA also showed potential stem-loop formation (TABLE 4) with the tip (GC) having nine silent and seven replacement mutations. Inverted repeats were not found, however, at all hot spots. Repetitive replacement mutations may also represent hot spots. The serine codon AGC at position 31, which has been shown to be a hot spot for murine V_H genes,[26] was heavily mutated with 12 replacement and five silent mutations.

The bias in the large number of silent mutations of the human V_{H6} gene reflects intrinsic specificity of the hypermutation process. Based on the asymmetry of A and T mutations, the mutation process appears to be targeted to one DNA strand. In the V_{H6} gene, mutations were targeted to the CDR. Some hot spots for mutation had the potential for stem-loop formation, transitions were favored over transversions and G → C, C → G, and A → C transversions were favored over G → T, C → A, and A → T transversions. These findings are most compatible with models that suggest mutation arises by alteration of fidelity of a polymerase or interference with proofreading. Future models will need to take account of preferential targeting to one DNA strand and targeting of transitions, some types of transversions, and to specific base sequences or hot spots.

INDUCTION OF HUMAN V GENE MUTATIONS

We do not know what antigen induced the high frequency of mutation observed in the human V_{H6} genes described above or the number of separate times these rearrangements underwent mutation. It is possible to immunize mice with specific antigens and characterize the mutations that arise in well-defined, restricted, dominant V genes that encode the antibody response to these antigens. No similar system exists for inducing and characterizing the rate and nature of mutations in the human system after immunization. We have developed an approach to induce mutations in human immunoglobulin V genes and characterize the rate of the mutation process.

The principal virulence determinant of the bacteria *Haemophilus influenzae* b (Hib) is its capsular polysaccharide (CP), a repetitive polymer of ribosyl-ribitol-phosphate. Vaccines composed of the HibCP have been developed to prevent this human infection.[31] The antibody response to the HibCP is one of the most extensively studied human antibody responses to date because of restricted diversity of the antibody response[32] and the development of these Hib vaccines.[31] We have shown that the L chain distribution of human antibody is predominantly κ L chain type, in contrast to the normal 60% κ and 40% L chain distribution in human immunoglobulin (Ig) and analytical IEF of postimmunization antibody shows a restricted repertoire with only one to four clones, or so-called clonotypes recruited into the antibody response in an individual.[32,33] Shared cross-reactive idiotypes were demonstrated on antibody, and one idiotype, HibId-1, was expressed on antibody of approximately 85% of immunized persons and on over 60% of their IgG antibody.[34] HibId-1 expression correlates with expression of a VκII subgroup antibody encoded by the VκII gene-A2.[34,35] Greater than 60% of adults immunized with the HibCP vaccines or CP generate a predominant A2-encoded response and greater than 90% of adults express some A2-encoded antibody. The CDR3 region of A2-encoded antibody, which arises from V_L-J_L joining, is unique in that it is one amino acid longer than the typical nine amino acid Vκ CDR3 with an invariant arginine at the Vκ-Jκ junction,[35] suggesting the importance of this region to antibody binding.

Vaccines in which the HibCP is covalently linked to proteins, so-called conjugate vaccines or conjugates, have the ability to induce protective antibody responses in young infants, the group with the highest incidence of Hib disease.[31] In contrast, unconjugated HibCP vaccines are not immunogenic in the infant. The HibCP conjugates show typical T-dependent (TD) properties of generating good IgG booster antibody responses to the HibCP in infants with second and third doses. With use of Hib conjugate vaccines, the incidence of disease in the United States has been decreased by over 90%.

We have cloned and sequenced the A2 repertoire in peripheral blood lymphocytes (PBL) before and after immunization with HibCP vaccines.[36] With this approach and analysis, we have taken advantage of A2 predominance and the use of the prototypic (10aa CDR3 with arginine) A2 sequence in the HibCP antibody response, the lack of polymorphism of the A2 gene in man, and our observation that a high proportion of circulating human PBL collected at seven or eight days after immunization with Hib CP vaccines secretes antibody to the CP.[37] A sense A2-specific leader primer and an antisense Jκ primer were used to PCR amplify cDNA prepared from PBL collected seven days after immunization of adults with the HibTITER conjugate vaccine, which is composed of the HibCP covalently linked to a nontoxic, immunogenic, mutant form of diphtheria toxin (CRM197).[31] The PCR product was cloned, and clones were randomly selected and sequenced. All of 19 randomly selected postimmunization clones arose from the same rearrangement event but were expressed with extensive diversification. This single rearrangement had the prototypic 10-amino-acid CDR3 with an arginine (CGT) at the Vκ-Jκ junction that arose by presumed N region addition. This rearrangement had been recruited into the repertoire by immunization as evidenced by failure to detect this rearrangement in a preimmunization PBL A2 library from the immunized subject. In addition, clones derived from nonimmunized donors showed multiple and not single A2 rearrangements, and the prototypic A2 L chain rearrangement was not detected.

Surprisingly, all but two of the 19 clones arising from this rearrangement were unique, and 18 of 19 showed evidence of somatic mutation. The isolation of a germline clone arising from the same rearrangement and the finding that no one single mutation was shared by all the clones indicate that this rearrangement had been recruited into the antibody repertoire as a germline sequence. Both shared and unique somatic mutations were apparent in the clones that allowed genealogical relationships among these clones to be defined. Among the 18 clones, 46 unique and 69 total mutations were introduced into this germline sequence by seven days after immunization with a wide range, from 0 to 8 mutations/clone. If all V gene mutations are considered, there was an average of 3.6 mutations/clone, or if only unique mutations were considered then there were 46 mutations for 17 mutated sequences or 2.7 unique mutations/A2 gene. These findings are similar to what has been observed in V gene sequences isolated from murine germinal centers.[6,7,28] It should be noted that the fidelity of PCR amplification of V genes quantitated in our laboratory is 5.4×10^{-4} bases, or only one PCR-induced error for approximately six clones.

Based on our finding of 45×10^{-6} antibody-secreting cells in the peripheral blood at eight days after immunization,[37] and a blood volume in an adult of approximately 5 liters, at least 18 cell divisions would have occurred starting with a single B cell. This calculation does not account for cells retained in the lymph node or that already reside in the bone marrow. This number is, however, compatible with a cell cycle time of 9 to 10 hours beginning immediately after immunization[28] or a cell cycle time of eight hours beginning day 2 postimmunization. It is possible to develop genealogical relationships of these clones, which showed stepwise accumulation of mutations occurring in at least nine different branches, which means that the original B cell with the germline sequence had to go through at least four generations before the onset of somatic mutation. This can be used to correct for the fact that mutation does not occur uniformly throughout all generations, so that $18 - 4 = 14$ generations should be considered in calculating a mutation rate. Based on 2.7 mutations/300 bp in 14 generations, then the mutation frequency is 6.4×10^{-4}/bp per generation, a figure very consistent with other estimates of mutation of murine V genes.[1-3,9,28]

This estimate assumes there is no selection of mutations. These PBL clones were, however, presumably derived from cells that escaped apoptosis in lymph nodes. Translation of the nt sequences demonstrated R/S ratios for the 46 unique mutations of total FR-17/7-2.4 and total CDR-13/9-1.4. This indicates some preservation of the germline CDR sequence, indicating that the germline A2 sequence, which has been conserved in the human population, may be ideal for antigen binding. If anything, negative selection of mutation has occurred. The dominant use of germline A2 suggests that, like the germline-encoded T-15 anti-phosphorylcholine (PC) Ab that dominates the 1° response and is optimally protective against *S. pneumoniae* in mice,[38] the A2 gene may have been conserved in human evolution because it provides optimal protection against Hib.

REGULATION OF THE MUTATION PROCESS

The X-linked HyperIgM syndrome (HIGM1-X) is associated with defective expression of the CD40 ligand, which is expressed on activated T cells and stimulates B cells to proliferate, differentiate, and respond to cytokines that induce isotype switching.[39,40] Patients with HIGM1-X fail to generate germinal centers, express IgG antibody, and demonstrate good 2° antibody responses, but they have the capacity to generate IgM antibody, even after immunization with TD antigen.[41,42] It is unclear what is the origin of the naturally occurring IgM level in these patients and whether immunoglobulin or IgM antibody of patients with HIGM1-X undergo somatic mutation and show increased antibody affinity. Targeted CD40 gene knockout mice also fail to generate germinal center and secondary antibody responses.[43] In these mice, IgM and IgG serum levels are normal, and IgM and IgG antibody responses to TI antigen and IgM antibody responses to TD antigen are preserved. The presence of somatic mutation in V genes of these mice has not been described.

It has been debated whether somatic mutation begins outside follicles before germinal center formation with B cells that undergo early somatic mutation migrating to form germinal centers and undergoing increased mutation versus whether somatic mutation initiates only in germinal centers.[1,2,44,45] It would be of interest to determine whether somatic mutation occurs in genetic defects associated with absence of germinal center formation. If somatic mutation does occur in the absence of germinal center formation in these patients, the selection pressure that occurs in germinal centers to select high-affinity antibody and delete Igs that have become self-reactive may not exist. This possibility could explain the high frequency of autoantibodies that arise in these patients.[41]

We have asked whether somatic mutation can be detected in V gene rearrangements isolated from the PBL of patients with the X-linked HyperIgM syndrome.[46] We have taken advantage of analysis of mutations in the nonpolymorphic human

V_{H6} sequence, as described above. Human V_{H6} rearrangements switched to IgM or IgG were cloned from cDNA from the PBL of two patients with X-HIM, sequenced, and compared to that of healthy persons. To identify rare rearrangements that have undergone somatic mutation, we analyzed the clones for single-stranded conformational polymorphism (SSCP), which is capable of detecting single-point mutations. SSCP was performed by PCR reamplifying cloned rearrangements with internal primers and labeled dNTP and running the product on a nondenaturing acrylamide gel. This technique has proven to be an efficient approach to screening multiple clones for somatic mutation in our laboratory.

The HIGM1-X IgG rearrangements were usually in germline form, which contrasted with the findings of extensive mutation in IgG rearrangements from normal persons. These IgG rearrangements may be arising from long-processed IgM transcripts and not from deletional isotype switching, a concept under investigation. Most HIGM1-X IgM rearrangements were also in germline form, which also contrasted with the V_{H6} IgM rearrangements of healthy persons, which, as we have shown, can be heavily mutated.[22] There were, however, 3 of 35 V_{H6} IgM and 1 of 5 V_{H6} IgG sequences that appeared mutated at a frequency far in excess of the frequency in our laboratory of PCR-induced error (5.4×10^{-4}). Up to nine mutations per clone with as many as five replacement mutations in the CDR were observed. The distribution and characteristics of their mutations were consistent with their arising by typical somatic mutation followed by antigen selection.

These mutations suggest that low-level mutation occurs in the absence of CD40-CD40 ligand interactions. Possibly, these somatic mutations are induced in extrafollicular locations through a CD40-independent pathway that induces low-level somatic mutation. In the normal situation, these activated cells may immediately traffic to germinal centers and undergo more extensive mutation and would, therefore, not be detected. The antigen selection of these mutations may have occurred in the periphery and not in the lymph nodes. Experiments to induce mutations in these patients should help further clarify the role of CD40-CD40 ligand interactions in the mutation process.

CONCLUSIONS

The results indicate that somatic mutation of human immunoglobulin V genes can be analyzed in a nonbiased manner to gain insights into its intrinsic properties. These properties include a bias toward generation of transitions and $C \rightarrow G$, $G \rightarrow C$, and $A \rightarrow C$ transversions. Mutation was targeted to the V region at sites with potential stem-loop formation and to adenine versus thymine nucleotides on the coding strand, suggesting strand polarity in the process. A mutation rate of the human A2 gene of 6.4×10^{-4}/bp per generation was observed. There was evidence that proliferation preceded mutation and that a stepwise introduction of point mutations occurred with development of a wide range of mutation events per clone, explained by the mutation process either ceasing in a stochastic manner or being reactivated in some clones. The ability to analyze mutations after immunization will offer a new approach to analyzing human antibody repertoires and evaluating vaccine immunogenicity. The finding of somatic mutation in patients with the X-linked HyperIgM syndrome suggests the mutation process can be initiated in the absence of CD40 ligand function. Ultimate understanding of mutation of immunoglobulin V genes will, however, require identification of the enzymes mediating and of the cellular interactions regulating the somatic mutation process.

REFERENCES

1. KOCKS, C. & K. RAJEWSKY. 1989. Stable expression and somatic hypermutation of antibody V regions in B cell developmental pathways. Ann. Rev. Immunol. **7:** 537.
2. FRENCH, D. L., R. LASKOV & M. D. SCHARFF. 1989. The role of somatic hypermutation in the generation of antibody diversity. Science **244:** 1152.
3. BEREK, C. & C. MILSTEIN. 1987. Mutation drift and repertoire shift in the maturation of the immune response. Immunol. Rev. **96:** 23.
4. LEVY, N. S., U. V. MALIPIERO, S. G. LEBECQUE & P. J. GEARHART. 1989. Early onset of somatic mutation in immunoglobulin VH genes during the primary immune response. J. Exp. Med. **169:** 2007.
5. WEISS, U., R. ZOEBELEIN & K. RAJEWSKY. 1992. Accumulation of somatic mutants in the B cell compartment after primary immunization with a T cell-dependent antigen. Eur. J. Immunol. **22:** 511.
6. BEREK, C., A. BERGER, & M. APEL. 1991. Maturation of the immune response in germinal centers. Cell **67:** 1121.
7. JACOB, J., G. KELSOE, K. RAJEWSKY & U. WEISS. 1991. Intraclonal generation of antibody mutants in germinal centres. Nature **354:** 389.
8. RADA, C., S. K. GUPTA, E. GHERARDI & C. MILSTEIN. 1991. Mutation and selection during the secondary response to 2-phenyloxazolone. Proc. Natl. Acad. Sci. USA **88:** 5508.
9. MCKEAN, D., K. HUPPI, M. BELL, L. STAUDT, W. GERHARD & M. WEIGERT. 1984. Generation of antibody diversity in the immune response of BALB/c mice to influenza virus hemagglutinin. Proc. Natl. Acad. Sci. USA **81:** 3180.
10. GEARHART, P. J. & D. F. BOGENHAGEN. 1983. Clusters of point mutations are found exclusively around rearranged antibody variable genes. Proc. Natl. Acad. Sci. USA **80:** 3439.
11. LEBECQUE, S. G. & P. J. GEARHART. 1990. Boundaries of somatic mutation in rearrangement immunoglobulin genes: 5′ boundary is near the promoter, and 3′ boundary is ≈1 kb from V(D)J gene. J. Exp. Med. **172:** 1717.
12. ROES, J., K. HUPPI, K. RAJEWSKY & F. SABLITZKY. 1989. V gene rearrangement is required to fully activate the hypermutation mechanism in B cells. J. Immunol. **142:** 1022.
13. BRENNER, S. & C. MILSTEIN. 1966. Origin of antibody variation. Nature **211:** 242.
14. MANSER, T. 1990. The efficiency of antibody affinity maturation: Can the rate of B-cell division be limiting? Immunol. Today **11:** 305.
15. ROGERSON, B., J. HACKETT, A. PETERS, D. HAASCH & U. STORB. 1991. Mutation pattern of immunoglobulin transgenes is compatible with a model of somatic hypermutation in which targeting of the mutator is linked to the direction of DNA replication. EMBO J. **10:** 4331.
16. UMAR, A., P. A. SCHWEITZER, N. S. LEVY, J. D. GEARHART & P. J. GEARHART. 1991. Mutation in a reporter gene depends on proximity to and transcription of immunoglobulin variable transgenes. Proc. Natl. Acad. Sci. USA **88:** 4902.
17. BETZ, A. G., C. MILSTEIN, A. GONZALEZ-FERNANDEZ, R. PANNELL, T. LARSON & M. S. NEUBERGER. 1994. Elements regulating somatic hypermutation of an immunoglobulin κ gene: Critical role for the intron enhancer/matrix attachment region. Cell **77:** 239–248.
18. MOTOYAMA, N., T. MIWA, Y. SUZUKI, H. OKADA & T. AZUMA. 1994. Comparison of somatic mutation frequency among immunoglobulin genes. J. Exp. Med. **179:** 395.
19. BERMAN, J. E., K. G. NICKERSON, R. R. POLLOCK, J. E. BARTH, R. K. B. SCHUURMAN, D. M. KNOWLES, L. CHESS & F. W. ALT. 1991. V_H gene usage in humans: Biased usage of the V_H6 gene in immature B lymphoid cells. Eur. J. Immunol. **21:** 1311.

20. VARADE, W. S., E. MARIN, A. M. KITTELBERGER & R. A. INSEL. 1993. Use of the most J_H-proximal human immunoglobulin heavy chain variable region gene, V_{H6}, in the expressed immune repertoire. J. Immunol. **150:** 4985.
21. VARADE, W. S. & R. A. INSEL. 1993. Isolation of germinal center-like vents from human spleen RNA: Somatic hypermutation of a clonally related $V_{H6}DJ_H$ rearrangement expressed with IgM, IgG, and IgA. J. Clin. Invest. **91:** 1838.
22. INSEL, R. A., W. S. VARADE & E. MARIN. 1994. Human splenic IgM immunoglobulin transcripts are mutated at high frequency. Mol. Immunol. **31:** 383–392.
23. INSEL, R. A. & W. S. VARADE. 1994. Bias in somatic hypermutation of human V_H genes. Intl. Immunol. **6:** 1437.
24. CHEN, C., V. A. ROBERTS & M. B. RITTENBERG. 1992. Generation and analysis of random point mutations in an antibody CDR2 sequence: Many mutated antibodies lose their ability to bind antigen. J. Exp. Med. **176:** 855.
25. MOTOYAMA, N., H. OKADA & T. AZUMA. 1991. Somatic mutation in constant regions of mouse λ1 light chains. Proc. Natl. Acad. Sci. USA **88:** 7933.
26. BETZ, A. G., C. RADA, R. PANNELL, C. MILSTEIN & M. S. NEUBERGER. 1993. Passenger transgenes reveal intrinsic specificity of the antibody hypermutation mechanism: Clustering, polarity, and specific hot spots. Proc. Natl. Acad. Sci. USA **90:** 2385.
27. MEEK, K., T. EVERSOLE & J. D. CAPRA. 1991. Conservation of the most J_H proximal Ig V_H gene segment (V_HVI) throughout primate evolution. J. Immunol. **146:** 2434.
28. JACOB, J., J. PRZYLEPA, C. MILLER & G. KELSOE. 1993. In situ studies of the primary immune response to (4-hydroxy-3-nitrophenyl)acetyl. III. The kinetics of V region mutation and selection in germinal center B cells. J. Exp. Med. **178:** 1293.
29. GOLDING, G. B., P. J. GEARHART & B. W. GLICKMAN. 1987. Patterns of somatic mutations in immunoglobulin variable genes. Genetics **115:** 169.
30. KOLCHANOV, N. A., V. V. SOLOVYOV & I. B. ROGOZIN. 1987. Peculiarities of immunoglobulin gene structures as a basis for somatic mutation emergence. FEBS Lett. **214:** 87.
31. ELLIS, R. W. & D. M. GRANOFF, Eds. 1994. Development and Clinical Uses of *Haemophilus* b Conjugate Vaccines. Marcel Dekker, Inc. New York, NY.
32. INSEL, R. A., E. E. ADDERSON & W. L. CARROLL. 1992. The repertoire of human antibody to the *Haemophilus influenzae* type b capsular polysaccharide. Intern. Rev. Immunol. **9:** 25.
33. INSEL, R. A., A. KITTELBERGER & P. ANDERSON. 1985. Isoelectric focusing of human antibody to the *Haemophilus influenzae* b capsular polysaccharide: Restricted and identical spectrotypes in adults. J. Immunol. **135:** 2810.
34. LUCAS, A. H., R. J. LANGLEY, D. M. GRANOFF, M. H. NAHM, M. Y. KITAMURA & M. G. SCOTT. 1991. An idiotypic marker associated with a germ-line encoded κ light chain V region that predominates the vaccine-induced human antibody response to the *Haemophilus influenzae* b polysaccharide. J. Clin. Invest. **88:** 1811.
35. SCOTT, M. G., J. J. TARRAND, D. L. CRIMMINS, D. W. MCCOURT, N. R. SIEGEL, C. E. SMITH & M. H. NAHM. 1989. Clonal characterization of the human IgG antibody repertoire to *Haemophilus influenzae* type b polysaccharide. II. IgG antibodies contain V_H genes from a single V_H family and V_L genes from at least four V_L families. J. Immunol. **143:** 293.
36. INSEL, R. A., E. MARIN & Y-W. CHU. 1995. Negative selection preserves the germline CDR of human antibody to a bacterial capsular polysaccharide. FASEB J. **9:** A495, Abstr. 2856.
37. MUNOZ, J. L. & R. A. INSEL. 1987. In vitro human antibody production to the *Haemophilus influenzae* type b capsular polysaccharide. J. Immunol. **139:** 2026.
38. BRILES, D. E., C. FORMAN, S. HUDAK & J. L. CLAFLIN. 1982. Antiphosphorylcholine antibodies of the T15 idiotype are optimally protective against *Streptococcus pneumoniae*. J. Exp. Med. **156:** 1177.
39. MARSHALL, L. S., A. ARUFFO, J. A. LEDBETTER & R. J. NOELLE. 1993. The molecular basis for T cell help in humoral immunity: CD40 and its ligand, gp39. J. Clin. Immunol. **13:** 165.

40. CALLARD, R. E., R. J. ARMITAGE, W. C. FANSLOW & M. K. SPRIGGS. 1993. CD40 ligand and its role in X-linked hyper-IgM syndrome. Immunol. Today **14:** 559–564.
41. NOTARANGELO, L. D., M. DUSE & A. G. UGAZIO. 1992. Immunodeficiency with hyper-IgM (HIM). Immunodef. Rev. **3:** 101.
42. NONOYAMA, S., D. HOLLENBAUGH, A. ARUFFO, *et al.* 1993. B cell activation via CD40 is required for specific antibody production by antigen-stimulated human B cells. J. Exp. Med. **178:** 1097.
43. KAWABE, T., T. NAKA, K. YOSHIDA, *et al.* 1994. The immune responses in CD40-deficient mice: Impaired immunoglobulin class switching and germinal center formation. Immunity **1:** 167.
44. MACLENNAN, I. C. M. 1994. Germinal centers. Annu. Rev. Immunol. **12:** 117.
45. BEREK, C. 1992. The development of B cells and the B-cell repertoire in the microenvironment of the germinal center. Immunol. Rev. **126:** 5.
46. CHU, Y-W., E. MARIN, R. FULEIHAN, N. RAMESH, F. S. ROSEN, R. S. GEHA & R. A. INSEL. Somatic mutation of human immunoglobulin V genes in the X-linked hyper IgM syndrome. J. Clin. Invest. **95:** 1389.

A Sequence Analysis of Human Germline Ig V_H and V_L Genes

The CDR1s of a Major Proportion of V_H, but Not V_L, Genes Display a High Inherent Susceptibility to Amino Acid Replacement[a]

BERNARD CHANG AND PAOLO CASALI

Division of Molecular Immunology
Department of Pathology
Cornell University Medical College
New York, New York 10021

Repeated antigenic challenge results in the production of antibodies with progressively higher affinity for antigen (affinity maturation).[1] Analysis of monoclonal antibodies (mAbs) generated at successive stages of various specific murine antibody responses[2–10] has shown that B-cell clones are originally recruited by virtue of the affinity of their unmutated surface immunoglobulin (Ig) receptor for antigen. Soon afterwards, the recruited B-cell clones begin to accumulate somatic point mutations in the expressed Ig-variable (V) segments and undergo selection by antigen. This process results in restriction of the response to those B cells with the best "fit" (highest affinity) for antigen.[11,12] High-affinity, antigen-selected antibodies exhibit a higher frequency of mutations resulting in amino acid replacement (R mutations) in the V gene complementarity-determining regions (CDRs) than in the framework regions (FRs), where the proportion of mutations not resulting in amino acid change (S mutations) may be greater. The high R : S mutation ratio characteristic of "affinity-mature" Ig V gene CDR sequences is consistent with these gene products' need to provide the best "fit" for antigen, while the low R : S mutation ratio characteristic of FR sequences is consistent with these gene products' role in maintaining structural integrity.

In an expressed antibody, the notion that high $R:S_{CDR}$ and low $R:S_{FR}$ mutation ratios are signs of antigen selection necessarily assumes that the CDRs and FRs of the template germline V gene possess an equal inherent potential for amino acid replacement given any nucleotide substitution. Random accumulation of nucleotide changes throughout the coding sequence of a germline Ig V gene segment comprising CDRs whose codons are inherently more susceptible to R mutations than the codons of the adjacent FRs would result in higher observed R : S mutation ratios in the CDRs than in the FRs, even in the absence of antigen selection.

We have recently demonstrated[13] that the CDR and FR sequences of a major proportion of human germline Ig V_H genes actually differ in their inherent susceptibility to amino acid replacement. This inherent susceptibility can be expressed

[a] This work was supported by U.S. Public Health Service Grants AR-40908 to P.C. P.C. is a Kaplan Cancer Scholar. B.C. was supported by U.S.P.H.S. Short-Term Training Grant 5T35 DK-07421. This is publication 3 from the Division of Molecular Immunology.

either as replacement frequency (*Rf*, or the proportion of total possible point mutations that result in amino acid replacement) or as inherent R : S mutation ratio (the ratio of total possible R mutations to total possible S mutations). We have shown that the CDR sequences of these germline V_H genes comprise codons more susceptible to R mutations (displaying higher inherent *Rf* and R : S mutation ratio values) than expected for a random sequence, whereas the FR sequences comprise codons less susceptible to R mutations (displaying lower inherent *Rf* and R : S mutation ratio values) than expected. Our analysis of human germline Ig V_L genes revealed a much smaller differential between CDR and FR sequences' inherent susceptibility to R mutations.

Here we have extended our analysis to additional three and four human germline V_H and V_L genes, respectively. The analysis was performed using the computer algorithm we devised to calculate inherent *Rf* and R : S mutation ratio values for any nucleotide sequence, as described.[13]

The sequence analysis of a total of 27 V_H germline genes[21–37] (TABLE 1), among those most frequently expressed in the early and adult B-cell repertoires, revealed that the inherent Rf_{CDR} values are significantly higher than the expected 0.7452 *Rf* value for a random sequence[38] ($p = 1.4 \times 10^{-16}$) and significantly higher than the respective Rf_{FR} values ($p = 4.1 \times 10^{-21}$). Conversely, the inherent Rf_{FR} values are significantly lower than the expected value ($p = 6.3 \times 10^{-3}$). Similarly, the $R:S_{CDR}$ mutation ratios are significantly higher than the 2.925 R : S mutation ratio expected for a random sequence ($p = 4.5 \times 10^{-14}$) and significantly higher than the respective $R:S_{FR}$ mutation ratios ($p = 6.1 \times 10^{-16}$). The $R:S_{FR}$ mutation ratios are significantly lower than the expected value ($p = 1.0 \times 10^{-2}$).

Inherent mutational profiles for the analyzed V_H genes were created as described elsewhere.[13] Overall, the CDR2s display somewhat higher inherent R : S mutation ratios than the expected random value and the corresponding values for the adjacent FRs, but this difference is not significant. In contrast, the CDR1 sequences on average display significantly higher inherent R : S mutation ratios, in 15 cases more than twofold higher, and in three cases (including the newly analyzed DP54 gene) about fivefold higher, than those of the adjacent FRs (TABLE 1, FIG. 1). We conclude that the CDRs, but not the FRs, of a major proportion of human germline Ig V_H genes display a significantly high structural susceptibility to R mutations and that this susceptibility is due mainly to the codon composition of the CDR1.

The analysis of four additional human germline V_L genes,[39–41] Humlv101 (VλI), Humlv318 (VλIII), Humlv418 (VλIV), and Vκ005 (VκII), the latter two being members of V_L subgroups not analyzed previously, further substantiated the conclusions derived from the original analysis of 12 V_L genes (TABLE 2).[42–53] In the 16 germline V_L genes, the Rf_{CDR} values are significantly higher than the expected 0.7452 *Rf* value ($p = 3.6 \times 10^{-8}$) and the respective Rf_{FR} values ($p = 2.9 \times 10^{-8}$), but they are significantly lower overall than those calculated for the V_H segment CDRs ($p = 1.6 \times 10^{-8}$). The V_L gene Rf_{FR} values do not significantly differ from the expected value ($p = 0.31$). The inherent $R:S_{CDR}$ mutation ratios are significantly higher than the theoretically expected 2.925 value ($p = 1.8 \times 10^{-7}$) and the respective $R:S_{FR}$ mutation ratios ($p = 6.9 \times 10^{-8}$), but are significantly lower than the R : S mutation ratios of the V_H segment CDRs ($p = 2.0 \times 10^{-9}$). The V_L gene $R:S_{FR}$ mutation ratios do not significantly differ from the expected value ($p = 0.38$). The inherent mutational profiles for the V_L gene segments show that inherent R : S mutation ratios are, on average, only moderately higher in the CDR3s than in the adjoining FRs and the theoretically expected value of 2.925, as exemplified in FIGURE 1. The V_L CDR1 sequences, with a narrower distribution

TABLE 1. Inherent *Rf* and R : S Mutation Ratio Values for Human Germline V_H Genes

			Mutations in CDRs			CDR1 Alone	Mutations in FRs		
Gene[a]	Percent Identity	Ref.	Total Possible Changes[b]	*Rf*[c]	R : S Mutation Ratio[c]	R : S Mutation Ratio	Total Possible Changes	*Rf*	R : S Mutation Ratio
V_H1									
51P1	100.0	21	191	0.7801	3.548	4.375	650	0.7477	2.963
hv1263	98.3	22	191	0.7801	3.548	4.375	650	0.7477	2.963
V_{I-2}	88.1	23	187	0.8128	4.343	5.833	650	0.7492	2.988
V_{I-3B}	87.4	23	186	0.8118	4.314	6.167	649	0.7504	3.006
V_H2									
V_H2-MC2	100.0	24	197	0.8020	4.051	4.083	635	0.7433	2.896
V_H2S12-8	92.7	25	197	0.7766	3.477	3.143	638	0.7414	2.867
V2-26	91.4	26	200	0.7950	3.878	3.769	669	0.7399	2.845
V_H3									
V_H26c	100.0	27	191	0.7801	3.548	6.167	652	0.7423	2.881
8-1B	92.5	28	180	0.7944	3.865	9.750	652	0.7408	2.858
WHG16G	91.2	29	194	0.8144	4.389	9.750	644	0.7376	2.811
22-2B	89.8	28	187	0.8128	4.343	9.250	650	0.7400	2.846
1.9III	89.1	28	185	0.7946	3.868	6.167	655	0.7404	2.853
GL-SJ2	89.1	30	185	0.7946	3.868	6.167	655	0.7404	2.853
hv3005	89.1	30	185	0.7946	4.162	6.167	655	0.7404	2.853
H11	88.4	31	191	0.8063	4.935	12.667	652	0.7362	2.791
HHG19	87.1	29	184	0.8315	4.743	12.667	652	0.7393	2.835
DP54	85.7	32	201	0.8259		12.667	652	0.7393	2.835
V_H4									
V_{71-2}	100.0	33	192	0.8177	4.486	6.125	654	0.7217	2.593
$V_{71.4}$	98.0	33	174	0.8218	4.613	8.750	654	0.7217	2.593
$V_H4.11$	97.6	34	174	0.8218	4.613	8.750	654	0.7232	2.613
4.33	95.3	35	190	0.8053	4.135	4.700	654	0.7278	2.674
$V_H4.18$	94.9	34	192	0.8177	4.486	6.167	657	0.7245	2.630
V79	92.6	33	188	0.8351	5.064	11.500	658	0.7264	2.656
$V_H4.21$	90.9	34	187	0.8182	4.500	5.857	642	0.7321	2.732
V_H5									
V_H251	100.0	36	188	0.7766	3.476	4.125	645	0.7659	3.272
V_H32	94.9	36	189	0.8095	4.150	7.200	647	0.7651	3.256
V_H6									
6-1G1	100.0	37	184	0.7935	3.842	4.778	678	0.7404	2.852
Mean value			188.5	0.8046	4.152	7.078	652.0	0.7394	2.845
Standard deviation			6.5	0.0170	0.449	2.885	8.4	0.0112	0.169

NOTE: Abbreviations: *Rf*, replacement frequency or the proportion of total possible point mutations that result in amino acid replacement; R : S, ratio of total possible replacement mutations to total possible silent mutations.

[a] Within each V_H family, genes are listed according to their sequence identity (percent) with an arbitrary prototypic sequence.

[b] Total possible mutations are calculated as the total number of possible single-base substitutions excluding those resulting in stop codons.

[c] For a random sequence, inherent *Rf* should be 0.7452 and inherent R : S should be 2.925 (Ref. 38).

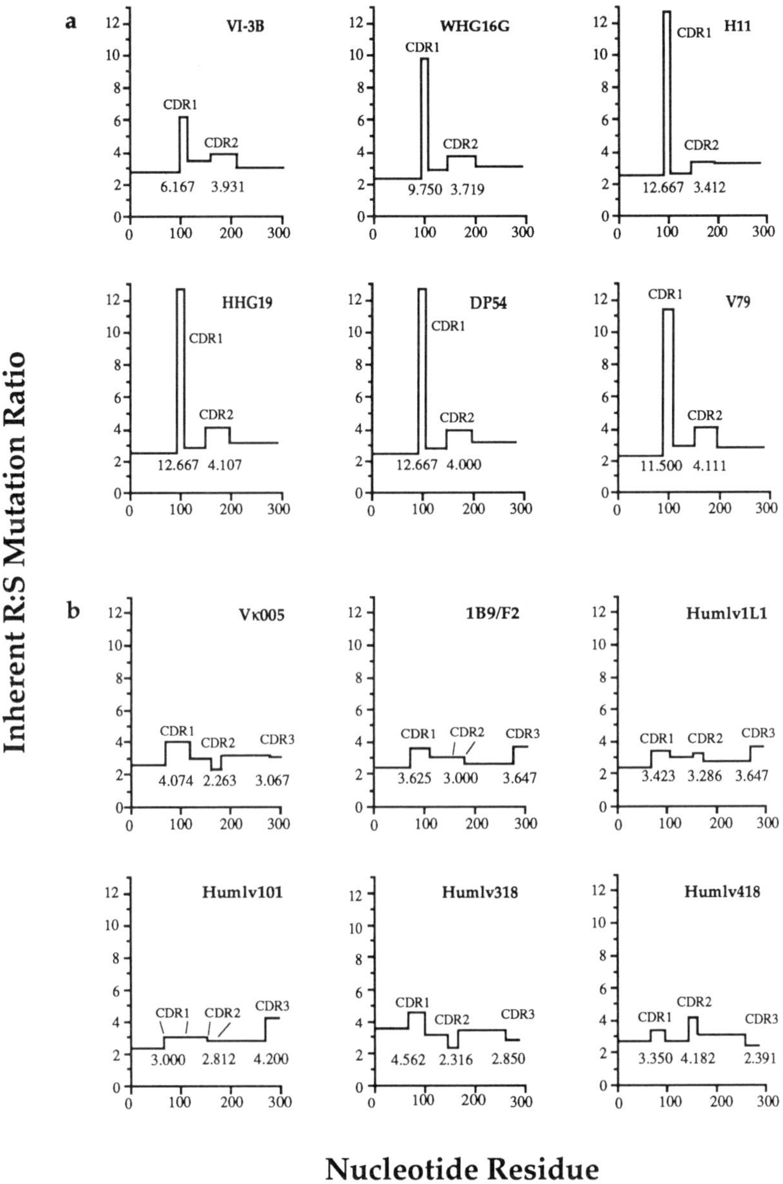

FIGURE 1. Inherent replacement susceptibility profiles of human germline V_H and V_L gene sequences. The CDR and FR sequences of the most commonly expressed human Ig V_H and V_L genes, as listed in TABLES 1 and 2, were analyzed using the computer algorithm that calculates the inherent susceptibility to amino acid replacement given any nucleotide change. Inherent R mutation susceptibility profiles were created by plotting the inherent R : S mutation ratio (ratio of total possible single-base substitutions resulting in amino acid replacement to total possible single-base substitutions not resulting in amino acid replacement) for each

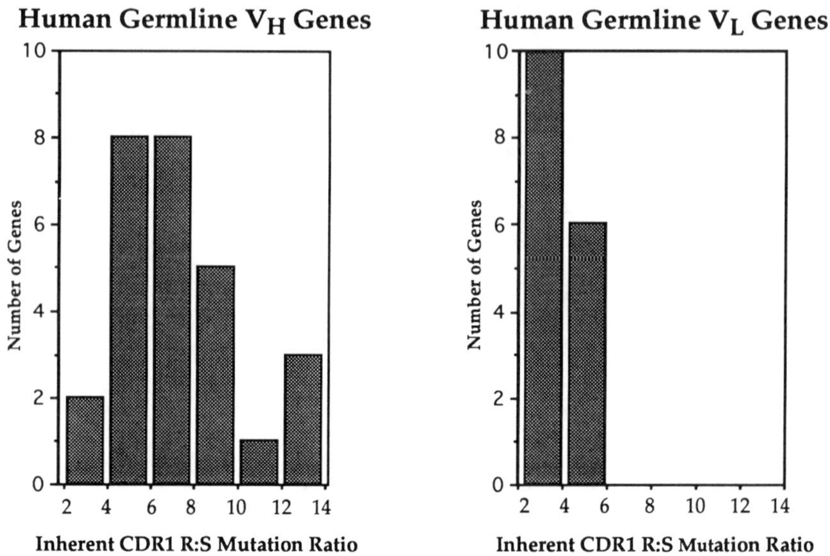

FIGURE 2. Distributions of the inherent R : S_{CDR1} mutation ratios of 27 germline V_H and 16 germline V_L genes selected from among those most commonly expressed in the human B-cell repertoire. The inherent R : S mutation ratio (ratio of total possible single-base substitutions resulting in amino acid replacement to total possible single-base substitutions not resulting in amino acid replacement) was calculated for each CDR1 sequence by applying the computer algorithm that calculates the inherent susceptibility to amino acid replacement to the CDRs and FRs of each of the human germline Ig V_H and V_L genes listed in TABLES 1 and 2. Histogram depicts the number of V_H or V_L genes (*ordinate*) with inherent R : S_{CDR1} mutation ratios within a given interval (*abscissa*).

and much lower mean value of inherent R : S_{CDR1} mutation ratios (FIG. 2, TABLE 2), do not display the striking susceptibility to amino acid replacement found in the V_H segment CDR1s. Furthermore, the average inherent R : S_{CDR2} mutation ratio is actually lower than those of the adjoining FRs. Thus, as compared with V_H genes, V_L genes exhibit a smaller overall difference between inherent R : S_{CDR} and R : S_{FR} mutation ratios and display virtually no codon clusters with high inherent susceptibility to replacement.

Our findings imply that when the observed R : S mutation ratios serve as the criterion for determining whether the somatic point mutations observed in an expressed Ig V gene have been selected by antigen, the inherent R : S mutation ratio for the germline gene template, not the theoretically expected figure of 2.925,

---

of the Ig V regions versus the respective V region nucleotide residue numbers. Numbers below each CDR represent the inherent R : S mutation ratio for that CDR. **(a)** Profiles of inherent susceptibility to amino acid replacement of the six germline V_H genes chosen as representative of those most commonly expressed in the human B-cell repertoire; **(b)** profiles of inherent susceptibility to amino acid replacement of the six germline V_L genes chosen as representative of those most commonly expressed in the human B-cell repertoire.

TABLE 2. Inherent *Rf* and R : S Mutation Ratios for Human Germline V_L Genes[a]

Gene	Ref.	Mutations in CDRs: Total Possible Changes	Mutations in CDRs: *Rf*	Mutations in CDRs: R : S Mutation Ratio	CDR1 Alone: R : S Mutation Ratio	Mutations in FRs: Total Possible Changes	Mutations in FRs: *Rf*	Mutations in FRs: R : S Mutation Ratio
Vκ1								
kalc6	42	195	0.8000	4.000	3.318	612	0.7549	3.080
Vκ2								
Vκ005	39	260	0.7654	3.262	4.074	604	0.7450	2.922
Vκ3								
κv3g	43	216	0.7917	3.800	3.947	606	0.7492	2.987
20G2	44	216	0.7824	3.596	4.053	607	0.7512	3.020
HumKv328h5	45	213	0.7606	3.176	4.053	607	0.7512	3.020
HumKv325	46	223	0.7713	3.373	4.150	605	0.7504	3.007
Vκ4								
Vκ4	47	268	0.7836	3.621	4.259	607	0.7529	3.047
Vλ1								
IB9/F2	48	250	0.7760	3.464	3.625	597	0.7236	2.618
YM-1	49	272	0.7831	3.610	3.567	613	0.7341	2.761
Humlv1L1	50	254	0.7756	3.456	3.423	593	0.7285	2.683
Humlv101	40	263	0.7642	3.242	3.000	592	0.7297	2.700
Humlv117	51	250	0.7720	3.386	3.385	590	0.7271	2.665
Vλ3								
Vλ3.1	52	211	0.7915	3.795	3.600	593	0.7420	2.876
Humlv318	41	229	0.7598	3.164	4.562	591	0.7733	3.410
Vλ4								
Humlv418	41	222	0.7568	3.111	3.350	589	0.7436	2.901
Vλ5								
p3C4λ5	53	238	0.7787	3.519	3.700	582	0.7406	2.854
Mean value		236.2	0.7758	3.473	3.754	599.2	0.7436	2.909
Standard deviation		23.6	0.0126	0.255	0.419	9.4	0.0129	0.201

[a] See TABLE 1 for abbreviations and footnotes.

must be used as the critical value for comparison. This principle is exemplified by the analysis of the V_H gene sequence of mAb57, which we have reported elsewhere.[20] The inherent $R:S_{CDR}$ mutation ratio of the germline gene template of mAb57, V_H hv1263, was calculated as 3.548, whereas the observed $R:S_{CDR}$ for mAb57 is 8.000 (eight R mutations and one S mutation). The inherent $R:S_{FR}$ of the germline gene hv1263 is 2.963, while the observed $R:S_{FR}$ for mAb57 is 2.500 (five R mutations and two S mutations). Thus, antigen selection of mAb57 results in observed $R:S_{CDR}$ and $R:S_{FR}$ mutation ratios significantly higher ($p = 2.1 \times 10^{-3}$) and lower ($p = 1.9 \times 10^{-2}$), respectively, than those expected from random mutation, even though the germline gene template hv1263 displays high inherent susceptibility to amino acid substitution in its CDRs and low inherent susceptibility in its FRs (*P* values were calculated according to the binomial distribution model[54] $p = \{n!/[k!(n - k)!]\} \times q^k \times (1 - q)^{n-k}$, where n = total number of observed mutations; k = number of observed R mutations in the CDRs or FRs; q = probability that an R mutation will localize to CDRs or FRs, or q = relative size of CDRs × inherent Rf_{CDR} or relative size of FRs × inherent Rf_{FR}).

The high inherent susceptibility to amino acid replacement of germline gene sequences encoding the Ig V_H segment CDRs, but not FRs, may stem from an evolutionary selection that reflects the different functions (antigen contact versus structural integrity) of these discrete portions of the Ig V_H gene product. Similarly, the significantly higher inherent R : S mutation ratios of the CDRs of V_H gene segments as compared with those of the CDRs of V_L gene segments might also reflect a function-related evolutionary selection, consistent with the evidence showing H chain V segments to be the major structural correlate of antigen binding.[55–58]

The inherently high susceptibility to replacement of certain V_H gene CDR1 sequences constitutes a source of amino acid hypervariability that is independent of the intrinsic "hotspot" mechanism of preferential targeting by the mutational machinery, as recently proposed by Betz and his colleagues.[59,60] The fact that the V_H segment CDR1, the sequence with maximal inherent susceptibility to replacement, may also be a preferentially targeted "hotspot" demonstrates that the two mechanisms can act synergistically to generate diversity in the CDRs and suggests that the concentration of R mutations observed within CDR1 and flanking areas in "affinity-mature" antibodies is not due solely to antigen selection. Further structural studies of the antibody molecule and increased understanding of the nucleotide substitution preferences of the enzymatic replicative machinery are necessary to determine the precise role of the hypervariability mechanism described here in antigen selection and affinity maturation.

REFERENCES

1. STEINER, L. A. & H. N. EISEN. 1967. Sequential changes in the relative affinity of antibodies synthesized during the immune response. J. Exp. Med. **126:** 1161–1183.
2. MCKEAN, D., K. HUPPI, M. BELL, L. STAUDT, W. GERHARD & M. WEIGERT. 1984. Generation of antibody diversity in the immune response of Balb/c mice to influenze virus hemagglutinin. Proc. Natl. Acad. Sci. USA **81:** 3180–3184.
3. BEREK, C., G. M. GRIFFITHS & C. MILSTEIN. 1985. Molecular events during maturation of the immune response to oxazolone. Nature **316:** 412–418.
4. SABLITZKY, F., G. WILDNER & K. RAJEWSKY. 1985. Somatic mutation and clonal expansion of B-cells in an antigen-driven immune response. EMBO J. **4:** 345–350.
5. WYSOCKI, L., T. MANSER & M. L. GEFTER. 1986. Somatic evolution of variable region structures during an immune response. Proc. Natl. Acad. Sci. USA **83:** 1847–1851.
6. MALIPIERO, U. V., N. S. LEVY & P. J. GEARHART. 1987. Somatic mutation in anti-phosphorylcholine antibodies. Immunol. Rev. **96:** 59–74.
7. CLAFLIN, J. L., J. BERRY, D. FLAHERTY & W. DUNNICK. 1987. Somatic evolution of diversity among anti-phosphorylcholine antibodies induced with *Proteus morganii*. J. Immunol. **138:** 3060–3068.
8. BEREK, C. & C. MILSTEIN. 1987. Mutation drift and repertoire shift in the maturation of the immune response. Immunol. Rev. **96:** 23–41.
9. BEREK, C. & C. MILSTEIN. 1988. The dynamic nature of the antibody repertoire. Immunol. Rev. **105:** 5–26.
10. FRENCH, D. L., R. LASKOV & M. D. SCHARFF. 1989. The role of somatic hypermutation in the generation of antibody diversity. Science **244:** 1152–1157.
11. MANSER, T. 1989. Evolution of antibody structure during the immune response. The differentiative potential of a single B lymphocyte. J. Exp. Med. **170:** 1211–1230.
12. WEISS, U. & K. RAJEWSKY. 1990. The repertoire of somatic mutants accumulating in

the memory compartment after primary immunization is restricted through affinity maturation and mirrors that expressed in the secondary response. J. Exp. Med. **172:** 1681–1689.

13. CHANG, B. & P. CASALI. 1994. The CDR1 sequences of a major proportion of human germline Ig V_H genes are inherently susceptible to amino acid replacement. Immunol. Today **15:** 367–373.
14. HARINDRANATH, N., I. S. GOLDFARB, H. IKEMATSU, S. E. BURASTERO, R. L. WILDER, A. L. NOTKINS & P. CASALI. 1991. Complete sequence of the genes encoding the V_H and V_L regions of low and high affinity monoclonal IgM and IgA1 rheumatoid factors produced by $Cd5^+$ B cells from a rheumatoid arthritis patient. Intl. Immunol. **3:** 865–875.
15. MANTOVANI, L., R. L. WILDER & P. CASALI. 1993. Human rheumatoid B-1a ($CD5^+$ B) cells make somatically hypermutated high affinity IgM rheumatoid factors. J. Immunol. **151:** 473–488.
16. IKEMATSU, H., M. T. KASAIAN, E. W. SCHETTINO & P. CASALI. 1993. Structural analysis of the V_H-D-J_H segments of human polyreactive IgG mAb: Evidence for somatic selection. J. Immunol. **151:** 3604–3616.
17. IKEMATSU, H., N. HARINDRANATH, Y. UEKI, A. L. NOTKINS & P. CASALI. 1993. Clonal analysis of a human antibody response. II. Sequences of the V_H genes of human IgM, IgG, and IgA to rabies reveal preferential utilization of V_HIII segments and somatic hypermutation. J. Immunol. **150:** 1325–1337.
18. KASAIAN, M. T., H. IKEMATSU, J. E. BALOW & P. CASALI. 1994. Structure of the V_H and V_L segments of monoreactive and polyreactive IgA autoantibodies to DNA in patients with systemic lupus erythematosus. J. Immunol. **152:** 3137–3151.
19. RIBOLDI, P., G. GAIDANO, E. W. SCHETTINO, T. G. STEGER, D. M. KNOWLES, R. DALLA-FAVERA & P. CASALI. 1994. Two AIDS-associated Burkitt's lymphomas produce specific anti-i IgM cold agglutinins utilizing somatically mutated V_H4–21 segments. Blood **83:** 2952–2961.
20. IKEMATSU, H., Y. ICHIYOSHI, E. W. SCHETTINO, M. NAKAMURA & P. CASALI. 1994. V_H and $V\kappa$ segment structure of anti-insulin IgG autoantibodies in patients with insulin-dependent diabetes mellitus. Evidence for somatic selection. J. Immunol. **152:** 1438–1441.
21. SCHROEDER, H. W., J. L. HILLSON & R. M. PERLMUTTER. 1987. Early restriction of the human antibody repertoire. Science **238:** 791–793.
22. CHEN, P. P., M.-F. LIU, C. A. GLASS, S. SINHA, T. J. KIPPS & D. A. CARSON. 1989. Characterization of two immunoglobulin V_H genes that are homologous to human rheumatoid factors. Arthritis Rheum. **32:** 72–76.
23. SHIN, E. K., F. MATSUDA, H. NAGAOKA, Y. FUKITA, T. IMAI, K. YOKOYAMA, E. SOEDA & T. HONJO. 1991. Physical map of the 3′ region of the human immunoglobulin heavy chain locus: Clustering of autoantibody-related variable segments in one haplotype. EMBO J. **10:** 3641–3645.
24. CAMPBELL, M. J., A. D. ZELENETZ, S. LEVY & R. LEVY. 1992. Use of family specific leader region primers for PCR amplification of the human heavy chain variable region gene repertoire. Mol. Immunol. **29:** 193–203.
25. ANDRIS, J. S., B. R. BRODEUR & J. D. CAPRA. 1993. Molecular characterization of human antibodies to bacterial antigens: Utilization of the less frequently expressed V_H2 and V_H6 heavy chain variable region gene families. Mol. Immunol. **30:** 1601–1616.
26. MATSUDA, F., E. K. SHIN, H. NAGAOKA, R. MATSUMURA, M. HAINO, Y. FUKITA, S. TAKA-ISHI, T. IMAI, J. H. RILEY, R. ANAND, E. SOEDA & T. HONJO. 1993. Structure and physical map of 64 variable segments in the 3′ 0.8-megabase region of the human immunoglobulin heavy-chain locus. Nature Genet. **3:** 88–93.
27. CHEN, P. P., M.-F. LIU, S. SINHA & D. A. CARSON. 1988. A 16/6 idiotype-positive anti-DNA antibody is encoded by a conserved V_H gene with no somatic mutation. Arthritis Rheum. **31:** 1429–1431.
28. BERMAN, J. E., S. J. MELLIS, R. POLLOCK, C. L. SMITH, H. SUH, B. HEINKE, C. KOWAL, U. SURTI, L. CHESS, C. R. CANTOR & F. W. ALT. 1988. Content and organization of the human IgV_H locus: Definition of three new V_H families and linkage to the Ig C_H locus. EMBO J. **7:** 727–738.

29. Kuppers, R., U. Fischer, K. Rajewsky & A. Gause. 1992. Immunoglobulin heavy and light chain gene sequences of a human CD5 positive immunocytoma and sequences of four novel V_HIII germline genes. Immunol. Lett. **34:** 57–62.
30. Chen, P. P. 1990. Structural analyses of human developmentally regulated V_H3 genes. Scand. J. Immunol. **31:** 257–267.
31. Rechavi, G., B. Bienz, D. Ram, Y. Ben-Neriah, J. B. Cohen, R. Zakut & D. Givol. 1982. Organization and evolution of immunoglobulin V_H gene subgroups. Proc. Natl. Acad. Sci. USA **79:** 4405–4409.
32. Tomlinson, I. M., G. Walter, J. D. Marks, M. B. Llewellyn & G. Winter. 1992. The repertoire of human germline V_H sequences reveals about fifty groups of V_H segments with different hypervariable loops. J. Mol. Biol. **227:** 776–798.
33. Lee, K. H., F. Matsuda, T. Kinashi, M. Kodaira & T. Honjo. 1987. A novel family of variable region genes of the human immunoglobulin heavy chain. J. Mol. Biol. **195:** 761–768.
34. Sanz, I., P. Kelly, C. Williams, S. Scholl, P. W. Tucker & J. D. Capra. 1989. The smaller human V_H gene families display remarkably little polymorphism. EMBO J. **8:** 3741–3748.
35. Weng, N., J. G. Snyder, L. Yu-Lee & D. M. Marcus. 1992. Polymorphism of human immunoglobulin V_H4 germline genes. Eur. J. Immunol. **22:** 1075–1082.
36. Humphries, C. G., A. Shen, W. A. Kuziel, J. D. Capra, F. R. Blattner & P. W. Tucker. 1988. A new human immunoglobulin V_H family preferentially rearranged in immature B-cell tumours. Nature **331:** 446–449.
37. Logtenberg, T., F. M. Young, J. H. van Es, F. H. J. Gmelig-Meyling & F. W. Alt. 1989. Autoantibodies encoded by the most J_H-proximal human immunoglobulin heavy chain variable region gene. J. Exp. Med. **170:** 1347–1355.
38. Jukes, T. H. & J. L. King. 1979. Evolutionary nucleotide replacements in DNA. Nature **281:** 605–606.
39. Straubinger, B., E. Huber, W. Lorenz, E. Osterholzer, W. Pargent, M. Pech, H. D. Pohlenz, F. J. Zimmer & H. G. Zachau. 1988. The human V_K locus: Characterization of a duplicated region encoding 28 different immunoglobulin genes. J. Mol. Biol. **199:** 23–34.
40. Daley, M. D., T. Olee, H.-Q. Peng, R. W. Soto-Gil, P. P. Chen & K. A. Siminovitch. 1992. Molecular characterization of the human immunoglobulin VλI germline gene repertoire. Mol. Immunol. **29:** 1031–1042.
41. Daley, M. D., H.-Q. Peng, V. Misener, X.-Y. Liu, P. P. Chen & K. A. Siminovitch. 1992. Molecular analysis of human immunoglobulin Vλ germline genes: Subgroups VλIII and VλIV. Mol. Immunol. **29:** 1515–1518.
42. Ermel, R. W., T. P. Kenny, P. P. Chen & D. L. Robbins. 1993. Molecular analysis of rheumatoid factors derived from rheumatoid synovium suggests an antigen-driven response in inflamed joints. Arthritis Rheum. **36:** 380–388.
43. Chen, P. P., K. Albrandt, T. J. Kipps, V. Radoux, F. Liu & D. A. Carson. 1987. Isolation and characterization of human VκIII germline genes. Implication for the molecular basis of human VκIII light chain diversity. J. Immunol. **139:** 1727–1733.
44. Silberstein, L. E., L. C. Jefferies, J. Goldman, D. Friedman, J. S. Moore, P. C. Nowell, D. Roelcke, W. Pruzanski, J. Roudier & G. J. Silverman. 1991. Variable region gene analysis of pathologic human autoantibodies to the related i and I red blood cell antigens. Blood **78:** 2372–2386.
45. Liu, M. F., D. L. Robbins, J. J. Crowley, S. Sinha, F. Kozin, T. J. Kipps, D. A. Carson & P. P. Chen. 1989. Characterization of four homologous L chain variable region genes that are related to 6B6.6 idiotype positive human rheumatoid factor L chain. J. Immunol. **142:** 688–694.
46. Kipps, T. J., E. Tomhave, P. P. Chen & D. A. Carson. 1988. Autoantibody-associated kappa light chain variable region gene expressed in chronic lymphocytic leukemia with little or no somatic mutation. Implications for etiology and immunotherapy. J. Exp. Med. **167:** 840–852.
47. Klobeck, H.-G., G. W. Bornkamm, G. Combriato, R. Mocikat, H.-D. Pohlenz & H. G. Zachau. 1985. Subgroup IV of human immunoglobulin κ light chain is encoded by a single germline gene. Nucl. Acid Res. **13:** 6515–6529.

48. CARROLL, W. L., M. YU, M. P. LINK & J. S. KORSMEYER. 1989. Absence of Ig V region gene somatic hypermutation in advanced Burkitt's lymphoma. J. Immunol. **143:** 692–698.

49. TSUNETSUSU-YOKOTA, Y., T. MINEKAWA, K. SHIGEMOTO, T. SHIRASAWA & T. TAKEMORI. 1992. Characterization of a new subgroup of human Ig Vλ cDNA clone and its expression. Mol. Immunol. **29:** 723–728.

50. OLEE, T., E. W. LIU, D.-F. HUANG, R. W. SOTO-GIL, M. DEFTOS, F. KOZIN, D. A. CARSON & P. P. CHEN. 1992. Genetic analysis of self-associating immunoglobulin G rheumatoid factors from two rheumatoid synovia implicates an antigen driven response. J. Exp. Med. **175:** 831–842.

51. SIMINOVITCH, K. A., V. MISENER, P. C. KWONG, Q.-L. SONG & P. P. CHEN. 1989. A natural antibody is encoded by germline heavy and lambda light chain variable region genes without somatic mutation. J. Clin. Invest. **84:** 1675–1678.

52. COMBRIATO, G. & H.-G. KLOBECK. 1991. Vλ and Jλ-Cλ gene segments of the human immunoglobulin λ light chain locus are separated by 14 kb and rearrange by a deletion mechanism. Eur. J. Immunol. **21:** 1513–1522.

53. YAMASAKI, N., S. KOMORI & T. WATANABE. 1987. Complementary DNA for a human subgroup IV immunoglobulin λ-chain. Mol. Immunol. **24;** 981–985.

54. SHLOMCHIK, M. J., A. H. AUCOIN, D. S. PISETSKY & M. G. WEIGERT. 1987. Structure and function of anti-DNA autoantibodies derived from a single autoimmune mouse. Proc. Natl. Acad. Sci. USA **84:** 9150–9154.

55. AMIT, A. G., R. A. MARIUZZA, S. E. V. PHILLIPS & R. J. POLJAK. 1986. Three-dimensional structure of an antigen–antibody complex at 2.8-Å resolution. Science **233:** 747–753.

56. WARD, E. S., D. GUSSOV, A. D. GRIFFITHS, P. T. JONES & G. WINTER. 1989. Binding activities of a repertoire of single immunoglobulin variable domains secreted from *Escherichia coli.* Nature **341:** 544–546.

57. RADIC, M. Z., M. A. MASCELLI, J. ERIKSON, H. SHAN & M. WEIGERT. 1991. Ig H and L chain contributions to autoimmune specificities. J. Immunol. **146:** 176–182.

58. KABAT, E. A. & T. T. WU. 1991. Identical V region amino acid sequences and segments of sequences in antibodies of different specificities: Relative contributions of V_H and V_L genes, minigenes, and complementarity-determining regions to binding of antibody-combining sites. J. Immunol. **147:** 1709–1719.

59. BETZ, A. G., R. CRISTINA, R. PANNELL, C. MILSTEIN & M. S. NEUBERGER. 1993. Passenger transgenes reveal intrinsic specificity of the antibody hypermutation mechanism: Clustering, polarity, and specific hotspots. Proc. Natl. Acad. Sci. USA **90:** 2385–2388.

60. BETZ, A. G., M. S. NEUBERGER & C. MILSTEIN. 1993. Discriminating intrinsic and antigen-selected mutational hotspots in immunoglobulin V genes. Immunol. Today **14:** 405–411.

Comparison of the Human Germline and Rearranged V_H Repertoire Reveals Complementarity between Germline Variability and Somatic Mutation

GERALD WALTER,[a] IAN M. TOMLINSON,[b,c]
PAUL H. DEAR,[b] ERIK L. L. SONNHAMMER,[b,d]
GRAHAM P. COOK,[b,c] AND GREG WINTER[b,c]

[a]*Cambridge Antibody Technology Ltd.*
Science Park
Melbourn SG8 6EJ
United Kingdom

[b]*MRC Laboratory of Molecular Biology*
Hills Road
Cambridge CB2 2QH
United Kingdom

[c]*MRC Centre for Protein Engineering*
Hills Road
Cambridge CB2 2QH
United Kingdom

[d]*Sanger Centre*
Hinxton Hall
Cambridge CB10 1RQ
United Kingdom

We have recently characterized 88 human germline V_H segments in a single individual[1] and completed the map of the human V_H locus (1.1 Mb) on chromosome 14q32.3.[2] We have now turned our attention to the somatic processes that shape the human V_H repertoire. We have sequenced 142 rearranged genes derived from peripheral blood lymphocyte cDNA of our individual (DP) to answer the following questions:

1. Is the usage of V_H segments in rearranged heavy chain genes biased according to their positions on the map of the locus?
2. Is the frequency of somatic mutations influenced by the isotype switch from IgM to IgG?
3. Is the variability created by somatic point mutations confined to the hypervariable loops H1 and H2[3] and does it coincide with the germline variability?

By sequencing germline and rearranged V_H segments from the same individual, we have excluded germline polymorphisms from our DP data. We have also used a specifically designed software package (Sonnhammer *et al.*, manuscript in preparation) for assessing questions (1) through (3) using the EMBL DNA sequence database.

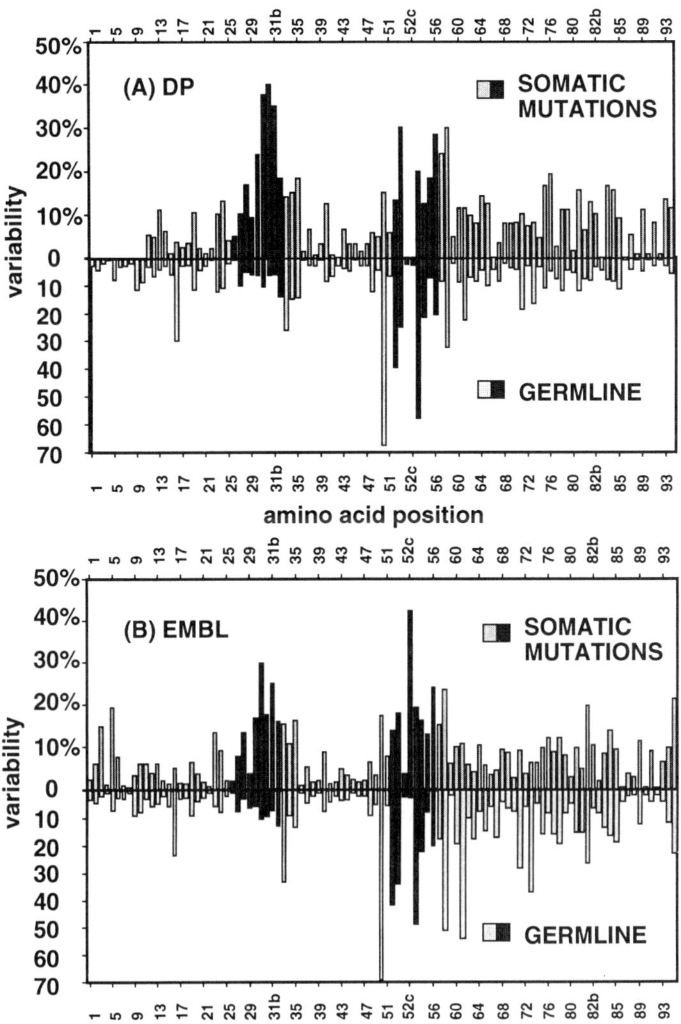

FIGURE 1. Variability of amino acid residues due to somatic point mutations compared to the variability in their germline counterparts scanned over the length of the V_H segments from our individual (DP) **(A)** and the EMBL sequence database **(B)**. The numbering is according to Kabat.[5] The dark-shaded residues form the loops H1 (position 26–32) and H2 (position 52–56),[3] which together with some neighboring residues are the regions of highest variability. *Methods*: Variability due to somatic mutations was determined for each position of the V_H segments as the number of changes per amino acid residue divided by the total number of relevant sequences at that position using purpose-written software (Dear, unpublished). For the germline variability, a score was calculated at each position as the number of different amino acids divided by the percentage frequency of occurrence of the most common amino acid.[5] The numbers of sequences scanned were 143 **(A)** and 1038 **(B)** for DP and EMBL, respectively.

CONCLUSIONS

Our data indicate the following:

1. The overall usage of V_H segments in rearranged heavy chain genes is not biased according to their positions on the map of the locus. Some V_H segments are preferentially expressed in DP (n = 143) and/or highly represented in the EMBL sequence database (n = 1038), as for example 6–1 (DP-74), 1–18 (DP-14), 3–23 (DP-47), and 1–69 (DP-10). Only one rearranged pseudogene (derived from DP-34 with one amino acid change at position 51) was found in the EMBL sequence database. V_H segments located on chromosome 15 or 16[4] were never found rearranged, nor were there any rearranged pseudogenes in DP.
2. The isotype switch from IgM to IgG has a profound influence on the overall frequency of somatic mutations as the average numbers of amino acid changes per V_H segment were calculated to be 2.7 for IgM and 10.1 for IgG.
3. The variability created by somatic point mutations clearly marks the extremities of the hypervariable loops H1 and H2 as well as some adjoining residues involved in shaping antigen-binding sites. It does coincide with the germline variability in H2 but not in H1 where the germline variability is low. That means that the pattern of somatic mutation is complementary to variability encoded by germline V_H segments, thereby maximizing the diversity required for antigen recognition (FIG. 1).

REFERENCES

1. TOMLINSON, I. M., G. WALTER, J. D. MARKS, M. B. LLEWELYN & G. WINTER. 1992. J. Mol. Biol. **227:** 776–798.
2. COOK, G. P., I. M. TOMLINSON, G. WALTER, H. RIETHMAN, N. P. CARTER, L. BULUWELA, G. WINTER & T. H. RABBITTS. 1994. Nature Genet. **7:** 162–168.
3. CHOTHIA, C., A. M. LESK, E. GHERARDI, I. M. TOMLINSON, G. WALTER, J. D. MARKS, M. B. LLEWELYN & G. WINTER. 1992. J. Mol. Biol. **227:** 799–817.
4. TOMLINSON, I. M., G. P. COOK, N. P. CARTER, R. ELASWARAPU, S. SMITH, G. WALTER, L. BULUWELA, T. H. RABBITTS & G. WINTER. 1994. Hum. Mol. Genet. **3:** 853–860.
5. KABAT, E. A., T. T. WU, H. M. PERRY, K. S. GOTTESMAN & C. FOELLER. 1991. Sequences of Proteins of Immunological Interest. 5th edit. USDHHS, NIH. Bethesda, MD.

Analysis of Somatic Mutation with a Reporter Gene Construct[a]

ESTHER M. CHUN, PATTY ZWOLLO, PHILLIP WONG, CHERI LOVE, ANTHONY DOFT, AND ROBERTA R. POLLOCK[b]

Department of Biology
Occidental College
Los Angeles, California 90041

Somatic mutation of immunoglobulin (Ig) variable region (V) genes plays an important role in the generation of antibody diversity. However, this process has been very difficult to study using *in vitro* approaches. One reason has been the difficulty in designing constructs that undergo "normal" somatic mutations; another is that very few cell lines show somatic mutation. Recent studies using animals expressing Ig transgenes have demonstrated that such mice provide excellent model systems for studying somatic mutation.

We are developing an assay to detect and isolate cells undergoing somatic mutation *in vivo* (in transgenic mice) and *in vitro* (in cell lines). Our assay uses a DNA construct in which somatic mutation in a rearranged Ig heavy (H) chain V gene results in expression of a *lac*Z reporter gene. The construct encodes a fusion protein that pairs a rearranged Ig H chain V region (VDJ), containing a stop codon, with the *lac*Z reporter gene.[1] The VDJ is 5′ of the *lac*Z; its stop codon therefore prevents translation of the *lac*Z. If the stop codon is reverted by somatic mutation, the entire fusion protein is translated, and *lac*Z expressed (FIG. 1**A**). The initial constructs contained the IgH chain intron enhancer, and new constructs have been made that contain both the intron and 3′ α enhancers (FIG. 1**B**). β-Galactosidase (β-gal), the *lac*Z gene product, can be readily detected by chromogenic or fluorescence assays.[2]

This construct has been transfected into the 18-8 pre-B-cell line, which undergoes somatic mutation *in vitro*. Integration of the construct was verified by Southern blot and polymerase chain reaction (PCR) analysis, and expression of the construct was verified by Northern blot and reverse transcriptase-PCR (RT-PCR). Expression of β-gal in cells expressing control constructs (in which the stop codon had been mutated by site-specific mutagenesis) was demonstrated by two assays.

The first assay uses the chromogenic substrate chloro-phenol-red-β-D-galactopyranoside (CPRG), which is converted by β-gal into a red product that is easily quantitated in an ELISA plate reader was a 570-nm filter. Sensitive detection of β-gal on the single-cell level can be done with the fluorogenic substrate, fluorescein di-β-D-galactoside (FDG), and a FACS (FIG. 2). The use of the FACS allows cells to be sorted on the basis of their *lac*Z expression. FDG is a vital stain, so positive cells can be recovered and grown in culture.

[a] This work was supported by National Institutes of Health Grant AI30020 and National Science Foundation Grant CAA MCB-9210186 to R.R.P.

[b] Address for correspondence: Roberta R. Pollock, Ph.D., Department of Biology, Occidental College, 1600 Campus Rd., Los Angeles, CA 90041-3314.

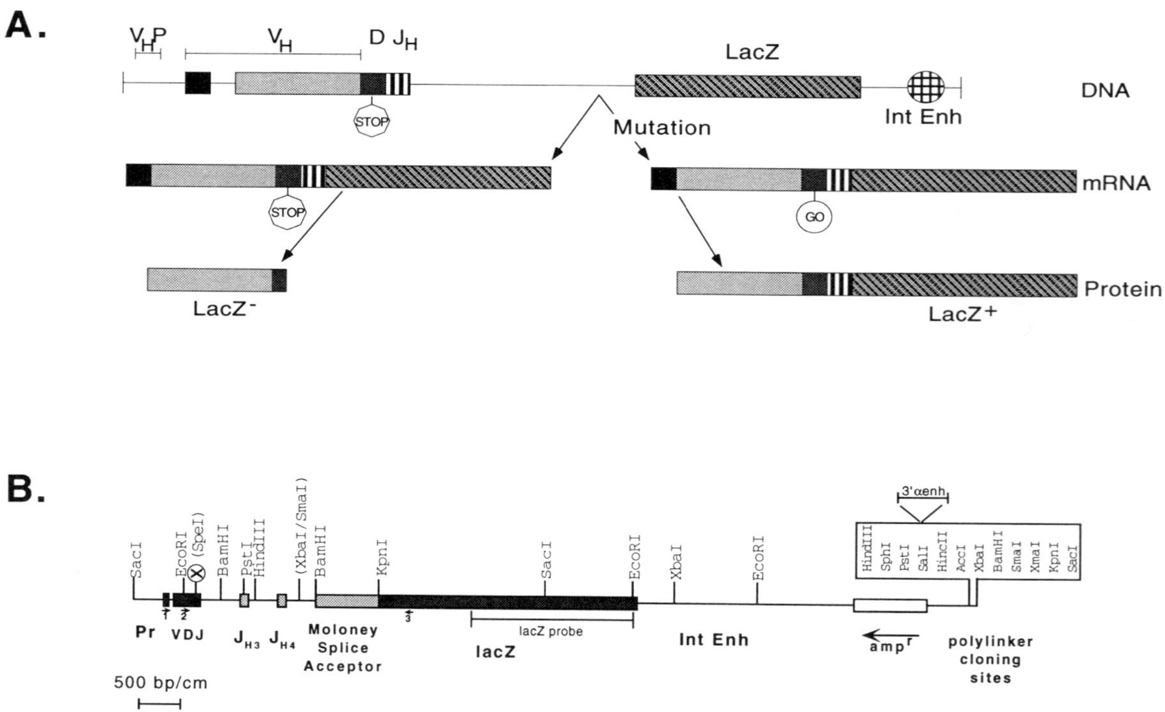

FIGURE 1. **(A)** Diagram of the overall approach to using *lac*Z as a reporter gene to study somatic mutation. **(B)** The pHIL experimental construct. The *Bsp*DI/*Eco*RI fragment used for probing blots is indicated. Primers used for PCR are indicated by small arrows and numbers. The sequences are as follows: Primer 1 (forward primer, leader exon): 5′TTGGTTTTCCTTGTCCTTATTTTA3′; Primer 2 (forward primer, VDJ exon): 5′GGTGGAGTCTGGGGGAGGCTTAGT3′; Primer 3 (reverse primer, *lac*Z): 5′AGCTGGCGAAAGGGGGATGTGCTG3′.

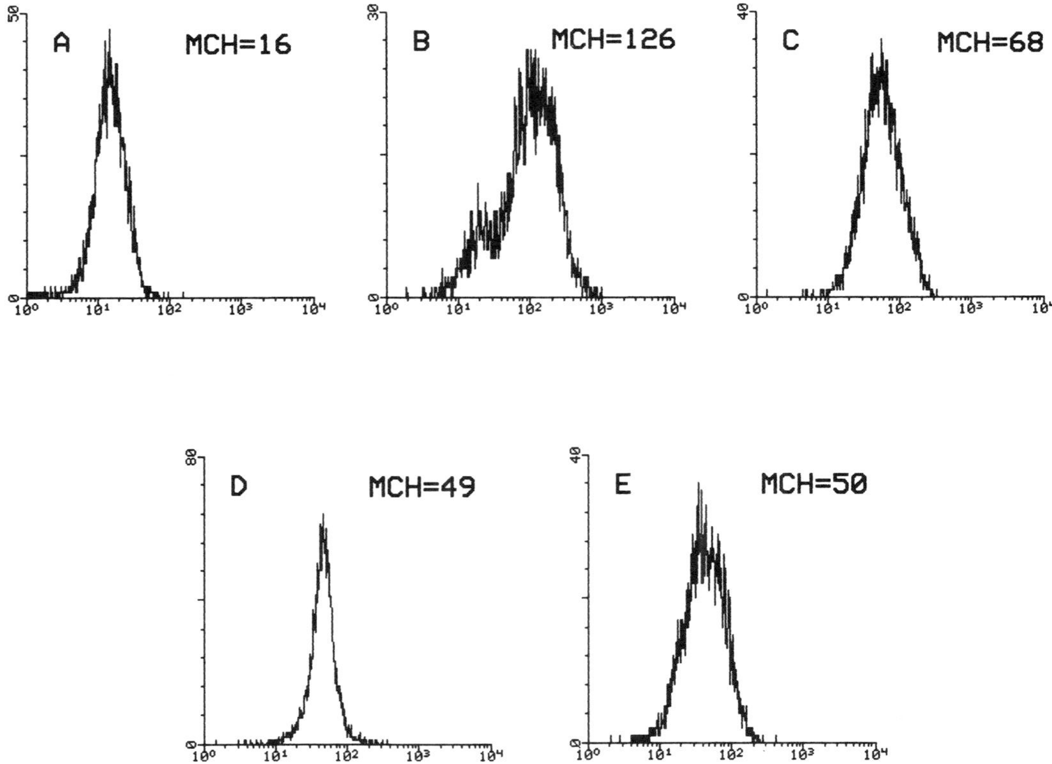

FIGURE 2. FDG labeling and FACS analysis of 18-8 cells transfected with *lac*Z constructs. FDG was loaded into cells by hypotonic shock, then cells were incubated for 30 min on ice. The reaction was stopped by adding an inhibitor, and the samples analyzed on a FACScan. MCH stands for mean channel of fluorescence. **(A)** 18-8 cells; **(B–D)** independently arising mutants from SSIII.1C10, 18-8 cells transfected with the pHIL experimental construct; **(E)** mutant arising from the SSIII.2D5, 18-8 cells transfected with the pHIL experimental construct.

We have used these assays to isolate several apparent mutants, from two different subclones, which now express *lac*Z (FIG. 2). Sequence analysis is under way to determine the nature of these mutations. Transgenic mice containing a control construct in which the stop codon has been reverted by site-specific mutagenesis have been made. When these transgenic lines are established, mice will be tested for *lac*Z expression. Once the assays are optimized in transgenic mice, we will make transgenic mice expressing the experimental construct.

ACKNOWLEDGMENTS

The authors would like to thank Drs. Bill Kerr and Len Herzenberg for providing the pACLAC and pcH110 plasmids and the FDG protocol.

REFERENCES

1. KERR, W. G., G. P. NOLAN, A. T. SERAFINI & L. A. HERZENBERG. 1989. Cold Spring Harbor Symp. Quant. Biol. **LIV:** 767–776.
2. NOLAN, G. P., S. FIERING, J.-F. NICOLAS & L. A. HERZENBERG. 1988. Proc. Natl. Acad. Sci. USA **85:** 2603–2607.

Mechanistic Properties of Immunoglobulin Somatic Mutation

JOHN E. FORD,[a] MICHAEL G. McHEYZER-WILLIAMS,[b] AND MICHAEL R. LIEBER[c]

Department of Pathology
Stanford University School of Medicine
Stanford, California 94305-5324
and
Department of Pathology
Washington University School of Medicine
St. Louis, Missouri 63110

Antibody diversity is generated through processes associated with variable (V), diversity (D), and joining (J) region recombination and subsequent peripheral antigen-driven somatic mutation. Although many aspects of V(D)J recombination are fully characterized, very little is known concerning the process of somatic mutation. Recent advances have been made in delineating the *cis*-directing sequences[1] and cellular microenvironment[2] necessary for this process. Nevertheless, the molecular mechanism responsible for specifically targeting nucleotide substitutions, deletions, and insertions to rearranged immunoglobulin (Ig) genes remains enigmatic.

Two major models have been proposed to explain the antigen-driven introduction of mutations into mammalian Ig genes. Gene conversion, involving the unidirectional, templated nucleotide transfer from homologous donor sequences (other V regions) has been proposed. A second model involves error-prone DNA synthesis as the underlying mechanism. To determine if gene conversion is the mechanism of murine somatic mutation, we examined the nucleotide changes at the lambda light chain locus from an early time point of a primary response (12 days post-immunization, to avoid biases in the frequency of certain mutations caused by antigenic selection) to the hapten (4-hydroxy-3-nitrophenyl)acetyl linked to a protein carrier. The murine lambda locus was chosen because the V-region repertoire is small, consisting of only three subexons, facilitating the identification of conversion events. Single germinal center cells (binding peanut agglutinin, expressing B220 and lambda light chain but low levels of surface IgD) were sorted from the spleens of immunized mice and individual lambda light chain genes were amplified and sequenced. Single-cell amplification was employed to avoid artifacts created when DNA from populations of cells are used as template.[3] None of the documented mutations were introduced through templated V-region conversion.[4] A recent study examining kappa light chain genes (another fully characterized Ig loci)

[a] J.E.F. was supported by a National Institutes of Health training grant (NIH 5T32AI07290). Present address: DNAK, Palo Alto, CA.

[b] Present address: Department of Microbiology and Immunology, Stanford University, Stanford, California 94305.

[c] Corresponding author.

from human splenic cDNA libraries also indicated that V-region gene conversion is not the basis for somatic mutation.[5]

Inferences about the nature of the somatic mutation machinery can only be made from unselected mutations. Unselected data sets have been obtained from passenger transgene studies[6] and by collating sets of silent mutations.[7] Mutations from early primary immune responses also demonstrate hallmark features of being unselected. No shared replacement mutations were observed in our study, which documents 38 mutations in 15 different V regions.[4] The replacement-to-silent ratio for these mutations is less than the ratio expected by randomly placed mutations, also indicating a lack of preferential expansion of certain variant clones by selective processes. Salient features of these unselected mutations include a transition to transversion ratio of 0.82 : 1 (resembling a value of 1 : 1, calculated for the eucaryotic DNA polymerase α[8]), a purine to pyrimidine ratio of 1.55 : 1 (which does not significantly differ by chi-squared analysis from a ratio of 1 : 1, therefore offering no evidence for a strand bias), and a propensity to accumulate unselected mutations in the third complementarity-determining region. The mutations were clustered; six out of seven V-regions containing more than one mutation harbor at least two mutations within ten codons of one another. Lastly, position 90 represents a putative hot spot, as two independent silent mutations occur at that position (the likelihood of this occurring by chance is 3.4×10^{-4}). The characteristics of these mutations are consistent with other unselected data sets[6,7] and error-prone DNA synthesis models.

REFERENCES

1. Betz, A. G., C. Milstein, A. González-Fernández, R. Pannell, T. Larson & M. Neuberger. 1994. Cell **77:** 239–248.
2. Küppers, R., M. Zhao, M.-L. Hansmann & K. Rajewsky. 1993. EMBO J. **12:** 4955–4967.
3. Ford, J. E., M. G. McHeyzer-Williams & M. R. Lieber. 1994. Gene **142(2):** 279–283.
4. Ford, J. E., M. G. McHeyzer-Williams & M. R. Lieber. 1994. Eur. J. Immunol. **24:** 1816–1822.
5. Klein, R., R. Jaenichen & H. G. Zachau. 1993. Eur. J. Immunol. **23:** 2348–2371.
6. Betz, A. G., C. Rada, R. Pannell, C. Milstein & M. S. Neuberger. 1993. Proc. Natl. Acad. Sci. USA **90:** 2385–2388.
7. Golding, G. B., P. J. Gearhart & B. W. Glickman. 1987. Genetics **115:** 169–176.
8. Steele, E. J. 1991. Somatic Hypermutation in V-Regions. CRC Press. Boca Raton, FL.

Variable-Region Gene Analysis of B-Cell Subsets Derived from a 4-Year-Old Child

Somatically Mutated Memory B Cells Accumulate in the Peripheral Blood Already at Young Age

ULF KLEIN, RALF KÜPPERS, AND KLAUS RAJEWSKY

Institute for Genetics
University of Cologne
Weyertal 121
50931 Cologne, Germany

Most of the previous studies on the B-cell repertoire of normal persons concentrated on the analysis of only a single B-cell subpopulation. We sought to characterize B-cell subsets differing in terms of cellular maturation derived from a single person at the level of rearranged V genes. The analysis comprised tonsillar "naive" IgM^+IgD^+ B cells and germinal center (GCC) B cells and IgG^- (memory) as well as IgM-expressing peripheral blood (PB) B cells derived from a 4-year-old child.[1] PB B cells were separated into conventional $CD5^-$ and $CD5^+$ B cells. Analyzed were rearranged V-region genes of the V_H4 gene family that has recently been extensively characterized.

RESULTS

Fifty V_H4 sequences could be assigned to one of eight published V_H4 germline genes (TABLE 1). Some of the sequences were more homologous to one another than to published germline genes and therefore most likely correspond to novel genes or alleles ([polymorphic] p3D279D and $V_H4.33r$ [related]). The V_H4 sequence of clone CD5/4 might correspond to a previously unidentified V_H4 gene and was therefore not considered in the mutation analysis. TABLE 1 summarizes the nucleotide differences in the rearranged V_H4 and J_H genes compared to the germline genes to which they showed highest homology. TABLE 2 shows the V_H4 gene mutation frequencies determined for the respective fractions.

DISCUSSION

In accordance with our earlier data,[2] both tonsillar IgM^+IgD^+ cells and the majority of IgM-expressing $CD5^+$ and $CD5^-$ PB B cells showed no or little somatic mutation. Tonsillar GCC and IgG-expressing PB B cells carried a high load of somatic mutations in their V-region genes. The extent of somatic mutation in IgG^+ cells from the child investigated in the present work and IgM^-IgD^- cells (mostly IgG^+ B cells) from a 67-year-old person[2] is in a similar range (3.2% versus 3.9%),

TABLE 1. Summary of the Sequence Analysis of V_H4-Encoded Transcripts Derived from Isolated Tonsillar and PB B-Cell Subpopulations of a 4-Year-Old Child

Tonsillar B Cell Subsets	Clones	V_H4	J_H	Sum of V_H & J_H Nucl. Diff.	PB B Cell Subsets	Clones	V_H4	J_H	Sum of V_H & J_H Nucl. Diff.
Tonsillar IgD^+ B cells [IGD] ($IgD^+/CD20^+$)	IGD1	4.21	3b	3	IgM^+, $CD5^+$ PB B cells [CD5] ($CD5^+/CD20^+$)	CD5/1	p3D279D	4b	0
	IGD2*	V71-4	4b	0		CD5/2	V71-4	5b	5
	IGD3	4.21	4b	1		CD5/3	4.21	3b	3
	IGD4	V71-4	3b	3		CD5/4	V71-4	5b	19
	IGD5	4.21	6b	0		CD5/5	4.21	6c	0
	IGD6	V71-4	3b	3		CD5/6	4.33	4b	5
	IGD8	4.21	3b	1		CD5/7	4.21	4b	0
	IGD9	4.21	6b	0		CD5/8	V2-1	4b	3
	IGD10	3D279D	4b	0					
Tonsillar IgM^+ GCC [GCM] ($PNA^+/CD20^{++}$)	GCM1	3D279D	6b	6	IgM^+, $CD5^-$ PB B cells [CBM] ($CD5^-/CD20^+$)	CBM1	4.21	4b	6
	GCM2	4.21	6c	3		CBM2	4.34	2	2
	GCM3	V71-4	4b	10		CBM3	4.21	4b	4
	GCM4*	V71-4	4b	2		CBM4	3D216D	3b	0
	GCM5	3D279D	3b	3		CBM5	V71-4	6b	1
	GCM6	4.21	4b	8		CBM6	VIV-4	5b	12
	GCM7	4.21	2	6		CBM7	V71-4	4b	0
	GCM8	p3D279D	3b	1		CBM8	V2-1	2	2
Tonsillar IgG^+ GCC [GCG] ($PNA^+/CD20^{++}$)	GCG1	VIV-4	6b	1	. IgG^+, $CD5^-$ PB B cells [CBG] ($CD5^-/CD20^+$)	CBG1	3D279D	4b	2
	GCG2	4.34	4b	3		CBG2	4.33r	2	6
	GCG3	4.33	6c	12		CBG3	3D279D	3b	11
	GCG4	p3D279D	6c	7		CBG4	V2-1	4b	10
	GCG5	4.21	4b	4		CBG5	4.21	4b	26
	GCG6	4.33r	3b	14		CBG6	V71-4	3b	12
	GCG7	3D216D	4b	6		CBG7	V2-1	2	2
	GCG8	p3D279D	4b	16		CBG8	p3D279D	3b	3
						CBG9	V71-4	3b	5

NOTE: The clones were designated according to the B-cell subset from which they were derived and, in the case of the GCC and conventional ($CD5^-$) B cells, also according to the immunoglobulin isotype expressed by the cells. The surface marker phenotype of the respective cell populations is shown in parentheses. Indicated are the V_H4 ($V_H4.33r$ and p3D279D see text) and J_H germline genes to which the V_H4 and J_H sequences showed highest homology and the sum of nucleotide differences of the combined V_H4 and J_H sequences. An asterisk indicates a nonproductive rearrangement. All sequences had unique V_H-D_H/DIR-J_H junctions. These sequence data are available from EMBO/GenBank/DDBJ under accession numbers Z30544-563, Z30573-581, Z30590-600, and Z30669-678.

TABLE 2. Summary of Mutations in Rearranged V_H4 Genes Expressed in Tonsillar and PB B-Cell Subsets Derived from a 4-Year-Old Child

Origin	Cell Population	n^a	V_H4 Range	V_H4 % Mutation
Tonsil	IGD	9	0–3	0.4
	GCM	8	1–10	2.0
	GCG	8	1–13	3.3
Peripheral blood	CD5 & CBM[b]	10	0–3	0.35
	CD5[c] & CBM[b]	5	3–11	2.6
	CBG	9	1–22	3.2

[a] Number of clones.
[b] V_H4 transcripts derived from IgM-expressing $CD5^+$ and $CD5^-$ PB B cells were analyzed together; mutation frequencies were determined for V_H4 transcripts that showed (i) three or less, and (ii) four or more mutations relative to the corresponding V_H4 and J_H germline genes.
[c] The V_H4 sequence of clone CD5/4 was excluded from the mutation analysis. Cell populations are designated as in TABLE 1.

indicating that the mutation frequency in PB memory B cells does not increase significantly with age. Furthermore, the V_H4 mutation frequencies determined for IgG-expressing GCC and IgG^+ PB B cells were in the same range (3.3% and 3.2%). These observations collectively suggest that long-lived memory cells do not acquire additional somatic mutations after they have left the GC.

A fraction of IgM-bearing PB B cells carried somatically mutated V-region genes. Somatically mutated μ chain transcripts have previously been found in PB B lymphocytes from adults[3,4] and are most likely to be derived from IgM^+IgD^- cells.[2] The mutation frequency determined for V_k3 transcripts of IgM^+IgD^- PB B cells[2] is similar to that determined for V_H4 transcripts of IgM^+IgD^- GCC (about 2.0% in both cases). These observations indicate that the IgM^+IgD^- PB B cell subpopulation represents GC-derived B cells that have left the GC at any early stage of the GC reaction without undergoing isotype switching.

In summary, the human PB B-cell repertoire is composed of at least three B-cell subsets that can be distinguished according to their isotype expression and according to the extent of somatic mutation found in their rearranged V genes.

REFERENCES

1. KLEIN, U., R. KÜPPERS & K. RAJEWSKY. 1994. J. Exp. Med. **180:** 1383–1393.
2. KLEIN, U., R. KÜPPERS & K. RAJEWSKY. 1993. Eur. J. Immunol. **23:** 3272.
3. VAN ES, J. H., F. H. J. GMELIG-MEYLING & T. LOGTENBERG. 1992. Eur. J. Immunol. **22:** 2761.
4. HUANG, C., A. K. STEWART, R. S. SCHWARTZ & B. D. STOLLAR. 1992. J. Clin. Invest. **89:** 1331.

Analysis of Hypermutation in a Mixed T-Cell Receptor β/Immunoglobulin Heavy Chain Passenger Transgene

J. M. JOHNSTON,[a] S. R. IHYER,[a] K-F. TAI,[a] R. S. SMITH,[a]
N. L. NADON,[b] AND W. L. CARROLL[a,c]

[a]*Departments of Pediatrics and Human Molecular Biology and Genetics University of Utah School of Medicine Salt Lake City, Utah 84112*

[b]*Department of Biology University of Tulsa Tulsa, Oklahoma 74104-3189*

The human B-cell antibody repertoire is estimated to contain 10^{10} or more distinct molecules. In mice and humans, antibody diversity and affinity are maximized by a process of somatic hypermutation, which targets rearranged Ig loci after antigen exposure.[1] Despite extensive study, the molecular mechanism responsible for Ig gene hypermutation is unknown. To characterize the *cis*-acting elements critical to the targeting of hypermutation, we have established a series of mouse lines transgenic for constructs containing different portions of an Ig heavy chain (IgH) locus. We have previously demonstrated that a rearranged VDJ segment and approximately 4 kb of immediately surrounding promoter and enhancer sequences constitute a target for a low but significant rate of mutation in hybridomas derived from hyperimmunized transgenic mice (Johnston *et al.*, submitted). Additional 3′ IgH sequences, perhaps including the 3′ enhancer, appear to be necessary for maximal mutation rates.[2]

We have now analyzed mutation in a transgene (FIG. 1**A**), consisting of a T cell receptor V(D)J embedded within an IgH minigene containing both the intron and 3′ enhancers. A nonsense mutation prevents translation of the transgenic antibody message. Five founder lines have been established, and this report describes preliminary analysis of a line containing three copies of the minigene.

In order to examine the transgene after circulation through the *in vivo* cell compartment associated with Ig hypermutation, we analyzed mutation in antigen-specific B cells rescued as hybridomas from mice hyperimmunized with the hapten 4-hydroxy-3-nitrophenyl acetyl (NP). Expression was analyzed by Northern blot and PCR using primers spanning the intron between the TCR region and Cα. Amplification of the transgene TCR variable region was performed with *Taq* or *Pfu* polymerase. Following 30 cycles of amplification, PCR products were cloned into BlueScript phagemid vectors and sequenced (Stratagene). Genuine mutations were distinguished from amplification errors by their repeated isolation from different PCRs.

[c] Address for correspondence: William L. Carroll, M.D., Departments of Pediatrics and the Eccles Program in Human Molecular Biology and Genetics, University of Utah, Room 3260, Building 533, Salt Lake City, UT 84112.

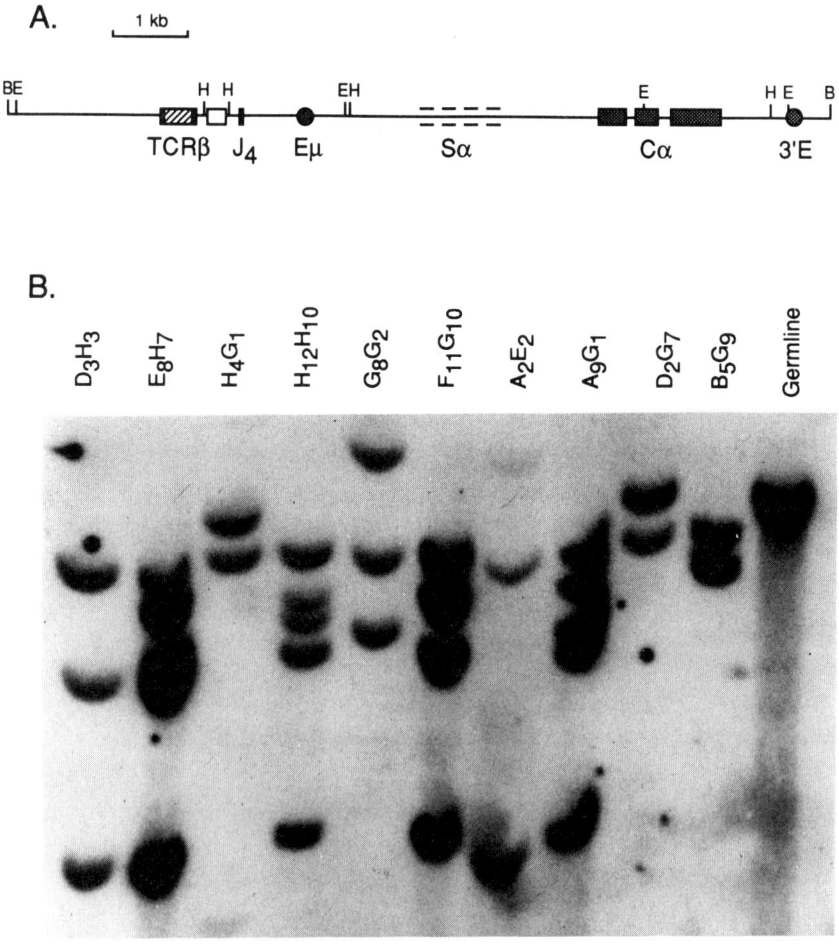

FIGURE 1. (A) The 2.1CE transgene. Restriction endonuclease sites are shown (B = *Bss*HII, E = *Eco*RI, H = *Hin*dIII). The TCR variable region is cross-hatched and was derived from cF3-42 (an alloreactive cytotoxic T lymphocyte clone). The TCR is flanked by approximately 2 kb of sequence 5′ of the murine IgH leader segment; approximately 1.6 kb of immediately 3′ IgH sequences, including a downstream J_4 segment, the intronic enhancer and 5′ matrix attachment region; the murine alpha switch/constant region; and the IgH 3′ enhancer, PCR amplified from Wistar rat genomic DNA. Important segments within the IgH locus are stippled; additional IgH sequences are shown as a horizontal line.

(B) Southern blot of hybridoma DNA. Genomic DNA from 10 hybridomas is shown, with normal mouse DNA in the right-hand lane, as a control. Only five hybridomas (D_3H_3, E_8H_7, $H_{12}H_{10}$, F_{11}/G_{10}, and A_9G_1) retain detectable levels of the transgene, as demonstrated by the two fastest migrating bands. Similarly, reverse transcription-PCR of hybridoma RNA revealed expression of the transgenic TCR-Cα message in only these same five hybridomas (data not shown). All hybridomas exhibit nongermline bands, corresponding to the expected rearranged anti-NP alleles.

Five of ten hybridomas contained the transgene (FIG. 1**B**). Expression from the transgene was limited, because Northern blot failed to identify a TCR transcript, whereas PCR detected a correctly spliced product. To date, we have examined subclones from four hybridomas, and we have failed to identify mutations other than those attributable to amplification error. Thus the TCR does not appear to be a target for hypermutation similar to endogenous Ig loci. This suggests that sequences within the Ig V(D)J itself or additional elements, not included in our minigene, may be essential to confer maximal mutation rates. However, ongoing analysis of additional hybridomas from this and other lines will be required to determine the true incidence of mutation within the passenger TCR transgene.

REFERENCES

1. FRENCH, D. L., R. LASKOV & M. D. SCHARFF. 1989. The role of somatic mutation in the generation of antibody diversity. Science **244:** 1152–1157.
2. SHARPE, M. J., C. MILSTEIN, J. M. JARVIS & M. S. NEUBERGER. 1991. Somatic mutation of immunoglobulin κ may depend on sequences 3′ of Cκ and occurs on passenger transgenes. EMBO J. **10:** 2139–2145.

Pattern of Usage of the V_H4-21 Gene by B Lymphocytes in a Patient with EBV Infection Indicates Ongoing Mutation and Class Switching[a]

CAROLINE CHAPMAN,[b] MYFANWY SPELLERBERG, TERRY HAMBLIN, AND FREDA STEVENSON

Molecular Immunology Group
Tenovus Laboratory
Southampton University Hospitals
Southampton SO16 6YD, United Kingdom

The V_H4-21 gene is a single member of the V_H4 family and is used to encode autoantibodies, particularly autoanti-red cell antibodies. In patients infected with EBV, there is a selective rise in V_H4-21-encoded Ig in serum. We have investigated usage of this V_H gene by circulating B cells from a typical post-infective patient. Evidence for activation of the unmutated germline gene, for intraclonal ongoing somatic mutation and for Ig class switching is presented.

METHODS

Blood lymphocytes were prepared from a patient with typical post-EBV infectious mononucleosis, and a cell line was established. Usage of the V_H4-21 gene was investigated by, first, using monoclonal anti-Id 9G4 to select and clone cells hybridized from the blood cells; the V_H gene was then analyzed by PCR. Second, the V_H4-21 gene was amplified by PCR directly from the line. Primers were for V_H4 or V_H4-21, with $C\mu$, $C\gamma$, or $C\alpha$, and for intraclonal analysis, a nested PCR with a CDR3-specific primer.

RESULTS

The patient had a high serum level of 9G4-positive idiotypic Ig, indicating V_H4-21 gene activation. Id-positive hybridomas were established, with eight out of nine secreting IgM, four out of six of which had anti-red cell (i) activity; one in nine of the hybridomas secreted IgG, and all used the V_H4-21 gene. Direct amplification of the cell line revealed an additional seven clonally distinct V_H4-21-encoded IgM sequences, two more IgG, but no IgA sequences. The major-

[a] These studies were supported by Tenovus and the Wessex Medical Trust, United Kingdom.

[b] Address for correspondence: Molecular Immunology Group, Tenovus Laboratory, Southampton University Hospitals, Tremona Road, Southampton SO16 6YD, U.K.

```
                                           |-CDR1--|                        |------CDR2------|                                                  |------CDR3-----|
VH4-21 GL       QVQLQQWGAGLLKPSETLSLTCAVYGGS FSGYYWS WIRQPPGKGLEWIG EINHSGSTNYNPSLKS RVTISVDTSKNQFSLKLSSVTAADTAVYYCAR
S10A+A2+B1+C3
E3+F8+G8+H7     ------------------s------D-- ------- -------------A e-S-T---T------- -----I---k-----k--s-------v----r PHYCSATTCSGPVE Y JH4
A9+B18           ----------------s------D-- ------- -------------A e-S-----T------- -----I---k-----k--s-------v----r -------------- - ---
A4+H2            ----------------s------D-- ------R -------------A e-S-T---T------- -----I---k-----k--s-------v----r -----------
E1+F1            ----------------s------D-- ------- -----------C-A e-S-T---T------- -----I---k-----k--s-------v----r -----------
G2+F10           ----------------s------D-- ------- -------------A e-S-T---T------- *----I---k-----k--s-------v----r -----------
```

FIGURE 1. Clonal heterogeneity of the V_H4-21 gene used by IgM-positive blood B lymphocytes of a patient with EBV infection. Deduced amino acid sequences of the gene in a hybridoma (S10A) and in the mixed cell line (A-H) are shown. A, B, and C denote separate PCRs; E through H denote nested PCR products; * indicates a stop codon.

```
                                      |-CDR1--|                |------CDR2------|                                      |---------CDR3-----------|
VH4.21  QVQLQQWGAGLLKPSETLSLTCAVYGGS FSGYYWS WIRQPPGKGLEWIG EINHSGSTNYNPSLKS RVTISVDTSKNQFSLKLSSVTAADTAVYYCAR
S13Ag   --H------g------------------ -sDy--G ---S---------- ----REG-K-----R- ----------------LT---V------yc-T APVCYLGWGA              FDS (JH4)
Ag5+Eg1   -------------------------- ----F-T -------------- ------VID----l-- -----------H---E-----V----V-F--- GDSMETAPSYDFWSGYYLDI aFDI (JH3)
Ag28      -------V-S-------------A- ---S--- -------------- ------N--------- ----Lv-----------T-------------r KRGYSYGYFKS                 (JH4/5)
```

FIGURE 2. Deduced amino acid sequences of the V_H4-21 gene used by IgG-positive blood B lymphocytes. Sequences in a hybridoma (S13Ag) and from the cell line by two separate PCRs (Ag, Eg) are shown.

ity of IgM-derived sequences were in germline configuration, but two were extensively mutated. One case showed intraclonal heterogeneity (FIG. 1), with amino acid differences and a stop codon. The IgG-derived sequences were all extensively mutated (FIG. 2) and were unrelated to the IgM sequences.

DISCUSSION

The V_H4-21 gene is commonly rearranged in B cells, but levels of encoded Ig in serum are low.[1] Following EBV infection, levels rise dramatically, and at least some of the secreted IgM has anti-red cell activity.[1] The majority of the activated B cells in blood use the V_H4-21 gene in germline configuration, but a minority of the IgM-positive cells display somatic hypermutation, with evidence for intraclonal heterogeneity. This pattern, and the presence of stop codons, is reminiscent of events in the germinal center, where cells with aberrant sequence would normally die. These findings, and the presence of IgG-switched sequences, indicate that the V_H4-21 gene can take part in normal B-cell maturation, and that EBV infection may promote survival of aberrant products.

REFERENCE

1. CHAPMAN, C. J., M. B. SPELLERBERG, G. A. SMITH, S. J. CARTER, T. J. HAMBLIN & F. K. STEVENSON. 1993. Autoanti-red cell antibodies synthesized by patients with infectious mononucleosis utilize the V_H4-21 gene segment. J. Immunol. **151:** 1051–1061.

Development of the Antibody Repertoire in Rabbits

KATHERINE L. KNIGHT AND MARY A. CRANE

Department of Microbiology and Immunology
Loyola University Chicago
Stritch School of Medicine
Maywood, Illinois 60153

Historically, immunologists were interested in rabbits because they develop both high-titer and high-affinity antibody responses to a wide variety of antigens. The heavy (H) chains of rabbit immunoglobulin (Ig) were of particular interest to geneticists because of the discovery of allotypes of the V_H regions. The V_H allotypes, a1, a2, and a3, were known to differ from each other by multiple amino acid residues, and they were known to be inherited in an allelic fashion. The a1, a2, and a3 allotypic specificities are found on approximately 80% of Ig; and because the germline contains multiple V_H genes, many of which we assumed were used in VDJ gene rearrangements, we could not understand how these allotypes could be inherited as if they were controlled by alleles. The discovery of a heavy chain mutant allele, *ali*, led to the solution of this problem and to the discovery that rabbits use predominantly one V_H gene in their VDJ gene rearrangements.[1–3]

PREFERENTIAL V_H GENE USAGE

Kelus and Weiss[1] discovered the mutant heavy chain allele, *ali*, when they found that offspring of an a2 allotype rabbit expressed exceedingly low levels of Ig molecules with the a2 allotype. Molecular genetic analysis of the V_H chromosomal region of *ali*/*ali* homozygous rabbits revealed that these rabbits were missing 10 kb of DNA at the D-proximal end of the heavy chain chromosomal region.[2,3] Included in this 10 kb of deleted DNA was the 3′-most V_H gene, V_H1, a gene that encodes a prototypic a2 allotype V_H region in a2 rabbits. Because the decrease in serum a2 Ig correlated with the loss of V_H1, we reasoned that most a2 Ig molecules in serum must be encoded by the V_H1 gene and that the V_H1 gene must therefore be used in the VDJ gene rearrangements in most B cells.

The V_H gene usage in VDJ genes was directly examined in leukemic B cells and in rabbit × mouse heterohybridomas.[4,5] In seven of eight VDJ genes cloned from B-cell leukemic rabbits, V_H1 was used in the VDJ gene rearrangement. The other VDJ gene rearrangement used V_Hy33, a gene that encodes some of the so-called V_Ha-negative molecules, the approximately 20% of Ig molecules that do not have V_Ha allotypic specificities.[6] Similarly, Raman *et al.* found that in eight of nine V_Ha-secreting rabbit × mouse heterohybridomas, V_H1 was used in the VDJ gene rearrangements.[5] These data indicate that nearly all V_Ha Ig is encoded by V_H1, and because approximately 80% of the Ig molecules have V_Ha allotypic specificities, we concluded that approximately 80% of the VDJ gene repertoire derives from the usage of V_H1. The preferential usage of the V_Ha allotype-encoding

TABLE 1. V_H Gene Usage in VDJ Genes Cloned from Spleen-Derived and Bone Marrow-Derived DNA and cDNA from Neonatal Rabbits[a]

V_H Gene	VDJ Genes Number	%
V_H1	50	74
V_Hx	5	7
V_Hy	5	7
V_Hz	2	3
Unidentified	6	9

[a] The utilized V_H genes of VDJ genes were identified by comparing their nucleotide sequences to the nucleotide sequences of known V_H1, V_Hx, V_Hy, and V_Hz genes. (During neonatal life the V_H genes are undiversified and therefore can be readily identified.)

V_H1 genes in VDJ gene rearrangements explained how the a1, a2, and a3 allotypic specificities are inherited in an allelic fashion.

V_H GENE USAGE IN NEONATES

To investigate whether the preferential usage of V_H1 in VDJ gene rearrangements was limited to the adult VDJ repertoire or whether it also dominated the neonatal repertoire, Friedman *et al.*[7] examined VDJ gene rearrangements in lymphoid tissues from newborn to 10-day-old rabbits. Approximately 75% of the VDJ gene rearrangements found in both DNA and mRNA had used V_H1, indicating that, like in the adult repertoire, V_H1 was used in most VDJ genes of the neonatal repertoire (TABLE 1). Most of the VDJ genes that did not use V_H1 used one of three other genes, V_Hx, V_Hy, and V_Hz, each of which appears to encode V_Ha-negative molecules.[7]

The finding that a limited number of V_H genes is used in VDJ gene rearrangements is striking because the germline appears to contain many functional V_H genes that could be used in VDJ gene rearrangements. The preferential usage of V_H1 may be due to preferential rearrangement of V_H1, or alternatively, it may be that many germline V_H genes are rearranged but that B cells expressing VDJ encoded by V_H1 are selectively expanded. To distinguish between these possibilities, it will be important both to examine VDJ gene rearrangements in pre-B cells, before surface expression of H chains, and to examine VDJ gene rearrangements early in fetal development.

GENERATION OF THE ANTIBODY REPERTOIRE

Rabbits are born immunoincompetent and with a limited neonatal VDJ repertoire.[8] This limited repertoire is due both to the use of the same V_H gene in most VDJ gene rearrangements and to the fact that these VDJ genes are not somatically diversified[2,7] (FIG. 1). We designate this somatically undiversified repertoire as the neonatal repertoire. The neonatal repertoire becomes highly diversified by seven to eight weeks of age, as determined by the nucleotide sequences of PCR-amplified VDJ genes from peripheral blood lymphocytes (PBL) (FIG. 2). We suggest that the repertoire at 7 to 8 weeks of age represents the primary antibody repertoire, a repertoire that allows the rabbit to become immunocompetent. This development of the primary antibody repertoire after birth is in contrast to the situation found in other species where the neonatal VDJ repertoire is reasonably diverse, due either to combinatorial joining of multiple V_H genes with one of several D and J segments, as occurs in mouse and man, or to somatic diversification during embryogenesis, as occurs in chicken and sheep.[9–11]

```
       FR1                                                                                 CDR1
       Q  S  V  K  E  S  E  G  G  L  F  K  P  T  D  T  L  T  L  T  C  T  V  S  G  F  S  L  S  S  N  A
VH1    CAGTCGGTGAAGGAGTCCGAGGGAGGTCTCTTCAAGCCAACGGATACCCTGACACTCACCTGCACAGTCTCTGGATTCTCCCTCAGTAGCAATGCA
675    ................................................................................................
676    ................................................................................................
677    ................................................................................................
678    ................................................................................................
679    ................................................................................................
681    ................................................................................................

           FR2                                   CDR2
       I  S  W  V  R  Q  A  P  G  N  G  L  E  W  I  G  A  I  G  S  S  G  S  A  Y  Y  A  S  W  A  K  S
VH1    ATAAGCTGGGTCCGCCAGGCTCCAGGGAACGGGCTGGAATGGATCGGAGCCATTGGTAGTAGTGGTAGCGCATACTACGCGAGCTGGGCGAAAAGC
675    ................................................................................................
676    ..................................................................G.............................
677    ................................................................................................
678    ................................................................................................
679    ................................................................................................
681    ................................................................................................

         FR3
       R  S  T  I  T  R  N  T  N  L  N  T  V  T  L  K  M  T  S  L  T  A  A  D  T  A  T  Y  F  C  A  R
VH1    CGATCCACCATCACCAGAAACACCAACCTGAACACGGTGACTCTGAAAATGACCAGTCTGACAGCCGCGGACACGGCCACCTATTTCTGTGCGAGA
675    ................................................................................................
676    ................................................................................................
677    ...............................................................................................T
678    ................................................................................................
679    ................................................................................................
681    ................................................................................................

           N          DH              N                         JH
675    TTGAGGGT ........................  AC    ...............................................  D5/J2
676             GAAA                                ...........................................  D?/J4
677             .............            TCCC        ..........................................  D4/J4
678             ..........................       ..................................................  D3/J4
679         TTA ...........................              ..........................................  D2b/J4
681         GGG  .............          CCCCCC        ..........................................  D1/J4
```

FIGURE 1. Comparison of the nucleotide sequence of germline V_H1 from the a^2 allele to the sequences of six VDJ gene rearrangements isolated from cDNA derived from spleen and bone marrow of newborn to 10-day-old rabbits. The germline D and J gene segments used in each VDJ gene rearrangement are indicated.[22] Dots indicate identity. Framework regions (FR) and complementarity-determining regions (CDR) are indicated. The nucleotide sequences were determined in one orientation only.

```
                     FR1
     G  V  Q  C  Q     S  V  K  E  S  E  G  G  L  F  K  P  T  D  T  L  T  L  T  C  25
VH1  GGTGTCCAGTGTCAG///TCGGTGAAGGAGTCCGAGGGAGGTCTCTTCAAGCCAACGGATACCCTGACACTCACCTGC 75
756  ..............................................................................
758  ..............................................................................
759  ...............GAGCA.C........................................................
760  ..............................................................................
761  ...................................G.....C..GG................................
762  ..............................................................................
763  ..............................................................................
764  ..............................................................................

                                CDR1               FR2
     T  V  S  G  F  S  L  S  S     N  A  I  S  W  V  R  Q  A  P  G  N  G  L  E  W  50
VH1  ACAGTCTCTGGATTCTCCCTCAGTAGC///AATGCAATAAGCTGGGTCCGCCAGGCTCCAGGGAACGGGCTGGAATGG 150
756  ..............................T.CAAC..GGC........................G............
758  ........C......A..A...................G.T.....................................
759  ..............................T..G.G.G........................................
760  .......................C......T.C.AC..G..........................G........G...
761  ....C..........A..T........AGCT.CTGG...T.........................G........G...
762  ..............................................................................
763  ..............................T.CTAC..G.....................................AC
764  ..............................T..G.G.G.T.........................G..........AC

           CDR2                                          FR3
     I  G  A  I  G        S  S  G  S  A  Y  Y  A  S  W  A  K  S  R  S  T  I  T  R  74
VH1  ATCGGAGCCATTGGT//////AGTAGTGGTAGCGCATACTACGCGAGCTGGGCGAAAAGCCGATCCACCATCACCAGA 222
756  ......AT....TA.......GC....A..A.A........................G......T.......T.....
758  ..............................................................................
759  ......................C.G........A............................................
760  ......TG....TA.GCTGGT............A...T...C...........T..TG......T...TC..T.....
761  ......TG....TA.ACTGGT............A..C................T..TG.T....T.T.....T.....
762  ..............................................................................
763  ......TT....A..........G......................................................
764  ......TA....A........TA.G...............................................T.....

     N  T  N  L  N  T  V  T  L  K  M  T  S  L  T  A  A  D  T  A  T  Y  F  C  A  R  100
VH1  AACACCAACCTGAACACGGTGACTCTGAAAATGACCAGTCTGACAGCCGCGGACACGGCCACCTATTTCTGTGCGAGA 300
756  .C.T.G//////.C....A..GA......................................................G
758  ..........................A...................................................
759  ............C.................G...............................................
760  G...T.G...A..G...A.GTTGC..AC..C...A......A....................TG...A..........
761  G.G.A..C..A..........T.....C......A..........C.T..............................
762  ...............................................T..............................
763  ..............................................................................
764  ..............................................................................
```

FIGURE 2. Comparison of nucleotide sequence of germline V_H1 to the sequences of eight VDJ gene rearrangements isolated from peripheral blood lymphocyte cDNA from 7- to 8-week-old rabbits. For other details, see legend to FIGURE 1.

Because the V_H region of VDJ genes is known to diversify by gene conversion and probably also by somatic mutation,[12,13] we examined the diversified VDJ genes for evidence that gene conversion is involved in developing the primary antibody repertoire. The nucleotide sequences of the diversified regions of the VDJ genes cloned from PBL of 7- to 8-week-old rabbits were compared with the nucleotide sequence of 5′ V_H genes in an effort to identify potential donor V_H genes that may have been used in gene conversion events. We found several diversified regions with nucleotide sequences identical to V_H genes 5′ of V_H1 (FIG. 3). For example, the nucleotide sequence of FR1 and CDR1 of the VDJ gene of clone 758 was identical to a region of V_H6 (FIG. 3A). We conclude that V_H6 or a gene very similar to it was used as donor in a gene conversion-like event. Similarly, two regions of clone 760 (FIG. 3B and 3C) had nucleotide sequences identical to V_Hz, indicating that V_Hz or a gene similar to it was used as donor in two separate gene conversion-like events. These data indicate that somatic gene conversion plays an important role in generating the primary antibody repertoire.

SEARCH FOR EVIDENCE OF UNDIVERSIFIED VDJ GENES IN ADULT RABBITS

We were surprised by the observation that all VDJ genes cloned from PBL of the 7- to 8-week-old rabbits were diversified (FIG. 2), because the rearranged V

```
A                           →
       25                  CDR1
VH1    TCTGGATTCTCCCTCAGTAGCAATGCAATAAGCTGGGTCCGC
758    --C------A--A----------------G-T----------
VH6    --C------A--A----------------G-T----------

B                               →
                               CDR2  52      53
VH1    GGGAACGGGCTGGAATGGATCGGAGCCATTGGT//////AGTAGT
760    -----G--------G---------TG----TA-GCTGGT------
VHz    -----G--------G---------TG----TA-GCTGGT------

C                    →
                    FR3        70                  80
VH1    TGGGCGAAAAGCCGATCCACCATCACCAGAAACACCAACCTGAACACGGTGACTCTG
760    ----T---TG------T---TC--T-----G---T-G---A--G---A-GTTGC--A
VHz    ----T---TG------T---TC--T-----G---T-G---A--G---A-GTTGC--A

            82A  B  C                    90
VH1    AAAATGACCAGTCTGACAGCCGCGGACACGGCCACCTATTTCTGTGCGAGA
760    C--C---A------A-------------------TG----A----------
VHz    C--C---A------A-------------------TG----A----------
```

FIGURE 3. Somatic gene conversion of V_H1 using VDJ genes. Comparison of nucleotide sequences of diversified regions of VDJ genes 758 and 760 (from FIG. 2) with germline V_H1 and with potential donor germline V_H genes. **A** = clone 758; **B** and **C** = clone 760. Numbers refer to codons. For other details, see legend to FIGURE 1.

genes of immature B cells of both human and mouse remain undiversified until after antigenic stimulation.[14,15] Because our data were from a relatively small number of VDJ genes, we tested for the presence of undiversified VDJ genes in bone marrow by testing, in an RNase protection assay, for the ability of RNA to protect a probe derived from undiversified V_H1. The RNase protection assay is highly sensitive and allowed us to examine a large pool of VDJ genes. We developed an RNA probe that spanned the leader region and most of the V_H region of undiversified V_H1[2] (FIG. 4A). Because the leader region is generally nonpolymorphic and rarely undergoes somatic diversification, the leader region serves as a control for total Ig H chain mRNA. The remainder of the probe will be protected by heavy chain RNA that has not undergone somatic diversification. If the V_H regions of all VDJ genes are diversified, then this region of the probe would not be protected from digestion by RNase. As shown in FIG. 4B, the V_H region of the probe is readily protected by RNA from bone marrow of a young rabbit, whereas little if any of the probe was protected by RNA from bone marrow from each of three adult rabbits. These data indicate that essentially all VDJ mRNA in bone marrow of adult rabbits is diversified. Assuming that newly generated B cells would use undiversified V_H1, then this finding suggests to us that the bone marrow of adult rabbits has few, if any, newly developed B-lineage cells and that limited B-lymphopoiesis occurs in bone marrow of the adult rabbit.

MODEL FOR B-CELL DEVELOPMENT AND GENERATION OF THE ANTIBODY REPERTOIRE

As immunologists study the immune systems of species other than mouse and human, we discover new pathways for generating the antibody repertoires. At one end of the spectrum lies mouse and human in which B cells develop throughout life and the primary repertoire develops by combinatorial joining of multiple V, (D), and J gene segments. These V(D)J genes then somatically diversify by hypermutation during an immune response following exposure to antigen. At the other end of the spectrum lies chicken, in which all B cells develop during embryogenesis and then migrate to the bursa where the V(D)J genes diversify by somatic gene conversion.[9] In this case, the primary repertoire develops not from combinatorial joining of multiple V(D) and J gene segments but from somatic diversification of V(D)J gene rearrangements in which the same V gene is rearranged in each B cell. As with human and mouse, the primary antibody repertoire develops during embryogenesis. Another variation on development of the antibody repertoire is provided by sheep. In this case B cells migrate to the ileal Peyers patches where the VJ genes undergo somatic hypermutation.[11] This hypermutation process occurs during gestation so that, as in humans, mice, and chickens, the sheep are born with an extensive primary antibody repertoire. Rabbits, on the other hand, seem to be different from any of the above-described species. We speculate that the B cells develop early in ontogeny and that little or no B-lymphopoiesis occurs in adults. Like chickens, rabbits use a limited number of V_H genes in their VDJ gene rearrangements and furthermore, they use a gene conversion-like mechanism to somatically diversify their VDJ genes. Unlike chickens, however, somatic diversification occurs in rabbits several weeks after birth rather than during embryogenesis. We do not yet know whether the initial diversification of rabbit VDJ genes is antigen driven, either by exogenous or endogenous antigen, or whether development of the primary antibody repertoire is a developmentally timed process.

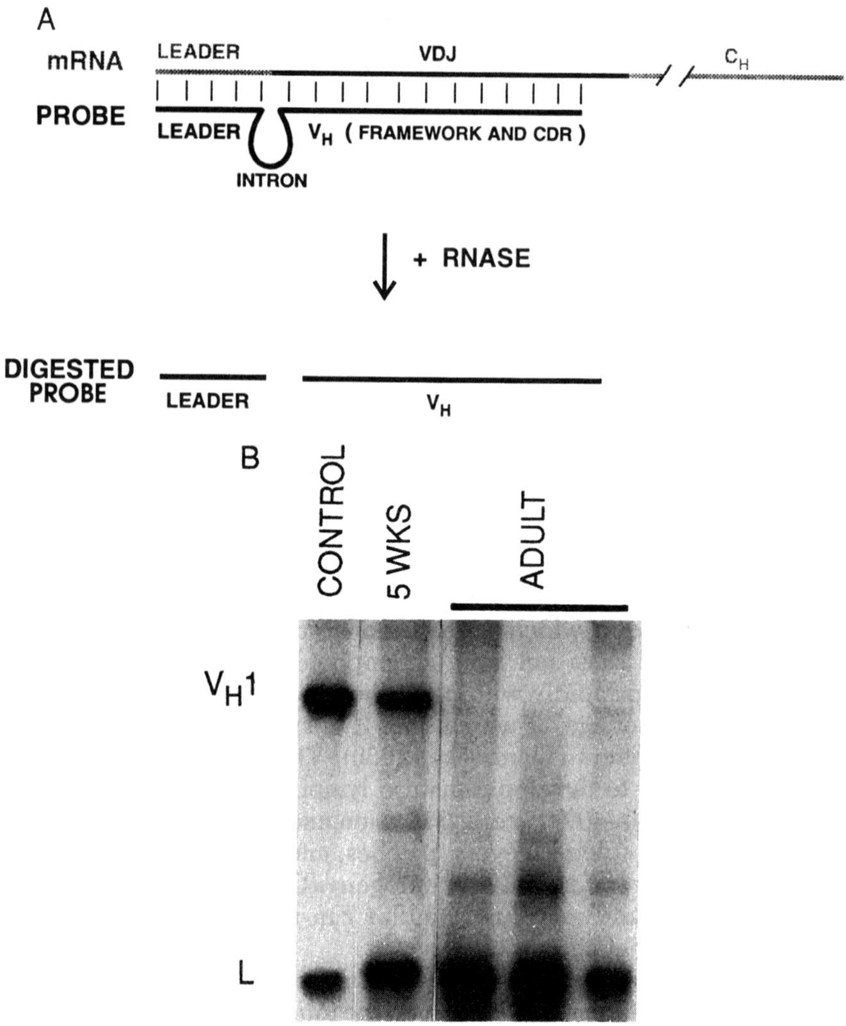

FIGURE 4. RNase protection assay for undiversified V_H1 mRNA in bone marrow of young and adult rabbits. **(A)** Method[23]: 10 μg RNA was hybridized to ^{32}P-antisense probe and subsequently digested with RNase and analyzed on 5% polyacrylamide gels. The V_H1 probe contains the 5′-untranslated region, leader, leader intron, and 280 bp of the V_H region. Two fragments, 120 bp and 280 bp, of the probe are protected by RNA from cells using undiversified V_H1; only the 5′-untranslated and leader regions of 120 bp are protected by RNA from cells using V_H1 that is diversified. **(B)** Autoradiogram of RNase-protected RNA from a positive control B-lineage cell line PBL-1 that expresses a VDJ gene that uses undiversified V_H1 (*lane 1, left*); bone marrow of 5-week-old rabbit (*lane 2*); bone marrow of three adult rabbits. L = protected leader region of probe; V_H1 = protected V_H coding region of probe.

Recently, we proposed a model for B-cell development and for generating the antibody repertoire in rabbits.[16] We suggested that B cells develop early in ontogeny and migrate to gut-associated lymphoid tissue (GALT) where the VDJ and VJ genes undergo somatic diversification by a somatic gene conversion-like mechanism and probably also by somatic mutation. We designate the repertoire developed in GALT as the primary antibody repertoire, and we propose that it is this repertoire that makes the rabbit immunocompetent. We suggest GALT as the site for development of the primary antibody repertoire for several reasons. One, the follicular structure of rabbit GALT, especially the appendix and sacculus rotundus,[17] is highly similar to that of the chicken bursa, the organ in which the primary antibody repertoire develops by somatic gene conversion.[9,18] Next, surgical removal of GALT from neonatal rabbits results in highly immunocompromised rabbits[19]; and finally, in germ-free rabbits the GALT does not develop, and these rabbits are immunoincompetent.[20,21] Because all rabbit B cells are $CD5^+$ and may be the functional homologue of self-renewing murine $CD5^+$ B cells, we suggest that once the primary antibody repertoire is formed, it is maintained by self-renewing B cells and that little, if any, B-lymphopoiesis occurs in the adult rabbit. Experiments are underway to test various aspects of this model.

REFERENCES

1. KELUS, A. S. & S. WEISS. 1986. Mutation affecting the expression of immunoglobulin variable regions in the rabbit. Proc. Natl. Acad. Sci. USA **83:** 4883–4886.
2. KNIGHT, K. L. & R. S. BECKER. 1990. Molecular basis of the allelic inheritance of rabbit immunoglobulin VH allotypes: Implications for the generation of antibody diversity. Cell **60:** 963–970.
3. ALLEGRUCCI, M., G. O. YOUNG-COOPER, C. B. ALEXANDER, B. A. NEWMAN & R. G. MAGE. 1991. Preferential rearrangement in normal rabbits of the 3′ V_Ha allotype gene that is deleted in *Alicia* mutants. Somatic hypermutation/conversion may play a major role in generating the heterogeneity of rabbit heavy chain variable region sequences. Eur. J. Immunol. **21:** 411–417.
4. BECKER, R. S., M. SUTER & K. L. KNIGHT. 1990. Restricted utilization of V_H and D_H genes in leukemic rabbit B cells. Eur. J. Immunol. **20:** 397–402.
5. RAMAN, C., H. SPIEKER-POLET, P.-C. YAM & K. L. KNIGHT. 1994. Preferential V_H gene usage in rabbit immunoglobulin-secreting heterohybridomas. J. Immunol. **152:** 3935–3945.
6. DRAY, S., G. O. YOUNG & A. NISONOFF. 1963. Distribution of allotypic specificities among rabbit γ-globulin molecules genetically defined at two loci. Nature **199:** 52–55.
7. FRIEDMAN, M. L., C. TUNYAPLIN, S. K. ZHAI & K. L. KNIGHT. 1994. Neonatal V_H, D and J_H gene usage in rabbit B-lineage cells. J. Immunol. **152:** 632–634.
8. THORBECKE, G. J. 1960. Gamma globulin and antibody formation in vitro: I. gamma globulin formation in tissues from immature and normal adult rabbits. J. Exp. Med. **112:** 279.
9. REYNAUD, C.-A., V. ANQUEZ, H. GRIMAL & J.-C. WEILL. 1987. A hyperconversion mechanism generates the chicken light chain preimmune repertoire. Cell **48:** 379–388.
10. REYNAUD, C.-A., A. DAHAN, V. ANQUEZ & J.-C. WEILL. 1989. Somatic hyperconversion diversifies the single V_H gene of the chicken with a high incidence in the D region. Cell **59:** 171–183.
11. REYNAUD, C.-A., C. R. MACKAY, R. G. MULLER & J.-C. WEILL. 1991. Somatic generation of diversity in a mammalian primary lymphoid organ: The sheep ileal Peyer's patches. Cell **64:** 995–1005.
12. BECKER, R. S. & K. L. KNIGHT. 1990. Somatic diversification of immunoglobulin heavy chain VDJ genes: Evidence for somatic gene conversion in rabbits. Cell **63:** 987–997.

13. SHORT, J. A., P. SETHUPATHI, S. K. ZHAI & K. L. KNIGHT. 1991. VDJ genes in V_Ha2 allotype-suppressed rabbits. Limited germline V_H gene usage and accumulation of somatic mutations in D regions. J. Immunol. **147:** 4014–4018.
14. KLEIN, U., R. RUPPERS & K. RAJEWSKY. 1993. Human IgM^+IgD^+ B cells, the major B cell subset in the peripheral blood, express V_κ genes with no or little somatic mutation throughout life. Eur. J. Immunol. **23:** 3272–3277.
15. GU, H., D. TARLINTON, W. MULLER, K. RAJWESKY & I. FORSTER. 1991. Most peripheral B cells in mice are ligand selected. J. Exp. Med. **173:** 1357–1371.
16. KNIGHT, K. L. & M. A. CRANE. 1994. Generating the antibody repertoire in rabbit. Adv. Immunol. **56:** 179–218.
17. ARCHER, O. K., D. E. R. SUTHERLAND & R. A. GOOD. 1963. Appendix of the rabbit: A homologue of the bursa in the chicken? Nature **200:** 337–339.
18. THOMPSON, C. B. & P. E. NEIMAN. 1987. Somatic diversification of the chicken immunoglobulin light chain gene is limited to the rearranged variable gene segment. Cell **48:** 369–378.
19. COOPER, M. D., D. Y. PEREY, A. E. GABRIELSEN, E. E. R. SUTHERLAND, M. F. MCKNEALLY & R. A. GOOD. 1968. Production of an antibody deficiency syndrome in rabbits by neonatal removal of organized intestinal lymphoid tissues. Intl. Arch. Allergy **33:** 65–88.
20. ŠTEPÁNKOVÁ, R., F. KOVÁŘŮ & J. KRUML. 1980. Lymphatic tissue of the intestinal tract of germ-free and conventional rabbits. Folia Microbiol. **25:** 491–495.
21. ŠTEPÁNKOVÁ, R. & F. KOVÁŘŮ. Immunoglobulin-producing cells in lymphatic tissues of germfree and conventional rabbits as detected by an immunofluorescence method. Folia Microbiol. **30:** 291–294.
22. BECKER, R. S., S. K. ZHAI, S. J. CURRIER & K. L. KNIGHT. 1989. Ig VH, DH, and JH germ-line gene segments linked by overlapping cosmid clones of rabbit DNA. J. Immunol. **142:** 1351–1355.
23. SPIEKER-POLET, H., P.-C. YAM & K. L. KNIGHT. 1993. Differential expression of 13 IgA-heavy chain genes in rabbit lymphoid tissues. J. Immunol. **150:** 5457–5465.

Development of the Mouse B-Cell Repertoire[a]

FLAVIUS MARTIN, XINJIAN CHEN, FENGYU SHU, AND JOHN F. KEARNEY[b]

Division of Developmental and Clinical Immunology
Department of Microbiology
University of Alabama at Birmingham
Birmingham, Alabama 35294

INTRODUCTION

The development of B lymphocyte precursors in mice occurs in multiple sites that change from embryonic to adult life. Early developmental sites include yolk sac, the paraaortic splanchnopleura, liver, and omentum.[1–4] A later site is the neonatal spleen, and finally development continues during adult life in the adult bone marrow.[5,6] The major steps that govern the development of the hematopoietic stem cell to B cells appear to be basically similar during these different ontogenetic stages. At each site of lymphopoiesis, B cells are generated that express a clonal surface immunoglobulin receptor as a result of the series of successive rearrangements within the immunoglobulin gene loci. The differentiation steps occurring during the progression from stem cell to the B lymphocyte are ordered and depend on the interaction of B-cell progenitors with specific microenvironmental factors.[7,8] Although the basic stages are similar, B cells generated during early versus adult B lymphopoiesis differ in several characteristics. Terminal deoxynucleotidyl transferase (Tdt) does not appear to be functionally active during early B-cell development,[9] and MHC class II antigens are not present on fetal pre-B and B cells.[10–12] The precursor lymphocyte-specific regulatory myosin light chain gene (PLRLC)[13] is also not expressed by fetal B cells. Except for the absence of Tdt activity, functional consequences of these developmental differences between early versus adult B cells have not yet been demonstrated. Apart from these qualitative differences between fetal and adult B-cell development, changes in the absolute numbers and frequencies of B-cell progenitors and acquisition of mature B-cell phenotypes have also been documented.[14,15]

There are extensive comparative data on the immunoglobulin repertoires of B cells from fetal and adult life,[16–20] and it is clear that the early B-cell repertoire is less diverse than that of the adult. Some of the molecular mechanisms responsible for these differences have been clarified, but the reasons for such a restriction in diversity and the role of B cells expressing this restricted repertoire during early development of the immune system are not understood.

[a] This work was supported in part by National Institutes of Health Grants AI 14782, AI 30879, and CA 13148.

[b] Address for correspondence: John F. Kearney, 378 WTI, University of Alabama at Birmingham, Birmingham, Alabama 35294-3300.

SAMPLING THE B-CELL REPERTOIRE

The Ig receptor expressed by mature B cells is an essential component in the recognition and triggering of effector cells in response to foreign antigens. It is also the receptor by which B-lineage cells receive clonally directed signals from the self-environment, which may permit them to become fully immunocompetent or, conversely, prevents them from developing further. It is not clear when this clonal receptor become functional, but if the variable portion of the μ chain in combination with the pseudo-light chain complex can bind to self-antigens, then clonally directed antigenic pressure begins at the pre-B-cell level. At this time, however, no ligand has yet been demonstrated for the pre-B-cell receptor complex. In contrast to the lack of evidence for μ-pseudo-light chain complex generated signals, there is ample evidence that after expression of the conventional μ-light chain molecule on young B cells, they are then susceptible to negative and also positive signals mediated through the Ig receptor. That B-cell clones can be negatively influenced by clonal deletion, anergy, editing, and antigenic competition is clearly shown in transgenic mouse models[21–25]; however, data supporting positive selection of developing B cells are less extensive and based mainly on indirect evidence.[26–30] The difficulties in demonstrating positive signaling effects via sIg arise from two directions. First the B-cell compartment is not homogenous, and positive signals may play a more important role in the development of subpopulations of B cells rather than all B cells as a whole. This has been shown in transgenic mouse models in which immunoglobulin receptors for antigens such as phosphorylcholine and phosphatidylcholine are clearly restricted to the B1 population.[29–31] Second, the Ig receptor-mediated signals might not result in either a negative or positive type of response on the fate of a B cell, but participate in a more complex effect, in which the fate of a newly generated B cell is determined by many factors apart from the actual signals delivered through sIg. The results of recent experiments in which Ig^- B-cell transgenic mice were created on $rag^{-/-}$ background[32,33] indirectly support the latter concept by showing variability in the B-cell development for different, presumably nonself-reactive, immunoglobulin transgenes.

Various stages in the acquisition of an adult repertoire can be delineated on both theoretical and experimental bases.[34] As reference points, the acquisition of a diverse repertoire can be divided into several stages: (i) the *potential* repertoire (the pool of receptors that could result from the theoretical combinations of the total unrearranged genomic immunoglobulin gene pool), (ii) the *emergent/nascent* repertoire (the sum of functional and nonfunctional immunoglobulin gene rearrangements), (iii) the *available* repertoire (the expressed clonal B-cell membrane receptor proteins on newly formed and long-lived populations), and finally there is (iv) the *actual* repertoire (the pool of secreted immunoglobulins). These levels of repertoire acquisition can be analyzed to some extent in terms of structure, which ranges from genomic DNA sequences to crystallized immunoglobulin, and of function—the reactivity against the self and foreign antigenic world. The stages of repertoire acquisition have been examined by several experimental methods, each with advantages and disadvantages. Nucleic acid libraries (genomic and complementary DNA) have been used to explore the molecular nature of Ig receptors. Antibody repertoire reactivity has been explored by a variety of detection techniques from ELISA to protein-blotting techniques.[36,39] A combination of the two can be used to analyze the molecular nature of antibody functions, but when used alone neither can do both. Hybridoma technology was the first method to permit the isolation of antibodies at different levels of repertoire expression

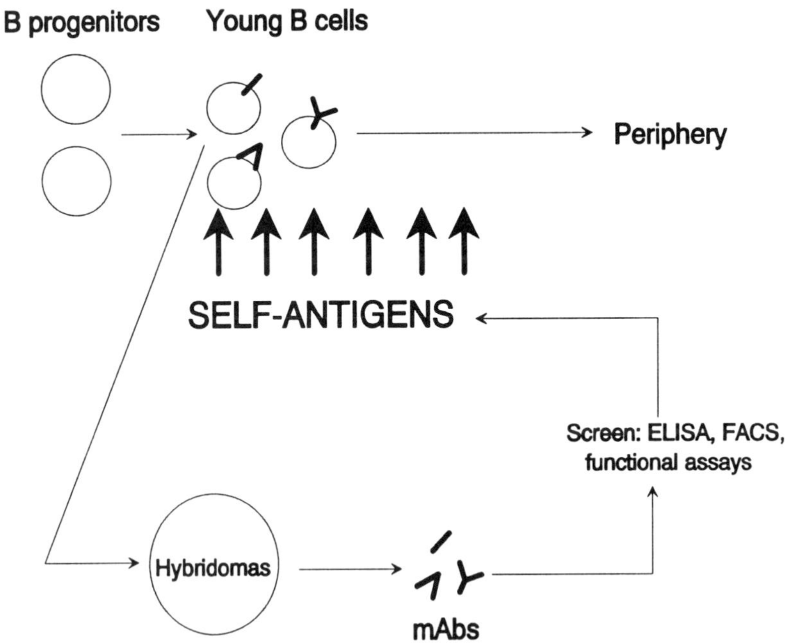

FIGURE 1. Rationale for the use of hybridomas to study the nature of the microenvironmental pressure on the nascent B-cell repertoire.

and facilitated assays for correlation of the molecular and function characteristics of antibodies from individual clones of desired populations of B cells. Recently, phage display libraries have provided an alternative,[37] but this approach does not at this stage of development facilitate the isolation of complete functional Ig receptors (paired heavy and light chains) from individual B cells.

We have used hybridoma technology to analyze functional characteristics of the fetal B-cell antibody repertoire and to obtain some information on the molecular nature of these antibodies. This approach not only permits insight into the molecular characteristics of the B-cell repertoire, but also facilitates the analysis of self-antigens that are involved in the clonal selection pressures on emerging newly formed B cells (FIG. 1).

CHARACTERISTICS OF THE MOUSE PERINATAL B-CELL REPERTOIRE

Restrictions in the Perinatal Repertoire

Extensive *in vitro* studies involving hybridoma technology and a variety of antibody-screening techniques have shown that the fetal repertoire is characterized by high connectivity, self-reactivity, and multireactivity.[38–41] These features of early B-cell-derived antibodies have been proposed to direct the subsequent development *in vivo* of the complete adult B-cell repertoire; however, because of the complex nature of the interactions involved, this function has been difficult to demonstrate conclusively. The ultimate role of mature B cells is to respond efficiently by the production of protective antibodies to a multitude of antigens expressed on potentially pathogenic environmental organisms. Because the fetal repertoire has the potential to generate at least a selection of the multitude of B-cell specificities in the adult B-cell repertoire, the special role of antibody connectivity, self-reactivity, and multireactivity that are characteristic of the fetal repertoire remain unclear.

Structural Base—the Germline-Restricted Component

Efforts of many groups have highlighted the differences in the molecular basis of Ig receptors in the fetal and adult B-cell repertoires. There is a clear restriction in the germline gene segments that are part of the emergent, available, and actual repertoire in fetal life. Proximity of the overexpressed V_H exons to the D locus appears to be a factor involved in the expression of certain V_H gene families.[42] This effect, attributed initially to the overexpression of 7183 and Q52 V_H families, is due in great part to the overexpression of the most proximal functional V_H exon − V_H81X (7183.1) a member of the V_H7183 family.[43] In addition to the overexpression of this particular V_H gene segment, one D region segment, DFL16.1, has been found in about half of the fetal liver-derived rearrangements both in DJ[44] and functional hybridoma VDJ rearrangements (unpublished data). Differences have also been found in the frequency of J_H and J_K usage and in V_κ-J_κ pairs between fetus and adult.[44–46]

Beside the restriction in the genomic exons used to generate immunoglobulin molecules during fetal life, the activity of the enzyme Tdt is also developmentally regulated and is involved in the generation of junctional diversity only after birth in spleen and then continues during adult B-cell lymphopoiesis in the bone marrow. The absence of N-region additions during fetal life results in a short average nucleotide length of the third complementarity-determining regions (CDR3) compared to those generated postnatally and in the adult. These shorter CDR3 regions have been hypothesized to generate a relatively flat antigen-binding site that may have the propensity to bind with low affinity to a variety of antigenic structures (Schroeder *et al.*, this volume). In addition, somatic mutation is very limited during early mouse B-cell development and also restricts the potential diversity. This combination of molecular constraints may be a partial explanation for the high frequency of multireactive antibodies evident in the early B-cell repertoire.

It has been shown that for several adult-derived autoantibodies, multireactive binding was dependent on the CDR3 regions of the binding site and particularly on the CDR3 of the heavy chain (Ref. 48 and this volume). In preliminary studies we have compared the amino acid composition of 23 heavy chain sequences derived from our perinatal B-cell hybridoma panel with 30 heavy chain sequences of adult mice (from GENEBANK), derived from high-affinity antibody responses to laboratory protein antigens. The composition of the full-length sequences was very similar between the two pools, but comparison of the CDR3 regions revealed a significantly higher percentage of aromatic residues in the fetal pool, which in the adult pool were replaced by neutral amino acids (FIG. 2). This difference was due mainly to the failure of germline-encoded tryptophan (W) residues to be preserved in the adult pool as a result of VDJ joining, N-region addition, and nucleotide nibbling. These differences also showed correlation with the experimental values

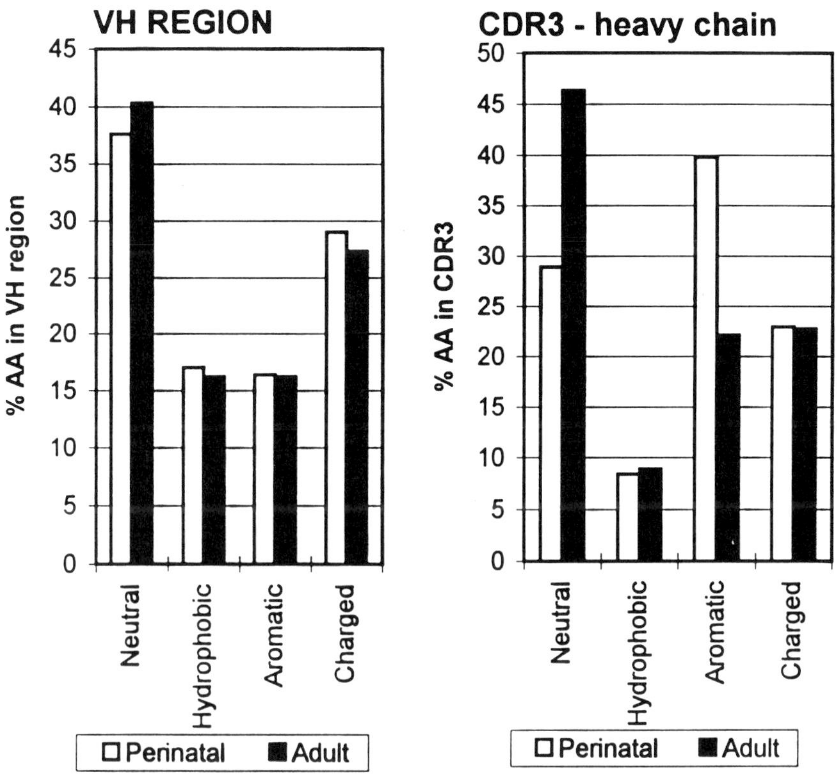

FIGURE 2. Comparison of the amino acid composition of individual V_H-D-J_H sequences obtained from our perinatal antibodies (*open bars*) with a representative panel of sequences derived from adult antibodies (*solid bars*).

for connectivity: antibodies with heavy chain CDR3 containing tryptophan were significantly higher in connectivity when compared with those without tryptophan in CDR3 (not shown). From a three-dimensional structural point of view, the presence of these large aromatic residues in the middle of a flat binding site might partially explain the characteristic functional behavior of fetal antibodies, but more detailed analysis at a molecular level is necessary to formally prove this hypothesis.

Functional Base—the Selective Component

Although there are molecular constraints in the formation of early mouse B-cell immunoglobulins, this does not automatically implicate these differences as being totally responsible for the observed restriction in the B-cell repertoire. There are selective processes occurring during the progression from the potential to the available repertoire. These pressures may result from *evolutionary selection* resulting from interaction of B cells with the inherited self-environment in which the immune system develops. There may also be a *developmental selection* resulting from identical active selection mechanisms within each organism that differ between individuals as a result of environmental experiences. The two alternatives can obviously coexist, and a genetically inherited set of regulatory elements could control the receptors expressed by each B cell. Superimposed on this genetic default, microenvironmental selection resulting from positive and negative signals from within the organism or from the external milieux may modify receptor expression.

To understand which of these alternatives plays a role in immunoglobulin V-region interactions and in repertoire development, we compared the average perinatal connectivity of antibodies derived from two strains of mice, BALB/c and C57BL/6. This was analyzed within a strain by studying the homologous combinations—BALB/c with BALB/c and C57BL/6 with C57BL/6—and also the heterologous combinations between strains—BALB/c with C57BL/6. We also performed the same experiments in F1(BALB/c × C57BL/6) mice where it was possible to isolate independent hybridomas expressing either of the parental Igh6a (BALB/c) or Igh6b (C57BL/6) allotypes that had in this case developed within the same host. The difference in connectivity in the panels of antibodies derived from the two genetically distinct mouse strains but from different origins (parent versus F1) were particularly obvious in the group of highly connected antibodies (connectivity > 20%). As shown in FIGURE 3, when the antibodies were derived from the single parental strains, there was much less connectivity in the heterologous than in the homologous combinations. Nevertheless, in the F1-derived panels the heterologous interactions were intermediate in their connectivity compared

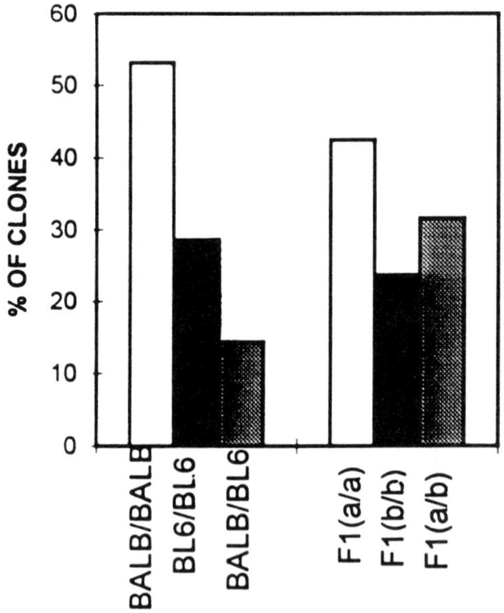

FIGURE 3. Clones with connectivity higher than 20%. Homologous and heterologous connectivity for BALB/c (BALB), C57BL/6 (BL6), and F1(BALB/c × C57BL/6) strains; "a" and "b" are the two parental allotypes in the F1.

to the homologous interactions, although the homologous combinations parallel the pattern of the single parental interactions.

These results, at face value, tell us that idiotypic connectivity in F1 mice is independent of the genetic origin of the heavy chain expressed by the antibodies. Connectivity within a strain, however, is characterized by a certian degree of strain-specific identity. B cells expressing IgH6a and IgH6b immunoglobulin that develop within the F1 mouse (even if derived from the same two sets of heavy chain germline genes as the parental panels) are more connected than the parental heterologous combinations when tested against each other. Therefore, the available B cells appear to be selected differently within the different strains from the same pool of immunoglobulin genes generated by the molecular mechanisms responsible for receptor generation.

Analyses of the association of light and heavy chain V regions with these reactivities show that antibodies using the D proximal V_H7183 family are more involved in interactions within the panel, both within a strain and across strains, compared with other more distal families (FIG. 4). The neonatal frequency of V_H7183 members is 1.5 to 2 times higher in BALB/c than in C57BL/6 and is part of the explanation for the difference in connectivity between the two strains. Nevertheless, it is clear that the connectivity does not depend solely on the particular member of a heavy chain family used in a particular antibody, because within a family of V_H genes there is a broad spectrum of connectivity of individual antibodies (not shown), suggesting that, as might be expected, light chains participating in the formation of the binding site also determine connectivity.

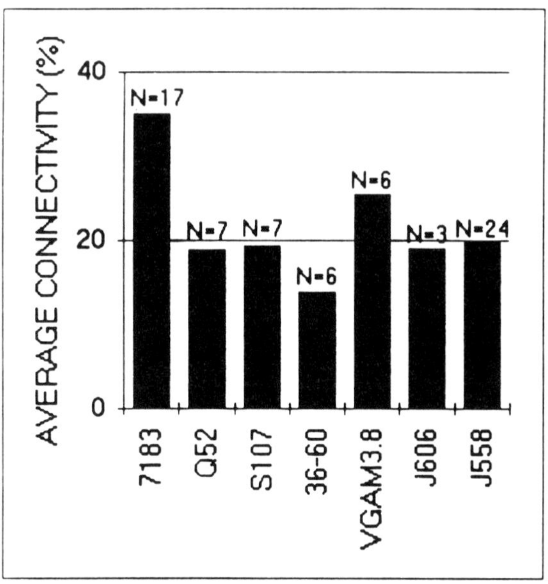

FIGURE 4. Average connectivity of neonatally derived antibodies and V_H family representation.

Relative Affinities of Perinatal-Derived Antibodies

Several groups have estimated that the relative binding affinity of perinatal or fetal-like antibodies in different systems are in the K_D range of 10^{-5} to 10^{-7}, values considered as low-affinity antibodies.[47–49] Some exceptions have been noted in which germline-encoded antibodies are of high affinity, but these exist at a very low frequency in the repertoire (germline anti-DNA antibodies[50]), or they are derived from restricted antibody response systems in mouse (phosphorylcholine[51]). We have compared the binding characteristics of idiotypic interactions between two natural fetal-type IgM antibodies with the binding of a conventional idiotype–antiidiotype pair generated by deliberate immunization. These preliminary results show that the relative affinity of a classic idiotype–antiidiotype pair of antibodies is two to three orders of magnitude higher than that of the naturally occurring pair (10^{-8} versus 10^{-5}–10^{-6}) (results not shown).

Many affinity estimations have been done with antigens that are frequently bound by antibodies expressed in the early repertoire (DNA, IgG, tubulin, fibronectin, collagen, etc.). If these perinatal antibody reactions are of low affinity, then the question remains as to the functional significance of these potential interactions *in vivo*. Are the antigens (used to demonstrate *in vitro* antibody reactions—foreign, self, idiotypes, etc.) part of the host environment that shapes the early repertoire? Until now Ig transgenic mouse models that have been developed demonstrating negative selection[21–25] have used Ig genes derived from antibodies of high affinity. Even when the affinity of interaction between transgenic Ig and the self-antigen (class I MHC) was experimentally reduced, the nature, abundance, and distribution of the antigen may not have been representative of the type of antigens contacted *in vitro* by emerging B cells. The main challenge in the field still remains to be determined: Are low-affinity self antigens (including immunoglobulin idiotypes) able to affect selection of the B-cell repertoire through a positive signal? The problem needs to be independently addressed for development of both the fetal and adult repertoires since they emerge in different microenvironments, and it may be found that there are different selective processes operating in fetal versus adult sites of B lymphopoiesis.

Specificities

Antibodies isolated from the early B-cell repertoire react *in vitro* with a broad range of self and nonself antigens.[47,52,53] The self antigens belong to a wide variety of components, including cell surface and intracellular glycoproteins, nucleic acids, extracellular matrix components, and soluble antigens. Identification by molecular approaches of new antigens isolated directly from tissues constituting the milieu in which the emerging B cells express their Ig receptors would permit the creation of *in vivo* (transgenic) models to determine the fate of reactive B cells. This would produce direct evidence to support the role of such antigens in hypothesized positive-selection mechanisms. Some support in this direction has come from species that have specialized organs for early B-cell development (chicken, duck, sheep, etc.) in which there is a strong indication for positive selection by self-antigens.[28,54,55] The parallels between B-cell development in these species versus man and mouse should be made with caution, because the molecular mechanisms for generation of clonal diversity of immunoglobulins are partially or totally different.

In conclusion, many antibodies constituting the early B-cell repertoire in mouse result from Ig gene rearrangements that are subjected to molecular restrictions compared to those of the adult and frequently have low affinity for a variety of foreign and self components. The transition of the fetal-type repertoire into an adult type is manifested by a change in the composition of the diversity of B cells. Two general questions can then be asked with respect to the fate of fetal-type B cells that develop in the adult organism: Can they compete with normal adult populations for a niche in the adult repertoire? Alternatively, if the final goal is to develop an adult repertoire, why it is not generated from the beginning, during the fetal life?

In order to approach these questions we have generated an Ig-transgenic mouse in which we have studied the fate of B cells expressing a rearranged V_H81X heavy chain gene, which is the most D-proximal functional V_H region and is expressed predominantly in fetal life, but is rarely detected in adult mice.

V_H81X-BEARING CELLS IN NORMAL AND TRANSGENIC MICE

The V_H81X exon is rearranged at a high frequency in fetal liver and adult bone marrow.[16,17] Although functional V_H81X rearrangements can be found in fetal liver and to some extent in adult bone marrow, B cells expressing functional rearrangements of this V_H gene are rarely, if ever, detected in adult peripheral lymphoid organs. In our panel of ~150 perinatal hybridomas, we have found four that secrete V_H81X-$V\kappa$ antibodies. So far as we know, these are the only four known hybridomas isolated from fetal liver of normal mice that secrete V_H81X-containing antibodies. In FIGURE 5 the junctions of these four heavy chains are compared with the only three known functional V_H81X sequences derived by PCR amplification from normal adult peripheral lymphoid organs (B cells) and also with representatives of functional rearrangements from adult bone marrow. The few adult peripherally derived sequences that have been described have a frequency of N region additions more similar to those derived from the perinatal period than those from adult bone marrow. This difference suggests that they arose either early in life or there was a strong selective process favoring the survival of perinatal type B cells. It is clear from these experiments that there is a strong negative influence on the abundant V_H81X-rearranging pre-B (and/or B) cells that prevents their development into mature peripheral surface immunoglobulin-bearing B cells. However, because a few cells are found in the periphery also, the selection process is not simply a yes/no phenomenon characteristic of the V_H81X segment. Since the frequency of the VH81X-bearing cells is so low, we expressed the rearranged heavy chain of the BC2 hybridoma (FIG. 5) under the μ enhancer and promoter control elements in a transgenic mouse.[56] Allelic exclusion was strong in this case and V_H81X surface immunoglobulin-expressing B cells produced were able to compete with the endogenously generated B cells and populate the peripheral lymphoid organs. The transgenic heavy chain paired with a variety of light chains, and the majority of the antibodies had a degree of reactivity toward self-tissues. Interestingly, a particular light chain rearrangement, V_K1A5-J_κ5, was detected by Southern blot in a large population of spleen B cells and hybridomas derived from these B cells (FIG. 6). This rearrangement has an identical V_κ-J_κ junction (giving rise to identical clonal Ig receptors when paired with the transgenic heavy chain) and is expressed by clonally independent B cells.[56] From the pool of hybridomas derived from this mouse, the antibody encoded by this combination is one of

```
                                    VH                 DH                        JH
                            ---------------><------------------><-------------------------
              VH81X(7183.1) TACTGTGCAAGACATG                    CTACTGGTACTTCGATGTCTGGGGC JH1
                            DFL16.1    TTTATTACTACGGTAGTAGCTAC       C-GG--T-C-TA-------- JH3
DAYS                                                            T----ATGCTA-G--CTA------T JH4
FROM
BIRTH   STRAIN         HYBRIDOMA

 -2     BALB/c (BA)    BC2          ----------------.......-------........----------------------
 +3     C57/BL6(BL)    F-3-1        ----------------..--------------...........C-GG--T-C-TA-------
 +2     F1(BA X BL)    31a          ----------------..----------...............-ATGCTA-G--CTA------T
 -3     BALB/c         16-165       ---------------.....---TA-G...............-ATGCTA-G--CTA------T


                      VH81X(7183.1)    N         D          N       JH
                      ---------------><------><------------------><------><--------------------

ADULT SPLEEN[1]       TACTGTGCAAGACA.....G...........TATAGG........................ATTACGCTATGG JH4

ADULT LN.*            ---------------T....A...........GTAACTA..........A............TTTGCTTACTGG JH3
                      ---------------TG...GG..........ACTGGGA..........GG......GCCTGGTTTGCTTACTGG JH3


ADULT B.M. B CELLS[2] ------------.......G.............AGG............C.....................CTGG JH?
                      -----------.......GCTA.....GACAGCTCGGG.........GT.......CCTGGTTTGCTTACTGG JH3
                      -------------.....GGGA..TTATTACTACGGTAGTAGC....AGG.........GCTATGGACTACTGG JH4
                      --------------....GGGA....ATTACTACGGCTAC......GGGGA...ACTGGTACTTCGATGTCTGG JH1
```

FIGURE 5. V_H-D-J_H junctional sequences of VH81X rearrangements derived from perinatal hybridomas (BC2, F-3-1, 31a, and 16-165) or from adult genomic DNA libraries (spleen, lymph nodes, bone marrow). * From two HA-immunized mice.[1] Carlsson, L., C. Overmo & D. Holmberg. 1992. Eur. J. Immunol. **22:** 71. [2] Huetz, F., L. Carlsson, U.-C. Tornberg & D. Holmberg. 1993. EMBO J. **12:** 1819.

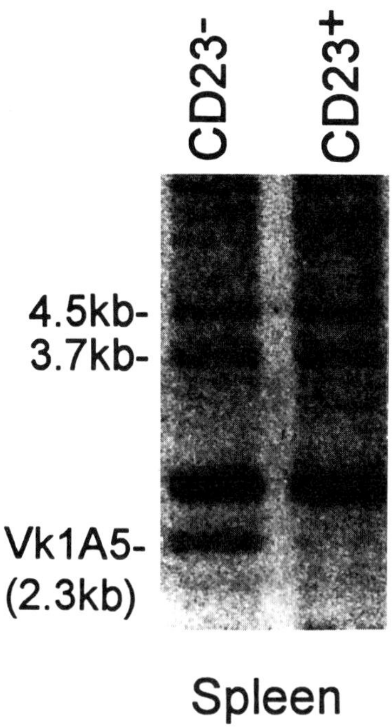

FIGURE 6. Southern blot analysis of the kappa light chain usage in splenic B cells derived from VH81X transgenic mouse. The cells were sorted as $CD23^+IgM^+$ or $CD23^-IgM^+$ B cells, and the Southern blots were hybridized with a J_κ genomic probe. The $V_\kappa 1A5$-$J_\kappa 5$ (2.3 kb) dominant rearrangement (confirmed by sequence analysis) is present in the $CD23^-IgM^+$ population and absent in the $CD23^+IgM^+$ cells.

the most self reactive, as shown by Western blot of tissue lysates (not shown). Phenotypically this predominant B-cell population is $CD5^-$, is found in the $CD23^-$ B cells from the spleen, and is absent from the $CD23^+$ B cells. The predominant $V_\kappa 1A5$-$J_\kappa 5$ rearrangement was not found in adult bone marrow or newborn liver and spleen. These data taken together suggest that there is a positive selection of nascent B cells bearing the $V_H 81X$-$V_\kappa 1A5$-$J_\kappa 5$ receptor into the splenic peripheral $CD23^-$ B cell compartment. Although the selection favoring this combination is ongoing during the life of the adult, it is followed paradoxically by negative selection from the $CD23^-$ to the $CD23^+$ B-cell compartment. The question arises as to why there appears to be a positive selection resulting in an enrichment of this B-cell population in the spleen followed by negative selection, which prevents B cells expressing this combination from entering the peripheral mature B-cell pool. Preferential combinatorial association of this particular light chain with the $V_H 81X$ heavy chain is unlikely to contribute to the high frequency of splenic B cells with this receptor, since there is not an overrepresentation of $V_\kappa 1A5$-$J_\kappa 5$ light chain rearrangement in bone marrow. Dominance is clearly not due to clonal proliferation of this set of B cells in the spleen because examination of hybridomas showed that these cells were multiclonal in origin and it is not likely that there is negative selection of all other combinations because of the wide variety of other κ chains detected in the transgenic mouse in addition to V_κ1A5.[56] It is most likely that there is a positive selection of this population by a signal through the surface IgM receptor and that a self-antigen is involved in this positive signal.[56]

In summary, our analysis of this transgenic mouse shows that B cells bearing rearranged fetal-type V_H genes can be generated in the adult and are positively selected into the compartment of $CD23^-$ cells in the periphery. However, this is followed by negative selection preventing the movement of these cells into the $CD23^+$ compartment. This second step involving negative selection may occur as a result of a newly described competition mechanism to exclude self-reactive cells from entering the long-lived B-cell pool.[25] Our experiments add to this new finding by clearly showing the existence of a positive selection step on a B-cell subpopulation at a stage that occurs between their generation in bone marrow and the migration into the $CD23^-$ splenic B compartment. This step occurs earlier and differs from the competition for the follicular niches that is clearly controlled by a negative selective step, the nature of which is unknown.[25]

SUMMARY

The mouse B-cell repertoire early in ontogeny contains B cells with receptor immunoglobulins exhibiting high connectivity, multi/self-reactivity, and generally low affinity. This is due structurally to extensive restriction in the germline components used to generate the B-cell receptor. The selective pressure acting on the nascent repetoire has both negative and positive components as seen in our *in vivo* models. V_H81X-bearing B cells from the V_H81X transgenic mice (and probably from normal mice) are subject to self-selective pressure with two components: a positive one favoring a certain (self-reactive) specificity in the $CD23^-IgM^+$ population and a negative one preventing the entry of B cells with this specificity into the $CD23^+IgM^+$ compartment.

ACKNOWLEDGMENTS

We wish to acknowledge the expert assistance of Ann Brookshire in the preparation of this manuscript. We would also like to acknowledge the expert technical assistance of Lisa Jia, Anne Elizabeth Ratcliffe, and Kathy Raper.

REFERENCES

1. Cumano, A., C. Furlonger & C. G. Paige. 1993. Differentiation and characterization of B-cell precursors detected in the yolk sac and embryo body of embryos beginning at the 10- to 12-somite stage. Proc. Natl. Acad. Sci. USA **90:** 6429–6433.
2. Godin, I. E., J. A. Garcia-Porrero, A. Coutinho, F. Dieterlen-Lievre & M. Marcos. 1993. Para-aortic splanchnopleura from early mouse embryos contains B1a cell progenitors. Nature **364:** 67–70.
3. Palacios, R. & B. A. Imhof. 1993. At day 8–8.5 of mouse development the yolk sac, not the embryo proper, has lymphoid precursor potential *in vivo* and *in vitro*. Proc. Natl. Acad. Sci. USA **90:** 6581–6585.
4. Solvason, N., A. Lehuen & J. F. Kearney. 1991. An embryonic source of Ly1 but not conventional B cells. Intl. Immunol. **3:** 543–550.

5. ROLINK, A. & F. MELCHERS. 1991. Molecular and cellular origins of B lymphocyte diversity. Cell **66:** 1081.
6. ROLINK, A. & F. MELCHERS. 1993. B lymphopoiesis in the mouse, Adv. Immun. **53:** 123–155.
7. HARDY, R. R., C. E. CARMACK, S. A. SHINTON, J. D. KEMP & K. HAYAKAWA. 1991. Resolution and characterization of pro-B and pre-pro-B cell stages in normal mouse bone marrow. J. Exp. Med. **173:** 1213.
8. KINCADE, P. W. 1991. Molecular interactions between stromal cells and B lymphocyte precursors. Semin. Immunol. **3:** 379–390.
9. LI, Y-S., K. HAYAKAWA & R. R. HARDY. 1993. The regulated expression of B lineage associated genes during B cell differentiation in bone marrow and fetal liver. J. Exp. Med. **178:** 951–960.
10. KEARNEY, J. F., M. D. COOPER, J. KLEIN, E. R. ABNEY, R. M. E. PARKHOUSE & A. R. LAWTON. 1977. Ontogeny of Ia and IgD on IgM—bearing B lymphocytes in mice. J. Exp. Med. **146:** 297–302.
11. HAYAKAWA, K., D. TARLINTON & R. R. HARDY. 1994. Absence of MHC class II expression distinguishes fetal from adult B lymphopoiesis in mice. J. Immunol. **152:** 4801–4807.
12. LAM, K.-P. & A. M. STALL. 1994. Major histocompatibility complex class II expression distinguishes two distinct B cell developmental pathways during ontogeny. J. Exp. Med. **180:** 507–511.
13. OLTZ, E. M., G. D. YANCOPOULOS, M. A. MORROW, A. ROLINK, G. LEE, F. WONG, K. KAPLAN, S. GILLIS, F. MELCHERS & F. W. ALT. 1992. A novel regulatory myosin light chain gene distinguishes pre-B cell subsets and is IL-7 inducible. EMBO J. **11:** 2759–2767.
14. ROLINK, A., D. HAASNER, S.-I. NISHIKAWA & F. MELCHERS. 1993. Changes in frequencies of clonable pre-B cells during life in different lymphoid organs in mice. Blood **81:** 2290–2300.
15. HAMILTON, A. M., A. LEHUEN & J. F. KEARNEY. 1994. Immunofluorescence analysis of B-1 B cell ontogeny in the mouse. Int. Immunol. **6:** 355–361.
16. YANCOPOULOS, G. D., S. V. DESIDERIO, M. PASKIND, J. F. KEARNEY, D. BALTIMORE & F. W. ALT. 1984. Preferential utilization of the most JH-proximal VH gene segments in pre-B cell lines. Nature **311:** 727–733.
17. CARLSSON, L. & D. HOLMBERG. 1990. Genetic basis of the neonatal antibody repertoire: Germline V-gene expression and limited N region diversity. Int. Immunol. **3:** 639–643.
18. FEENEY, A. J. 1990. Lack of N regions in fetal and neonatal mouse immunoglobulin V-D-J junctional sequences. J. Exp. Med. **172:** 1377–1390.
19. MEDINA, C. A. & J. M. TEALE. 1993. Restricted k chain expression in early ontogeny: Biased utilization of Vk exons and preferential Vk-Jk recombinations. J. Exp. Med. **177:** 1313–1330.
20. KEARNEY, J. F., J. BARTELS, A. M. HAMILTON, A. LEHUEN, N. SOLVASON & M. VAKIL. 1992. Development and function of the early B cell repertoire. Int. Rev. Immunol. **8:** 247–257.
21. HARTLEY, S. B., J. CROSBIE, R. BRINK, A. B. KANTOR, A. BASTEN & C. C. GOODNOW. 1991. Elimination from peripheral lymphoid tissues of self-reactive B lymphocytes recognizing membrane-bound antigens. Nature (London) **353:** 765.
22. GOODNOW, C. C., J. CROSBIE, H. JORGENSEN, R. A. BRINK & A. BASTEN. 1989. Induction of self-tolerance in mature peripheral B lymphocytes. Nature **342:** 385–391.
23. TIEGS, S. L., D. M. RUSSELL & D. NEMAZEE. 1993. Receptor editing in self-reactive bone marrow B cells. J. Exp. Med. **177:** 1009–1020, 1993.
24. GAY, D., T. SAUNDERS, S. CAMPER & M. WEIGERT. 1993. Receptor editing: An approach by autoreactive B cells to escape tolerance. J. Exp. Med. **177:** 999–1008.
25. CYSTER, J. G., S. B. HARTLEY & C. C. GOODNOW. 1994. Competition for follicular niches excludes self-reactive cells from the recirculating B-cell repertoire. Nature **371:** 389–395.
26. GU, H., I. FORSTER & K. RAJEWSKY. 1990. Sequence homologies, N sequence insertion and JH gene utilization in VHDJH joining: Implications for the joining mechanism

and the ontogenetic timing of Ly1 B cell and C-CLL progenitor generation. EMBO J. **9:** 2133–2140.

27. GU, H., D. TARLINTON, W. MULLER, K. RAJEWSKY & I. FOSTER. 1991. Most peripheral B cells in mice are ligand selected. J. Exp. Med. **173:** 1357–1371.
28. PANDEY, A., L. W. TJOELKER & C. B. THOMPSON. 1993. Restricted immunoglobulin junctional diversity in neonatal B cells results from developmental selection rather than homology-based V(D)J joining. J. Exp. Med. **177:** 329–337.
29. KENNY, J. J., C. O'CONNEL, D. G. SIECKMANN, R. T. FISHER & D. L. LONGO. 1991. Selection of antigen-specific, idiotype-positive B cells in transgenic mice expressing a rearranged M167-μ heavy chain gene. J. Exp. Med. **174:** 1189–1201.
30. ARNOLD, L. W., C. A. PENNELL, S. K. MCCRAY & S. H. CLARKE. 1994. Development of B-1 cells: Segregation of phosphatidyl choline-specific B cells to the B-1 population occurs after immunoglobulin gene expression. J. Exp. Med. **179:** 1585–1595.
31. HAYAKAWA, K., C. E. CARMACK, S. A. SHINTON & R. R. HARDY. 1992. Selection of autoantibody specificities in the Ly-1 B subset. Ann. N.Y. Acad. Sci. **651:** 346.
32. SPANOPOULOU, E., C. A. ROMAN, L. M. CORCORAN, M. S. SCHLISSEL, D. P. SILVER, D. NEMAZEE, M. C. NUSSENZWEIG, S. A. SHINTON, R. R. HARDY & D. BALTIMORE. 1994. Functional immunoglobulin transgenes guide ordered B-cell differentiation in rag-1-deficient mice. Genes Dev. **8:** 1030–1042.
33. YOUNG, F., B. ARDMAN, Y. SHINKAI, R. LANSFORD, T. K. BLACKWELL, M. MENDELSOHN, A. ROLINK, F. MELCHERS & F. W. ALT. 1994. Influence of immunoglobulin heavy and light-chain expression on B-cell differentiation. Genes Dev. **8:** 1043–1057.
34. LUNDKVIST, I., A. COUTINHO, F. VARELA & D. HOLMBERG. 1989. Evidence for a functional idiotypic network among natural antibodies in normal mice. Proc. Natl. Acad. Sci. USA **86:** 5074–5078.
35. HAURY, M., A. GRANDIEN, A. SUNDBLAD, A. COUTINHO & A. NOBREGA. 1994. Global analysis of antibody repertoires. I. An immunoblot method for quantitative screening of a large number of reactivities. Scand. J. Immunol. **39:** 79–87.
36. NOBREGA, A., M. HAURY, A. GRANDIEN, E. MALANCHERE, A. SUNDBLAD & A. COUTINHO. 1993. Global analysis of antibody repertoires. II. Evidence for specificity, self-selection and the immunological "homunculus" of antibodies in normal serum. Eur. J. Immunol. **23:** 2851–2859.
37. WINTER, G., A. D. GRIFFITHS, R. E. HAWKINS & H. R. HOOGENBOOM. 1994. Making antibodies by phage display technology. Ann. Rev. Immunol. **12:** 433–445.
38. HOLMBERG, D., S. FORSGREN, F. IVARS & A. COUTINHO. 1984. Reactions amongst IgM antibodies isolated from the early B cell repertoire of BALB/c mice. Eur. J. Immunol. **14:** 435–442.
39. ELLIOTT, M. & J. F. KEARNEY. 1992. Idiotypic regulation of development of the B cell repertoire. Ann. N.Y. Acad. Sci. **651:** 336–345.
40. KEARNEY, J. F. 1993. Idiotypic networks. *In* Fundamental Immunology. Vol. 3. W. E. Paul, Ed.: 887–902. Raven Press. New York.
41. AVRAMEAS, S. & T. TERNYNCK. 1993. The natural autoantibodies system: Between hypotheses and facts. Mol. Immunol. **30:** 1133–1142.
42. MALYNN, B. A., G. D. YANCOPOULOS, J. E. BARTH, C. A. BONA & F. W. ALT. 1990. Biased expression of JH-proximal VH genes occurs in the newly generated repertoire of neonatal and adult mice. J. Exp. Med. **171:** 843–859.
43. CARLSSON, L., C. OVERMO & D. HOLMBERG. 1992. Selection against N-region diversity in immunoglobulin heavy chain variable regions during the development of preimmune B cell repertoires. Int. Immunol. **4:** 549–553.
44. CHANG, Y., C. J. PAIGE & G. WU. 1992. Enumeration and characterization of DJH structures in mouse fetal liver. EMBO J. **11:** 1891–1899.
45. HUETZ, F., L. CARLSSON, U.-C. TORNBERG & D. HOLMBERG. 1993. V-region directed selection in differentiating B lymphocytes. EMBO J. **12:** 1819–1826.
46. TEALE, J. M. & C. A. MEDINA. 1992. Comparative expression of adult and fetal V gene repertoires. Int. Rev. Immunol. **8:** 95–111.
47. TERNYNCK, T. & S. AVRAMEAS. 1986. Murine natural monoclonal autoantibodies: A study of their polyspecificities and their affinities. Immun. Rev. **94:** 98–112.

48. ICHIYOSHI, Y. & P. CASALI. 1994. Analysis of the structural correlates for antibody oligoreactivity by multiple reassortments of chimeric human immunoglobulin heavy and light chain V segments. J. Exp. Med. **180:** 885–895.
49. HOLMBERG, D., G. WENNERSTROM, L. ANDRADE & A. COUTINHO. 1986. The high idiotypic connectivity of "natural" newborn antibodies is not found in adult mitogen-reactive B cell repertoires. Eur. J. Immunol. **16:** 82–87.
50. SHEFNER, R., G. KLEINER, A. TURKEN, L. PAPAZIAN & B. DIAMOND. 1991. A novel class of anti-DNA antibodies identified in BALB/c mice. J. Exp. Med. **173:** 287–296.
51. CHEN, C., V. A. ROBERTS & M. B. RITTENBERG. 1992. Generation and analysis of random point mutations in an antibody CDR2 sequence: Many mutated antibodies lose their ability to bind antigen. J. Exp. Med. **176:** 855–866.
52. LEHUEN, A., J. BARTELS & J. F. KEARNEY. 1992. Characterization, specificity and Ig V gene usage of anti-lymphocyte monoclonal antibodies from perinatal mice. Int. Immunol. **4:** 1073–1084.
53. KEARNEY, J. F. 1992. Early B cell repertoire. *In* Current Topics in Microbiology and Immunology. Springer-Verlag. Berlin. pp. 81–84.
54. MCCORMACK, W. T., L. W. TJOELKER, C. F. BARTH, L. M. CARLSON, B. PETRYNIAK, E. H. HUMPHRIES & C. B. THOMPSON. 1989. Selection for B cells with productive IgL gene rearrangements occurs in the bursa of *Fabritius* during chicken embryonic development. Genes Dev. **3:** 838–847.
55. REYNAUD, C.-A., C. R. MACKAY, R. G. MULLER & J.-C. WEILL. 1991. Somatic generation of diversity in a mammalian primary lymphoid organ: The sheep ileal Peyer's patches. Cell **64:** 995–1005.
56. CHEN, X. & J. F. KEARNEY. Positive selection of spontaneous auto-reactive B cells. Submitted.

Analysis of Immunoglobulin Gene Rearrangement in Fetal B-Cell Progenitors Developing *in Vivo* or *in Vitro*

AARON J. MARSHALL, BARBARA L. KEE, CHRISTOPHER J. PAIGE, AND GILLIAN E. WU

The Wellesley Hospital Research Institute and Department of Immunology University of Toronto Toronto, Ontario, Canada M4Y 1J3

The mature B-cell repertoire has been well studied; nevertheless, the molecular and cellular mechanisms that shape the repertoire are still poorly understood. We are currently using both *in vivo* and *in vitro* models of B-cell development, in conjunction with PCR assays specific for rearranged immunoglobulin (Ig) loci, to study the mechanisms that influence repertoire formation during B-cell development. One approach that we are using is to examine immunoglobulin gene rearrangements in mouse fetal liver at various time points during a period of ontogeny spanning the initial onset of rearrangement to the appearance of surface Ig-positive B cells.[1,2] This analysis allows us to follow the course of Ig repertoire formation from the emergent repertoire to the mature repertoire. We have also begun to examine Ig rearrangements in an *in vitro* system in which lineage-uncommitted progenitor cells from early fetal liver mature into mitogen-responsive, Ig-secreting B cells during culture.[3,4] This approach also allows us to follow the course of Ig repertoire formation, but has all of the experimental advantages of an *in vitro* system.

In this study we developed and characterized a DNA-based PCR assay that can proportionally amplify the large majority of VDJ_H rearrangements and used this assay to investigate repertoire formation *in vivo* and *in vitro*. The assay uses a degenerate V_H primer that is homologous to the framework 1 region of all V_H genes in conjunction with a single sequence primer homologous to a region just 3′ of J_H4. In order to verify the ability of this primer pair to amplify a large diversity of VDJ_H rearrangements and provide an accurate representation of the VDJ_H repertoire in experimental samples, several control experiments were performed. First, when DNA samples from a panel of B-cell lines were amplified with these primers, it was found that one or more VDJ_H rearrangements could be amplified from 39 out of the 40 lines that were screened. Second, using defined mixtures of VDJ_H rearrangements from various B-cell lines, we determined that the proportion of VDJ_H structures in the amplified products is approximately the same as in the original DNA mixtures. Third, we determined the utilization of four V_H families in VDJ_H rearrangements amplified from Balb/c or C57BL6 splenic LPS blasts and found that the pattern of V_H usage was very similar to that seen using B-cell colony assays.[5]

We used this assay to study the kinetics and repertoire of VDJ_H rearrangement in mouse fetal liver. Quantitative analysis indicated that VDJ_H rearrangement begins at about day 13 of gestation, in contrast to DJ_H rearrangement, which begins at day 12, and $VJ\kappa$ rearrangement, which begins at day 14. Fetal liver VDJ_H rearrangements were analyzed for V_H usage, and it was found that a single V_H segment called V_H81_x[6] predominates on early days of gestation. However, this predominance varies among mouse strains and declines sharply during ontogeny.

We have also begun to examine Ig rearrangement in fetal B-cell progenitors developing *in vitro*. In our assay, day 12 fetal liver cells are enriched for lineage-uncommitted B-cell progenitors and cultured under conditions that allow maturation to mature Ig-secreting B cells.[3,4] Using PCR analysis, we found that DJ_H rearrangements were detected on day 3 of culture, whereas VDJ_H rearrangements were detected at day 6 of culture. Rearrangement was reduced below levels of reliable detection when the pre-B-cell growth factor interleukin 7 was omitted from the culture. Analysis of V_H81_x utilization demonstrated that V_H81_x was also highly used in the repertoire that develops *in vitro*, both in the presence and the absence of stromal cells. Upon extended culture, however, both in conditions that select for cells that are able to mature and in conditions that maintain immature cells, V_H81_x utilization drops by approximately fourfold. This may reflect the same selection events that are responsible for the decline in V_H81_x utilization *in vivo*.

REFERENCES

1. CHANG, Y., C. J. PAIGE & G. E. WU. 1992. Embo J. **11:** 1891–1899.
2. RAMSDEN, D., C. PAIGE & G. WU. 1994. J. Immunol. **153:** 1150–1160.
3. CUMANO, A. & C. PAIGE. 1992. EMBO J. **11:** 593–601.
4. KEE, B. L., A. CUMANO, N. N. ISCOVE & C. J. PAIGE. 1994. Int. Immunol. **6:** 401–407.
5. WU, G. & C. PAIGE. 1986. EMBO J. **5:** 3475–3481.
6. YANCOPOULOUS, G. D., S. V. DESIDERO, M. PASKIND, J. F. KEARNEY, D. BALTIMORE & F. W. ALT. 1984. Nature **311:** 727–733.

Development of the Antibody Repertoire as Revealed by Single-Cell PCR of FACS-Sorted B-Cell Subsets[a]

AARON B. KANTOR,[b,c] CYNTHIA E. MERRILL,[d]
JOHN D. MACKENZIE,[c] LEONORE A. HERZENBERG,[c]
AND JAN L. HILLSON[d]

[c]*Department of Genetics*
Beckman Center B007
Stanford University Medical Center
Stanford, California 94305-5125

[d]*Department of Medicine*
University of Washington Medical Center
Seattle, Washington 98195

We distinguish three murine B-cell lineages[1-4]: conventional B cells (B-2 cells), which develop late and are continually replenished throughout life from progenitors in adult bone marrow; B-1a cells (Ly-1 or CD5 B cells), which develop early and maintain their numbers by self-replenishment; and B-1b cells (Ly-1 B "sister" cells), which share many of the properties of Ly-1 B cells, including self-replenishment and feedback regulation of development, but which can also readily develop from progenitors in adult bone marrow. Several studies indicate that the expressed antibody repertoire of mature B-1a and conventional B cells differ with respect to V_H gene usage and antigen specificity. There is no information on B-1b cells. Earlier methods of repertoire analysis, which include generation of immortalized cell lines (hybridomas and transformations), *in situ* hybridization with V_H probes, and the amplification of cDNA from polyclonal cells, are not necessarily representative of the populations studied. For example, transformation and hybridoma production are very inefficient and may select cells in favored states of activation, and bulk cDNA libraries favor cells with the most message.

We have developed a method for making cDNA from individual, unstimulated, FACS-sorted murine and human B cells that permits the recovery and amplification of transcripts from up to 90% of the cells and will allow the analysis of multiple genes from a single cell. Cells are sorted at one cell per tube, lysed in hypotonic medium containing RNAse inhibitors, and cDNA is prepared by using random hexanucleotide primers. No purification of the RNA or cDNA is needed. The murine cDNA is amplified with primers homologous to the 5′ portion of $C\mu$ and to a conserved consensus region in framework I of V_H. We have recovered transcripts from hybridomas and sorted B cells that represent the 14 murine V_H families. This result, along with our high recovery rate, demonstrates that the procedure is void of bias for particular rearrangements.

[a] This work is supported by National Institutes of Health Grants HD-01287 and LM-04836 to Leonore Herzenberg and NRSA awards AI-07937 and AI-07290 to A. B. Kantor.

[b] Address for correspondence: A. B. Kantor, AmCell Corporation, 1190 Bordeaux Drive, Sunnyvale, CA 94089.

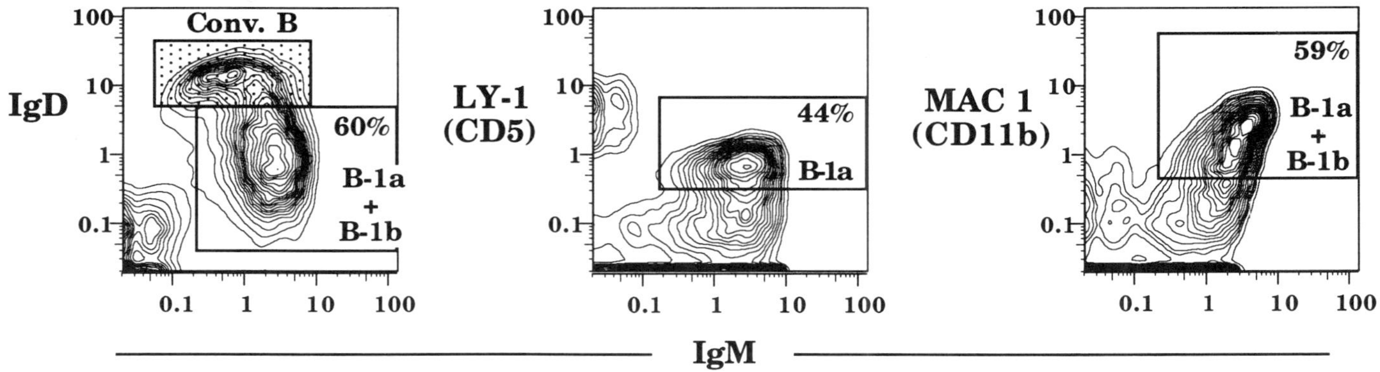

FIGURE 1. FACS analysis of peritoneal B-cell populations. Conventional B cells are identified by a broad, positive IgM and tight, bright IgD FACS profile. They are negative for Ly-1 and Mac 1. All B-1 cells are IgM bright and low to moderate for IgD. B-1 cells are also MAC 1 positive in the peritoneum. B-1 cells are divided into B-1a cells, which are $CD5^+$, and B-1b cells, which are $CD5^-$.

The method allows us to analyze V gene expression from any population of B cells that we can define phenotypically by FACS. We begin with B-1a, B-1b, and conventional B cells from the adult Balb/c peritoneum. The phenotype is shown in FIGURE 1. Conventional B cells are IgM^{lo}, IgD^{hi}, and $CD5^-$. Both B-1a and B-1b cells are IgM^{hi}, IgD^{lo}; however, B-1a cells are $CD5^+$, and B-1b cells are $CD5^-$. Each population is bulk sorted first, based on size, viability, and expression of IgM, IgD, and Ly-1. After reanalysis, single cells are sorted into the lysis solution and snap-frozen for future analysis. The data described below is from 170 transcripts distributed among the three populations.

The three B-cell populations show some similarities in V_H gene usage with respect to the V_H families. In all cases, the J558 family is used most frequently, and the Q52 family is used the second most frequently. Interestingly, the V_H 10 and 3660 families are overrepresented among B-1b cells. In order to evaluate the use of individual V_H gene segments, we developed a database of known and putative germline genes. The most striking finding from this analysis is that a single Q52 V_H element is used in 22% of all B-1a cells, 13% of all B-1b cells, but only 3% of B-2 cells. Some of these cells are among the phosphatidylcholine liposome-binding cells (anti-bromelein-treated mouse red blood cells). Cells with this specificity also commonly use V_H11 and V_H12, as previously shown by others.[5–8]

Does the repertoire of B-1a cells found in the adult animal reflect any features of fetal/neonatal B-cell development? It has been well established that fetal and neonatal B cells rarely have N-region insertions at the VD and DJ junctions, whereas most such junctions recovered from adults have N-regions.[9–13] Contrary to a previous study that evaluated B-1a hybridomas,[12,14] we find that B-1a cells isolated from the adult peritoneum have a significantly larger proportion of V_H transcripts with no (or few) N bases at both the V-D and D-J junctions than either B-1b or conventional B cells. In fact 25% of the B-1a cell transcripts, but only 5% of the conventional B-cell transcripts, lack N-region insertions. This result is consistent with the model that many of the B-1a cells present in the adult, which contain no or few N-region nucleotides, arose very early in ontogeny when TdT activity is low[15,16] and survived via self-replenishment into adulthood. However, TdT expression, as monitored by the incidence of N-region insertions, clearly does not define B-1a cells, since most V_H transcripts contain N-regions, consistent with B-1a cell VDJ rearrangement occurring through the first six weeks of life.[2]

We anticipate that the approach presented here will be extremely powerful in analyzing the developmental and selective events that produce the expressed antibody repertoire.

ACKNOWLEDGMENTS

We thank Ometa Herman for technical assistance.

REFERENCES

1. STALL, A. M., S. ADAMS, L. A. HERZENBERG & A. B. KANTOR. 1992. Characteristics and development of the murine B-1b (Ly-1 B sister) cell population. Ann. N.Y. Acad. Sci. **651:** 33.
2. KANTOR, A. B. & L. A. HERZENBERG. 1993. Origin of murine B cell lineages. Annu. Rev. Immunol. **11:** 501.

3. KANTOR, A. B., A. M. STALL, S. ADAMS, L. A. HERZENBERG & L. A. HERZENBERG. 1992. Differential development of progenitor activity for three B-cell lineages. Proc. Natl. Acad. Sci. USA **89:** 3320.
4. KANTOR, A. B., A. M. STALL, S. ADAMS, K. WATANABE & L. A. HERZENBERG. 1994. De novo development and self-replenishment of B cells. Intl. Immunol. **7:** 55–68.
5. HARDY, R. R., C. E. CARMACK, S. A. SHINTON, R. J. RIBLET & K. HAYAKAWA. 1989. A single VH gene is utilized predominantly in anti-BrMRBC hybridomas derived from purified Ly-1 B cells. Definition of the VH11 family. J. Immunol. **142:** 3643.
6. MERCOLINO, T. J., L. W. ARNOLD, L. A. HAWKINS & G. HAUGHTON. 1988. Normal mouse peritoneum contains a large population of Ly-1$^+$ (CD5) B cells that recognize phosphatidyl choline. Relationship to cells that secrete hemolytic antibody specific for autologous erythrocytes. J. Exp. Med. **168:** 687.
7. MERCOLINO, T. J., A. L. LOCKE, A. AFSHARI, D. SASSER, W. W. TRAVIS, L. W. ARNOLD & G. HAUGHTON. 1989. Restricted immunoglobulin variable region gene usage by normal Ly-1 (CD5$^+$) B cells that recognize phosphatidyl choline. J. Exp. Med. **169:** 1869.
8. PENNELL, C. A., T. J. MERCOLINO, T. A. GRDINA, L. W. ARNOLD, G. HAUGHTON & S. H. CLARKE. 1989. Biased immunoglobulin variable region gene expression by Ly-1 B cells due to clonal selection. Eur. J. Immunol. **19:** 1289.
9. BANGS, L. A., I. E. SANZ & J. M. TEALE. 1991. Comparison of D, JH, and junctional diversity in the fetal, adult, and aged B cell repertoires. J. Immunol. **146:** 1996.
10. CARLSSON, L. & D. HOLMBERG. 1990. Genetic basis of the neonatal antibody repertoire: Germline V-gene expression and limited N-region diversity. Int. Immunol. **2:** 639.
11. FEENEY, A. J. 1990. Lack of N regions in fetal and neonatal mouse immunoglobulin V-D-J junctional sequences. J. Exp. Med. **172:** 1377.
12. GU, H., I. FORSTER & K. RAJEWSKY. 1990. Sequence homologies, N sequence insertion and JH gene utilization in VHDJH joining: Implications for the joining mechanism and the ontogenetic timing of Ly1 B cell and B-CLL progenitor generation. EMBO J. **9:** 2133.
13. MEEK, K. 1990. Analysis of junctional diversity during B lymphocyte development. Science **250:** 820.
14. GU, H., I. FORSTER & K. RAJEWSKY. 1992. Study of murine B-cell development through analysis of immunoglobulin variable region genes. Ann. N.Y. Acad. Sci. **651:** 304.
15. DESIDERIO, S. V., G. D. YANCOPOULOS, M. PASKIND, E. THOMAS, M. A. BOSS, N. LANDAU, F. W. ALT & D. BALTIMORE. 1984. Insertion of N-regions into heavy chain genes is correlated with the expression of terminal deoxytransferase in B cells. Nature **311:** 752.
16. GREGOIRE, K. E., I. GOLDSCHNEIDER, R. W. BARTON & F. J. BOLLUM. 1979. Ontogeny of terminal deoxynucleotide transferase-positive cells in lymphohemopoietic tissues of rat and mouse. J. Immunol. **123:** 1347.

Analysis of DHQ52 Gene Segment Transcription and Rearrangement During B-Cell Development in Human Fetal Bone Marrow[a]

F. E. BERTRAND III,[b] L. G. BILLIPS,[b]
AND H. W. SCHROEDER, JR.[b,c]

University of Alabama at Birmingham
Division of Developmental and Clinical Immunology
Departments of [b]Microbiology and [c]Medicine
Birmingham, Alabama 35294-3300

Human B lymphopoiesis is a highly regulated process that occurs in the bone marrow where a pluripotent stem cell matures into an immunoglobulin-expressing B cell. The hallmark of this developmental pathway is somatic DNA rearrangement of the gene segments that encode the heavy and light chains of the antibody molecule. Rearrangement usually begins in the immunoglobulin heavy chain (IgH) locus with germline transcription of IgH gene segments. Next, one of 30 D_H segments is joined to one of six J_H elements. Subsequently, the DJ join undergoes rearrangement with one of about functional 50 V_H segments, creating a complete heavy chain variable domain.[1–4]

Although this process can potentially generate more than 10^8 random combinations, the rearrangement process is regulated such that specific IgH gene segment combinations are preferred at different stages of human ontogeny. In particular, the DHQ52 gene segment is present in nearly 40% of VDJ and DJ IgH gene segment rearrangements in human fetal bone marrow. In contrast, DHQ52 is found in fewer than 1% of VDJ and DJ rearrangements in adult bone marrow and peripheral blood.[5–8] Therefore, DJ rearrangement in fetal bone marrow may be regulated differently than in adult bone marrow.

In order to precisely define the stages during which DHQ52 transcription and rearrangement may be regulated during human fetal B lymphopoiesis, fetal bone marrow cells were sorted into subpopulations representative of stem cell and progenitor, pro-B, pre-B I and II, and B-cell populations on the basis of surface expression of the hematopoietic markers CD34, CD19, and IgM (FIG. 1). RT-PCR followed by Southern analysis was used to detect germline and rearranged DHQ52 DJ transcripts in each subpopulation.

Germline transcription of the DHQ52 gene segment was detected in all of the B-cell subpopulations tested. Furthermore, a rearranged DHQ52 transcript was detected at all stages of B-cell differentiation, even before expression of the presumed pan-B-cell marker CD19 (FIG. 1). The presence of DHQ52 to J_H rearrangements was confirmed by analysis of genomic DNA isolated from the subpopulations.

[a] This work was supported, in part, by AI33621, AI07051, AI34568, and AI30879.

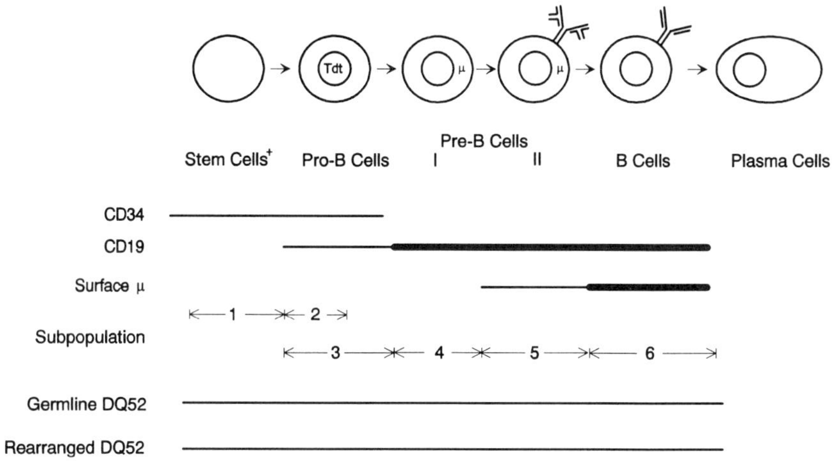

FIGURE 1. Model of human B-cell development based on CD34, CD19, and IgM surface expression and DHQ52 transcription. Fetal bone marrow cells were sorted into the following subpopulations: (1) $CD34^+$ $CD19^-$ (stem cells and committed progenitors); (2) $CD34^+$ $CD19^+$ (pro-B); (3) $CD19^{lo}$ IgM^- (pro-B); (4) $CD19^{hi}$ IgM^- (pre-B I); (5) $CD19^+$ IgM^{lo} (pre-B II); and (6) $CD19^+$ IgM^+ (B cells). This figure also correlates DHQ52 transcription and rearrangement with B lymphopoiesis.

These data suggest a model of human fetal B-cell development in which IgH rearrangement begins with the activation of the DHQ52 locus, which is phenotypically distinguished by germline transcription of DHQ52. Subsequently, the DHQ52 exon can undergo rearrangement with one of the six J_H gene segments to form a DHQ52-J_H join. Initial activation of the DHQ52 locus, followed by DJ rearrangement, occurs before the expression of CD19, excluding this molecule from being involved in the signaling pathway for IgH activation in this population of human fetal B lineage precursors.

ACKNOWLEDGMENTS

The authors thank Dr. Larry Gartland for help in sorting B-cell subpopulations and Drs. Peter Burrows, Hiromi Kubagawa, and Max Cooper for their helpful discussions.

REFERENCES

1. Uckun, F. M. 1990. Blood **76:** 1908–1923.
2. Alt, F. W., K. T. Blackwell, R. A. DePinho, M. G. Reth & G. D. Yancopoulos. 1986. Immunol. Rev. **89:** 5–30.
3. Matsuda, F., E. K. Shin, H. Nagaoka, R. Matsumura, M. Haino, Y. Fukita, S. Taka-ishi, T. Imai, J. H. Riley, R. Anand, E. Soeda & T. Honjo. 1993. Nature Genet. **3:** 88–94.

4. TOMLINSON, I. M., G. WALTER, J. D. MARKS, M. B. LLEWELYN & G. WINTER. 1992. J. Mol. Biol. **227:** 776–798.
5. SCHROEDER, H. W. & J. Y. WANG. 1990. Proc. Natl. Acad. Sci. USA **87:** 6146–6150.
6. SHIOKAWA, S., F. MORTARI, C. NUNEZ & H. W. SCHROEDER, JR. FASEB J. **8:** 2701.
7. BANGS, L. A., I. E. SANZ & J. M. TEALE. 1991. J. Immunol. **146:** 1996–2004.
8. YAMADA, M., R. WASSERMAN, B. A. REICHARD, S. SHANE, A. H. CANTON & G. ROVERA. 1991. J. Exp. Med. **173:** 395–407.

Fetal versus Adult PreB or B Cells: The Human VH Repertoire[a]

CÉCILE TONNELLE,[b] ANNE MARIE CUISINIER,[b]
LAURENT GAUTHIER,[b] VALÉRIE GUELPA-FONLUPT,[b]
MICHÈLE MILILI,[b] CLAUDINE SCHIFF,[b]
AND MICHEL FOUGEREAU[b,c]

[b]*Centre d'Immunologie Marseille Luminy*
Case 906
13288 Marseille Cedex 9, France

[c]*CIML and Institut Universitaire de France*
France

INTRODUCTION

The large diversity of the expressed human VH repertoire results from several events that include combinatorial recombinations between *ca.* 50 different VH genes, at least 30 DH, and 6 JH segments, followed by somatic mutations that occur during the maturation of B lymphocytes.

The human IGH locus, located at 14q32.3, covers approximately 1100 kb, with some variations according to the haplotypes. The most recent reports indicate that it contains about 100 variable genes, of which 40 to 50 appear functional.[1–3] Seven families could be defined on the basis of sequence homology, each containing a variable number of genes, ranging from one for VH6 to more than 50 for VH3. The complete VH locus has been sequenced, confirming the interspersed organization of VH families[1,3,4] as opposed to most mouse situations.

The steps of differentiation of the B lineage are characterized by the successive events of recombination and expression that occur first on heavy then on light chain genes of immunoglobulin and by sequential expression of several surface antigens such as CD10, CD19, CD20, and CD22.[5] At the preB stage, the rearranged μ chain is expressed in association with two pseudo-light (ΨL) chains, encoded by the λ-like and VpreB genes, which are directly expressed without rearrangement.[6,7] Gene targeting of either μm[8] or λ5[9] in the mouse has been shown to result in the arrest of differentiation at the preB step. This makes the μ-ΨL complex a good candidate for regulating critical steps in the transition between preB- and B-cell differentiation, presumably through transducing signal(s) initiated upon specific triggering at the cell surface. This raises the question of a particular VH repertoire and the selection of this repertoire at this step of differentiation.

Using a monoclonal antibody directed against a peptide of the COOH-terminal part of VpreB, discrete steps of B-cell differentiation could be defined in normal adult bone marrow, in coordination with the expression of certain membrane

[a] This work was supported by CNRS (Centre National de la Recherche Scientifique), INSERM (Institut National de la Santé et de la Recherche Médicale), Ministère de la Recherche et de l'Enseignement Supérieur, and ARC (Association de la Recherche contre le Cancer).

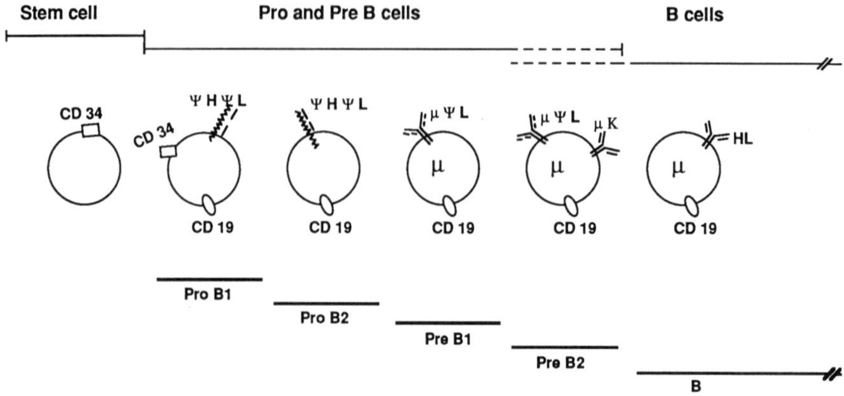

FIGURE 1. The different steps of VpreB expression in human adult bone marrow[10] are indicated on the basis of the model of early B-cell differentiation.

markers.[10] The VpreB marker is present only on a very small percentage of CD34$^+$ cells and appears partially expressed on CD10$^+$ cells, already indicating that it is not expressed until the differentiation is indeed engaged in the B pathway on CD19$^+$ cells (19%), just before the occurrence of CD20, CD21, and CD22. Relative proportions of three VpreB-positive subpopulations could be defined as 23% (pro B1 + B2) without μ chain, 4% (preB1) $\mu\Psi$L, and 73% (preB2) $\mu\Psi$L/μLC (FIG. 1). The pre-B2 population expresses essentially κ (54%) or λ (19%). Somewhat different conclusions were obtained by the group of M. D. Cooper, who identified ΨL at the cell surface of a μ + ΨL + preB-cell population only.[11] We have recently analyzed the VH3 repertoire in a VpreB-positive /κ + λ-negative cell population isolated by FACS from adult bone marrow (Milili *et al.*, in preparation). The VH3-containing transcripts were amplified and sequenced. The characteristics of this repertoire were compared with those of the fetal repertoire,[12–18] of X-linked agammaglobulinemia (XLA), characterized by a blockade at the preB stage,[19] and with B cells from PBL.[20–22]

RESULTS AND DISCUSSION

VH Germline Gene Usage

The first crude approach indicated that only VH5 and VH6 family transcripts were detected in a fetal liver at seven weeks of gestation,[12] whereas all families could be observed at eight weeks either in fetal liver or in bone marrow, suggesting a very rapid expansion of the VH repertoire. Furthermore, when looking at the VH family usage frequencies, it occurred that the "adult" pattern of EBV-transformed B cells derived from various bone marrows between 8 and 13 weeks of gestation was rapidly acquired with a major expression of the VH3 subgroup and an expression of a pattern of VH usage correlating with the estimated number of VH genes.[13] This overexpression of VH3 was not found in a 13-week-old liver,[14] where VH4 accounted for over 40% of the transcripts differing from previous analyses performed on human fetal liver at 15 and 18 weeks of gestation that emphasized a major expression of VH3,[15] or at 16 to 24 weeks where a bias of utilization of VH6 was reported.[23] These differences might be due to the discrete technical approaches (anchored PCR and direct screening on colonies or cDNA clone analysis or RNA quantification on Northern) or to the variation of the individual repertoire with time, as it was described in adult peripheral blood.[22]

A detailed analysis of the VH gene usage was then undertaken, using a classical sequencing approach on materials derived from human fetal liver at 8 and 13 weeks of gestation. A clear pattern of preferential gene emerged, characterized by a restricted gene usage. No clear evidence of a preferential expression of the most 3′ VH genes can be generalized.[14]

The approach used with the total RNA extracts from fetal liver did not provide information as to the nature, preB or B, of the transcripts. We thus turned to the analysis of the repertoire expressed in human normal bone marrow preB cells sorted on the basis of surface VpreB chain expression, with negative selection on L chain expression. In this study, we focused on the VH3 family, as representing the most widely expressed. This provided a basis for a comparison of VH3 genes used in fetal liver,[14–18] adult preB cells (Milili *et al.*, in preparation), XLA,[19] and adult B cells,[22] in connection with the gene localization described by Matsuda *et al.*[1] and Cook *et al.*[3] (FIG. 2). The first remark was the common usage of certain genes, mostly 3.30 and 3.23, consistently overrepresented in all situations, whereas other genes were not found expressed. This bias of expression could be due to the number of copies of the same gene, as 3.30 appeared in triplicate[3] and 3.23 occasionally in duplicate.[24] The existence of multiple copies might of course reflect an evolutionary pressure to retain and amplify these sequences in the genome. The particular strength of promoters, as the accessibility of certain V genes to

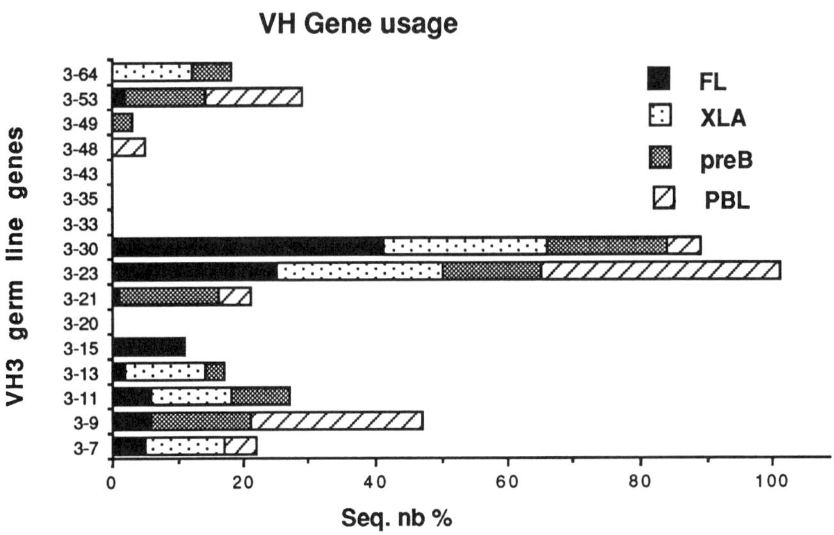

FIGURE 2. Comparison of VH3 gene usage between fetal liver (FL),[14–17] XLA,[19] preB (Milili *et al.*, in preparation), and B cells PBL[22] following the localization and gene number described by Matsuda *et al.*[1]

recombinases, could be other types of explanations. These genes participated in building autoimmune repertoire[25] and probably could be involved in the first step of selection.

VH Germline Transcripts

Identification of germline VH transcripts in mouse preB cells has led to the proposals that transcription might be a prerequisite to gene rearrangement, in connection with DNA accessibility to the recombinase machinery.[26] In the mouse, a possible restriction of the VH germline transcripts to some members of the VH J558 family has been reported,[27] and similar observations have indicated that transcripts of the VH5 family, a paucigene subgroup, could be identified in the human.[28] A systematic screening, performed on an 8-week-old fetal liver, has allowed us to extend this observation to VH germline transcripts pertaining to all six families and including pseudogenes.[14] Furthermore, sequencing of these transcripts revealed that this population profoundly differed from the VH repertoire that was used in rearranged genes. Restriction of germline transcripts to one given family is clearly not the rule, and our data rather suggest that a large portion of the IGVH locus, if not the entire locus, is indeed accessible at this stage, suggesting that the gene usage selection that is observed is not directly linked to a preferential rearrangement resulting from a targeted accessibility. Surprisingly, four VH3 transcripts (TABLE 1) resulted from the expression of orphons localized on chromosome 16,[29] thus indicating a clear dissociation between the transcription of any particular VH gene and its immediate involvement in V-D-J rearrangement. Transcription would thus be most likely considered a passive witness of DNA accessibility.

Somatic Mutations

All sequences analyzed, whatever their origin (fetal, preB, or XLA), contained only occasional evidence of somatic mutations, with the exception of a series of VH6 clones derived from an 8-week-old fetal liver, derived from the same VDJ

TABLE 1. Fetal Germline Transcripts

Clones	Germline Genes	Difference Number
VHGL1.8	DP10[a]/1.69	2
VHGL1.2	DP8[a]/1.2[b]	5
VHGL1.9	?	
VHGL3.1	DP38[a]/3.15[b]	0
VHGL3.4	HC16-15[c]	1
VHGL3.5	HC16-15/HC16-16[c]	3
VHGL3.6	HC16-15[c]	1
VHGL3.9	HC16-16[c]	1
VHGL3.7	?	
VHGL3.0	?	

[a] Cook *et al.*[3]
[b] Matsuda *et al.*[1]
[c] Tomlinson *et al.*[29]

rearrangement,[14] where extensive mutations were accumulated in the CDR3. This observation is puzzling, and probably rare. It deserves further investigation, as it would suggest that the specialized enzymatic machinery that generates somatic mutation may be at hand very early during human development.

JH Gene Usage

We had already noticed a relatively increased usage of JH3 and JH2 in the fetal repertoire,[14] but the predominance of JH4 expression, also reported in other studies,[18,20–22] was consistently observed as well (FIG. 3). The predominance of JH4 could be mechanistic, as was demonstrated in human Ig transgenic minilocus in mouse,[30] where a predominant rearrangement of a VH pseudogene occurred with DQ52 and JH4. Such a bias could not be explained by any protein selection, but only by some mechanistic facilities: DNA accessibility to recombinases and sequences for recombination. If the predominance of JH4 were mechanistic, however, a necessary selective step would have occurred after this first rearrangement. In fact, only the fetal repertoire expressed a relatively high percentage of DQ52 compared to the other VH repertoire. Therefore, elimination of such clones or secondary rearrangements could occur in the adult repertoire.

D Gene Usage

All D gene families were expressed, with a general predominance of DXP (considered the largest D family) and contributing between 24 to 45% of DH genes

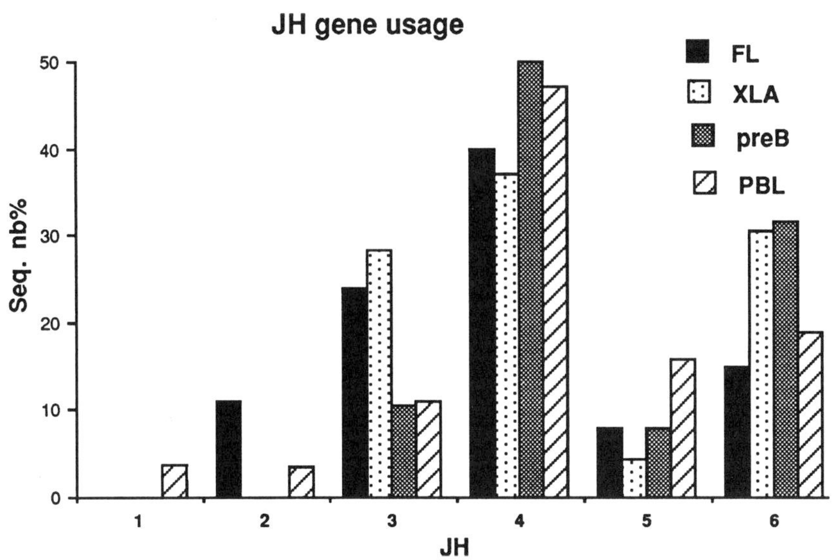

FIGURE 3. Comparison of JH gene usage between fetal liver (FL),[14] XLA,[19] preB (Milili *et al.*, in preparation), and B cells from PBL.[20]

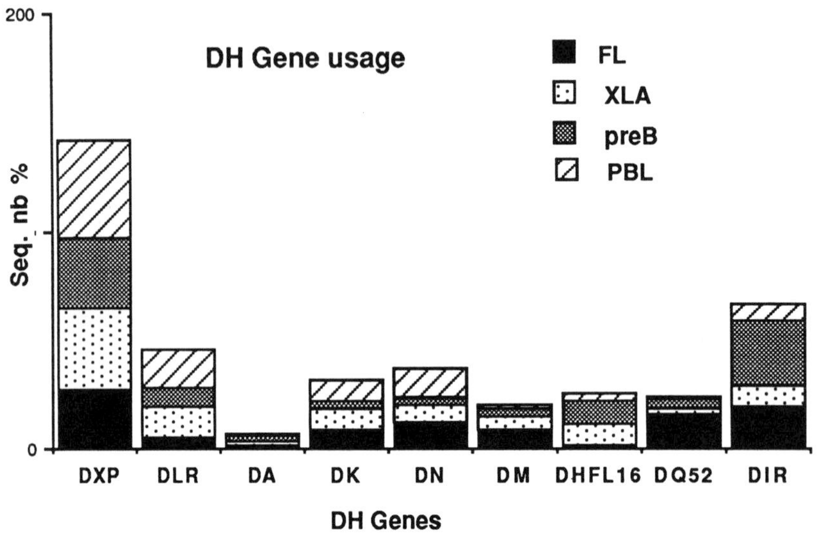

FIGURE 4. Comparison of DH gene usage between fetal liver (FL),[14] XLA,[19] preB (Milili *et al.*, in preparation), and B cells from PBL.[21]

used in embryo,[14] XLA,[19] and PBL[21] (FIG. 4). Preferential usage of DQ52 seemed limited to fetal samples. DIR, composed of 151 nucleotides and thus appearing much longer than the other DH segments, was part of CDR3 in 30% of the sequences in the preB repertoire, in which it appeared largely overexpressed compared with other situations. The relatively low level of expression of DQ52 in our preB samples as compared to the fetal repertoire could be a mark of selection at this stage.

CDR3 Characteristics

N diversity, reported as very limited in the fetal repertoire although already present as early as the eighth week of gestation, remained low in the preB and XLA repertoires (35% and 73%, respectively, at the VD junction and 67% and 47%, respectively, for DJ). Classical mechanisms of rearrangement by using heptamer and nonamer adjacent regions could explain only very few CDR3 sequences. On many occasions, several D fusions were used to generate either repertoire. Some sequences might be explained on the basis of general recombination,[31] and the majority of CDR3 sequences cannot be simply explained by known somatic events, but appear rather to result from imprecision in the recombination, gene conversion, random addition of nucleotides, or action of exonucleases.[20,21]

One hallmark of fetal repertoire was the shorter length of CDR3. The comparison between fetal, preB, XLA, or PBL CDR3 sequences revealed a gradient of CDR3 length from fetal to PBL repertoire (FIG. 5). A close analysis suggested a bimodal distribution of the CDR3 lengths. Clearly, active depletion of the $\kappa^+\lambda^+$ population in the preB-cell preparation (sample BEL) considerably diminished

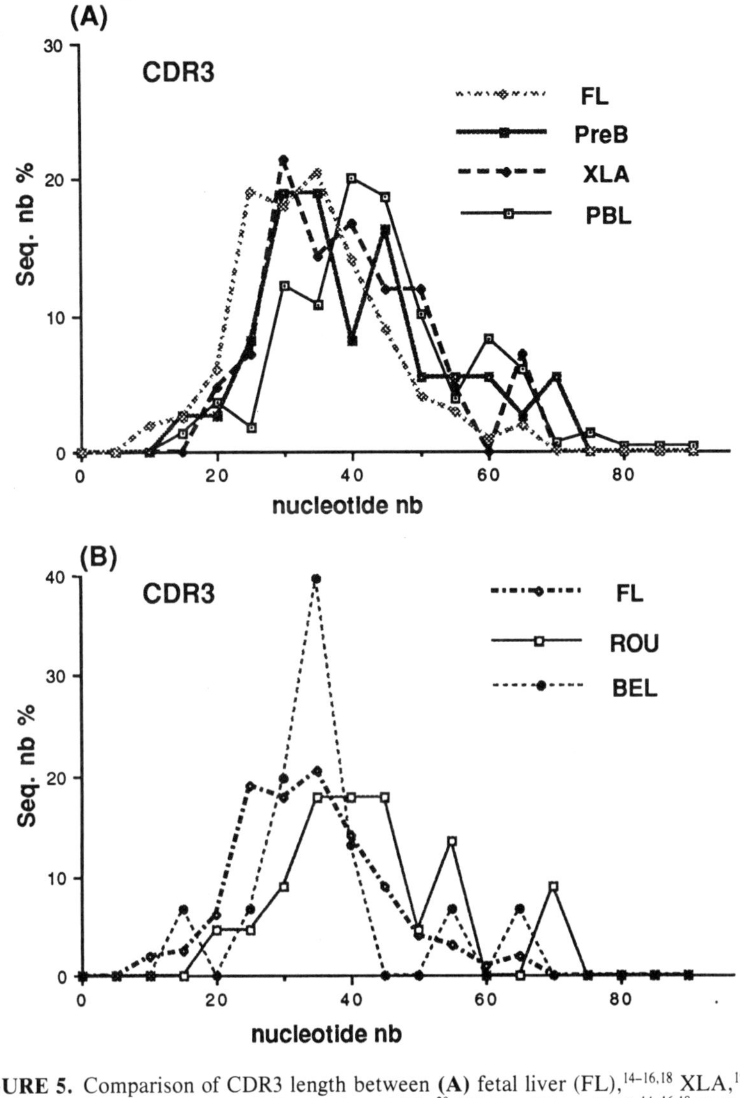

FIGURE 5. Comparison of CDR3 length between **(A)** fetal liver (FL),[14–16,18] XLA,[19] preB (Milili *et al.*, in preparation), and B cells from PBL.[20] **(B)** Fetal liver (FL),[14–16,18] BEL: preB sample depleted of the $\kappa + \lambda$ population; ROU: preB sample undepleted of the $\kappa + \lambda$ population (Milili *et al.*, in preparation).

the contribution of transcripts with the longest CDR3s, thus making it close to the fetal repertoire. PBL contained mostly transcripts of the second, longest CDR3 population, whereas XLA appeared bimodal, with a major population in the shortest portion, in relation with the leakiness of the defect. In the absence of active depletion of the $\kappa^+\lambda^+$ cells, the "preB" population also appears bimodal, which presumably suggests contamination with B cells (ROU sample).

TABLE 2. Gly and Tyr Residues in CDR3

Residue	Percent Gly CDR3	Percent Tyr CDR3
PreB[a]	96	87
XLA[b]	88	80
F.L.(a)[c]	69	64
F.L.(b)[d]	84	84

[a] Milili *et al.*, in preparation.
[b] Milili *et al.*[19]
[c] Cuisinier *et al.*[14]
[d] Pascual *et al.*[18]

We did not find in human repertoire the biased reading frames noted in the mouse DH usage,[32] although a slight preferential usage of the second reading frame for the direct D usage or the first one for the reverse was observed. This reading frame correlated with the presence of glycine codons. It was striking to note that, in either repertoire, more than 80% of CDR3s contained glycine codons, expected for the fetal samples, in which this value dropped to less than 70% (TABLE 2). Abergel and Claverie,[33] with a computation of murine and human IgH and TcRb D segments, confirmed a strong bias in the expression of Ig H D segments from a preferred sequence, rating 90% in humans, and containing at least one glycine and one tyrosine residue. The presence of these residues could be necessary to form a flexible loop within this region generally considered important in antigen binding. Alternatively, residues of the D-J-encoded region have been shown to be involved in the interaction with the L chain V domains.[34] Similarly, this region could interact with the conserved homologous regions of the pseudo light chains: VpreB and $\lambda 5$[35] in preB cells. Altogether, the composition in amino acid and the length could be involved in a first selection, as a structure necessary to the association between these three chains. This does not exclude another step of selection due to a binding to some ligand or autoantigens, in this preB stage.

Previous studies on fetal repertoire were always performed on total B population, except in the liver at eight weeks of gestation which did not present any light chain transcripts. Nevertheless, we could not extrapolate these characteristics to the first steps of differentiation, as we could not distinguish the role of earliness in development or in differentiation. Our approach by a cell selection based on the detection of a pseudo light chain at the cell surface gave a definitive answer on the step of selection that occurred on heavy chains associated with the pseudo light chains.

In summary, our results permit us to conclude that fetal repertoire remained distinct, with a higher expression of DQ52 and a shorter length of CDR3. PreB and XLA repertoires appeared very close if not identical and shared some characteristics: (i) There is no more restriction of expression of VH germline genes at the preB stage, when μ chains are expressed in association with pseudo light chains; (ii) the low average of DQ52 DH gene usage was compatible with selection pressure at this step of differentiation; (iii) the almost constant presence of Gly residues in the CDR3 could not be a random event; and (iv) the shortest length of CDR3 should be a characteristic of this first selection step. However, we cannot decide whether this selection occurred when the μ chains were present at the preB-cell surface in association with pseudo light chains, and whether it was dependent on interaction with a hypothetical ligand. If there were some evidence of the pseudo Ig complex at the surface of preB-cell lines or in human bone marrow,[11,36] the window of expression was very narrow[10] or even not demonstrated in mouse.[37,38] Alternatively, this step could be too short to be identified in normal bone marrow. In the mouse, it was suggested that the complex could be internalized as soon as it appears at the cell surface, eventually after binding with a putative ligand on the stroma cells.[37] Another hypothesis could be that an intracellular selection signal exists in the cytoplasm. Whatever the mechanism, the VH repertoire at this early stage of differentiation seems already selected. Emergence of a more mature B-cell repertoire is thus likely to require a strong selection, implying elimination of a large number of cells and/or alternative escape pathways, such as secondary rearrangements, that might be induced on signals that remain to be elucidated.

SUMMARY

At the preB stage, when only the IGH locus has rearranged, μ chains become expressed in association with the ΨL chains, λ-like and VpreB, thus forming the preB receptor. By the use of a monoclonal anti VpreB antibody, preB cells were isolated from two adult bone marrow samples, and the VH repertoire was analyzed and compared to fetal, XLA (X-linked agammaglobulinemia), and adult B repertoires. Most VH genes identified were also expressed in fetal liver, XLA bone marrow, and adult PBLs, with similar predominant usage of certain germline genes. Multiple D/D fusions, limited N diversity, and preferential use of JH4 with a low level of DQ52 usage were also identified. Few mutations could be observed, not specifically localized in CDR regions, that could be interpreted as not positively selected. Conversely, a shorter length of CDR3 appeared to be the hallmark of the preB step. Thus, the association of ΨL chains with μ does not bring about a bias in the VH gene usage, but a first selection on the CDR3 region could be the result of recognition by given autoantigens or ligands different for preB cells and B cells.

ACKNOWLEGMENTS

We would like to thank Léon Boubli from Hopital de la Conception, Marseille; Didier Blaise and the "Marrow Transplant Unit" staff from the Centre Anti-Cancereux Paoli Calmettes, Marseille; Joel Ladec from the Centre Regional de Transfusion Sanguine, Marseille; and Alain Fisher and his collaborators from INSERM, U132, Hopital des Enfants Malades, Paris for providing physiological samples.

REFERENCES

1. MATSUDA, F., E. K. SHIN, H. NAGAOKA, R. MATSUMURA, M. HAINO, Y. FUKITA, S. TAKA-ISHI, T. IMAI, J. H. RILEY, R. ANAND, E. SOEDA & T. HONJO. 1993. Structure and physical map of 64 variable segments in the 3′ 0.8-megabase region of the human immunoglobulin heavy-chain locus. Nature Genet. **3:** 88–94.
2. WALTER, M. A., U. SURTI, M. H. HOFKER & D. W. COX. 1990. The physical organization of the human immunoglobulin heavy chain gene complex. EMBO J. **9:** 3303–3313.

3. Cook, P. G., I. M. Tomlinson, G. Walter, H. Riethman, N. P. Carter, L. Buluwela, G. Winter & T. H. Rabbitts. 1994. A map of the human immunoglobulin VH locus completed by analysis of the telomeric region of chromosome 14q. Nature Genet. **7:** 162–168.
4. Tomlinson, I. M., G. Walter, J. D. Marks, M. B. Llewerlyn & G. Winter. 1992. The repertoire of human germline VH sequences reveals about fifty groups of VH segments with different hypervariable loops. J. Mol. Biol. **227:** 776–798.
5. Uckun, F. M. 1990. Regulation of human B cell ontogeny. Blood **76:** 1908–1923.
6. Melchers, F., H. Karasuyama, D. Haasner, S. R. Bauer, A. Kudo, N. Sakaguchi, B. Jameson & A. Rolink. 1993. The surrogate light chain in B-cell development. Immunol. Today **14:** 60–68.
7. Schiff, C., M. Milili, D. Bossy & M. Fougereau. 1992. Organization and expression of the pseudo-light chain genes in human B-cell ontogeny. Intern. Rev. Immunol. **8:** 135–145.
8. Kitamura, D., J. Roes, R. Kühn & K. Rajewsky. 1991. A B cell-deficient mouse by targeted disruption of the membrane exon of the immunoglobulin m chain gene. Nature **350:** 423–426.
9. Kitamura, D., A. Kudo, S. Schaal, W. Müller, F. Melchers & K. Rajewsky. 1992. A critical role of lambda5 protein in B cell development. Cell **69:** 823–831.
10. Guelpa-Fonlupt, V., C. Tonnelle, D. Blaise, M. Fougereau & F. Fumoux. 1994. Discrete early pro-B and pre-B stages in normal human bone marrow as defined by surface pseudo-light chain expression. Eur. J. Immunol. **24:** 257–264.
11. Lassoued, K., C. A. Nuñez, L. Billips, H. Kubagawa, R. C. Monteiro, T. W. LeBien & M. D. Cooper. 1993. Expression of surrogate light chain receptors is restricted to a late stage in pre-B cell differentiation. Cell **73:** 73–86.
12. Cuisinier, A. M., V. Guigou, L. Boubli, M. Fougereau & C. Tonnelle. 1989. Preferential expression of VH5 and VH6 immunoglobulin genes in early human B cell ontogeny. Scand. J. Immunol. **30:** 493–497.
13. Cuisinier, A. M., F. Fumoux, D. Moinier, L. Boubli, V. Guigou, M. Milili, C. Schiff, M. Fougereau & C. Tonnelle. 1990. Rapid expansion of human immunoglobulin repertoire (VH, Vk, Vl) expressed in early fetal bone marrow. New Biol. **2:** 689–699.
14. Cuisinier, A. M., L. Gauthier, L. Boubli, M. Fougereau & C. Tonnelle. 1993. Mechanisms that generate human immunoglobulin diversity operate from the 8th week of gestation in fetal liver. Eur. J. Immunol. **23:** 110–118.
15. Schroeder, H. W., Jr. & J. Y. Wang. 1990. Preferential utilization of conserved immunoglobulin heavy chain variable gene segments during human fetal life. Proc. Natl. Acad. Sci. USA **87:** 6146–6150.
16. Hillson, J. L., I. R. Oppliger, E. H. Sasso, E. C. B. Milner & W. H. Wener. 1992. Emerging human B cell repertoire: Influence of developmental stage and interindividual variation. J. Immunol. **149:** 3741–3752.
17. Raaphorst, F. M., E. Timmers, M. J. H. Kenter, M. J. D. Van Tol, J. M. Vossen & R. K, B. Schuurman. 1992. Restricted utilization of germ-line V_H3 genes and short diverse third complementarity-determining regions (CDR3) in human fetal B lymphocyte immunoglobulin heavy chain rearrangements. Eur. J. Immunol. **22:** 247–251.
18. Pascual, V., L. Verkruyse, M. L. Casey & J. D. Capra. 1993. Analysis of IgH chain segment utilization in human fetal liver. J. Immunol. **151:** 4164–4172.
19. Milili, M., F. Le Deist, G. De Saint-Basile, A. Fischer, M. Fougereau & C. Schiff. 1993. Bone marrow cells in X-linked agammaglobulinemia express pre-B-specific genes (lambda-like and V pre-B) and present immunoglobulin V-D-J gene usage strongly biased to a fetal-like repertoire. J. Clin. Invest. **91:** 1616–1629.
20. Yamada, M., R. Wasserman, B. A. Reichard, S. Shane, A. J. Caton & G. Rovera. 1991. Preferential utilization of specific immunoglobulin heavy chain diversity and joining segments in adult human peripheral blood B lymphocytes. J. Exp. Med. **173:** 395–407.
21. Sanz, I. 1991. Multiple mechanisms participate in the generation of diversity of human H chain CDR3 regions. J. Immunol. **147:** 1720–1729.

22. Huang, C. & B. D. Stollar. 1993. A majority of Ig H chain cDNA of normal human adult blood lymphocytes resembles cDNA for fetal Ig and natural autoantibodies. J. Immunol. **151:** 5290–5300.
23. Berman, J. E., K. Nickserson, J. R. Pöllock, J. Barth, R. Schuurman, D. Knowles, L. Chess & F. Alt. 1991. VH gene usage in humans: Biased usage of the VH6 gene in immature B lymphoid cells. Eur. J. Immunol. **21:** 1311–1314.
24. Rubinstein, D., M. Symann, A. Stewart & T. Guillaume. 1993. Restriction fragment length polymorphisms and single germline coding region sequence in VH18/2, a duplicated gene encoding autoantibody. Mol. Immunol. **30:** 403–412.
25. Stewart, A. K., C. Huang, A. A. Long, B. D. Stollar & R. S. Schwartz. 1992. VH-gene representation in auto antibodies reflects the normal human B-cell repertoire. Immunol. Rev. **128:** 101–122.
26. Alt, F. W., T. K. Blackwell, R. A. de Pinho, M. G. Reth & M. D. Yancopoulos. 1986. Regulation of genome rearrangement events during lymphocyte differentiation. Immunol. Rev. **89:** 5–30.
27. Yancopoulos, G. D. & F. W. Alt. 1985. Developmentally controlled and tissue-specific expression of unrearranged VH gene segments. Cell **40:** 271–281.
28. Berman, J. E., C. G. Humphries, J. Barth, F. W. Alt & P. W. Tucker. 1991. Structure and expression of human germline VH transcripts. J. Exp. Med. **173:** 1529–1535.
29. Tomlinson, I. M., G. P. Cook, N. P. Carter, R. Elaswarapu, S. Smith, W. Walter, L. Buluwela, T. H. Rabbitts & G. Winter. 1994. Human immunoglobulin VH and D segments on chromosomes 15q11.2 and 16p11.2. Hum. Mol. Genet. **3:** 853–860.
30. Tuaillon, N., A. B. Miller, N. Lonberg, P. W. Tucker & J. D. Capra. 1994. Biased utilization of DQ52 and JH4 gene segments in a human Ig transgenic minilocus is independent of antigenic selection. J. Immunol. **152:** 2912–2920.
31. Hochtl, J., C. R. Muller & H. G. Zachau. 1982. Recombined flanks of the variable and joining segments of immunoglobulin genes. Proc. Natl. Acad. Sci. USA **79:** 1383–1387.
32. Gu, H., D. Kitamura & K. Rajewsky. 1991. B cell development regulated by gene rearrangement: Arrest of maturation by membrane-bound Dm protein and selection of D_H element reading frames. Cell **65:** 47–54.
33. Abergel, C. & J. M. Claverie. 1991. A strong propensity toward loop formation characterizes the expressed reading frames of the D segments at the Ig H and T cell receptor loci. Eur. J. Immunol. **21:** 3021–3025.
34. Segal, D. M., E. A. Padlan, G. H. Cohen, S. Rudikoff, M. Potter & D. R. Davies. 1974. The three-dimensional structure of a phosphorylcholin-binding mouse immunoglobulin Fab and the nature of the antigen binding site. Proc. Natl. Acad. Sci. USA **71:** 4298–4302.
35. Köhler, G., A. Iglesias, R. Lamers, M. Kopf, B. Bülher & U. Fritzsche. 1989. Regulation of immunoglobulin gene rearrangement. *In* Progress in Immunology. F. Melchers *et al.*, Eds. **7:** 324–330. Springer-Verlag. Berlin.
36. Nishimoto, N., H. Kubagawa, T. Ohno, G. L. Gartland, A. K. Stankovic & M. D. Cooper. 1991. Normal pre-B cells express a receptor complex of m heavy chains and surrogate light-chain proteins. Proc. Natl. Acad. Sci. USA **88:** 6284–6288.
37. Karasuyama, H., A. Rolink, Y. Shinkai, F. Young, F. W. Alt & F. Melchers. 1994. The expression of $V_{pre\text{-}B}$/lambda5 surrogate light chain in early bone marrow precursor B cells of normal and B cell-deficient mutant mice. Cell **77:** 133–143.
38. Young, F., B. Ardman, Y. Shinkai, R. Lansford, T. K. Blackwell, M. Mendelsohn, A. Rolink, F. Melchers & F. W. Alt. 1994. Influence of immunoglobulin heavy- and light-chain expression on B-cell differentiation. Genes Dev. **8:** 1043–1057.

Developmental Regulation of the Human Antibody Repertoire[a]

HARRY W. SCHROEDER, JR.,[b,c,d] FRANK MORTARI,[c]
SATOSHI SHIOKAWA,[c] PERRY M. KIRKHAM,[d]
ROTEM A. ELGAVISH,[c] AND F. E. BERTRAND III[d]

Division of Developmental and Clinical Immunology
[c]Department of Medicine
[d]Department of Microbiology
University of Alabama at Birmingham
Birmingham, Alabama 35294-3300

GENERATION OF IMMUNOGLOBULIN VARIABLE DOMAINS

Specific immunity in vertebrates is dependent on the individual's ability to generate a heterogeneous repertoire of antigen-binding structures that are displayed on the surface of lymphocytes. These antigen receptors are encoded by a superfamily of gene segments that undergo rearrangements exclusively in lymphoid cells.[1] The strategies responsible for diversification of antigen recognition structures have been most completely dissected in the case of antibody genes. Starting with a germline repertoire that includes fewer than 300 elements, combinatorial joining of VH, DH, and JH gene segments (for heavy chains) and of VL and JL gene segments (for light chains) coupled with combinatorial association of the resulting heavy and light polypeptides, can yield more than 10^6 different antibody molecules. Every codon added by N region addition increases the potential diversity of the repertoire 20-fold. These N regions can be inserted both between the V and the D, between Ds in D-D fusions, and between the D and the J.[2,3] Thus, more than 10^9 different H chain VDJ junctions, or CDR 3-encoding intervals, can be generated at the time of gene segment rearrangement, yielding a potential preimmune antibody repertoire in excess of 10^{15} different immunoglobulins. Maturation of the primary response is in many cases characterized by the appearance of higher affinity variants generated by somatic mutation of germline-encoded sequence.[4]

Comparison of antibody sequences and studies of antibody crystals has allowed definition of the structure of the antigen-binding site.[5,6] Each immunoglobulin variable domain contains three intervals of sequence hypervariability (termed complementarity-determining regions or CDRs) that are separated from each other by four intervals of relatively constant sequence (termed frameworks or FRs). The juxtaposed CDR domains create an antigen-binding surface of approximately 2500 to 3000Å.[6] CDR 1 and 2 form the outside border of the binding site and are

[a] This work was supported, in part, by National Institutes of Health Grants AI07051, AI33621, AI34568, AR20216, and AR30555.

[b] Address for correspondence: Harry W. Schroeder, Jr., M.D., Ph.D., Division of Developmental and Clinical Immunology, Departments of Medicine and Microbiology, The University of Alabama at Birmingham, 378 Wallace Tumor Institute, VAB Station, Birmingham, AL 35294-3300.

completely encoded by their respective V gene segments. The CDR 3 domains are generated by V(D)J joining and are found in the center of the antigen-binding site. It has been estimated that an average protein antigen interacts with about a third of this surface.[6] The central location of CDR 3 in the antigen binding site attests to the critical role that the sequence and structure of this region plays in defining antigen specificity.[6]

THE ABILITY TO RESPOND TO SPECIFIC ANTIGENS DEVELOPS IN A PROGRAMMATIC FASHION

Immunologic competence arises very early in embryonic development.[7] Premature infants weighing as little as 1500 grams can respond to vaccination with bacteriophage $\phi\chi$174.[8] Fetuses can respond to syphillis and toxoplasmosis by the 29th week of gestation,[7] and up to 0.6% of the cord blood IgG can derive from the infant.[9] Nevertheless numerous studies have shown that the human neonate is unable to mount satisfactory responses to many vaccines.[10] The delay in achieving adequate responses can extend beyond infancy. For example, 95% of neonates are able to respond to tetanus toxoid before the first six months of life, but the 95% response level to diphtheria toxin is not achieved until after the first six months of life.[11] Similarly, children under the age of two are often unable to respond to unmodified bacterial polysaccharide vaccines.[10,12] Studies in mice and lambs reveal striking similarities to the human system in the ability to respond to specific antigens during fetal life and infancy. For example, the response to bacteriophage $\phi\chi$174 develops early in sheep, mice and man, whereas both sheep and man are unable to mount a humoral response to diphtheria toxoid until after birth.[7]

ONTOGENY OF THE MOUSE ANTIBODY REPERTOIRE

These restrictions in antigen specificities might seem paradoxical, given the stochastic mechanisms that appear to underlie the generation of the antibody repertoire. This paradox was in part resolved by the demonstration that the composition of the antibody repertoire also follows an ordered developmental program (reviewed in Schroeder and Perlmutter[13]). The mouse system is the most thoroughly studied. The VH7183 family contributes disproportionately to the mouse fetal and neonatal repertoire, whereas the VHJ558 family, which comprises more than 50% of the germline repertoire, is minimally represented in the fetal repertoire but contributes to the majority of the postnatal splenic preimmune antibody repertoire. The mechanisms controlling the differential use of VH gene segments are poorly understood. Linear proximity to the JH locus and the CH enhancer regions appears to play a role,[14,15] with the most proximal functional gene segment, VH81X of the VH7183 family, present in more than 50% of pre-B-cell rearrangements. Studies of L chain utilization also indicate nonrandom use of Vκ gene segments. For example, members of the Vκ 8 family are used.[16]

Fetal CDR 3 diversity is further limited because of the lack of N regions in fetal VDJ joins.[17,18] Thus, fetal H chains containing a given VH gene segment are more likely to share similar germline-encoded CDR 3 domains.

These observations suggest that fetal restrictions in the repertoire of antigen-binding sites play a role in limiting the ability of the mouse fetus to respond to specific antigens. In order to gain insight into the mechanisms underlying the acquisition of immunocompetence in the human, we have sought to understand which of these observations are fundamental properties by determining which of these fetal restrictions are preserved in both mouse and human.

THE ONTOGENY OF HUMAN B-CELL DEVELOPMENT

Experimentally, immunofluorescent staining of fetal liver has shown the presence of pre-B cells (cytoplasmic $C\mu^+$, $sIgM^-$) by the seventh to eighth week of gestation, immature B lymphocytes ($sIgM^+$, $sIgD^-$) by the ninth week of gestation, and mature B lymphocytes ($sIgM^+$, $sIgD^+$) by the 12th week of gestation.[19] Surface IgG^+ ($sIgG^+$) and $sIgA^+$ cells are first apparent in fetal liver by 12 and 13 weeks' gestation, yet plasma cells secreting these antibodies cannot be detected until 20 and 30 weeks' gestation, respectively.[20] Active transport of maternal IgG across the human placenta begins around 20 to 21 weeks,[21] but does not reach significant levels until 27 to 28 weeks. Late in the second trimester, production of B-lineage cells shifts to the bone marrow where it remains in adult life. At birth, B cells expressing surface IgM ($sIgM^+$) comprise approximately 18% of cord blood mononuclear cells, whereas $sIgG^+$ and $sIgA^+$ cells comprise <5% and <1%, respectively (G. Silverman, personal communication). Adult serum IgM levels typically are not reached until 2 years of age, until 6 years of age for IgG, and until puberty for IgA.[22] Although endogenous serum IgG and IgA is difficult to detect in cord blood, infants responding to prenatal infection can produce IgG and IgA in conjunction with an elevation in the serum level of IgM.[23,24]

HUMAN IMMUNOGLOBULIN GENES

The approximately 90 (~50 functional) VH gene segments in human haploid genomes can be grouped into seven families of related sequence.[25–27] Unlike the mouse, gene segments of the different families are dispersed throughout the locus. The most JH proximal VH gene segment (V6-1) belongs to the single-member VH 6 family. The DH locus contains four 9-kb repeats that each include one or more members of seven of the eight known DH families (FIG. 1). These repeats begin 20 kb telomeric to the six functional JH gene segments (reviewed in Schroeder and Perlmutter[13]). The single exception to this rule is DHQ52, which is less than 100 base pairs 5′ of JH 1.

The κ locus, on chromosome 2, contains approximately 85 Vκ (35 potentially functional), five Jκ gene segments[28] and a single Cκ gene. Four major Vκ families have been recognized in man, ranging in size from more than 30 members (Vκ 1) to a family with a single member (Vκ 4).[29] Vκ utilization in the peripheral B-cell repertoire appears nonrandom. One of the most frequently utilized Vκ gene segments is Vκ 4,[30] which is located immediately adjacent to the Jκ-Cκ locus, in close proximity to the Cκ enhancers.[29] The λ locus, on chromosome 22, contains four functional Cλ constant domains, each with its own Jλ. Upstream of this constant locus are more than 50 Vλ gene segments that can be grouped into at least seven families.[31]

DEVELOPMENT OF THE HUMAN VH ANTIBODY REPERTOIRE

Plasma cells are relatively rare in fetal liver, and pre-B cells outnumber B cells.[32] By generating cDNA libraries from human fetal liver mononuclear cells,

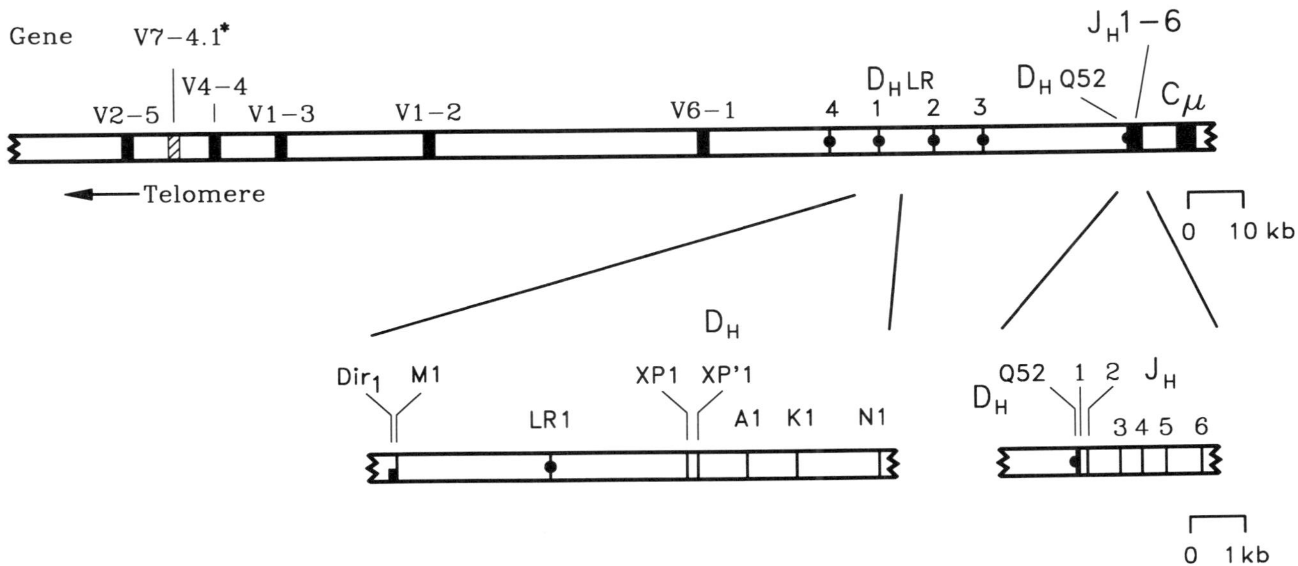

FIGURE 1. Map of the VH, DH, JH region proximal to the Cμ locus on chromosome 14q32. Shown in the top panel are the six most JH-proximal VH gene segments in the human genome (reviewed in Schroeder and Perlmutter[13]). An *asterisk* (*) marks the V7-4.1 gene segment, which is absent in some haploid genomes. The *bottom panel* contains an expanded view of one of the four 9-kb repeating structures of the DH locus, identified by the corresponding DH LR gene segment.

we hoped to sample the early fetal antibody repertoire in a relatively unbiased way. Because of VH polymorphism in the human heavy-chain locus, we elected to generate libraries using individual fetal samples. Oligo(dT)-primed cDNA libraries were independently generated from mononuclear cell RNA derived from two unrelated fetuses of 104 and 130 days' gestation.[13,33,34] These stages were chosen because early second-trimester fetal liver was of sufficient quantity to allow generation of a cDNA library by traditional means.

One-third of the $C\mu^+$ transcripts were germline $C\mu$ transcripts that initiated immediately upstream of the CH1 exon and were transcribed without an intervening intronic sequence (FIG. 2). These transcripts had identical 5′ sequences that contained numerous stop codons in all three reading frames. Two additional sterile transcripts contained sequence derived from the CH intronic enhancer ($I\mu$) between JH and $C\mu$. The presence of these "sterile" transcripts is common in early lymphoid cells, indicating that the cDNA libraries were indeed enriched for the products of pre-B and immature B cells.[35]

Examination of the nucleotide sequences of 32 VH-DH-JH-$C\mu^+$ transcripts revealed that each derived from an independent VDJ joining event; hence, the libraries were not biased by the presence of a few clones expressing large amounts of particular mRNAs. Nevertheless, only 15 different VH gene segments were used to generate the 32 VH-containing sequences. Six of these belonged to the VH 3 family. As shown graphically in FIGURE 3, nine of the VH gene segments identified were used more than once, with six represented in both libraries. Five of the sequences (16%) contained out-of-frame VDJ joins. Representation of the same limited population of VH gene segments in both translatable and nontranslatable transcripts indicates that the preference for these VH gene segments reflects B-cell events before antigen selection.

In order to determine if the antibody repertoire expressed at birth differed from that expressed in the fetus, we performed sequence analysis of VDJ-containing $C\mu$ cDNA transcripts from unselected cord blood mononuclear cells.[36] Only two of the 11 $C\mu^+$ transcripts with identifiable VH exons contained VH gene segments previously observed in the second-trimester fetal life cDNA libraries. These studies suggested that this infant had accessed a broader array of VH elements than those commonly used in second-trimester fetal liver.

In the fetal mouse, DJC-proximal VH81X is used in approximately 50% of VDJ rearrangements, whereas distal gene segments, such as members of the J558 family, are underrepresented (reviewed in Schroeder and Perlmutter[13]). Use of J558 family members increases with advancing age, and the ratio of VH7183 to J558 family utilization serves as a rough measurement of maturation of the repertoire.

In second-trimester fetal liver, V3-30.3 (56p1) was found in 22% of the VDJ transcripts (FIG. 3). However, more than one-third of the functional VH repertoire separates this gene segment from the DJC locus. V6-1, the most JH proximal VH gene segment, was present in only 6% of the 32 rearrangements. At first glance, these findings would suggest that linear proximity to the DJC locus might play a limited role in the diversification of the human VH repertoire. However, examination of the distribution of VH utilization in FIGURE 3 reveals that 40% of the VH gene segments expressed in 130-day fetal liver were distal to V5-51, the most telomeric VH gene segment detected in the $C\mu^+$ transcripts from 104-day fetal liver ($p < 0.02$, chi square). Most notably V1-69 (51p1) was not detected in the 104-day sample.

The above observations raise the possibility that human VH utilization at the earliest stages of fetal development could, like mouse, be influenced by linear proximity to the DJC locus. Extrapolating from mouse, the greatest limitations

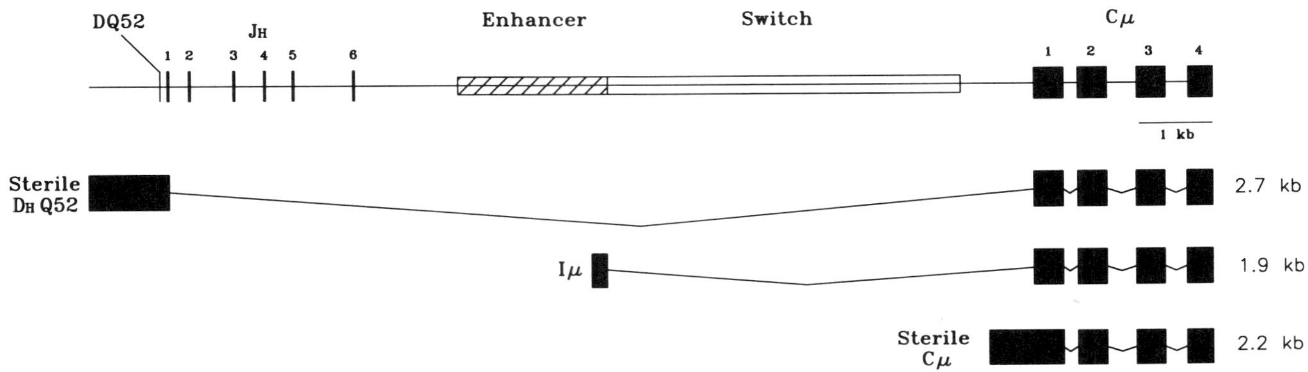

FIGURE 2. Unrearranged transcripts from the IgH locus in fetal liver B-lineage cells. The majority of untranslatable ("sterile") transcripts include germline sequences immediately 5′ to the CH 1 domain.[33,34] Iμ sterile transcripts initiate from within the CH enhancer region.[60] Also depicted are DHQ52-JH 1 containing $C\mu^+$ transcripts that include the intervening DNA between DH and JH.[34]

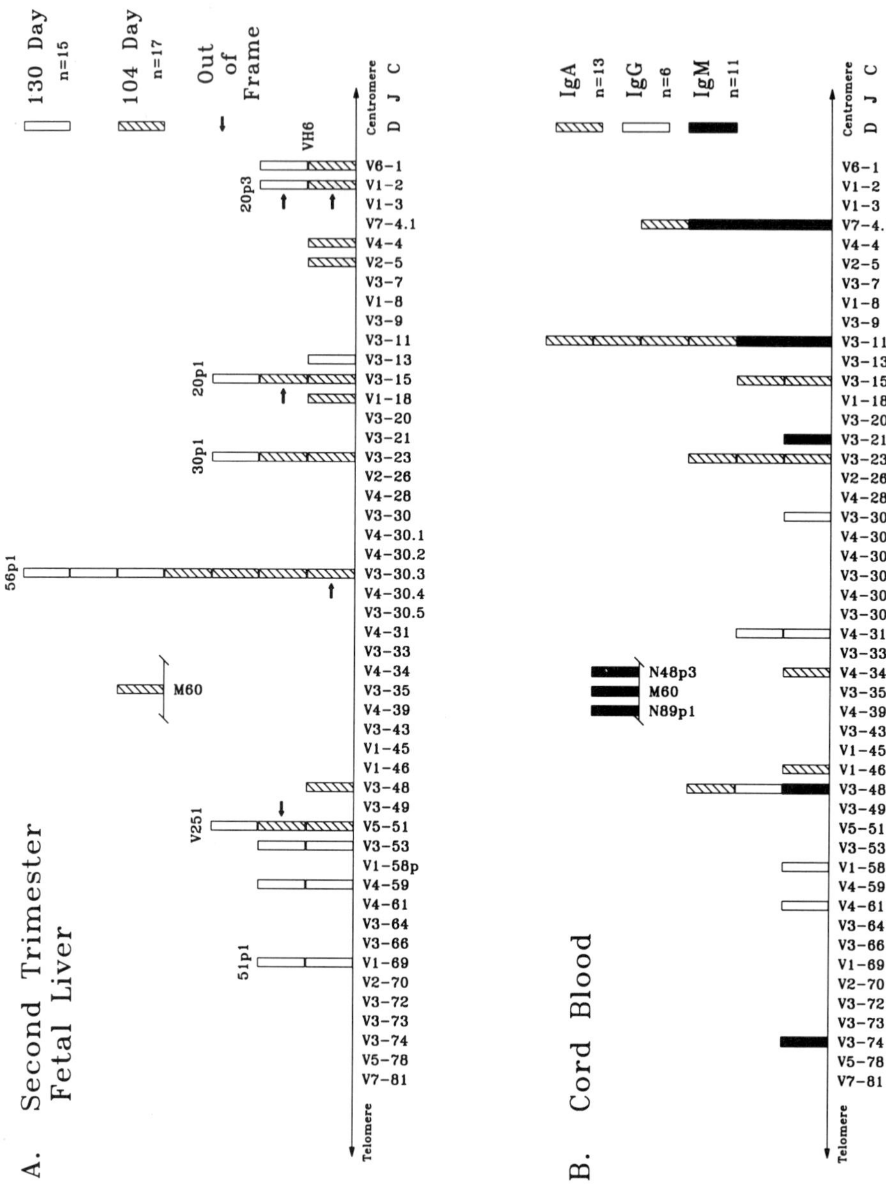
A. Second Trimester Fetal Liver
130 Day n=15
104 Day n=17
Out of Frame
B. Cord Blood
IgA n=13
IgG n=6
IgM n=11
Centromere
D J C
Telomere
VH6
20p3
20p1
30p1
56p1
M60
V251
51p1
N48p3
M60
N89p1
V6-1
V1-2
V1-3
V7-4.1
V4-4
V2-5
V3-7
V1-8
V3-9
V3-11
V3-13
V3-15
V1-18
V3-20
V3-21
V3-23
V2-26
V4-28
V3-30
V4-30.1
V4-30.2
V3-30.3
V4-30.4
V3-30.5
V4-31
V3-33
V4-34
V3-35
V4-39
V3-43
V1-45
V1-46
V3-48
V3-49
V5-51
V3-53
V1-58p
V4-59
V4-61
V3-64
V3-66
V1-69
V2-70
V3-72
V3-73
V3-74
V5-78
V7-81

in VH utilization would be expected to be present at the time of the first appearance of B-lineage cells, that is, 8- to 10-week fetal liver. Using a cocktail of VH leader and 5′ Cμ PCR primers,[37] we cloned 12 VH1/7 family containing μ transcripts from five separate fetal liver samples of 54 to 76 days' gestation. Of these, 75% used V1-2, the second most DJC-proximal VH gene segment, whereas none used V1-69 (FIG. 4). These observations support the hypothesis that chromosomal position plays a role in preferential V rearrangement early in fetal life.[14]

SELF-REACTIVITY, A CONSISTENT PROPERTY OF THE EARLY REPERTOIRE

In mice, the early repertoire is enriched for multireactive antibodies that exhibit low-affinity binding to many self antigens (reviewed in Schroeder and Perlmutter[13]). These self-reactive antibodies often use members of JH proximal VH families, including VH7183 (reviewed in Schroeder and Perlmutter[13]). The multireactivity of the early antibody repertoire has been hypothesized to help prime the development of B cells (reviewed in Kearney *et al.*[38]). It has been proposed that interactions between these antibodies and self-components are essential for the acquisition of a mature B-cell repertoire.[39] A corollary of this hypothesis holds that the early repertoire is following a genetic program which, if disrupted by environmental factors or adverse mutations, can have a major impact on the immunocompetence of the individual.

Self-reactivity is not always benign. A number of B-cell lines have been generated that express antibodies with the self-reactive specificities typical of autoantibodies expressed in human autoimmune diseases,[40] for example, anti-DNA, rheumatoid factor, and anti-Sm antibodies, among others. Of the first 26 self-reactive human antibodies whose sequences were published, sixteen shared greatest sequence similarity with members of the fetal repertoire (reviewed in Schroeder and Perlmutter[13]). Two sets of sequences deserve special note. V6-1 is the most JH proximal VH gene segment in man (see FIG. 1 in Schroeder *et al.*[41]) and is more frequently represented in fetal B-lineage cells than expected by random chance alone.[33,34,42] We are aware of the antigen reactivities of four functional antibodies from VH 6-expressing cell lines,[43] and all four bind DNA. It is possible that the decreased representation of VH 6 in cord blood and later life reflects selection against antibodies with a high potential for pathogenic self-reactivity. The V3-23 gene segment is a major component of both the fetal and the adult repertoires (reviewed in Schroeder and Perlmutter[13] and Schwartz and Stollar[39]). However, the V3-23 (30p1) sequence is also frequently found in anti-DNA antibodies, including those that express the 16/6 idiotype commonly found in patients with systemic

←

FIGURE 3. VH utilization in fetal liver and cord blood H chain transcripts. The X-axis depicts the location of the potentially functional human VH gene segments that are telomeric to the DJC locus on the right.[25,26] Each rectangle represents a VH-containing H chain transcript cloned from oligo d(T)-primed cDNA libraries from **(A)** 104- and 130-day fetal liver[13,33,34] and **(B)** cord blood.[36] Only Cμ$^+$ transcripts were studied in fetal liver, whereas Cγ$^+$ and Cα$^+$ transcripts were also studied in cord blood. Transcripts derived from germline gene segments that are likely polymorphic in the population and whose germline position is unknown are shown above (e.g., VH 2 M60 in the fetal liver and cord blood, and VH 3 N48p3 and N89p1 in cord blood).

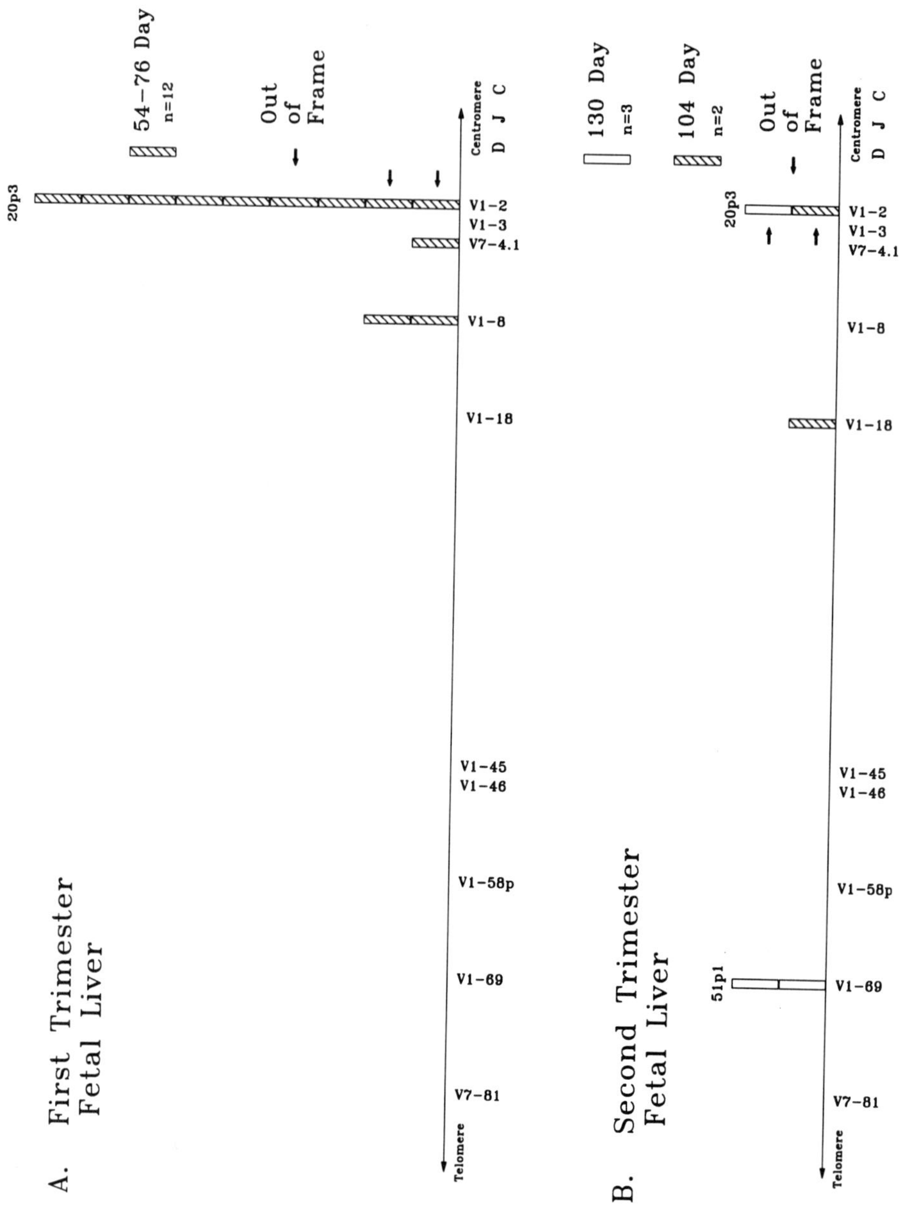
A. First Trimester Fetal Liver
54–76 Day
n=12
Out of Frame
20p3
Centromere
D J C
V1–2
V1–3
V7–4.1
V1–8
V1–18
V1–45
V1–46
V1–58p
V1–69
V7–81
Telomere
B. Second Trimester Fetal Liver
130 Day
n=3
104 Day
n=2
51p1

lupus erythematosus.[44,45] Conservation in the germline of VH gene segments that can directly encode autoreactive specificities suggests that their potential for pathogenicity is outweighed by their beneficial, developmentally regulated functions.

V3-23 (30p1), A PROTOTYPIC ANTIBODY GENE VARIABLE DOMAIN

Both fetal mice and premature human infants are able to respond to similar types of antigens, such as bacteriophage $\phi\chi$174. These observations would suggest that both species are able to generate similar antigen-binding sites at the earliest stages of ontogeny. Remarkably, there is extensive sequence homology between the human VH 3 and murine VH7183 gene families. The two most similar mouse and human VH gene segments are both used in fetal life: V3-23 and murine VH283[46] (fetal E415[14]) (see FIG. 5). The extensive diversity of the germline repertoire of VH gene segments makes it difficult to interpret the significance of the similarity, or differences, between two gene segments drawn at random from the germline pool. Because gene conversion within a family appears to play a major role in the evolution of VH gene segments,[47] the confounding effects of individual gene segment variation can be minimized by treating each family as a group (reviewed in Kirkham and Schroeder[27]). Using the sequence analysis program SAW,[48] family-specific consensus sequences for the known human and mouse families were generated by identifying the most commonly used nucleotide at each base pair position for all the germline elements available at the time of analysis. These consensus sequences compose the group norm for each family. V3-23 and VH283 differ by no more than three base pairs from their respective family consensus sequences. Remarkably, the fetal VH 4 gene segment V4-59a (58p2) and VH1 gene segment V1-2 (20p3) are also >98% identical to their respective family consensus sequence.[49] These findings suggest that the fetal repertoire contains a set of prototypical, phylogenetically conserved gene segments capable of generating a plastic, polyreactive antibody repertoire that can provide a low-affinity defense against a wide array of antigens.[50]

Analysis of the amino acids specified by mutated codons can yield insight into the forces that have shaped the evolution of a gene. Due to redundancy in the genetic code, base pair changes in a codon may result in either a replacement (R) of one amino acid by another or preservation of the same residue (a silent [S] mutation). Codons undergoing random mutation are predicted to yield an R to S substitution ratio (R/S ratio) of 2.9.[51,52] Positive selection for conservation of protein sequence yields R/S ratios of less than 2.9, whereas ratios significantly greater than 2.9 can only be achieved through selection for diversity.

Although the nucleotide sequences of V3-23 and VH283 are similar, they are not identical. Pairwise comparisons have shown that FR 2 is conserved in all VH gene segments and that sequence intervals within FR 1 and FR 3 are conserved in all members of a given VH family. Both V3-23 and VH283 share extensive sequence similarity in FR 1 and 3 and are likely derived from a common progenitor.[49] The FR 3 subdomain is immediately adjacent to CDR 1 and 2 and helps

←

FIGURE 4. Distribution of VH 1 and seven gene segments in $C\mu^+$ transcripts from fetal liver. See FIGURE 3 and text for further details. Nonfunctional (out-of-frame) joins are marked with an arrow.

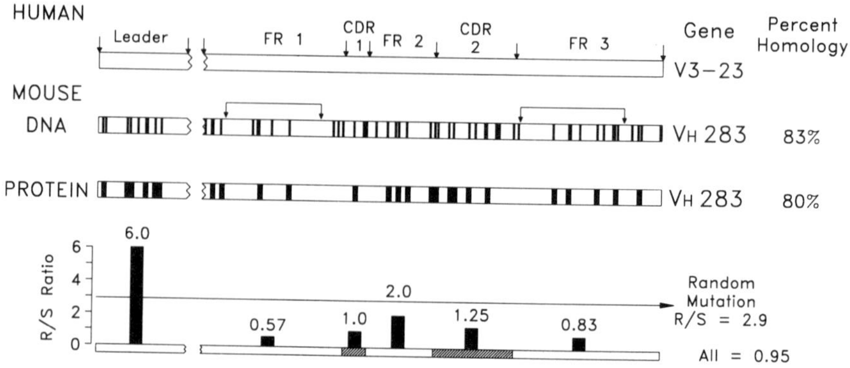

FIGURE 5. Two VH gene segments used in human and murine fetal life share extensive sequence homology. The functional domains of human V3-23 and mouse VH7183 VH283[5,27,46] are listed above. The arrows above the mouse DNA sequence demonstrate the location of the clan-specific region within FR 1 and family-specific region within FR 3.[49] Each line represents a nucleotide or peptide replacement mutation, respectively. Shown below are R/S ratios for each domain. The average R/S ratio for the entire mouse sequence, in comparison to human 30p1, is 0.95.

define the general shape of the H chain component of the antibody-binding site (reviewed in Kirkham *et al.*[49]). VH gene segments with an FR 3 interval homologous to human VH 3 can be found in *Xenopus,* rabbit, and even shark.[27,49] Thus, the human V3-23 and mouse VH283 gene segments contain a highly conserved scaffold that surrounds and shapes the antigen-binding site.

In addition to the framework regions, V3-23 and VH283 have conserved the sequence and structure of CDR 1 and 2. The length and predicted conformations of these domains are similar (data not shown), and R/S ratio analysis confirms that the protein sequence of these gene segments has been preserved (R/S ratios of 1.0 and 1.25 for CDR 1 and 2, respectively, see FIG. 5). The striking conservation of these two fetal sequences across species barriers provides further support for the hypothesis that antibodies that bear antigen-binding sites encoded by these genes play an important role in the development of the antibody repertoire. In this context, it is remarkable that VH283 can also contribute to an anti-DNA response.[53]

"FETAL" VH GENE SEGMENTS CAN BE USED IN ANTI-POLYSACCHARIDE ANTIBODIES

One possible biologic role for V3-23 is suggested by the studies of Adderson and colleagues.[54] They generated heterohybridomas secreting human antibodies directed against the capsular polysaccharide of *Haemophilus influenzae* type b (Hib PS), a developmentally controlled antigen specificity. A number of these antibodies contain sequences highly similar to V3-23. Intriguingly, VH283 can also be used to generate antibodies against polysaccharides.[46] The fact that V3-23 is a major component of the fetal repertoire and encodes antibodies with specificities that a normal infant will not respond to until the second year of life suggests that factors other than VH utilization must play a role in the programmed development of the humoral immune response.

AN EVOLUTIONARILY CONSERVED Vκ GENE SEGMENT IS USED BY FETAL B-LINEAGE CELLS

Regulation of VL utilization could also contribute to restrictions in the composition of the fetal antibody repertoire. In preliminary studies of the fetal Vκ repertoire, six Vκ light chains from the 104- and 130-day fetal liver cDNA libraries have been sequenced (Hillson, Schroeder, and Perlmutter, unpublished data). Two of the sequences use the single member Vκ 4 gene family, the most Cκ-proximal Vκ gene segment.[55] Like V3-23, the human Vκ 4 family has a mouse homologue, Vκ 8 (Elgavish, Kirkham, and Schroeder, unpublished observations). Pairwise comparisons of Vκ 4 and a Vκ 8 family consensus sequence reveal that both FR 3 and CDR 1, 2, and 3 demonstrate the low R/S ratios indicative of preservation of protein sequence (FIG. 6). Thus, as in the case of the VH locus, highly conserved Vκ gene segments also appear to be expressed in the fetal repertoire. (See above and Medina and Teale.[16]) The evolutionary preservation of the CDRs and the CDR-proximal FR3 region supports the hypothesis that V gene segments used in fetal life have been preserved in order to generate antibodies with essential antigen specificities.

However, the Vκ 4 gene segment, like V3-23, is also a common component of the adult repertoire,[30] making it less likely that developmental control of Vκ utilization controls the programmed development of antigen specificities during ontogeny. There remains the possibility that there are developmental differences in pairing of VH and Vκ. Even less is known regarding Vλ utilization in fetal and adult life. However, due to the lack of a D gene segment, the

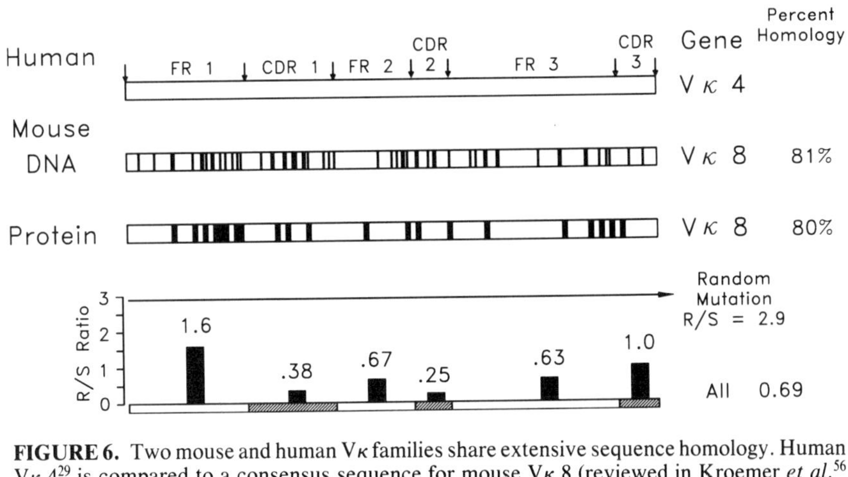

FIGURE 6. Two mouse and human Vκ families share extensive sequence homology. Human Vκ 4[29] is compared to a consensus sequence for mouse Vκ 8 (reviewed in Kroemer *et al.*[56] and D'Hoostelaere and Klinman[57]). The functional domains of the sequences are listed above. Comparisons of DNA and protein sequences are shown as in FIGURE 5.

diversity of the κ and λ CDR 3 domain is inherently more limited than that of the H chain CDR 3.

DEVELOPMENTAL CONTROL OF THE DIVERSITY OF THE H CHAIN CDR 3

The H chain CDR 3 forms the center of the antigen-binding site. This is the site of greatest somatic diversification in the preimmune repertoire. Although DH and JH utilization appears similar at different stages of ontogeny,[17,58] the structure and sequence of CDR 3 in the fetal and neonatal mouse is severely restricted as a result of the lack of N region addition. We found that the diversity of the human fetal CDR 3 repertoire was also limited, but some of the mechanisms leading to these restrictions differ from those seen in mouse.

Shown in FIGURE 7 is a summary of DH utilization, JH utilization, and the distribution of CDR 3 lengths in a representative sample of first- and second-

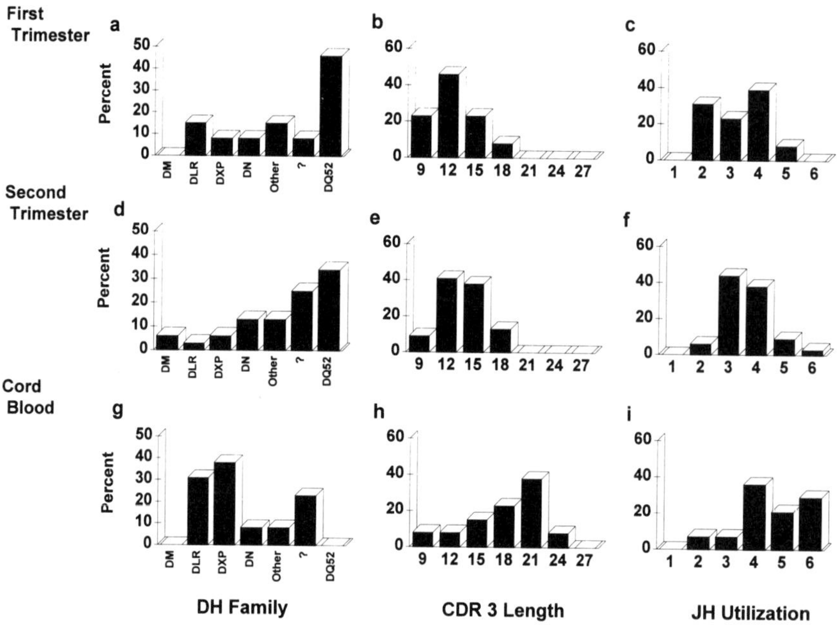

FIGURE 7. DH and JH utilization and CDR 3 length distributions during human ontogeny. Depicted is DH and JH and the distribution of H chain CDR 3 lengths as found in 13 $C\mu^+$ VDJ transcripts from five fetal liver samples ranging in age from 54 to 76 days' gestation, 32 $C\mu^+$ VDJ transcripts from 104- and 130-day fetal liver, and 14 cord mononuclear cell VDJ $C\mu^+$ transcripts from one sample of cord blood mononuclear cells. On the left are the percentage of transcripts that use members of the designated DH families, and on the right are the percentage of transcripts that use the designated JH gene segment. In the middle is the distribution of the lengths of the CDR 3 intervals of these transcripts (residues 93–102[5,27]) divided into three residue intervals (e.g., ≤9, 10–12, 13–15, 16–18, 19–21, 22–24, and ≥25 codons).

trimester fetal liver μ transcripts in comparison to those observed in cord blood. Unlike the mouse, DH and JH utilization differed as a function of developmental age. DHQ52 contributed to approximately 40% of all fetal transcripts, but was present in none of the cord blood transcripts. DHDXP- and DLR-containing transcripts were frequent in cord blood, but rare in fetal liver. Similarly, the fetal repertoire was enriched for JH 3 and 4, whereas cord blood $C\mu^+$ transcripts were enriched for use of JH 4, 5, and 6. Finally, D-D fusion was rare in fetal liver, but common in the cord blood (data not shown).

The three DH-JH-containing transcripts from the second-trimester libraries all contained DHQ52 in association with JH 1, 2, and 3, respectively. Intriguingly, one of these transcripts contained unrearranged DHQ52 and JH 1 gene segments, complete with the intervening intron (FIG. 2).[34] This transcript may represent an initial activation product of the DH-JH locus.[34,59] The preferential use of DHQ52 in the incomplete products of H chain rearrangement as well as in complete VDJ transcripts again indicates that nonrandom gene segment expression reflects a regulatory process intrinsic to the B cell itself.

In the mouse, fetal and neonatal VDJ junctions lack N region addition. Although N region addition at the V → D junction was present in the majority of first- and second-trimester μ transcripts, N region addition at the D → J junction varied depending on the identity of the DH gene segment (FIG. 8). Nine of eleven DHQ52 → JH rearrangements had N additions with an average of six nucleotides per join (i.e., two codons), whereas only four of nine non-DHQ52 → JH rearrangements had N additions, averaging only one nucleotide per join (FIG. 8). In contrast, N region addition in non-DHQ52 → JH cord blood transcripts averaged seven nucleotides per join. DHQ52 is the shortest human DH, encoding at most four codons, whereas DH DXP and DLR gene segments can encode up to 10 codons. JH 3 and 4 are the shortest JH gene segments, JH 5 and 6 the longest. The lengths of H chain CDR 3s in fetal transcripts averaged 11 to 12 codons and were all less than 18 codons long. Through increased use of the longer DH and JH gene segments, the increased number of D-D fusions, and the increased extent of N region addition, cord blood transcripts contained an average of 17 codons, ranging between 6 and 24 codons in length. Thus in human, as in mouse, the fetal repertoire lacks long CDR 3 domains. However, the mechanisms used to achieve this goal differ.

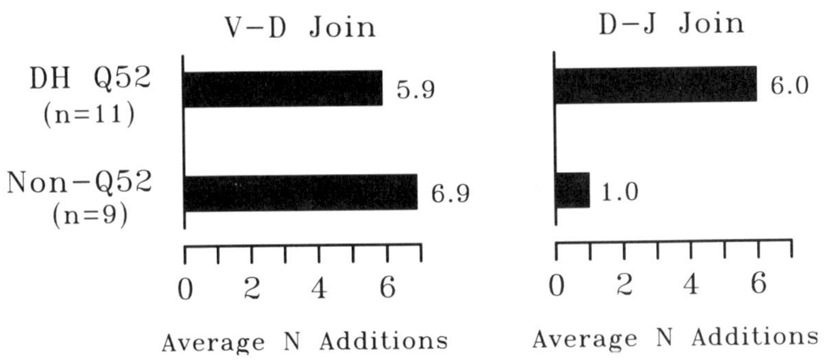

FIGURE 8. Gene-specific differences in the extent of N region addition in second-trimester fetal liver VDJ $C\mu$ transcripts.

CORD BLOOD IgG TRANSCRIPTS CAN EXPRESS "FETAL" CDR 3 DIVERSITY

We sought to determine if the neonatal IgG and IgA response would be associated with restrictions in the expressed antibody repertoire that are evident in the IgM response. $C\gamma^+$ and $C\alpha^+$ transcripts from cord blood lymphocytes were cloned and sequenced and then compared to the μ transcripts from the same cord blood cDNA library. Virtually all of the γ and α transcripts contained secretory termini, whereas the μ transcripts were evenly divided between membrane and secretory forms. Moreover, almost half of the α transcripts contained VH gene segments also found in the μ transcripts (FIG. 3). These findings are compatible with the expectation that the class-switched repertoire in cord blood reflects antigen receptor-influenced selection.

Although the repertoire of VH gene segments employed by the γ and α transcripts was more diverse than that in fetal liver μ, the average CDR 3 lengths of the fetal liver μ and cord blood γ sequences were found to be statistically similar (FIG. 9), whereas both of these CDR 3 populations differed from the cord blood

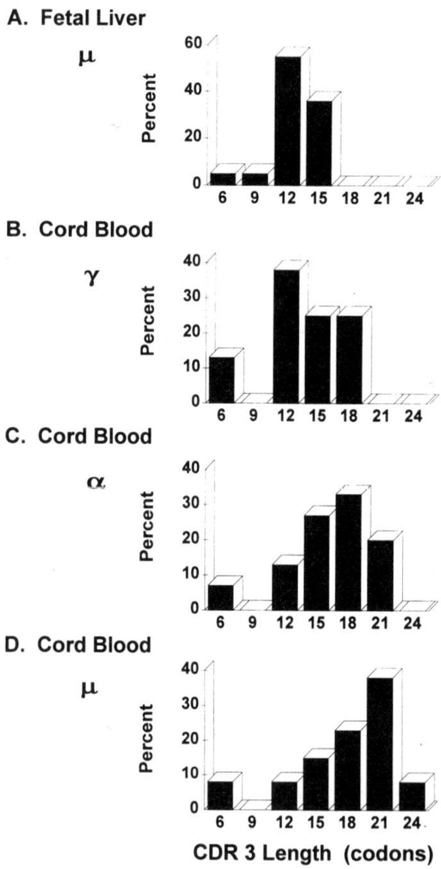

FIGURE 9. H chain CDR 3 length distributions in fetal liver and cord blood H chain transcripts. Shown is a comparison of CDR 3 length distribution in **(A)** 24 fetal liver $C\mu^+$ transcripts,[33,34] **(B)** eight $C\gamma^+$ cord blood mononuclear cell transcripts, **(C)** 15 $C\alpha^+$ cord blood mononuclear cell transcripts, and **(D)** 13 $C\mu^+$ cord blood mononuclear cell transcripts from the same cord blood donor.[36] The deduced CDR 3 intervals (residues 93 to 102[5,27]) have been divided into three amino acid residue intervals (≤6 codons, 7–9 codons, 10–12 codons, 13–15 codons, 18–21 codons, and 22–24 codons).

α and μ clones at the $p = 0.05$ level (Duncan's multiple range test). Our findings suggest that in this neonate, cord blood B cells generating secretory IgG could have been derived from an earlier population of progenitor B cells than those that gave rise to B cells generating secretory IgA or B cells expressing surface IgM. These results support the hypothesis that differences exist in the maturation of separate B-cell compartments.

FORM FOLLOWS FUNCTION: THE FLAT ANTIBODY-BINDING SITE HYPOTHESIS

Our preliminary results suggest that the fetal repertoires of mouse and human share similar VH and Vκ gene segment sequences. Moreover, although the mechanisms used to regulate CDR 3 diversity differ between the species, both mouse and human limit the length of H chain CDR 3. The fetal repertoire has been shown to be enriched for polyreactive sequences of low affinity. We propose that these sequence and structural limitations are the basis for the characteristic features of the fetal antibody repertoire specificities. Limitations in the size of CDR 3 results in a "flat" antibody-binding site. Such a "flat" site would maximize the number of different interactions possible between the residues of the CDR 3 and potential antigens, resulting in polyspecificity. However, the lack of topography would prevent a tight conformational fit, correlating with the observed low affinity of the fetal repertoire.

SUMMARY

The ability to respond to antigen develops in a programmed fashion during ontogeny. In human, "fetal" immunoglobulin gene segment utilization appears biased towards a small set of evolutionarily conserved V gene segments. Many of these gene segments are also used in antibodies with antigen specificities that do not arise until after infancy. The human fetus primarily regulates the diversity of the antibody repertoire through control of the H (heavy) chain CDR 3, which is generated by VDJ joining and forms the center of the antigen-binding site. Molecular modeling suggests that limitations in the length and composition of fetal CDR 3 intervals result in antibodies that contain a relatively "flat" antigen-binding surface that could serve to maximize the number of different interactions possible between the antibody and potential antigens. We propose that these limitations in the sequence and structure of H chain CDR 3 contribute to the low affinity and multireactivity of fetal antibody repertoires. The specific mechanisms used to generate a restricted fetal repertoire appear to differ between human and mouse. Nevertheless, included in the final products of both human and mouse fetal B cells will be antibodies that are quite homologous in composition and structure. The precise role that these antibodies play in the development of immunocompetence remains to be elucidated.

REFERENCES

1. YANCOPOULOS, G. D. & F. W. ALT. 1986. Regulation of the assembly and expression of variable-region genes. Ann. Rev. Immunol. **4:** 339–368.

2. YAMADA, M., R. WASSERMAN, B. A. REICHARD, S. SHANE, A. H. CATON & G. ROVERA. 1991. Preferential utilization of specific immunoglobulin heavy chain diversity and joining segments in adult human peripheral blood B lymphocytes. J. Exp. Med. **173:** 395–407.
3. SANZ, I. 1991. Multiple mechanisms participate in the generation of diversity of human H chain CDR3 regions. J. Immunol. **147:** 1720–1729.
4. KOCKS, C. & K. RAJEWSKY. 1989. Stable expression and somatic hypermutation of antibody V regions in B-cell developmental pathways. Ann. Rev. Immunol. **7:** 537–559.
5. KABAT, E. A., T. T. WU, H. M. PERRY, K. S. GOTTESMAN & C. FOELLER. 1991. Sequences of Proteins of Immunological Interest. Vol. **5:** 1–2387. USPHS. Bethesda, MD.
6. PADLAN, E. A. 1994. Anatomy of the antibody molecule. Mol. Immunol. **31:** 169–217.
7. SILVERSTEIN, A. M. 1977. Ontogeny of the immune response: A perspective. *In* Development of Host Defense. M. D. Cooper, Ed. Vol. **1:** 1–10. Raven Press. New York.
8. UHR, J. W., J. DANCIS, E. C. FRANKLIN, M. S. FINKELSTEIN & E. W. LEWIS. 1962. The antibody response to bacteriophage $\phi\chi$174 in newborn premature infants. J. Clin. Invest. **41:** 1509.
9. MARETENSSON, L. & H. H. FUDENBERG. 1965. Gm genes and gamma-G-globulin synthesis in the human fetus. J. Immunol. **94:** 514–520.
10. STEIN, K. E. 1992. Thymus-independent and thymus-dependent responses to polysaccharide antigens. J. Inf. Dis. **165:** S49–S52.
11. COOKE, J. V., J. HOLOWACH, J. E. ATKINS & J. R. POWERS. 1948. Antibody formation in early infancy against diphtheria and tetanus toxoids. J. Pediatr. **33:** 141–146.
12. PATON, J. C., I. R. TOOGOOD, R. A. COCKINGTON & D. HANSMAN. 1986. Antibody response to pneumococcal vaccine in children aged 5 to 15 years. Am. J. Dis. Child. **140:** 135–138.
13. SCHROEDER, H. W., JR. & R. M. PERLMUTTER. 1993. Development of the human antibody repertoire. *In* New Concepts in Immunodeficiency Diseases. S. Gupta & C. Griscelli, Eds.: 1–20. Chichester, United Kingdom.
14. YANCOPOULOS, G. D., S. V. DESIDERIO, M. PASKIND, J. F. KEARNEY, D. BALTIMORE & F. W. ALT. 1984. Preferential utilization of the most JH-proximal VH gene segments in pre-B cell lines. Nature **311:** 727–733.
15. YANCOPOULOS, G. D., B. A. MALYNN & F. W. ALT. 1988. Developmentally regulated and strain-specific expression of murine VH gene families. J. Exp. Med. **168:** 417.
16. MEDINA, C. A. & J. M. TEALE. 1993. Restricted κ chain expression in early ontogeny: Biased utilization of Vκ exons and preferential Vκ-Jκ recombinations. J. Exp. Med. **177:** 1317–1330.
17. GU, H., I. FORSTER & K. RAJEWSKY. 1990. Sequence homologies, N sequence insertion and JH gene utilization in VH-D-JH joining: Implications for the joining mechanism and the ontogenetic timing of Ly1 B cell and B-CLL progenitor generation. EMBO J. **9:** 2133–2140.
18. FEENEY, A. J. 1990. Lack of N regions in fetal and neonatal mouse immunoglobulin V-D-J junctional sequences. J. Exp. Med. **172:** 1377–1390.
19. OWEN, J. J. T., M C. RAFF & M. D. COOPER. 1975. Studies on the generation of B lymphocytes in the mouse embryo. Eur. J. Immunol. **5:** 468–473.
20. GATHINGS, W. E., H. KUBAGAWA & M. D. COOPER. 1981. A distinctive pattern of B cell immaturity in perinatal humans. Immunol. Rev. **57:** 107–126.
21. KOHLER, P. F. & R. S. FARR. 1966. Elevation of cord over maternal IgG immunoglobulin: Evidence for an active placental IgG transport. Nature **210:** 1070–1071.
22. COOPER, M. D. 1987. B lymphocytes: Normal development and function. N. Engl. J. Med. **317(23):** 1452–1456.
23. STIEHM, E. R., A. J. AMMANN & J. D. CHERRY. 1966. Elevated cord macroglubulins in the diagnosis of intrauterine infections. N. Engl. J. Med. **275:** 971–977.
24. COSTA, T. E. F., H. SUH & M. C. NUSSENZWEIG. 1992. Chromosomal position of rearranging gene segments influences allelic exclusion in transgenic mice. Proc. Natl. Acad. Sci. USA **89:** 2205–2208.

25. MATSUDA, F., E. K. SHIN, H. NAGAOKA, *et al.* 1993. Structure and physical map of 64 variable segments in the 3′ 0.8-megabase region of the human immunoglobulin heavy-chain locus. Nat. Genet. **3:** 88–94.
26. TOMLINSON, I. M., G. WALTER, J. D. MARKS, M. B. LLEWELYN & G. WINTER. 1992. The repertoire of human germline VH sequences reveals about fifty groups of VH segments with different hypervariable loops. J. Mol. Biol. **227:** 776–798.
27. KIRKHAM, P. M. & H. W. SCHROEDER, JR. 1994. Antibody structure and the evolution of immunoglobulin V gene segments. Semin. Immunol. **6:** 1–13.
28. MEINDL, A., H.-G. KLOBECK, R. OHNHEISER & H. G. ZACHAU. 1990. The Vkappa gene repertoire in the human germline. Eur. J. Immunol. **20:** 1855–1863.
29. ZACHAU, H. G. 1989. Immunoglobulin light-chain genes of the kappa type in man and mouse. *In* Immunoglobulin Genes. T. Honjo, F. W. Alt & T. H. Rabbitts, Eds. Vol. **1:** 91–110. Academic Press. London.
30. KLEIN, R., R. JAEHICHEN & H. G. ZACHAU. 1993. Expressed human immunoglobulin κ genes and their hypermutation. Eur. J. Immunol. **23:** 3248–3271.
31. SELSING, E., J. DURDIK, M. W. MOORE & D. M. PERSIANI. 1989. Immunoglobulin lambda genes. *In* Immunoglobulin Genes. T. Honjo, F. W. Alt & T. H. Rabbitts, Eds. Vol. **1:** 111–122. Academic Press. London.
32. KUBAGAWA, H., W. E. GATHINGS, D. LEVITT, J. F. KEARNEY & M. D. COOPER. 1982. Immunoglobulin isotype expression of normal pre-B cells as determined by immunofluorescence. J. Clin. Immunol. **2:** 264–269.
33. SCHROEDER, H. W., JR., J. L. HILLSON & R. M. PERLMUTTER. 1987. Early restriction of the human antibody repertoire. Science **238:** 791–793.
34. SCHROEDER, H. W., JR. & J. Y. WANG. 1990. Preferential utilization of conserved immunoglobulin heavy chain variable gene segments during human fetal life. Proc. Natl. Acad. Sci. USA **87:** 6146–6150.
35. KELLEY, D. E. & R. P. PERRY. 1986. Transcriptional and posttranscriptional control of immunoglobulin mRNA production during B lymphocyte development. Nucleic Acids Res. **14:** 5431–5447.
36. MORTARI, F., J. A. NEWTON, J. Y. WANG & H. W. SCHROEDER, JR. 1992. The human cord blood antibody repertoire. Frequent use of the VH7 gene family. Eur. J. Immunol. **22:** 241–245.
37. LARRICK, J. W., L. DANIELSSON, C. A. BRENNER, M. ABRAHAMSON, K. E. FRY & C. A. K. BORREBAECK. 1989. Rapid cloning of rearranged immunoglobulin genes from human hybridoma cells using mixed primers and the polymerase chain reaction. Biochem. Biophys. Res. Commun. **160:** 1250–1255.
38. KEARNEY, J. F., M. VAKIL & N. SOLVASON. 1989. The role of idiotypic interactions and B-cell subsets in development of the B-cell repertoire. Cold Sping Harbor Symp. Quant. Biol. **54:** 203–207.
39. SCHWARTZ, R. S. & B. D. STOLLAR. 1994. Heavy-chain directed B-cell maturation: Continuous clonal selection beginning at the pre-B cell stage. Immunol. Today **15:** 27–32.
40. CARSON, D. A., P. P. CHEN & T. J. KIPPS. 1991. New roles for rheumatoid factor. J. Clin. Invest. **87:** 379–383.
41. SCHROEDER, H. W., JR., M. A. WALTER, M. H. HOSKER, *et al.* 1988. Physical linkage of a unique human immunoglobulin VH gene segment (VH6) to DH and JH gene segments. Proc. Natl. Acad. Sci. USA **85:** 8196–8200.
42. BERMAN, J. E., K. G. NICKERSON, R. R. POLLOCK, *et al.* 1991. VH gene usage in humans: Biased usage of the VH6 gene in immature B lymphoid cells. Eur. J. Immunol. **21:** 1311–1314.
43. LOGTENBERG, T., F. M. YOUNG, J. H. VAN ES, F. H. J. GMELIG-MEYLING & F. W. ALT. 1989. Autoantibodies encoded by the most JH-proximal human immunoglobulin heavy chain variable region gene. J. Exp. Med. **170:** 1347–1355.
44. SANZ, I., P. CASALI, J. W. THOMAS, A. L. NOTKINS & J. D. CAPRA. 1989. Nucleotide sequences of eight human natural autoantibody VH regions reveals apparent restricted use of VH families. J. Immunol. **142:** 4054–4061.
45. DERSIMONIAN, H., R. S. SCHWARTZ, K. J. BARRETT & B. D. STOLLAR. 1987. Relation-

ship of human variable region heavy chain germ-line genes to genes encoding anti-DNA autoantibodies. J. Immunol. **139:** 2496–2501.

46. OLLO, R., J.-L. SIKORAV & F. ROUGEON. 1983. Structural relationships among mouse and human immunoglobulin VH genes in the subgroup III. Nucleic Acids Res. **11:** 7887–7897.
47. PERLMUTTER, R. M., B. BERSON, J. A. GRIFFIN & L. E. HOOD. 1985. Diversity in the germline antibody repertoire: Molecular evolution of the T15 VH gene family. J. Exp. Med. **162:** 1998–2016.
48. ELGAVISH, R. A. & H. W. SCHROEDER, JR. 1993. SAW: A graphical user interface for the analysis of immunoglobulin variable domain sequences. BioTechniques **15:** 1066–1071.
49. KIRKHAM, P. M., F. MORTARI, J. A. NEWTON & H. W. SCHROEDER, JR. 1992. Immunoglobulin VH clan and family identity predicts variable domain structure and may influence antigen binding. EMBO J. **11:** 603–609.
50. DIGHIERO, G., P. LYMBERI, J.-C. MAZIE, *et al.* 1983. Murine hybridomas secreting natural monoclonal antibodies reacting with self antigens. J. Immunol. **131**(5): 2267.
51. JUKES, T. H. & J. L. KING. 1979. Evolutionary nucleotide replacements in DNA. Nature **281:** 605–606.
52. SHLOMCHIK, M. J., A. MARSHAK-ROTHSTEIN, C. B. WOLFOWICZ, T. L. ROTHSTEIN & M. G. WEIGERT. 1987. The role of clonal selection and somatic mutation in autoimmunity. Nature **328:** 805–811.
53. SHLOMCHIK, M. J., M. MASCELLI, H. SHAN, *et al.* 1990. Anti-DNA antibodies from autoimmune mice arise by clonal expansion and somatic mutation. J. Exp. Med. **171:** 265–297.
54. ADDERSON, E. E., P. G. SHACKELFORD, A. QUINN & W. L. CARROLL. 1991. Restricted Ig H chain V gene usage in the human antibody response to *Haemophilus influenzae* type b capsular polysaccharide. J. Immunol. **147:** 1667–1674.
55. KLOBECK, H.-G., G. W. BORNKAMM, G. COMBRIATO, R. MOCKIKAT, H.-D. POHLENZ & H. G. ZACHAU. 1985. Subgroup IV of human immunoglobulin k light chains is encoded by a single germline gene. Nucleic Acids Res. **13:** 6515–6529.
56. KROEMER, G., A. HELMBERG, A. BERNOT, C. AUFFRAY & R. KOFLER. 1991. Evolutionary relationships between human and mouse immunoglobulin kappa light chain variable region genes. Immunogenetics **33:** 42–49.
57. D'HOOSTELAERE, L. A. & D. M. KLINMAN. 1990. Characterization of new mouse V kappa groups. J. Immunol. **145:** 2706–2712.
58. FEENEY, A. J. 1992. Predominance of VH-D-JH junctions occurring at sites of short sequence homology results in limited junctional diversity in neonatal antibodies. J. Immunol. **149:** 222–229.
59. BERTRAND, F. E. III, L. G. BILLIPS & H. W. SCHROEDER, JR. 1995. Analysis of DHQ52 gene segment transcription and rearrangement during B cell development in human fetal bone marrow. Ann. N.Y. Acad. Sci. This volume.
60. NEALE, G. A. & G. R. KITCHINGMAN. 1991. mRNA transcripts initiating within the human immunoglobulin mu heavy chain enhancer region contain a nontranslatable exon and are extremely heterogeneous at the 5′ end. Nucleic Acids Res. **19:** 2427–2433.

Human IgH-V Gene Repertoire in Neonatal Cord Blood, Adult Peripheral Blood, and EBV-Transformed Cells[a]

YOUNGNIM CHOI,[b,c] MICHAEL H. RICKERT,[b]
MARK BALLOW,[d] AND STEVEN J. GREENBERG[b,c,e,f]

[b]*Department of Neurology*
Roswell Park Cancer Institute
Buffalo, New York 14263

[c]*Department of Microbiology/Immunology*
RPCI Division
State University of New York at Buffalo
Buffalo, New York 14263

[d]*Division of Allergy, Immunology and Rheumatology*
Department of Pediatrics
Children's Hospital of Buffalo
Buffalo, New York 14263

[e]*State University of New York at Buffalo*
Buffalo, New York 14263

The repertoire of completed human V_H gene rearrangements demonstrates restricted usage of certain V_H genes among fetal cells,[1] CD5$^+$ cells,[2] and autoantibody-producing cells,[3] whereas the normal adult B cell is assumed to rearrange V_H genes randomly from among those in the available gene pool.[4] Nevertheless, there is disagreement concerning the onset of randomization and the full degree of randomness. For example, Cuisnier *et al.*[5] suggested that randomization of repertoire usage is evident after eight weeks of gestation, whereas Zouali and Theze[6] reported biased V_H family usage in adults. To gain insight into the process of B-cell repertoire maturation and to determine the degree of influence that EBV transformation has on IgH-V gene repertoire recruitment, a novel DNA-based PCR strategy was used that detects all rearranged V_HDJ_H gene segments independent of transcriptional or translational expression. This strategy, therefore, avoids potential biases introduced by *in vitro* stimulation and cell selection. This novel method permitted the analysis of IgH-V gene repertoires from among four different polyclonal B-cell populations *ex vivo*. This approach has proven to be a very useful tool for profiling IgH-V gene family usage among mono-, oligo-, or polyclonal B-cell populations in normal or pathologic disease states.[7]

[a] This study was supported in part by Grant No. 857-0514 C from Endowment for the Neurosciences and Grant No. 857-0593 A from The Association for Research of Childhood Cancer.

[f] Address for correspondence: Dr. Steven J. Greenberg, Laboratory of Neuroimmunology and Neurovirology, Department of Neurology, Roswell Park Cancer Institute, Elm and Carlton Streets, Buffalo, New York 14263.

RESULTS

The hierarchy and frequency of IgH-V gene family usage was found to be V1 (31 ± 3.7%) > V3 (24 ± 6.8%) ≈ V4 (24 ± 2.6%) > V5 (15 ± 3.8%) > V2 (4 ± 1.0%) > V6 (2 ± 0.9%) > V7 (0.3 ± 0.8%) in cord blood (n = 8) and was V3 (44 ± 5.6%) > V4 (23 ± 3.2%) > V1 (18 ± 4.3%) > V5 (8 ± 2.3%) > V2 (5 ± 2.5%) > V6 (2 ± 0.7%) > V7 (0.3 ± 0.8%) in adult peripheral blood (n = 9) (FIG. 1). These observations reflect a significant increase in V3 usage ($p < 0.0001$) and decreases in V1 ($p < 0.0001$) and V5 usage ($p < 0.0002$) during the process of B-cell repertoire evolution from neonate to adulthood. IgH-V5 gene family was overrepresented in cord blood compared to germline complexity[8] ($p < 0.001$), but was less used and was not significantly different from the germline V_H gene composition in adult peripheral blood. Although usage of V3, V4, and V7 in cord blood and usage of V1 and V7 in adult peripheral blood also showed substantial differences from gene complexity, these were not statistically significant.

To investigate the influence of EBV transformation on B-cell repertoire, mononuclear cells from cord and adult peripheral blood were transformed by EBV *in vitro,* and the resultant IgH-V gene family repertoires subsequently profiled three

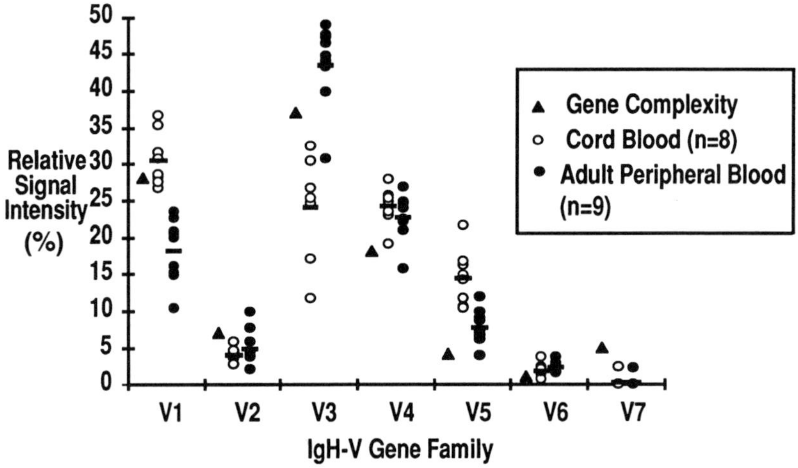

FIGURE 1. IgH-V gene family usage in cord blood and adult peripheral blood. The immunoglobulin heavy chain variable (IgH-V) gene usage in cord blood (n = 8) and adult peripheral blood (n = 9) were compared to germline complexity. The scatter graph represents the comparison of IgH-V gene family usage between cord blood and normal adult peripheral blood in relationship to the germline complexity. gDNA (1 μg) was subjected to PCR amplification, directed by a ^{32}P end-labeled composite IgH-J generic primer and a complement of IgH-V family-specific primers.[7] Amplicons were electrophoresed through native polyacrylamide gels and autoradiographed. Autoradiograms were scanned densitometrically, and variable IgH-V gene family usage was assessed according to relative signal intensity. There was a significant increase of V3 usage and decrease of V1 and V5 usage in the transition from neonate to adulthood. In cord blood, V5 was overrepresented compared to germline complexity, but the IgH-V gene repertoire in adult peripheral blood was not significantly different from the germline V_H gene composition.

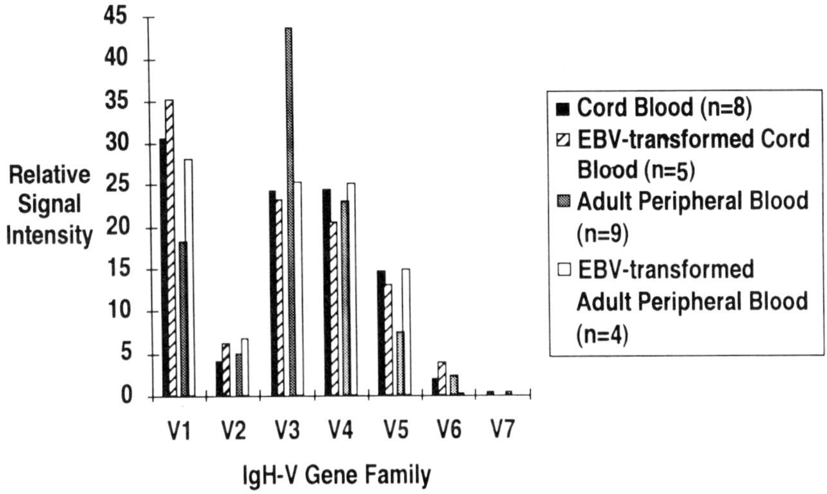

FIGURE 2. The influence of EBV-induced transformation on IgH-V gene repertoire. Mononuclear cells derived from cord blood (n = 5) and adult peripheral blood (n = 4) were infected and transformed by EBV. All adult specimens were obtained from seropositive persons, suggesting prior exposure to EBV. After a three-week culture, IgH-V family gene usage was subsequently analyzed using the same strategy as depicted in FIGURE 1. The resulting IgH-V family repertoire profiles were compared with the original, uninfected IgH-V gene profiles. The bar graph represents mean values of a variable number of persons. The number of persons observed for each population is shown in parentheses. EBV transformation of cord-blood-derived B cells did not alter the original IgH-V gene profile. However, when adult peripheral blood-derived B cells were transformed by EBV, the resulting polyclonal cells displayed increased usage of V1 and V5 and decreased usage of V3 and V6.

weeks post EBV infection. All adult donors were EBV immune positive. EBV transformation of cord blood-derived B cells (n = 5) did not result in any skewing of the original IgH-V gene profile. However, when adult peripheral blood-derived B-cells were transformed by EBV (n = 4), the population of transformed polyclonal cells displayed increased usage of V1 and V5 and decreased usage of V3 and V6 (all, $p < 0.001$; FIG. 2).

DISCUSSION

A significant increase of V3 usage ($p < 0.0001$) and decreases of V1 ($p < 0.0001$) and V5 usage ($p < 0.0002$) were observed during the process of B-cell repertoire evolution from neonate to adulthood. The IgH-V gene repertoire in adult peripheral blood was not significantly different from the germline V_H gene composition. In cord blood, however, V5 was overrepresented compared to germline complexity. EBV transformation of cord blood-derived B cells produced polyclonal populations of IgH-V gene repertoires that did not differ from that of the original cord blood IgH-V gene profile. Nevertheless, when adult peripheral blood-derived B-cells (all obtained from seropositive individuals) were transformed by EBV, the resulting polyclonal cells displayed increased usage of V1 and V5 and decreased usage of V3 and V6. This result suggests that EBV may preferentially transform B cells of certain IgH-V gene families in EBV-immune positive individuals.

REFERENCES

1. SCHROEDER, J. H. W., J. L. HILLSON & R. M. PERLMUTTER. 1987. Science **238:** 791–793.
2. MAGEED, R. A., L. E. MACKENZIE, F. K. STEVENSON, B. YUKSEL, F. SHOKRI, B. R. MAZIAK, R. JEFFERIS & P. M. LYDYARD. 1991. J. Exp. Med. **174:** 109–113.
3. PASCUAL, V. & J. D. CAPRA. 1992. Scand. J. Immunol. **35:** 11–18.
4. GUIGOU, V., A.-M. CUISINIER, C. TONNELLE, D. MOINIER, M. FOUGEREAU & F. FUMOUX. 1990. Mol. Immunol. **27:** 935–940.
5. CUISINIER, A.-M., F. FUMOUX, D. MOINIER, L. BOUBLI, V. GUIGOU, M. MILILI, C. SCHIFF, M. FOUGEREAU & C. TONNELLE. 1990. New Biol. **2:** 689–699.
6. ZOUALI, M. & J. THEZE. 1991. J. Immunol. **146:** 2855–2864.
7. GREENBERG, S. J., Y. CHOI, M. BALLOW, T-L. DU, P. WARD, M. H. RICKERT, S. FRANKEL, S. BERNSTEIN & M. BRECHER. 1995. J. Leuk. Biol. In press (May).
8. TOMLINSON, I. M., G. WALTER, J. D. MARKS, M. B. LLEWELYN & G. WINTER. 1992. J. Mol. Biol. **227:** 776–799.

The Expressed Heavy Chain V Gene Repertoire of Circulating B Cells in Normal Adults[a]

B. DAVID STOLLAR

Department of Biochemistry
Tufts University School of Medicine
Boston, Massachusetts 02111

Analyses of the expressed human Ig repertoire are undertaken for several purposes. Comparisons of fetal, neonatal, and adult repertoires explore the developmental program of the immune system, asking how the capacity for diverse immune responses develops in an individual and how it varies among different persons. How are choices made among the vast numbers of potential combinations of gene segments with or without mutations? How do particular antibody responses to viral or bacterial antigens arise from a primary repertoire formed before exposure to such environmental antigens? How does one evaluate restricted patterns of V gene segment usage seen in malignancy and autoimmunity? What is the relationship between disease-associated and natural autoantibodies that appear in a primary repertoire?

Early information reflecting repertoire expression was present in primary structure data for myeloma proteins.[1] Subsequently, there has been an accumulation of sequence data for rearranged genomic DNA, and even more for Ig cDNA from human hybridomas, EBV-transformed B cells, and leukemic B cells.[2-4] Many hybridomas and virus-transformed B cells were chosen for study mainly because their products were of interest—either autoantibodies or antibodies to bacterial or viral antigens. In addition, cDNAs from fetal and neonatal B cells were analyzed in studies of repertoire development.[5-9] Limited numbers of individual hybridomas and EBV-transformed cells can be analyzed. For examination of larger numbers of cells, *in situ* hybridization with specific probes[10-12] and PCR[8,13-15] have been applied to both fetal and adult B-cell populations and to both H and L chains. Simultaneous advances in defining the genomic content of V region segments[16-21] have been of paramount importance for interpretation of sequence data obtained at the levels of rearranged DNA, mRNA, and protein.

ANTIBODY 18/2 AND GERMLINE V_H GENE SEGMENTS

Analysis of the expressed human V gene repertoire in my laboratory grew out of collaborative studies with the laboratory of Dr. Robert Schwartz on monoclonal anti-DNA autoantibodies. We began with analysis of cDNA for anti-DNA autoantibodies produced by human hybridomas from patients with autoimmune disease.[22,23] The first human autoantibody V region we sequenced, antibody 18/2, was

[a] Research in the author's laboratory is supported by Grants AI33180 and GM32375 from the National Institutes of Health.

an IgM molecule that bound ssDNA, cardiolipin, and platelet surface glycolipid. Antibody 18/2 cDNA had an unmutated V_H segment, identical to the corrected sequence of germline V_H26[22,24] (now also known as DP47[18] or V_H3-23[17]). The H chain CDR3 and FR4 of 18/2 cDNA had a long portion of an unmutated Dxp'1 segment and a J_H5 segment with one or no mutation (a single, probably polymorphic, variation from the first published J_H5 sequence). The 18/2 L chain cDNA had a $V\kappa$I segment with more than 98% sequence identity to the germline $V\kappa$A30, and an unmutated $J\kappa4$ segment. Antibody 18/2 was an early example of an autoantibody coded by V region gene segments with few or no mutations.

Subsequent studies focused on H chains related to antibody 18/2 because their V_H26 segment is associated with a disease-related public idiotype first identified as Id16/6.[25] When a panel of 65 EBV-transformed human B cells was examined in the Schwartz lab, strong reactivity of their IgM products with anti-Id was closely correlated with use of V_H26 even when very different κ or λ chains were used.[26,27] Polyclonal anti-Id serum identified corresponding immunoglobulin in lesions and sera of patients with SLE[28,29] and their family members[30] and recognized features shared by both human and mouse immunoglobulins.[31] Cultured B cells of a patient with SLE spontaneously synthesized Id-positive Ig with anti-DNA activity.[32] Low levels of Id-positive immunoglobulins occur in normal sera.[33]

The V_H segment found in antibody 18/2 was of interest also because it is expressed frequently in several other contexts. V_H segments with 98% or higher identity to V_H26 recurred in: other antibodies that bind DNA[27,34,35]; rheumatoid factor (Genbank accession No. L26907); fetal libraries[5,7,8,36,37]; antibodies to bacterial,[38] viral,[39] and tumor[40] antigens; a recombinant phage single chain Fv with anti-B blood group autoreactivity[41]; and EBV-transformed B cells without known binding activity.[27,42]

Several other V_H segments, such as V_H56P1, V_H4.21, and V_H1.9III, also occur frequently in various contexts. As with V_H26, they are found, with few or no mutations, both in IgM antibodies that were chosen for study because of their reactions with autoantigens and in libraries prepared without the criterion of antigen-binding activity. Many of these IgM molecules belong to a population of autoreactive and polyreactive natural autoantibodies.[43] Natural autoantibodies bind to self antigens such as ssDNA, IgG, albumin, actin, and cell surfaces, but they also bind to a variety of bacterial and viral antigens.[43–45] They comprise a substantial fraction of the fetal and neonatal repertoire in both mice and humans.[46–48]

Growing experience has revealed a nonrandom frequency of expression of certain V_H gene segments, D_H segments and J_H segments in special sets of antibodies, such as fetal immunoglobulins, autoantibodies,[3,49,50] and leukemic B-cell products.[4,51] These findings led us and others to ask what is the pattern of V segment expression in normal adults. Approaches to this question have involved analyses of products of EBV-transformed or mitogen-stimulated peripheral blood cells, either individually or, on a larger scale, by *in situ* hybridization[11,12] or PCR.[8,13] In our studies, we tested mRNA expression as it occurs in circulating B cells at a level that allows us to capture it through PCR amplification without *in vitro* stimulation.

PREPARATION OF REPRESENTATIVE LIBRARIES FROM NORMAL SUBJECTS

In order to prepare a representative amplified cDNA library without introducing bias from the choice of primers (as can occur with degenerate or consensus

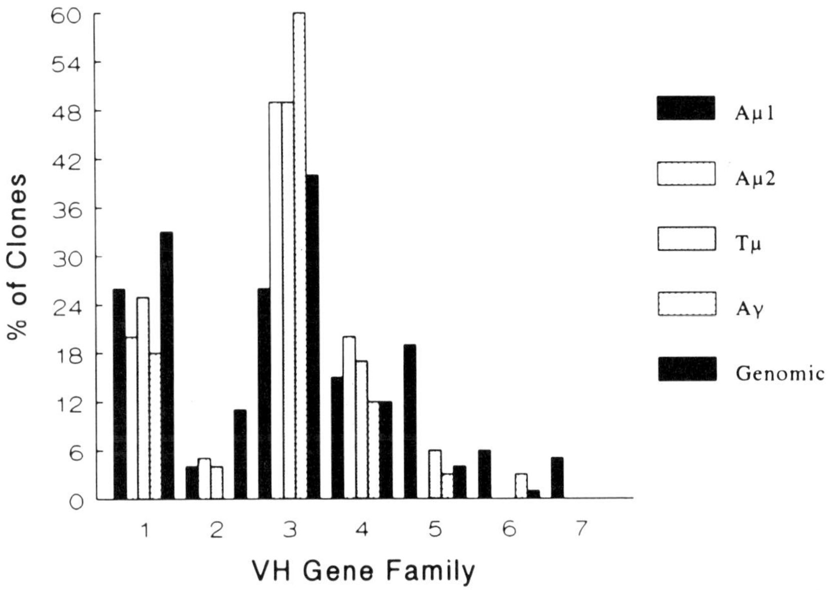

FIGURE 1. V_H gene segment family usage in four peripheral blood B-cell cDNA libraries from two subjects, A and T. The Aμ1 (n = 54) and Aμ2 (n = 34) libraries were prepared from one subject at two times separated by 11 months. For the Tμ library, n = 49. The Aγ (n = 34) library was prepared from the same RNA sample as Aμ1. The relative genomic complexity is taken from Berman *et al.*[16] V_H family assignments are based on sequencing from the J_H-Cμ junction at least through FR3.

sequences or primer mixtures), Chichi Huang devised a two-step PCR procedure that does not use any V region sequence in the primers.[52] No step in the procedure favors any particular mRNA. He prepared IgM and IgG libraries by use of γ or μ constant region primers in the reverse transcriptase and PCR reactions. He studied two subjects and, for one of them, prepared libraries at two times, separated by 11 months. Libraries were analyzed by hybridization with family-specific FR probes and individual gene-specific CDR probes. Partial cDNA sequences, through the J_H, D_H, and at least FR3 of the V_H segment, were obtained for 165 distinct clones. Sequences through all or nearly all of the V region were obtained for 105 of these clones.[53,54]

A MAJORITY OF IgM V_H SEGMENTS IN cDNA OF CIRCULATING B CELLS HAVE FEW OR NO MUTATIONS AND USE THE SAME V_H SEGMENTS AS FETAL Ig

Averaging of data combined from all libraries indicated that gene family usage was generally parallel to family complexity.[53–55] V_H usage in individual libraries, however, sometimes differed from that pattern (FIG. 1). One of the IgM libraries contained more than the expected numbers of gene segments from the small V_H5, V_H6, and V_H7 families. The gene family distribution changed to resemble the

TABLE 1. A Small Number of V_H Gene Segments Account for a Majority of IgM Clones

Germline V_H Gene	V_H Families	No. of IgM Clones >99% $n = 48$	No. of IgM Clones 95–98% $n = 21$
1-18; DP-14;	V_H1	3	2
1-02; DP-75; 20P3;		2	3
L33		2	0
1-08; 1-08		2	0
2-05	V_H2	2	1
3-23; DP-47; V_H26;18/2;30P1	V_H3	5	0
3-09; DP-31; HHG19		5	3
3-53; DP-42; 60P2; 63P1		3	1
4-39; $V_H4.18$	V_H4	5	0
4-59; DP-71; 58P2; 4.11		5	0
4-34; DP-63; 4.21		2	0
		36	10

genomic complexity in a second library from the same subject 11 months later. The skewed distribution in the first library may have reflected a recent immunological stimulus. The analysis provides a time-framed picture of Ig expression.

The V_H segments of 48 (approximately two-thirds) of the 77 fully sequenced IgM cDNA clones in three libraries had more than 99% identity with known germline V_H gene segments. This number is similar to the fraction of circulating B cells that are IgM^+/IgD^+,[56] a population described as naive, as distinct from memory cells.[56] Consistent with this assignment, there were also very few mutations in the CDR3 portion of the J_H segments in these clones, and the D_H of many could be accounted for by known D_H segments with few or no mutations.[54] In two cases "n" nucleotides were absent at both the VD or the DJ junctions, and in 12 more they were absent from one or the other junction. Thus, properties of the fetal or neonatal repertoire persist in a substantial fraction of cDNAs from unmanipulated circulating B cells of normal adults. This finding is also consistent with the reported presence in adults of B cells bearing a cross-reactive idiotype characteristic of fetal cells.[12]

Also as in fetal libraries, there was recurrent expression of a restricted number of germline sequences, so that 36 of the 48 clones with nearly unmutated V_H gene segments (75%) could be accounted for by 11 V_H gene segments (TABLE 1). The highly represented segments were the same as those found previously in autoantibodies and fetal libraries. One of them was V_H26.

A second population of IgM cDNA clones had V_H segments with multiple mutations. This observation could be made with confidence for members of the small V_H6 or V_H5 families that have only one or few members.[53] In most of these clones, there were also mutations in the CDR3 portion of J_H sequences. These clones may represent circulating memory B cells that have been subject to previous antigen exposure.

In contrast to the IgM clones, only 2 of 28 IgG cDNA clones had V_H segments with 99% or higher identity and another 6 had 95 to 98% identity to known germline V_H segments. V_H26 was used in both of the nearly unmutated clones and in two of the six with 95 to 98% germline identity.

The high frequency of V_H26 expression among randomly picked and sequenced clones was consistent with results of surveying much higher numbers of clones of these libraries with oligonucleotide probe hybridization; from 4 to 10% of clones in different libraries hybridized with V_H26-specific probes.[55] Combining results of all three IgM and the IgG libraries, a total of 171 of 2540 clones (6.8%) hybridized with V_H26-specific oligonucleotide probes, whereas 27 (1%) were positive with 56P1-specific probes and only 14 (0.6%) were positive with probes for the V_H1 family member that occurs in antibody 21/28 (V_H1-3b, DP25).

Because V_H26 recurred frequently in these libraries, Keith Stewart further tested its representation in V_H3 family-specific libraries from other subjects.[55] V_H26 was found in 19%, 20%, and 28% of V_H3 cDNA clones in three different subjects, but was totally absent from the V_H3 library of another person. Another frequently expressed segment, V_H56P1, was found in 6%, 15%, and 15% of V_H3 family clones in these three subjects. The hybridization data supported the conclusion that nonrandom use of V_H gene segments is characteristic of a normal adult repertoire, as it is of the populations of natural autoantibodies, fetal Ig, and leukemic B-cell products.

CDR3H DIVERSITY: EACH CLONE IS UNIQUE

There were nonrandom features in the use of D_H and J_H segments as well as in the V_H segments. Consistent with results of others,[3,13,57,58] we found a high frequency of J_H4 and J_H5 and much less use of J_H1 and J_H2.[54] Certain D_H families and individual members recurred more frequently than others. The most marked feature, however, again consistent with experience of others,[13,58] was the extensive variability of the CDR3H segments. Each of the CDR3 sequences was unique. When the "n" nucleotides, D_H segments, and sites of D and J segment joining were considered, no CDR3 was closely related to any other in overall structure, even though several clones used a particular D_H segment. Furthermore, even though major loci of D gene segments have been identified, cDNA libraries (including ours) regularly contain a significant number of examples for which a related germline sequence cannot be assigned. The CDR3H variability in both length and sequence is evident in a comparison of V regions in which the V_H segment has 99% or more identity with V_H26 (TABLE 2). On the basis of PCR studies focused on the CDR3H, Yamada *et al.*[13] estimated that the frequency of any given B-cell clone in peripheral blood of normal subjects was no higher than 1/20,000.

IMPLICATIONS FOR SELECTION OF THE B-CELL REPERTOIRE

The starting point for development of an individual's repertoire is the inherited genomic set of V region gene segments. There are polymorphic features in the human genomic V gene set—most notably in the presence or absence or the copy number of certain V_H segments.[59–62] Nevertheless, the sequence conservation among humans of widely differing ethnic origin is more striking than these differences.[3,63] Futhermore, there is substantial primary structure conservation among V_H gene segments of different species[64]; some V_H segments of mouse and human immunoglobulin DNA have fewer differences than C domains.[65]

A high degree of conservation of V domain primary structure among species and among diverse human populations may reflect various kinds of selective

TABLE 2. Heavy Chain CDR3H Amino Acid Sequences in cDNA Clones with V_H26 Segments

Genbank Accession	Clone	. . . V_H	CDR3H	. . J_H. .
U00491	Aμ1c3-3	tavyyca	m v g a s h	dywgq
yU00495	Aμ1d11-3	tavyycak	c s e l r y f d w s v	dywgq
U00510	Aμ1g5-3	tavyyca	n d y y g s g r y f t y a t	dvwgq
U00525	Aγ34-1	tavyyca	s p s k d f h y y g m	dvwgq
U00536	Aγ6-3	tavyycar	d r e r n h y a d y e v r i y y y g l	dvwgk
U00537	Aγ62-3	tavyycak	w s v s m v p s l d d a f	diwgq
U00541	Aμ10-2	tavyycak	d r r n y d f w s g y y y y y y g m	dvwgq
U00549	Aμ35-2	tavyycak	d i e d t a m f p y y y g m	dvwgq
M17746	18/2	tavyycak	g q v l y y g s g s y h w f	dpwgq
M57940	V-B19.7	tavyycak	r a p k l t t k g y y f	dywg
M57941	V-A77	tavyycak	y p i y g y s v g a t y f	dywg
M57942	V-A73	tagyyca	l r w r w s y p p	dywg
M57943	V-GL16	tavyycak	g d s s g y p r y f	dlwg
M57944	V-A95	tavyycak	g v g s s g w	dhwg

NOTE: Clones beginning U00 are from Huang *et al.*[54] Clone M17746 (18/2) is from Dersimonian *et al.*[22] Clones beginning M57 are from Guillaume *et al.*[26] Antibodies 18/2 and B19.7 bind ssDNA.[22,27] The CDR3H portion of JH is underlined.

pressures acting on an evolutionary time scale. Some features of primary structure are stable because they are required for the fundamental antibody architecture, that is, the formation of the beta sheet framework, disulfide loops, and interchain faces. Other patches of conserved sequence exposed on the surface of FR structure have been suggested to play a role in regulating gene usage.[64] However, ligand selection of the repertoire may also act on this time scale and may be reflected in two properties of antibodies in the natural autoantibody and fetal repertoires of both mice and humans: autoreactivity to highly conserved self antigens[46,65] and polyreactivity[66,67] that includes the ability to bind to bacterial surface antigens[45] and viruses.[68]

The pattern of nonrandom V segment expression in both fetal and adult repertoires also suggests the action of selective processes within a lifetime. Structural factors may dictate some of this pattern: the presence of multiple copies of some segments,[69,70] varying efficiency of switch sequences for rearrangement, and accessibility in chromatin structure. The repertoire is also shaped by negative selection based on strong interaction with self antigen,[71,72] and it is probable that many emerging B cells die for lack of any interaction.

Positive ligand selection by low-affinity interaction with self antigens , including idiotopes of other immunoglobulins,[73,74] is also considered to shape the primary repertoire, which includes natural autoantibodies. Although V_H segment usage may be influenced by the ability of V_H segments to facilitate certain ligand interactions, the CDR3H is likely to play a major role in specific ligand interaction and repertoire development.[75–77] What appears to be a restricted repertoire from the point of view of germline segment composition and even V_H and J_H combinations is in fact much more highly diverse because of CDR3H variability. CDR3H diversity is particularly significant because of the important contribution of this region to the antigen-binding specificity of many antibodies. Thus, there are many examples in which completely identical V_H segments occur in antibodies with very different antigen-binding properties.

In the case of CDR3, it is functional analysis rather than inspection of sequences that provides evidence for selection. Mutagenesis experiments show the importance of particular CDR3H residues for natural autoantibody activity.[76] Mutagenesis and domain grafting of V region segments of a mouse antibody to Z-DNA also illustrate how one to four amino acids in the H chain CDR3 can determine whether the antibody is a highly selective antibody to Z-DNA; an autoantibody-like molecule that binds ssDNA, dsDNA, and Z-DNA; or an immunoglobulin that does not bind any of these nucleic acids.[78]

Polyreactivity of the diverse primary repertoire provides an ability to make a primary response to diverse organisms. Even though the intrinsic affinity of a single polyreactive binding site may be relatively low, magnification by multivalent binding of IgM to multivalent bacterial surfaces can yield very stable complexes. Thus, antibody of the primary repertoire of circulating B cells can be effective in combining with organisms and delivering antigen to central lymphoid organs where mutation and antigen selection can lead to production of a more specific antibody of higher affinity. The CDR3H diversity allows for many different modes of interaction with either self or foreign antigen.

Positive selection by autoantigen may act on early stages of B-cell development, at a time when H chain has been expressed but before L chains have been rearranged[79]; at this stage, H chain is associated with other proteins, notably surrogate L chain, both within the cells and on the cell surface.[80] This selection begins in fetal development before exposure to foreign antigen and may continue through adult life to act on all emerging B cells at an early stage in their development.[80,81] A continuing process of positive selection by low-affinity interaction with self antigen could account for the shared features of fetal and adult IgM repertoires, that is, the substantial populations of IgM cDNA with unmutated V_H and J_H segments but highly variable CDR3 sequences. This repertoire can also provide the substrate for development of disease-related autoantibodies through continued selection by autoantigen[82] if there is a failure in appropriate elimination of cells making high-affinity autoantibody.

REFERENCES

1. Kabat, E. A., T. T. Wu, M. Reid-Miller, H. M. Perry & K. S. Gottesman. 1993. Sequences of Proteins of Immunological Interest, 4th edit. U.S. Public Health Service. National Institutes of Health. Bethesda, MD.
2. Casali, P. & A. L. Notkins. 1989. Annu. Rev. Immunol. **7:** 513–535.
3. Pascual, V. & J. D. Capra. 1991. Adv. Immunol. **49:** 1–74.
4. Logtenberg, T., M. E. M. Schutte, G. Inghirami, J. E. Berman, F. H. J. Gmelig-Meyling, R. A. Insel, D. M. Knowles & F. W. Alt. 1989. Int. Immunol. **1:** 362–366.
5. Schroeder, H. W., Jr., J. L. Hillson & R. M. Perlmutter. 1987. Science **238:** 791–793.
6. Mortari, F., J. A. Newton, J. Y. Wang & H. W. Schroeder, Jr. 1992. Eur. J. Immunol. **22:** 241–245.
7. Raaphorst, F. M., E. Timmers, M. J. H. Kenter, M. J. D. Van Tol, J. M. Vossen & R. K. B. Schuurman. 1992. Eur. J. Immunol. **22:** 247–251.
8. Cuisinier, A. M., L. Gauthier, L. Boubli, M. Fougereau & C. Tonnelle. 1993. Eur. J. Immunol. **23:** 110–118.
9. Hillson, J. L. & R. M. Perlmutter. 1990. Int. Rev. Immunol. **5:** 215–229.

10. Guigou, V., A. M. Cuisinier, C. Tonnelle, D. Moinier, M. Fougereau & F. Fumoux. 1990. Mol. Immunol. **27:** 935–940.
11. Zouali, M. & J. Theze. 1991. J. Immunol. **146:** 2885–2864.
12. Long, A. A., J. Mueller, J. André-Schwartz, K. J. Barrett, R. Schwartz & H. Wolfe. 1992. Diag. Mol. Pathol. **1:** 45–57.
13. Yamada, M., R. Wasserman, B. A. Reichard, S. Shane, A. J. Caton & G. Rovera. 1991. J. Exp. Med. **173:** 395–407.
14. Campbell, M. J., A. D. Zelenetz, S. Levy & R. Levy. 1992. Mol. Immunol. 193–198.
15. Klein, R., R. Jaenichen & H. G. Zachau. 1993. Eur. J. Immunol. **23:** 3248–3271.
16. Berman, J. E., S. J. Mellis, R. Pollock, C. L. Smith, H. Suh, B. Heinke, C. Kowai, U. Surti, L. Chess, C. R. Cantor & F. W. Alt. 1988. EMBO J. **7:** 727–738.
17. Matsuda, F., E. K. Shin, H. Nagaoka, R. Matsumura, M. Haino, Y. Fukita, S. Taka-ishi, T. Imai, J. H. Riley, R. Anand, E. Soeda & T. Honjo. 1993. Nature Genet. **3:** 88–94.
18. Tomlinson, I. M., G. Walter, J. D. Marks, M. B. Llewelyn & G. Winter. 1992. J. Mol. Biol. **227:** 776–798.
19. Chothia, C., A. M. Lesk, E. Gherardi, I. M. Tomlinson, G. Walter, J. D. Marks, M. B. Llewelyn & G. Winter. 1992. J. Mol. Biol. **227:** 799–817.
20. Huber, C., K. F. Schable, E. Huber, R. Klein, A. Meindl, R. Thiebe, R. Lamm & H. G. Zachau. 1993. Eur. J. Immunol. **23:** 2868–2875.
21. Daley, M. D., T. Olee, H. Peng, R. W. Soto-gil, P. P. Chen & K. A. Siminovitch. 1992. Mol. Immunol. **29:** 1031–1042.
22. Dersimonian, H., R. S. Schwartz, K. J. Barrett & B. D. Stollar. 1987. J. Immunol. **139:** 2496–2501.
23. Dersimonian, H., K. P. W. J. McAdam, C. Mackworth-Young & B. D. Stollar. 1989. J. Immunol. **142:** 4027–4033.
24. Chen, P. P., M. F. Liu, S. Sinha & D. A. Carson. 1988. Arthritis Rheum. **31:** 1429–1431.
25. Shoenfeld, Y., D. A. Isenberg, J. Rauch, M. Madaio, B. D. Stollar & R. S. Schwartz. 1983. J. Exp. Med. **158:** 718–730.
26. Guillaume, T., D. B. Rubinstein, F. Young, L. Tucker, T. Logtenberg, R. S. Schwartz & K. Barrett. 1990. J. Immunol. **145:** 1934–1945.
27. Young, F., L. Tucker, D. Rubinstein, T. Guillaume, J. Andre-Schwartz, K. J. Barrett, R. S. Schwartz & T. Logtenberg. 1990. J. Immunol. **145:** 2545–2553.
28. Isenberg, D. A. & C. Collins. 1985. J. Clin. Invest. **76:** 287–294.
29. Watts, R. A., C. T. Ravirajan, L. S. Wilkinson, W. Williams, M. Griffiths, D. Butcher, A. T. Horsfall, N. A. Staines & D. A. Isenberg. 1991. Clin. Exp. Immunol. **83:** 267–273.
30. Isenberg, D. A., Y. Shoenfeld, M. Walport, C. Mackworth-Young, C. Dudeney, A. Todd-Pokropek, S. Brill & A. Weinberger. 1985. Arthritis Rheum. **28:** 999–1007.
31. Watts, R. A., C. T. Ravirajan, N. A. Staines & D. A. Isenberg. 1990. Immunology **69:** 348–354.
32. Datta, S. K., Y. Naparstek & R. S. Schwartz. 1986. Clin. Immunol. Immunopathol. **38:** 302–318.
33. Madaio, M. P., A. Schattner, M. Schattner & R. S. Schwartz. 1986. J. Immunol. **137:** 2535–2540.
34. Lampman, G. W., B. Furie, R. S. Schwartz, B. D. Stollar & B. C. Furie. 1989. Blood **74:** 262–269.
35. Winkler, T. H., H. Fehr & J. R. Kalden. 1992. Eur. J. Immunol. **22:** 1719–1728.
36. Hillson, J. L., I. R. Oppliger, E. H. Sasso, E. C. Milner & M. H. Wener. 1992. J. Immunol. **149:** 3741–3752.
37. Milili, M., F. Le Deist, G. de Saint-Basile, A. Fischer, M. Fougereau & C. Schiff. 1993. J. Clin. Invest. **91:** 1616–1629.
38. Adderson, E. E., P. G. Shackelford, A. Quinn, P. M. Wilson, M. W. Cunningham, R. A. Insel & W. L. Carroll. 1993. J. Clin. Invest. **91:** 2734–2743.

39. IKEMATSU, H., N. HARINDRANATH, Y. UEKI, A. L. NOTKINS & P. CASALI. 1993. J. Immunol. **150:** 1325–1337.
40. KISHIMOTO, T., H. OKAJIMA, T. OKUMOTO & M. TANIGUCHI. 1989. Nucleic Acids Res. **17:** 4385.
41. MARKS, J. D., W. H. OUWENHAND, J. M. BYE, R. FINNERN, B. D. GORICK, D. VOAK, S. J. THORPE, N. C. HUGHES-JONES & G. WINTER. 1993. BioTechnology **11:** 1145–1149.
42. KNIGHT, G. B., V. AGNELLO, V. BONAGURA, J. L. BARNES, D. J. PANKA & Q. X. ZHANG. 1993. J. Exp. Med. **178:** 1903–1911.
43. DIGHIERO, G., B. GUILBERT & S. AVRAMEAS. 1982. J. Immunol. **128:** 2788–2797.
44. PRABHAKAR, B. S., J. SAEGUSA, T. ONODERA & A. L. NOTKINS. 1984. J. Immunol. **133:** 2815–2817.
45. KABAT, E. A., K. G. NICKERSON, J. LIAO, L. GROSSBARD, E. OSSERMAN, E. GLICKMAN, L. CHESS, J. B. ROBBINS, R. SCHNEERSON & Y. YANG. 1986. J. Exp. Med. **164:** 642–654.
46. DIGHIERO, G., P. LYMBERI, D. HOLMBERG, I. LUNDQUIST, A. COUTINHO & S. AVRAMEAS. 1985. J. Immunol. **134:** 765–771.
47. SOUROUJON, M., M. E. WHITE-SCHARF, J. ANDRÉ-SCHWARTZ, M. L. GEFTER & R. S. SCHWARTZ. 1988. J. Immunol. **140:** 4173–4179.
48. GUIGOU, V., B. GUILBERT, D. MOINIER, C. TONNELLE, L. BOUBLI, S. AVRAMEAS, M. FOUGEREAU & F. FUMOUX. 1991. J. Immunol. **146:** 1368–1374.
49. SANZ, I., H. Y. DANG. M. TAKAI, N. TALAL & J. D. CAPRA. 1989. J. Immunol. **142:** 883–887.
50. PASCUAL, V., I. RANDEN, K. THOMPSON, M. SIOUD, O. FORRE, J. NATVIG & J. D. CAPRA. 1990. J. Clin. Invest. **86:** 1320–1328.
51. KIPPS, T. J., E. TOMHAVE, L. F. PRATT, S. DUFFY, P. P. CHEN & D. A. CARSON. 1989. Proc. Natl. Acad. Sci. USA **86:** 5913–5917.
52. HUANG, C. & B. D. STOLLAR. 1991. J. Immunol. Methods **141:** 227–236.
53. HUANG, C., A. K. STEWART, R. S. SCHWARTZ & B. D. STOLLAR. 1992. J. Clin. Invest. **89:** 1331–1343.
54. HUANG, C. & B. D. STOLLAR. 1993. J. Immunol. **151:** 5290–5300.
55. STEWART, A. K., C. HUANG, B. D. STOLLAR & R. S. SCHWARTZ. 1993. J. Exp. Med. **177:** 409–418.
56. KLEIN, U., R. KÜPPERS & K. RAJEWSKY. 1993. Eur. J. Immunol. **23:** 3272–3277.
57. BRAUN, J., L. BERBERIAN, L. KING, I. SANZ & H. L. GOVAN III. 1992. J. Clin. Invest. **89:** 1395–1402.
58. SANZ, I. 1991. J. Immunol. **147:** 1720–1729.
59. SASSO, E. H., K. W. VAN DIJK & E. C. B. MILNER. 1990. J. Immunol. **145:** 2751–2757.
60. WILLEMS VAN DIJK, K., H. W. SCHROEDER, JR., R. M. PERLMUTTER & E. C. MILNER. 1989. J. Immunol. **142:** 2547–2554.
61. CHEN, P. P., K. A. SIMINOVITCH, N. J. OLSEN, R. A. ERGER & D. A. CARSON. 1989. J. Clin. Invest. **84:** 706.
62. ADDERSON, E. E., F. H. AZMI, P. M. WILSON, P. G. SHACKELFORD & W. L. CARROLL. 1993. J. Immunol. **151:** 800–809.
63. SANZ, I., P. KELLY, C. WILLIAMS, S. SCHOLL, P. TUCKER & J. D. CAPRA. 1989. EMBO J. **8:** 3741–3748.
64. KIRKHAM, P. M., F. MORTARI, J. A. NEWTON & H. W. SCHROEDER, JR. 1992. EMBO J. **11:** 603–609.
65. STOLLAR, B. D. 1991. Mol. Immunol. **28:** 1399–1412.
66. SHOENFELD, Y., J. RAUCH, H. MASSICOTTE, S. K. DATTA, J. ANDRÉ-SCHWARTZ, B. D. STOLLAR & R. S. SCHWARTZ. 1983. N. Engl. J. Med. **308:** 414–420.
67. LAFER, E. M., J. RAUCH, C. ANDRZEJEWSKI, JR., D. MUDD, B. FURIE, R. S. SCHWARTZ & B. D. STOLLAR. 1981. J. Exp. Med. **153:** 897–909.
68. UEKI, Y., I. S. GOLDFARB, N. HARINDRANATH, M. GORE, H. KOPROWSKI, A. L. NOTKINS & P. CASALI. 1990. J. Exp. Med. **171:** 19–34.
69. RUBINSTEIN, D. B., M. SYMANN, A. K. STEWART & T. GUILLAUME. 1993. Mol. Immunol. **30:** 403–412.

70. Van Dijk, K. W., L. A. Milner, E. H. Sasso & E. C. B. Milner. 1992. Proc. Natl. Acad. Sci. USA **89:** 10430–10434.
71. Russell, D. M., Z. Dembic, G. Morahan, J. F. A. P. Miller, K. Bürki & D. Nemazee. 1991. Nature **354:** 308–311.
72. Goodnow, C. C. 1992. Annu. Rev. Immunol. **10:** 489–518.
73. Kearney, J. F. & M. Vakil. 1986. Immunol. Rev. **94:** 39–50.
74. Freitas, A. A., O. Burlen & A. Coutinho. 1988. J. Immunol. **140:** 4097–4102.
75. Martin, T., S. F. Duffy, D. A. Carson & T. J. Kipps. 1992. J. Exp. Med. **175:** 983–991.
76. Martin, T., R. Crouzier, J-C. Weber, T. J. Kipps & J-L. Pasquali. 1994. J. Immunol. **152:** 5988–5996.
77. Clarke, S. H. & S. K. McCray. 1993. Eur. J. Immunol. **23:** 3327–3334.
78. Polymenis, M. & B. D. Stollar. 1994. J. Immunol. **152:** 5319–5329.
79. Schwartz, R. S. & B. D. Stollar. 1994. Immunol. Today **15:** 27–32.
80. Melchers, F., D. Haasner, U. Grawunder, C. Kalberer, H. Karasuyama, T. Winkler & A. G. Rolink. 1994. Annu. Rev. Immunol. **12:** 209–225.
81. Freitas, A. A., L. Andrade, M. P. Lembezat & A. Coutinho. 1990. Int. Immunol. **2:** 15–23.
82. Radic, M. Z. & M. Weigert. 1994. Annu. Rev. Immunol. **12:** 487–520.

Molecular Analysis of V_H Regions Expressed in the Adult B-Cell Repertoire

NIENKE VAN DER STOEP,[a] SASKIA B. EBELING,[b] AND TON LOGTENBERG[a]

[a]*Department of Immunology*
University Hospital Utrecht
Heidelberglaan 100
3508 GA Utrecht, the Netherlands

[b]*Red Cross Blood Transfusion Center*
Plesmanlaan 125
1066 CX Amsterdam, the Netherlands

INTRODUCTION

Immunoglobulin (Ig) molecules are composed of two identical heavy (H) chain and two identical light (L) chain polypeptides. Functionally, Ig molecules can be divided into two portions: a constant region that determines the biological properties of the molecule and a variable (V) region involved in antigen binding. During precursor B-cell development, the V regions of Ig molecules are assembled from three multigene families of DNA segments: V_H, D, and J_H genes for the H chain and V_L and J_L genes for the L chain. The combinatorial assortment of the various gene segments present in the germline and the variability generated at the junctions during somatic recombination results in the generation of a large diversity of H and L polypeptide chains. The net result of these processes is the continuous generation of B lymphocytes that each express a unique cell surface-bound Ig receptor (reviewed in Tonegawa[1] and Alt *et al.*[2])

Once expressed on the cell surface of B lymphocytes, the repertoire of Ig molecules is shaped by V-region-mediated selection processes. Selection may already operate on newly formed B cells in primary lymphoid organs and, with intervals, may continue throughout the life of a B lymphocyte.[3,4] Several factors affect the selection process, including the nature of the selecting ligand, ontogenic stage, anatomical location, and B-cell subpopulation. Depending on these variables, Ig-mediated interactions of a B lymphocyte with components of its environment may result in (functional) deletion or clonal expansion of B cells and activation of B-cell-specific processes such as isotype switching and somatic hypermutation, often associated with commitment of a B cell to a particular differentiation pathway.[5]

We have studied aspects of human B-cell repertoire generation by studying the expression of individual V_H genes in peripheral adult lymphoid organs. In particular, we analyzed the occurrence of somatic mutations in $CD5^+/IgM^+$ and $CD5^-/IgM^+$ B cells in peripheral blood, tonsil, and spleen of healthy persons and determined the molecular structure of V regions encoding IgE molecules in patients with atopic dermatitis as a model of a human antibody response to clinically relevant antigens. To facilitate molecular analysis, V_H genes were selected that

TABLE 1. Summary of Mutations in the V Regions Expressed in IgM-Bearing Human B Cells

Source		n^a	Number GL/MUT	Mutations	Range	Average
Spleen 1	V_H5	8	1/7	27	2–10	3.9
	V_H6	8	0/8	53	1–14	6.6
Spleen 2	V_H5	8	0/8	77	2–17	9.6
	V_H6	9	0/9	61	2–16	6.8
PB 1	V_H5	11	3/8	47	2–14	5.9
PB 2/CD5$^+$	V_H5	11	4/7	23	1–7	3.3
	V_H6	9	3/6	30	1–19	5.0
PB 2/CD5$^-$	V_H5	12	7/4	16	1–9	4.0
TO/CD5$^+$	V_H6	3	0/3	40	11–17	13.3
TO/CD5$^-$	V_H6	3	0/3	15	3–8	5.0

[a] *n*: number of cDNA sequences analyzed. Calculations are based on mutated V regions only. GL: germline; MUT: mutated.

are easily distinguishable from other V_H genes in the germline and that appear very well conserved in the outbred human population: the V_H6 gene, the single member of this family, and V_H5-1R1 and V_H5-2R1, the two functional members of the small V_H5 gene family.

SOMATICALLY MUTATED IgM RECEPTORS IN PERIPHERAL CD5$^+$ AND CD5$^-$ B CELLS

Using a cDNA/PCR-based approach, we analyzed the nucleotide sequences of >80 μ transcripts from adult peripheral blood (PB), tonsil, and spleen B-cell populations.[5] The analysis of somatic mutations was restricted to the portion of the V region encoded by the V_H gene segments because germline D and J_H segments from the persons studied were not available for comparison.

The analysis of 11 V_H5-encoded μ transcripts from unseparated PB B cells of a healthy person showed that three cDNAs were 100% identical in sequence to the published germline V_H5-1R1 gene. In eight cDNAs, we detected a total of 47 mutations (average 5.9 mutations/V_H gene; range 2–14 mutations/V_H gene). The analysis of splenic μ transcripts encoded by the V_H5 and V_H6 gene segment showed comparable results. (The sequences for V_H5 are shown in FIG. 1.) Among 33 transcripts from two spleen samples analyzed, we detected a single germline-encoded transcript, whereas the remaining 32 cDNAs contained a total of 218 somatic mutations (average 6.8 mutations/V_H gene; range 1–17 mutations/V_H gene). The results of these experiments are summarized in TABLE 1.

We next analyzed μ transcripts from FACS-sorted CD5$^+$ and CD5$^-$ PB and tonsillar B cells to determine whether somatic mutations in μ transcripts segregated with either B-cell population. The sorted CD5$^+$ and CD5$^-$ populations were >98% pure as determined by rerun FACS analysis (results not shown). Comparing the extent and distribution of somatic mutations, we detected no significant differences in μ transcripts from CD5$^+$ and CD5$^-$ B cell populations in either tonsil or PB. (The results for the V_H5 transcripts from PB B cells are shown in FIG. 1.)

Comparison of the total set of V_H5 and V_H6 cDNA clones reveals the presence of multiple recurrent replacement (R) and substitution (S) mutations within sequences from one person and between sequences of unrelated persons. Some of these may be attributed to polymorphisms. The silent C → G transition at position 42 in of the V_H5-encoded transcripts has also been reported in a germline V_H5 gene and likely represents a polymorphism (FIG. 1). Considering that members of the V_H5 and V_H6 gene families are extremely well conserved, most of the recurrent substitutions cannot be accounted for by polymorphisms. This notion is further supported by the occurrence of three different nucleotides at particular positions in cDNAs from a single person (an example is indicated with an arrow in FIG. 1). Recurrent mutations leading to R mutations may have been selected by antigen or reflect "hot spots" of mutation, whereas recurrent S mutations may denote "hotspots" of mutation. TABLE 2 summarizes the distribution of R and S mutations over complementary-determining (CDR) and framework (FR) regions, considering all mutations within the collections of V_H5 and V_H6-encoded cDNAs as two separate pools. The overall ratio of R/S mutations is compatible with an antigen-driven selection process. In particular, the CDR1 region contains a high number of R mutations. In individual cases, both random and skewed distributions of R/S mutations are noticeable.

In the mouse, molecular studies of Ig V regions have used hybridomas, unselected pre-B and B-cell populations, and B cells separated on the basis of Ly1 or high level of membrane IgD expression as a starting material. Importantly, PCR-based strategies for V_H region analysis of polyclonal murine B-cell populations have employed both cDNA and genomic DNA libraries. In the former case, selection may occur for transcripts from B lymphocytes with a high copy number of μ H chain mRNA molecules (e.g. plasma cells), whereas in genomic libraries, each B-cell clone in the starting sample is represented as a single functional rearrangement. Independent of the experimental strategy or starting B-cell population, the peripheral repertoire of IgM-bearing murine B cells comprises predominantly unmutated Ig V regions, although some μ transcripts from B cells isolated early after primary immunization may contain low numbers of mutations. In this context, the high incidence of mutated V regions in μ transcripts from adult human peripheral lymphoid tissues, containing up to 18 mutations in individual transcripts, is surprising. Several, not mutually exclusive, explanations may account for the apparent difference between mice and humans. Although we previously found little evidence for spontaneous IgM-secreting cells in freshly isolated PB B-cell samples, it is conceivable that our cDNA/PCR-based cloning strategy selects for "activated" B cells containing relatively high levels of μ mRNA. Alternatively, the peripheral IgM-bearing B lymphocyte pool in adult humans may comprise a relatively large population of long-lived B cells that have accumulated mutations in their V regions over extended periods of time. Independent support for this scheme comes from the analysis of human monoclonal, IgM-secreting EBV-transformed cell lines from adult blood that reportedly express mutated V_H5 and V_H6 genes (8–10). It has been suggested that this B-lymphotrophic virus immortalizes a wide spectrum of B-lineage cells.[11,12]

The distribution and types of mutations in the V region of a B-cell clone have been suggested to reflect the nature of the activating and selecting forces causing its expansion.[13] Under restraint of preserving the Ig molecule's structure and in the absence of selection, R/S values have been estimated to be 1.5 in the FR and 2.9 in the CDR.[14] Substantially higher R/S values in the CDR have been interpreted to reflect Ig-mediated positive selection. The degree and distribution of mutations in individual μ transcripts in our collection varies extensively, indicating that they derived from a heterogeneous population of B cells. Several cDNA clones in our collection appear to be the product of antigen-driven activation and selection. For

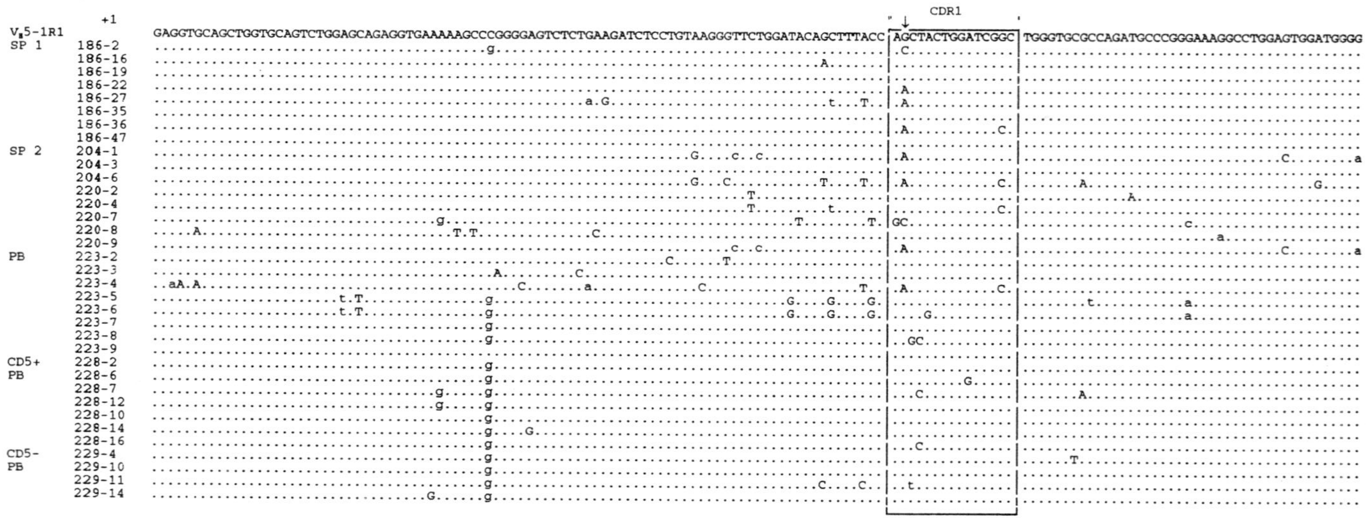

FIGURE 1. Nucleotide sequences of V_H5-encoded μ transcripts. The sequences are compared to the germline V_H5-1R1 gene. The sequences of 15 cDNAs that were 100% identical to V_H5-1R1 are not shown. Dots indicate nucleotide identity. Lower case letters represent S mutations. Upper case letters represent R mutations. Arrows indicate examples of mutational "hotspots." SP: spleen.

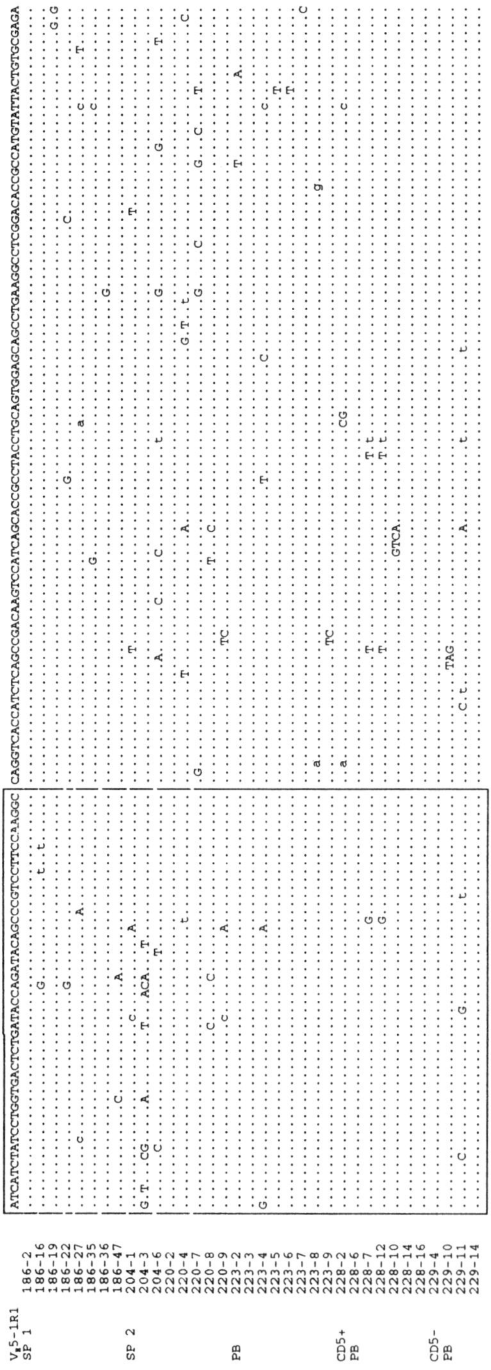

FIGURE 1 *(Continued).*

TABLE 2. Distribution of R and S Mutations over FR and CDR Regions of μ Transcripts

V_H	Region	S	R	R/S	V_H	Region	S	R	R/S
V_H5	FR1	11	27	2.5	V_H6	FR1	7	13	1.9
	CDR1	1	25	25.0		CDR1	4	21	5.3
	FR2	5	6	1.2		FR2	3	10	3.3
	CDR2	6	25	4.2		CDR2	13	33	2.5
	FR3	15	51	3.4		FR3	26	41	1.6

example, cDNA clone 204-3 contains a minimum of five R mutations, all located in CDR2, and no S mutations. In contrast, the distribution of R and S mutations in the V regions expressed in other B-cell clones appears to be at variance with the "classical" scheme associated with antigen selection. For example, cDNA 204-6 harbors four R and no S mutations in the combined CDR1 and CDR2 regions and 12 R and 1 S mutation in the FR regions. Recent data suggest that the "classical" model of antigen-induced indication and selection of R and S mutations in V regions requires extension. The analysis of "passenger" V_H genes in transgenic mouse models that are not the targets of antigenic selection has revealed site- (or sequence-) specific mutational preferences of the hypermutation mechanism proper.[15] In addition, it has been suggested that V regions in long-lived B cells may accumulate "selectively neutral" mutations that do not affect the affinity or structural integrity of the Ig molecule, thus altering the distribution of initially antigen-selected mutations. Thus, because we do not know the antigen specificity and affinity of the antibodies encoded by these μ transcripts and in light of the complexity of antigen-independent factors affecting somatic mutations in V regions, it is difficult to evaluate the life history of the corresponding B cell clones.

MOLECULAR EVOLUTION OF THE IgE RESPONSE IN PATIENTS WITH ATOPIC DERMATITIS

As a model for a clinically relevant immune response to a recurrent antigenic stimulation, we have analyzed the molecular structure of V_H regions associated with the IgE response in patients with atopic dermatitis.[6] To investigate the molecular structure of ε V_H5 regions, first-strand cDNA was synthesized from total RNA of PB mononuclear cells and subsequently amplified in the PCR using a Cε and V_H5-specific primer pair. In FIGURE 2, we have compared the nucleotide sequences of the expressed and germline V_H5 genes of patients P1, P2, and P3. TABLE 3 summarizes the distribution of R and S mutations over framework and CDR regions.

cDNA clones V_HP2-57 and V_HP1-58 derive from the germline V_H5-2R1 gene, whereas the other 17 cDNA clones descend from the germline V_H5-1R1 gene.[16] All ε V_H5 sequences contain somatic mutations, ranging from 6 to 37 in the V_H portion of individual transcripts. Nucleotide sequences P2-51, P2-52, P2-53, and P2-54 are not identical (between 95.7 and 99.2% sequence homology) but use the same D_H and J_H gene segments and are indistinguishable at the V_H/D_H and D_H/J_H junctions (results not shown). This strongly suggests that the corresponding B-cell clones are the progeny of a common precursor. By the same criteria, the

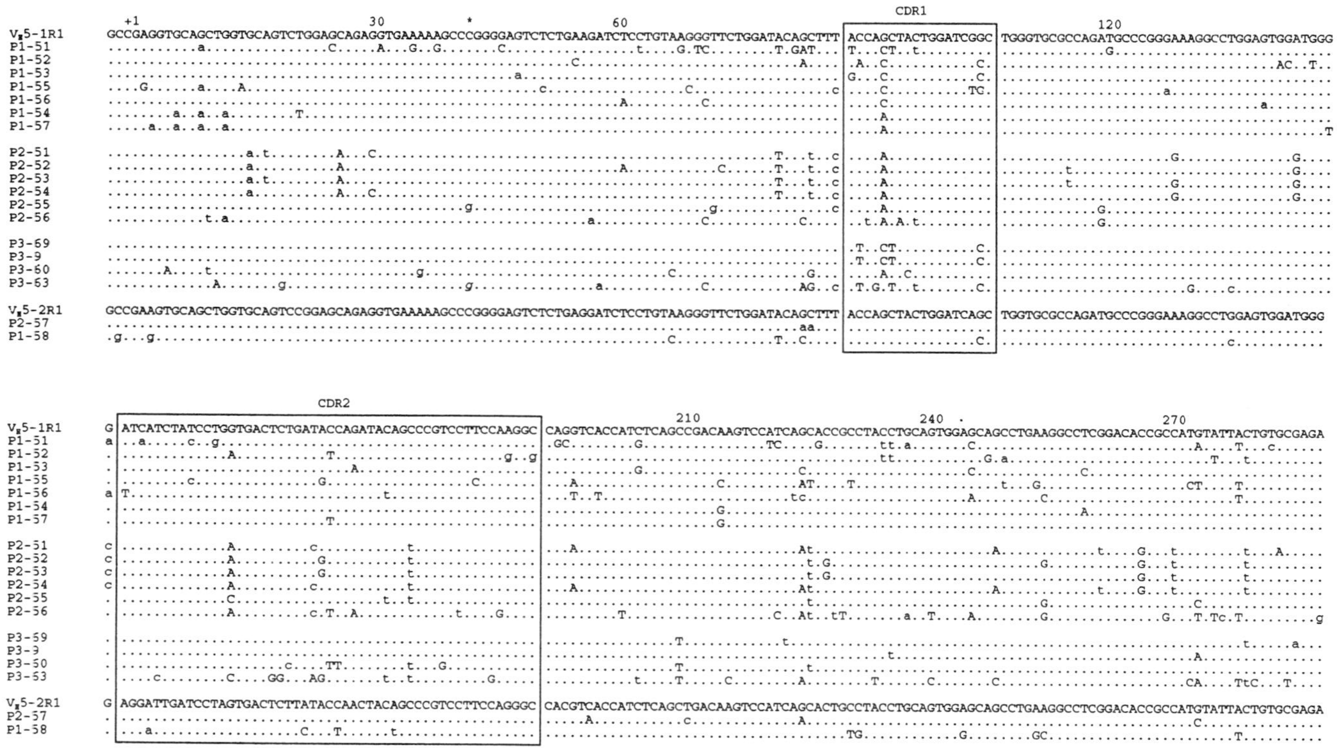

FIGURE 2. Nucleotide sequences of ε V_H5 transcripts from atopic dermatitis patients P1, P2, and P3. Nucleotide sequences are compared to the rearranged V_H5-1R1 and V_H5-1R2 genes that represent the two functional V_H5 genes in germline configuration. The nucleotide sequences of the corresponding germline V_H5 gene segments of patients P1, P2, and P3 were 100% identical, except for a single difference at position 42 in one of the alleles of patient P3 (denoted by an * and a *g*). Dashes indicate nucleotide identity. Capital and lower case letters refer to R and S mutations, respectively.

TABLE 3. Distribution of Replacement and Silent Mutations in the V_H5 Regions of ε-Transcripts from Three Patients with Atopic Dermatitis

Seq.	Total m[a]	Total R/S	CDR R/S	FR R/S	Seq.	Total m[a]	Total R/S	CDR R/S	FR R/S
P1-51	36	2.2	0.6	3.3	P2-51	23	1.1	1	1.1
P1-52	18	2.0	2.0	2.0	P2-52	20	1.2	3	1.0
P1-53	9	8.0	≥3.0	5.0	P2-53	19	0.9	3	0.7
P1-54	7	0.7	≥1.0	≤0.2	P2-54	21	1.0	1	1.1
P1-55	23	2.8	5.0	2.4	P2-55	11	0.8	1	0.8
P1-56	14	1.8	2.0	1.7	P2-56	32	1.6	2	1.5
P1-57	8	1.0	≥2.0	0.5	P2-57	6	1.0	—	1.0
P1-58	17	2.4	1.5	3.0					
					P3-9	6	5.0	≥4.0	1.0
					P3-60	14	1.8	2.5	1.3
					P3-63	36	2.6	2.5	2.7
					P3-69	8	1.7	≥4.0	0.3

NOTE: For each ε transcript, the total number of mutations (columns 2 and 7), the R/S ratio in the overall V_H5 region (columns 3 and 8), the CDR region (columns 4 and 9), and the FR regions (columns 5 and 10) are depicted.

[a] Total number of mutations.

collection of eight sequences from patient P1 contains two clonally related IgE transcripts, P1-54 and P1-57, that differ by four nucleotides in the V_H segment.

Comparison of the nucleotide sequences of all ε V_H5 transcripts revealed that numerous mutations are shared among clonally related and unrelated sequences within one person and among several persons. These shared mutations are present in both CDR and FR regions and are silent or result in amino acid replacements. Among the most extreme examples are mutations that occur in multiple independent ε transcripts derived from all three donors (compare, e.g., the C → T substitution at position 171, the G → A at position 228 and the C → T substitution at position 178). Notably, these mutations involve descendants from both the V_H5-1R1 and V_H5-2R1 germline gene segments.

The R/S ratios in the CDR and FR regions of the ε V_H5 transcripts do not exhibit a consistent pattern (TABLE 3). In fact, all combinations of high and low R/S ratios in FR regions with high and low R/S rations in CDR regions occur (TABLE 3). Although this appears unusual for an antigen-driven selection process, several notions argue in favor of a role for antigen. Activation of the somatic hypermutation mechanism is associated with antigen stimulation and does not occur in polyclonally stimulated B cells.[17] Moreover, the allergen-specificity of the IgE antibodies in the sera of these patients suggests an important role for allergen. Finally, the occurrence of clonally related B cells, as observed in two out of three patients, is characteristic for an antigen-driven selection process. Several explanations, not mutually exclusive, for the heterogeneous R/S patterns in ε V_H5 transcripts are conceivable. The FR regions, especially FR3, may contribute to antigen binding, as has been argued on the basis of experimental data and theoretical considerations. In this respect, it is important to realize that antigen determines the shape of the antigen-binding portion of V regions and that current concepts concerning R/S distributions have largely been derived from analysis involving immunization with haptenic groups; the antigen-combining site of hapten-binding antibodies may differ profoundly from that of protein-binding antibodies. Examples of nonhaptenic immune responses generating high-affinity antibodies with "aberrant" R/S mutations have been reported in the literature. An alternative interpretation is an extension of a model proposed by Manser.[17] It may be envisaged that, initially, somatic mutations are efficiently selected by antigen, resulting in affinity maturation of the immune response. In case of recurrent or prolonged antigenic stimulation, the extent to which somatic mutations lead to higher binding affinity in an already selected B-cell clone becomes limited and finally saturated. At this point persistent antigenic stimulation may result in the accumulation of "neutral" R and S mutations, that is, mutations that do not affect affinity or the structural integrity of the Ig molecule. Such mutations are not selected for or against and may shift the R/S ratio towards a pattern resembling random mutations. It may be envisaged that such a pattern of mutations of V regions is associated with antigens that chronically and/or recurrently stimulate the immune system as is the case with allergens and perhaps particular autoantigens.[18] In the above scenario, mutations continue to occur in secondary and higher order immune responses as a result of repeated or chronic exposure to the same antigen. In this context it is interesting to note that the majority of circulating IgE^+ B cells from these atopic patients express CD20 and CD38 cell surface markers, reminiscent of the phenotype of germinal center centroblasts and centrocytes in peripheral lymphoid organs.[19] These populations of B cells have been shown to actively accumulate somatic mutations. From the above considerations and the observation that ε-V_H5 transcripts contain on the average significantly higher numbers of mutations than V_H5-encoded μ or γ transcripts,[19] we hypothesize that these B cells have undergone multiple rounds of selection in germinal centers.

REFERENCES

1. Tonegawa, S. 1983. Somatic generation of antibody diversity. Nature **302:** 575–581.
2. Alt, F. W., T. K. Blackwell & G. D. Yancopoulos. 1987. Development of the primary antibody repertoire. Science **238:** 1079–1087.
3. Freitas, A. A. & B. Rocha. 1993. Lymphocyte lifespans: Homeostasis, selection, and competition. Immunol. Today **14:** 25–29.
4. Gu, H., D. Tarlington, W. Muller, K. Rajewsky & I. Forster. 1991. Most peripheral B cells in mice are ligand selected. J. Exp. Med. **173:** 1357–1371.
5. Kocks, C. & K. Rajewsky. 1989. Stable expression and somatic hypermutation of antibody V regions in B-cell developmental pathways. Ann. Rev. Immunol. **7:** 537–559.
6. van der Stoep, N., J. van der Linden & T. Logtenberg. 1993. Molecular evolution of the human immunoglobulin E response: High incidence of shared mutations and clonal relatedness among V_H5 transcripts from three unrelated patients with atopic dermatitis. J. Exp. Med. **177:** 99–107.
7. Ebeling, S. B., M. E. M. Schutte & T. Logtenberg. 1993. Peripheral human $CD5^+$ and $CD5^-$ B cells may express somatically mutated V_H5- and V_H6-encoded receptors. J. Immunol. **151:** 6891–6899.
8. van der Heijden, R. W. J., H. Bunschoten, V. Pascual, F. G. C. M. Uytdehaag, A. D. M. E. Osterhaus & J. D. Capra. 1990. Nucleotide sequence of a human monoclonal anti-idiotypic antibody specific for a rabies virus-neutralizing monoclonal antibody reveals extensive somatic variability suggestive of an antigen-driven immune response. J. Immunol. **144:** 2835–2839.
9. Cai, J., C. Humphries, A. Richardson & P. W. Tucker. 1992. Extensive and selective mutation of a rearranged V_H5 gene in B cell chronic lymphocytic leukemia. J. Exp. Med. **176:** 1073–1081.
10. Logtenberg, T., F. M. Young, J. H. van Es, F. H. J. Gmelig-Meyling & F. W.

ALT. 1989. Autoantibodies encoded by the most J_H-proximal human immunoglobulin heavy chain variable region gene. J. Exp. Med. **170:** 1347–1355.

11. MARTINEZ-MAZA, O. & S. BRITTON. 1983. Frequencies of separate human B cell subsets activatable to Ig secretion by Epstein-Barr virus and pokeweed mitogen. J. Exp. Med. **157:** 1808–1814.
12. AMAN, P., B. EHLIN-ERIKSSON & G. KLEIN. 1984. Epstein-Barr virus susceptibility of normal human B lymphocyte populations. J. Exp. Med. **159:** 208–220.
13. SHLOMCHIK, M. J., A. MARSHAK-ROTHSTEIN, C. B. WOLFOWICZ, T. L. ROTHSTEIN & M. G. WEIGERT. 1987. The role of clonal selection and somatic mutation in autoimmunity. Nature **328:** 805–811.
14. SHLOMCHIK, M. J., A. H. AUCOIN, D. S. PISETSKY & M. G. WEIGERT. 1987. Structure and function of anti-DNA autoantibodies derived from a single autoimmune mouse. Proc. Natl. Acad. Sci. USA **84:** 9150–9154.
15. BETZ, A. G., C. RADA, R. PANNELL, C. MILSTEIN & M. S. NEUBERGER. 1993. Passenger transgenes reveal intrinsic specificity of the antibody hypermutation mechanism: Clustering, polarity, and specific hot spots. Proc. Natl. Acad. Sci. USA **90:** 2385–2388.
16. SANZ, I., P. KELLY, C. WILLIAMS, S. SCHOLL, P. TUCKER & J. C. CAPRA. 1989. The smaller human V_H gene families display remarkably little polymorphism. EMBO J. **8:** 3741–3748.
17. MANSER, T., L. J. WYSOCKY, M. N. MARGOLIES & M. L. GEFTER. 1987. Evolution of antibody variable region structure during the immune response. Immunol. Rev. **96:** 141–162.
18. RANDEN, I., D. BROWN, K. M. THOMPSON, N. HUGHES-JONES, V. PASCUAL, K. VICTOR, J. D. CAPRA, Ø, FØRRE & J. B. NATVIG. 1992. Clonally related IgM rheumatoid factors undergo affinity maturation in the rheumatoid synovial tissue. J. Immunol. **148:** 3296–3301.
19. PASCUAL, V., Y-J. LIU, A. MAGALSKI, O. DE BOUTEILLER, J. BANCHEREAU & D. CAPRA. 1994. Analysis of somatic mutation in five B cell subsets of human tonsil. J. Exp. Med. **180:** 329–339.

V_H4.21 Expression in the Normal Human B-Cell Repertoire

DAVID F. FRIEDMAN,[a] PIOTR KRAJ,[b]
AND LESLIE E. SILBERSTEIN[b]

[a]*Departments of Pediatrics and Pathology*
Children's Hospital of Philadelphia
[b]*Department of Pathology and Laboratory Medicine*
Hospital of the University of Pennsylvania
University of Pennsylvania School of Medicine
Philadelphia, Pennsylvania 19104

INTRODUCTION

The human immunoglobulin heavy chain locus contains over 100 variable region genes and pseudogenes distributed over a large region of chromosome 14.[1] The number of these variable (V_H) genes available to form functional VDJ rearrangements is an important element of the overall diversity of the immunoglobulin repertoire. The mechanisms that govern the relative levels of expression of individual V_H genes in various B lymphocyte subpopulations are not completely understood and may be influenced by several factors. The position of individual V_H genes relative to the heavy chain joining (J_H) regions and constant regions that make up the 3′ end of the heavy chain locus and untranslated regulatory elements flanking the coding region of individual V_H genes might influence the relative frequencies at which individual V_H genes are rearranged. In addition to these "mechanistic" factors affecting V_H gene rearrangement, cellular selection processes mediated by antigen or ligand binding to antigen receptors containing the Ig heavy chain might affect the frequencies at which individual V_H genes are represented in samples of various B-cell populations. Selection by exogenous antigens or by autoantigens may influence the final distribution of V_H genes expressed by B cells in peripheral blood or spleen. These issues have been reviewed previously.[2,3]

The V genes expressed in B-cell neoplasms and in human monoclonal autoantibodies appear to be restricted to a small subset of the available variable region genome. Specific individual V_H (and $V\kappa$) gene segments have been found repeatedly in antibodies with anti-DNA activity[4] and in CD5 positive neoplasms such as B-CLL.[5] This apparent restriction of V gene utilization has been interpreted as evidence that selection and expansion of B cells mediated by autoantigens may play a significant role in the pathogenesis of some autoimmune diseases and in some B-cell malignancies.

Although the expression of V_H genes in normal adult B-cell repertoires was initially thought to be a "randomized" sample of the V_H genome, studies of the expressed repertoire of V_H genes early in development,[6,7] in newborns,[8,9] and in adults[10] have suggested that certain V_H genes are expressed preferentially in normal B-cell populations. For example, the V_H3 gene VH26 has been found to be disproportionately expressed in normal fetal and adult repertoires and in autoantibodies with anti-DNA activity.[10,11]

The most consistent correlation of a particular V_H gene segment with a defined autospecificity has been found with the $V_H4.21$ gene, which is highly correlated with cold agglutinin specificity directed against the i/I carbohydrate antigens found on erythrocytes. Nearly all pathological cold agglutinins with this specificity have been found to bear the 9G4 idiotype[12] and, of those for which the V_H gene nucleotide sequence has been determined, to contain heavy chains encoded by the $V_H4.21$ gene.[13,14] The structural basis of the 9G4 idiotype appears to be in the first framework of the $V_H4.21$ heavy chain.[15] In addition to anti-i/I cold agglutinins, the $V_H4.21$ gene has also been shown to encode anti-DNA antibodies, rheumatoid factors, antibodies directed against bacterial lipids,[16] and alloantibodies against red cell antigens.[17]

We wished to define the frequency of V_H4 genes, and of $V_H4.21$ in particular, in normal human expressed repertoires as a basis for understanding the overrepresentation of these genes in autoreactive antibodies. Previous studies employing the 9G4 idiotype reagent suggested that the $V_H4.21$ gene might be found at high frequency in normal adult peripheral B cells and in fetal repertoires as well. However, repertoire studies employing nucleic acid sequencing to identify individual gene segments expressed in the adult have not generally focused on the V_H4 family.[10] At early developmental stages, expression of V_H4 family genes has been reported, but the gene segments have not been fully identified.[8,9]

METHODS

Lymphocytes were isolated by density gradient centrifugation from peripheral blood from normal healthy adults and from umbilical cord blood from normal newborns. Cells were used immediately or in some cases frozen in DMSO until use. Total RNA was extracted from 10^7 to 10^8 unstimulated lymphocytes by a method using guanidinium isothiocyanate and enriched for poly-A RNA by oligo-dT column chromatography. An amplified but unbiased cDNA library of the Ig V_H genes expressed in the peripheral B cells was then prepared by a method based on that described by Huang and Stollar.[18] Double-stranded cDNA is synthesized using oligonucleotide primers specific for the immunoglobulin constant regions, $C\mu$ and $C\gamma$, producing the cDNA basis for separate, isotype-specific V_H gene libraries. The cDNA is amplified with two rounds of polymerase chain reactions (PCR), the first using nonspecific linker primers and the second using a 3′ primer specific for the constant region and a 5′ primer derived from the intermediate vector into which the product of the first amplification is ligated. Thus, the method employs amplification primers specific for Ig constant regions, but no consensus primers from V_H genes. Such a library contains the V_H genes that were expressed as mRNA transcripts in the original sample, represented in the same relative proportions as in the original sample. The method avoids introducing bias in the V_H gene representation because it does not rely on consensus primers or mixtures of degenerate primers for PCR amplification. Huang and Stollar have shown that this approach preserves the relative proportions of V_H genes originally present in a sample.

We have modified the procedure to permit the library to be cloned into a lambda phage vector (λ Zap II), to increase the cloning efficiency and permit screening of larger numbers of clones. Plaque lifts are screened with radiolabeled oligonucleotide probes specific for $C\mu$ and $C\gamma$ constant regions, and for each V_H family, with oligonucleotide probes in the first framework regions, to determine overall V_H gene representation and to identify plaques for subsequent rescue of inserts into plasmids (pBluescript II) for sequence analysis. Plaques that hybridized with the V_H4 oligonucleotide probe were sampled at random, rescued, and sequenced, either by the dideoxynucleotide termination method with [^{35}S]dCTP and Sequenase or by using fluorochrome-labeled universal primers, Taq polymerase cycle-sequencing, and an automated sequence reader.

Two- and three-color flow cytometry analysis was performed using fluorochrome-conjugated markers for surface IgM, IgG, CD19 or CD20, CD3, CD5, and controls. The rat antibody 9G4, which identifies an idiotope highly correlated with the $V_H4.21$ heavy chain, and an irrelevant rat IgG2a antibody as an isotype control for 9G4 were biotinylated and stained with strepavidin-phycoerythrin.

Serum from the same samples was used to measure total IgM, IgG, and 9G4 idiotype by ELISA using rabbit anti-human μ or γ heavy chain on the plate, and developing with either alkaline phosphatase-conjugated anti-human IgM or IgG, or with the biotinylated 9G4 or isotype control and strepavidin-alkaline phosphatase. Standard curves were constructed using polyclonal human IgM, IgG, or purified antibody from the supernatant of a human EBV-transformed cell line known to express the $V_H4.21$ heavy chain as IgM or IgG.

RESULTS

The distribution of V_H families in two adult and three neonatal expressed V_H-$C\mu$ gene libraries, as determined by screening plaques with oligonucleotide probes, is shown in FIGURE 1. For the purpose of comparison, the distribution of V_H families in the genomic DNA as reported by Matsuda *et al.*,[1] and the distribution of V_H families in a panel of EBV-transformed cell lines as reported by Schutte and Logtenberg[19,20] are shown in the figure. In each library 200 to 700 plaques containing a V_H gene were screened on one or two filters. The screening results were consistent between filters from the same library and were reproducible when a library was replated and rescreened. There were no $C\gamma$-positive plaques in the $C\mu$ libraries or vice versa, and fewer than 5% of plaques hybridized with more than one screening probe.

The distribution of V_H families was consistent among the $C\mu$ libraries from adult and neonate. The $C\gamma$ libraries (not shown) contained the same overall proportions of V_H genes, but there was greater variation between individuals. In both $C\mu$ and $C\gamma$ libraries, the V_H4 family accounted for about 30% of expressed genes, with the exception of one neonatal $C\gamma$ library that contained fewer than 5% V_H4-positive plaques. The figure 30% for percentage of V_H4 genes in the expressed $C\mu$ repertoire is somewhat higher, and the percentage of V_H3 genes somewhat lower, than the proportion of V_H4 genes in the genomic DNA and in the panel of EBV cell lines. This figure agrees with the percentage of V_H4 genes among sequenced V_H genes from expressed libraries in adults as reported by Stewart *et al.*[2,10]

Sequence analysis of V_H4-positive plaques was carried out to identify the V_H4 gene segments represented in the libraries and to assess the clonal diversity of each library. A series of 13 to 30 clones was sequenced in both directions from each library to obtain a sample size roughly comparable to the number of V_H4 genes in the genome.[21] In the $C\mu$ libraries from both neonates and adults, and in the $C\gamma$ libraries from adults, the $V_H4.21$ gene was consistently the most frequent V_H4 gene, accounting for 23 to 40% of V_H4 clones. $V_H4.21$ was not detected among 14 and 15 clones sequenced from the neonatal $C\gamma$ libraries. The $V_H4.21$ gene was usually expressed at 98 to 100% homology (one clone at 95%) with the

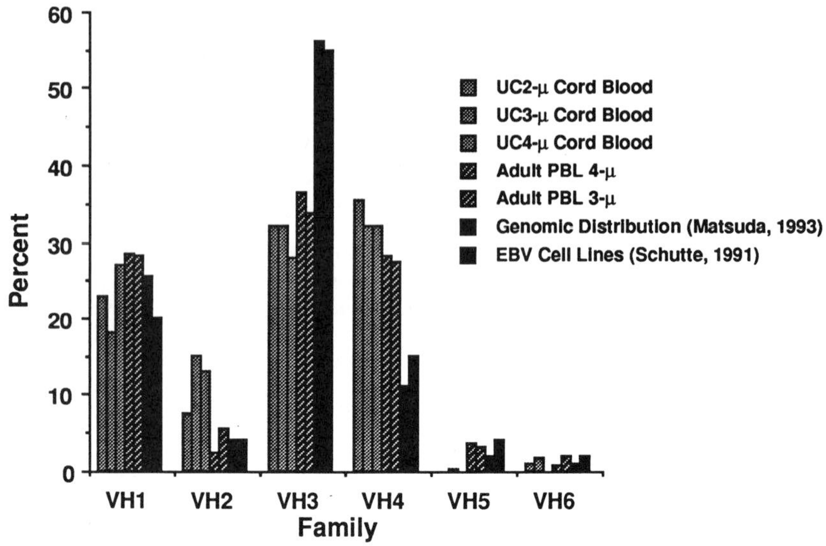

FIGURE 1. Distribution of V genes by family: oligonucleotide screening. The distribution of V_H family expression in three umbilical cord V_H-$C\mu$ libraries (*shaded bars*) and two adult V_H-$C\mu$ libraries (*hatched bars*) is shown. Family-specific oligonucleotide probes in the first framework region of the V_H gene were radiolabeled and hybridized to replicate plaque lifts from the libraries. For comparison, the *black bars* show the distribution of V_H genes by family in the genome as determined by Matsuda *et al.*[1] from mapping with yeast artificial chromosomes and the distribution of V_H family usage in a large panel of EBV-transformed B cell lines as reported by Logtenberg and Schutte.[19,20]

germline sequence in neonatal and adult $C\mu$ libraries, and 96 to 97% germline homology in the adult $C\gamma$ libraries. In FIGURE 2, these data are combined with the screening results to show that the V_H4.21 gene accounts for 6 to 12% of the total expressed V_H gene repertoire of peripheral blood $C\mu$-expressing B cells.

The second most frequent V_H4 gene expressed in these libraries was 58P2 in the neonatal $C\mu$ libraries and V71-4 in the adult $C\mu$ libraries. 58P2, first described as a rearranged gene in the fetal library,[7] and V71-4 differ by a single nucleotide and may be allelic forms of the same germline gene. Only five to eight genes were isolated from any single library, and a total of 11 different V_H4 genes were isolated among all the clones sequenced from adult and neonatal $C\mu$ and $C\gamma$ libraries.

Flow cytometry using the 9G4 idiotype reagent was employed as an independent measurement of the prevalence of the V_H4.21 gene product among peripheral B cells, using the same samples from which the libraries were derived and additional samples as well. An example of the three-color analysis of the umbilical cord sample UC4 is shown in FIGURE 3. The right contour plot shows staining with 9G4 on the vertical axis against IgM on the horizontal. The left panel shows staining with the isotype control rat IgG2a against IgM. The plots are gated on all lymphocytes based on forward and side scatter and on CD20 as a B-cell marker. Subtracting 1.7% nonspecific staining by IgG2a from 6.5% total 9G4 staining gives the result that 4.8% of IgM-positive B cells in this sample bear the 9G4 idiotype. In all cases, the 9G4-positive population was brightly stained and clearly distinct

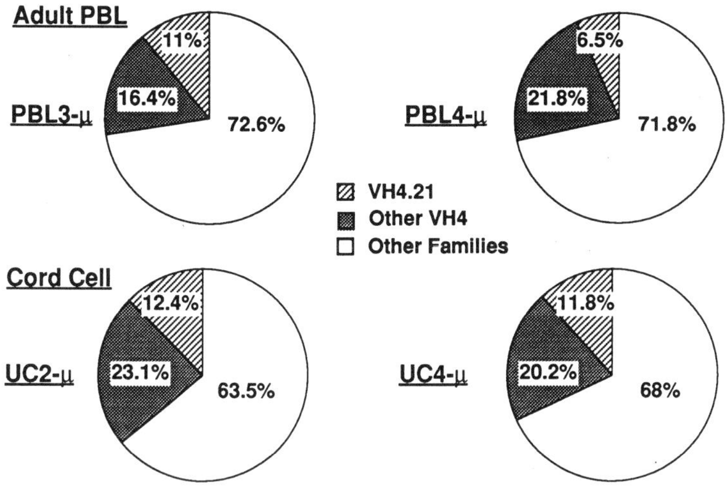

FIGURE 2. V_H4.21 gene as a percentage of the total $C\mu$ library. Pie diagrams illustrate the high frequency of expression of the V_H4.21 gene (*hatched*) relative to other V_H4 genes (*shaded*) in adult and neonatal V_H-$C\mu$ libraries. The numbers represent percentages of the entire expressed library.

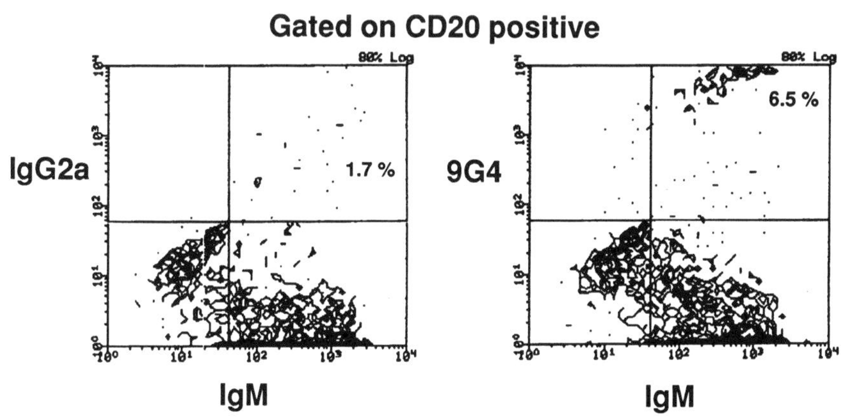

FIGURE 3. Expression of idiotype 9G4 in umbilical cord B cells: 4.8% bear 9G4 idiotype. Three-color flow cytometry analysis of umbilical cord lymphocytes from the UC4 specimen shows a distinct population of IgM-positive B cells that bear the 9G4 idiotype, which is highly correlated with the V_H4.21 gene. The IgM and CD20 monoclonal antibodies were directly conjugated to FITC and PerCP, respectively; the 9G4 and isotype control rat IgG2a antibodies were biotinylated and developed with strepavidin-PE. The plots shown are gated on CD20-positive lymphocytes as a B-cell marker. This brightly staining, distinct population was typical of all experiments.

as seen in this example. In the additional umbilical cord and adult samples, the percentage of 9G4-positive B cells ranged from 3.7% to 16.3%. IgG-positive cells bearing 9G4 were detectable in the adult peripheral blood but not the umbilical cord lymphocytes.

Finally, the serum antibody from each sample was assayed for 9G4 idiotype in an isotype-specific ELISA assay. The total amount of IgM and of 9G4-IgM was determined based on standard curves as described in METHODS. Total IgG and 9G4-IgG was also determined in the adult sera. Consistent with data that has been published, the percentage of 9G4-positive IgM was 0.5 to 1.8% of the total IgM in both cord blood serum and adult; no 9G4 idiotype was detectable in the adult serum by IgG ELISA.

DISCUSSION

Several reports have suggested that the V_H4 family, and in particular, the $V_H4.21$ gene, are overrepresented in the human "autoimmune repertoire"[3,13,22] and in B-cell neoplasms.[20,23] This apparent restriction of V_H gene usage might evolve as part of the pathologic process in these diseases, or alternatively, might be a characteristic of the normal B-cell populations from which disregulated B-cell expansions arise. In the first model, selection mediated by antigens or autoantigens might influence the evolution of the transformed or disregulated clones, while in the second, antigenic selection might be responsible for expanding certain clones of normal B cells, possibly making them more likely to undergo transforming events or escape regulatory mechanisms. The purpose of studying the V_H gene repertoire in normal B-cell populations is to provide a basis on which to understand the pathologic B-cell populations, and to help distinguish between these two models.

The overrepresentation of the V_H4 family and of the $V_H4.21$ gene segment in the normal B-cell populations reported here adds to the evidence of preferential expression of individual V_H genes in normal adults,[10] neonates,[8,9] and in the fetus.[6,7] These studies, which employ unbiased library methods, avoid the confounding variables that may be introduced by the use of EBV cell lines or hybridoma technology, LPS stimulation of B-cell populations for *in situ* hybridization, or V_H gene consensus primers for amplification, while still permitting assessment of the frequency of individual V_H gene segments by nucleotide sequence analysis. Our studies suggest that the $V_H4.21$ gene, like the VH26 (30p1), is preferentially expressed in normal peripheral resting (IgM^+) B cells. This gene segment may account for as much as 10% or more of functionally rearranged V_H genes.

This estimate of the frequency of $V_H4.21$ expression, from the screening and sequence analysis of our libraries, is in good agreement with published studies of the prevalence of 9G4 idiotype in normal adult tissues (summarized in Pascual and Capra[13]), as well as our own flow cytometry analysis of the samples from which the libraries were derived. Also in agreement with the idiotype-based findings is the high degree of homology with the germline sequence that we found in the $C\mu$ libraries. The preferential expression of $V_H4.21$ appears to be characteristic of both the neonate and the adult; interestingly, the 58P2 gene, but not the $V_H4.21$ gene, was found in normal fetal libraries. This may have been a problem of sample size, since the 9G4 idiotype has been detected in the fetus.[13]

Although the conservation of the $V_H4.21$ gene among individuals and the high level of expression of this gene is established, the function of the $V_H4.21$ gene product in immunologic defense is not well understood. The $V_H4.21$ heavy chain has been found in polyspecific antibodies and antibodies that bind bacterial lipids as well as the anti-i/I cold agglutinins. $V_H4.21$ has also been found in alloimmune responses to red cell antigens,[17] and our adult libraries demonstrate that isotype-switched B cells may express $V_H4.21$. It has been speculated that the $V_H4.21$ gene may play an important role in defense against bacterial pathogens and in other common alloimmune responses, explaining its conservation and frequent expression. Alternatively, $V_H4.21$-encoded immunoglobulins have also been reported to bind to i/I carbohydrate moieties on lymphocytes,[24] suggesting that $V_H4.21$-encoded antibodies may play a regulatory role.

The discrepancy between the frequency of $V_H4.21$ expression in circulating B cells and the fraction of 9G4 idiotype in serum has not been explained. It is possible that cells expressing $V_H4.21$ become anergic or fail to differentiate into secretory cells because of regulatory signals mediated by binding of surface $V_H4.21$-IgM with autoantigens. Alternatively, soluble $V_H4.21$-encoded antibody may be selectively removed from the serum because of its binding specificities.

The preferential expression of $V_H4.21$ in normal B-cell populations suggests that restriction of V_H gene usage is a feature of normal repertoires. The high frequency of this gene in autoimmune clones and in B-cell neoplasms may therefore be a reflection of the B cells from which those diseases arise rather than a characteristic unique to the disease processes.

REFERENCES

1. MATSUDA, F., E. K. SHIN, H. NAGAOKA, R. MATSUMURA, M. HAINO, Y. FUKITA, S. TAKA-ISHI, T. IMAI, J. H. RILEY, R. ANAND, E. SOEDA & T. HONJO. 1993. Structure and physical map of 64 variable segments in the 3′ 0.8 megabase region of the human immunoglobulin heavy-chain locus. Nature Genet. **3:** 88–94.
2. STEWART, A. K., C. HUANG, A. A. LONG, B. D. STOLLAR & R. S. SCHWARTZ. 1992. VH-Gene representation in autoantibodies reflects the normal human B cell repertoire. Immunol. Rev. **128:** 101–122.
3. PASCUAL, V., G. WIDHOPF & J. D. CAPRA. 1992. The human VH repertoire: A restricted set of Vh genes may be the target of immune regulation. Int. Rev. Immunol. **8:** 147–157.
4. DERSIMONIAN, H., R. S. SCHWARTZ, K. J. BARRETT & B. D. STOLLAR. 1987. Relationship of human variable region heavy chain germline genes to genes encoding anti-DNA autoantibodies. J. Immunol. **139:** 2496–2501.
5. KIPPS, T. J., B. A. ROBBINS, A. TEFFERI, G. MEISENHOLDER, P. M. BANKS & D. A. CARSON. 1990. CD5-positive B-cell malignancies frequently express cross-reactive idiotypes associated with IgM autoantibodies. Am. J. Pathol. **136:** 809–816.
6. SCHROEDER, H. W. & J. Y. WANG. 1990. Preferential utilization of conserved immunoglobulin heavy chain variable gene segments during human fetal life. Proc. Natl. Acad. Sci. USA **87:** 6146–6150.
7. SCHROEDER, H. W., J. I. HILLSON & R. M. PERLMUTTER. 1987. Early restriction of the human antibody repertoire. Science **238:** 791–793.
8. MORTARI, F., J. A. NEWTON, J. Y. WANG & H. W. SCHROEDER. 1992. The Human Cord Blood Antibody Repertoire: Frequent Usage of the VH7 Gene Family. Eur. J. Immunol. **22:** 241–245.
9. MORTARI, F., J.-Y. WANG & H. W. SCHROEDER. 1993. Human cord blood antibody repertoire: Mixed population of VH gene segments and CDR3 distribution in the expressed $C\alpha$ and $C\gamma$ repertoires. J. Immunol. **150:** 1348–1357.
10. STEWART, A. K., C. HUANG, B. D. STOLLAR & R. S. SCHWARTZ. 1993. High-frequency representation of a single VH gene in the expressed human B cell repertoire. J. Exp. Med. **177:** 409–418.

11. CHEN, P. P., N. J. OLSEN, P.-M. YANG, R. W. SOTO-GIL, T. OLEE, K. A. SIMINOVITCH & D. A. CARSON. 1990. From human autoantibodies to the fetal antibody repertoire to B cell malignancy: It's a small world after all. Int. Rev. Immunol. **5:** 239–251.
12. STEVENSON, F. K., M. WRIGHTHAM, M. J. GLENNIE, A. R. CATTAN, T. FEIZI, T. J. HAMBLIN & G. T. STEVENSON. 1986. Antibodies to shared idiotypes as agents for analysis and therapy of human B cell tumors. Blood **68:** 430–436.
13. PASCUAL, V. & J. D. CAPRA. 1992. VH4-21, a human VH gene segment overrepresented in the autoimmune repertoire. Arthritis Rheum. **35:** 11–18.
14. PASCUAL, V., K. VICTOR, M. SPELLERBERG, T. J. HAMBLIN, F. K. STEVENSON & J. D. CAPRA. 1992. VH restriction among human cold agglutinins. The VH4-21 gene segment is required to encode anti-I and anti-i specificities. J. Immunol. **149:** 2337–2344.
15. POTTER, K. N., Y. LI, V. PASCUAL, R. C. J. WILLIAMS, L. C. BYRES, M. SPELLERBERG, F. K. STEVENSON & J. D. CAPRA. 1993. Molecular characterization of a cross-reactive idiotope on human immunoglobulins utilizing the VH4-21 gene segment. J. Exp. Med. **178:** 1419–1428.
16. BHAT, N. M., M. M. BIEBER, C. J. CHAPMAN, F. K. STEVENSON & N. N. TENG. 1993. Human antilipid A monoclonal antibodies bind to human B cells and the i antigen on cord red blood cells. J. Immunol. **151:** 5011–5021.
17. THOMPSON, K. M., J. SUTHERLAND, G. BARDEN, M. D. MELAMED, I. RANDEN, J. B. NATVIG, V. PASCUAL, J. D. CAPRA & F. K. STEVENSON. 1991. Human monoclonal antibodies against blood group antigens preferentially express a VH4-21 variable region gene-associated epitope. Scand. J. Immunol. **34:** 509–518.
18. HUANG, C. & B. D. STOLLAR. 1991. Construction of representative immunoglobulin variable region cDNA libraries from human peripheral blood lymphocytes without in vitro stimulation. J. Immunol. Methods **141:** 227–236.
19. SCHUTTE, M. E. M., S. B. EBELING, K. E. AKKERMANS, F. H. J. GMELIG-MEYLING & T. LOGTENBERG. 1991. Antibody specificity and immunoglobulin V_H gene utilization of human monoclonal $CD5^+$ B cell lines. Eur. J. Immunol. **21:** 1115–1121.
20. LOGTENBERG, T., M. E. M. SCHUTTE, G. INGHIRAMI, J. E. BERMAN, F. H. J. GMELIG-MEYLING, R. A. INSEL, D. M. KNOWLES & F. W. ALT. 1989. Immunoglobulin V_H gene expression in human B cell lines and tumors: Biased V_H gene expression in chronic lymphocytic leukemia. Int. Immunol. **1:** 360–364.
21. VAN DER MAAREL, S., K. W. VAN DIJK, C. M. ALEXANDER, E. H. SASSO, A. BULL & E. C. B. MILNER. 1993. Chromosomal organization of the human VH4 gene family. J. Immunol. **150:** 2858–2868.
22. PASCUAL, V. & J. D. CAPRA. 1991. Human immunoglobulin heavy-chain variable region genes: Organization, polymorphism, and expression. Adv. Immunol. **49:** 1–74.
23. MAYER, R., T. LOGTENBERG, J. STRAUCHEN, A. DIMITRIU-BONA, L. MAYER, S. MECHANIC, N. CHIORAZZI, L. BORCHE, G. DIGHIERO, A. MANNHEIMER-LORY, B. DIAMOND, F. ALT & C. BONA. 1990. CD5 and immunoglobulin V gene expression in B-cell lymphomas and chronic lymphocytic leukemia. Blood **75:** 1518–1524.
24. GRILLOT-COURVALIN, C., J.-C. BROUET, F. PILLER, L. Z. RASSENTI, S. LABAUME, G. J. SILVERMAN, L. SILBERSTEIN & T. J. KIPPS. 1992. An anti-B cell autoantibody from Wiskott-Aldrich syndrome which recognizes i blood group specificity on normal human B cells. Eur. J. Immunol. **22:** 1781–1788.

A Complete Preimmune Human V_H3 Repertoire[a]

WENDY O. HUFNAGLE, SHU-CAI HUANG, IVY SUZUKI, AND ERIC C. B. MILNER[b]

Virginia Mason Research Center and
University of Washington
Seattle, Washington 98101

Recent evidence in humans suggests that the adult repertoire is skewed toward overexpression of a few V_H genes[1-4] and, it can be inferred, the concomitant underexpression of other genes. We endeavored to determine the complete set of V_H3 genes that contribute to the preimmune antibody repertoire of a single individual. We surmised that it would be possible to identify a significant number of the rearranged genes from IgD^+ B cells using sequence-specific oligonucleotide probes because IgD^+ cells are thought to have recently emerged from the bone marrow (BM) and probably have not yet accumulated many somatic mutations.[5]

The germline source and frequency of occurrence of all the rearranged V_H3 genes in a library from a single individual were assessed by oligonucleotide hybridization and sequence analysis. V_H3 rearrangements were amplified quantitatively from IgD^+ B cells, and the resulting PCR products were cloned into a phagemid vector. Of the 260 clones examined, approximately 92% contained rearranged V_H3 genes that were identifiable using gene-specific oligonucleotide probes. The remaining 8% were identified by sequence analysis. In each case where a gene that was identified by sequence analysis failed to hybridize with a probe, at least one mutation was detected in the sequence in the region of the probe (CDR1 or CDR2). Two anomalous clones hybridized to a probe, but no V_H gene was found by sequence analysis.

As few as 16 V_H3 genes composed this library from the preimmune repertoire of this subject (FIG. 1). One V_H3 element, V3-23,[6,7] accounted for 29% of all the V_H3 rearrangements examined. Seven other genes—V3-33, V3-30b, V3-30.3, V3-30, V3-15, V3-11, and V3-74—each made up 5 to 10% of the V_H3 rearrangements. The remaining eight V_H3 genes—V3-53, V3-20, V3-73, V3-72, V3-13, V3-43, V3-64, and 23d11 (a previously unidentified V_H3b gene)—each made up $\leq$4% of the total V_H3 rearrangements. Of the 16 genes identified, 15 are known to be functional.[7] It remains to be determined if the remaining gene, 23d11, is fully functional. No correlation between location on the chromosome and rearrangement frequency was observed.

These results corroborate and extend our earlier findings that the rearranged V_H3 repertoire consists of fewer elements than previously thought.[4] In this analysis, the repertoire was skewed toward the overexpression of one V_H3 segment (V3-23) and the concomitant underexpression of eight other V_H genes. Differences

[a] This work was supported in part by National Institutes of Health Grants AI31241 and AR39918, and American Cancer Society Grant IM485.

[b] Address for correspondence: E. C. B. Milner, Virginia Mason Research Center, 1000 Seneca St., Seattle, WA 98101.

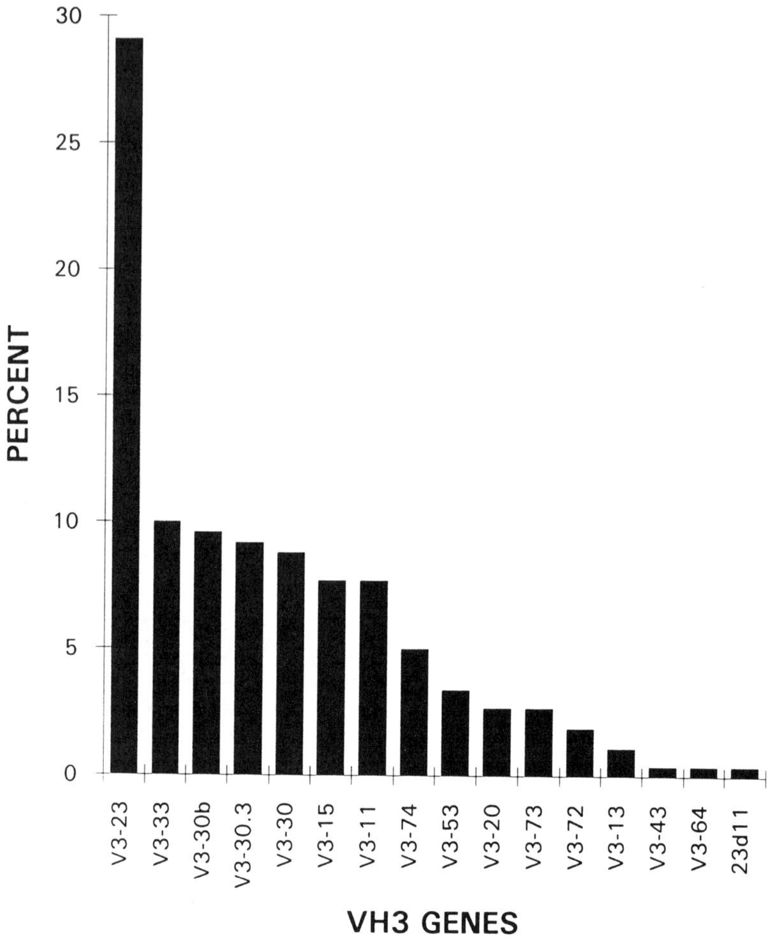

FIGURE 1. V_H3 Rearrangements in a preimmune repertoire. Occurrence (%) of each specific V_H gene segment was calculated by dividing the number of clones hybridizing with an oligonucleotide probe or detected by sequence analysis by the total number of clones hybridizing with the family-specific probe. Genes are named according to Matsuda *et al.*[6] and Cook *et al.*[7] An unidentified member of the V_H3b subfamily is designated by its clone name, 23d11.

in levels of expression may, in part, be explained by copy number in the germline. Three copies of V3-23 occur in the germline of this subject and may contribute to the overexpression of this gene. However, V3-13 is present in two copies in the germline but is rearranged less frequently than V3-15, V3-11, or V3-33, which are also present in two copies. Thus, copy number cannot account for all the differences in expression levels. Perhaps the genes that are rearranged more frequently (≥5%) have been selected evolutionarily for the antibodies they encode. On the other hand, the genes that are rearranged at low frequencies may provide a means for quickly diversifying the repertoire upon an encounter with a novel foreign antigen. It remains to be determined whether the expressed repertoire that is driven by antigen selection is limited to these 16 V_H3 genes or whether one, some, or all of the six remaining functional V_H3 genes[7] could also be used at low frequencies by this subject.

REFERENCES

1. HUANG, C., A. K. STEWART, R. S. SCHWARTZ & B. D. STOLLAR. 1992. J. Clin. Invest. **89:** 1331–1343.
2. STEWART, A. K., C. HUANG, B. D. STOLLAR & R. S. SCHWARTZ. 1993. J. Exp. Med. **177:** 409–418.
3. HUANG, C. & B. D. STOLLAR. 1993. J. Immunol. **151:** 5290–5300.
4. SUZUKI, I., L. PFISTER, A. GLAS, C. NOTTENBURG & E. C. B. MILNER. 1994. Biased representation of rearranged V_H gene segments in the human adult antibody repertoire. J. Immunol. In press.
5. KLEIN, U., R. KUPPERS & K. RAJEWSKY. 1993. Eur. J. Immunol. **23:** 3272–3277.
6. MATSUDA, F., E. K. SHIN, H. NAGAOKA, R. MATSUMURA, M. HAINO, Y. FUKITA, S. TAKA-ISHI, T. IMAI, J. H. RILEY, R. ANAND, E. SOEDA & T. HONJO. 1993. Nature Genet. **3:** 88–94.
7. COOK, G. P., I. M. TOMLINSON, G. WALTER, H. RIETHMAN, N. P. CARTER, L. BULUWELA, G. WINTER & T. H. RABBITTS. 1994. Nature Genet. **7:** 162–168.

Contribution of the V_λ Light Chain to the Development of the Primary Antibody Repertoire[a]

A. NANGPAL,[b] J. KNOTT,[b] V. LIPSANEN,[b] C. BONA,[c] AND A. KAUSHIK[b]

[b]*Department of Veterinary Microbiology and Immunology*
University of Guelph
Guelph, Ontario, Canada

[c]*Department of Microbiology*
Mount Sinai School of Medicine
New York, New York

The development of the primary antibody repertoire is influenced by mechanistic processes that generate germline diversity and endogenous selection of immunoglobulin (Ig) receptors expressed on B lymphocytes. Most mammalian Ig molecules bear either κ or λ light chains,[1] and their usage varies with different species, for example, horse (5% κ, 95% λ), man (60% κ, 40% λ), and mouse (95% κ, 5% λ). In the mouse, 100 to 500 V_H genes have been grouped into 14 families based on their nucleotide similarity.[2] As for the L chain, about 300 V_κ genes are classified into 16 families,[3] whereas three V_λ genes provide four recombination units.[4] The contribution of V_κ and V_λ genes to the development of a primary antibody repertoire is not well understood. We earlier noted stochastic V_H (Q52, X24, and S107), and V_κ (V_κ1 and V_κ8) gene family pairings among mitogen-activated C57BL/6 lymphocyte clones.[5] Some nonstochastic associations between certain V_H and V_λ gene families have been observed.[6,7] In order to assess the contribution of the V_λ L chain to the development of primary antibody repertoire, we have examined V_H gene expression and self-reactive antibodies in κ-deficient mice, generated by targeted deletion of J_κ, κ-intron enhancer, and C_κ locus.[8]

MATERIALS AND METHODS

Mice

Healthy three-month-old homozygous κ-deficient[8] and Ig haplotype-matched normal 129/Sv mice were bred and maintained in the animal facility of Mount Sinai School of Medicine in New York. The κ-deficient mice lacked κ light chain in their serum and exclusively expressed λ light chain (data not shown).

[a] This work was supported by grants from NSERC, Canada (OGP0121465) and the Canadian Arthritis Society.

[d] Address for correspondence: A. Kaushik, D.Sc., Dept. of Vet. Microbiology and Immunology, University of Guelph, Guelph, Ontario N1G 2W1, Canada.

DNA Probes

DNA probes specific for V_H J558, J606, S107, 36-60, 36-09, V_H11, 7183, and Q52 gene families and the C_μ region have previously been described.[5,9] These probes were routinely ^{32}P-labeled by random priming (Boehringer Mannheim, Mannheim, Germany).

In Situ *Hybridization*

Specific V_H and V_κ gene family expression was determined by sequential hybridizations of B lymphocyte clones grown on Whatman paper discs with V-gene family specific DNA probes followed by a probe specific for C_μ region.[5,10] Between successive hybridizations >95% bound ^{32}P was stripped off by stringent washing. Percent frequencies of V_H gene family were determined as:

$$\frac{\text{Number of } V_H{}^+ \text{ clones per disc}}{\text{Number of } C_\mu{}^+ \text{ clones per disc}} \times 100$$

Antibody Specificity Analysis

A solid-phase enzyme immunoassay (ELISA) was developed to determine antibody specificities to conserved self antigens (albumin, thyroglobulin, ubiquitin, hsp70, and hsp65) as described earlier.[11] Briefly, antigens were coated on immulon4 ELISA plates (Fisher Scientific) at a concentration of 2.0 to 2.5 μg/ml. Nonspecific sites were blocked, and serial serum dilutions were added followed by enzyme-labeled antibody and goat anti-mouse Ig (H + L) coupled to alkaline phosphatase (Southern Biotechnology). Appropriate methodological and positive and negative controls were included in each test, and optical densities were measured at 405 nm.

RESULTS AND DISCUSSION

Nonrandom V_H Gene Expression Occurs in κ-Deficient Mice

All nine V_H gene families examined in this study were observed to be expressed in κ-deficient mice (FIG. 1), and therefore, availability of only four V_λ recombination units of the L chain did not altogether obliterate expression of individual V_H gene families in the available LPS-activated B-cell repertoire. Nevertheless, expression of 36-60 (21% versus 12%) and 36-09 (5% versus 2.5%) V_H gene families was significantly increased (χ^2, $p < 0.0001$), while V_HJ558 (29% versus 39%), J606 (1.7% versus 8.0%), and S107 (0.2% versus 4.1%) gene families were expressed at significantly lower frequencies (χ^2, $p < 0.0001$) in comparison to normal 129/Sv mice. Certain V_H gene families, for example, 7183, Q52, and V_H11, were expressed at similar levels both in κ-deficient and normal 129/Sv mice (FIG. 1). These data suggest nonrandom expression of V_H gene families in the absence of κ L chains as a result of restricted available V_λ L-chain diversity. Such a nonrandom expression of V_H gene families in the primary antibody repertoire of κ-deficient mice probably reflects the impact of a few amino acid substitutions between predominantly expressed V_λ1 and V_λ2 L-chain genes and defines restrictions imposed by

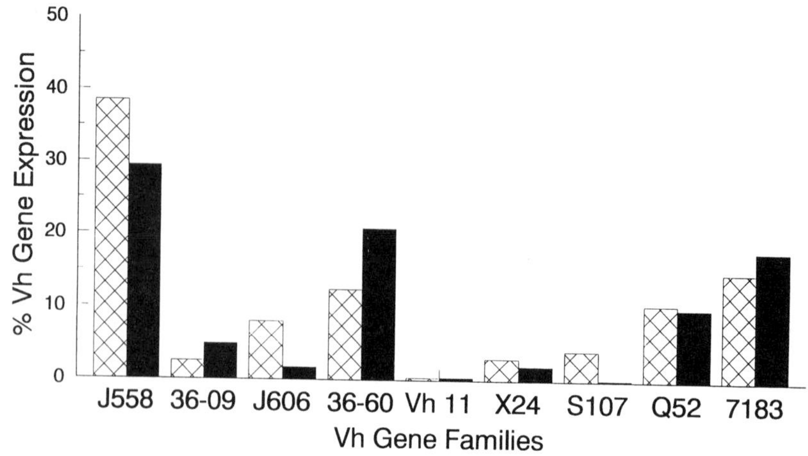

FIGURE 1. V_H gene expression in normal 129/Sv and κ-deficient congenic 129/Sv mice. Normal mice are represented by cross-hatched bars and the κ-deficient mice by black bars.

V_λ L chains. Further studies are in progress to determine if preferential $V_H + V_\lambda$ gene family pairings occur in the mitogen-responsive B-cell repertoire of κ-deficient mice.

Self-Reactive Antibodies Are Increased in κ-Deficient Mice

In contrast to normal 129/Sv mice, serum antibody titers to conserved self antigens like albumin, thyroglobulin, hsp65, and hsp70 were increased, at least two to three times in κ-deficient 129/Sv mice (FIG. 2). Similarly, antibodies to ubiquitin were noted at reasonable titers in κ-deficient mice, while these were undetectable in normal 129/Sv mice. These data suggest that exclusive association of three V_λ genes with various V_H gene families results in generation of self-reactive antigen-binding sites, which is intrinsic to the development of B-cell repertoire. In fact, the number of conventional B lymphocytes in κ-deficient mice is decreased by 50%, while B1 cell numbers remain unaffected.[8] Therefore, the high titers of natural antibodies observed here may be related to B1 cells and, also, reveal the available back-up mechanisms in the preimmune repertoire as a result of available restricted V_λ gene diversity.

SUMMARY

The experiments outlined here provide evidence that most V_H gene families are expressed in association with either of three V_λ genes available in κ-deficient mice, though their expression is nonrandom. An exclusive use of V_λ L-chain genes, where predominantly expressed $V_\lambda 1$ and $V_\lambda 2$ differ by only seven amino acid substitutions in the germline, results in a higher degree of self-reactive antibodies to conserved natural antigens like albumin, thyroglobulin, and stress proteins. The

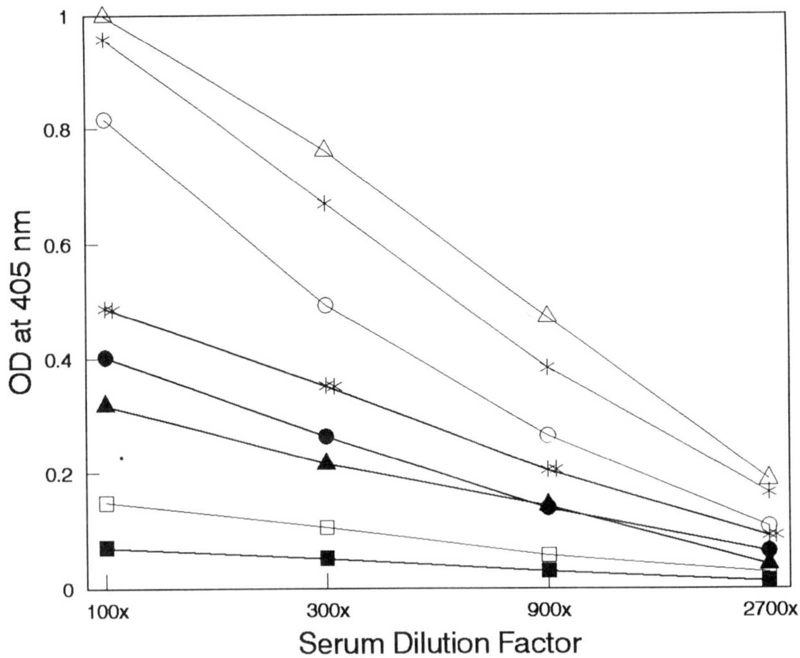

FIGURE 2. Serum antibody titers to conserved natural antigens, albumin, thyroglobulin, ubiquitin, and hsp70, in κ-deficient and normal 129/Sv mice. Symbols: △, BSA ($\kappa-$); ▲, BSA (129/Sv); ○, thyroglobulin ($\kappa-$); ●, thyroglobulin (129/Sv); □, ubiquitin ($\kappa-$); ■, ubiquitin (129/Sv); *, hsp70 ($\kappa-$); **, hsp70 (129/Sv).

higher levels of self-reactive antibodies, characteristic of preimmune B-cell repertoire, are likely to be a product of nonrandom V_H gene family expression and their restricted pairing with three V_λ L-chain genes. These data suggest that a low λ L-chain expression in normal mice is due, in part, to endogenous selection against λ L chain in order to restrict self reactivity within the homeostatic threshold.

ACKNOWLEDGMENT

The authors would like to thank Dr. Dennis Huszar, Genpharm International, Mountain View, California, for providing κ-deficient mice.

REFERENCES

1. TONEGAWA, S. 1983. Somatic generation of antibody diversity. Nature **302:** 575.
2. BRODEUR, P. H. & R. RIBLET. 1984. The heavy chain variable region (IgH-V) in the mouse. I. 100IgH-V genes comprise 7 families of homologous genes. Eur. J. Immunol. **14:** 922.
3. KOFLER, R., S. GELEY, H. KOFLER & A. HELMBERG. 1992. Mouse variable region

gene families: Complexity, polymorphism and use in non-autoimmune responses. Immunol. Rev. **128:** 5.

4. Sanchez, P., B. Nadel & P.-A. Cazenave. 1991. Vλ-Jλ rearrangements are restricted within a V-J-C recombination unit in the mouse. Eur. J. Immunol. **14:** 1159.
5. Kaushik, A., D. H. Schulze, F. A. Bonilla, C. Bona & G. Kelsoe. 1990. Stochastic pairing of heavy chain and κ-light chain variable region families occurs in polyclonally activated B cells. Proc. Natl. Acad. Sci. USA **87:** 4932.
6. Yurovsky, V. V. & G. Kelsoe. 1993. Pairing of V_H gene families with the λ1 light chain: Evidence for a non-stochastic association. Eur. J. Immunol. **23:** 1975.
7. Nadel, B., A.-M. Drapier, L. Andrade, P.-A. Cazenave & P. Sanchez. 1992. Preferential V_H/Vλ pairings occur in the available B cell repertoire of adult BALB/c mice. Intl. Immunol. **4:** 501.
8. Chen, J., M. Trounstine, C. Kurahara, F. Young, C. C. Kuo, Y. Xu, J. F. Loring, F. W. Alt & D. Huszar. 1993. B cell development in mice that lack one or both immunoglobulin κ light chain genes. EMBO J. **12:** 821.
9. Painter, C. J., M. Monestier, A. Chew, A. Dimitrui-Bona, K. N. Kasturi, C. Baily, V. E. Scott, C. L. Sidman & C. Bona. 1988. Specificities and V genes encoding monoclonal autoantibodies from viable motheaten mice. J. Exp. Med. **167:** 1137.
10. Schulze, D. H. & G. Kelsoe. 1987. Genotypic analysis of B cell colonies by in situ hybridization. J. Exp. Med. **166:** 163.
11. Kaushik, A., A. Lim, P. Poncet, X. R. Ge & G. Dighiero. 1988. Comparative analysis of natural antibody specificities among hybridomas originating from spleen and peritoneal cavity of adult NZB and BALB/c mice. Scand. J. Immunol. **27:** 461.

Analysis of V_κI and V_λII Light Chain Genes in the Expressed B-Cell Repertoire[a]

AUDREY MANHEIMER-LORY, MACARENA IRIGOYEN, BRUCE GAYNOR, RASHEL MONHIAN, ADAM SPLAVER, AND BETTY DIAMOND

Department of Microbiology and Immunology
Albert Einstein College of Medicine
of Yeshiva University
Bronx, New York 10461-1602

INTRODUCTION

Idiotypic analyses have constituted an important first approach to the elucidation of the molecular genetics of a variety of antibody responses.[1-7] Coupled with a knowledge of the immunoglobulin germline gene repertoire used in forming idiotypic antibodies, one can define the origins and mechanisms of diversification of antibodies of a particular antigenic specificity. Over the years our laboratory has characterized the idiotypic profile of human anti-double-stranded (ds) DNA autoantibodies from patients with systemic lupus erythematosus (SLE). We have used idiotypic analysis to address the following issues: Do autoimmune and normal persons have the same or different germline variable (V) gene repertoires? Do anti-foreign and anti-self antibodies use the same V gene pool? Is the subset of V genes used for anti-self antibodies a small or large subset of the available germline genes? Do autoantibodies use conventional mechanisms of diversification?

Two human anti-dsDNA idiotypic systems, 3I and 8.12, have been well characterized.[2,4] 3I recognizes a determinant present on kappa light chains of anti-dsDNA antibodies. It is present in elevated titers in about 80% of patients with SLE whose serum is dsDNA reactive and in clinically unaffected family members. Idiotypic antibodies can be found in both the kidney and skin of these patients[2] and in the anti-pneumococcal response of nonautoimmune persons.[8] The 3I idiotype is primarily encoded by members of the V_κI gene family.[2]

The 8.12 idiotype is present on lambda light chains.[9,10] About 50% of SLE patients have elevated titers of 8.12^+ antibodies; 8.12^+ antibodies, like $3I^+$ antibodies, can be found in the glomeruli of lupus patients. Unlike 3I, titers of 8.12-reactive antibodies are not elevated in family members (unpublished). 8.12^+ antibodies are, however, also part of the protective antibody repertoire; vaccination with pneumocococcal polysaccharide leads to the production of 8.12^+ anti-pneumococcal antibodies.[11] 8.12^+ antibodies are exclusively encoded by V_λII genes.[10] Therefore, the V_λII and V_κI gene families contribute significantly to the anti-DNA response in patients with SLE.[2-4]

[a] This work was supported by grants from the National Institutes of Health. Audrey Manheimer-Lory is supported by a Career Development Award from the SLE Foundation.

The human V_κ light chain locus has been extensively studied over the last decade.[12–25] This locus consists of 75 to 80 genes including about 25 pseudogenes.[23] Zachau and colleagues[13–18,20,21,23,24] have systematically cloned and sequenced the kappa locus and have identified five gene clusters: A, B, L, O, and W. These are generally present in duplicate copies. The largest V_κ family (approximately 30 genes including pseudogenes) is the V_κI subgroup, whose members are interspersed and can be found within the A, O, and L regions.[15,17,18,24] The V_λ locus is similar in complexity to V_κ and has been estimated to contain about 70 germline genes that have been grouped into 10 distinct families.[26–28] There is approximately the same percentage of nonfunctional V_λ genes as V_κ genes.[29] The V_λII family has 10 to 15 members.[30,31]

To better understand the V_κI and V_λII genes involved in generating idiotype-positive anti-DNA antibodies and the contribution of somatic mutation to DNA binding, we have used the polymerase chain reaction (PCR)[32] and genomic libraries to clone V_κI and V_λII germline genes from patients with SLE. An analysis of the germline repertoire can help show whether autoreactivity results from the expression of specific germline genes and/or whether particular gene rearrangements or mutations contribute to the production of these disease-related antibodies.

MATERIALS AND METHODS

PCR Amplification

Genomic DNA from spleen was amplified (using standard protocol from Perkin-Elmer Cetus) from either the $3I^+$ or 8.12^+ lupus patient, using 5′ primers corresponding to the conserved leader regions and 3′ primers corresponding to an area of the 3′ untranslated region including the nonamer recombination signal sequence (RSS). Oligomers are described below. Products of the appropriate size were cloned directly using the TA cloning system (Invitrogen) and screened with either V_κI (400 bp *Sac*I/*Pst*I) or V_λII (255bp *Hin*f/*Pst*I) coding region probes. Sequencing was performed using Sequenase (U.S. Biochemical, Cleveland, OH). The University of Wisconsin Genetics Computer Group (GCG) program version 7.3 was used to analyze the nucleotide, amino acid, and isoelectric point data.

Oligonucleotides

The oligonucleotide sequences are as follows: for V_κI: LV_κI: 5′TGGGGCTTC-TGCTGCTCTGG3′ (the mid-region of the leader sequence) and 3′ V_κI: 5′CTGGG-GGTTTATGTTCGGGT3′ (RSS). Oligonucleotides used to clone V_λII genes have been previously described.[4]

Construction of the Genomic Library

The method that was used has been described elsewhere.[4,33]

RESULTS

We analyzed the V_κI and V_λII nucleotide sequences and the deduced amino acid sequences of PCR products derived from two persons with SLE. The results are summarized in TABLE 1. Twenty V_κI PCR products were generated. Three clones represent germline genes that differ from previously reported genes by two or more nucleotides; these are probably alleles of reported genes, inasmuch as there is little room in the locus for novel genes. Six clones are identical to sequences of three previously published V_κI genes, and the remaining 11 clones differ from previously published V_κI genes by a single nucleotide. These clones are likely to be alleles or Taq polymerase errors (TABLE 1).[33]

For V_λII, a gene family that had not been extensively characterized, seven new germline genes were isolated from the genomic and PCR libraries (TABLE 1).[4] From the genomic library, two novel genes were identified, one a pseudogene and one a potentially functional gene, as well as two previously reported genes. From the PCR library, 21 clones were analyzed: five were novel, three were previously reported, seven were identical to each other, and six differed from other genes by a single nucleotide.[4] It is not easy to distinguish between alleles and new genes in this family, because the V_λ locus has not been sufficiently well mapped to identify the number of genes in the V_λ locus, although, as mentioned above, Southern blot analysis of V_λII genes suggests approximately 10 to 15 members.

To facilitate our understanding of the potential for V_κ and V_λ genes to encode altered amino acid sequences, we calculated random replacement/silent (R/S) mutation ratios over the framework regions (FRs) and complementarity-determining regions (CDRs) of all currently reported germline genes (TABLE 2). This R/S ratio can be defined as the inherent susceptibility of a random nucleotide change within the gene to encode for an altered amino acid sequence.[4] As seen in TABLE 2, the V_κI and V_λII gene families are not more likely to generate replacement mutations than other light chain families.

An analysis of expressed V_κI and V_λII genes confirms the previously published reports that anti-DNA antibodies are encoded by somatically mutated genes. When all expressed antibodies encoded by V_κI and V_λII genes are analyzed (TABLE 3), the average frequency of R mutations of V_κI autoantibodies is 3.0% in FRs and 8.4% in CDRs. This is very similar to what is observed for V_λII autoantibodies (1.0% in FRs and 8.7% in CDRs). Molecular studies of the human protective antibody repertoire have lagged behind studies of the pathogenic repertoire. Therefore, only a few V_κI and V_λII antibodies to foreign antigens are available for comparison. The average frequency of R mutations in these antibodies is 1.8% in FRs and 4.2% in CDRs for V_κI and 0.3% in FRs and 6.9% in CDRs for V_λII (TABLE 3). For both autoantibodies and protective antibodies, the higher frequency of replacement amino acids in CDRs is consistent with antigen selection.

TABLE 1. Summary of Cloned V_κI and V_λII Genes

	Library	N[a]	Novel Genes	Published Genes	Alleles[b]
V_κI	PCR	20	3	6	11
V_λII	PCR	21	5[c]	10	6
	Genomic	4	2[d]	2	—

[a] N = number of clones.

[b] These sequences differ by one nucleotide from other reported sequences. They may represent Taq polymerase errors but are more likely to represent allelic forms of those genes.

[c] These probably represent alleles of previously reported V_κI genes, but differ by two or more nucleotides.

[d] One novel gene is a pseudogene.

TABLE 2. Random R/S Ratios of Germline V_κ and V_λ Genes

Germline[a]	FR1	CDR1	FR2	CDR2	FR3	CDR3
V_κI	2.97	3.57	3.46	3.13	3.11	4.59
V_κII	2.48	3.67	3.12	2.76	3.30	3.89
V_κIII	2.54	4.13	3.28	2.23	3.36	4.19
V_κIV	2.82	3.59	3.69	2.33	3.17	3.53
V_λI	2.39	2.99	3.01	2.93	2.64	3.86
V_λII	2.18	2.92	3.71	2.65	2.74	3.85
V_λIII	2.30	3.76	3.14	3.36	2.85	5.68
V_λIV[b]	2.81	4.41	3.28	2.00	2.85	3.50
V_λVI[b]	2.86	3.80	3.07	3.67	2.79	5.67
V_λVII	2.18	2.58	3.31	4.40	2.55	2.51
V_λVIII	2.23	3.02	3.84	3.54	3.04	3.51
V_λIX[b]	2.03	4.15	3.20	2.31	3.40	3.74
V_λX[b]	2.39	2.67	3.25	2.07	2.65	4.83

[a] Pseudogenes were excluded from this analysis.
[b] Analysis of only one gene.

A more extensive analysis of the frequency of R substitutions for the variable region of autoantibodies encoded by all V_κ and V_λ gene families is summarized in TABLE 4. Both IgM and IgG antibodies display somatic mutation. The frequency of amino acid substitutions for V_κI is higher in FRs (3.0%) than for V_κII, V_κIII, and V_κIV antibodies (0.4%, 0.9%, and 1.0%, respectively). The frequency of altered amino acids in CDRs (8.4%) is also significantly higher for the V_κI autoantibodies compared to the other V_κ families.[33] This is of interest, because the random R/S ratios of the V_κI germline genes do not show them to be more likely to encode altered amino acid sequences than other V_κ genes. The frequency of altered amino acids in expressed V_λII antibodies is higher than in antibodies encoded by most other V_λ families. V_λII-encoded autoantibodies also have a higher frequency of replacement mutations in both FRs (1.0%) and CDRs (8.7%).

TABLE 3. Actual %R Mutation in Expressed Autoantibodies and Anti-Foreign Antibodies

Antibodies	N	%R FRs	%R CDRs
A. V_κI			
Autoantibodies[a]	42	3.0	8.4
Anti-foreign antibodies[b]	3	1.8	4.2
B. V_λII			
Autoantibodies[c]	6	1.0	8.7
Anti-foreign antibodies[d]	3	0.3	6.9

[a] Seven anti-DNA[2,34]; three RF[35,36]; 25 anti-TPO,[6] anti-AchR,[37] anti-U1 SnRNP[EMBL], anti-platelet integrin IIB,[38] anti-striated muscle,[39] and anti-cardiolipin[EMBL]; and two anti-ganglioside.[40]
[b] Anti-hepatitis B,[41] anti-*N. meningitidis* (EMBL Ayala *et al.*), anti-gp41 of HIV.[42]
[c] Two anti-RBC,[43] three anti-DNA[10,44] and RF.[45]
[d] Anti-Hib polysaccharide,[46] anti-Id (EMBL Van der Heijden, R.W.J. *et al.*, 1993), anti-hepatitis A.[47]

TABLE 4. %R Mutations in Expressed Autoantibodies

Family	%R FR	%R CDR
$V_{\kappa}I$[2,34–40,EMBL]	3.0	8.4
$V_{\kappa}II$[6,34,40,48,49]	0.4	1.3
$V_{\kappa}III$[2,7,37,39,40,44,50,51]	0.9	3.2
$V_{\kappa}IV$[2]	1.0	1.0
$V_{\lambda}I$[44,45,52–54]	0.4	1.6
$V_{\lambda}II$[10,43–45]	1.0	8.7
$V_{\lambda}III$[EMBL]	0.5	1.2
$V_{\lambda}IV$[49]	0.0	0.0
$V_{\lambda}VIII$[44,55,56]	0.8	4.9
$V_{\lambda}IX$[44]	0.0	0.0

Somatic mutation of both $V_{\kappa}I$ and $V_{\lambda}II$ genes appears to proceed by point mutation. There is no evidence at present to suggest that gene conversion is occurring. Studies of somatic mutation of mouse V region genes provide evidence for the presence of "hot spots," which may more readily undergo base substitutions.[57,58] $V_{\kappa}I$ and $V_{\lambda}II$ gene sequences also contain these "hot spot" motifs, and some mutations in expressed antibodies occur at these sites (TABLE 5).

Serum studies of 3I and 8.12 anti-DNA idiotypic antibodies have shown that they are cationic.[4,9] HK137/L1 and Vd/L8 genes used in the production of anti-DNA antibodies both encode cationic sequences. Four of six reported monoclonal anti-DNA antibodies encoded by these $V_{\kappa}I$ genes have light chains that are even more cationic than the protein encoded by the unmutated gene (data not shown). Two $V_{\kappa}I$ antibodies with anionic sequences are, however, derived from the same germline genes and the two genetically characterized $V_{\lambda}II$ anti-DNA antibodies also have anionic light chains, though they are derived from germline genes that without mutation encode anionic protein sequences (data not shown). These four light chains, by virtue of their associated heavy chains, form cationic antibodies. Thus, the reason that these light chain families are preferentially used in an anti-DNA response is not because they encode a cationic sequence in their germline configuration.

TABLE 5. Mutations in "Hot Spots" of anti-DNA $V_{\kappa}I$ and $V_{\lambda}II$ Antibodies

	Antibody	Germline Gene	FRs	CDRs	No. of Mutations at Hot Spots / Total No. of Mutations
A. $V_{\kappa}I$	I-2A[2]	Vd/L8[23]	2	2	4/11
	2A4[2]		1	11	12/40
	III-3R[2]	HK137/L1[23]	2	3	5/25
	III-2R[2]		1	2	3/19
	IC4[2]		3	5	8/31
	1X7RG1[2]		1	0	1/15
B. $V_{\lambda}II$	KS3[10]	hslv215.23/DPL11[4]	4	3	7/11
	SD6[10]		1	5	6/11
	33.H11[44]		1	0	1/5

DISCUSSION

Because the majority of genes obtained from V gene libraries derived from patients with an autoimmune disease, SLE, are identical or one nucleotide different from previously reported genes, it is clear that the germline repertoire of an autoimmune person is highly homologous to that of a healthy person. Furthermore, although the human protective antibody repertoire is just now beginning to be characterized[4,59–64] and only a small number of antibody sequences have been reported, an analysis of the V_κI- and V_λII-encoded antibodies shows that the same V_L genes are used to encode both autoantibodies and, presumably, protective antibodies to foreign antigens. Current data also show that the expressed antibody repertoire is derived from a small number of the apparently available germline genes. A recent report by Cox, Tomlinson, and Winter[23] similarly demonstrates that the normal expressed V_κ repertoire is derived from a restricted number of V gene segments.

Despite the finding that the genes used in the presumed protective repertoire are the same as those used to construct autoantibodies, the relative frequency of use of each gene may be different between the two. Cox *et al.*[23] found that only a small percent of rearranged light chain sequences use V_κ genes located in the J_κ distal region, whereas over 95% of sequences are encoded by J_κ proximal genes. Similarly, for V_κIII autoantibodies, the majority are encoded by J_κ proximal genes. This is in striking contrast to the V_κI genes used to encode autoantibodies of which two-thirds are from J_κ distal germline genes. It is not clear if autoantibodies, in general, use J_κ distal V_κ genes or if this is only the case for V_κI-encoded autoantibodies.

Although most V_κI- or V_λII-expressed light chains are encoded by somatically mutated genes and all anti-DNA antibodies encoded by these gene families display somatic mutation, only a small number of V_κI- or V_λII-encoded antibodies have an R/S ratio greater than random. An R/S ratio greater than the random R/S ratio has been cited as evidence for antigen selection.[65] It is clear, however, that antibodies without high R/S ratios can also be the products of antigen selection.[66] Specific mutations in CDR and in FR of antibodies with a low R/S ratio may cause biologically significant changes in antigen binding.

V_κI-encoded antibodies have a higher frequency of amino acid substitutions than antibodies encoded by the other V_κ families. Furthermore, among V_κI-encoded antibodies, anti-DNA antibodies have the highest number of replacement mutations, which may be a contributing factor in the use of V_κI genes in the anti-DNA response. V_λII-encoded autoantibodies also show a higher frequency of amino acid substitutions compared to antibodies encoded by other V_λ families. The reasons for this higher frequency of replacement mutations are not known.

Recent reports demonstrate that mutations that induce a cationic shift in the charge of the antibody are prevalent in murine lupus nephritis.[67] Such mutations might not be necessary if in fact the germline genes used to encode an anti-DNA antibody encode a cationic sequence. An analysis of the isoelectric points of a panel of V_κI-encoded anti-DNA antibodies has indicated that in four of six cases a definite cationic shift from the germline-encoded sequence to the expressed V_κI light chain has occurred. In contrast, two other V_κI-encoded anti-DNA antibodies and two V_λII-encoded anti-DNA antibodies do not have cationic light chains, although the heavy and light chains together form a cationic protein.[33,68] Thus, an analysis of any particular amino acid substitution and its importance in antigen binding may be incomplete if the associated heavy or light chain is not similarly examined.

Our data do not suggest that there is a special subset of germline V_κI or V_λII sequences used exclusively to encode DNA-binding antibodies, but rather that certain kinds of mutations and/or heavy and light chain associations can lead to DNA binding. On the basis of data derived from what is still a small number of anti-DNA antibodies, we speculate there may be subtle differences in V gene usage in autoreactive B cells. Whether these differences reflect a difference in the process of V gene rearrangement that exists in the naive B-cell repertoire of autoimmune persons or whether these differences reflect antigen selection in an autoimmune person is not known at this time.

SUMMARY

Our previous studies of anti-DNA antibodies in SLE have demonstrated a preferential use of V_κI and V_λII gene families to encode light chains of antibodies that express the anti-DNA-associated 3I and 8.12 idiotypes, respectively. In this study, we employed PCR to obtain V_κI and V_λII germline genes from lupus patients in order to compare the germline genes to genes encoding expressed V_κI and V_λII light chains and to analyze the extent of somatic mutation among autoantibodies that derive from these light chain families.

Our analysis shows that the germline repertoire among all persons (autoimmune and healthy) is comparable and that somatic mutation is used to diversify autoantibodies as well as anti-microbial antibodies. We have observed that autoantibodies encoded by V_κI and V_λII genes have a higher number of amino acid replacements in CDRs than autoantibodies encoded by other V_L gene families. In addition, there may be subtle differences in V gene usage that distinguish the V_κI-encoded light chains from other expressed V_κ light chains.

REFERENCES

1. Davidson, A., A. Manheimer-Lory, C. Aranow, R. Peterson, N. Hannigan & D. Diamond. 1990. Molecular characterization of a somatically mutated anti-DNA antibody bearing two systemic lupus erythematosus-related idiotypes. J. Clin. Invest. **85:** 1401–1409.
2. Manheimer-Lory, A., J. B. Katz, M. Pillinger, C. Ghossein, A. Smith & B. Diamond. 1991. Molecular characteristics of antibodies bearing an anti-DNA-associated idiotype. J. Exp. Med. **174:** 1639–1652.
3. Manheimer-Lory, A. J., A. Davidson, D. Watkins, N. R. Hannigan & B. Diamond. 1990. Generation and analysis of clonal IgM- and IgG-producing human B cells line expressing an anti-DNA-associated idiotype. J. Clin. Invest. **87:** 1519–1525.
4. Irigoyen, M., A. Manheimer-Lory, B. Gaynor & B. Diamond. 1994. Molecular analysis of the human immunoglobulin V_λII family. J. Clin. Invest. **94:** 532–538.
5. Liu, M.-F., D. L. Robbins, J. J. Crowley, S. Sinha, F. Kozin, T. J. Kpps, D. A. Carson & P. P. Chen. 1989. Characterization of four homologous L chain variable region genes that are related to 6B6.6 idiotype positive human rheumatoid factor L chains. J. Immunol. **142:** 688–694.
6. Chazenbalk, G. D., S. Portolano, D. Russo, J. S. Hutchison, B. Rapoport & S. McLachlan. 1993. Human organ-specific autoimmune disease. Molecular cloning and expression of an autoantibody gene repertoire for a major autoantigen reveals an antigenic immunodominant region and restricted immunoglobulin gene usage in the target organ. J. Clin. Invest. **92:** 62–74.
7. Kipps, T. J., E. Tomhave, P. P. Chen & R. I. Fox. 1989. Molecular characterization

of a major autoantibody-associated cross-reactive idiotype in Sjögren's syndrome. J. Immunol. **142;** 4261–4268.
8. Grayzel, A., A. Solomon, C. Aranow & B. Diamond. 1991. Antibodies elicited by pneumococcal antigens bear an anti-DNA-associated idiotype. J. Clin. Invest. **87:** 842–846.
9. Livneh, A., A. Halpern, D. Perkins, A. Lazo, R. Halpern & B. Diamond. 1987. A monoclonal antibody to a cross-reactive idiotype on cationic human anti-DNA antibodies expressing λ light chains: A new reagent to identify a potentially differential pathogenic subset. J. Immunol. **138:** 123–127.
10. Paul, E., A. Iliev, A. Livneh & B. Diamond. 1992. The anti-DNA-associated idiotype 8.12 is encoded by the V_λII gene family and maps to the vicinity of L chain CDR1. J. Immunol. **149:** 3588–3595.
11. Livneh, A., E. Gazit & B. Diamond. 1994. The preferential expression of the anti-DNA-associated idiotype 8.12 in lupus is not genetically controlled. Autoimmunity, **18:** 1–6.
12. Bentley, D. L. & T. H. Rabbitts. 1980. Human immunoglobulin variable region genes: DNA sequences of two V_κ genes and a pseudogene. Nature (London) **288:** 730–733.
13. Pech, M. & H. G. Zachau. 1984. Immunoglobulin genes of different subgroups are interdigitated within the V_κ locus. Nucleic Acids Res. **12:** 9229–9236.
14. Pohlenz, H.-D., B. Straubinger, R. Thiebe, M. Pech, F.-J. Zimmer & H. G. Zachau. 1987. The human V_κ locus: Characterization of extended immunoglobulin gene regions by cosmid cloning. J. Mol. Biol. **193:** 241–253.
15. Straubinger, B., E. Huber, W. Lorenz, E. Osterholzer, W. Pargent, M. Pech, H-D. Pohlenz, F.-J. Zimmer & H. G. Zachau. 1988. The human V_κ locus: Characterization of a duplicated region encoding 28 different immunoglobulin genes. J. Mol. Biol. **199:** 23–34.
16. Pargent, W., K. F. Schäble & H. G. Zachau. 1991. Polymorphisms and haplotypes in the human K locus. Eur. J. Immunol. **21:** 1829–1835.
17. Lautner-Rieske, A., C. Huber, A. Meindl, W. Pargent, K.-F. Schäble, R. Thiebe, I. Zocher & H. G. Zachau. 1992. The human immunoglobulin K locus: Characterization of the duplicated A regions. Eur. J. Immunol. **22:** 1023–1029.
18. Pargent, W., A. Meindl, R. Thiebe, S. Metzel & H. G. Zachau. 1991. The human immunoglobulin K locus. Characterization of the duplicated 0 regions. Eur. J. Immunol. **21:** 1821–1827.
19. Kurth, J. H., J. L. Mountain & L. L. Cavalli-Sforza. 1993. Subclustering of human immunoglobulin kappa light chain variable region genes. Genomics **16:** 69–77.
20. Pech, M., H. R. Jaenichen, H.-D. Pohlenz, P. S. Neumaier, H. G. Klobeck & H. G. Zachau. 1984. Organization and evolution of a gene cluster for human immunoglobulin variable regions of the kappa type. J. Mol. Biol. **176:** 189–204.
21. Klobeck, H.-G., A. Meindl, G. Cambriato, A. Solomon & H. G. Zachau. 1985. Human immunoglobulin kappa light chain genes of subgroups II and III. Nucleic Acids Res. **13:** 6499–6513.
22. Bentley, D. L. & T. H. Rabbitts. 1983. Evolution of immunoglobulin V genes: Evidence indicating that recently duplicated human V-kappa sequences have diverged by gene converison. Cell **32:** 181–189.
23. Cox, J. P. L., I. M. Tomlinson & G. Winter. 1994. A directory of human germline V_κ segments reveals a strong bias in their usage. Eur. J. Immunol. **24:** 827–836.
24. Huber, C., E. Huber, A. Lautner-Rieske, K.-F. Schäble & H. G. Zachau. 1994. The human immunoglobulin K locus. Characterization of the partially duplicated L region. Eur. Immunol. **23:** 2860–2867.
25. Klobeck, H.-G., G. W. Bornkamm, G. Combriato, R. Mocikat, H.-D. Pohlenz & H. G. Zachau. 1985. Subgroup IV of human immunoglobulin K light chains is encoded by a single germline gene. Nucleic Acids Res. **13:** 6515–6529.
26. Chuchana, P., A. Blancher, F. Brockly, D. Alexander, G. Lefranc & M-P. Lefranc. 1990. Definition of the human immunoglobulin variable lambda (IGLV) gene subgroups. Eur. J. Immunol. **20:** 1317–1325.

27. WILLIAMS, S. C. & G. WINTER. 1993. Cloning and sequencing of human immunoglobulin V_λ gene segments. Eur. J. Immunol. **23:** 1456–1461.
28. STIERNHOLM, N. B. J., B. KUZNIAS & N. L. BERINSTEIN. 1994. Identification of a new human V_λ gene family—V_λX. J. Immunol. **152:** 4969–4975.
29. DALEY, M. D., T. OLEE, H.-Q PENG, R. W. SOTO-GIL, P. P. CHEN & K. A. SIMINOVITCH. 1992. Molecular characterization of the human V_λI germline gene repertoire. Mol. Immunol. **29:** 1031–1042.
30. PAUL, E., A. LIVNEH, A. J. MANHEIMER-LORY & B. DIAMOND. 1991. Characterization of the human Ig V_λII gene family and analysis of V_λII and C_λ polymorphism in systemic lupus erythematosus. J. Immunol. **147:** 2771–2776.
31. COMBRIATO, G. & H.-G. KLOBECK. 1991. V_λ and J_λ-C_λ gene segments of the human immunoglobulin λ light chain locus are separated by 14 kb and rearrange by a deletion mechanism. Eur. J. Immunol. **21:** 1513–1522.
32. ECKERT, K. A. & T. A. KUNKEL. 1990. High fidelity DNA synthesis by the *Thermus aquaticus* DNA polymerase. Nucleic Acids Res. **18:** 3739–3744.
33. MANHEIMER-LORY, A. J., R. MONHIAN, A. SPLAVER, B. GAYNOR & B. DIAMOND. 1994. Analysis of the V_κI family: Germline genes and expressed antibodies. Submitted.
34. VAN ES, J. H., H. AANSTOOT, F. H. J. GMELIG-MEYLING, R. H. W. M. DERKSEN & T. LOGTENBERG. 1992. A human systemic lupus erythematosus-related anti-cardiolipin/single stranded DNA autoantibody is encoded by a somatically mutated variant of the developmentally restricted 51P1 V_H gene. J. Immunol. **149:** 2234–2240.
35. CHEN, P. P., K. ALBRANDT, T. J. KIPPS, V. RADOUX, F. T. LIU & D. A. CARSON. 1987. Isolation and characterization of human V_κIII germ-line genes. Implications for the molecular basis of human V_κIII light chain diversity. J. Immunol. **139:** 1727–1733.
36. ERMEL, R., T. P. KENNY, P. P. CHEN & D. ROBBINS. 1993. Molecular analysis of rheumatoid factors derived from rheumatoid synovium suggests an antigen-driven response in inflamed joints. Arthritis Rheum. **36:** 380–388.
37. VICTOR, K. D., V. PASCUAL, A. K. LEFVERT & J. D. CAPRA. 1992. Human anti-acetylcholine receptor antibodies use variable region gene segments analogous to those used in autoantibodies of various specificities. Molec. Immunol. **29:** 1501–1506.
38. KUNICKI, T. J., D. S. ANNIS, J. GORSKI & D. J. NUGENT. 1991. Nucleotide sequence of the human autoantibody 2E7 specific for the platelet integrin IIb heavy chain. J. Autoimmunity **4:** 433–446.
39. VICTOR, K. D., V. PASCUAL, C. L. WILLIAMS, V. A. LENNON & J. D. CAPRA. 1992. Human monoclonal striational autoantibodies isolated from thymic B lymphocytes of patients with myasthenia gravis use V_H and V_L gene segments associated with the autoimmune repertoire. Eur. J. Immunol. **22:** 2231–2236.
40. WENG, N.-P., L. Y. YU-LEE, I. SANZ, B. M. PATTEN & D. M. MARCUS. 1992. Structure and specificities of antiganglioside autoantibodies associated with motor neuropathies. J. Immunol. **149:** 2518–2529.
41. ANDRIS, J. S., P. H. EHRLICH, L. OSTBERG & J. D. CAPRA. 1992. Probing the human antibody repertoire to exogenous antigens. Antiganglioside autoantibodies associated with motor characterization of the H and L chain V region gene segments from anti-Hepatitis B virus antibodies. J. Immunol. **149:** 4053–4059.
42. FELGENHAUER, M. N., J. KOHL & F. RUEKER. 1990. Nucleotide sequences of the cDNAs encoding the V regions of H and L chains of a human monoclonal antibody specific to HIV-1 gp41. Nucleic Acids Res. **18:** 4927.
43. CHAPMAN, C. J., M. B. SPELLERBERG, G. A. SMITH, S. J. CARTER, T. J. HAMBLIN & F. STEVENSON. 1993. Auto anti-red cell antibodies synthesized by patients with infectious mononucleosis utilize the V_H4-21 gene segment. J. Immunol. **151:** 1051–1061.
44. WINKLER, T. H., H. FEHR & J. R. KALDEN. 1992. Analysis of immunoglobulin variable region genes from human IgG anti-DNA hybridomas. Eur. J. Immunol. **22:** 1719–1728.
45. VICTOR, K. D., I. RANDEN, K. THOMPSON, O. FORRE, J. B. NATVIG, S. M. FU & J. D. CAPRA. 1991. Rheumatoid factors isolated from patients with autoimmune disorders are derived from germline genes distinct from those encoding the Wa, Po, and Bla cross-reactive idiotypes. J. Clin. Invest. **987:** 1603–1613.

46. CARROLL, W. L., P. M. WILSON, A. QUINN, R. A. INSEL, P. G. SHACKELFORD & E. ANDERSON. 1992. Immunoglobulin light chain variable region gene sequences for human antibodies to *Haemophilus influenzae* type b capsular polysaccharide are dominated by a limited number of V kappa and V lambda segments and VJ combinations. J. Clin. Invest. **89:** 729–738.
47. LEWIS, A. P., N. PARRY, T. S. PEAKMAN & J. S. CROWE. 1992. Rescue and expression of human immunoglobulin genes to generate functional human monoclonal antibodies. Hum. Antib. Hybrid. **3:** 146–152.
48. WEISBART, R. H., A. L. WONG, D. NORITAKE, A. KACENA, G. CHAN, C. RULAND, E. CHIN, I. S. Y. CHEN & J. D. ROSENBLATT. 1991. The RF reactivity of a human IgG monoclonal autoantibody is encoded by a variant V_κII L chain. J. Immunol. **147:** 2795–2801.
49. PASCUAL, V., K. VICTOR, I. RANDEN, K. THOMPSON, M. STEINITZ, O. FORRE, S. M. FU, J. B. NATVIG & J. D. CAPRA. 1992. Nucleotide sequence analysis of rheumatoid factors and polyreactive antibodies derived from patients with rheumatoid arthritis reveals diverse use of V_H and V_L gene segments and extensive variability in CDR3. Scand. J. Immunol. **36:** 349–362.
50. YOUNGBLOOD, K., L. FRUCHTER, G. DING, J. LOPEZ, V. BONAGURA & A. DAVIDSON. 1994. Rheumatoid factors from the peripheral blood of two patients with rheumatoid arthritis are genetically heterogeneous and somatically mutated. J. Clin. Invest. **93:** 852–861.
51. SPATZ, L., K. K. WONG, M. WILLIAMS, R. DESAI, J. GOLIER, J. E. BERMAN, F. W. ALT & N. LATOV. 1990. Cloning and sequence analysis of the V_κ and V_L regions of an anti-myelin/DNA antibody from a patient with peripheral neuropathy and chronic lymphocytic leukemia. J. Immunol. **144:** 2821–2828.
52. HUGHES-JONES, N. C., J. M. BYE, D. BEALE & J. COADWELL. 1990. Nucleotide sequences and three-dimensional modeling of the V_H and V_L domains of two human monoclonal antibodies specific for the D antigen of the human Rh-blood group system. Biochem. J. **268:** 135–140.
53. BERINSTEIN, N., S., LEVY & R. LEVY. 1989. Activation of an excluded immunoglobulin allele in a human B lymphoma line. Science **244:** 337–339.
54. CAIRNS, E., P. C. KWONG, V. IP, P. MIESNER, D. A. BILL & K. A. SIMINOVITCH. 1989. Analysis of variable region genes encoding a human anti-DNA antibody of normal origin. Implications for the molecular basis of human autoimmune responses. J. Immunol. **143:** 685–691.
55. SPATZ, L. A., M. WILLIAMS, B. BRENDER, R. DESAI & N. LATOV. 1992. DNA sequence analysis and comparison of the variable heavy and light chain regions of the IgM, monoclonal, anti-myelin associated glycoprotein antibodies. J. Neuroimmunol. **36:** 29–39.
56. DEFTOS, M., R. SOTO-GIL, M. QUAN, T. OLEE & P. P. CHEN. 1994. Utilization of a potentially universal downstream primer in the rapid identification and characterization of V_λ genes from two new human V_λ gene families. Scand. J. Immunol. **39:** 95–103.
57. ROGOZIN, I. B. & N. A. KOLCHANOV. 1992. Somatic hypermutagenesis in immunoglobulin genes. II. Influence of neighboring base sequences on mutagenesis. Biochem. Biophys. Acta **1171:** 11–18.
58. BETZ, A. G., M. S. NEUBERGER & C. MILSTEIN. 1993. Discriminating intrinsic and antigen-selected mutational hotspots in immunoglobulin V genes. Immunol. Today **14:** 405–411.
59. SCOTT, M. G., D. L. CRIMMINS, D. W. MCCOURT, G. CHUNG, K. F. SCHÄBLE, R. THIEBE, E. M. QUENZEL, H. G. ZACHAU & M. H. NAHM. 1991. Clonal characterization of the human IgG antibody repertoire to *Haemophilus influenzae* Type b polysaccharide. IV. The less frequently expressed V_L are heterogenous. J. Immunol. **147:** 4007–4013.
60. FERNÁNDEZ DE COSSIÓ, M. E., M. OHLIN, M. LLANO, B. SELANDER, S. CRUZ, J. DEL VALLE & C. A. K. BORREBAECK. 1992. Human monoclonal antibodies against an epitope on the class 5C outer membrane protein common to many pathogenic strains of *Neisseria meningitidis*. J. Infect. Dis. **166:** 1322–1328.

61. CHUNG, G.-H., M. G. SCOTT, K. H. KIM, J. KEARNY, G. R. SIBER, D. M. AMBROSINO & M. H. NAHM. 1993. Clonal characterization of the human IgG antibody repertoire of *Haemophilus influenzae* Type b polysaccharide. J. Immunol. **151:** 4352–4361.
62. SCOTT, M. G., D. L. CRIMMINS, D. W. MCCOURT, I. ZOCHER, R. THIEBE, H. G. ZACHAU & M. H. NAHM. 1989. Clonal characterization of the human IgG antibody repertoire to *Haemophilus influenzae* Type b polysaccharide. III. A single V_{κ}II gene and one of several J_{κ} genes are joined by an invariant arginine to form the most common L chain V region. J. Immunol. **143:** 4110–4116.
63. ANDRIS, J. S., S. JOHNSON, S. ZOLLA-PAZNER & J. D. CAPRA. 1991. Molecular characterization of five human anti-human immunodeficiency virus type 1 antibody heavy chains reveals extensive somatic mutation typical of an antigen-driven immune response. Proc. Natl. Acad. Sci. USA **88:** 7783–7787.
64. RAFF, H. V., C. BRADLEY, W. BRADY, K. DONALDSON, L. LIPSICH, G. MALONEY, W. SHUFORD, M. WALLS, P. WARD, E. WOLFF & L. J. HARRIS. 1990. Comparison of functional activities between IgG1 and IgM class switched human monoclonal antibodies reactive with Group B. *Streptococci* or *Escherichia coli* K1. J. Infect. Dis. **163:** 346–354.
65. SHLOMCHIK, M. J., A. MARSHAK-ROTHSTEIN, C. B. WOLFOWICZ, T. L. ROTHSTEIN & M. G. WEIGERT. 1987. The role of clonal selection and somatic mutation and autoimmunity. Nature **328:** 805–811.
66. DIAMOND, B. & M. D. SCHARFF. 1984. Somatic mutation of T15 heavy chain gives rise to an antibody with autoantibody specificity. Proc. Natl. Acad. Sci. USA **81:** 5841–5844.
67. O'KEEFE, T. L., S. K. DATTA & T. IMANISHI-KARI. 1992. Cationic residues in pathogenic anti-DNA autoantibodies arise by mutations of a germ-line gene that belongs to a large V_H gene subfamily. Eur. J. Immunol. **22:** 619–624.
68. PAUL, B. & B. DIAMOND. 1993. Characterization of two human anti-DNA antibodies bearing the pathogenic idiotype 8.12. Autoimmunity **16:** 13–21.

Anomalous Diversification of the Antibody Repertoire following Bone Marrow Transplantation[a]

ANNUSKA M. GLAS,[b] WENDY O. HUFNAGLE,[b]
IVY SUZUKI,[b] LAURIE PFISTER,[c] P. SAMBASIVA RAO,[b]
CAROL NOTTENBURG,[c] AND ERIC C. B. MILNER[b,d]

[b]*Virginia Mason Research Center*
Seattle, Washington 98101

[c]*The Fred Hutchinson Cancer Research Center*
Seattle, Washington 98104

[d]*Department of Immunology*
University of Washington
Seattle, Washington 98195

Bone marrow transplantation (BMT) is a recognized treatment for certain leukemias, other blood diseases, and some inborn errors of metabolism, and it has potential as a vehicle for gene therapy. Nevertheless, patients undergoing BMT may remain immunodeficient for several years despite the return to normal IgM and IgG levels.[1,2] These observations suggest continued defects in both T- and B-cell compartments. To determine whether abnormal usage of V_H genes may contribute to this immunodeficiency, the relative usage of V_H genes after marrow transplantation has been determined. As a measure of V_H gene utilization, rearrangements of eight individual V_H3 genes were assessed in peripheral blood B cells of four BMT recipients and compared to similarly obtained rearrangements from two healthy subjects. In addition, rearrangements of 14 individual V_H4 genes were assessed in peripheral blood B cells of one of the BMT recipients and compared to similarly obtained rearrangements from one of the healthy subjects. Heavy chain rearrangements were amplified quantitatively, and the resulting PCR products were cloned into a phagemid vector. Identification of the V_H gene present in an individual clone was established by hybridization with a sequence-specific oligonucleotide probe. More than 700 independent rearrangements from each person were analyzed.

The pattern of usage of V_H genes assessed at 90 days and approximately one year after transplant was similar to that observed in healthy subjects and was marked by overuse of two elements (V3-23 and V4-34).[3] However, the repertoires of the BMT recipients were less diversified than the repertoires of the healthy subjects. Sequence-specific oligonucleotide probes for 14 individual V_H4 genes, composing 100% of the germline repertoire, identified more than 99% of the rearrangements in the BMT recipient, but only 80% of the rearrangements in the healthy subject. Similarly, oligonucleotide probes specific for eight V_H3 genes detected 75% of rearrangements in the BMT recipients, but only 55% of the

[a] This work was supported in part by National Institutes of Health Grant A131241 and ACS Grant IM485.

rearranged V_H3 genes in the healthy subjects. The difference in the percentage of identified rearrangements between healthy subjects and BMT patients is presumed to be the result of somatic mutations in the regions of the probe that abrogate hybridization. To assess more directly differences in the occurrence of somatic mutations, we determined the nucleotide sequence of 20 randomly picked V_H3 and 20 randomly picked V_H4 rearrangements from one of the BMT recipients and from one healthy subject. The BMT recipient exhibited fewer mutations in both V_H3 gene segments (FIG. 1) and V_H4 gene segments (FIG. 2) than did the healthy subject. The ratio of replacement to silent substitutions (R/S ratio) in the complementarity-determining regions (CDRs) from the BMT recipient was lower than in the healthy subject. Somewhat unexpectedly, although the absolute number of substitutions was lower in the BMT recipient, the R/S ratio in the framework regions in the BMT recipient was much higher than in the healthy subject. The significance of this high ratio is not certain, but suggests that the mechanism of somatic mutation favors amino acid replacement.

A high R/S ratio and the selective accumulation of somatic mutations in the CDRs is an indicator of antigen-driven responses. The failure in the BMT recipients to accumulate somatic mutations is consistent with a defect in antigen-driven B-cell responses. This is also consistent with the observation that germinal center

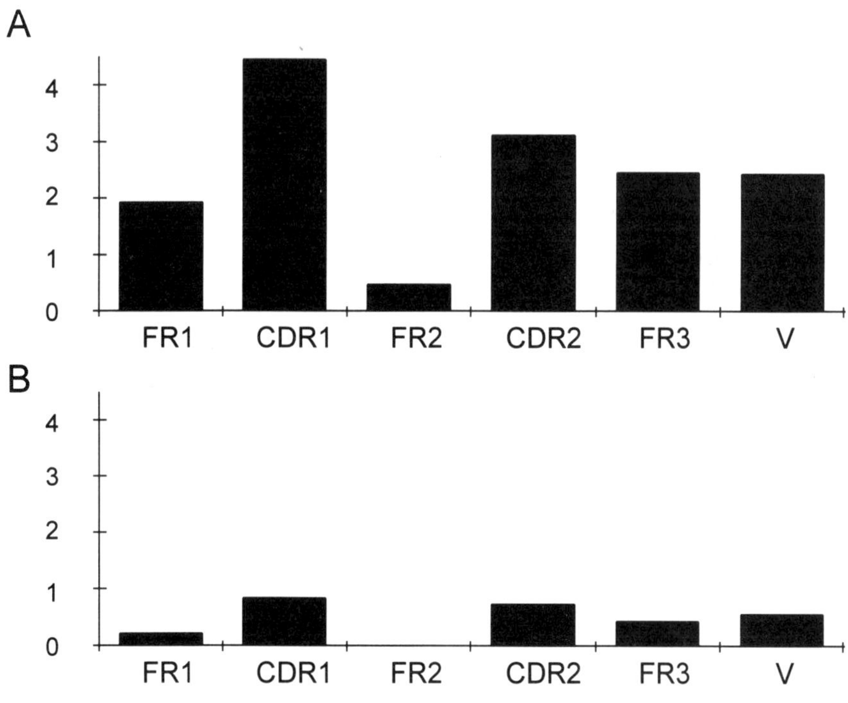

FIGURE 1. Nucleotide substitutions per 100 bp in randomly picked V_H3 rearrangements. Sequences were determined in both orientations to assess differences in the occurrence of somatic mutations. **Panel A,** healthy subject (VMRC3116); **Panel B,** BMT recipient (UPN5012).

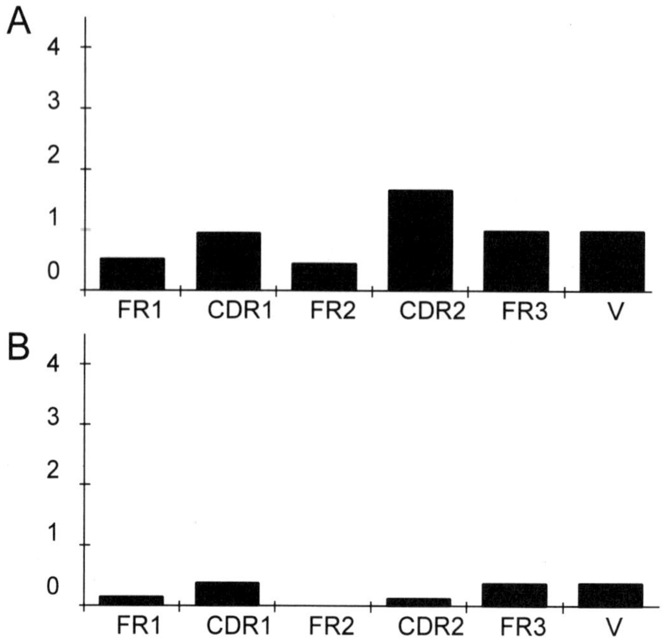

FIGURE 2. Nucleotide substitutions per 100 bp in randomly picked V_H4 rearrangements. Sequences were determined in both orientations to assess differences in the occurrence of somatic mutations. **Panel A,** healthy subject (VMRC3116); **Panel B,** BMT recipient (UPN5012).

formation is defective in BMT recipients months to years post transplant.[4] Together, these data indicate that, although the V_H gene content of the repertoire has normalized by 90 days post transplant, a defect in B-cell differentiation associated with antigen activation persists for at least one year after BMT.

REFERENCES

1. Lum, L. G., M. C. Seigneuret, R. F. Storb, R. P. Witherspoon & E. D. Thomas. 1981. Blood **58:** 431–439.
2. Witherspoon, R. P., H. J. Deeg, L. G. Lum, H. D. Ochs, J. A. Hansen, E. D. Thomas & R. Storb. 1984. Transplantation **37:** 456–461.
3. Suzuki, I., L. Pfister, A. M. Glas, C. Nottenburg & E. C. B. Milner. 1995. J. Immunol. In press.
4. Dilly, S. A., J. P. Sloane & I. S. M. Psalti. 1986. J. Pathol. **150:** 213–221.

Structure and Thermodynamics of Antigen Recognition by Antibodies[a]

BRADFORD C. BRADEN,[b] ANA CAUERHFF,[b]
WILLIAM DALL'ACQUA,[b] BARRY A. FIELDS,[b]
FERNANDO A. GOLDBAUM,[b] EMILIO L. MALCHIODI,[b]
ROY A. MARIUZZA,[b] ROBERTO J. POLJAK,[b,c]
FREDERICK P. SCHWARZ,[b] XAVIER YSERN,[d]
AND T. N. BHAT[e]

[b]*Center for Advanced Research in Biotechnology*
University of Maryland Biotechnology Institute and
National Institute of Standards & Technology
Rockville, Maryland 20850

[d]*Center for Drug Evaluation and Research*
Food and Drug Administration
Rockville, Maryland 20857

[e]*Structural Biochemistry Program*
Biomedical Supercomputing Center, NCI-FCRFDC
Frederick, Maryland 21202

The molecular basis of antigen recognition by antibodies has been approached by the X-ray diffraction study of crystalline complexes between protein antigens and Fab or Fv fragments. The antigens in these complexes were hen egg-white lysozyme (HEL),[1-7] pheasant lysozyme,[8] the influenza virus neuraminidase,[9,10] the anti-HEL FabD1.3,[11] a phosphocarrier protein from *Escherichia coli,*[12] and the Fab730.1.4.[13] These studies showed that antigenic determinants consist of discontinuous sequence segments of the antigen, bringing about 15 amino acid residues in contact with a similar number of antibody residues, mostly from the complementarity-determining regions (CDRs). Surfaces with areas ranging from about 1300 $Å^2$ to 1900 $Å^2$ are buried upon complex formation in these complexes. The antigen and the antibody may undergo small conformational changes upon complex formation. Chemical contacts between them include a variable number of hydrogen bonds, extensive van der Waals interactions, and, in some complexes, a few ion pairs. In agreement with the fact that the CDRs of antibodies contain many arginines and aromatic residues, a large proportion of the contacts with antigens are made by these residues.

[a] This research was supported by grants from the National Institute of Standards and Technology, the Human Frontiers Science Program, and the W. E. Elkins Professorship (to R.J.P.). Certain commercial equipment or materials are identified in this paper in order to specify adequately the experimental procedure. Such identification does not imply recommendation or endorsement by the National Institute of Standards and Technology, nor does it imply that the materials or equipment identified are necessarily the best available for this purpose.

[c] Address for correspondence: Roberto J. Poljak, CARB, 9600 Gudelsky Drive, Rockville, MD 20850-3479.

Although these studies have provided a detailed description of the three-dimensional structures of antigenic determinants and of the combining site of antibodies, a number of fundamental problems relating to the recognition process remain to be solved. For example, the relative contributions of entropic factors, such as hydrophobicity, versus enthalpic ones, such as van der Waals interactions, hydrogen bonding, and ion pairs, to the energetics of binding and the stability of complexes has not been elucidated. In addition, the binding of water molecules may be important, beyond the solvent entropy effect, by optimizing contacts between antigen and antibody. Thus, shape and chemical complementarity would arise not only from direct interactions between amino acid residues but also by the participation of solvent molecules. If this were the case, water molecules would be expected to make a favorable enthalpic contribution to antigen binding, compensated in part by an unfavorable loss of entropy. These issues are important in understanding the molecular basis of immunological specificity and, in general, protein–protein association processes.

To address such questions at the molecular level requires (1) higher resolution structures of antigen–Fab complexes than have generally been attained (3.0–2.8 Å) and (2) structure determination of the free antigen and the free antibody to examine conformational changes and their states of hydration before and after complex formation.

Although the three-dimensional structures of complexes involving avian lysozymes with Fabs or Fvs from six different monoclonal antibodies (mAbs) have been reported,[1–8] only in the case of two of these complexes, those derived from mAbs D1.3 and D44.1, has the structure of the free Fv (D1.3),[4,5] or free Fab (D44.1)[8] also been determined, permitting a comparison of the free and antigen-bound combining sites and an analysis of the conformational changes that take place upon formation of the complex.

The Fv fragment of D1.3 has been cloned and expressed in *Escherichia coli,* and subsequently crystallized as a free Fv and in a complex with the antigen, HEL.[14] Their three-dimensional structures have been determined to 1.8-Å resolution,[4,5] permitting a high-resolution description of an antibody-combining site in its free and antigen-bound states. In addition, the structure of the Fab from mAb D44.1 has been determined in its free form and in a complex with HEL at 2.1-Å and 2.5-Å resolution, respectively, also allowing a description of the free and the antigen-bound combining site.[7]

Structural information, in particular that concerning D1.3 and D44.1, has been used in the interpretation of the thermodynamic and physicochemical data obtained in studies of the reactions of those mAbs with the antigen HEL. This communication will analyze thermodynamic data obtained by titration calorimetry and will correlate it with crystallographic analyses of the antigen–antibody complexes, including that of a site-directed mutant of FvD1.3.

STRUCTURES OF THE FREE AND THE ANTIGEN-BOUND FabD44.1

The three-dimensional structures of the free and antigen-complexed Fab from the mouse monoclonal anti-HEL antibody D44.1 have been determined by X-ray crystallographic techniques.[7] The crystals of the free and lysozyme-bound Fab were grown under nearly identical conditions, and their X-ray diffraction data were collected to 2.1- and 2.5-Å resolution, respectively. The antigenic determinant of lysozyme in the FabD44.1–HEL complex includes residues Gln41, Arg45,

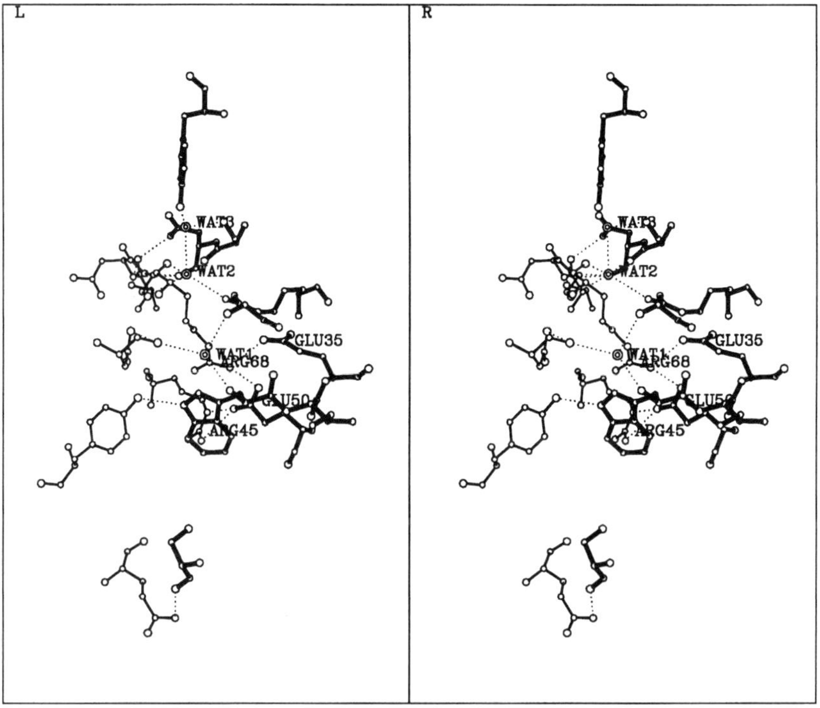

FIGURE 1. Hydrogen bonding scheme between FabD44.1 (*heavy trace*) and HEL (*lighter trace*). A stacking interaction between HEL residue Arg68 and the indole ring of V_H Trp33, at the bottom of the figure, can be seen. Water molecules, labeled WAT 1, 2, and 3, are shielded from the solvent continuum by the protein atoms of the interface.

Asn46, Thr47, Asp48, Gly49, Tyr53, Gly67, Arg68, Pro70-Ser81, and Leu84. This epitope is extremely similar to that recognized by HyHEL5: Gln41, Thr43, Asn44, Arg45, Asn46, Thr47, Asp48, Gly49, Tyr53, Gly67, Arg68, Thr69, Pro70, and Leu84.[2] The antibody residues that make contacts with HEL are Asn32 (V_L CDR1); Tyr50 (V_L CDR2); Ser91, Asn92, Trp94 (V_L CDR3); Tyr32, Trp33, Glu35 (V_H CDR1); Glu50, Ser55, Ser57, Tyr59 (V_H CDR2); and Gly99, Asp100, Gly101 (V_H CDR3). V_H-contacting residues Trp33, Glu35 (CDR1), Glu50, Ser55, Ser57 (CDR2), and 99Gly (CDR3) are common to D44.1 and HyHEL-5.

A review of the hydrogen-bonding scheme in the FabD44.1–HEL complex is shown in FIGURE 1. There are three water molecules totally buried in the interface, giving a clear example of complementarity achieved by the participation of solvent molecules. At the position labeled WAT3 (FIG. 1), a water molecule is also bound in the free FabD44.1 structure (to V_L Tyr50-OH and to V_L Gly94 O), thus making only a partial contribution to the stabilizing enthalpy in the complex; in compensation, the negative entropy contribution of this molecule will be smaller. WAT2 (FIG. 1) is found in two of the HEL crystal structures for which solvent has been located.[15,16] Consequently, the hydrogen bonds between WAT2 and HEL Gly49 O and Arg68 O may not make a full contribution to the stabilization of the Fab-

D44.1–HEL complex; in compensation the negative entropic contribution of WAT2 will be negligible. In contrast, the third buried water molecule, WAT1, is not present in the HEL nor in the free FabD44.1 crystal structures; all bonds to it should be important to the formation and stability of the complex, compensated in part by a negative entropic contribution. Thus, only seven of the 11 hydrogen bonds formed by water molecules contribute fully to the stabilization of the complex. The temperature factors for these buried water molecules are in the range 23 to 25 Å^2, slightly higher than those of protein atoms at the antibody–antigen interface but fully compatible with firmly bound water molecules.

The complex includes five direct antibody–antigen hydrogen bonds: HEL Gln41 Nε2-V_H Ser55 O, HEL Arg45 Nε-V_H Glu50 Oε1, HEL Tyr53 OH-V_H Trp33 Nε1, HEL Thr47 O-V_L Asn92 Nδ2, and HEL Gly49 N-V_L Asn92 O, as well as a stacking interaction between HEL Arg68 and V_H Trp33. The remaining polar interactions are three salt links involving HEL residues Arg45 and Arg68 and Fab residues V_L Asp 35 and V_H Asp 50. The surface areas buried by the antibody–antigen interaction, including the three buried water molecules, are 1,240 Å^2 (626 Å^2 for D44.1, 614 Å^2 for HEL) based on an atomic probe of 1.7-Å radius (these values decreased by about 5% if measured with a 1.4-Å probe). There are two buried "holes" in the interface, one approximately 3 Å in diameter situated in a hydrophobic environment defined by the side chains of V_H Trp33, V_H Leu52, HEL Tyr53, and HEL Leu84 and a second larger hole bounded by the main and side chains of V_H Thr31 (CDR1) and V_H Leu52 (CDR2) and HEL residues Tyr53, Asp66, and Ser81. Probing of these holes with a test water molecule reveals that whenever the water molecule is in position for a hydrogen bond to a proton donor or acceptor it is also too close to a hydrophobic atom, a feature that may explain why no water molecules are found in these holes.

Superposition of the free and the complexed FabD44.1 provides evidence of conformational changes upon binding of HEL, which are particularly visible in the V_H domain. Alterations in atomic positions of the FabD44.1 fall into two categories: (1) concerted movements of the peptide backbone and (2) changes in the orientations of side chains.

Examples of concerted movements are those of V_L CDR3 residues Ser93 to Arg96 away from HEL, apparently from contact with HEL Arg45. Likewise, V_H CDR3 is also displaced by contact with HEL residue Arg68. A radical difference between the free and complexed FabD44.1 occurs in the V_H CDR2 loop where the Pro53 backbone torsion angle ψ changes by 180°. The residues of V_H CDR2 exhibit the highest mobility (thermal values) of all Fv residues in the free Fab, and this mobility is greatly reduced in the Fab–HEL complex. Additional stability of V_H CDR2 in the Fab–HEL complex is afforded by the formation of an intraloop hydrogen bond between Leu52 O and Ser55 Oγ. Reduced mobility of V_L CDR1 and CDR3 in the Fab–HEL complex is also quite evident from a comparison of thermal values. The lower temperature factors for the V_L CDR2 loop of the free Fab is due to three hydrogen bonds from Fab atoms Gly57 N, Ser60 N, and Ser60 O to a neighboring free Fab molecule in the crystal.

Side chain rotations occur in V_L Asn92, which forms a hydrogen bond to HEL Thr47 O, and in the V_H Glu35 and Glu50, which form the salt links to HEL Arg45 and Arg68. By this rearrangement a hydrogen bond between V_H Glu50 Oε1 and V_L Trp94 Nε1 is broken and a close contact between the acid moieties of the free Fab V_H Glu35 and Glu50 (2.8 Å) is alleviated.

Comparison of the FabD44.1-bound HEL with "uncomplexed" HEL structures reveals little or no change in main chain conformation.

mAbs D44.1 and HyHEL-5 both exhibit reduced association with bobwhite

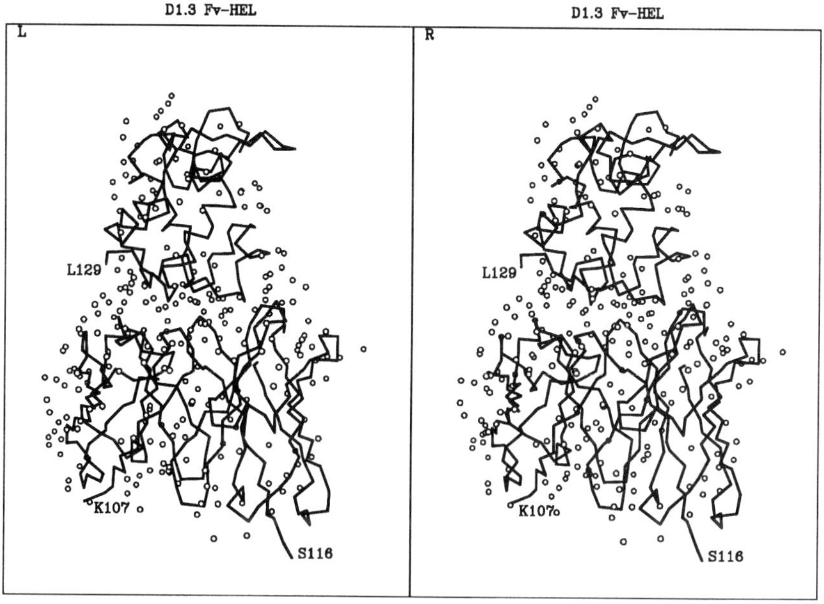

FIGURE 2. Stereo view of the FvD1.3–HEL complex. The α-C backbone of the structure is shown, with water molecules represented as circles. The COOH-termini of the V_L, V_H, and HEL polypeptide chains are indicated.

quail lysozyme where residue 68 is Lys rather than Arg as in HEL.[17,18] The reduced association with bobwhite quail lysozyme certainly indicates that HEL Arg68 plays a role in complex formation. The orientation of the salt links in the FabD44.1 and FabHyHel-5 complexes with HEL are different as a result of differences in the orientation of HEL Arg45. In the HyHEL-5-HEL complex, three salt links are formed between the HEL arginines 45 and 68 and V_H residue Glu50. The D44.1–HEL complex includes one bridge each from HEL Arg68 and Arg45 to V_H Glu50 and a third bridge between HEL Arg68 and V_H Glu35. Although the strength of the ionic interactions made by Arg45 and Arg68 in the HyHEL-5 and D44.1 complexes cannot be evaluated, they seem qualitatively similar, suggesting that they do not determine the high association constant observed for HyHEL-5 (4×10^{10} M^{-1}),[19] since that of D44.1 is only about 1.4×10^7 M^{-1}.[20,21] Thus, mAbs D44.1 and HyHEL-5 bind to a very similar epitope on HEL, although only the salt links and two hydrogen bonds (HEL Tyr53-V_H Trp33 and HEL Gln41-V_H Ser55) are common to the two complexes. This illustrates, at the molecular level, the degeneracy and diversity of antibody immune responses.

WATER MOLECULES IN THE FvD1.3–HEL COMPLEX

The 1.8-Å resolution structures of the free and the complexed FvD1.3 permit an analysis of water molecules bound by the antibody and by the antigen–antibody complex.[5] Many water molecules are observed at the antigen–antibody interface in the Fv–HEL complex (FIGS. 2 and 3). In general, these molecules make several

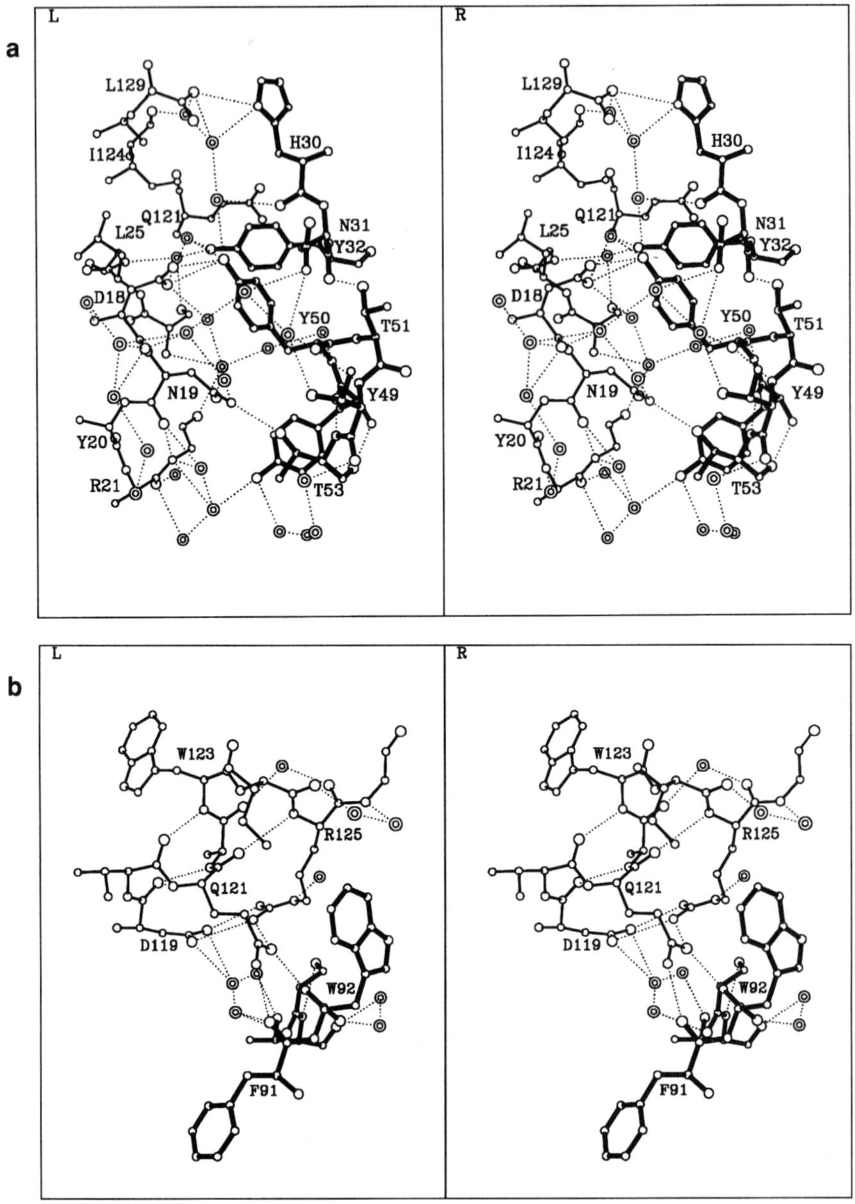

FIGURE 3. Stereo view of the V_L and V_H CDRs (*thick bonds*) and neighboring solvent and HEL atoms (*thin bonds*). Hydrogen bonds are indicated with broken lines. **(a)** V_L CDR1 and CDR2; **(b)** V_L CDR3; **(c)** V_H CDR1 and CDR2; **(d)** V_H CDR3.

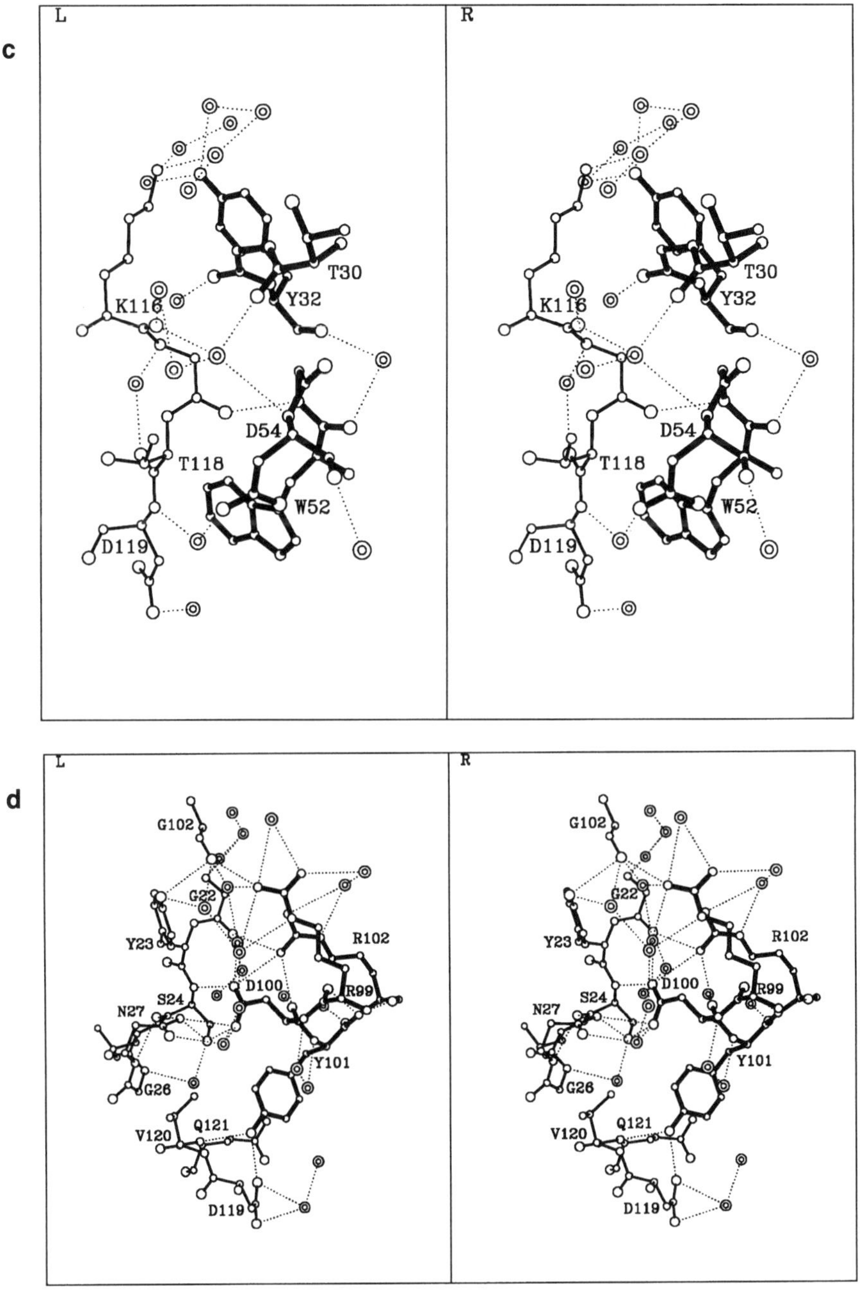

FIGURE 3 (*Continued*).

hydrogen bonds and have low temperature factors, sometimes comparable to those of adjacent protein atoms, and can therefore be considered an integral part of the complex.

More than 20 water molecules were located in the free Fv combining site, and about 50 in the complex, in the region of the antigen–antibody interface. Their positions were compared by aligning the V_H $C\alpha$ of the complexed and the free Fv. Several of the water molecules superpose or are very close to each other and may therefore be considered common to both the free and the complexed antibody-combining site. Furthermore, the comparison shows that antigen residues generally do not displace water molecules of the Fv, with the exception of a small displacement of two molecules by Asn19 and the COOH-terminal carboxyl of Leu129.

Several types of interactions involving hydrogen-bonded water molecules are seen in the FvD1.3–HEL complex. Some of these are mediated by only one, others require additional bridging water molecules (FIG. 3). Also, framework residues of the antibody interact with antigen by water bridges, and thus, they are also important in stabilizing antigen–antibody complexes.

THE MUTANT V_L TRP92ASP

Several mutants of the bacterially expressed FvD1.3 were obtained to explore the role of individual amino acids in the stability of the complex. Among these, oligonucleotide site-directed mutagenesis[22] was carried out to introduce the V_LTrp92 → Asp (W92D) mutation. The mutant Fv was prepared from the culture supernatants of transformed *Escherichia coli* cells and complexed to HEL. The complex gave crystals isomorphous with those obtained using wild-type FvD1.3, thus facilitating the structural analysis by X-ray diffraction, which was carried out to a nominal resolution of 1.8 Å.[23]

The major change observed in the structure of the W92DFv–HEL complex relative to that of the wild-type Fv is a rearrangement of solvent at the antigen–antibody interface so that two water molecules occupy space taken by the Trp side chain in the wild-type complex (FIG. 4). These two water molecules have temperature factor values (B, in Å^2) of 37 and 40, compared to the Asp92 atoms, which have B values ranging from 22 to 42. The B values for atoms of the wild-type Trp92 range from 14 to 17. In the vicinity of Asp92, V_L His30 retains position, although its ring atoms have rotated around the $C\beta$-$C\gamma$ bond (FIG. 4). There also is a movement of Leu129, which is not surprising because this COOH-terminal residue has a high average temperature factor in the mutant (B = 59 Å^2) and in the wild type (B = 54 Å^2) complexes. The side chain of HEL Ile124 also moves slightly, away from the solvent molecules (FIG. 4). All these small changes in conformation occur around the site of the mutation without affecting the rest of the antigen–antibody interface.

There is a difference between wild-type Fv and the W92DFv mutant in the surface excluded from solvent after complex formation. Using a radius of 1.4 Å for a probing water molecule, the solvent-excluded area or buried surface for the wild-type complex is 1386 Å^2 (675 Å^2 from the Fv, 711 Å^2 from HEL), whereas for the W92DFv complex with HEL it is 1238 Å^2 (604 Å^2 from the Fv, 634 Å^2 from HEL). The difference in the buried surface area between the wild-type complex and that with the mutated Fv is thus 148 Å^2. If a radius of 1.7 Å is used for the probing water molecule, the buried surface area is 1589 Å^2 (Fv 773 Å^2, HEL 816 Å^2) for the wild-type complex and 1444 Å^2 for the mutant (Fv 702 Å^2,

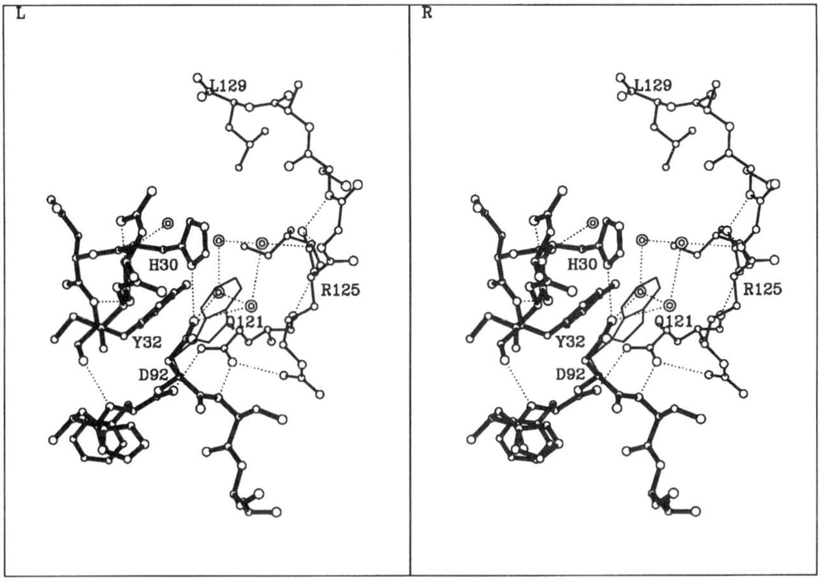

FIGURE 4. Stereo view of the HEL–FvD1.3 mutant complex in the area of the mutated residue V_L 92. The wild-type V_L Trp92 is superposed (*in thinner trace*) with Asp92. HEL residues are shown in the thinner trace, and its Gln121, Arg125, and Leu129, as well as V_L residues His30, Tyr32, and Asp92 are labeled. Hydrogen bonds are shown by *broken lines,* water molecules as *double circles*. Superposition of the wild-type and the mutant Fv structures gives an rms deviation of 0.19 Å for all backbone atoms and of 0.33 Å for the side-chain atoms.

HEL 742 Å^2), giving a difference between wild-type and mutant complexes of 145 Å^2. This figure is essentially the same as that calculated using a radius of 1.4 Å for the probing water molecule. The differences in the number and type of contacts made by the wild-type and the W92D mutant with antigen amount to a loss of one C-N and seven C-C atomic contacts.

The crystallographic study of the FvD1.3–HEL complex[5] has shown that water molecules occupy channels resulting from the association of antigen and antibody. In the structure of the W92DFv–HEL complex, a cavity has been introduced by the replacement of Trp92 by the smaller Asp. This cavity is partially filled by two water molecules, further supporting the proposal that solvent contributes to the stability of the complex.

THERMODYNAMIC ANALYSES OF LYSOZYME–ANTIBODY REACTIONS

The results in TABLE 1 show that the binding reactions displayed by mAbs D1.3, D11.15, D44.1, F9.13.7, and HyHEL-5 are under enthalpic control, with some opposition from a negative entropy contribution.[20,21,24] In the case of antibody

TABLE 1. Thermodynamic Parameters of the Binding of Antibodies to Lysozyme

mAb	Temp. °C	ΔH kJ mol^{-1}	K_a M^{-1}	ΔG kJ mol^{-1}	$T \Delta S$ kJ mol^{-1}
D1.3	23.4	-90.0 ± 0.5	$2.7 \pm 1.0 \times 10^8$	-47.9 ± 0.9	-42.1 ± 2.0
FvD1.3	24.0	-90.0 ± 1.8	$2.7 \pm 1.0 \times 10^8$	-47.9 ± 0.9	-42.1 ± 2.0
FvW92D	24.0	-74.2 ± 4.0	$2.6 \pm 1.0 \times 10^5$	-30.8 ± 0.5	-43.4 ± 3.3
D44.1	24.2	-43.3 ± 1.9	$1.4 \pm 0.1 \times 10^7$	-40.7 ± 0.4	$+2.6 \pm 1.9$
F9.13.7	24.2	-42.7 ± 0.4	$3.4 \pm 1.0 \times 10^8$	-48.6 ± 0.7	-5.9 ± 0.8
F10.6.6	24.2	-76.4 ± 3.3			
D11.15	24.0	-79.1 ± 0.2	$1.1 \pm 0.2 \times 10^8$	-45.7 ± 0.4	-33.4 ± 1.5
HyHEL-5[a]	25.0	-94.6 ± 1.8	4×10^{10}	-70.7	-23.9

NOTE: The values given for D44.1 and F10.6.6 are for their Fab fragments; those of ΔH and K_a were obtained using titration calorimetry.[20,21] See text for more details.

[a] Values taken from Hibbitts *et al.*,[24] except for K_a, which is from Benjamin *et al.*[19]

D44.1, the entropic factor is small and positive at 24°C. For antibody F10.6.6, the binding constant is too high for a measurement by titration calorimetry. For values ranging from 10^9 M^{-1} to 10^{12} M^{-1}, however, the binding entropy would be on the order of -25 to -8 kJ mol^{-1}, indicating that enthalpic factors also drive the reaction of F10.6.6 with antigen.

The results of titration calorimetry study of the reaction of the W92DFv mutant with HEL (TABLE 1), compared to that of the wild-type, indicate differences in the enthalpy of reaction and in the equilibrium binding constant K_a, but no significant difference in the entropy. Because the structures of the complexes of HEL with wild-type Fv and with the W92DFv mutant are nearly identical, the major entropic factors (loss of rotational, translational, and conformational freedom) in their reactions with antigen should be essentially the same. Contacts made by the apolar part of Trp92 do not seem to contribute to entropic factors, or alternatively, the loss of conformational entropy and the hydrophobic effects cancel each other. The decrease in the binding constant from 2.3×10^8 M^{-1} (wild-type) to 2.6×10^5 M^{-1} (W92D mutant) corresponds to a $\Delta\Delta G$ of 17.1 kJ mol^{-1} (4.1 kcal mol^{-1}). Neglecting the possible contributions to entropic and enthalpic factors by the solvent and the small amino acid side-chain rearrangements, one can calculate a mean "hydrophobic" energy of 117 J $mol^{-1} \cdot Å^{-2}$ (28 cal $\cdot$ mol^{-1} $Å^{-2}$) of buried surface. This agrees with the figure of 25 to 50 cal $\cdot$ mol^{-1} $Å^{-2}$ proposed for carbon–carbon interactions.[25–27] Because eight van der Waals antigen–antibody interactions are lost (7 C-C; 1 C-N) as a result of the mutation, the average van der Waals interaction from Trp92 contributes -2.1 kJ mol^{-1} (-0.5 kcal mol^{-1}) to the stability of the complex.

DISCUSSION

The results presented in TABLE 1, showing that enthalpic factors (mainly hydrogen bonds and van der Waals interactions) drive the reactions of antibodies with antigen, are not in agreement with the widespread notion that solvent entropy is the dominant factor in protein–protein association.

In reviewing the thermodynamic parameters characterizing protein association reactions, Ross and Subramanian[28] concluded that the hydrophobic effect alone cannot account for the observed changes in enthalpy and entropy, since in most cases the reactions are enthalpically driven. The reactions between the anti-lysozyme mAbs and HEL reviewed here fall in this category. These results can be correlated with the three-dimensional structures of the complexes of the FabD1.3, the FvD1.3, the FvD11.15, the FabF9.13.67, the FabD44.1, and the FabHyHEL-5 with lysozyme. An important conclusion from the structural studies of these and other antigen–antibody complexes is that their contacts are predominantly hydrogen bonds and van der Waals interactions, extending over a large area, about 700 to 900 $Å^2$. Thus, the negative binding enthalpies can be correlated with these chemical bonds, which are a dominant feature of the complexes.

A second observation of interest for the interpretation of the thermodynamic parameters is that of the conformational changes occurring upon antigen binding in the mAbs D1.3 and D44.1, the only antibodies in TABLE 1 for which it has been possible to compare the structure of the free with that of the antigen-bound site. These comparisons indicate a stabilization of the relative orientations of the V_H and V_L domains upon binding, which should translate into a negative entropy component opposing the enthalpic forces that drive the reactions and the positive contribution from solvent entropy. TABLE 1 indicates that in most reactions there is a negative entropy component that should arise, in part, from the rearrangement of the V_H and V_L domains. Rearrangements and stabilization of the CDRs, as seen in D44.1 and D1.3, also contribute to this negative entropy.

A third observation of interest in interpreting the results of TABLE 1 and immunological specificity in general is that of the bound water molecules at the antigen–antibody interface. The importance of bound water in enzyme catalysis, allosteric regulation, molecular assembly, and conformational transitions has been recently reviewed.[29] Several crystallographic studies have demonstrated the essential role of well-ordered water molecules in the structure and activity of proteins.[30–33] Bridging water molecules have also been localized in the contacting regions of several protease–protease inhibitor complexes[31–33] and of protein–DNA complexes[34] whose structures have been determined to high resolution. Clearly, the issue of water mediating a bridging role in protein–protein interactions is a general one.[35]

The arrangement of water molecules in and around the antigen–antibody combining sites may be compared with that at the larger V_H-V_L interface where, in the case of the mAb D1.3, only one buried water molecule is found. Presumably, the residues making the V_H-V_L contacts have been optimized over evolutionary time to reduce spatial voids. As antibodies are essentially derived from a preexisting genetic repertoire and have been subjected to different selection pressures to optimize their fit with antigens, some imperfections in complementarity might be expected and are in fact observed at their interfaces with antigens. It should be noted, however, that evolution does not necessarily act to completely eliminate buried waters from protein–protein interfaces. This is illustrated by the high-resolution structures of the complexes between proteinases and inhibitors,[31–33] in which a number of ordered solvent molecules mediate interactions between two protein surfaces that presumably have co-evolved to optimize complementarity. Furthermore, the presence of buried water molecules is not necessarily incompatible with high affinity, as the association constant of mAbs D1.3 (3×10^8 M^{-1}) and D44.1 (1.4×10^7 M^{-1}) for lysozyme are greater than that reported[36] for V_H-V_L dimers in general (about 10^6 M^{-1}).

REFERENCES

1. Amit, A. G., R. A. Mariuzza, S. E. V. Phillips & R. J. Poljak. 1986. Science **233:** 747–753.
2. Sheriff, S., E. W. Silverton, E. A. Padlan, G. H. Cohen, S. J. Smith-Gill, B. C. Finzel, & D. R. Davies. 1987. Proc. Natl. Acad. Sci. USA **84:** 8075–8079.
3. Padlan, E. A., E. W. Silverton, S. Sheriff, G. H. Cohen, S. J. Smith-Gill & D. R. Davies. 1989. Proc. Natl. Acad. Sci. USA **86:** 5938–5942.
4. Bhat, T. N., G. A. Bentley, T. O. Fischmann, G. Boulot & R. J. Poljak. 1990. Nature **347:** 483–485.
5. Bhat, T. N., G. A. Bentley, G. Boulot, M. I. Greene, D. Tello, W. Dall'Acqua, H. Souchon, F. P. Schwarz, R. A. Mariuzza & R. J. Poljak. 1994. Proc. Natl. Acad. Sci. USA **91:** 1089–1093.
6. Chitarra, V., P. M. Alzari, G. A. Bentley, T. N. Bhat, J.-L. Eiselé, A. Houdusse, J. Lescar, H. Souchon & R. J. Poljak. 1993. Proc. Natl. Acad. Sci. USA **90:** 7711–7715.
7. Braden, B. C., H. Souchon, J.-L. Eiselé, G. A. Bentley, T. N. Bhat, J. Navaza & R. J. Poljak. 1994. J. Mol. Biol. **243:** 767–781.
8. Lescar, J., M.-M. Riottot, H. Souchon, V. Chitarra, G. A. Bentley, J. Navaza, P. M. Alzari & R. J. Poljak. 1993. Proteins: Struct. Funct. Genet. **15:** 209–212.
9. Colman, P. M., W. G. Laver, J. N. Varghese, A. T. Baker, P. A. Tulloch, G. M. Air & R. G. Webster. 1987. Nature **326:** 358–363.
10. Tulip, W. R., J. N. Varghese, W. G. Laver, R. G. Webster & P. M. Colman. 1992. J. Mol. Biol. **227:** 122–148.
11. Bentley, G. A., G. Boulot, M.-M. Riottot & R. J. Poljak. 1990. Nature **348:** 254–257.
12. Prasad, L., S. Sharma, M. Vandonselaar, J. W. Quail, J. S. Lee, E. B. Waygood, K. S. Wilson, Z. Dauter & L. T. J. Delbaere. 1993. J. Biol. Chem. **268:** 10705–10708.
13. Ban, N., C. Escobar, R. Garcia, K. Hasel, J. Day, A. Greenwood & A. McPherson. 1994. Proc. Natl. Acad. Sci. USA **91:** 1604–1608.
14. Boulot, G., J.-L. Eiselé, G. A. Bentley, T. N. Bhat, E. S. Ward, G. Winter & R. J. Poljak. 1990. J. Mol. Biol. **213:** 617–619.
15. Madhusudan, R. Kodanapani & M. Vijayan. 1993. Acta Cryst. **D49:** 234–245.
16. Ramanadham, M., L. C. Sieker & L. H. C. Jensen. 1990. Acta Cryst. **B46:** 63–69.
17. Smith-Gill, S. J., A. C. Wilson, M. Potter, E. M. Prager, R. J. Feldmann & C. R. Mainhart. 1982. J. Immunol. **128:** 314–322.
18. Harper, M., F. Lema, G. Boulot & R. J. Poljak. 1987. Mol. Immunol. **2:** 97–108.
19. Benjamin, D. C., D. C. Williams, S. J. Smith-Gill & G. S. Rule. 1992. Biochemistry **31:** 9539–9545.
20. Tello, D., F. A. Goldbaum, R. A. Mariuzza, X. Ysern, F. P. Schwarz & R. J. Poljak. 1993. Biochem. Soc. Trans. **21:** 943–946.
21. Schwarz, F. P., D. Tello, F. A. Goldbaum, R. A. Mariuzza & R. J. Poljak. 1995. Eur. J. Biochem. In press.
22. Kunkel, T. A., J. D. Roberts & R. A. Zakour. 1987. Methods Enzymol. **154:** 367–382.
23. Ysern, X., B. A. Fields, T. N. Bhat, F. A. Goldbaum, W. Dall'Acqua, F. P. Schwarz, R. J. Poljak & R. A. Mariuzza. 1994. J. Mol. Biol. **238:** 496–500.
24. Hibbits, K. A., D. S. Gill & R. C. Willson. 1994. Biochemistry **33:** 3584–3590.
25. Chothia, C. 1974. Nature **248:** 338–339.
26. Eisenberg, D. & A. D. McLachlan. 1986. Nature **319:** 199–203.
27. Nicholls, A., K. A. Sharp & B. Honig. 1991. Proteins **11:** 281–296.
28. Ross, P. D. & S. Subramanian. 1981. Biochemistry **20:** 3096–3102.
29. Rand, R. P. 1992. Science **256:** 618.
30. Quiocho, F. A., D. K. Wilson & N. K. Vyas. 1989. Nature **340:** 404–407.
31. James, M. N. G., A. R. Sielecke, G. D. Brayer, L. T. J. Delbaere & C.-A. Bauer. 1980. J. Mol. Biol. **144:** 43–88.

32. Bolognesi, M., G. Gatti, E. Menegatti, M. Guarneri, M. Marquart, E. Papamokos & R. Huber. 1982. J. Mol. Biol. **162:** 839–868.
33. Fujinaga, M. & M. N. G. James. 1987. J. Mol. Biol. **195:** 373–396.
34. Joachimiak, A., T. E. Haran & P. B. Sigler. 1994. EMBO J. **13:** 367–372.
35. Kornblatt, J., M. J. Kornblatt, G. Hui Bon Hoa & A. G. Mauk. 1993. Biophys. J. **65:** 1059–1065.
36. Dorrington, K. J. 1978. Can. J. Biochem. **56:** 1087–1096.

Analysis of the Structural Correlates for Self-Antigen Binding by Natural and Disease-Related Autoantibodies

In Vitro Expression of Recombinant and/or Mutagenized Human IgG[a]

YUJI ICHIYOSHI[b] AND PAOLO CASALI[c]

Division of Molecular Immunology
Department of Pathology
Cornell University Medical College
New York, New York 10021

INTRODUCTION

Analysis of the immunoglobulin (Ig) V gene segment sequences of various human autoantibodies has delineated some of the structural features of the antigen (Ag)-binding sites and has allowed some of the mechanisms underlying autoantibody emergence to be inferred.[1–8] Those studies, however, are retrospective and do not provide direct evidence for a structure–function relationship. For instance, (i) natural autoantibodies have been thought to rely on the use of selected V_H segments, possibly in unmutated configuration, for the recognition of multiple Ags,[1–3] but direct evidence of consistent binding of different Ags by given unmutated V segments has not been provided; and (ii) some human autoimmune disease-related autoantibodies display point-mutations in the V_H and V_L segments that are consistent in nature and distribution with a clonal selection process driven by Ag,[4–8] but no evidence is available indicating that self-Ags are actually responsible for clonal selection of these autoantibodies.

Functional analysis of recombinant and/or mutagenized antibodies expressed *in vitro* enables us to measure directly the influence of structural changes on Ag binding, thereby pointing at the role played by different structures in that function. Using a system for the *in vitro* expression of human IgG1 κ, we have performed two series of recombinatorial experiments to analyze (i) the structural correlates for polyreactivity of a natural antibody, mAb55; and (ii) the structural changes underlying the affinity maturation of a specific anti-insulin autoantibody, mAb13.

[a] This work was supported by the U.S. Public Health Service Grant AR40908. This is publication #2 from the Division of Molecular Immunology.

[b] Present address: Department of General Medicine, Kyushu University Hospital, Fukuoka 812, Japan.

[c] Address correspondence and reprint requests to: Paolo Casali, M.D., Division of Molecular Immunology, Department of Pathology, Cornell University Medical College, 1300 York Avenue, New York, NY 10021-4896.

THE *IN VITRO* HUMAN IgG1 κ EXPRESSION SYSTEM

We used the pcDNAIG and pSXRDIG plasmid vectors[9–11] for the *in vitro* expression of human Ig V_H-D-J_H and V_κ-J_κ gene segments, respectively. pcDNAIG encodes a whole human IgG1 H chain. The mouse H chain leader sequence, the V_H-D-J_H gene segment, and the Cγ1 gene are accommodated in a 2.3-kb segment between *Hin*d III and *Xba* I sites and are driven by a human CMV promoter (FIG. 1A). The vector contains a neomycin resistance gene for selection. pSXRDIG encodes a whole human Ig κL chain. The mouse κ chain leader sequence, the Ig V_κ-J_κ gene segment, and the human C_κ gene are accommodated in a 1.3-kb segment between *Hin*d III and *Xba* I sites and are driven by a SV40 promoter (FIG. 1B). The vector contains the DHFR gene for selection by methotrexate. In both pcDNAIG and pSXRDIG, the unique *Hin*d III and *Xho* I sites allow for the substitution of the vector V(D)J gene segment (FIG. 2).

The "recombinant PCR" technique we used to recombine different human Ig V gene regions and segments, is illustrated in FIGURE 2. The mouse leader sequence of pcDNAIG (or pSXRDIG) and the V(D)J (V_HDJ_H or $V_\kappa J_\kappa$) gene segment of a

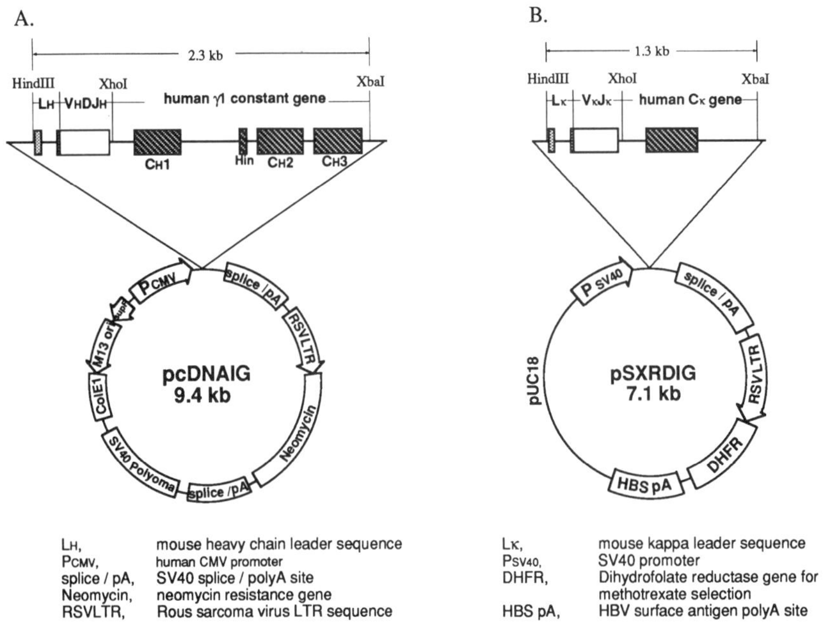

FIGURE 1. Structure of the pcDNAIG **(A)** and pSXRDIG **(B)** expression vectors. The segment between *Hin*d III and *Xba* I sites, comprising the Ig H or κ chain gene is detailed: Boxes indicate exons (*shaded boxes,* mouse leader exons; *open boxes,* human Ig variable chain gene exons; and *hatched boxes,* human Ig constant chain gene exons), and *solid lines* indicate introns and noncoding sequences. Abbreviations at the bottom of the figure are used for gene segments related to mammalian expression. Other abbreviations are as follows: supF, supF suppressor tRNA gene; M13 ori, M13 origin bases; ColE1, ColE1 origin bases; SV40 Polyoma ori, SV40 origin bases and polyoma origin bases.

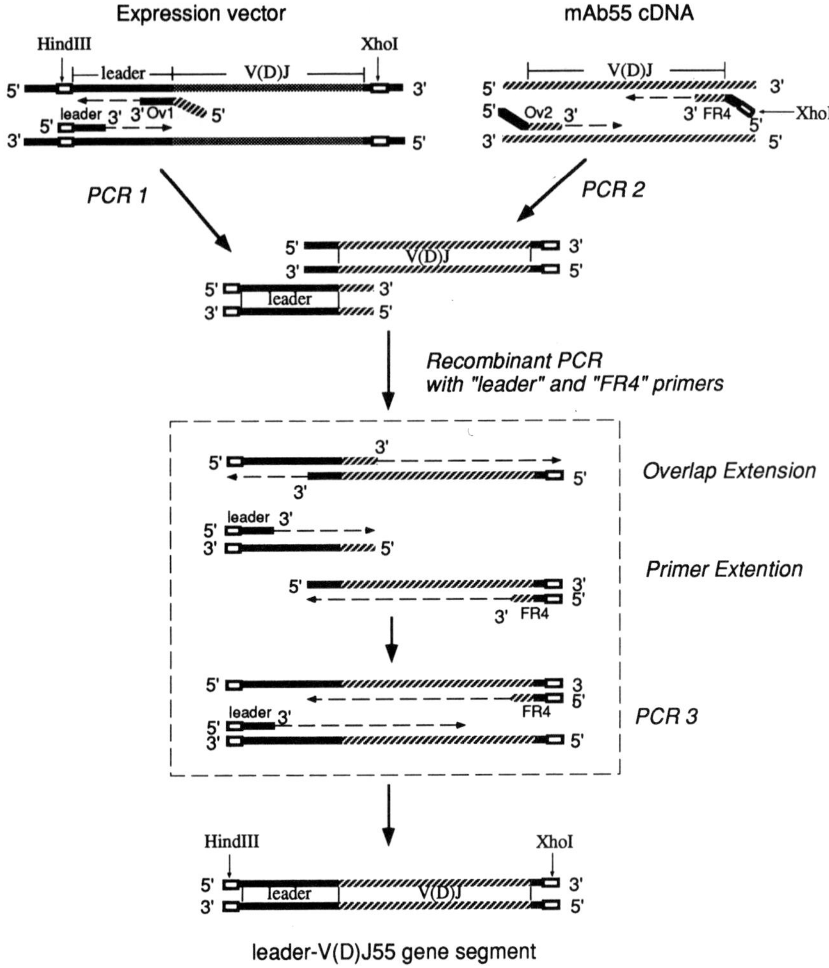

FIGURE 2. Construction of the recombinant leader V(D)J55 gene segment (leader-V_HDJ_H55 or leader-$V_\kappa J_\kappa$55). Solid (▬) and hatched (////) lines depict the sequences of the expression vectors (pcDNAIG or pSXRDIG), and that of the V(D)J gene segment of mAb55, respectively. *Open boxes* depict restriction sites. *Broken arrows* indicate nucleotide chain elongation by DNA polymerase. The mouse leader IgV gene segment in the expression vector and the V_H-D-J_H or V_κ-J_κ gene segment of mAb13 were amplified in separate PCR (PCR 1 and PCR 2), and joined by recombinant PCR.

given mAb, for example, mAb55, are amplified in separate PCRs (PCR 1 and PCR 2) and joined by overlap extension. The primers used at the ends to be joined (Ov1 and Ov2) are made complementary to one another by including in one primer a sequence complementary to the 3′ portion of the other primer. This allows for the products of the PCR 1 and PCR 2 to overlap at the end to be joined. These products are mixed with excess "leader" and "FR4" (framework region) primers, and subject to repeated rounds of denaturation, annealing, and primer extension by Taq DNA polymerase (PCR 3). Reactions in the mixture are shown in the shaded box. One strand from each fragment contains the overlap sequence at the 3′ end, and these strands can serve as primers for one another. Extension of this overlap by DNA polymerase yields the recombinant product (overlap extension). The other strands, which have the overlap at their 5′ ends, are not capable of priming each other, but they can act as templates for the "leader" and "FR4" primers and generate more of the original PCR products (primer extension). The recombinant product is amplified with "leader" and "FR4" primers by PCR as soon as it is formed. The recombinant fragment "leader-V(D)J55" is digested with *Hin*d III and *Xho* I, and then ligated into expression vector.

Recombinant Ig V gene segment pcDNAIG (4 μg) and pSXRDIG (4 μg) vectors are transfected in F3B6 cells by electroporation. F3B6 is the Ig nonsecretor human-mouse heterohybridoma we used for the generation of somatic cell hybrids.[12–14] After electroporation and a 48-hr culture in RPMI1640 complete medium containing 10% FCS and 1% L-glutamine, cells were distributed into 96-well flat-bottom plates in selective medium containing 0.4 mg/ml of G418. Neomycin only was used as selecting agent because in transfectants expressing only pcDNAIG (H chain), the accumulation of unsecretable H chain molecules leads to cell death. Clumps of transformant cells are detected within 7 to 10 days. After two weeks, culture fluids are tested by specific ELISAs using plates coated with goat $F(ab')_2$ fragment to human Igγ H chain. Double γ1 and κ chain producer cells are identified by developing separate ELISA plates with peroxidase-conjugated affinity-purified goat-anti-human Ig γ and κ chain.[10,11] In each transfection, up to 12 clones secreting IgG κ are generated. The three most efficient secretors are expanded and frozen. Antibodies are purified by affinity chromatography on a protein G-Sepharose column.[10,11]

We used Ig V gene recombination and expression techniques similar to those reported above to generate chimeric and/or site-directed mutagenized V gene segments described below.

ANALYSIS OF THE STRUCTURAL CORRELATES FOR MULTIPLE SELF-ANTIGEN BINDING IN A HUMAN NATURAL POLYREACTIVE IgM AUTOANTIBODY

Polyreactive natural antibodies constitute a major proportion of the early antibody repertoire and are an important component of the natural defense mechanisms against infections.[1] They are primarily IgM, although polyreactive IgG and IgA also exist and bind a variety of structurally dissimilar self and exogenous, mainly microbial, Ags with moderate affinity.[1] The characteristic spectrum of Ags bound by polyreactive antibodies presumably reflects fundamental differences in the structure of their Ag-binding sites, as compared to that of Ag-induced (monoreactive) specific antibodies.[1–3,12,15]

To define the structural correlates for Ig polyreactivity, we generated a variety of chimeric Ig molecules by replacing different portions of the polyreactive mAb55 H and κ chain V segments with the corresponding portions of monoreactive mAb13.[10] These mAbs were chosen because they use the same V_H and V_L gene segments, and, accordingly, display highly conserved FR structures (TABLE 1). mAb55 is a "prototypic" natural human antibody; that is, it is a polyreactive and moderate affinity IgM κ and was derived from a healthy subject.[13,16] The anti-insulin mAb13 is a "prototypic" Ag-induced "mature" human autoantibody; that is, it is a monoreactive and highly specific IgG κ and was derived from the insulin-specific autoantibody response in a patient with insulin-dependent diabetes mellitus (IDDM).[8,14] Deduced amino acid sequences of mAb55 and mAb13 V_H-D-J_H genes were 94.9% identical throughout the V_H area, differing in four amino acids in the complementarity-determining region 1 (CDR1) and CDR2 (residues 31, 51, 56, and 58) and one amino acid in the FR3 (residue 94), but were highly divergent in length and composition of the somatically generated CDR3 (FIG. 3A). The two mAb H chain FR4 sequences differed only in two amino acids. The two mAb V_κIII segment sequences were 87.6% identical, differing in four amino acids (including one deletion) in the CDR1 and CDR2 and seven in the FRs (FIG. 3B). The κ chain CDR3s were identical in length, but differed in five of the nine amino acids. The κ chain FR4 sequences were identical. Thus, mAb55 and mAb13 were highly similar or identical in their V_H, H chain FR4, V_κ and κ chain FR4 segments, but differed substantially in the CDR3 of their H chains and, to a lesser extent, κ chains. The mAb55 and mAb13 dose-saturable binding to human insulin, human IL-1β, ssDNA, human IgG Fc fragment, tetanus toxoid, and BSA are depicted in FIGURE 4A and C, respectively. The relative avidity (Av_{rel}) for human recombinant insulin, used as a measure of the mAb ability to bind Ag in the fluid phase, were 5.0×10^{-7} and 4.9×10^{-9} g/μl for mAb55 and mAb13, respectively.

The decavalency of the Ag-binding sites has been claimed to be an important factor in the polyreactivity of natural IgM antibodies.[1] To verify whether the pentameric structure of mAb55 is, in fact, a major factor in the ability of this antibody to bind multiple Ags, we expressed *in vitro* the mAb55 V_H-D-J_H and V_κ-J_κ genes as V segments of a human IgG1 κ molecule (recIgG55). The bivalent recombinant IgG antibody bound multiple Ags with an efficiency comparable to that of the original decavalent IgM mAb55 (FIG. 4A and B). The functional similarity of the recIgG55 to its IgM counterpart was further emphasized by the comparable Av_{rel} for insulin of the two molecules (8.6×10^{-7} and 5.0×10^{-7} g/μl, respectively). These experiments show that class switch from IgM to IgG does not significantly affect the ability of mAb55 to bind multiple Ags.

Having established that polyreactivity is a function inherent to the V portion of the Ig molecule, we evaluated the relative contribution of the whole V_H-D-J_H and V_κ-J_κ segments to the polyreactivity of recIgG55 by reassorting the mAb55-derived H or κ chain with the mAb13-derived κ or H chain, respectively. The hybrid IgG antibody bearing the mAb55-derived H chain V segment paired with the mAb13-derived κ V segment (recV_H55-55/V_κ13-13), but not that bearing the mAb13-derived H chain V segment paired with the mAb55-derived κ V segment (recV_H13-13/V_κ55-55), bound multiple Ags (FIG. 4E and F). Thus, the mAb55 V_H-D-J_H segment is necessary and sufficient for multiple Ag binding.

To determine whether mAb55 polyreactivity rests in the V_H segment itself (FR1 through FR3 sequence) or the somatically generated H chain CDR3, we shuffled the FR1-FR3 and CDR3 regions of the H and κ chain V segments of the mAb55-derived IgG molecule with the corresponding regions of the monoreactive IgG mAb13. The chimeric molecules satisfied the criteria set by Chothia and Lesk,[17] that is, identity in those canonical V_H and V_κ segment FR sequences that are adjacent to the CDRs, as defined by Kabat and Wu[18] and critical for the full preservation of the CDR loop structure. Thus, our choice of antibodies and experimental design for the shuffling of the FR and CDR sequences should have prevented the structural deleterious effects observed when CDR sequences are inserted into FR contexts different from those of the original antibody molecule.[19,20] The chimeric mAb55-derived IgG molecule lost polyreactivity when the H chain

TABLE 1. Properties and Structure of Human mAb55 and mAb13

Clone	Source	Chains H	Chains L	Reactivity	Av_{rel} for insulin (g/μl)	V_H segment: V_H gene[a]	V_H segment: D gene[b]	V_H segment: J_H gene[c]	V_κ segment: V_κ gene[d]	V_κ segment: J_κ gene[e]
mAb55	Healthy subject	μ	κ	Polyreactive	5.0×10^{-7}	H11	DXP'1	J_H6	kv325 (V_κIII)	$J_\kappa1$
mAb13	IDDM patient	$\gamma1$	κ	Monoreactive (anti-insulin)	4.9×10^{-9}	H11	DXP'1	J_H2	kv3g (V_κIII)	$J_\kappa1$

[a] The sequence of the germline H11 gene has been reported by Rechavi *et al.*[30]
[b] D genes have been reported by Ichihara *et al.*[31]
[c] J_H genes have been reported by Ravetch *et al.*[32]
[d] The sequence of germline kv325 gene has been reported by Kipps *et al.*[33]; the sequence of germline kv3g gene has been reported by Pech *et al.*[34]
[e] J_κ genes have been reported by Heiter *et al.*[35]

TABLE 2. Binding of Insulin by Different Recombinant IgG1s Consisting of the "Wild-Type" V_κ Chain and Variably Mutated V_H Chains, as Assessed by Competitive Inhibition and Kinetic Analyses

Recombinant IgG1	Residues in the V_H Segment: CDR1 31	CDR2 52	CDR2 56	CDR2 58	Competitive Inhibition Av_{rel} (g/μl)	Kinetic Analysis: k_{on} (10^4 $M^{-1} \cdot sec^{-1}$)	k_{off} (10^{-3} sec^{-1})	K_D(nM)
$V_H13G12\text{-}DJ_H13/V_\kappa J_\kappa 13$	S	N	S	S	1.44×10^{-8}	2.60	4.92	189
$V_HDJ_H13/V_\kappa J_\kappa 13$	G	—	T	R	4.91×10^{-9}	2.73	1.35	49
Gly31/$V_\kappa J_\kappa 13$	G	—	—	—	1.62×10^{-8}	2.98	4.35	148
Thr56, Arg58/$V_\kappa J_\kappa 13$	—	—	T	R	4.62×10^{-9}	2.91	1.12	38
Thr56/$V_\kappa J_\kappa 13$	—	—	T	—	1.87×10^{-8}	2.56	4.22	164
Arg58/$V_\kappa J_\kappa 13$	—	—	—	R	5.14×10^{-9}	2.61	1.54	59
Lys52/$V_\kappa J_\kappa 13$	—	K	—	—	No binding	ND	ND	ND

```
A
                     FR1               CDR1     FR2            CDR2                      FR3                               CDR3                         FR4
H11          EVQLVESGGGLVQPGGSLRLSCAASGFTFSSYWMHWVRQAPGKGLVWVSRINSDGSSTTYADSVKGRFTISRDNAKNTLYLQMNSLRAEDTAVYYCAR
VHDJH55      ------------------------------R--------------------K-----S---------------------------------------V LKVGFGEPEGWVLWGQAKIPREDYYYYGMDV WGQGTTVTVSS
VHDJH13      ------------------------------G------------------------T-R---------------------------------------- DDRGITMVRGIILFGFLDL             --R--L-----

B
                     FR1              CDR1          FR2          CDR2                  FR3                  CDR3       FR4
kv325        EIVLTQSPGTLSLSPGERATLSCRASQSVSSSYLAWYQQKPGQAPRLLIYGASSRATGIPDRFSGSGSGTDFTLTISRLEPEDFAVYYC QQYGSSP
VκJκ55       ----------------------------------------------------------------------------------------- -------ET FGQGTKVEIK
kv3g         --------A----------------------.-----------------D--N------A----------------S------------ --RSNW-
VκJκ13       --------AI---C------------N---.-----------------SD--N------A---------------NN------------ --RSNW-Q- ----------

C
                   FR1              CDR1     FR2            CDR2                      FR3                        CDR3              FR4
VH13G12-DJH13 EVQLVESGGGLVQPGGSLRLSCAASGFTFSSYWMHWVRQAPGKGLVWVSRINSDGSSTSYADSVKGRFTISRDNAKNTLYLQMNSLRAEDTAVYYCAR  DDRGITMVRGIILFGFLDLWGRGTLVTVSS
VHDJH13       ------------------------------G------------------------T-R----------------------------------------  ------------------------------
Gly31         ------------------------------G--------------------------------------------------------------------  ------------------------------
Thr56Arg58    -------------------------------------------------------T-R----------------------------------------  ------------------------------
Thr56         -------------------------------------------------------T------------------------------------------  ------------------------------
Arg58         ---------------------------------------------------------R----------------------------------------  ------------------------------
Lys52         ---------------------------------------------------K----------------------------------------------  ------------------------------

D
                    FR1             CDR1          FR2          CDR2                  FR3                    CDR3          FR4
Vκ13K20-Jκ13  EIVLTQSPATLSLSPGERATLSCRASQSVSSYLAWYQQKPGQAPRLLIYDASNRATGIPARFSGSGSGTDFTLTISSLEPEDFAVYYCQQRSNWP  QTFGQGTKVEIK
VκJκ13        ---------I---C------------N--------------------S--------------------------NN------------------  ------------
```

FIGURE 3. Deduced amino acid sequences of the V_H-D-J_H **(A)** and V_κ-J_κ **(B)** gene segments of mAb55 and mAb13, and those of the recombinant V_H-D-J_H **(C)** and V_κ-J_κ **(D)** gene segments. *Dashes* depict identity to each top sequence. *Solid lines* on the top of each cluster depict CDRs.

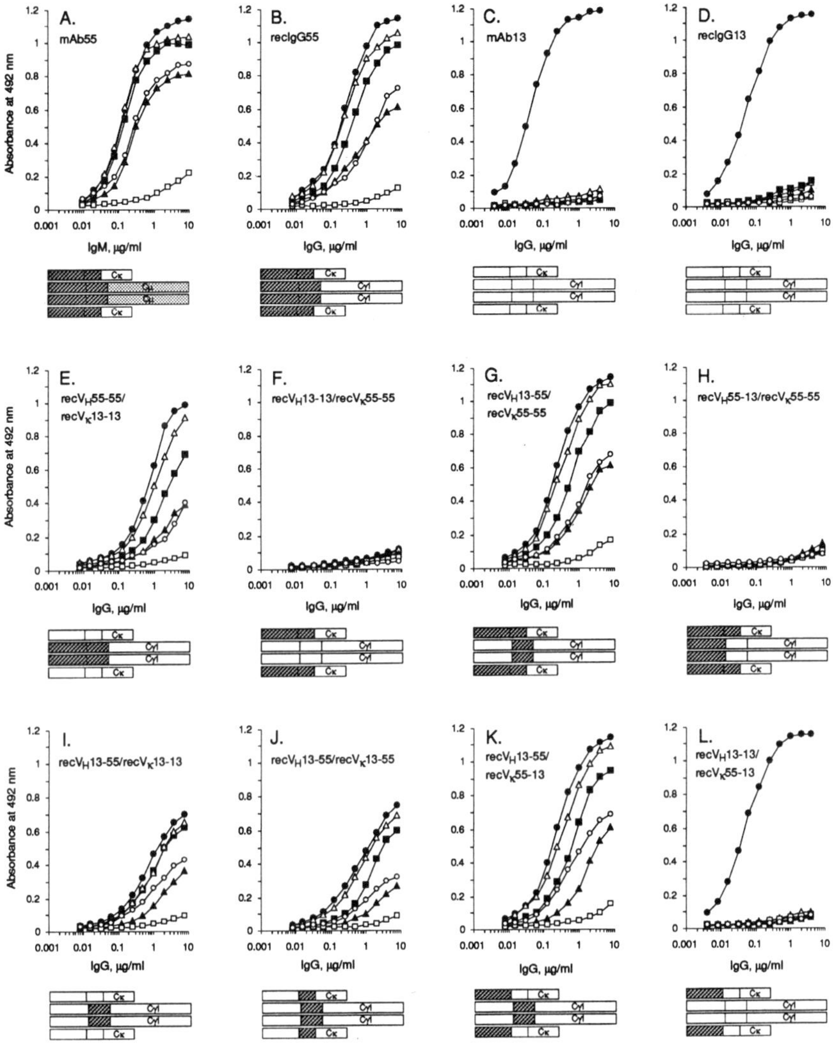

FIGURE 4. Binding of the original IgM κ mAb55 **(A)**, recIgG55 **(B)**, original IgG κ mAb13 **(C)**, recIgG13 **(D)**, recV$_H$55-55/recV$_\kappa$13-13 **(E)**, recV$_H$13-13/recV$_\kappa$55-55 **(F)**, recV$_H$13-55/recV$_\kappa$55-55 **(G)**, rec V$_H$55-13/recV$_\kappa$55-55 **(H)**, recV$_H$13-55/recV$_\kappa$13-13 **(I)**, recV$_H$13-55/recV$_\kappa$13-55 **(J)**, recV$_H$13-55/recV$_\kappa$55-13 **(K)**, and recV$_H$13-13/recV$_\kappa$55-13 **(L)** antibodies to human recombinant insulin (●), ssDNA (△), tetanus toxoid (■), human recombinant IL-1β (△), human IgG Fc fragment (○), and BSA (□). Bars below each panel schematize the composition of each original or recombinant antibody. Human Cγ and Cκ segments are depicted as *open boxes,* and human Cμ segments as *shaded boxes*. The V segments derived from the polyreactive mAb55 are depicted as *hatched boxes,* and those derived from the monoreactive anti-insulin IgG mAb13 are depicted as *open boxes*. The Ag-binding activity of each antibody is expressed as optical absorbance at 492 mm.

CDR3, but not the FR1-FR3 sequence, was replaced by the corresponding region of mAb13 (FIG. 4, H and G), suggesting that the H chain CDR3 provides the major structural correlate for multiple Ag binding. This was formally proved by the multiple Ag binding of the originally monoreactive mAb13-derived IgG molecule grafted with the mAb55-derived H chain CDR3 (FIG. 4I). The polyreactivity of this chimeric IgG was further enhanced by grafting the mAb55-derived κ chain FR1-FR3, but not the κ chain CDR3 sequence (FIG. 4J and K). The mAb55-derived κ chain FR1-FR3, however, failed to yield a polyreactive antibody when grafted onto an IgG that was in all other parts identical with mAb13. Rather, this chimeric antibody, recV_H13-13/V_κ55-13, bound specifically to insulin but to none of the other Ags (FIG. 4L). Its exquisite specificity for insulin (Av_{rel}, 6.0×10^{-9} g/μl) was comparable to that measured for mAb13 and rec IgG13 (FIG. 4, C and D).

In these studies, we investigated the contribution of Ig decavalency and that of the primary structure of V segments to the multiple Ag binding of a polyreactive human IgM antibody, mAb55. The recombinant Ig V_H and V_L gene expression system we adopted was designed to produce a complete human IgG1 κ molecule. Accordingly, the IgM mAb55-derived V_H-D-J_H and V_κ-J_κ gene segments were expressed and secreted as a recombinant bivalent monomer (designated recIgG55), effectively "switching" the polyreactive IgM to IgG. On a weight-to-weight basis, the multiple Ag binding of the recIgG55 molecule closely mimicked that of the "wild-type" IgM mAb55, suggesting that IgM to IgG class switch does not necessarily affect the characteristic feature of a polyreactive antibody and that recognition of multiple Ags is a function that is absolutely restricted to the Ig V domains. Thus, the role of natural polyreactive IgM may extend far beyond that of providing the structural elements of the idiotypic regulatory network in early ontogeny and that of enhancing opsonization and/or complement activation after binding invading microorganisms in the early phases of infection. Natural polyreactive IgM-producing cells likely provide the precursors of cells producing high-affinity polyreactive IgG or IgA autoantibodies or antibodies to exogenous microbial Ags, after undergoing an Ag-dependent process of clonal activation and expansion entailing class switch and, possibly, somatic mutation and Ag selection.[5,21,22] The experiments in which the presence or absence of the mAb55-derived H chain CDR3 sequence resulted in acquisition and abolition, respectively, of polyreactivity in the mAb55 or mAb13-like structural context indicated that the CDR3 provides the critical structural correlate for the prominent multiple Ag-binding activity of mAb55. Although limited to a single antibody pair, the present findings suggest that polyreactivity can be preserved after Ig class switch, and it depends primarily on the structures that are generated through somatic rearrangement events (H chain CDR3) and structures that represent the expression of an unmutated gene (kv325 gene).

ANALYSIS OF THE STRUCTURAL CORRELATES FOR SELF-ANTIGEN BINDING AND AFFINITY MATURATION IN A HUMAN MONOREACTIVE ANTI-INSULIN IgG AUTOANTIBODY

We have shown that, like other human autoimmune disease-related autoantibodies, anti-insulin IgG autoantibodies in IDDM patients display point mutations in the V_H and V_L segments that are consistent in nature and distribution with selection by Ag.[8] As mentioned earlier in this paper, however, the structural features, specificity, and high affinity for insulin of these autoantibodies do not necessarily point at insulin itself as the Ag responsible for the clonal selection process. In fact, the ability of self-Ags to induce affinity maturation in autoantibodies has been questioned, and it has been suggested that autoantibodies might be selected by unrelated, perhaps foreign, Ags, such as those on microbial agents.[23] Stronger circumstantial evidence that an autoantibody-producing cell precursor might have been selected by a self-Ag would entail the demonstration that the germline template Ig progenitor of the somatically mutated autoantibody is capable of specifically binding the relevant self-Ag, and that some of the point mutations observed in the "affinity mature" autoantibody are crucial in enhancing this binding.

To investigate the possibility that insulin was the Ag originally responsible for the selection of the mAb13-producing cell clone, we cloned the autologous germline genes that putatively gave rise to the expressed V_H and V_κ segments and used them to generate the (unmutated) "germline revertant" of the (somatically mutated) "wild-type" mAb13.[11] The deduced amino acid sequences of the full "germline revertant" V_H and V_κ segments differed in three and six amino acids, respectively, from those of the "wild-type" counterparts (FIG. 3C and D). The full "germline revertant" bound insulin specifically and in a dose-saturable fashion (not shown), but with a relative avidity (Av_{rel}) more than threefold lower than that of its "wild-type" counterpart (Av_{rel}, 1.6×10^{-8} vs. 4.91×10^{-9} g/μl; inhibition curves are shown in FIG. 5). Thus, the unmutated Ig gene product expressed by a B cell that was the putative progenitor of the somatically mutated mAb13-producing cell could specifically bind insulin.

To analyze the contribution of the mutated V_H and/or V_κ segments to the increased mAb13 Av_{rel} for insulin, we reassorted the "wild-type" and "germline revertant" forms of the mAb13 V_H and V_κ segments. The competitive inhibition curves derived from the experiments involving the V_H13G12-DJ_H13/$V_\kappa J_\kappa$13 IgG1 (unmutated V_H segment paired with the mutated V_κ segment) was similar to that derived from the analysis of the V_H13G12-DJ_H13/V_κ13K20-J_κ13 IgG1 (full germline-revertant Ig), whereas the curves derived from the analysis of the V_H-DJ_H13/V_κ13K20-J_κ13 IgG1 (mutated V_H segment paired with the unmutated V_κ segment) was similar to that derived from the analysis of the $V_H DJ_H$13/$V_\kappa J_\kappa$13 IgG1 (wild-type Ig) (FIG. 5). The Av_{rel} for insulin of the V_H13G12-DJ_H13/$V_\kappa J_\kappa$13 IgG1 and $V_H DJ_H$13/V_κ13K20-J_κ13 IgG1 were 1.44×10^{-8} (g/μl) and 5.7×10^{-9} (g/μl), respectively (TABLE 2). These experiments suggest that the somatic mutations in the V_H segment, not those in the V_κ segment, dominantly contribute to the increased affinity of mAb13 for insulin.

To determine which of the three mutations in the mAb13 V_H segment (Ser to Gly, Ser to Thr, and Ser to Arg at position 31, 56, and 58, respectively) was or were critical in increasing the mAb13 binding affinity for insulin, we constructed, by site-directed mutagenesis, partial "germline revertants" of the mAb13 V_H segment. Recombinant molecules consisting of a "germline revertant" V_H segment bearing Gly31 or Thr56 and individually paired with a "wild-type" V_κ segment did not differ from their Ig counterpart consisting of the full "germline revertant" V_H segment in Av_{rel} for insulin or any other of the kinetic parameters tested, including association and dissociation rate constants (k_{on} and k_{off}, respectively) (TABLE 2). In contrast, replacement of Ser58 by Arg in the unmutated V_H segment resulted in an at least threefold increase in avidity (Av_{rel}, 5.14×10^{-9} vs. 1.44×10^{-8} g/μl) and affinity (K_D, 59 vs. 189 nM), which directly reflected an about threefold decrease in k_{off} (1.54×10^{-3} vs. 4.92×10^{-3} sec^{-1}) not k_{on} (2.61×10^4 vs. 2.60×10^4 M^{-1} sec^{-1}). The affinity enhancement mediated by the V_H segment Arg58 residue contrasted with the complete loss of insulin binding resulting from

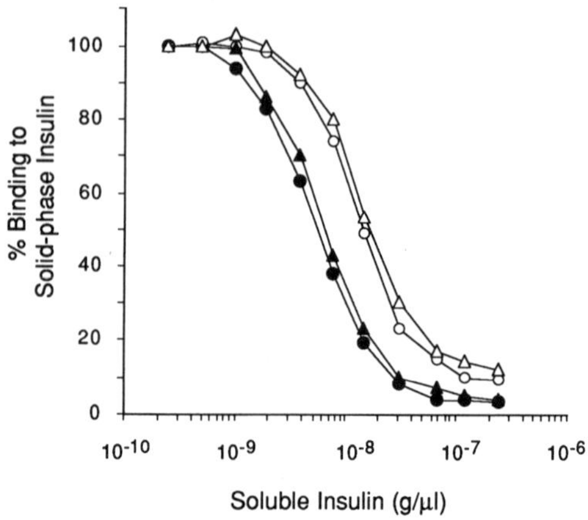

FIGURE 5. The inhibition of the binding of recombinant V_HDJ_H13/V_κ-$J_\kappa13$ (●), V_H13G12-DJ_H13/V_κ-$J_\kappa13$ (○), $V_HDJ_H13/V_\kappa13K20$-$J\kappa13$ (△), and V_H13G12-$DJ_H13/V_\kappa13K20$-$J_\kappa13$ (△) to solid-phase insulin by increasing the amount of soluble insulin.

the substitution of Asn52 by Lys that was accidentally obtained by PCR misincorporation.

In the present studies, we generated the full "germline V_H and V_κ revertant" of "somatically mutated" anti-insulin IgG mAb13 using the putative autologous germline gene sequences. Our experimental system thus enabled us to generate the unmutated Ig gene product of a putative B-cell progenitor of the mAb13-producing cell clone, although we could not identify somatic mutations that might have occurred in the junctional V_H-D-J_H sequence. This full "germline revertant" recombinant antibody, "V_H13G12-$DJ_H13/V_\kappa13K20$-$J_\kappa13$," bound specifically to insulin, suggesting that naturally occurring monoreactive autoantibodies encoded by unmutated V genes to insulin can exist in the normal B-cell repertoire, and possibly provide the templates for the specific high-affinity autoantibodies elicited by homologous insulin. Because somatic hypermutation occurs predominantly in the Ag-dependent phase of B-cell differentiation,[24] unmutated Ig receptors need to bind efficiently to the selecting Ag to be recruited into affinity maturation.[25] Thus, the ability of the full germline revertant of mAb13 to bind insulin specifically, as demonstrated here, constitutes strong circumstantial evidence that insulin was the Ag responsible for the selection of the specific anti-insulin mAb13-producing cell progenitor clone. Once B cells enter the Ag-dependent phase of differentiation, they are considered to undergo strong selective pressure, entailing the loss of a high proportion of mutants with low or no affinity for Ag.[26–29] The present demonstration that only one replacement in the V_H CDR2 (Asn to Lys at position 52) totally abolished the ability of mAb13 to bind insulin emphasizes the antibody susceptibility to binding loss resulting from somatic mutation. This fact and the demonstration that the mAb13 insulin affinity was enhanced by another V_H segment single amino acid replacement (Arg58) further suggests that the positive selection of the somatically mutated Ab13-producing cell clone occurred in concomitance with the purging of a number of binding loss mutants.

We used the *in vitro* human IgG1 κ expression system and addressed two major issues related to the structural correlates for self-Ag binding by autoantibodies: Recombinant IgGs with various chimeric V segments derived from monoreactive mAb13 and polyreactive mAb55 were instrumental in identifying the structures responsible for polyreactivity. The partial germline revertants of the anti-insulin mAb13 were instrumental in simulating the structural changes brought about by self-Ag-mediated pressure on the putative germline progenitor clone. The present findings provide direct evidence for the crucial role played by unique sequences, that is, H chain CDR3, in human antibody polyreactivity and unique amino acid residue changes in the recruitment of autoantibody-producing cell precursor by a given self-Ag. Further studies are needed to generalize the principles unveiled by the analysis of the single pair of mAb55 and mAb13.

REFERENCES

1. RIBOLDI, P., M. T. KASAIAN, L. MANTOVANI, H. IKEMATSU & P. CASALI. 1993. Natural antibodies. *In* The Molecular Pathology of Autoimmune Disease. C. A. Bona, K. Siminovitch, M. Zanetti & A. N. Theofilopoulos, Eds.: 45–64. Harwood Academic Publishers GmbH. Chur.
2. CHEN, P. P., M.-F. LIU, S. SINHA & D. A. CARSON. 1988. A 16/6 idiotype-positive anti-DNA antibody is encoded by a conserved V_H gene with no somatic mutation. Arthritis Rheum. **31:** 1429–1431.
3. SIMINOVITCH, K. A., V. MESINER, P. C. KWONG, Q.-L. SONG & P. P. CHEN. 1989. A natural autoantibody is encoded by germline heavy and lambda light chain variable region genes without somatic mutation. J. Clin. Invest. **84:** 1675–1678.
4. VAN ES, J. H., F. H. J. G. MEYLING, W. R. M. VAN DE AKKER, H. AANSTOOT, R. H. W. M. DERKSEN & T. LOGTENBERG. 1991. Somatic mutations in the variable regions of a human IgG anti-double-strand DNA autoantibody suggest a role for antigen in the induction of system lupus erythematosus. J. Exp. Med. **173:** 461–470.
5. KASAIAN, M. T., H. IKEMATSU, J. E. BALOW & P. CASALI. 1994. Structure of the V_H and V_L segments of monoreactive and polyreactive IgA autoantibodies to DNA in patients with SLE. J. Immunol. **152:** 3137–3151.
6. OLEE, T., E. W. LU, D.-F. HUANG, R. W. SOTO-GIL, M. DEFTOS, F. KOZIN, D. A. CARSON & P. P. CHEN. 1992. Genetic analysis of self-associating immunoglobulin G rheumatoid factors from two rheumatoid synovia implicates an antigen-driven response. J. Exp. Med. **175:** 831–842.
7. MANTOVANI, L., R. L. WILDER & P. CASALI. 1993. Human rheumatoid B-1a ($CD5^+$ B) cells make somatically hypermutated high-affinity IgM rheumatoid factors. J. Immunol. **151:** 473–488.
8. IKEMATSU, H., Y. ICHIYOSHI, E. W. SCHETTINO, M. NAKAMURA & P. CASALI. 1994. V_H and V_κ segment structure of anti-insulin IgG autoantibodies in patients with insulin-dependent diabetes mellitus. Evidence for somatic selection. J. Immunol. **152:** 1430–1441.
9. LARRICK, J. W., E. F. WALLACE, M. J. COLOMA, U. BRUDERER, A. B. LANG & K. FRY. 1992. Therapeutic human antibodies derived from PCR amplification of B-cell variable regions. Immunol. Rev. **130:** 69–85.
10. ICHIYOSHI, Y. & P. CASALI. 1994. Analysis of the structural correlates for antibody polyreactivity by multiple reassortments of chimeric human Ig H and L chain V segments. J. Exp. Med. **180:** 885–895.
11. ICHIYOSHI, Y., M. ZHOU & P. CASALI. 1995. A human anti-insulin IgG autoantibody apparently arises through clonal selection from an insulin specific "germ-line" natural antibody template. Analysis by V gene reassortment and site-directed mutagenesis. J. Immunol. **154:** 226–238.
12. HARINDRANATH, N., I. S. GOLDFARB, H. IKEMATSU, S. E. BURASTERO, R. L. WILDER, A. L. NOTKINS & P. CASALI. 1991. Complete sequence of the genes encoding the

V_H and V_L regions of low and high affinity monoclonal IgM and IgA1 rheumatoid factors produced by $CD5^+$ B cells from a rheumatoid arthritis patient. Intl. Immunol. **3:** 865–875.

13. Ueki, Y., I. S. Goldfarb, N. Harindranath, M. Gore, H. Koprowski, A. L. Notkins & P. Casali. 1990. Clonal analysis of a human antibody response: Quantitation of precursors of antibody-producing cells and generation and characterization of monoclonal IgM, IgG, and IgA to rabies virus. J. Exp. Med. **171:** 19–34.
14. Casali, P., M. Nakamura, F. Ginsberg-Fellner & A. L. Notkins. 1990. Frequency of B cells committed to the production of antibodies to insulin in newly diagnosed patients with insulin-dependent diabetes mellitus and generation of high affinity monoclonal IgG to insulin. J. Immunol. **144:** 3741–3747.
15. Sanz, I., P. Casali, J. W. Thomas, A. L. Notkins & J. D. Capra. 1989. Nucleotide sequence of eight human natural autoantibodies V_H region reveal apparent restricted use of V_H families. J. Immunol. **142:** 4054–4061.
16. Ikematsu, H., N. Harindranath, Y. Ueki, A. L. Notkins & P. Casali. 1993. Clonal analysis of a human antibody response II. Sequence of the V_H gene of human IgM, IgG, and IgA to rabies virus reveal preferential utilization of V_HIII segments and somatic hypermutation. J. Immunol. **150:** 1325–1337.
17. Chothia, C. & A. M. Lesk. 1987. Canonical structures for the hypervariable regions of immunoglobulins. J. Mol. Biol. **196:** 901–907.
18. Kabat, E. A. & T. E. Wu. 1991. Identical V region amino acid sequences and segments of sequences in antibodies of different specificities: Relative contributions of V_H and V_L genes, minigenes, and complementarity-determining regions to binding of antibody-combining sites. J. Immunol. **147:** 1709–1719.
19. Reichmann, L., M. Clark, H. Waldmann & G. Winter. 1988. Reshaping human antibodies for therapy. Nature **332:** 323–327.
20. Kao, C.-Y. Y. & J. Sharon. 1993. Chimeric antibodies with anti-dextran-derived complementarity-determining regions and anti-*p*-azophenylarsonate-derived framework regions. J. Immunol. **151:** 1968–1979.
21. Ikematsu, H., M. T. Kasaian, E. W. Schettino & P. Casali. 1993. Structural analysis of the V_H-D-J_H segments of human polyreactive IgG mAb: Evidence for somatic selection. J. Immunol. **151:** 3604–3616.
22. Harindranath, N., H. Ikematsu, A. L. Notkins & P. Casali. 1993. Structure of the V_H and V_L segments of polyreactive and monoreactive human natural antibodies to HIV-1 and *E. coli* β-galactosidase. Int. Immunol. **5:** 1523–1533.
23. Cunningham, M. 1993. Bacterial antigen mimicry. *In* The Molecular Pathology of Autoimmune Disease. C. A. Bona, K. Simminovitch, M. Zanetti & A. N. Teofilopoulos, Eds.: 245–256. Harwood Academic Publishers GmbH. Chur.
24. Manser, T. & M. L. Gefter. 1986. The molecular evolution of the immune response: Idiotype-specific suppression indicates that B cells express germ-line-encoded V genes prior to antigenic stimulation. Eur. J. Immunol. **16:** 1439–1444.
25. Kocks, C. & K. Rajewsky 1988. Stepwise intraclonal maturation of antibody affinity through somatic hypermutation. Proc. Natl. Acad. Sci. USA **85:** 8206–8210.
26. Berek, C. & C. Milstein. 1987. Mutation drift and repertoire shift in the maturation of the immune response. Immunol. Rev. **96:** 23–41.
27. Manser, T., B. Parhami-Seren, M. N. Margolies & M. L. Gefter. 1987. Somatically mutated forms of a major anti-*p*-azophenylarsonate antibody variable region with drastically reduced affinity for *p*-azophenylarsonate: By-products of an antigen-driven immune response? J. Exp. Med. **166:** 1456–1463.
28. Shlomchik, M. J., A. H. Aucoin, D. S. Pisetsky & M. G. Weigert. 1987. Structure and function of anti-DNA autoantibodies derived from a single autoimmune mouse. Proc. Natl. Acad. Sci. USA **84:** 9150–9154.
29. Chen, C., V. A. Roberts & M. B. Rittenberg. 1992. Generation and analysis of random point mutations in an antibody CDR2 sequence: Many mutated antibodies lose their ability to bind antigen. J. Exp. Med. **176:** 855–866.
30. Rechavi, G., B. Bienz, D. Ram, Y. Ben-Neriah, J. B. Cohen, R. Zakut & D. Givol. 1982. Organization and evolution of immunoglobulin V_H gene subgroups. Proc. Natl. Acad. Sci. USA **79:** 4405–4409.

31. ICHIHARA, Y., H. MATSUOKA & Y. KUROSAWA. 1988. Organization of human immunoglobulin heavy chain diversity gene loci. EMBO J. **7:** 4141–4150.
32. RAVETCH, J. V., U. SIEBENLIST, S. J. KORSMEYER, T. WALDMANN & P. LEDER. 1981. Structure of the human immunoglobulin μ locus: Characterization of embryonic and rearranged J and D genes. Cell **27:** 583–591.
33. KIPPS, T. J., E. TOMHAVE, P. P. CHEN & D. A. CARSON. 1988. Autoantibody-associated κ light chain variable region gene expressed in chronic lymphocytic leukemia with little or no somatic mutation: Implications for etiology and immunotherapy. J. Exp. Med. **167:** 840–852.
34. PECH, M. & H. ZACHAU. 1984. Immunoglobulin genes of different subgroups are interdigitated within the Vκ locus. Nucl. Acids Res. **12:** 9229–9236.
35. HEITER, P. A., J. V. MAIZEL, JR. & P. LEDER. 1982. Evolution of human immunoglobulin κJ region gene. J. Biol. Chem. **257:** 1516–1522.

Unconventional B-Cell Antigens and Human Immune Repertoires[a]

GREGG J. SILVERMAN

The Sam and Rose Stein Institute for Research on Aging and Department of Medicine
University of California, San Diego
La Jolla, California 92093-0663

INTRODUCTION

Within the immune system, the lymphocyte antigen receptor is responsible for the central function of distinguishing between foreign and self. In order to achieve this goal, a sequential series of somatic molecular processes are superimposed to mold a fixed set of germline gene segments into a virtually unlimited range of receptors. This system enables exquisite specificity and high-affinity recognition of novel antigens for which the individual is both personally and evolutionarily naive. In addition, because of the existence of inherent functional potentials associated with certain receptor genes and biases in their use, certain binding specificities are ensured to be highly represented in the preimmune repertoire.

The immunoglobulin (Ig) receptors of human B lymphocytes are heterodimers composed of heavy (H) and light (L) chains. Each chain is functionally divided into a constant domain that specifies effector function and a variable (V) domain that provides Ag binding specificity. To create the V domain, two or more minigenes are joined by splicing events. Thus, the V_H domain is derived from one of about 100 V_H gene segments, one or two of about 30 D_H segments, and one of six J_H gene segments. Diversity is further increased by insertion at the splice sites of nongermline-encoded nucleotides (N regions). Differential sites of splicing, D_H-D_H fusions, multiple D_H reading frames, and the N region additions increase the potential diversity of the CDR3 of the H chain, ensuring a vast spectrum of potential Ag binding sites. The V domains of L chains are similarly created beginning with splicing of V_L and J_L gene segments. A hypermutational mechanism further enhances diversity, such that a binding site for an Ag can theoretically be created from any pair of antibody (Ab) V gene segments.

Within each V domain there are three intervals of sequence hypervariability (termed complementarity-determining regions, CDR) that are separated by four intervals of relatively conserved sequence (termed framework regions, FRs) (FIG. 1). The CDR1 and CDR2 are contributed by V gene segments, which are themselves grouped into families based on conserved sequence homologies primarily within the FRs. These FRs have specific amino acid compositions and sequences that contribute the scaffolding responsible for conservation of the overall three-dimensional beta barrel structure called the "immunoglobulin fold."[1] On these beta strands, the CDRs are positioned at the spatially adjacent bends that are

[a] This work was supported in part by National Institutes of Health Grant AI34001. GJS is a recipient of a Biomedical Sciences Award from the Arthritis Foundation and is a Scholar of the Lupus Foundation of America, Inc.

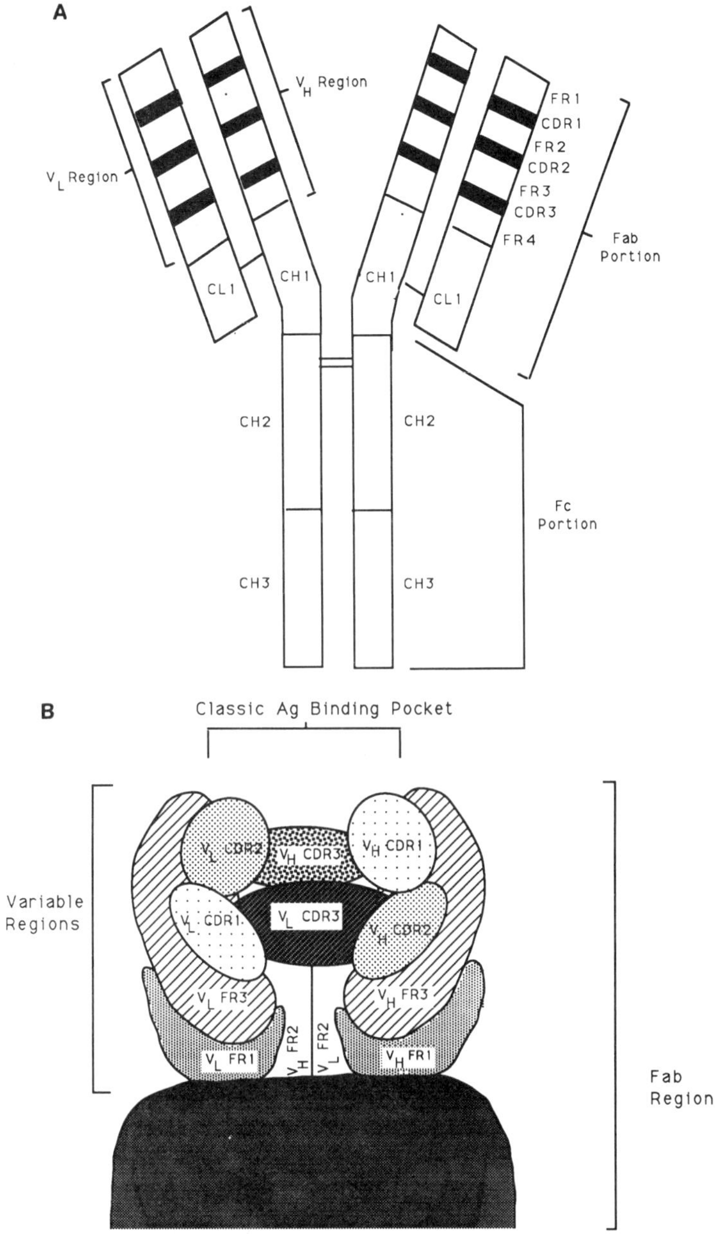

FIGURE 1. Classic and unconventional antigen-binding sites on antibodies. (A) Schematic diagram of primary sequences of H and L chain rearrangements that create the V regions of an antibody. **(B)** Idealized space filling model of the relative locations of the subdomains within the V regions of an antibody. In this cartoon, portions of framework 2 subdomains are visible, but in the folded Fab they are almost completely internalized. Figure adapted from Silverman.[21]

most distant from the constant regions. Cumulatively the CDRs create the classic antigen-binding pocket (FIG. 1). At the center of this conventional Ag-binding site are the CDR3s of the H and L chains, which are created by V_H-D_H-J_H and V_L-J_L joinings, respectively.[2] The walls of the classic Ag-binding pocket are composed of the CDR1 and CDR2 of the V gene segments. Immediately circumjacent to this are the FR3 and subsequently the FR1 from the V gene segments. Crystallographic studies have confirmed that conventional antigens generally interact with a central position within the classic antigen-binding pocket and have contacts with amino acids in CDR of both chains that always include a CDR3.[3]

In contrast to Ag-binding capacities for conventional Ag, with each specificity generally representing less than 0.1% of the total expressed B-cell repertoire, recent studies have described a distinct type of V region-dependent binding specificity that can be expressed by 5 to 50% of the B-cell pool. Each of these naturally occurring Ag-binding sites appears to be recurrently created with a single type of V_H gene segment with little conservation of other elements in the V regions of the Ag receptor. The prevalence of such Ag-binding sites no doubt derives from the fact that expression of this type of functional binding activity generally is dependent only on the use of an appropriate V_H gene segment. This represents an exception from the more common pattern in which the CDRs in both H and L chains contribute to the creation of an Ag-binding site. Recent investigations of these unconventional Ags and their Ig receptors suggest that they have much in common with the binding interaction between known superantigens and their corresponding T lymphocyte sets.

STAPHYLOCOCCAL PROTEIN A: THE PROTOTYPIC B-CELL SUPERANTIGEN

Of all known Ig-binding proteins, staphylococcal protein A (SpA) is the best studied. This 45-kDa membrane protein is produced by greater than 95% of clinical isolates of *Staphylococcus aureus*,[4] which may indicate a role in pathogenicity. In addition to the well-known capacity of SpA to bind to Fcγ, a constant region portion of IgG, SpA has two or more sites that facilitate binding interactions with certain Fabs, the V region containing fragments of Ig.[5,6] Through this interaction, 15 to 50% of human polyclonal IgM, IgA, and IgG $F(ab')_2$ can bind SpA.[7,8] The sites on SpA bound by Fab appear to be structurally separate from the Fcγ-binding sites, because chemical modification of SpA can eliminate the Fc-binding activity without altering the ability to be bound by Fab.[9,10] Binding of SpA does not generally block binding of a conventional ligand,[11] which suggests that the site on Fab responsible for SpA binding is functionally distinct from the conventional Ag-binding pocket. This Fab–SpA binding interaction is critical for the B-cell mitogenicity of *S. aureus* extracts.[12,13] Fab-mediated SpA binding is also highly conserved in mammalian and more primitive species.[14,15,16]

Evidence that Fab-mediated SpA binding capacity correlates with specific V region utilization was first obtained by use of a panel of V family-specific serologic reagents that enabled identification of the V family origins of binding and nonbinding Ig populations.[7,8,17] Subsets of polyclonal IgM, IgA, and IgG $F(ab')_2$ that bind SpA were found to be highly enriched in the products of the V_H3 family.[7,8] Studies of monoclonal Ig confirmed that Fab-mediated SpA binding was common among V_H3 antibodies but absent in monoclonal IgM that derived from the other V_H families.[17–20]

To further evaluate the molecular requirements for Fab-mediated SpA binding, we have used combinatorial Ig expression libraries created using the phage-display vector, pComb3.[20] From peripheral blood cells of an adult donor, a large ($>10^7$ members) Ig library was constructed that included gamma chains, which were anticipated to better display the influence of somatic hypermutation on the SpA-binding capacity. All V_H families were represented in the library, and about 17% of Ab-expressing clones in the library bound SpA. Significantly, SpA binding was completely restricted to Fab with V_H3 H chains, and about 60% of V_H3 IgG Fab in the unselected library had SpA-binding capacity. Competition experiments demonstrated that all Fab binding is directed against the same site(s) on SpA.[10,20]

In contrast to the conserved V_H3 usage by SpA binders, sequence analysis detected diverse L chains. To explicitly test the hypothesis that L chains are relatively passive in Fab-mediated SpA binding, H and L chains were differentially paired. These studies demonstrated that the global capacity to bind (or not bind) SpA was dictated by specific V_H3 usage, and different L chain usage caused only minor changes in SpA-binding affinity, accounting for at most a fourfold change.[20]

Analyses of the DNA sequences of the V_H regions of over 40 monoclonal V_H3 antibodies from combinatorial libraries[20] (also P. Roben and G. J. S., manuscript in preparation) have demonstrated that SpA binders are encoded by V_H gene segments with great heterogeneity in the CDR1 and CDR2, and there is also no conservation of V_H CDR3 or J_H usage. These analyses, taken together with data from antibodies obtained by conventional cellular methods, indicate that at least 15 of the 30 estimated germline (and perhaps all) V_H3 genes can potentially encode an SpA-binding site (reviewed in Silverman[21]) (FIG. 2) (G. J. S., unpublished observation). Significantly, studies of fetal B-cell lines from Hillson *et al.*[19] indicate that to date, all antibodies encoded by a V_H3 gene segment in germline configuration have been found to bind SpA.

Identifying the portions of the V_H3 region responsible for SpA binding has been complicated by the observation that SpA binders generally display conserved sequences throughout the FR1, the 5′ portion of CDR2, and the majority of the FR3 subdomains (FIG. 2). Crystallographic studies have demonstrated that these discontinuous subdomains create a large solvent-exposed surface of the Fab that is separate from the classical Ag-binding site[22] (FIG. 3). Sequence comparisons also indicate that minor amino acid variations from the V_H3 consensus are generally associated with retention of SpA-binding capacity.[17–21] Although single point mutations have not generally been correlated with functional differences in SpA-binding capacity, Randen *et al.* have reported a notable exception in which a single nonconservative amino acid change at position 57 in the V_H CDR2 of a hv3005-encoded V_H3 rheumatoid factor can destroy SpA binding.[23] However, this correlation is not universally observed (see PRSJ1 and SpA1-17 in FIG. 2). Discordance from this rule could at times be due to the cumulative subtle effects of variations in other portions of the V_H sequence (see 0-17 in FIG. 2), and it remains to be determined whether the replacement of the lysine at position 57 for a glutamic acid actually results in the abolition of a contact site or if it instigates remote conformational changes. In our studies, we have speculated that SpA binding correlates with conservation of a highly conserved V_H3 family-specific sequence (Ala/Ser-Lys-Asn-Ser/Thr) in the tip of the FR3 loop of the V_H domain (positions 74 to 77)[8] (FIGS. 2 and 3). Variations from this germline-encoded four amino acid sequence in FR3 are usually associated with an inability to bind SpA (FIG. 2). Taken together, available data indicate that the V_H3 requirements for SpA binding largely reiterate specific V_β requirements for binding of the MIs and staphylococcal

T-cell superantigens.[24–27] Mutational analyses of V_H regions will be required to further define the sites critical for SpA binding.

The construction of a more accurate molecular model of this unconventional Ag–Ab binding interaction should also be aided by the identification of antibodies with the highest SpA-binding affinity, as these antibodies should contain complementary surfaces with the best fit for SpA. The pComb3 phage-display vector is designed to enable efficient selection of antibody species based on relative affinity.[28] For these studies, Fab-expressing phages are added to microtiter wells coated with SpA, then by sequential rounds of specific elution, library expansion, and repeat panning, antibodies with the highest SpA-binding affinity are selected. By comparison with SpA binders in the unselected library, this approach has enabled identification of V_H3 IgG Fab with a spectrum of SpA-binding affinities.[20] In one experiment, after three rounds of selection eight of nine V_H regions sampled from the resultant library appeared to be derived from the same V_H-D_H-J_H rearrangement and differed only by one or a few single mutations.[20] Surface plasmon resonance measurements of affinity showed that one of these IgG Fabs, termed JGSpA3-08, binds SpA with a K_a of 6×10^7 M^{-1} (G. J. S., manuscript in preparation), which is in agreement with our earlier relative affinity values obtained from competitive inhibition ELISA.[20] The Fabs with the highest SpA-binding affinity selected in these, and more recent biopanning studies,[20] have all been found to use the V_H26c gene segment. In support of this, among a panel of monoclonal V_H3 antibodies, an IgM directly encoded by the V_H26 gene, 18/2, had the highest binding affinity (G. J. S., manuscript in preparation). The prominence of V_H26c rearrangements in the V_H genes obtained after *in vitro* stimulation with SpA further supports this notion.[12] The V_H26c gene segment has also been shown to be preferentially rearranged in the early fetal repertoire and to be overexpressed in the B-cell repertoire throughout development.[29,30] Even though SpA is not normally present during early development, it is appealing to speculate that the association of

---

FIGURE 2. SpA binding and V_H utilization. Antibody sequences and relative reactivity with the Fab-mediated binding site of SpA are compiled. V_H regions are grouped by V_H gene segment family and SpA reactivity. [a]V_H regions with a V_H3 gene segment in germline configuration are indicated below the heading "Germline." Certain H chains were obtained from fetal tissues and are presumed to contain V_H segments in germline configuration,[19] although the germline source has not been independently isolated. [b]SpA-binding activity is frequently expressed by V_H3 antibodies that also have a conventional antigen-binding activity, which is indicated below the heading "Ag." [c]The immunoglobulin isotype is indicated below the heading "Isotype." The IgG Fabs were derived from a combinatorial antibody expression library[20]; all others were derived via conventional cellular approaches. Sequences are compiled from previous reports by Silverman[17] designated [d]; by Sasano *et al.*[20] designated [e]; by Ibrahim *et al.*[18] designated [f]; by Randen *et al.*[23] designated [g]; by Hillson *et al.*[19] designated [h]; and by Sasso *et al.*[7] designated [i]. The V_H sequence for CA4/PRP12 is previously unreported[10,20] (A. H. Lucas, personal communication). Amino acid residues are indicated by the single letter code. **X** is an indeterminate residue; - indicates identity with the amino acid in the same position in the 18/2 sequence; * marks the amino acid positions in V_H3 FR3[20] and CDR2[23] that have been implicated as critical to expression of SpA-binding activity. From mutagenesis experiments with a V_H3-encoded IgM RF, Randen *et al.*[23] have reported that at the indicated position in CDR2, **K** endows SpA-binding capacity, whereas **E** abolishes SpA binding. However, the absence of SpA binding by PRSJ1 and SpA1-17 indicates that this is not a universal rule. (The antibody 0-19 has *very* weak SpA binding activity.[20]) Figure adapted from Silverman.[21]

	Germline[a]	Ag[b]	Isotype[c]	Framework1	CDR1		Framework2	CDR2		Framework3	CDR3	Framework4
SpA Binders									*	****		
18/2[d]	V_H26	DNA	IgM	EVQLLESGGGLVQPGGSLRLSCAASGFTFS	S	YAMS	WVRQAPGKGLEWVS	AISG	SGGSTYYADSVKG	RFTISRDNSKNTLYLQMNSLRAEDTAVYYC	TKGQVLYYGSGSYHWFDP	WGQGTLVTVSS
SpA2-08[e]			IgGFab	Q-K------T--------------------	-			----	-------------	L-------A------L--------------	ARAELGGLMVIVTPSEY	-----------
ED8.4[f]		HibPS	IgM	------------------------------	-	----	--L-P---------	----	---R---------	------E-----V-----V-----------	TKGYGMDV	-----T-----
SpA3-08[e]			IgGFab	Q-K---------------------------	-	H---	--------------	D--A	-------------	--------------------------L---	ASSGAGWGLPSLDY	-----------
SpA3-02[e]			IgGFab	Q-K---------------------------	-	H---	--------------	D--A	-------------	--------------------------L---	ASNGAGWGLPSLDY	-----------
SpA2-13[e]			IgGFab	Q-K---------------------------	-	H---	-------R------	D--A	-------------	--------------------------L---	ASNGAGWGLPSLDY	-----------
SpA3-39[e]			IgGFab	VKL-E-------------------------	-	H---	-------R------	D--A	-------------	--------------------------L---	ASNGAGWGLPSLDY	-----------
SpA1-30[e]			IgGFab	VKL-EQ------------------------	-	H---	-------R------	D--A	-------------	--------------------------L---	ANSGAGWGLPSLDY	-----------
SpA3-37[e]			IgGFab	Q-K---------------------------	-	H---	-------R------	D--A	-------------	--------------------------L---	ASNGAGWGLPSLDY	L----------
0-16[e]			IgGFab	VKL-E----------------A-G------	-	H---	-------R------	D--A	M------------	--------------------------L---	ASNGAGWGLPSLDY	-----------
KIM4.6[d]	1-9III	DNA	IgM	Q---V-----V----R--------------	-	-G-H	-------------A	V--Y	--SNK--------	------------------------------	AKVREYYYGSGWYYYYGMDV	-----T-----
RF-TS1[g]		RF	IgM	Q---V-----V----R--------------	-	---H	------------YA	V--Y	D-SNK--------	------------------------------	AREDPYGDYVANPFDI	-----M--
PR-SJ2[g]		RF	IgM	Q---V-----V----R--------------	-	---H	------------YA	V--Y	D-SNK--------	-------------S----------------	ARGRYCSSISCWYFDL	--R-----
2E8[h]	FETAL		IgM	------R--------------	-	-W--	-------------A	N-KQ	D-SEK--V-----	--------A--S------------------	AREDYYYGMDV	-----T-----
0-19[e]			IgGFab	Q-K------------R-------T------	T	F--N	-------------A	VVWY	D-TTK------Q-	---------E----------V---------	AKGYQLLHGQ	-------
SFL[d]		RF	IgG	----V------------------------T	X	YX--	-------------A	F-YR	--INK------N-	--------X---XXX-----------L---	AKDAGLKVEKSV	
RIV[i]		RF	IgM	----V-----V----S--------------	-	F--H	-------------A	VM-Y	--DNK--V-----	--------------------------L---	AKLSTAASYFTFDTYGMD	-----T-----
SpA1-14[e]			IgGFab	Q-K---------------------------	-	-W-H	----------V---	R-NT	D-SR-S-------	--------A------L--------------	ARAELRRPNGYANPPPEY	-----------
RF-KL1[g]		RF	IgM	------------------------------	-	----	------------Y-	---D	N--I--T-Y----	-----------P------------------	AKLRSGLVPYYFDS	-------N
3G11[h]	FETAL		IgM	--V---R--------------	-	---H	-----------Y--	---S	I--R---N-----	-----------------G------M-----	ARVFTSCFDY	-----------
4D5[h]	13-2		IgM	--V------------------	-	-P--	-----T--------	--G	TA-D--PG-----	-------EA--S---------G--------	AAARVAAAGPYQLYYYYGMGV	-----T-----
RF-TS5[g]		RF	IgM	Q---V-----V-K-----------------	D	-Y--	-I------------	Y--S	--SII--------	--------A--S------------------	GTPFIDY	--------
5D4[h]	22-2B		IgM	----K----------------	D	-Y--	-I---T--------	Y--S	--STI--PG----	--------A--S------------------	ARDPINWYFDL	--R--------
2E5[h]	FETAL		IgM	--VL-----------------	D	-Y--	--------------	Y--S	-SSY-N-------	--------A--S------------------	AGYHWYFDL	--R--------
6H9[h]	FETAL		IgM	--V-R--R------------D	D	-G--	--------------	G-NW	I----S-------	--------A--S--------------L-H-	ARDDLIIMPYFDY	-----------
SpA3-33[e]			IgGFab	VKL-E-----------VP------E--P--	N	-G--	--------------	S---	--D----------	--------A--S------------------	ARDAWDAFDI	-----M-----
SpA1-29[e]			IgGFab	Q-K---------------------------	N	----	--------------	G--A	--DT-------R-	--A-----F---------------------	GRGYSYPV	-----T-----
LAY[i]		RF	IgM	A-----------------------------	A	S---	-------------A	WKYE	N-NDKH-----N-	------ND-------------Q--S-I---	ARDAGPLVSPTFFAH	-----------
POM[i]		RF	IgM	------------------------------	-	S---	-------------A	WKYE	N-NDKH-----N-	------ND-------L-----Q----L---	ARDAGPLVSPTFFAHY	-----------
4B4[d]	9-1	Sm	IgM	----V-------K-----------------	N	AW--	-------------G	R-KSKTD	--T-D--AP---	-------D----------KT----------	TTDSLPPHRV	-----------
SB5/D6[f]		HibPS	IgA1	----V-------K-----------------	N	AW-N	----P--------G	R-KTKTD	--T-D--AP---	------ND----------KT----------	TTGGGVG	-----------
VH3 SpA Non-binders									*	****		
RF-SJ3[g]		RF	IgM	Q--MV-----V----R--------------	-	-G-H	------------YA	V--Y	D-SNE--------	------------------------------	AKWGGYCTNGVCYRGGYGMDV	--K-ST--
PR-SJ1[g]		RF	IgM	Q---V-----V----R-----------S--	-	-G-H	----C-------YA	V--D	D-SNK--------	----------K-S---S----D--------	AKEAGWGLPSLGYFGRRGYFDY	--------
PR-TS2[g]		RF	IgM	Q---V-----V----R--------------	-	-G-H	------------YA	V--Y	D-SNE-F------	---------N---M----------------	ARGRRGAARLHYYYYMDV	--K-ST--
SpA1-17[e]			IgGFab	Q-K-------------------V-------	-	-W-R	-------------A	S-KQ	D-SEK--V-----	--------AE-S--------------I---	ARSYNWASDY	-C---------
0-17[e]			IgGFab	Q-K---------------------------	T	HS-N	--------------	SV-S	-SSYI--------	--------A--S------------------	ASLYFWANGMDV	-----T-I---
0-12[e]			IgGFab	Q-K------------------------S--	-	HW--	L------------A	T-NQ	G-SAK--V-----	-----F--VQDS---L----V---------	ASGYSDSSGYRGSYFYY	-C---------
CA4/PRP-12		HibPS	IgG2	----V-------T----------------T	F	NTY-MN	--------------	S--R	-SDYI--P-----	-------TP--PV---T--D----------	AK	
LSF2[f]		HibPS	IgA1	----V------IK-----T-------I---	N	AW-N	-------R-----G	R-KKKTD	--T-D--AP---	---V---D-QS-V---S---T---------	GTRDY	-----------
Other SpA Non-binders												
BOR[d]	V_H1	RF	IgM	---VQ--AEVKK--S-VKVT-K---D---	-	SAI-	-------Q----MG	G-IP	IF-TPN--QKFQ-	-V--TT-E-TS-A-MEVR---S----L---	AREGRRMAINPFDY	-----------
KAS[d]	V_H1	RF	IgM	---VQ--AEVKK--S-VKVT-K---G---	-	--I-	-------Q----MG	G-IP	IF-QAN--QKFQ-	-V--TA-E-T--A-MELR---SD---M---	AKEGYGDYGRPFDF	-----------
CESS[d]	V_H2		IgM	Q-N-R---PA--KATHT-T-T-TF--LSVN	T	RG--VS	-I--P---A---LA	R-D	WDDDK--GT-LET	-L---K-T---QVV-KVTNMDPA---T---	ARMQVTMVREVMITSNAFDI	-----M-----
LES[d]	V_H4	RF	IgM	QVQ-QQW-A--LK-S-T-S-T--VY-GP--	G	-YW-	----P----P--IG	E-N	HS-R-T-NP-LTS	-V-M-L-T---QFS-KLT-VT-A---V--L	ARGPCEAYCTDDAPQAYFQH	-----------
A224[d]	V_H5	Poly	IgM	----VQ--AEVKK--E--KI--KT--YS-T	N	-WIG	----M-------MG	L-YP	GDSG-R-SP-F--	QV---A-K-IS-A---WS----S---M---	AR	
L16[d]	V_H6	Poly	IgM	Q---QQ--P---K-SQT-S-T--I--DSV-	-	NSAAWN	-I--S-SR----LG	RTYYR	-KWYND--V---S	-I--NP-T---QFS--LN-VTP----V---	ARELGDAFDI	-----M-----
				EVQLLESGGGLVQPGGSLRLSCAASGFTFS	S	YAMS	WVRQAPGKGLEWVS	AISG	SGGSTYYADSVKG	RFTISRDNSKNTLYLQMNSLRAEDTAVYYC	TKGQVLYYGSGSYHWFDP	WGQGTLVTVSS

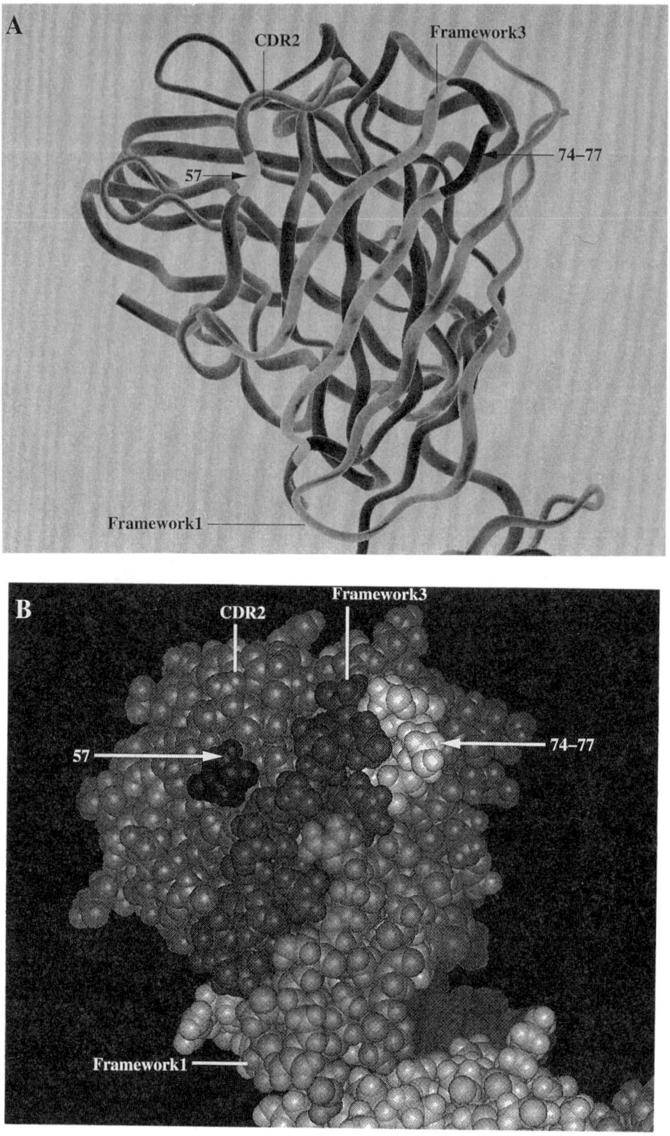

FIGURE 3. The structure of the V_H3 region. In **panel A,** the α-carbon backbone is displayed, and the beta bends of the CDR2, FR3, and FR1 loops are indicated. Positions 57 in CDR2 and 74–77 in the FR3, which have been correlated with SpA binding, are indicated. In **panel B** is the same view of a space-filling model that includes the side chains. The model is based on coordinates for the Kol, V_H3 Fab (obtained from the Brookhaven Data Base, Upton, NY), and software of the Biosym Corp. (San Diego, CA) running on a Silicon Graphics (Mountain View, CA) workstation. Of importance, the sequence within the CDR2 of Kol differs from the V_H3 IgM RF studies by Randen *et al.*,[23] and here position 57 is an aspartic acid.

this gene segment with high affinity for another unconventional ligand, that interacts at a related nonclassical site, may contribute to the *in vivo* selection of V_H26c rearrangements.

To better understand the expression of an unconventional antigen binding specificity with the physiologic B-cell repertoire, we evaluated Fab-mediated SpA binding within different B lymphocyte populations. Multiparameter flow cytometry studies were performed using a form of SpA that was chemically modified to remove Fc binding (Mod SpA).[10] As anticipated, Mod SpA binding activity was completely absent from T cells and non-B, non-T mononuclear cells, but was highly prevalent among B cells (14–54%). Significantly, the proportion of peripheral blood B lymphocytes with this V_H3-restricted binding capacity was found to undergo an age-associated change (FIGS. 4 and 5). In neonates, a large proportion (mean ± SD, 46.0 ± 2.9%) of peripheral B cells bind this site on SpA, but by 10 months of age there is a significant decrease to a lower binding level (30.2 ± 3.3%) that appears to stay roughly constant into adulthood (FIG. 5).[10]

In studies of B lymphocyte subsets, we found that Fab-mediated binding of SpA is always expressed by a greater proportion of CD5-positive B cells than CD5-negative B cells.[10] However, the above-described age-associated decrease in the peripheral blood level of SpA binders is not due to the well-known drop from the high neonatal level of CD5-positive B cells that occurs during infancy and childhood. In both the CD5-positive and the CD5-negative peripheral blood B cell subsets, the proportion of Mod SpA binders undergo comparable decreases between 6 and 10 months of age (FIGS. 4 and 5).[10] Thus, the age-associated decrease in B cell binding of Mod SpA is not simply due to a decrease in the proportion of B cells that express the CD5 phenotype. Fab-mediated SpA binding is also more common within the repertoire of IgM-bearing B cells than the repertoire of IgG-bearing B cells. On average, IgM-bearing B cells also have stronger binding interactions with Mod SpA than do IgG-bearing B cells.[10]

Taken together with the above immunochemical and molecular analyses, these data further support the impression that, although commonly retained in the IgG repertoire, Fab-mediated SpA binding is still more frequent in the IgM repertoire. It is relevant to note that the developmental interval during which the level of Fab-mediated SpA binding decreases is nearly identical with the period that serum IgG levels nadir. This occurs because maternal antibodies wane, and endogenous IgG levels are only beginning to increase. This part of the first year of life is the time when environmental Ags first exert their full, unbuffered force in selecting the B-cell repertoire. Therefore, the decrement in the germline-encoded binding capacity for SpA is likely a side effect of the somatic process of B lymphocyte clonal selection and hypermutation that occurs following exposures to conventional environmental ligands.

PROTEIN Fv—A SELF GLYCOPROTEIN WITH SUPERANTIGEN-LIKE B-CELL BINDING PROPERTIES

Investigations of the Fab-binding specificity of staphylococcal protein A have clearly demonstrated that there is an antibody site, which likely involves the FRs, that is responsible for a highly specific binding capacity shared by most V_H3 antibodies. Although the interaction with SpA has aided the characterization of this unconventional antibody site, it does not necessarily imply that this site has been retained in evolution solely to interact with a single bacterial protein.

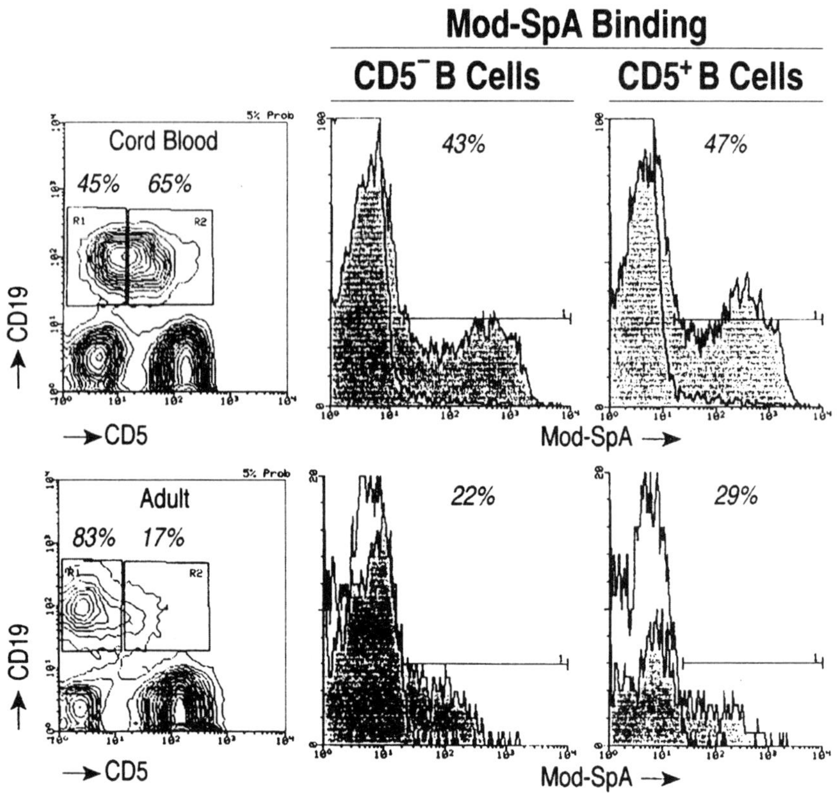

FIGURE 4. Proportion of CD5-negative and CD5-positive peripheral B lymphocytes that bind Mod SpA. A representative three-color flow cytometric study of human B lymphocytes that bind Mod SpA (SpA that has been chemically modified to remove Fc binding but retains V_H3 Fab-mediated binding activity). Mod SpA is bound by a subset of both the CD5-negative (R1) and CD5-positive (R2) B cell populations at each age. The proportion of peripheral blood B cells that are CD5-positive decreases with aging, from neonate (cord blood, 65%) to adult (17%). In both B-cell subpopulations, the proportion of B cells that bind Mod SpA is higher in the neonate than in the adult. A greater proportion of CD5-positive B cells bind Mod SpA relative to CD5-negative B cells in both individuals. (Reproduced from Silverman *et al.*[10] with permission; copyright 1993, *The Journal of Immunology*.)

Superantigens have been shown at times to act as virulence factors,[31] and in one possible scenario the SpA membrane protein of *S. aureus* has acquired this ability as a means of circumventing or incapacitating the host's immune defense. A large proportion of B cells that encounter secreted SpA are activated, and this may interfere with the acquisition of a more selective antibody response to critical conventional bacterial antigens required for immune defense. With this in mind, we sought evidence that unconventional binding interactions might also be part of normal immune regulation.

In recent publications, Bouvet and coworkers have demonstrated that a 175-kDa dimeric sialoprotein produced in the human liver, termed protein Fv (pFv), possesses Fab-binding properties with similarities to SpA.[32–35] A single molecule of pFv can bind six $F(ab')_2$ fragments,[35] and binding to pFv has been clearly shown to be mediated by the V_H region.[34] In earlier reports, virtually all human IgM paraproteins have some pFv binding capacity. A large proportion of Ig of other isotypes also bind pFv, but the pFv binding capacities of different antibodies vary greatly. pFv binding capacity is also expressed by Ig from a variety of mammalian, avian, reptilian, amphibian, and fish species.[3] Also reminiscent of SpA, the interaction of an antibody with pFv does not generally block binding of conventional ligands, which suggests that the pFv-binding site is outside of the conventional antigen-binding pocket of Fab.[34,35]

In collaboration with Dr. Bouvet, we have begun to investigate the molecular basis for antibody binding of pFv by application of approaches and insights obtained from molecular studies of Fab-mediated SpA binding. The pFv binding capacity of a panel of over 50 human monoclonal IgMs have been evaluated in a direct-binding ELISA, in which about 40% of IgMs were found to bind pFv.

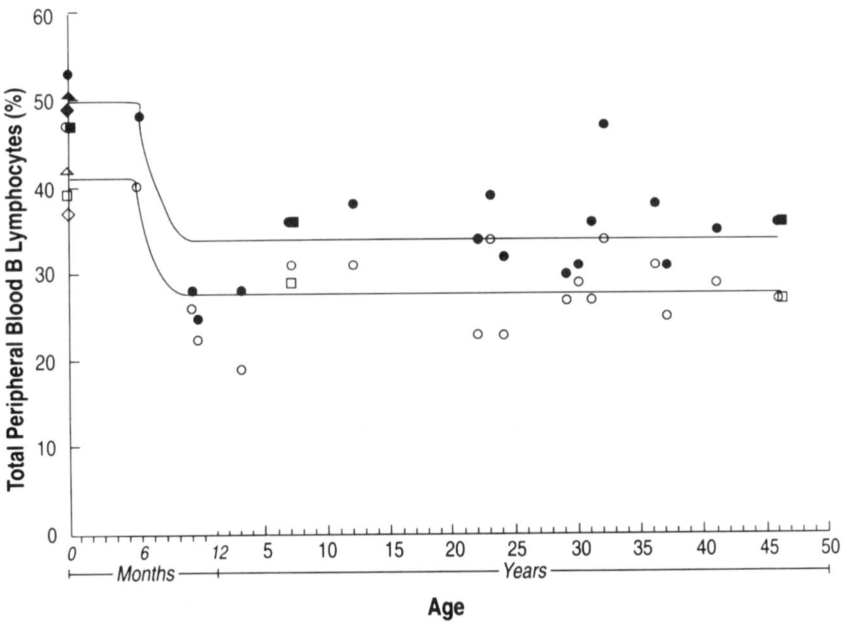

FIGURE 5. Age-associated changes in binding of Mod SpA by circulating CD5-positive and CD5-negative B lymphocytes. Mod SpA binding by CD5-positive and CD5-negative B-cell subsets from the peripheral blood of 23 healthy persons. The proportion of CD5-positive B cells that bind Mod SpA is displayed with a solid marker, while the proportion of CD5-negative B cells that bind Mod SpA in the sample is displayed with an open marker. The lines represent mean values for these B-cell subsets in neonates and in older persons. In all samples, a greater proportion of CD5-positive B cells bound Mod SpA relative to the CD5-negative B-cell subpopulation. There is an age-associated decrease in the binding of Mod SpA between 6 and 10 months of age in both the CD5-negative and CD5-positive peripheral B lymphocyte subpopulations. (Reproduced from Silverman *et al.*[10] with permission; copyright 1993, *The Journal of Immunology*.)

Competitive inhibition studies confirmed that the binding of structurally diverse IgMs are coinhibitable, suggesting that all IgM binding is directed to the same limited site(s) on pFv. The affinity of pFv binding by certain IgMs has also been evaluated, and was roughly in the same range detected for the binding of most V_H3 IgMs to SpA.[36]

Evaluations of variable-region usage have indicated a correlation of V_H usage with pFv binding with similarities to SpA binding but without the same rigid V_H family restriction. In preliminary studies, less than 20% of V_H1 and V_H4 IgMs bind pFv, whereas greater than 90% of V_H3 IgMs and 75% of V_H6 IgMs bind pFv, among IgMs tested to date. An equivalent V_H gene correlation has been detected among IgM Fab binders from an unselected combinatorial expression library made from the PBL of a healthy donor. In addition to corroborating results from V-region analysis of IgM binders, these data also indicate that this binding interaction does not involve a lectin-like activity of pFv inasmuch as bacterially produced Fabs are not glycosylated.

By use of the phage-display form of the IgM Fab library, we have also performed selection experiments with pFv. Colony hybridization studies using V_H family-specific probes also clearly indicate that pFv binding preferentially enriches for the representation of V_H3 and V_H6 IgM Fabs. Sequence analyses indicate that pFv binding clearly does not correlate with V_L usage or with particular V_H CDR3 sequence or J_H usage.

UNCONVENTIONAL B-CELL ANTIGENS AND HUMAN IMMUNE DEVELOPMENT

On the basis of differential reliance on the several somatic mechanisms responsible for the creation of antibody diversity, Klinman and coworkers have suggested that during B-cell ontogeny the developing repertoire can be divided into three sequential phases or tiers, and within the mature repertoire remnants of earlier tiers persist.[37] Each tier is likely an ontogenetic representation of different approaches to creating diversity within the immune repertoire that have been adopted during evolution. The first tier is created earliest *in utero*. It contains H chain rearrangements that have little or no N additions,[38] which results in limited diversity and skewing towards expression of a limited set of preferentially rearranged V_H gene segments[29] with recurrent usage of certain CDR3s that are created by homology-directed rearrangements.[39] Within this tier are a limited diversity of clonotypes, many of which are evolutionarily conserved. The second tier contains H chains that exhibit a greater influence of N-linked diversity with a resultant expansion in the range of V_H CDR3s represented in the repertoire. The third tier appears following birth and exposure to environmental ligands. Exogenous antigens begin to play a dominant role in B-cell clonal selection, and V regions of antibodies are highly affected by the hypermutation mechanism that is coupled to affinity maturation within the germinal centers of secondary lymphoid tissue.

The first tier has therefore been suggested to include a preimmune repertoire with biased representation of ligand-binding specificities useful for the survival of the otherwise immunologically naive neonate. As part of this theoretical model, it appears that the binding specificities for pFv and SpA are represented at high frequency, because they are linked to the basic design of many V_H gene segments. Thus, unconventional antigen-binding specificities likely are part of the foundation of the first tier. In support, pFv has recently been shown to aid in the agglutination of bacterial and viral pathogens,[35] facilitate activation of complement by immune complexes,[41] and aid cellular triggering via membrane-bound Ig.[42] Future studies will investigate the contribution of unconventional antigens to the acquisition of an effective immune defense.

REFERENCES

1. Poljak, R. J., L. M. Amzel, H. P. Avery, B. L. Chen, R. P. Phizackerley & F. Saul. 1973. The three-dimensional structure of the Fab′ fragment of a human immunoglobulin at 2.8 Å resolution. Proc. Natl. Acad. Sci. USA **70:** 3305–3309.
2. Kirkham, P. M., F. Mortari, J. A. Newton & H. W. Schroeder Jr. 1992. Immunoglobulin V_H clan and family identity predicts variable domain structure and may influence antigen binding. EMBO J. **11:** 603–609.
3. Padlan, E. A. 1994. Anatomy of the antibody molecule. Mol. Immunol. **31:** 169–217.
4. Langone, J. J. 1982. Protein A of *Staphylococcus aureus* and related immunoglobulin receptors produced by streptococci and pneumococci. Adv. Immunol. **32:** 157–252.
5. Inganas, M. 1981. Comparison of mechanism of interaction between protein A from *Staphylococcus aureus* and human monoclonal IgG, IgA and IgM in relation to the classical Fc and the alternative $F(ab')_2$ protein A interactions. Scand. J. Immunol. **13:** 343–352.
6. Harboe, M. & I. Folling. 1974. Recognition of two distinct groups of human IgM and IgA based on different binding of staphylococci. Scand. J. Immunol. **3:** 471–482.
7. Sasso, E. H., G. J. Silverman & M. Mannik. 1989. Human IgM molecules that bind staphylococcal protein A contain VHIII heavy chains. J. Immunol. **142:** 2778–2783.
8. Sasso, E. H., G. J. Silverman & M. Mannik. 1991. Human IgA and IgG $F(ab')_2$ that bind to staphylococcal protein A belong to the VHIII subgroup. J. Immunol. **147:** 1877–1883.
9. Romagnani, S., M. G. Guidizi, G. del Prete, E. Maggi, R. Biagiotti, F. Almerigogna & M. Ricci. 1982. Demonstration on protein A of two distinct immunoglobulin-binding sites and their role in the mitogenic activity of *Staphylococcus aureus* Cowan I on human B cells. J. Immunol. **129:** 595–602.
10. Silverman, G. J., M. Sasano & S. B. Wormsley. 1993. Age-associated changes in binding of human B lymphocytes to a VH3 restricted unconventional bacterial antigen. J. Immunol. **151:** 5840–5855.
11. Young, W. W., Y. Tamura, D. M. Wolock & J. W. Fox. 1984. Staphylococcal protein A binding to the Fab fragments of mouse monoclonal antibodies. J. Immunol. **133:** 3163–3166.
12. Vasquez-Kristiansen, S., V. Pascual & P. E. Lipsky. 1994. Staphylococcal protein A induces biased production of immunoglobulin by VH3 expressing B lymphocytes. J. Immunol. **153:** 2974–2984.
13. Levinson, A. I., Z. Yi, L. Wheatley, L. Kozlowski & S. Zorn. 1993. IgM rheumatoid factor produced in response to *S. aureus* Cowan I: Role of protein A and preferential utilization of VH3 heavy chains. J. Immunol. **150:** 237A (abstract).
14. Myhre, E. B. 1990. Interaction of bacterial immunoglobulin receptors with sites in the Fab region. *In* Bacterial Immunoglobulin-Binding Proteins. M. D. P. Boyle, Ed. Vol. **1:** 244–256.
15. Seppala, I., M. Kaartnen, S. Ibrahim & O. Makela. 1990. Mouse Ig coded by V_H families S107 and J606 bind to protein A. J. Immunol. **145:** 2989–2993.
16. Silverman, G. J., M. Sasano & S. B. Wormsley. 1993. The variable region specificity of bacterial Fab binding proteins: The search for B-cell superantigens. Immunomethods **2:** 17–32.
17. Silverman, G. J. 1992. Human antibody responses to bacterial antigens: Studies of a model conventional antigen and a proposed model B cell superantigen. Int. Rev. Immunol. **9:** 57–78.
18. Ibrahim, S., I. Seppala & O. Makela. 1993. V-region-mediated binding of human Ig by protein A. J. Immunol. **151:** 3597–3603.

19. HILLSON, J. L., N. S. KARR, I. R. OPPLIGER, M. MANNIK & E. H. SASSO. 1993. The structural basis of germline-encoded V_H3 immunoglobulin binding to staphylococcal protein A. J. Exp. Med. **178:** 331–336.
20. SASANO, M., D. R. BURTON & G. J. SILVERMAN. 1993. Molecular selection of human antibodies with an unconventional bacterial B-cell antigen. J. Immunol. **151:** 5822–5839.
21. SILVERMAN, G. J. 1994. Superantigens and the spectrum of unconventional B cell antigens. Immunologist **2:** 57–64.
22. SATOW, Y., G. H. COHEN, E. A. PADLAN & D. R. DAVIES. 1986. Phosphocholine binding immunoglobulin Fab McPc603: An X-ray diffraction study at 2.7 Å. J. Mol. Biol. **190:** 593–604.
23. RANDEN, I., K. N. POTTER, Y. LI, K. M. THOMPSON, V. PASCUAL, O. FORRE, J. B. NATVIG & J. D. CAPRA. 1993. Complementarity-determining region 2 is implicated in the binding of staphylococcal protein A to human immunoglobulin VHIII variable regions. Eur. J. Immunol. **23:** 2682–2686.
24. PULLEN, A. M., T. WADE, P. MARRACK & J. W. KAPPLER. 1990. Identification of the region of T cell receptor beta chain that interacts with the self-superantigen MIs-1a. Cell **61:** 1365–1374.
25. CAZENAVE, P., P. N. MARCHE, E. JOUVIN-MARCHE, D. VOEGTLE, F. BONHOMME, A. BANDEIRA & A. COUTINHO. 1990. $V_\beta17$ gene polymorphism in wild-derived mouse strains: Two amino acid substitutions in the $V_\beta17$ region greatly alter T cell receptor specificity. Cell **63:** 717–728.
26. PULLEN, A. M., J. BILL, R. T. KUBO, P. MARRACK & J. W. KAPPLER. 1991. Analysis of the interaction site for the self superantigen $MIs\text{-}1^a$ on T cell receptor V_β. J. Exp. Med. **173:** 1183–1192.
27. WHITE, J., A. PULLEN, K. CHOI, P. MARRACK & J. W. KAPPLER. 1993. Antigen recognition properties of mutant $V_\beta3^+$ T cell receptors are consistent with an immunoglobulin-like structure for the receptor. J. Exp. Med. **177:** 119–125.
28. BARBAS, C. F., A. S. KANG, R. A. LERNER & S. J. BENKOVIC. 1991. Assembly of combinatorial antibody libraries on phage surfaces: The gene III site. Proc. Natl. Acad. Sci. USA **88:** 7978–7982.
29. SCHROEDER, H. W., JR., J. L. HILLSON & R. M. PERLMUTTER. 1987. Early restriction of the human antibody repertoire. Science **238:** 791–793.
30. STEWART, A. K., C. HUANG, B. D. STOLLAR & R. S. SCHWARTZ. 1993. High-frequency representation of a single V_H gene in the expressed human B cell repertoire. J. Exp. Med. **177:** 409–418.
31. ROTT, I. & B. FLEISCHER. 1994. A superantigen as virulence factor in an acute bacterial infection. J. Infect. Dis. **169:** 1142–1146.
32. BOUVET, J.-P., R. PIRÈS, F. LUNEL-FABIANI, B. CRESCENZO-CHAIGNE, P. MAILLARD, D. VALLA & P. OPOLON. 1990. A novel F(ab)-binding factor, present in normal liver, and largely released in the digestive tract during hepatitis. J. Immunol. **145:** 1176–1180.
33. BOUVET, J.-P., R. PIRÈS, J. CHARLEMAGNE, J. PILLOT & S. ISCAKI. 1991. Non-immune binding of human protein Fv to immunoglobulins from various mammalian and nonmammalian species. Scand. J. Immunol. **34:** 491–496.
34. BOUVET, J.-P., R. PIRÈS, C. QUAN, S. ISCAKI & J. PILLOT. 1991. Non-immune V_H-binding specificity of human protein Fv. Scand. J. Immunol. **33:** 381–386.
35. BOUVET, J.-P., R. PIRÈS, S. ISCAKI & J. PILLOT. 1993. Non-immune macromolecular complexes of Ig in human gut lumen. J. Immunol. **151:** 5685–5698.
36. SILVERMAN, G. J., P. ROBEN, J.P. BOUVET & M. SASANO. Superantigen properties of a human sialoprotein involved in gut-associated immunity. Submitted.
37. DECKER, D. J. & N. R. KLINMAN. 1992. Interrelating B cell subpopulations and environmental regulation with the expression of three tiers of repertoire diversity. Intern. Rev. Immunol. **8:** 159–171.
38. SANZ, I. 1991. Multiple mechanisms participate in the generation of diversity of human H chain CDR3 region. J. Immunol. **147:** 1720–1729.
39. FEENEY, A. J. 1990. Lack of N regions in fetal and neonatal mouse immunoglobulin V-D-J junction sequences. J. Exp. Med. **172:** 1377–1390.

40. FEENEY, A. J. 1992. Predominance of VHDJH junctions occurring at sites of short sequence homology results in limited junctional diversity in neonatal antibodies. J. Immunol. **149:** 222–229.
41. RUFFET, E., R. PIRÈS, J. PILLOT & J.-P. BOUVET. 1994. Activation of the classical pathway of complement by non-immune complexes of immunoglobulins with human protein Fv (Fv fragment-binding protein). Scand. J. Immunol. **40:** 359–362.
42. PATELLA, V., J.-P. BOUVET & G. MARONE. 1993. Protein Fv produced during viral hepatitis is a novel activator of human basophils and mast cells. J. Immunol. **151:** 5685.

Effect of a Putative B Cell Superantigen on Complement

LISA M. KOZLOWSKI,[a] JOHN D. LAMBRIS,[b] AND ARNOLD I. LEVINSON[a,c]

[a]*Section of Allergy/Immunology*
[b]*Department of Pathology and Laboratory Medicine*
University of Pennsylvania School of Medicine
Philadelphia, Pennsylvania 19104

Staphylococcal protein A (SPA), a cell wall component of *S. aureus,* binds not only to the Fc fragment of IgG, but also via an alternative site to the Fab fragments of immunoglobulins (Igs) independent of heavy chain isotype.[1,2] In man, this binding is restricted to Igs with V_H3 heavy chains and occurs at a site(s) outside the conventional antigen-binding region.[2,3]

An increasing body of evidence has indicated that SPA shares many analogous properties with T cell superantigens, and it has thus been characterized as a B cell superantigen.[3,4] Unlike a T cell superantigen, however, a B cell superantigen has the ability to react with potentially large amounts of soluble B cell antigen receptors in the serum. Encountering such a large reservoir of reactive Igs could have deleterious effects on the host, particularly if this interaction leads to activation of the complement cascade with resultant tissue inflammation. The addition of SPA to serum has been shown to cause activation of the complement cascade. This activation has been attributed to the binding of the classical site on SPA to the Fc region of IgG.[5] It is also possible, however, that complement activation is caused by the interaction of SPA with Fab regions of V_H3^+ Igs. The generation of such complement-activating "immune complexes" by SPA may represent a novel and important biologic activity of a B cell superantigen.

Therefore, we sought to determine *in vitro* if the interaction of the alternative site on SPA with the Fab region of Ig molecules causes complement activation. Using a total hemolytic complement assay, we demonstrated that SPA, abrogated of its IgG Fc-binding activity by hyperiodination (mod-SPA), causes complement consumption when incubated with human serum (TABLE 1). To further test this hypothesis, we determined whether the interaction of SPA with polyclonal IgM or a panel of monoclonal IgM proteins (representative of human V_H gene families) led to C1q binding in an ELISA. These proteins were first analyzed in an SPA-binding ELISA to determine which bind to SPA. Our results (data not shown) demonstrated that polyclonal IgM bound SPA as well as four out of six V_H3^+ IgM proteins. No proteins from other V_H gene families had binding activity. Because only a subset of V_H3^+ Ig molecules is known to bind SPA, it was anticipated that some of our V_H3^+ IgM monoclonals would also not bind SPA.[3] To determine if the interaction of the IgM proteins with SPA led to binding of C1q, biotinylated SPA was incubated with either human polyclonal IgM or the monoclonal IgM

[c] Address for correspondence: Dr. Arnold I. Levinson, Section of Allergy/Immunology, University of Pennsylvania School of Medicine, 726 Clinical Research Building, 415 Curie Boulevard, Philadelphia, PA 19104.

TABLE 1. The Effect of mod-SPA on the Total Hemolytic Complement Activity of Human Serum

	Protein Added (240 μg/ml)	% Hemolysis Inhibition
Experiment #1	mod-SPA	76%
Experiment #2	mod-SPA	71%

NOTE: Normal human serum was incubated in the presence or absence of hyperiodinated SPA (mod-SPA) that was devoid of Fc IgG-binding activity. Hemolytic complement activity was then determined in a total hemolytic complement assay.

proteins (FIG. 1). Aliquots of the incubation mixtures were added to wells coated with C1q or the control protein BSA. After incubation and washing, streptavidin-peroxidase was added to the wells, which were subsequently developed with OPD substrate. In control experiments, no binding was observed when (1) biotinylated SPA alone was added to C1q-coated wells and (2) IgM was added to C1q-coated wells before incubation with biotinylated SPA. In addition, no binding to BSA-coated wells was detected under any of the above conditions. The interaction of SPA with polyclonal IgM led to binding of C1q. Because SPA will bind only V_H3^+ IgMs, we anticipated that C1q binding would be limited to V_H3^+ IgM/SPA

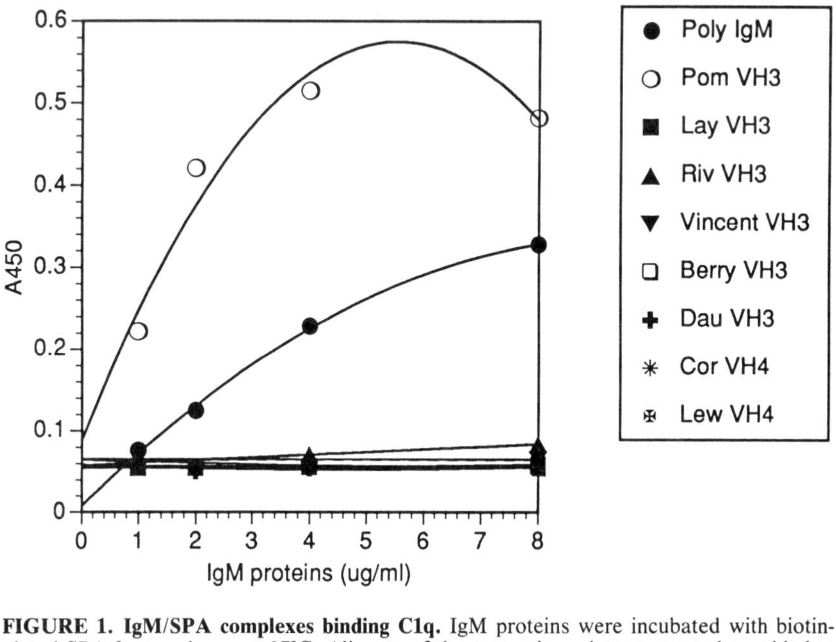

FIGURE 1. IgM/SPA complexes binding C1q. IgM proteins were incubated with biotinylated SPA for two hours at 37°C. Aliquots of these reaction mixtures were then added to C1q-coated wells (5 μg/ml) of an ELISA plate for two hours at 37°C. After washing, streptavidin-HRP was added for 1 hour at 37°C, and the wells were then washed with BSA/PBS. Wells were developed by addition of OPD substrate. Controls are as described in the text.

complexes. This hypothesis proved to be correct, although surprisingly, not all V_H3^+ IgM/SPA complexes bound C1q. We are currently evaluating these V_H3^+ proteins to determine the structural basis for their inability to bind C1q following interaction with SPA.

In summary, because addition of modified SPA to human serum leads to complement consumption and because the interaction of SPA with either polyclonal IgM or some V_H3^+ IgM proteins leads to binding of C1q, our results suggest that the interaction of IgM with the Fab-binding site on SPA leads to complement activation. Accordingly, the *in vivo* interaction of a B cell superantigen with secreted Igs may lead to complement activation and ensuing tissue inflammation.

REFERENCES

1. INGANAS, M. 1981. Comparison of mechanisms of interaction between protein A from *Staphylococcus aureus* and human monoclonal IgG, IgA, and IgM in relation to the classical Fcγ and the alternative F(ab')$_2\varepsilon$ protein A interactions. Scand. J. Immunol. **13:** 343–352.
2. ROMAGNANI, S., M. G. GUIDIZI, G. DEL PRETE, E. MAGGI, R. BIAGIOTTI, F. ALMERIGOGNA & M. RICCI. 1982. Demonstration on protein A of two distinct immunoglobulin-binding sites and their role in the mitogenic activity of *Staphylococcus aureus*. J. Immunol. **129:** 596–602.
3. SASSO, E. H., G. J. SILVERMAN & M. MANNIK. 1989. Human IgM molecules that bind staphylococcal protein A contain VHIII H chains. J. Immunol. **142:** 2778–2783.
4. SILVERMAN, G. J. 1992. Human antibody response to bacterial antigens: Studies of a model conventional antigen and a proposed model B cell superantigen. Int. Rev. Immunol. **9:** 57–78.
5. KRONVALL, G. & H. GEWURZ. 1970. Activation and inhibition of IgG-mediated complement fixation by staphylococcal protein A. Clin. Exp. Immunol. **7:** 211.

Structural Correspondences between Mouse and Human Immunoglobulin V_H Genes

Application to the Humanization of Mouse Monoclonal Antibodies

MARIE-ALIX POUL AND MARIE-PAULE LEFRANC

Laboratoire d'ImmunoGénétique Moléculaire
LIGM, IGMM, UMR 9942
1919, route de Mende, BP 5051
34033 Montpellier CEDEX 1, France

The use of murine monoclonal antibodies (mAbs) in immunotherapy is impaired by their limited half-life in circulation because patients raise human-anti-mouse antibodies. A solution for circumventing immunogenicity of murine mAbs *in vivo* is to genetically engineer "humanized" antibodies.[1] The hypervariable regions (complementarity-determining regions or CDR) of the light (L) and the heavy (H) chain of the mAb of interest are grafted on human framework regions (FR) to obtain humanized variable V_L and V_H regions that conserve high affinity and specificity for the antigen. Furthermore, humanized antibodies comprise constant C_L and C_H regions. The immunogenicity of the humanized mAb is reduced, and the effector functions depend on the selected C_H isotype. The choice of the human FRs for the construction of the humanized V_L and V_H regions is crucial for conservation of the antigen-binding site.[2] The antigen-binding site is formed by six loops: three from V_L (L1, L2, and L3) and three from V_H (H1, H2, and H3). Binding site loops, except for the H3 loop, have one of only a small number of main-chain conformations, designated as canonical structures (CS).[3] The CS of a particular loop is determined by its size and the presence of certain residues at key sites in both the loop and the FR regions. In an attempt to humanize the murine mAb K20, an anti-human CD29 antibody, known for its *in vitro* and *in vivo* immunosuppressive properties,[4] we elaborated a strategy based on structural correspondences between mouse and human V regions. We illustrate here the choice of the human V_H region used for the humanization of K20 mAb.

The K20 V_H gene belongs to the mouse VHIIB subgroup. The canonical structure of the K20 H1 loop of K20 V_H is class 1, whereas that of the H2 loop is class 2 (CS 1-2, as defined by Chothia *et al.*,[5]) (TABLE 1). The NEW V_H gene (from human subgroup V_HII) has been successfully used,[6,7] with additional modifications of its FR regions,[8] for the graft of the binding site of various mouse mAbs. However, the FR regions of NEW have only a low homology with those of K20 V_H gene (55%); furthermore, the CS class of NEW V_H is 1-1 (compared to 1-2 for K20 V_H) (TABLE 1). Because the genes belonging to human subgroup V_HI are more homologous to K20 V_H (>65%),[9] we selected, in that subgroup, the F1 V_H gene, whose FR regions present 71% identity with those of K20 V_H. This allows us to obtain, for the humanized K20-F1 V_H, first an H1 loop CS class identical to that of the murine K20 V_H (a deletion of R94 was introduced at the

TABLE 1. Canonical Structures of H1 and H2 Loops from K20[a] (Mouse Subgroup IIA), Newm (Human Subgroup II), F1 (Human Subgroup I) and Humanized K20-F1 V_H Genes

	H1 loop			H2 loop		Canonical structure class
Position	2 3 3 6 lab 5	2 4	9 4	5 5 5 0 2abc 8	7 1	
	* * * *	*	*	* *	*	
K20 V_H	GYTFTD--YYIS	A	—	RIYP--GSGNTF	A	1-2
	* * * *	*	*	*	*	
Newm V_H	GSTFSN--DYYT	V	R	YVF---YHGTSD	V	1-1
	* * * *	*	*	*	*	
F1 V_H	GYTYTL--YAMH	A	R	WING--GNGETK	R	1-3
	* * * *	*	*	* *	*	
K20-F1 V_H	GYTYTD--YYIS	A	—	RIYP--GSGNTF	R	1-2[b]

[a] H1 loop of K20 V_H is of CS class 1: The size of the loop is 7 residues (26 to 32). G26 produces a sharp turn, and F29, a hydrophobic residue, packs deep in the interior between hydrophobic A and I residues at positions 24 and 34, respectively. The conformation of this loop is also influenced by the residue at position 27 (Y), which packs into a surface cavity. Residue 94, which usually packs against H1 residues and influences the H1 loop conformation, is absent in the case of K20 V_H. H2 loop of K20 V_H is of CS class 2. The H2 loop is made of six residues (52 to 56), with small- or medium-sized hydrophobic residues at position 52a (P) and 71 (A).

[b] The H2 loop of canonical structure of class 2 generally has P at position 52a and G at position 55a, and packs against a small or medium hydrophobic residue at position 71.

* Indicates the residues at the key sites responsible for the canonical structure class.

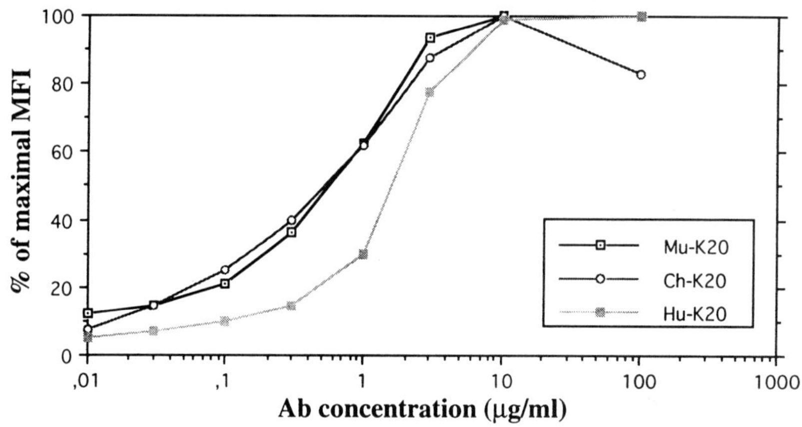

FIGURE 1. Fluorescence of CD29+ cells (Jurkat) stained with the mouse K20 (Mu-K20), the chimeric K20 (Ch-K20), or the humanized K20 (Hu-K20) mAbs. Jurkat cells (10^7/ml in PBS 0.1% sodium azide) were incubated for 15 min at 4°C with various concentrations of Hu-K20, Ch-K20, or irrelevant human IgG1 Abs, diluted in staining medium (0.1% sodium azide in PBS) then revealed with FITC-anti-human IgG Abs (15 min at 4°C). Cells were washed three times between each step and fixed in 0.1% paraformaldehyde before analysis on a FACScan (Becton Dickinson, Mountain View, CA). Forward and right-angle scatter were set in order to exclude dead cells and debris. As a reference, Mu-K20 binding on Jurkat cells was analyzed with FITC-Mu-K20. Negative control (irrelevant human IgG1) gave no staining (not shown). Results are expressed as percent relative to maximal MFI.

FR3 end of F1 V_H) and second an H2 loop that is almost identical for both genes (R71 in humanized K20-F1 V_H instead of A71 in murine K20 V_H) (TABLE 1). K20 Vκ gene, which belongs to the murine VκV subgroup, was humanized with a Vκ gene from the human VκI subgroup. The completely humanized (human Cκ/Cγ1) K20 mAb (Hu-K20) was constructed and produced in myeloma NS0 cells as described elsewhere.[8,10] The avidity of Hu-K20 for CD29 is slightly decreased compared to the avidity of a chimeric K20 mAb (composed of K20 V_H/V_L regions and human Cκ/Cγ1 regions, and is designated Ch-K20) (FIG. 1). Another version of the humanized K20-F1 V_H, with R71 changed in A71, will be produced in an attempt to check the influence of that residue, at that position, in the antigen-binding site.[11]

Recently, Chothia *et al.*[11] described the CS of a large number of human germline V_H genes in terms of structural repertoire. We compared the structural repertoire of mouse germline and rearranged V_H genes from the Kabat compendium.[9] Except for the 2-2 CS class, any combination of H1 and H2 loop CS classes found in mouse V_H genes (1-1, 1-2, 1-3, 1-4, 2-1, 3-1) can be found in the human V_H genes. The choice of human V_H genes for humanization of mouse mAbs should be facilitated by taking into account the structural homology of the V_H loop CS classes.

REFERENCES

1. WINTER, G. & C. MILSTEIN. 1991. Nature **349:** 293–299.
2. GORMAN, S. D., M. R. CLARK, E. G. ROUTLEDGE, S. P. COBBOLD & H. WALDMANN. 1991. Proc. Natl. Acad. Sci. USA **88:** 4181–4185.
3. GROUX, H., S. HUET, H. VALENTIN, D. PHAM & A. BERNARD. 1989. Nature **339:** 152–154.
4. CHOTHIA, C., A. M. LESK, A. TRAMONTANO, M. LEVITT, G. S. SMITH, G. AIR, S. SHERIFF, E. A. PADLAN, D. DAVIES, W. R. TULIP, P. M. COLMAN, S. SPINELLI, P. M. ALZARI & R. POLJAK. 1989. Nature **342:** 877–883.
5. CHOTHIA, C., A. M. LESK, E. GHERARDI, I. M. TOMLINSON, G. WALTER, J. D. MARKS, M. B. LLEWELYN & G. WINTER. 1992. J. Mol. Biol. **227:** 799–817.
6. JONES, P. T., P. H. DEAR, J. FOOTE, M. S. NEUBERGER & G. WINTER. 1986. Nature **321:** 522–325.
7. VERHOEYEN, M., C. MILSTEIN & G. WINTER. 1988. Science **239:** 1534–1536.
8. RIECHMANN, L., M. CLARK, H. WALDMANN & G. WINTER. 1988. Nature **332:** 323–327.
9. KABAT, E. A., T. T. WU, H. M. PERRY, K. S. GOTTESMAN & C. FOELLER. 1991. Sequences of Proteins of Immunological Interest. United States Department of Health and Human Services. Bethesda, MD.
10. POUL, M.-A., M. TICCHIONI, A. BERNARD & M.-P. LEFRANC. 1995. Mol. Immunol. **32:** 101–116.
11. TRAMONTANO, A., C. CHOTHIA & A. M. LESK. 1990. J. Mol. Biol. **215:** 175–182.

Structure of the V_H-D-J_H Segments of Human Natural Polyreactive IgM and IgG Antibodies[a]

Use of Germline or Somatically Mutated Forms of Commonly Expressed V_H Genes

HIDEYUKI IKEMATSU,[b] EDWARD W. SCHETTINO,[c] AND PAOLO CASALI[c]

[b]*Department of General Medicine*
Kyushu University Hospital
Fukuoka 813, Japan

[c]*Division of Molecular Immunology*
Department of Pathology
Cornell University Medical College
New York, New York 10021

It has been shown that sera of normal humans and normal (nonimmunized) animals contain antibodies that react with multiple antigens (Ags) including self-Ags. These antibodies have been termed "natural autoantibodies."[1–4] Most natural monoclonal autoantibodies generated from humans and mice react with multiple and dissimilar self- and foreign antigens, such as proteins, polysaccharides, nucleic acids, and haptens, including structural cellular and tissue components, soluble hormones, and other proteins. The relative avidities of natural polyreactive autoantibodies are different for different antigens, but are often of the same order of magnitude as those of monoreactive antibodies induced by antigen or those of monoreactive antibodies generated from patients with autoimmune diseases. Because of their broad range of reactivity, natural antibodies have been postulated to be involved in the elimination of cellular debris and, possibly, toxic substances. In addition, they may participate in the primary defense mechanisms against bacterial or viral aggressions before the specific immune response arises. It is unclear whether natural polyreactive autoantibodies can provide the templates for specific high-affinity autoantibodies or antibodies induced by antigen, as found, for instance, in autoimmune patients. If natural polyreactive autoantibodies provide the templates on which the pressure of an antigen selection process is exerted, they must (i) use V_H, D, and J_H gene segments similar to those used by high-affinity (auto)antibodies and (ii) be able to accumulate somatic mutations of characteristic nature and distribution.

To address the above issues, we cloned and sequenced the V_H-D-J_H genes of six IgM and five IgG human natural polyreactive antibodies. Their dose-saturable bindings to various Ags have been reported elsewhere.[5–7] Although all mAbs bind

[a] This work was supported by the U.S. Public Health Service Grant AR-40908. This is publication #4 from the Division of Molecular Immunology.

the same panel of multiple self and exogenous Ags, each mAb displays unique fine specificities.

The use of V_H, D, and J_H gene elements by the 11 polyreactive mAbs is summarized in TABLE 1. The expressed V_H gene sequences were 93.2 to 100% identical to those of reported germline V_H sequences.[8–14] Members of the V_H4 and V_H3 gene families were used by six (55%) and four (36%) mAbs, respectively. Members of these two gene families are frequently expressed in the early and adult human B cell repertoires. Overuse of V_H4 genes by the present polyreactive antibodies is consistent with their putative B-1a cell origin. Overuse of V_H4 family genes has been reported in human umbilical cord blood, where almost all B cells are $CD5^+$, and in adult $CD5^+$ B cells and $CD5^+$ chronic B cell leukemia.[15–17] Gene sequences identical or virtually identical to those of the V_H gene of the 11 polyreactive mAbs are also used by specific antibodies induced by foreign Ags in healthy subjects and/or autoantibodies in autoimmune patients. The polyreactive mAb D genes contained stretches of sequence that were identical to those of reported germline D genes.[18–21] The genes of the DXP family were used by seven mAbs. DLR, DK, and DN genes were used by two, two, and one mAb, respectively. The DXP family is prevalent in the fetal liver and adult B cell repertoires, comprising about 30% of the expressed D gene elements,[20,21] suggesting that D gene utilization by polyreactive mAbs may be comparable to that in the fetal and adult B cell repertoires. The J_H4 gene was used by six mAbs (54%) and the J_H6 by three (27%). The J_H1 and the J_H3 genes were each used by a single mAb (9%). Reported frequencies of J_H4, J_H6, J_H1, and J_H3 gene expression in the adult B cell repertoire are 53%, 22%, 1%, and 9%, respectively.[20] Thus, the distribution of J_H gene usage by the 11 polyreactive mAbs reflected that of the adult B cell repertoire. These findings suggested that the pool of V_H, D, and J_H genes used by the present polyreactive IgM and IgG autoantibodies overlaps with that of V_H, D, and J_H gene elements commonly expressed in the early and late human B cell repertoires and recruited in high-affinity immune response to foreign Ags in healthy subjects or to self-Ag in autoimmune patients.

The results of the comparisons of the expressed V_H gene sequences to those of the closest germline V_H genes are given in TABLE 1. The high degree of conservation of human V_H genes would suggest that the majority of the nucleotide differences displayed by the expressed V_H gene sequences compared with the closest germline V_H gene sequences results from somatic mutation. The sequence of the mAb103 V_H gene was that of a perfect germline V_H26c gene. Three mAb V_H genes (mAb102, mAb104, and mAb410.7F91) had no replacement mutation in the CDRs. One mAb (mAb426.4.2F20) showed only one putative replacement mutation in the CDRs. The total putative mutations in these five mAb FRs were fewer than three. The frequencies of putative mutations in the CDRs and the FRs were 0.9×10^{-2} and 0.9×10^{-2} per base, respectively. The putative replacement to silent (R/S) mutation ratios were comparable in the CDRs (0.5) and in the FRs (0.7). Three mAbs (mAb55, mAb52, and mAb59) contained three or four putative replacement mutations in the V_H segment CDRs. They also displayed three or four silent mutations in the FRs. Overall putative mutations in these three mAb V_H segment sequences were 21 in the CDRs and 24 in the FRs, yielding 1.1×10^{-1} and 3.5×10^{-2} changes per base, respectively. The putative R/S mutation ratios were 0.9 and 1.0 in the CDRs and the FRs, respectively. Three mAbs (mAb426.12.3F1.4, mAb410.30F305, and mAb10) contained three or four R but few S putative mutations in the CDRs. Putative mutations in these three mAb V_H segment sequences were 12 in the CDRs and 31 in the FRs, yielding 6.1×10^{-2} and 4.5×10^{-2} changes per base, respectively. The putative R/S mutation ratio

TABLE 1. Utilization of V_H, D, and J_H Gene Elements by Human Polyreactive IgM and IgG mAbs

Clone	Origin	Chains H, L	V_H Gene Family	Closest V_H Gene	Nucleotide Identity (%)	Nucleotide Differences CDR R S	Nucleotide Differences FR R S	D Gene	J_H Gene
mAb103	HS[a]	IgM, λ	V_H3	V_H26c	100	0:0	0:0	DN2	J_H4
mAb102	HS[a]	IgM, λ	V_H4	4.33	99.0	0:0	1:2	D21/10	J_H4
mAb104	HS[a]	IgM, λ	V_H4	$V_H4.21$	98.7	0:1	1:2	DXP′1	J_H1
mAb410.7F91	SLE[b]	IgG1, κ	V_H4	$V_H4.21$	99.0	0:0	1:2	DXP′1	J_H4
mAb426.4.2F20	IDDM[c]	IgG1, κ	V_H4	$V_H4.11$	99.0	1:1	1:0	DXP′4	J_H6
mAb55	HS[a]	IgM, κ	V_H3	H11	96.9	3:3	2:3	DXP′1	J_H6
mAb52	HS[a]	IgM, κ	V_H3	22-2B	93.2	3:4	5:8	DLR4	J_H6
mAb59	HS[a]	IgM, λ	V_H3	V_H26c	95.2	4:4	5:1	D21/9	J_H4
mAb426.12.3F1.4	IDDM[c]	IgG1, κ	V_H4	$V_H4.11$	95.9	3:0	2:7	DK4	J_H3
mAb410.30F305	SLE[b]	IgG3, κ	V_H1	hv1263	95.9	4:0	2:6	DK4	J_H4
mAb10	HS[a]	IgG3, λ	V_H4	$V_H4.18$	93.6	4:1	6:8	DXP′4	J_H4

[a] HS, healthy subject.
[b] SLE, systemic lupus erythematosus.
[c] IDDM, insulin-dependent diabetes mellitus.

was significantly higher in the CDRs (11.0) than in the FRs (0.5). The features (distribution and nature) of the mutations observed in the last three mAb V_H segments are consistent with those resulting from a process of Ag-driven clonal selection.[22–25] Thus, according to the configuration of the V_H gene segments, the present human polyreactive autoantibodies can be divided into three groups: (i) those with V_H gene segments in germline or virtual germline configuration, (ii) those with mutated but not Ag-selected V_H gene segments, and (iii) those with mutated and Ag-selected V_H gene segments. Whereas the first and second groups contained mainly IgMs, the third group consisted of IgGs only.

To verify the somatic mutational nature of the differences found in the V_H segment sequences of polyreactive autoantibodies when compared with the sequences of the closest germline V_H gene segments, we performed experiments according to the strategy outlined in FIGURE 1A and involving the use of *ad hoc* synthesized PCR primer as well genomic DNA extracted from mAb-producing hybridoma cells and autologous PMNs. We synthesized the sense primer 58-5, encompassing a FR1 sequence shared by the V_H4.11 and mAb426.12.3F1.4 V_H genes. We also synthesized the antisense primer 12.3-3, the sequence of which was identical with the reverse complement of a 5′ portion of the FR3 region of mAb 426.12-3F1.3 V_H sequence, but differed in four nucleotides from the reverse complement of the corresponding V_H4.11 sequence. We chose this FR3 area because it was the only one throughout the mAb426.12.3F1.4 V_H gene that contained at least four nucleotide differences when compared with the V_H4.11 gene sequence. The PCR involving the 58-5 and 12.3-3 primers resulted in amplification of DNA from the mAb426.12.3F1.4-producing B cell hybridoma DNA, but not from autologous PMN DNA. The molecular size of the amplified product (~160 bp) was consistent with that of the sequence-spanning residues 65 to 226 (161 bp) of the expressed V_H gene (FIG. 1B). The PCR amplified product was cloned and sequenced, and the sequence of this DNA fragment (58-5/12.3-3 sequence) was identical with the corresponding sequencing of the mAb426.12.3F1.4 V_H gene (FIG. 1A). These findings suggested that the expressed V_H gene was somatically mutated, and were confirmed by the results of the subsequent experiments. We synthesized the antisense primer 58-3, the sequence of which was identical with the reverse complement of a V_H4.11 FR3 sequence and differed in two bases from the reverse complement of the corresponding mAb426.12.3F1.4 V_H gene sequence. PCR involving the 58-5 and 58-3 primer pair yielded products of the expected and identical size (~200 bp) when applied to the amplification of not only B cell hybridoma, but also autologous PMN DNA (FIG. 1C). These findings demonstrated that the failure to amplify any DNA from autologous PMN DNA using the 58-5 and 12.3-3 primers was not due to flaws inherent to the DNA preparation. The PCR products shown in FIG. 1C were then hybridized with the 12.3-3 oligonucleotide as a probe. The 12.3-3 nucleotide hybridized with DNA amplified from B cell hybridoma, but not with that amplified from PMN (FIG. 1D), also suggesting that the expressed V_H gene was somatically mutated. The DNA fragment PCR amplified from the PMN DNA was used to identify, by cloning and sequencing, the putative germline gene that gave rise to the expressed mAb426.12.3F1.4 V_H gene. Five independent clones were analyzed and found to contain an identical V_H gene sequence. This sequence G411 was identical with the sequence of the V_H4.11 gene throughout the overlapping area (169 bp) (FIG. 1A). These experiments proved that the expressed mAb 426.12.3F1.4 V_H gene was somatically mutated and strongly suggested that V_H4.11 was the germline gene that gave rise to it. They also indicated that somatic mutations can accumulate in the V_H segments of polyreactive autoantibodies, as they do in high-affinity monoreactive antibodies.

The deduced amino acid sequences of the D and J_H genes of the polyreactive IgM and IgG mAbs are depicted in FIGURE 2. The deduced amino acid sequences of the expressed V_H-D-J_H joining regions were segregated into CDR3 and FR4 according to Kabat *et al.*[26] The lengths and the deduced amino acid sequences of the CDR3s were highly divergent, whereas those of the FR4s were highly conserved. The CDR3 of two mAbs (mAb426.4.2F20 and mAb10) displayed an identical stretch of five amino acids (RFLEW) that derived from the use of the DXP′4 gene in the same reading frame. No other characteristic common amino acid

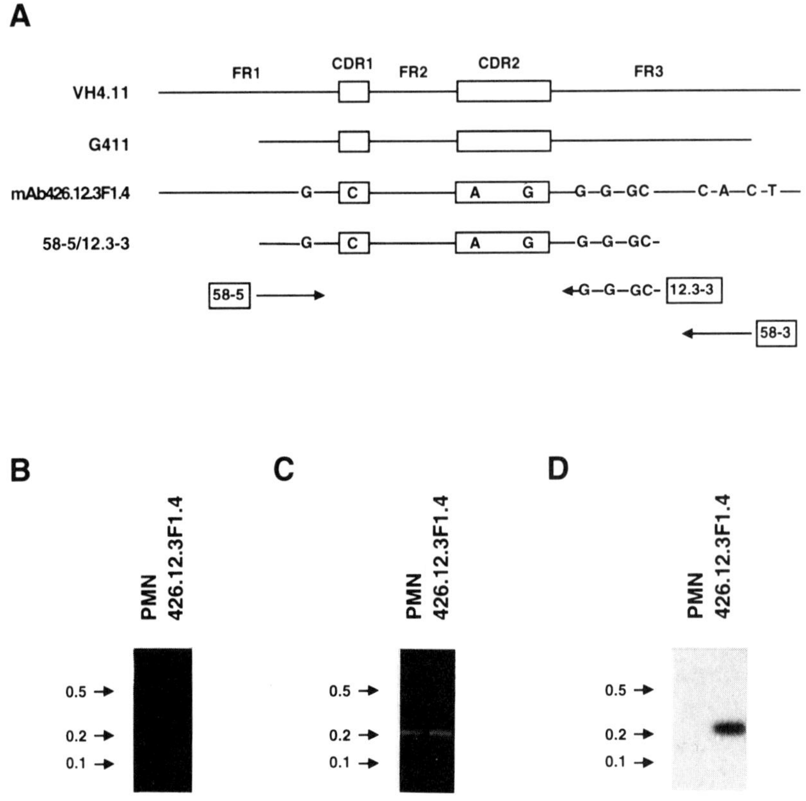

FIGURE 1. Analysis of somatic mutations in the expressed mAb426.12.3F1.4 V_H gene. **(A)** Nucleotide sequence of mAb426.12.3F1.4 V_H gene and the location of the primers used in PCR amplification experiments. *Horizontal arrows* indicate the location of primers and *arrowheads* indicate the direction for PCR amplification. Letters in the line indicate the nucleotide differences. **(B)** Ethidium bromide staining of PCR-amplified products using the 58-5 and 12.3-3 primers and genomic DNA extracted from autologous PMNs (*left lane*) or the mAb426.12.3F1.4-producing B cell hybridoma (*right lane*). **(C)** Ethidium bromide staining of the DNA PCR amplified using the 58-5 and 58-3 primers and genomic DNA extracted from autologous PMNs (*left lane*) or the mAb426.12.3F1.4-producing B cell hybridoma (*right lane*). **(D)** Southern blot hybridization of the PCR-amplified products shown in C with the ^{32}P-labeled 12.3-3 oligonucleotide probe.

Antibody	CDR3	FR4
mAb103	SSVGWYHTVDY	WGQGTLVTVSS
mAb102	TLGDYVWGSYRKSYFDY	-----------
mAb104	SKTFGTGSYSPLRRGPPYFQH	-----------
mAb410.7F91	GRYYHGSGRQPFQY	-----------
mAb426.4.2F20	TISGGRLRFLEWTRGGNYYYYMDV	--K--T-----
mAb55	LKVGFGEPEGWVLWGQAKIPREYYYYGMDV	-----T-----
mAb52	YPHCSSTSCYVYYYYGMDV	-----T-----
mAb59	VVGGTRRRGLTLMIVVYYFDY	-----------
mAb426.12.3F1.4	RRWAMVGKKYAFDV	-----M-----
mAb410.30F305	DTSIYGDY	-----------
mAb10	HFNARTPTVRFLEWLSGGEIDY	----I--S---

FIGURE 2. Deduced amino acid sequences of the V_H-D-J_H joining region of the 11 polyreactive mAbs. The deduced amino acid sequences of the expressed V_H-D-J_H regions were segregated into CDR3 and FR4 according to Kabat *et al.*[26] Dashes indicate amino acid identities.

sequence motif could be detected in the eleven CDR3s. The H chain CDR3 may provide the major structural correlate for polyreactivity,[27] but its heterogeneity in length and deduced amino acid sequence possibly accounts for the unique fine specificity of each polyreactive mAb.

In conclusion, our results suggest that V_H regions of the polyreactive autoantibodies are encoded by commonly used V_H, D, and J_H genes. The vast variation of V_H-D-J_H recombination and the highly divergent CDR3 structure are consistent with the distinctive fine specificities of different polyreactive autoantibodies. Somatic mutations can accumulate in the V_H segments of polyreactive antibodies and be selected by Ag, as they do in high-affinity monoreactive antibodies, without loss of polyreactivity. The nature of the Ag or Ags initially recruiting polyreactive germline clonotypes, and subsequently responsible for applying the selective pressure necessary for affinity maturation, remains to be determined.

REFERENCES

1. VINCENT, C. & J. P. REVILLARD. 1983. Autoantibodies specific for β2 microglobulin in normal human serum. Mol. Immunol. **20:** 877–884.
2. BENDTZEN, K., M. SVENSON, V. JONSSON & E. HIPPE. 1990. Autoantibodies to cytokines—friends of foes? Immunol. Today **11:** 167–169.
3. AVRAMEAS, S. 1991. Natural autoantibodies. Immunol. Today **12:** 154–159.
4. RIBOLDI, P., M. T. KASAIAN, L. MANTOVANI, H. IKEMATSU & P. CASALI. 1993. Natural antibodies. *In* The Molecular Pathology of Autoimmune Disease. C. A. Bona, K. Siminovitch, M. Zanetti & A. N. Theofilopoulos, Eds.: 45–64. Harwood Academic Publishers GmbH. Chur.
5. UEKI, Y., I. S. GOLDFARB, N. HARINDRANATH, M. GORE, H. KOPROWSKI, A. L. NOTKINS & P. CASALI. 1990. Clonal analysis of a human antibody response: Quantitation of precursors of antibody-producing cells and generation and characterization of monoclonal IgM, IgG, and IgA to rabies virus. J. Exp. Med. **171:** 19–34.
6. HARINDRANATH, N., H. IKEMATSU, A. L. NOTKINS & P. CASALI. 1993. Structure of the V_H and V_L segments of polyreactive and monoreactive human natural antibodies to HIV-1 and *Escherichia coli* β-galactosidase. Intl. Immunol. **5:** 1523–1533.

7. IKEMATSU, H., M. T. KASAIAN, E. W. SCHETTINO & P. CASALI. 1993. Structural analysis of the V_H-D-J_H segments of human polyreactive IgG mAb. J. Immunol. **151:** 3604–3616.
8. WENG, N., J. G. SNYDER, L. YU-LEE & D. M. MARCUS. 1992. Polymorphism of human immunoglobulin V_H4 germ-line genes. Eur. J. Immunol. **22:** 1075–1082.
9. VAN ES, J. H., M. HEUTININK, H. AANSTOOT & T. LOGTENBERG. 1992. Sequence analysis of members of the human Ig V_H4 gene family derived from a single V_H locus. J. Immunol. **149:** 492–497.
10. CHEN, P. P., M.-F. LIU, S. SINHA & D. A. CARSON. 1988. A 16/6 idiotype-positive anti-DNA antibody is encoded by a conserved V_H gene with no somatic mutation. Arthritis Rheum. **31:** 1429–1431.
11. SANZ, I., P. KELLY, C. WILLIAMS, S. SCHOOL, P. TUCKER & J. D. CAPRA. 1989. The smaller human V_H gene families display remarkably little polymorphism. EMBO J. **8:** 3741–3748.
12. BERMAN, J. E., S. J. MELLIS, R. POLLOCK, C. L. SMITH, H. SUH, B. HEINKE, C. KOWAL, U. SURTI, L. CHESS, C. R. CANTOR & F. W. ALT. 1988. Content and organization of the human Ig V_H locus: Definition of three new V_H families and linkage to the Ig C_H locus. EMBO J. **7:** 727–738.
13. RECHAVI, G., B. BIENZ, D. RAM, Y. BEN-NERIAH, J. B. COHEN, R. ZAKUT & D. GIVOL. 1982. Organization and evolution of immunoglobulin V_H gene subgroups. Proc. Natl. Acad. Sci. USA **79:** 4405–4409.
14. CHEN, P. P., M.-F. LIU, C. A. GLASS, S. SINHA, T. J. KIPPS & D. CARSON. 1989. Characterization of two immunoglobulin V_H genes that are homologous to human rheumatoid factors. Arthritis Rheum. **32:** 72–76.
15. MAGEED, R. A., L. E. MACKENZIE, F. K. STEVENSON, B. YUKSEL, F. SHOKRI, B. R. MAZIAK, R. JEFFERIS & P. M. LYDYARD. 1991. Selective expression of a V_HIV subfamily of immunoglobulin genes in human $CD5^+$ B lymphocytes from cord blood. J. Exp. Med. **174:** 109–113.
16. SANZ, I., P. CASALI, J. W. THOMAS, A. L. NOTKINS & J. D. CAPRA. 1989. Nucleotide sequences of eight human natural autoantibodies V_H region reveals apparent restricted use of V_H families. J. Immunol. **142:** 4054–4061.
17. IKEMATSU, W., H. IKEMATSU, S. OKAMURA, T. OTSUKA, M. HARADA & Y. NIHO. 1994. Surface phenotype and Ig heavy-chain gene usage in chronic B-cell leukemias: Expression of myelomonocytic surface markers in $CD5^-$ chronic B-cell leukemia. Blood **83:** 2602–2610.
18. ICHIHARA, Y., H. MATSUOKA & Y. KUROSAWA. 1988. Organization of human immunoglobulin heavy chain diversity gene loci. EMBO J. **7:** 4141–4150.
19. RAVETCH, J., U. SIEBENLIST, S. KORSMEYER, T. WALDMANN & P. LEDER. 1981. Structure of the human immunoglobulin m locus: Characterization of embryonic and rearranged J and D genes. Cell **27:** 583–591.
20. YAMADA, M., R. WASSERMAN, B. A. RICHARD, S. SHANE, A. J. CATON & G. ROVERA. 1991. Preferential utilization of specific immunoglobulin heavy chain diversity and joining segments in adult human peripheral blood B lymphocytes. J. Exp. Med. **173:** 395–407.
21. SANZ, I. 1991. Multiple mechanism participate in the generation of diversity of human H chain CDR3 regions. J. Immunol. **147:** 1720–1729.
22. WEIGERT, M. 1986. The influence of somatic mutation on the immune response. *In* Progress in Immunology VI. B. Cinader & R. G. Miller, Eds.: 138–144. Academic Press, Inc. New York.
23. KOCKS, C. & K. RAJEWSKY. 1989. Stable expression and somatic hypermutation of antibody V regions in B-cell developmental pathways. Ann. Rev. Immunol. **7:** 537–559.
24. FRENCH, D. L., R. LASKOV & M. D. SCHARFF. 1989. The role of somatic hypermutation in the generation of antibody diversity. Science **244:** 1152–1157.
25. IKEMATSU, H., N. HARINDRANATH, U. UEKI, A. L. NOTKINS & P. CASALI. 1993. Clonal analysis of a human antibody response. II. Sequences of the V_H genes of human monoclonal IgM, IgG, and IgA to rabies virus reveal preferential utilization of V_HIII segments and somatic hypermutation. J. Immunol. **150:** 1325–1337.

26. KABAT, E., T. WU, H. BILOFSKY, M. REID-MILNER & H. PERRY. 1991. Sequences of Proteins of Immunological Interest. Fifth Ed. U.S. Department of Health and Human Services. Washington D.C.

27. ICHIYOSHI, Y. & P. CASALI. 1994. Analysis of the structural correlates for antibody polyreactivity by multiple reassortments of chimeric human IgH and L chain V segments. J. Exp. Med. **180:** 885–895.

Structure–Function Studies of Human Monoclonal Antibodies to Pneumococcus Type 3 Polysaccharide

DENISE R. SHAW,[a,b] PERRY KIRKHAM,[b]
HARRY W. SCHROEDER, JR.,[b] PAUL ROBEN,[c]
AND GREGG J. SILVERMAN[c]

[b]*Division of Hematology and Oncology*
University of Alabama at Birmingham
Birmingham, Alabama 35294

[c]*Department of Medicine*
University of California at San Diego
La Jolla, California

Antibodies to *Streptococcus pneumoniae* type-specific capsular polysaccharide (PPS) are important for resistance to pneumococcal disease.[1,2] Although PPS-specific humoral immunity has been studied for over a century, virtually nothing is known about the immunoglobulin V genes associated with protective versus nonprotective antibodies in either man or experimental animals.

We have characterized two human monoclonal antibodies (designated HUAB14-3 and HUAB2-3) with specificity for type 3 PPS (PPS-3) derived from two immunized young adults. PPS-3 is of note because (a) the antigen has a simple disaccharide repeat structure, (b) type 3 organisms are commonly isolated from patients with pneumococcal disease, and (c) this PPS type has been widely used in both basic and clinical investigations.[3]

Monoclonal antibodies HUAB14-3 and HUAB2-3 are both IgM and exhibit similar binding specificity and avidity when tested *in vitro* with purified carbohydrate antigens. Yet the two antibodies exhibit strikingly divergent activities in assays with intact type 3 pneumococci (TABLE 1).[4–8] Additionally, antigen polyreactivity was readily detected with HUAB2-3, but not the more biologically active HUAB14-3, when tested against nonprotein antigens, including nitrophenyl haptens conjugated with protein carriers (e.g., DNP-OVA, NP-BSA, NIP-BSA, and ssDNA in FIG. 1). Of interest was the ability of soluble, unconjugated NIP hapten to competitively inhibit binding of HUAB2-3 to the nominal PPS-3 antigen, by >50% at 0.1 mM NIP, suggesting that polyreactivity of HUAB2-3 is mediated by antibody structures near or identical to the PPS-3 binding site.

Sequencing of antibody-variable region cDNA (GenBank accession numbers L06101, L06102, L19893, L19894) and three-dimensional modeling of each antigen-

[a] Address for correspondence: Dr. Shaw, Division of Hematology and Oncology, University of Alabama at Birmingham, 690 WTI, UAB Station, Birmingham, AL 35294-3300.

TABLE 1. Summary of Monoclonal Antibody Properties

Characteristic	HUAB14-3	HUAB2-3
Antibody isotype	μ, κ	μ, λ
V_H region	V_H3 DXP′1 J_H5b	V_H3 DHFL16 J_H5b
V-J germline homology	97%	99%
CDR3 length (amino acids)	7	15
	SpA binder	SpA binder
V_L region	$V\kappa4$ $J\kappa4$	$V\lambda7$ $J\lambda2\&3$
V-J germline homology	96%	100%
Polysaccharide reactivity		
PPS (20 types tested)	PPS-3 only	PPS-3 only
Phosphocholine	Neg	Neg
α-1,3-Dextran	Neg	Neg
H. influenzae type b	Neg	Neg
PPS-3 binding constants with purified antigens		
k_{on} ($M^{-1}sec^{-1}$)	5.1×10^4	7.4×10^4
k_{off} (sec^{-1})	1.0×10^{-4}	2.5×10^{-4}
K (k_{on}/k_{off}, M^{-1})	5.0×10^8	3.0×10^8
Assays with type 3 pneumococcal bacteria		
Flow cytometry (MFI at 100 μg/ml)	3,190	52
Opsonophagocytosis (killing at 0.1 μg/ml)	100%	3%
Mouse protection (alive at 10 μg/mouse i.p.)	85%	17%
Polyreactivity	None detected	NIP, NP, DNP, ssDNA

ABBREVIATIONS: V_H, V_L, immunoglobulin variable region heavy and light chain genes, respectively; J_H, immunoglobulin joining gene; CDR, complementarity determining region; SpA, staphylococcal protein A; PPS, pneumococcal type-specific capsular polysaccharide; k or K, binding constants determined by BIAcore™ surface plasmon resonance analysis; MFI, mean fluorescence intensity; NP, nitrophenylacetic acid; NIP, 3-iodo-4-hydroxyl-5-nitrophenylacetic acid; DNP, dinitrophenyl; ssDNA, single-stranded DNA.

binding domain[9,10] revealed differences between the two antibodies that may relate to the observed biologic activities. HUAB2-3 has relatively long CDR regions that in modeling produced a relatively flat antigen-binding surface. In contrast, HUAB14-3 has shorter CDR regions and modeled with an antigen-binding surface groove of size and charge characteristics compatible with binding to PPS-3 antigen. The more active HUAB14-3 was also more somatically divergent by homology with reported germline-variable gene sequences (TABLE 1).

The data generated by this study provide a basis for further investigations into the molecular/genetic characteristics of protective antibody responses to type 3 pneumococci. The recent emergence of antibiotic-resistant *S. pneumoniae* clinical isolates, and the failure of currently approved PPS vaccines to elicit effective immunity among many patients at risk for pneumococcal disease, suggests that these studies might have a significant impact on the control of pneumococcal disease.

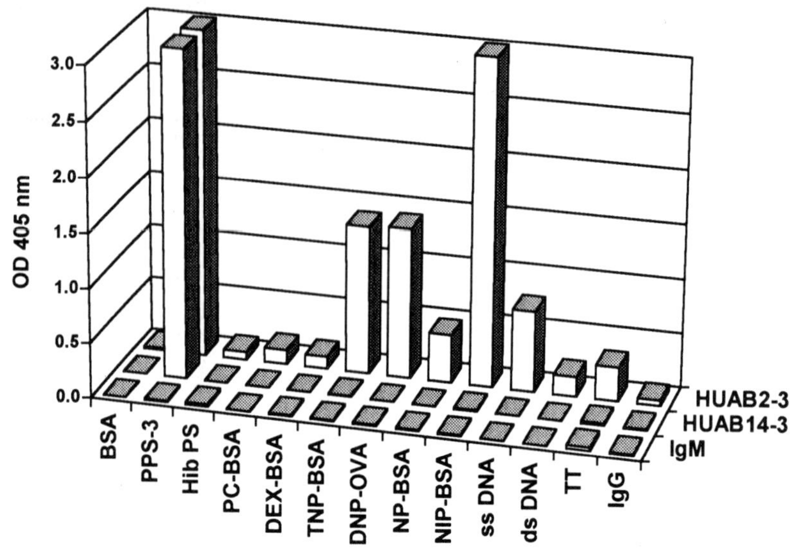

FIGURE 1. Binding of HUAB14-3, HUAB2-3, and control polyclonal human IgM to antigens as determined by ELISA. Binding to solid-phase antigens was determined with alkaline-phosphatase-conjugated goat anti-human IgM and quantitated by absorbance at 405 nm following addition of alkaline phosphatase substrate. Antigens: BSA, bovine serum albumin; PPS-3, type 3 pneumococcal capsular polysaccharide; HibPS, *Haemophilus influenzae* type b capsular saccharide; PC-BSA, phosphocholine conjugated to BSA; DEX-BSA, α-1,3-dextran BSA conjugate; TNP-BSA, trinitrophenyl hapten BSA conjugate; DNP-OVA, dinitrophenyl hapten ovalbumin conjugate; NP-BSA, nitrophenylacetic acid hapten BSA conjugate; NIP-BSA, 3-iodo-4-hydroxyl-5-nitrophenylacetic acid hapten BSA conjugate; ssDNA, single-stranded DNA; dsDNA, double-stranded DNA; TT, tetanus toxoid; IgG, polyclonal human serum IgG.

REFERENCES

1. Austrian, R. 1981. Pneumococcus: The first one hundred years. Rev. Infect. Dis. **3:** 183–189.
2. Shapiro, E. D., A. T. Berg, R. Austrian, D. Schroeder, V. Parcells, A. Margolis, R. K. Adair & J. D. Clemmens. 1991. Protective efficacy of polyvalent pneumococcal polysaccharide vaccine. N. Engl. J. Med. **325:** 1453–1460.
3. Jennings, H. J. 1983. Capsular polysaccharides as human vaccines. Adv. Carbohydr. Chem. Biochem. **41:** 155–208.
4. Gray, B. M. 1979. ELISA methodology for polysaccharide antigens: Protein coupling of polysaccharides for adsorption to plastic tubes. J. Immunol. Methods **28:** 187–192.
5. Gray, B. M. & D. R. Shaw. 1993. Artifacts with the thiocyanate elution method for estimating relative antibody avidity. J. Immunol. Methods **157:** 269–271. [Published erratum appears in J. Immunol. Methods **161:** 277, 1993.]
6. Karlsson, R., A. Michaelsson & L. Mattsson. 1991. Kinetic analysis of monoclonal antibody–antigen interactions with a new biosensor analytical system. J. Immunol. Methods **145:** 229–240.
7. Gray, B. M. 1990. Opsonophagocidal activity in sera from infants and children immunized with *Haemophilus influenzae* type b conjugate vaccine (meningoccoccal protein conjugate). Pediatrics **85:** 694–697.

8. BRILES, D. E., C. FORMAN & M. CRAIN. 1992. Mouse antibody to phosphocholine can protect mice from infection with mouse-virulent human isolates of *Streptococcus pneumoniae*. Infect. Immun. **60:** 1957–1962.
9. MARQUAT, M., J. DEISENHOFER, R. HUBER & W. PALM. 1980. Crystallographic refinement and atomic models of the intact immunoglobulin molecule Kol and its antigen-binding fragment at 3.0 angstroms and 1.9 angstroms resolution. J. Mol. Biol. **141:** 369–391.
10. SATOW, Y., G. H. COHEN, E. A. PADLAN & D. R. DAVIES. 1986. Phosphocholine binding immunoglobulin Fab McPC603: An X-ray diffraction study at 2.7 angstroms. J. Mol. Biol. **190:** 593–604.

The Immunoglobulin Gene Repertoire to *Haemophilus influenzae* Type b

WILLIAM L. CARROLL,[a,b]
PENELOPE G. SHACKELFORD,[c]
AND ELISABETH E. ADDERSON[a]

[a]*Eccles Program in Human Molecular Biology and Genetics and Department of Pediatrics University of Utah Salt Lake City, Utah 84107*

[c]*Division of Infectious Diseases Mallinckrodt Department of Pediatrics Washington University School of Medicine St. Louis, Missouri 63110*

To examine the human antibody repertoire generated against a biologically significant antigen, we have obtained sequences of heavy and light chain immunoglobulin variable region genes from monoclonal antibodies specific for the capsular polysaccharide (PS) of the important bacterial pathogen *Haemophilus influenzae* type b (Hib).[1,2]

Variable (V) region gene usage in the anti-Hib PS heavy chain variable region response is notable for its marked restriction and relative paucity of somatic mutation. All V_H segments are members of the V_H3 gene family. Approximately half of these rearranged genes are members of the V_H3b subfamily and are 94 to 99% homologous to the germline genes 9.1, LSG6.1, and LSG12.1 (FIG. 1).[2] The remainder are V_H3a subfamily members that are 85 to 98% homologous to the germline gene V_H26. Using a polymerase chain reaction assay optimized to detect base pair differences between genomic DNA and rearranged genes, we have demonstrated that at least a portion of expressed anti-Hib PS V_H genes have undergone somatic mutation, although at a low rate. Both the 9.1 and V_H26 germline genes are preferentially expressed in early development, implying that the poor response of young infants to this polysaccharide antigen is not due to the inability to rearrange critical V_H gene segments.[3]

Heavy chain CDR-3 regions are also restricted in this immune response. In several instances heavy chain variable region genes obtained from unrelated subjects share identical V(D)J joinings.

Although more diverse than that of heavy chains, light chain variable region gene usage in the human anti-Hib PS antibody response is also restricted. Quantita-

[b] Address for correspondence: Dr. William L. Carroll, Departments of Pediatrics and Pathology and the Eccles Program in Human Molecular Biology and Genetics, University of Utah, Rm. 3260, Building 533, Salt Lake City, UT 84107.

→

FIGURE 1. Nucleotide sequences of anti-Hib PS V_H segments. Shown are the nucleotide sequences of 15 human monoclonal anti-Hib PS antibodies. Also shown are the nucleotide sequences of the germline V_H3 genes LSG6.1, 9.1, LSG12.1, and V_H26. Complementarity-determining regions are boxed. (From Adderson *et al.*,[2] with permission.)

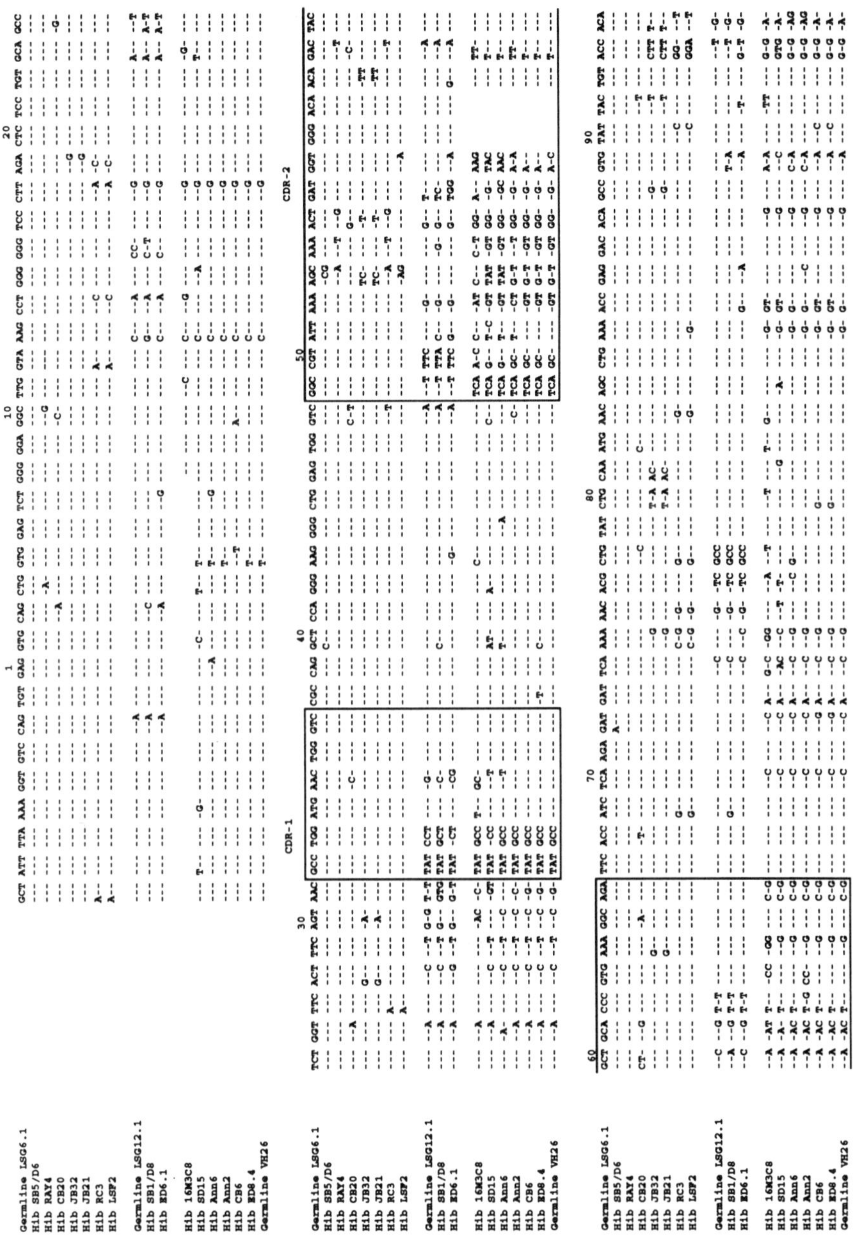

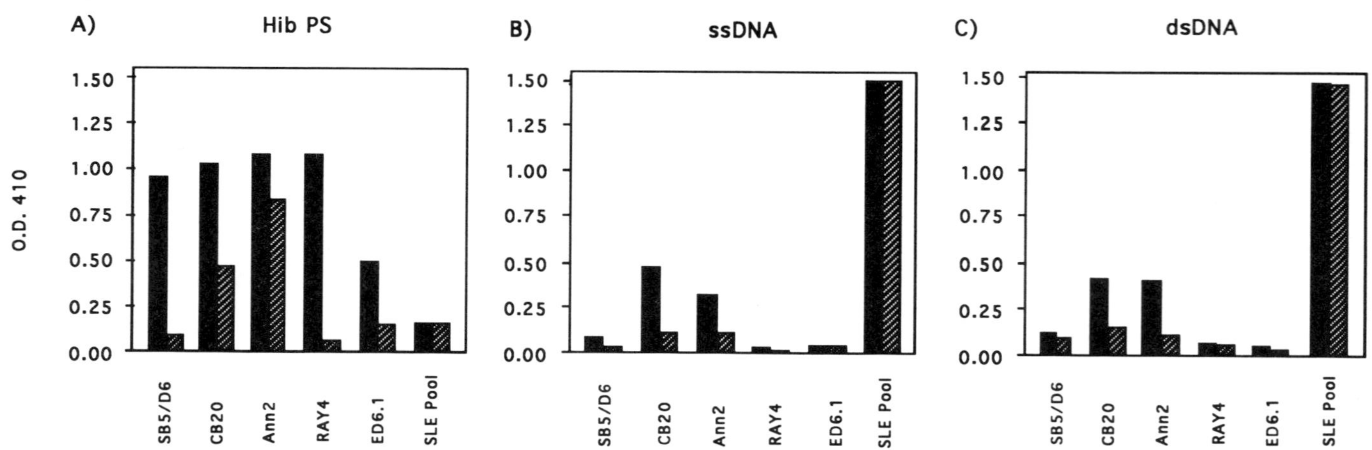

FIGURE 2. Reactivity of human monoclonal anti-Hib PS antibodies. Shown is the binding of five human monoclonal antibodies to **(A)** Hib PS, **(B)** single-stranded DNA, and **(C)** double-stranded DNA in the presence (*stippled bars*) or absence (*solid bars*) of preincubation with 10 μg/ml of soluble Hib PS. Also shown is the reactivity of pooled sera from patients with systemic lupus erythematosis. (From Adderson *et al.*,[2] with permission.)

tively, most antibody is encoded by a V_{κ}II gene segment, A2. A2-encoded variable segments from two unrelated subjects are 98 to 100% homologous to the A2 germline segment.[1,4] Four lambda antibodies, all cross-reactive with the structurally related antigen *E. coli* K100 PS, use V_{λ}VII gene segments that are 96 to 98% homologous to one another and appear to be encoded by the same germline variable segment.

Heavy and light chain variable segments are homologous to segments encoding autoantibodies. Despite having high affinity for Hib PS, some monoclonal anti-Hib PS antibodies cross react with double-stranded and single-stranded DNA (FIG. 2).[2] Comparison of V_H and V_L regions between self-reactive and monospecific anti-Hib PS antibodies does not demonstrate a consistent structural feature correlating with fine antigen specificity. Considerable data associates the immune response to bacterial polysaccharides with the generation of anti-DNA antibodies. It is possible that under certain conditions, the antigen-driven expansion of autoreactive anti-Hib PS antibody secreting B cells may stimulate or perpetuate autoimmune disease.

REFERENCES

1. ADDERSON, E. E., P. G. SHACKELFORD, R. A. INSEL, A. QUINN, P. M. WILSON & W. L. CARROLL. 1992. Immunoglobulin light chain variable region gene sequences for human antibodies to *Haemophilus influenzae* type b capsular polysaccharide are dominated by a limited number of V_{κ} and V_{λ} segments and VJ combinations. J. Clin. Invest. **89:** 729–738.
2. ADDERSON, E. E., P. G. SHACKELFORD, A. QUINN, P. M. WILSON, M. W. CUNNINGHAM, R. A. INSEL & W. L. CARROLL. 1993. Restricted immunoglobulin V_H usage and VDJ combinations in the human response to *Haemophilus influenzae* type b capsular polysaccharide—nucleotide sequences of monospecific anti-*Haemophilus* antibodies and polyspecific antibodies cross-reacting with self-antigens. J. Clin. Invest. **91:** 2734–2743.
3. SCHROEDER, H. W. & J. Y. WANG. 1990. Preferential utilization of conserved immunoglobulin heavy chain segments during human fetal life. Proc. Natl. Acad. Sci. USA **87:** 6146–6150.
4. ADDERSON, E. E., P. G. SHACKELFORD, A. QUINN, P. M. WILSON & W. L. CARROLL. 1993. Diversity of immunoglobulin light chain usage in the human immune response to *Haemophilus influenzae* type b capsular polysaccharide. Ped. Res. **33:** 307–311.

Patterns of V-Region Gene Segment Association in the Human Antibody Response to *Haemophilus influenzae* Type B Capsular Polysaccharide[a]

GEORGE V. PINCHUK,[b,c,d] CAROL NOTTENBURG,[e] AND ERIC C. B. MILNER[c,f]

[c]*Virginia Mason Research Center*
Seattle, Washington 98101

[d]*A. A. Bogomoletz Institute of Physiology*
Kiev, Ukraine

[e]*Division of Clinical Sciences*
Fred Hutchinson Cancer Research Center
Seattle, Washington 98104

[f]*Department of Immunology*
University of Washington
Seattle, Washington 98195

Haemophilus influenzae type b (Hib) is a major pathogen in young children, natives of some ethnic groups, and long-term survivors of bone marrow transplantation. Antibodies to the Hib capsular polysaccharide (PS) protect against invasive disease. The Hib PS antibody gene repertoire in healthy subjects involves a small subset of V_H and a larger but still restricted subset of V_L genes.[1-3] Therefore, the diversity of Hib PS antibodies and subsequent selection of antibodies with good protective properties may rely on mechanisms other than random assortment of various germline elements. For precise characterization of these mechanisms, additional sequencing data are needed.

We have obtained the complete V_H and V_L gene sequences of four human monoclonal Hib PS antibodies derived from four unrelated adult subjects immunized with Hib PS-diphtheria toxoid conjugate.[3] The heavy chains of two antibodies are encoded by gene segments related to the V3-23 gene, whereas the other two heavy chains are encoded by gene segments that belong to the V_H3b subfamily and are most similar to the V3-15 gene.[4] The light chains of V3-23-encoded antibodies are products of genes that are very similar to V_κA2,

[a] Supported in part by National Institutes of Health Grants AR39918 (E.M.), AI31241 (E.M.), and CA18221 (C.N.) and American Cancer Society Grant IM485 (E.M.)

[b] Address for correspondence to G. V. Pinchuk, Virginia Mason Research Center, 1000 Seneca St., Seattle, WA 98101.

TABLE 1. Association of V3-23-Related Heavy Chains Expressed by Anti-Hib PS Hybridomas

Hybridoma[a]	D Segment (bp)	J_H	V_L
D3F3[3]	3	6	$V\kappa$A2
Gar6E8[3]	3	6	$V\kappa$A2
Ann2[1,2]	1	4	$V\kappa$A2
16M3c8[1]	34	4	$V\kappa$1 (non-A2)
SD15[1]	1	6	Not known
CB61[1]	1	6	Not known
ED8.4[1]	1	6	Not known
Ann6[1]	1	4	Not known

[a] Superscript numbers refer to references.

the latter element being thought to dominate the Hib PS antibody light chain repertoire.[2] The light chains of V_H3b-encoded antibodies are, on the contrary, distinct from V_κA2. Both V3-23-, V_κA2-encoded antibodies have extraordinary short D segments and use the J_H6 segment, whereas both V_H3b-, non-V_κA2-encoded antibodies have D segments of conventional length and use the J_H4 segment. As detailed in TABLES 1 and 2, the two V-region gene configurations (V3-23, short D, J_H6 and V_κA2 or V_H3b, conventional D, non-J_H6 and non-V_κA2) are highly predominant in a panel of Hib PS antibody-producing hybridomas obtained in our laboratory and elsewhere and derived from a total of 14 subjects. The first configuration encodes antibodies with "fetal" CDR3.[5] Possibly, a B-cell population with features of fetal B cells that may represent human homologues of murine B1 cells, and may be selected by PS antigens, substantially contributes to the Hib PS antibody response of adult humans. It would be interesting to determine if the two described gene configurations in fact correspond to different B-cell lineages and if both of them are equally protective against the infection.

TABLE 2. Association of V_H3b Heavy Chains Expressed by Anti-Hib PS Hybridomas

Hybridoma[a]	D Segment (bp)	J_H	V_L
BA3[3]	24	4	$V\kappa$1 (non-A2)
ED6.1[1,2]	18	4	$V\kappa$1 (non-A2)
SC6[3]	33	4	$V\kappa$2 (non-A2)
RC3[1,2]	3	4	$V\kappa$A2
SB5/D6[1]	10	4	$V\lambda$ (non-A2)
RAY4[1]	10	4	$V\lambda$ (non-A2)
JB32[1]	14	5	$V\lambda$ (non-A2)
JB21[1]	14	5	$V\lambda$ (non-A2)
CB20[1]	28	1	$V\lambda$ (non-A2)
LSF2[1,2]	3	4	$V\lambda$ (non-A2)
SB1/D8[1]	34	3	Not known

[a] Superscript numbers refer to references.

REFERENCES

1. ADDERSON, E. E., P. G. SHACKELFORD, A. QUINN, P. M. WILSON, M. W. CUNNINGHAM, R. A. INSEL & W. L. CARROLL. 1993. J. Clin. Invest. **91:** 2734–2743.
2. SCOTT, M. G., H. G. ZACHAU & M. H. NAHM. 1992. Int. Rev. Immunol. **9:** 45–55.
3. PINCHUK, G. V., C. NOTTENBURG & E. C. B. MILNER. 1995. Scand. J. Immunol. In press.
4. MATSUDA, F., E. K. SHIN, H. NAGAOKA, R. MATSUMURA, M. HAINO, Y. FUKITA, S. TAKA-ISHI, T. IMAI, J. H. RILEY, R. ANAND, E. SOEDA & T. HONJO. 1993. Nature Genet. **3:** 88–94.
5. WORTIS, H. H. 1992. Int. Rev. Immunol. **8:** 235–246.

Generation of Diversity in a Human Anti-Viral Response

JOHN D. RIOUX,[a] BERNARD R. BRODEUR,[b]
YOLANDE LAROSE,[b] MATS OHLIN,[c]
CARL A. K. BORREBAECK,[c]
AND MARIANNA M. NEWKIRK[a,d]

[a]*Montreal General Hospital Research Institute*
McGill University
Montreal, Quebec, Canada H3G 1A4

[b]*The National Laboratory for Immunology*
Ottawa, Ontario, Canada K1A 0L2

[c]*Department of Immunotechnology*
Lund University
Lund, Sweden

Cytomegalovirus (CMV), a DNA virus of the herpes family, can establish a latent infection in as much as 50% of populations in developed countries and 100% in the Third World. Reactivation of the virus can occur and is associated with severe consequences in immunosuppressed persons (organ transplantation or HIV infection). Studies of the antibody response to this virus have indicated that one of the major targets is a phosphoprotein of molecular weight 65 kDa (pp65). Although the function has not been fully elucidated, pp65 appears to be a structural matrix protein. CMV dense bodies are predominantly composed of pp65. We studied the structure of human monoclonal antibodies (mAbs) specific for CMV pp65 in order to determine whether any restrictions in antibody gene elements could be identified.

METHODS

Eight IgG mAbs were generated by human hybridoma technology from peripheral blood or lymph nodes of four persons who were seropositive for CMV infection.[1,2] These mAbs bound specifically to pp65 and were cloned and sequenced.[3,4]

RESULTS AND DISCUSSION

In TABLES 1 and 2, the rearranged variable (V) region gene elements used in the construction of the heavy (HC) and light chains (LC) of the eight mAbs and

[d] Address for correspondence: Marianna M. Newkirk, Ph.D., Montreal General Hospital Research Institute, 1650 Cedar Avenue, Montreal, Quebec, Canada H3G 1A4.

TABLE 1. Light Chains of Human Monoclonal Anti-CMV Antibodies That Have Specificity for pp65

mAb	Source of B Cells	V_L	Gene Germline	Percent Identity to GL	J Region
HCV-2	PBL	KIIIa	ka3d1	97	JK-1
HCV-3	PBL	LI	1.2	94	JL-2
HCV-63	LN	LI	1.2	95	JL-2
HCV-65	LN	LIII	cML13[a]	79	JL-2
MO53	PBL	KI	KLVA20	95	JK-3
MO58	PBL	KI	KL11A	97	JK-1
MO61	PBL	LI	LBE	98	JL-2
MO81	PBL	KI	HK102	94	JK-1

ABBREVIATIONS: GL, germline; PBL, peripheral blood lymphocyte; LN, lymph node.
[a] GL of highest homology.

their closest germline counterparts are identified. Interestingly, only two mAbs, HCV-3 and -63 appeared to derive LCs from the same V_LI 1.2 ancestral gene. All other gene elements, except J-region genes (six of eight used J_H4, three of four kappa mAbs used $J_\kappa 1$, all of the four lambda mAbs used $J_\lambda 1$) were unique. In the generation of diversity, although one mAb, MO61, was essentially germline encoded, many of the V gene elements showed evidence of antigen-selected somatic mutation events. A "hot spot" in the HC CDR2 could be identified. The most divergent gene element was the CDR3 of the HC. This varied in length and in composition. N segment addition, different D segments, and D-D fusion contributed to the variability of the HC CDR3. It is likely that these mAbs each recognize different epitopes, possibly with some overlap, on the immunodominant CMV pp65 molecule, and the diversity is a reflection of the highly antigenic nature of this viral protein.

TABLE 2. Heavy Chains of Human Anti-pp65 Antibodies

mAb	V_H	Gene Germline	Percent Identity to GL	CDR3 Length	D-D Fusion[a]	J Region
HCV-2	3	V_H26	92	22	3	4
HCV-3	3	WHG26	94	19	2	4
HCV-63	4	WHG16	96	11	3	5
HCV-65	3	V71-4	93	22	1	6
MO53	3	Hv3005	88	14	2	4
MO58	1	HG3	93	13	1	4
MO61	3	WHG16	98	9	2	4
MO81	4	V71-2	88	10	2	4

ABBREVIATION: GL, germline.
[a] Number of D region genes, and likely D-D fusion.

REFERENCES

1. LAROSE, Y., E. S. TACKABERRY & B. R. BRODEUR. 1991. Human monoclonal antibodies to cytomegalovirus recognize viral epitopes on the surface of virus-infected cells. Hum. Antibod. Hybridomas **2:** 67–73.
2. OHLIN, M., V.-A. SUNQVIST, G. GILLJAM, U. RUDÉN, F. O. GOMBERT, B. WAHREN & C. A. K. BORREBAECK. 1991. Characterization of human monoclonal antibodies directed against the pp65-kD matrix antigen of human cytomegalovirus. Clin. Exp. Immunol. **84:** 508–514.
3. RIOUX, J. D., Y. LAROSE, B. R. BRODEUR, D. RADZIOCH & M. M. NEWKIRK. 1994. Structural characteristics of four human hybridoma antibodies specific for the pp65 protein of the human cytomegalovirus and their relationship to human rheumatoid factors. Mol. Immunol. **8:** 585–597.
4. RIOUX, J. D., M. OHLIN, C. A. K. BORREBAECK & M. M. NEWKIRK. 1995. Molecular characterization of human monoclonal antibodies specific for the human cytomegalovirus: Relationship of variable region sequence to antigen specificity and rheumatoid factor-associated idiotype expression. Immunol. Infect. Dis. In press.

Origins of Anti-DNA Antibodies and Their Implications for B-Cell Tolerance[a]

MARKO Z. RADIC[b] AND MARTIN WEIGERT[c]

[b]*Department of Microbiology and Immunology*
Medical College of Pennsylvania
Philadelphia, Pennsylvania 19129

[c]*Department of Molecular Biology*
Princeton University
Princeton, New Jersey 08544

INTRODUCTION

Antibodies to DNA and histones are the common denominators of systemic lupus erythematosus (SLE) autoantibodies.[1] A likely reason for the high frequency of these specificities is that they may be the most common and/or immunogenic self antigens. This idea depends on whether autoimmunity is in fact an antigen-driven process, and if so, whether DNA, RNA, and/or their associated proteins are the driving antigens.

The argument for antigen activation of autoantibodies in SLE is based on the observed hierarchy of specificities. As pointed out by Hardin,[2] even though most persons with SLE express anti-DNA and antihistone antibodies, a person will express a unique spectrum of antibodies to the less popular targets. Because unique spectra of antibodies to the epitopes on a complex antigen are a feature of the immune response to foreign antigens such as influenza virus hemagglutinin (HA),[3] the idiosyncratic patterns of autoantibodies suggest that they too arise by immunization. The parallel to the immune response to HA extends to features of specific autoantibodies. Autoantibodies directed to a particular auto-epitope are oligoclonal and display patterns of somatic mutation that indicate positive selection.[4–6] None of these properties is predicted by models that explain autoimmunity by nonspecific mechanisms of B-cell activation.

THE NATURE OF THE ANTIGEN

The antigens that elicit the common lupus autoantibodies are likely to be nucleoprotein particles. As noted by Hardin,[2] the set of anti-nuclear antibodies

[a] Support for this work was provided to M. Z. Radic by the National Institutes of Health (AI34881), the Investigator Award from the Arthritis Foundation, and the Julius Erwing Lupus Research Award from the Lupus Foundation of Delaware Valley and to M. Weigert by the National Institutes of Health (GM-20964) and the Sheryl N. Hirsch Award from the Lupus Foundation of Philadelphia.

(ANA) from an individual tend to include antibodies to physically linked molecules such as DNA and histones. That a foreign antigen could mimic a nucleoprotein in this way is unlikely. In addition, several nucleoprotein autoepitopes have been demonstrated to be conformation dependent.[7,8] Again, it is unlikely that an unrelated antigen could independently arrive at such complex structures. Shlomchik *et al.* have used similar logic to argue that DNA is involved in activating anti-DNA antibodies.[6] They observed that selected mutants modify the specificity for DNA. This is seen as increased affinity for ssDNA and/or dsDNA or the acquisition of binding to new epitopes on DNA. Once again, it is unlikely that one antigen (non-DNA) would select upon mutations that modify the specificity for another (DNA).

Individual components of nucleoproteins such as DNA fragments are certainly available to the immune system, but recently a delivery system has been described that explains the linked specificities in systemic autoimmunity. Casciola-Rosen and colleagues[9] have shown that cells undergoing apoptosis parcel nucleoproteins into structures called blebs. These include many, perhaps most, of the targets of the autoantibodies in lupus and are conveniently presented on the surface of dying cells. This remarkable discovery not only provides evidence for the availability of autoantigens, but also it connects the presentation of autoantigens to apoptosis. Thus these autoantigens may be focused at sites where tolerance is exerted. Indeed, as Casciola-Rosen *et al.*[9] point out, a breakdown in mechanisms leading to cell death may "deprive" these sites of self antigen and prevent tolerance induction.

THE STRUCTURAL BASIS FOR ANTI-DNA SPECIFICITY

A structural basis for anti-DNA specificity has emerged from the observation that certain somatic mutations correlate with altered binding properties of clonally related anti-DNA antibodies.[5,6,10,11] Several clones studied in detail showed intraclonal differences in affinity and/or specificity for DNA. For example, the 3H9 clone has one member (3H9) with a higher affinity for ssDNA than its siblings. This clone member also acquired the ability to bind dsDNA.[5] Reversion analysis (FIG. 1) pinpoints the mutation responsible for these properties; one that leads to an arginine substitution in V_HCDR2.[12] The pleiotropic effect of this mutation to arginine is understandable if one considers the nature of the arginine side chain.

The structure and chemistry of arginine endow it with the capacity to participate in multiple and diverse binding interactions. The positively charged guanidino group at the tip of arginine's side chain is planar and has aromatic character due to π electrons that are shared by the three nitrogen atoms. Each of the five hydrogens that are part of the guanidinium group can be brought into position to donate H bonds by the long and flexible stem of the arginine side chain. Thus, the multifunctionality of arginine should ensure its frequent use in antibody-combining sites.

The utility of the functional groups of arginine is evident from an inspection of protein–DNA interactions. Arginine can interact with phosphate groups or donate H bonds to atoms located in the major or minor grooves of DNA with lone electron pairs. As the distribution of H-bond acceptors is different for each nucleotide in the major groove of B-form DNA, arginine can contribute to DNA binding in a sequence-specific manner. Moreover, the potential for multiple H bonds allows arginine to interact with adjacent bases on the same or opposite

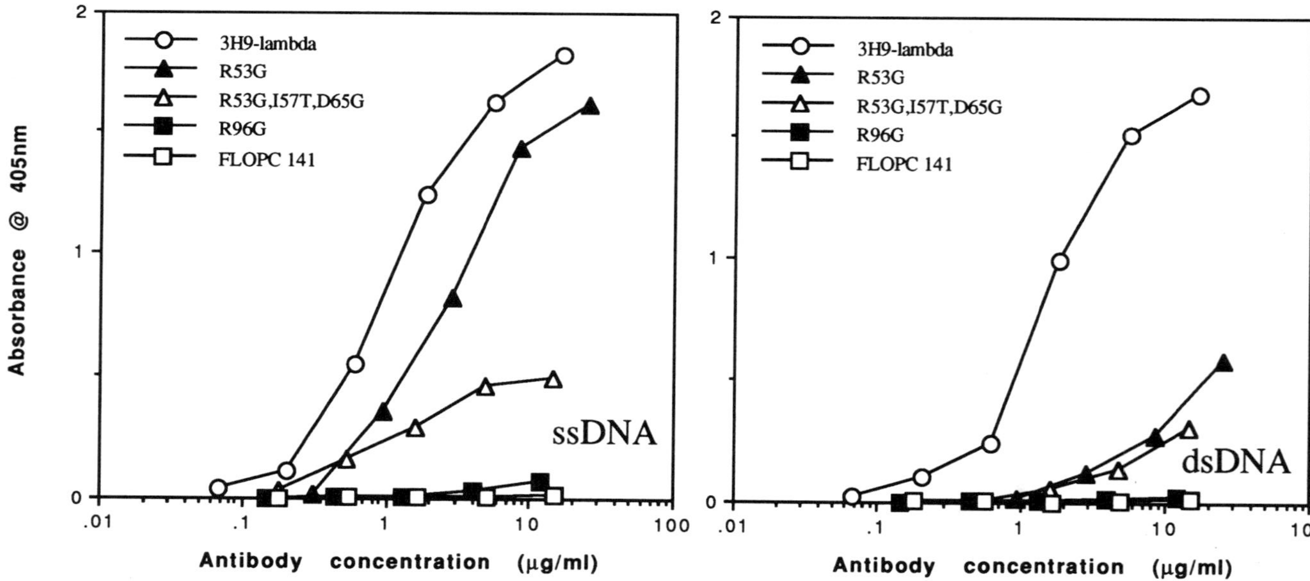

FIGURE 1. Reversions in V_H3H9. Solution-phase DNA-binding assay of 3H9-λ and revertants of V_H3H9. Relative affinities for ssDNA and dsDNA were measured as previously described[31] and were plotted as OD readings at 405 nm for each serial antibody dilution. The antibodies are identified by different symbols listed in the figure.

strands of DNA. Even though the number of protein–DNA complexes that have been analyzed by X-ray crystallography is still limited, each of these different interactions of arginine with DNA have been demonstrated.[13-18] Therefore, it stands to reason that arginine could increase an antibody's affinity for DNA by recognizing one of several suitable epitopes on this polymer.

Detailed studies of antibody–antigen interactions have revealed that the specificity and affinity of binding depend in large part on H bonds and aromatic interactions.[19] Indeed, tyrosine and tryptophan, amino acids with large aromatic side chains and the capability of donating H bonds, are overrepresented in antibody-combining sites.[20] Arginine, however, is underrepresented in conventional antibodies. The preponderance of arginine in anti-DNA antibodies presents a vivid contrast to this general situation.[21] One possible reason for the frequent use of arginine in anti-DNA antibodies but not in antibodies directed against exogenous antigens may be arginine's versatility in mediating interactions with DNA, a versatility that readily gives rise to self-reactivity.

ARGININE IN ANTI-DNA ANTIBODIES

To investigate the contribution of arginine to anti-DNA specificity, we compiled sequences from over 250 spontaneous murine anti-DNA antibodies.[21] By assigning sequences to homology groups (sets of three or more sequences differing from each other by no more than four nucleotides), we defined 13 V_H and 11 V_K recurrent germline genes that code for disease-associated anti-DNA antibodies. A partial list of these recurrent V_H sequences shows that 10 of the 13 V_H genes belong to the J558 H chain family (TABLE 1). Although these 10 genes differ from each other by as many as 11 amino acid residues in CDRs 1 and 2, they must share properties that make them more suitable for DNA binding than other V_H genes from this family of more than 60 genes. One indication for why these genes may be favored by anti-DNA antibodies is provided by their pattern of somatic mutations.

Although we only include V_H genes with a maximum of four point mutations, it is striking that several sequences have two replacement mutations yielding arginine codons (TABLE 1). Overall, mutations to arginine are frequent, yet they involve only a limited number of positions in the combining site. In fact, 14 of 16 independent mutations to arginine cluster at five positions; position 31 of CDR1, positions 53, 56, and 58 of CDR2, and position 76 of FWR3 (TABLE 1). The high frequency and nonrandom distribution of replacement mutations to arginine suggest that arginine is used to establish contacts with DNA from key positions in the combining site. A second implication of arginine's recurrence is that different anti-DNA V_H domains appear to use a similar binding mechanism despite their germline-encoded differences.

TESTING ARGININE'S ROLE IN ANTI-DNA ANTIBODIES

To test the role of arginine introduced by somatic mutations in anti-DNA V_H genes, we transfected a λ1 L chain-only plasmacytoma cell line with an IgG2b expression vector.[31] The expression vector contained the 3H9 H chain (TABLE 1) and the coding segments for the γ2b constant region. By site-directed mutagenesis, we reverted the R53 in CDR2 of 3H9 either by itself or along with the other two

TABLE 1. Comparison of Recurrent Anti-DNA J558 H Chains

Gene:	Hybridoma:	CDR1: 31	32	33	34	35	CDR2: 50	51	52	52a	53	54	55	56	57	58	59	60	61	62	63	64	65	76	CDR3:	Vk:	Isotype:	Specificity:	Strain:	Ref.:
1	D20	S	S	W	M	N	R	I	Y	P	G	D	G	D	T	N	Y	N	G	K	F	K	G	S	SGLRQLGLPFAY	**8B**	IgG2a	DNA	M/1	6
	108-10	-	-	-	-	-	-	-	-	-	-	-	-	-	-	-	-	-	-	-	-	-	-	-	GDYFWY-DV	**23**	IgM	DNA	M/1	12
	DP7	(R)	-	-	-	-	-	-	-	-	-	-	-	-	-	-	-	-	-	-	-	-	-	(R)	GGNRW---	9	IgG3	DNA	M/1	6
	DP11	(N)	-	-	-	-	-	-	-	-	-	-	-	-	-	-	-	Y	-	-	-	-	-	(R)	GGNRW---					
	3H9	-	-	-	-	-	-	-	-	-	(R)	-	-	-	I	-	-	-	-	-	-	-	D	-	ARSKYSYVMD-	4/5	IgG2b	DNA, H3	M/1	5
	DNA5	-	Y	-	-	-	-	-	-	-	-	(N)	-	-	-	-	-	-	-	-	-	-	-	-	EKT-TG---	**8A**	IgG2a	DNA	B/W	11
2	1-42	-	-	-	-	-	W	-	-	-	-	(N)	-	-	-	-	-	-	-	-	-	-	-	-	-RRNYYGD-	4/5	IgG3	DNA, SmD	M/1	22
	17s-c4	(R)	-	-	-	-	W	-	-	-	-	-	-	-	I	(K)	-	-	-	-	-	-	-	-	GGNYGGSFYALD-	**21**	IgG2a	DNA	B/W	23
	17s.13	-	-	-	-	-	W	-	x	-	(R)	-	-	-	-	-	-	-	-	-	-	-	D	(N)	SRGYYFGSSRF-DV	1	IgM	DNA	B/W	23
3	MRL DNA22	-	Y	-	I	-	N	-	-	-	-	S	S	S	-	-	-	-	E	-	-	-	S	-	RLYRYYAMD-	4/5	IgM	DNA	M/1	24
	bfd03	(R)	Y	-	I	(K)	N	-	-	-	-	S	S	(R)	-	-	-	-	E	-	-	-	S	-	-HYYGSSYGx--	**9**	IgG2a	DNA	B/c	25
4	82-3	G	Y	(K)	-	-	N	-	N	-	Y	Y	-	S	-	S	-	-	Q	-	-	-	-	-	RDGLYYAMD-	4/5	IgM	DNA	M/1	10
	A52	G	Y	N	-	-	(K)	-	N	-	Y	Y	-	S	-	S	-	-	Q	-	-	-	-	-	-RLRR-GY-D-	8	IgG2b	DNA	B/W	26
5	H130	G	Y	F	-	-	-	-	N	-	Y	N	-	-	-	F	-	-	Q	-	-	-	-	-	-RAYSNYFAMD-	?	IgM	DNA, CL	M/1	27
	SECF5/9	G	Y	F	-	H	-	-	N	-	Y	N	-	-	-	F	-	-	-	-	-	-	-	-	RGSYAMD-	21	IgG1	DNA (E.coli)	B/c	28

6	DP13	D	Y	Y	-	-	-	V	N	-	S	N	-	G	-	S	-	-	Q	-	-	-	-	-	SGTSGYPAW---	**8A**	IgG1	DNA	M/1	6
	23-9	D	Y	Y	-	-	-	V	N	-	S	N	-	(R)	-	T	-	-	Q	-	-	(R)	-	(N)	-YY-D-	3	IgG3	DNA	M/1	10
	25-3	D	Y	(N)	-	-	-	V	N	-	S	N	-	G	-	(R)	-	D	Q	-	-	-	-	-	RDYAMD-	**23**	IgG2a	DNA	M/1	10
	CAL-11	D	Y	Y	-	-	-	V	N	-	S	Y	-	G	-	(K)	-	-	Q	-	-	-	-	-	--GWLHPAW---	?	IgG2a	**DNA-**, CL	M/1	29
7	25.12m	D	Y	(N)	-	D	Y	-	-	-	N	N	-	G	-	G	-	-	Q	-	-	-	S	-	GRYT---	1	IgM	DNA	B/W	23
	18F7	D	Y	Y	-	D	Y	-	-	-	N	N	-	D	I	(R)	-	-	Q	-	-	-	-	-	YGAYVMD-	**1A**	IgG2a	**DNA-**, SmB	M/1	22
8	BXW DNA7	-	Y	V	-	H	Y	-	N	-	Y	N	D	G	-	(N)	-	-	E	-	-	-	-	-	RAYYGSSPHWY-DV	?	IgM	DNA	B/W	24
	8-1	-	Y	V	-	H	Y	-	N	-	Y	N	D	G	-	K	-	-	E	-	-	-	A	-	CRYD_GYYYAMD-	**10**	IgG3	DNA	M/1	10
	111-c2	(R)	Y	V	-	H	Y	-	N	-	Y	N	D	G	-	K	-	-	E	-	-	-	-	-	GTVIGDYYAMD-	2	IgG2a	DNA	B/W	23
	3-7	-	Y	V	-	H	Y	-	N	-	Y	N	D	G	I	K	-	-	E	-	-	-	S	(R)	GLRRSYAMD-	9	IgG2a	DNA, SmD	M/1	22
9	S57	D	Y	Y	I	-	W	-	-	S	-	S	-	N	-	K	-	-	E	-	-	-	D	-	RGTLGKGY-D-	8	IgG3	DNA	M/1	6
	DP9	D	Y	Y	I	-	W	-	-	S	-	S	D	N	-	K	-	-	E	-	-	E	-	G	NGLRRRAW---	**19**	IgG1	DNA	M/1	6
10	05	(R)	Y	-	I	H	E	-	N	-	S	N	-	R	-	-	F	-	E	-	-	-	T	-	-TILNY-NYLD-	**21**	IgG2b	DNA	S/N	30
	564	-	Y	-	-	H	E	-	N	-	S	N	-	R	-	-	-	-	E	-	-	-	(R)	-	SGGFYYGMD-	4/5	IgG2b	DNA	S/N	30

NOTE: Sequences were grouped according to criteria described elsewhere.[21] Sequences presumed to derive from the same gene are separated by a horizontal line. Residues of CDR1 (31–35), CDR2 (50–65), FWR3 residue 76, and CDR3 (variable length) are numbered according to Kabat *et al.*[37] and are shown in the single letter code. Identities with the sequence at the top of the comparison are indicated by a dash, whereas germline differences and replacement mutations are shown, and those resulting in arginine, K, or asparagine are circled. The L chain used by each antibody is indicated by a number that signifies the L chain group. If the L chain is itself recurrent,[21] this fact is highlighted by bold print. The isotype of each antibody is indicated, as well as its specificity. Specificity for DNA, lack of DNA binding (bold), and binding to cardiolipin (CL), histone H3, or ribonucleoproteins (SmB or SmD) are listed. The antibodies were derived from the following mouse strains: MRL/lpr (M/1), BALB/c (B/c), NZBxNZW)F_1 (B/W), and SWRxNZB)F_1 (S/N). All sequences are taken from the sources listed.

mutations, isoleucine 57 (I57) and aspartic acid 65 (D65), and assayed the effect of these changes in a solution-phase DNA binding assay (FIG. 1). Of the three replacement mutations that were fixed in the 3H9 H chain, the mutation that gave rise to R53 was found to be most important in determining the unique characteristics of this clone member: high affinity for ssDNA and specificity for dsDNA. That a single mutation can accomplish both is plausible considering the multiple ways in which arginine can approach DNA. An analogy is to be found in the structure of the GCN4 homodimer bound to DNA.[15] R243 of one monomer contacts phosphates, while the same residue in the other monomer binds to guanine.

The fact that R53 plays such a crucial role in the binding of 3H9 to DNA encouraged us to test the contribution of other arginines that were the product of somatic mutations. We tested arginines at positions 31, 56, 64, and 76 (TABLE 1) by introducing them as single mutations into the H chain of 3H9. Each of the mutant H chains raised the affinity of ss and dsDNA antibodies formed with the λ1 L chain by approximately 10-fold,[12] an increment that is consistent with the formation of direct protein–DNA contacts. Although the simultaneous reversion of D65 to the germline glycine (G) was required to observe the effect of R64 (mutant K64R/D65G), arginines at the other positions showed uniform affinity increments in the 3H9 tester H chain. We suggest that arginines at these positions contribute to DNA binding of the original anti-DNA antibodies as they do in 3H9.

If in fact the different anti-DNA H chains use a similar mechanism of DNA binding, then different arginines may contribute independently to binding, provided that they do not sterically hinder each other's access to suitable DNA epitopes. This prediction was tested by constructing double arginine mutants. One such pair, D56R/S76R, was assayed as above and compared to the two single mutants, D56R and S76R (FIG. 2). The improvement in binding affinity of the double mutant was equal to the product of the two single arginine mutants, suggesting that R56 and R76 can, along with the R53 mutation, contribute independently to binding. A possible scenario consistent with these results is that each of the three arginines is engaged in a direct contact with DNA, a possibility supported by the pleiotropic potential of arginine's interactions with DNA.

GENETIC CONTROL OF ANTI-DNA

Arginine is introduced into antibody V regions in three ways, the first by being part of the germline V sequence. Although there is no known case in which a germline V sequence *per se* determines anti-DNA specificity, certain germline-encoded structures may be more favorable substrates for somatic mechanisms that introduce arginines. Repeated Vs (TABLE 1) suggest this may be the case, but the structural features that lead to selection of these preferred Vs are not obvious.[21]

In addition, somatic mutations to arginine influence affinity for DNA. This is likely to be an extremely potent way of developing anti-DNA specificity because, as shown above, arginine can interact with DNA in an autonomous fashion. Apart from fixing the position of the α carbon, the immunoglobulin backbone may play only a minor role in determining the contact of this residue with antigen, and mutations to arginine at any one of numerous sites may lead to DNA binding. Clone E[6] is a case in point (TABLE 2). The V_H of this clone has accumulated five independent mutations (possibly six depending on the origin of the CDR3 arginine). The independent origin of the mutations has been established because different

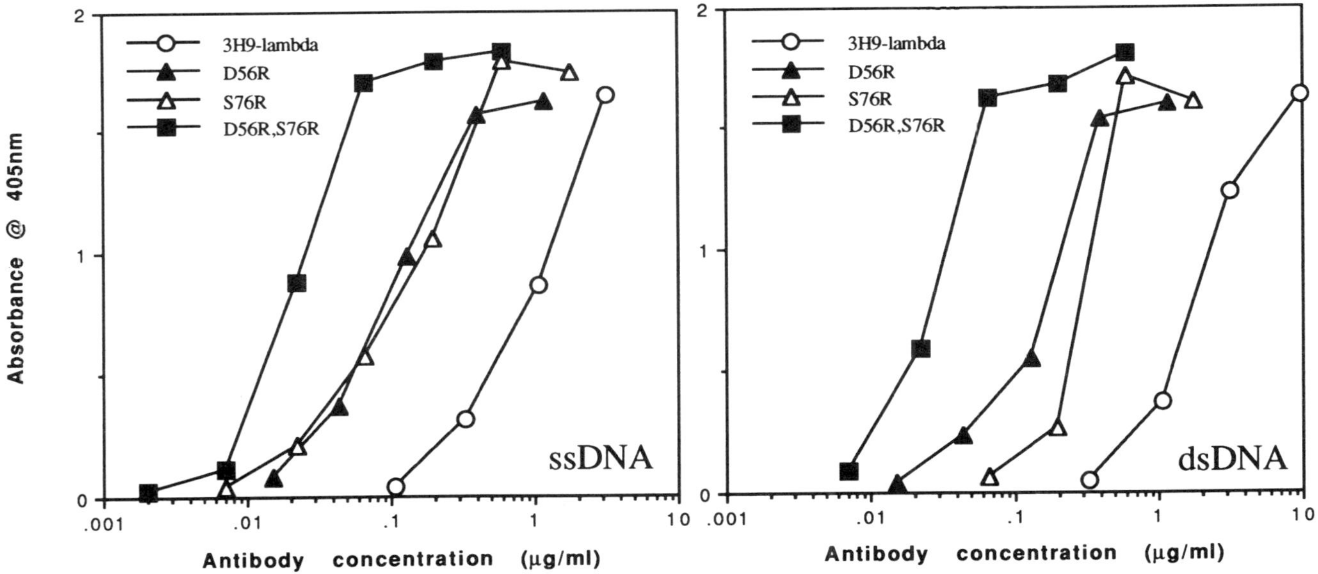

FIGURE 2. Forward mutations in V_H3H9. Solution-phase DNA-binding assay of 3H9-λ and higher affinity mutants generated by addition of arginine to V_H3H9. Relative affinities for ssDNA and dsDNA were measured as previously described[31] and were plotted as OD readings at 405 nm for each serial antibody dilution. The antibodies are identified by different symbols listed in the figure.

TABLE 2. Codon Degeneracy and Selection of Mutations

Codon	Result of Single Base Change	Mutations in Clone E V_H
SER(AGY)	ARG(CCY)	SER31(AGC) → ARG(AGG)
	ARG(AGR) × 2	SER31(AGC) → ASN(ACC)
	ASN(AAY)	SER76(AGC) → ARG(AGA)
	CYS(TGY)	SER76(AGC) → ARG(CGC)
	ILE(ATY)	ASN60(AAT) → TYR(TAT)
	GLY(GGY)	
	THR(ACY)	

base changes were involved in creating each new codon. The effect of codon degeneracy on the use of arginine can be seen by inspecting the codons generated from the serine codon by single base changes. As shown in TABLE 2, three of eight replacement mutations lead to arginine. Strikingly, all four mutations in serine codons of clone E are to arginine or asparagine, strongly suggesting selection upon mutants by DNA. Codon biases in CDRs also increase the frequency of mutations to these DNA-binding residues. One-half of the mutations at the serine codon, AGY, yield arginine or asparagine (TABLE 2), and CDRs show a marked bias toward this serine codon over its synonym, TCN. At the estimated somatic mutation rate of 10^{-3}/bp/div, an antibody gene with 10 CDR AGY codons will give rise to a CDR arginine mutant on average once every five or six cell divisions.

Arginine codons can also arise through VDJ rearrangement. D genes can code for arginine, depending on their reading frame and the orientation in which they are rearranged. For example, the reading frame 2 of D_H SP2.4 gives rise to an arginine codon (FIG. 3). The use of reading frame 2 accounts for many of the CDR3 arginine codons, as originally noted by Eilat.[32] The use of inverted D_H can also yield arginine codons (orientation B; FIG. 3) and indeed many CDR3 arginine(s) are encoded by D_H genes in such a fashion (TABLE 1). Because arginine is underrepresented in CDR3 of conventional antibodies,[6] it follows that the reading frame and orientation that yield arginine codons will be infrequent. Indeed, as noted by Makela *et al.*,[33] this is the case. Reading frame 1 is used in the vast majority of antibodies, and orientation A (FIG. 3) is also preferred.

We have observed an inverse relationship between the presence of arginines introduced by VDJ rearrangement and the presence of arginines that are the product of somatic mutations in CDRs 1 and 2 of anti-DNA V_H regions (TABLE 1). Two-thirds of the sequences have arginines that are generated by either one or the other mechanism. Thus, anti-DNA antibodies with CDR3 arginine(s) often lack mutations to arginine, whereas antibodies with mutations to arginine need not have CDR3 arginine. How do we interpret this finding? Somatically generated arginines have implications for the generation of autoreactivity in the peripheral lymphoid system, as discussed above. Conversely, CDR3 arginines imply that anti-DNA specificity can also be established as part of the primary repertoire.

One indication that arginine in CDR3 may determine whether an antibody will be autoreactive or not was provided by an *in vitro* generated mutant of V_H3H9.[12] Substitution of R96 with glycine (R96G) completely eliminated DNA binding, whereas retention of R96 in a variant lacking all somatic mutations (R53G/I57T/D65G) resulted in low but detectable binding (FIG. 1). Clearly, the CDR3 arginine of 3H9 was required to establish the anti-DNA phenotype of this B-cell clone, thus determining anti-DNA specificity at the earliest stage of Ig expression in the

```
VH                 Orientation A                  JH
       ------------------------------------->

   T C T A C T A T G G T T A C G A C

   Ser    Thr    Met    Val    Thr        = RF1
     Leu    Leu    Trp    Leu    Arg      = RF2
       Tyr    Tyr    Gly    Tyr    Asp    = RF3


VH                 Orientation B                  JH
       <-------------------------------------

   G T C G T A A C C A T A G T A G A

   Val    Val    Thr    Ile    Val        = RF1
     Ser    Trm    Pro    Trm    Trm      = RF2
       Arg    Asn    His    Ser    Arg    = RF3
```

FIGURE 3. Six reading frames for DSP2.4.

bone marrow. Subsequent mutations further improved upon this initial specificity. The fact that most anti-DNA H chains in TABLE 1 have either one or the other type of arginine implies that autoreactive B-cell clones may be generated either in the bone marrow or in the periphery, and the example of 3H9 shows that continued antigen selection can further improve antigen binding and result in the addition of arginine.

IMPLICATIONS FOR THE REGULATION OF ANTI-DNA

The processes by which anti-DNA specificity arises are universal to antibody diversification: random somatic mutation is central to creating the antibody repertoire to foreign antigens, and rearrangement flexibility leads to enormous CDR3 diversity. Nevertheless, mutations to arginine are rare in conventional antibodies. Arginine is also underrepresented in the CDR3 of conventional antibodies even though certain reading frames and orientation of D_H genes include arginine codons. According to the "rules of recombination,"[34] these reading frames and orientation should be used as often as any other, yet they are not.

Melvin Cohn has explained these biases under the rubric, "D for disaster." In this model[35,36] sequences coded for by certain D gene reading frames or orientation are selected against because they lead to a structurally unsound antibody or an autoantibody. There is now abundant evidence for negative regulation of B cells expressing autoantibodies[38,39] and regulation of anti-DNAs has been inferred from the phenotypes of anti-DNA transgenic mice.[40,41,42]

The mechanisms by which anti-DNAs are formed point to the levels of B-cell development at which regulation must act. There are multiple sites at which arginine can be introduced. Antibodies with CDR3 arginines will be part of the emerging repertoire in the bone marrow. Antibodies with arginine(s) due to somatic mutation will arise in the periphery. Thus, regulation of these antibodies must occur both in "central" and peripheral lymphoid organs.

REFERENCES

1. TAN, E. M. 1982. Autoantibodies to nuclear antigens (ANA): Their immunobiology and medicine. Adv. Immunol. **33:** 167–240.
2. HARDIN, J. A. 1986. The lupus autoantigens and the pathogenesis of systemic lupus erythematosus. Arthritis Rheumatol. **29:** 457–460.
3. STAUDT, L. M. & W. GERHARD. 1983. Generation of antibody diversity in the immune response of BALB/c mice to influenza virus hemagglutinin. I. Significant variation in repertoire expression between individual mice. J. Exp. Med. **157:** 687.
4. SHLOMCHIK, M. J., A. MARSHAK-ROTHSTEIN, C. B. WOLFOWICZ, T. ROTHSTEIN & M. G. WEIGERT. 1987. The role of clonal selection and somatic mutation in autoimmunity. Nature **328:** 805–811.
5. SHLOMCHIK, M. J., A. H. AUCOIN, D. S. PISETSKY & M. G. WEIGERT. 1987. The structure and function of anti-DNA autoantibodies derived from a single autoimmune mouse. Proc. Natl. Acad. Sci. USA **84:** 9150–9154.
6. SHLOMCHIK, M. J., M. A. MASCELLI, H. SHAN, M. Z. RADIC, D. S. PISETSKY, A. MARSHAK-ROTHSTEIN & M. G. WEIGERT. 1989. Anti-DNA antibodies from autoimmune mice arise by clonal expansion and somatic mutation. J. Exp. Med. **171:** 265–297.
7. FATENEJAD, S., M. J. MAMULA & J. CRAFT. 1993. Role of intermolecular/intrastructural B- and T-cell determinants in the diversification of autoantibodies to ribonucleoprotein particles. Proc. Natl. Acad. Sci. USA **90:** 12010–12014.
8. LOSMAN, J. A., T. M. FASY, K. E. NOVICK, M. MASSA & M. MONESTIER. 1993. Nucleosome-specific antibody from an autoimmune MRL/Mp-lpr/lpr mouse. Arthritis Rheumatol. **36:** 552–560.
9. CASCIOLA-ROSEN, L. A., G. ANHALT & A. ROSEN. 1994. Autoantigens targeted in systemic lupus erythematosus are clustered in two populations of surface structures on apoptotic keratinocytes. J. Exp. Med. **179:** 1317–1330.
10. RADIC, M. Z., M. A. MASCELLI, J. ERIKSON, H. SHAN, M. J. SHLOMCHIK & M. G. WEIGERT. 1989. Structural patterns in anti-DNA antibodies from MRL/lpr mice. Cold Spring Harbor Symposium Quant. Biol. **54:** 933–946.
11. MARION, T. N., D. M. TILLMAN & N.-T. JOU. 1990. Interclonal and intraclonal diversity among anti-DNA antibodies from an (NZBxNZW)F_1 mouse. J. Immunol. **145:** 2322–2332.
12. RADIC, M. Z., J. J. MACKLE, J. ERIKSON, C. MOL, W. F. ANDERSON & M. G. WEIGERT. 1993. Residues that mediate DNA binding of autoimmune antibodies. J. Immunol. **150:** 4966–4977.
13. PAVLETICH, N. P. & C. O. PABO. 1993. Crystal structure of a five-finger GLI-DNA complex: New perspectives on zinc fingers. Science **261:** 1701–1707.
14. FERRE-D'AMARE, A. R., G. C. PRENDERGAST, E. B. ZIFF & S. K. BURLEY. 1993. Recognition by Max of its cognate DNA through a dimeric b/HLH/z domain. Nature **363:** 38–45.
15. ELLENBERGER, T. E., C. S. BRANDL, K. STRUHL & S. C. HARRISON. 1992. The GCN4 basic region leucine zipper binds DNA as a dimer of uninterrupted α helices: Crystal structure of the protein–DNA complex. Cell **71:** 1223–1237.

16. PAVLETICH, N. P. & C. O. PABO. 1991. Zinc finger–DNA recognition: Crystal structure of a Zif268-DNA complex at 2.1 Å. Science **252:** 809–817.
17. KIM, Y., J. C. GRABLE, R. LOVE, P. J. GREENE & J. M. ROSENBERG. 1990. Refinement of *Eco*R1 endonuclease crystal structures: A revised protein chain tracing. Science **249:** 1307–1309.
18. CHO, Y., S. GORINA, P. D. JEFFREY & N. P. PAVLETICH. 1994. Crystal structure of a p53 tumor suppressor-DNA complex: Understanding tumorigenic mutations. Science **265:** 346–355.
19. ALZARI, P. M., M.-B. LASCOMBE & R. J. POLJAK. 1988. Three-dimensional structure of antibodies. Ann. Rev. Immunol. **6:** 555–580.
20. MIAN, I. S., A. R. BRADWELL & A. J. OLSON. 1991. Structure, function and properties of antibody binding sites. J. Mol. Biol. **217:** 133–151.
21. RADIC, M. Z. & M. G. WEIGERT. 1994. Genetic and structural evidence for antigen selection of anti-DNA antibodies. Annu. Rev. Immunol. **12:** 487–520.
22. BLOOM, D. D., J.-L. DAVIGNON, M. W. RETIER, M. J. SHLOMCHIK, D. S. PISETSKY, P. L. COHEN, R. A. EISENBERG & S. H. CLARKE. 1993. V region gene analysis of anti-Sm hybridomas from MRL/Mp-lpr/lpr mice. J. Immunol. **150:** 1591–1610.
23. TILLMAN, D. M., N.-T. JOU, R. J. HILL & T. N. MARION. 1992. Both IgM and IgG anti-DNA antibodies are the products of clonally selective B cell stimulation in (NZBxNZW)F_1 mice. J. Exp. Med. **176:** 761–789.
24. KOFLER, R., R. STROHAL, R. S. BALDERAS, M. E. JOHNSON, D. J. NOONAN, M. A. DUCHOSAL, F. J. DIXON & A. N. THEOFILOPOULOS. 1988. Immunoglobulin k light chain variable region gene complex organization and immunoglobulin genes encoding anti-DNA autoantibodies in lupus mice. J. Clin. Invest. **82:** 852–860.
25. KRISHNAN, M. R. & T. N. MARION. 1993. Structural similarity of antibody variable regions from immune and autoimmune anti-DNA antibodies. J. Immunol. **150:** 4948–4957.
26. EILAT, D., D. M. WEBSTER & A. R. REES. 1988. V region sequences of anti-DNA and anti-RNA autoantibodies from NZB/NZW F_1 mice. J. Immunol. **141:** 1745–1753.
27. TREPICCHIO, W., A. MANYA & K. J. BARRETT. 1987. The heavy chain genes of a lupus anti-DNA autoantibody are encoded in the germline of a nonautoimmune strain of mouse and conserved in strains of mice polymorphic for this gene locus. J. Immunol. **139:** 3139–3145.
28. GILKESON, G. S., D. D. BLOOM, D. S. PISETSKY & S. H. CLARKE. 1993. Molecular characterization of anti-DNA antibodies induced in normal mice by immunization with bacterial DNA. J. Immunol. **151:** 1353–1364.
29. KITA, Y., T. SUMIDA, K. ICHIKAWA, T. MAEDA, F. YORAHA, I. IWANOTO, S. YOSHIDA & T. KOIKE. 1993. V gene analysis of anticardiolipin antibodies from MRL/lpr/lpr mice. J. Immunol. **151:** 849–856.
30. O'KEEFE, T. L., S. BANDYOPADHYAY, S. K. DATTA & T. IMANISHI-KARI. 1990. V region sequences of an idiotypically connected family of pathogenic anti-DNA autoantibodies. J. Immunol. **144:** 4275–4283.
31. RADIC, M. Z., M. A. MASCELLI, J. ERIKSON, H. SHAN & M. G. WEIGERT. 1991. Immunoglobulin heavy and light chain contributions to autoimmune specificities. J. Immunol. **146:** 176–182.
32. EILAT, D. 1990. The role of germline gene expression and somatic mutation in the generation of autoantibodies to DNA. Mol. Immunol. **27:** 203–210.
33. KAARTINEN, M. & O. MAKELA. 1985. Reading of D-genes in variable frames as a source of antibody diversity. Immunol. Today **6:** 324–327.
34. EARLY, P., H. HUANG, M. DAVIS, K. CALAME & L. HOOD. 1980. An immunoglobulin heavy chain variable region is generated from three segments of DNA: V_H, D, and J_H. Cell **19:** 981–992.
35. SAKANO, H., K. HUPPI, G. HEINRICH & S. TONEGAWA. 1979. Sequences at the somatic recombination sites of immunoglobulin light-chain genes. Nature **280:** 288–294.
36. LANGMAN, R. E. & M. COHN. 1987. The E-T (elephant-tadpole) paradox necessitates the concept of a unit of B-cell function: The protecton. Mol. Immunol. **24:** 675–697.
37. KABAT, E. A., T. W. TAI, H. M. PERRY, K. S. GOTTESMAN & C. FOELLER. 1991.

Sequences of Proteins of Immunological Interest. U.S. Govt. Printing Off. Bethesda, MD.

38. BASTEN, A., R. BRINK, P. PEAKE, E. ADAMS, J. CROSBIE, S. HARTLEY & C. C. GOODNOW. 1991. Self tolerance in the B-cell repertoire. Immunol. Rev. **122:** 5–19.
39. NEMAZEE, D., D. RUSSELL, B. ARNOLD, G. HAEMMERLING, J. ALLISON, J. F. A. P. MILLER, G. MORAHAN & H. BUERKI. 1991. Clonal deletion of autospecific B lymphocytes. Immunol. Rev. **122:** 117–132.
40. ERIKSON, J., M. Z. RADIC, S. CAMPER, R. HARDY, C. CARMACK & M. WEIGERT. 1991. Expression of anti-DNA immunoglobulin transgenes in non-autoimmune mice. Nature (London) **349:** 331–334.
41. CHEN, C., M. Z. RADIC, J. ERIKSON, S. A. CAMPER, S. LITWIN, R. R. HARDY & M. WEIGERT. 1994. Deletion and editing of B cells that express antibodies to DNA. J. Immunol. **152:** 1970–1982.
42. CHEN, C., Z. NAGY, M. Z. RADIC, R. R. HARDY, D. HUSZAR, S. A. CAMPER & M. WEIGERT. 1995. The site and stage of anti-DNA B cell deletion. Nature (London) **373:** 252–255.

Does Immunological Tolerance Explain the Waste in the B-Lymphocyte Immune System?

Experiment and Theory[a]

DAVID NEMAZEE[b]

Division of Basic Sciences
Department of Pediatrics
National Jewish Center for Immunology and Respiratory Medicine
Denver, Colorado 80206

The increasing acceptance of the notion that autospecific B lymphocytes are inactivated, either by clonal deletion or anergy (reviewed by Nossal[1]), now permits a new look into some of the implications of self-tolerance. In this review we discuss the implications of a newly discovered form of B-cell tolerance and speculate on the possibility that an immune system that generates in its primary repertoire a very high proportion of self-reactive B cells may in fact be pursuing the most efficient strategy for building a repertoire capable of optimally recognizing non-self.

Our recent work with transgenic mice[2] along with the results of Weigert's group[3–5] has indicated a novel mechanism of B-cell tolerance called receptor editing. This mechanism involves secondary (nested) immunoglobulin light chain gene rearrangement in immature, bone marrow B cells whose receptors are cross-linked by autoantigen. Such rearrangements are mediated by ongoing or newly induced recombinase activity, which allows the cell to change its specificity and effectively results in deletion of the autoreactive receptor while retaining the cell that originally expressed it. These conclusions are based on the observed increase in recombination activator gene (RAG) expression and endogenous light chain gene rearrangement and expression in cells that should otherwise be constrained from further rearrangement by the mechanisms of allelic exclusion. Although the validity of the receptor-editing model has yet to be proved in a nontransgenic system, information from a number of sources suggests, albeit indirectly, that it may well be operative in normal cells.[6–9]

Why should one need the efficiency afforded by receptor editing? There is already massive waste in B-cell development, both at early stages of development in the bone marrow[10] and later, after potentially responsive B-cells have been generated and exported to the periphery.[11] One possible explanation is that self reactivity is actually far more likely than commonly thought, that is, that a far higher proportion of self-reactive B cells is generated in the repertoire than one

[a] This work was supported by grants from the Arthritis Foundation and the National Institutes of Health (RO1GM44809, KO4AI01161, AI33608-02).

[b] Address for correspondence: Division of Basic Sciences, Department of Pediatrics, National Jewish Center for Immunology and Respiratory Medicine, 1400 Jackson St., Denver, CO 80206.

might expect. Because the generated repertoire is highly diverse and there is no compelling reason to suppose, at least in the mouse system, that the repertoire should be prebiased to recognize self, the concept that there is extensive self-reactivity in the generated repertoire must imply that B-cell tolerance is extraordinarily sensitive, having a low affinity/avidity threshold. We have explored this possibility experimentally using our anti-major histocompatibility complex (MHC) class I-specific immunoglobulin gene transgenic mice. The antibody genes used to generate these transgenic mice come from hybridoma 3-83, which was derived from a fusion of spleen cells from an H-2^d strain mouse immunized with H-2^k spleen cells. The 3-83 antibody binds with moderate affinity to K^k and only very weakly to a number of other MHC class I molecules. The 3-83 cross-reactive interaction with K^{bm3} and D^k molecules is particularly weak, and even at the highest concentrations of antibody it is difficult to demonstrate any binding to cell lines bearing these antigens. Nevertheless, when 3-83 transgenic mice are bred onto the D^k or K^{bm3} background, deletion of these weakly autoreactive B cells is complete (J. Lang and D.N., manuscript in preparation). One possible conclusion that could be drawn from this result is that without receptor editing an even more dramatic depletion of B cells would occur at the pre-B-to-B-cell transition in the bone marrow.

Another aspect of self-reactivity that is of potential theoretical interest emerges if one considers the well-known concepts of degeneracy and cross-reactivity.[12] An individual antibody or B cell must bind to many different antigens, which can be related or unrelated to one another. This property is probably a function of both the size and the flexibility of the antigen combining site.[13] Can one come up with a model that could describe the amount of self-reactivity in the preselected repertoire that would be optimal to provide the most efficient use of the cells that are not deleted? I discuss below the interrelationship between two key numbers: (1) the fraction of B cells that must be inactivated, and (2) the degeneracy of the antibody combining site.

THE MODEL

Consider FIGURE 1, in which the box **A** represents the total number of potential antigen epitopes, and the oval **S** represents the subset of self epitopes to which tolerance is induced. Circles numbered 1 to 4 represent collections of antigenic epitopes recognized by individual B cells. Note that some these epitopes may overlap **S**. In the example shown in FIGURE 1, cells 2 and 4 would be deleted, while cells 1 and 3 would be permitted to function. The number of epitopes recognized by an individual receptor is defined as **c**, the *capacity* of an immunoglobulin receptor (IgR). According to this model, the circles of epitopes recognized by the individual B cells in FIGURE 1 is really a random distribution of dots of number **c**, but is shown as a circle to make the point that a fraction of the epitopes recognized by a single B cell may or may not bind self at a "deletable" affinity. Assuming that **A** is large, but finite, and that each B cell has the same value for **c**, what would be the best strategy for the immune system to adopt in order to maximize the "coverage" of the foreign universe of antigens? As the size of **c** increases, the potential coverage that each permitted B cell affords will increase linearly, but the probability that it will be eliminated by tolerance, and that its potential coverage of the foreign antigens will be lost, will increase exponentially. This means that there should be an optimal value of **c**. Incorporating these ideas

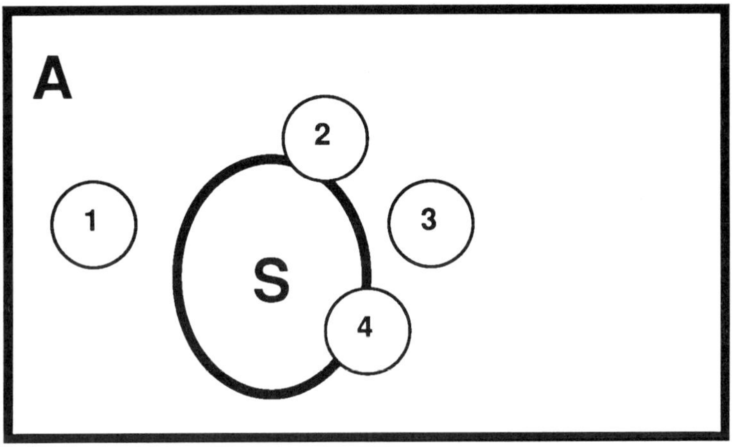

FIGURE 1

into a simple mathematical model, the fraction, **F**, of the foreign antigenic universe recognized by some functional B cell in the system is

$$F = 1 - e^{-m} \tag{1}$$

Equation 1 expresses the value of one minus the probability that a portion of the foreign epitopes will fail to be seen by the produced lymphocytes either because of chance or because the cells are disallowed as the result of self tolerance. This second term of Equation 1 is the first term of the Poisson series (representing the class with zero events) and the exponent m represents the mean number of "events" per epitope, that is, the number of cells recognizing a particular epitope. This value is seen to be $nc/(A - S)$ multiplied by a correction factor to include only the fraction of clones that are non-self-reactive, which is $[(A - S)/A]^c$. Thus,

$$m = nc[(A - S)/A]^c/(A - S)$$

where n is the total number of cells (clones of independent specificities) in the system before selection. We assume that the generated lymphocyte receptor specificities are essentially random, but of fixed c. Differentiating F with respect to c and setting to zero, one can determine the value of c that maximizes the coverage of the foreign epitopic universe. We find an optimal c:

$$c = -\{\ln [(A - S)/A]\}^{-1}$$

The probability P that a particular B cell will be autoreactive (deleted) is

$$P = 1 - [(A - S)/A]^c$$

What then is the probability P that a particular B cell will be autoreactive (deleted) at optimal c? Substituting for c, this value is

$$P = 1 - e^{-1}$$

Surprisingly, at optimal c the *probability that a particular lymphocyte will be tolerized* is $P = 1 - e^{-1}$ and is independent of the absolute or relative values of n, A, and S. However, optimal c is sensitive to the *ratio* of self to non-self. Thus, the system should provide the most complete coverage or protection from foreign antigens when most ($1 - e^{-1} = \sim 63.2\%$) of the B cells are forbidden. If c was suboptimal, the immune response would be more specific and delete a smaller percentage of cells, but would require far more cells to avoid intolerably large "holes" in the repertoire. If c was greater than optimal, an even larger fraction of produced cells would be eliminated. In either case, the apparent waste in the system would be greater, given the assumptions of the model. We have argued elsewhere that the B lymphoid system could partly compensate for the potential waste caused by self tolerance by using a mechanism of receptor editing to rapidly screen out useless receptors at an early developmental stage.[2]

On the basis of the few assumptions mentioned, the model suggests that the nature of the immunoglobulin binding site, which is the biochemical expression of c, may be under pressure to evolve to reflect the *fraction* of the antigenic universe that is self. It also provides a rationale for receptor editing, which would overcome to some extent this, theoretical, difficulty—namely that even an efficient immune system must generate useless receptors.

ACKNOWLEDGMENTS

The author thanks Dr. P. Archer of the University of Colorado Health Sciences Center for checking the mathematical model.

[*Note added in proof*: Since submitting this paper, we have become aware of a paper (R. J. De Boer and A. S. Perelson, 1993. Proc. R. Soc. London. B **252:** 171–175) that comes to a conclusion similar to the one presented here.]

REFERENCES

1. Nossal, G. V. J. 1992. Cellular and molecular mechanisms of B lymphocyte tolerance. Adv. Immunol. **52:** 283–331.
2. Tiegs, S. L., D. M. Russell & D. Nemazee. 1993. Receptor editing in self-reactive bone marrow B-cells. J. Exp. Med. **177:** 1009–1020.
3. Gay, D., T. Saunders, S. Camper & M. Weigert. 1993. Receptor editing: An approach by autoreactive B cells to escape tolerance. J. Exp. Med. **177:** 999–1008.
4. Radic, M. Z., J. Erikson, S. Litwin & M. Weigert. 1993. B lymphocytes may escape tolerance by revising their antigen receptors. J. Exp. Med. **177:** 1165–1173.
5. Chen, C., M. Z. Radic, J. Erikson, S. A. Camper, S. Litwin, R. R. Hardy & M. Weigert. 1994. Deletion and editing of B cells that express antibodies to DNA. J. Immunol. **152:** 1970–1982.
6. Feddersen, R. M. & B. G. Van Ness. 1985. Double recombination of a single immunoglobulin κ-chain allele: Implications for the mechanism of rearrangement. Proc. Natl. Acad. Sci. USA **82:** 4793–4797.
7. Shapiro, M. A. & M. Weigert. 1987. How immunoglobulin V_κ genes rearrange. J. Immunol. **139:** 3834–3839.
8. Harada, K. & H. Yamagishi. 1991. Lack of feedback inhibition of V_κ gene rearrangement by productively rearranged alleles. J. Exp. Med. **173:** 409–415.
9. Ma, A., P. Fisher, R. Dildrop, E. Oltz, G. Rathbun, P. Achacoso, A. Stall & F. W. Alt. 1992. Surface IgM mediated regulation of RAG gene expression in Eμ-N-myc B cell lines. EMBO J. **11:** 2727–2734.

10. OSMOND, D. G. 1990. B cell development in the bone marrow. Semin. Immunol. **2:** 173–180.
11. FÖRSTER, I. & K. RAJEWSKY. 1990. The bulk of the peripheral B-cell pool in mice is stable and not rapidly renewed from the bone marrow. Proc. Natl. Acad. Sci. USA **87:** 4781–4784.
12. YOSHIOKA, M., J. LIFTER, C. L. HEW, C. A. CONVERSE, M. Y. ARMSTRONG, W. H. KONIGSBERG & F. F. RICHARDS. 1973. Studies on the combining region of protein 460, a mouse gamma A immunoglobulin which binds several haptens. Binding and reactivity of two types of photoaffinity labeling reagents. Biochemistry **12:** 4679–4685.
13. FOOTE, J. & C. MILSTEIN. 1994. Conformational isomerism and the diversity of antibodies. Proc. Natl. Acad. Sci. USA **91:** 10370–10374.

B-1 Cells and Autoimmunity[a]

MASAO MURAKAMI AND TASUKU HONJO

Department of Medical Chemistry
Kyoto University
Yoshida, Sakyo-ku
Kyoto 606, Japan

B-1 cells are distinguished from conventional B cells by anatomical localization, surface phenotypes, and functional characteristics.[1,2] B-1 cells are virtually absent from lymph nodes and present at a low frequency in spleen; however, they constitute the major fraction of B cells in the peritoneal cavity. Several important questions, such as the origin, function, and development of B-1 cells, have to be settled. Although adoptive transfer studies suggest that B-1 cells and conventional B cells have different developmental pathways through distinct progenitor cells,[3–6] *in vitro* studies showed that B-1 cells may derive from conventional B cells.[7] It is also unknown why the peritoneal cavity contains a large number of B-1 cells. They produce the majority of IgM and particularly most of IgM class autoantibodies (Abs), including those against thymocytes/T cells, single-stranded (ss) DNA, and bromelain-treated mouse erythrocytes, which appear to be cross-reactive to bacterial components (e.g., dextrane and phosphatidylcoline).[8] Although several observations (including motheaten mice) suggest a possible role for B-1 cells in autoimmune diseases,[9–11] it remains to be tested whether B-1 cells are directly involved in the pathogenesis of autoimmune diseases. In this review, we would like to clarify these questions by using anti-red blood cell (RBC) autoantibody (autoAb) transgenic (Tg) mice and classical lupus mice (NZB and NZB/WF1).

ANTI-RBC AUTOAB TG MICE

We established Tg mouse lines in which all B cells express anti-RBC autoAb on the surface by introducing the Ig genes derived from the anti-RBC autoAb-producing hybridoma into fertilized eggs.[14–16] In these mice, almost all B cells are deleted in the periphery, but a normal number of B-1 cells survived in the peritoneal cavity and in lamina propria of the gut. About 50% of the animals of the transgenic line suffer from autoimmune hemolytic anemia. Several lines of evidence have demonstrated that production of the autoantibody by B-1 cells is responsible for autoimmune hemolytic anemia in the Tg mice. First, intraperitoneal injection of autoAg (RBC) killed almost all B-1 cells and cured autoimmune symptoms by reduction of serum autoAb levels. Second, adoptive transfer of peritoneal B cells of the Tg mice to SCID mice induced hemolytic anemia. Finally, male anti-RBC Tg mice crossed with the *xid* mice, in which B-1 cells are congenitally absent, did not suffer from autoimmune hemolytic anemia. These results suggest a senario in which B-1 cells in the peritoneal cavity and gut, surviving from clonal deletion, may be activated by some environmental factors, produce autoAbs, and induce

[a] This work was supported in part by grants from the Ministry of Education, Science and Culture and the Dr. Shimizu Foundation for the Promotion of Immunology Research Grant for 1995.

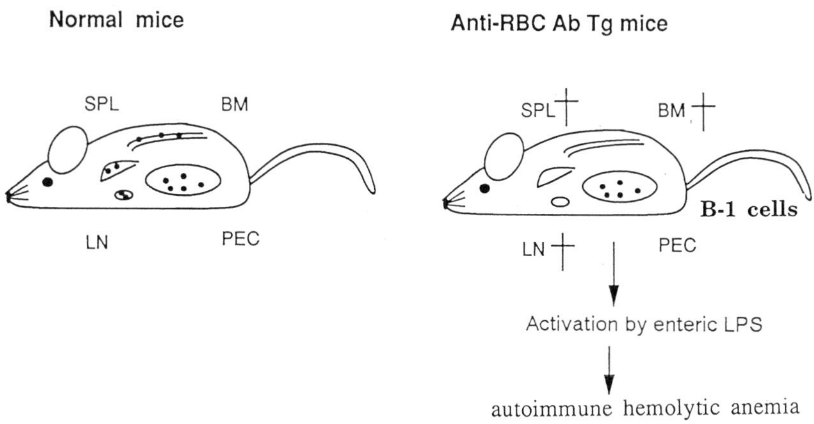

FIGURE 1. Distribution of B cells in anti-RBC Ab Tg mice. Abbreviations: SPL, spleen; BM, bone marrow; LN, lymph node; PEC, peritoneal cavity.

the autoimmune disease (FIG. 1). The anti-RBC autoAb Tg mice have provided the first autoimmune animal model in which B-1 cells are directly shown to be involved in the occurrence of autoimmune diseases.

MECHANISM OF ACTIVATION OF B-1 CELLS

Analyses of cell cycle and cytokine response[15–17] suggest that B-1 cells are in a pre-activated state and produce IgM in response to IL-2, IL-5, and IL10. How are B-1 cells activated *in vivo*? This question can be answered in our Tg mouse system in which only B-1 cells survive *in vivo* and the activation of B-1 cells can be monitored by checking the occurrence of autoimmune hemolytic anemia. The reduced frequency of hemolytic anemia in Tg mice with the same genetic background and the same number of B-1 cells, but maintained in cleaner conditions (semi-SPF), suggests that some environmental factors such as infection may be involved in the activation of B-1 cells and the subsequent onset of the autoimmune disease. B-1 cells in the peritoneal cavity and gut were activated by oral administration of lipopolysaccharides (LPS) but not by other routes of systemic administration such as intramuscular or intraperitoneal injection of LPS,[14] indicating that B-1 cells surviving in the peritoneal cavity and lamina propria of the gut can be activated by enteric bacteria LPS and induced autoimmune hemolytic anemia. This conclusion was confirmed by the fact that Tg mice breeding in germ-free conditions did not suffer from autoimmune hemolytic anemia (unpublished data). The fact that B-1 cells can be strongly activated by enteric bacteria or enteric flora suggests that B-1 cells may be involved in mucosal immunity.

B-1 CELLS AS A COMPONENT OF GALT

Anatomical location and functional properties distinguish mucosal immunity from the systemic immune system. The mucosal immune system is also referred to as the gut-associated lymphoid tissues (GALT), which include Peyer's patches, mesenteric lymph nodes, lamina propria lymphocytes, and intraepithelial lymphocytes.[18] A large fraction of intraepithelial T cells are generated independently from thymus and express $\gamma\delta$T cell receptors.[19–21] B cells in GALT preferentially produce and transport IgA into the gut lumen. Moreover, lymphocytes in GALT are generally known to have a unique homing pattern; committed cells leave the Peyer's patches after antigenic stimulation, migrate to mesenteric lymph nodes, go through the thoracic duct into the blood circulation arriving at the spleen, and finally return to GALT[22] (FIG. 2). According to this pathway, peritoneal lymphocytes seem to be related to GALT. Several studies, on the other hand, showed the possibility that peritoneal B-1 cells communicate with the gut and serve as progenitors of Ig-producing cells in the lamina propria of gut.[23] It remains to be seen whether or how peritoneal B cells communicate with GALT, where Peyer's patches and mesenteric lymph nodes contain few B-1 cells and abundant conventional B cells in normal mice. We have demonstrated that B-1 cells preferentially exist in the lamina propria of the gut as well as in the peritoneal cavity in Tg mice. B-1 cells exist in the lamina propria and in the peritoneal cavity in spite of an almost complete absence of conventional B cells in every lymphoid organ including GALT in anti-RBC autoAb Tg mice.[16] The B-1 cells in these two locations were activated by enteric bacterial antigens such as LPS (TABLE 1). These results strongly suggest

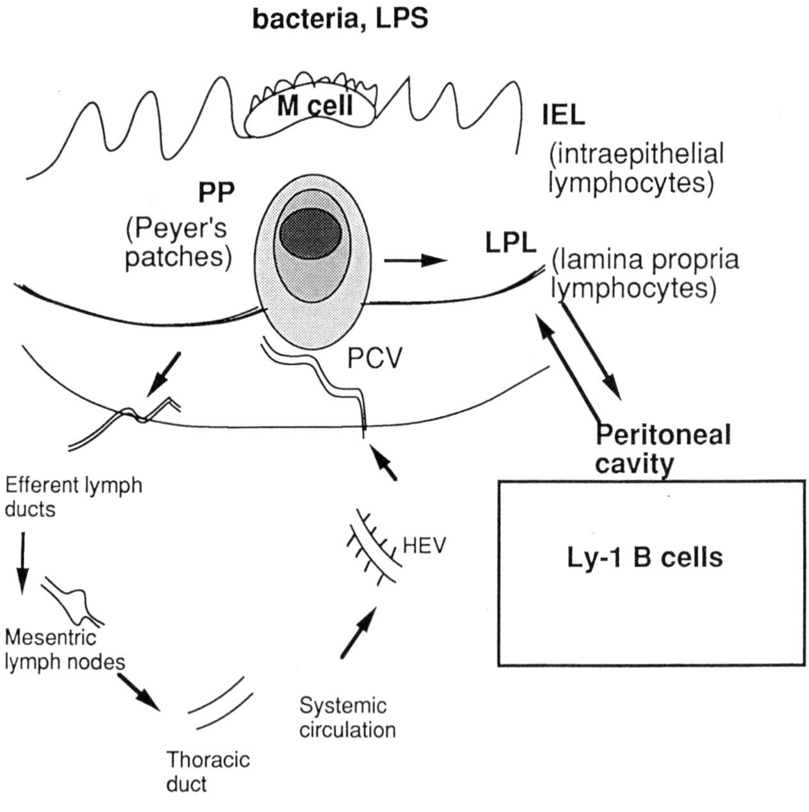

FIGURE 2. Communication of peritoneal Ly-1 B cells with LPL.

TABLE 1. Distribution of B Cells in GALT[a]

	Normal Mice		Anti-RBC Tg Mice	
Tissue[a]	Conventional	B-1	Conventional	B-1
IEL	–	–	–	–
P.P.	+	–	–	–
LPL	+	+	–	+
PEC	+	+	–	+

[a] Abbreviations: IEL, intraepithelial lymphocytes; P.P., Peyer's patches; LPL, lamina propria lymphocytes; PEC, peritoneal cavity.

that peritoneal B-1 cells may communicate directly with GALT (not with Peyer's patches but directly with lamina propria of the gut). B-1 cells in the peritoneal cavity possibly make a common pool with B cells in lamina propria of the gut and serve as one component of GALT for mucosal immunity. These results also clearly demonstrate that conventional B cells in Peyer's patches are separate from and not linked with peritoneal and lamina propria B-1 cells and that GALT contains two separate lineages of B cells. One is conventional B cells whose supply is probably dependent on Peyer's patches. The other is lamina propria B-1 cells, which communicate with B-1 cells in the peritoneal cavity and grow independently of Peyer's patches (FIG. 2).

DUAL SYSTEM IN MUCOSAL IMMUNITY

Intraepithelial T cells, frequently reactive to enteric bacteria with strong killer activities, partly originate independently of the thymus and partly have $\gamma\delta$ T cell receptors.[20,21] B-1 cells and $\gamma\delta$ T cells share many common characteristics: predominant localization ouside lymphoid tissues of the gut and restricted antigen specificity to enteric and self antigens. These primitive immune systems may have played an important role for the first line of defense against invading pathogens in primordial vertebrates until more sophisticated immune systems, that is, conventional B cells and $\alpha\beta$ T cells, evolved later. Upon the invasion of bacteria, antibacterial IgM antibodies, albeit with low affinity, may be immediately produced by B-1 cells in the peritoneal cavity and the lamina propria. Subsequently, after somatic mutation and class switching of conventional B cells in Peyer's patches, IgG or IgA class antibodies with high affinity will be produced. Inasmuch as the gut mucosa is continuously attacked by numerous microorganisms, GALT may require a two-layered line of defense, namely, the ancestral immune system of $\gamma\delta$T and B-1 cells, and the modern immune system of $\alpha\beta$T and conventional B cells (FIG. 3).

B-1 CELLS AND AUTOIMMUNE DISEASES

Several observations[8–11] suggest a possible role for B-1 cells in autoimmune diseases. Moreover, we directly dcmonstrated that B-1 cells were involved in the pathogenesis of autoimmune hemolytic anemia in anti-RBC autoAb Tg mice. However, because there are several arguments against the pathogenic role of B-1 cells, particularly in system lupus erythematosus (SLE) mice and humans,[24,25] whether B-1 cells are really involved in the development of autoimmune disease remains to be tested. To answer this question, elimination of B-1 cells *in vivo* would be valuable. So far, there have been several reports to eliminate B-1 cells *in vivo*. In NZB/*xid* and NZB/W F1/*xid* mice, which congenitally lack B-1 cells as a result of the X-linked immunodeficiency (*xid*) gene, the production of autoAbs was prevented, and the occurrence of autoimmune symptoms was inhibited.[26] The elimination of peritoneal B-1 cells by repeated injection of anti-IL10 Ab *in vivo* reduced the autoimmune symptoms in NZB/W F1 mice.[27] Unfortunately, however, the effects of these methods are more complex because of the change of the genetic background or effects on T cells, and therefore the potential for interpretation is limited. To clarify the pathogenesis of B-1 cells, the simpler method of eliminating B-1 cells *in vivo* was used.

ELIMINATION OF B-1 CELLS BY INTRAPERITONEAL INJECTION OF DISTILLED WATER

We used simple hypotonic shock to eliminate peritoneal B-1 cells. Injection of 1 ml distilled water into the peritoneal cavity of 6-week-old normal BALB/c mice

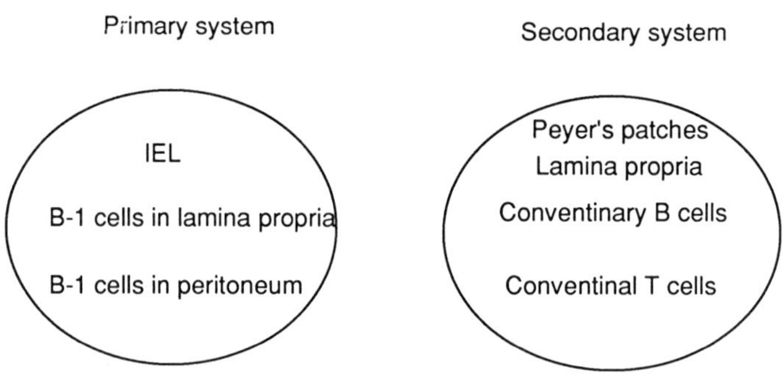

	Primary system		Secondary system
T cells	extrathymic origin	T cells	intrathymic origin
B cells	Bone marrow-derived self-replenishing capacity(+) Peyer's patches independent	Bcells	Bone marrow- independent self -replenshing capacity (-) Peyer's paches dependent
antigens	Sugar, phospholipid (bacterial antigens)		Peptides
reactivity to antigens	Low-specificity Low affinity		high-specificity high affinity

FIGURE 3. Dual immune system in gut.

TABLE 2. Effect of Elimination of B-1 Cells *in Vivo* in Lupus Mice

NZB (12 months old)	NZB/WF1 (6 months old)
Anti-RBC Abs IgM ↓ IgG ↓	Anti-DNA Abs IgM ↓ IgG ↓
Occurrence of autoimmune hemolytic anemia ↓	Occurrence of lupus nephritis ↓

could in a few minutes almost completely extinguish peritoneal cells, including T cells, B cells, and macrophages, whereas the number and relative proportion of lymphocyte subsets did not change in the spleen and bone marrow even after repeated injection of water into the peritoneal cavity. Because conventional B cells, T cells, and macrophages are constantly supplied from bone marrow, whereas B-1 cells are derived from peritoneal cells and self-replenish in the peritoneal cavity, repeated intraperitoneal injection of water decreased the number of B-1 cells *in vivo* but not other lymphoid cells, including conventional B cells, T cells, and macrophages. To eliminate peritoneal B-1 cells completely, we continued to inject water intraperitoneally every 7 days into neonatal mice of NZB and NZB/W F1 and examined whether B-1 cells could affect the onset of autoimmune diseases in the autoimmune-prone mice. The results were surprising: In water-treated NZB mice, the production of anti-RBC Abs including IgM and IgG decreased, and the occurrence of autoimmune hemolytic anemia was inhibited. In water-treated NZB/WF1 mice, the titer of serum IgG anti-DNA Abs decreased, and pathological signs of lupus nephritis were reduced (TABLE. 2). These results have clearly shown that B-1 cells are deeply involved in the production of IgG-class autoAbs and the pathogenesis of autoimmune diseases.

CAN B-1 CELLS CHANGE INTO CONVENTIONAL B CELLS?

In contrast to our reports, there have been several reports that conventional B cells, but not B-1 cells, produce IgG autoAbs in SLE in humans and mice.[24,25] This apparent discrepancy may be explained by the possibility that pathogenic IgG-class anti-DNA Ab-producing cells that do not express Ly-1 are derived from B-1 cells.[28] IgM-class autoAbs produced by B-1 cells are not necessarily pathogenic. To develop an autoimmune disease like SLE, T cells should be involved in isotype switching and somatic mutation of variable regions of immunoglobulin genes.[29–31] Indeed, some IgM-producing B-1 cells became IgG-producing cells *in vitro* upon stimulation by cytokines and/or CD40. The striking similarity of the immunoglobulin variable region sequences of IgM and IgG anti-DNA Abs derived from B/W F1 mice also suggested that anti-DNA Abs of the two classes might originate from the same variable gene and thus the same B cell.[33,34] It would be interesting to test whether the elimination of B-1 cells can decrease the production of pathogenic autoAbs and cure autoimmune symptoms in MRL/*lpr* mice.

ACKNOWLEDGMENT

We thank Drs. M. Okamoto, T. Tsubata, and S. Nishitani for their collaboration in analyzing anti-RBC autoAb Tg mice and Drs. T. Ushi, H. Yoshioka, and T. Shirai for their technical assistance and fruitful suggestions.

REFERENCES

1. Herzenberg, L. A., A. M. Stall, P. A. Lalor, C. Sidman, W. A. Moore, D. R. Parks & L. A. Herzenberg. 1986. The Ly-1 B cell lineage. Immunol. Rev. **93:** 81–102.
2. Hardy, R. R. & K. Hayakawa. 1986. Development and physiology of Ly-1 B and its human homolog, Leu-1 B. Immunol. Rev. **93:** 53–80.
3. Hardy, R. & K. Hayakawa. 1992. Generation of Ly-1B cells from developmentally distinct precursors. Ann. N.Y. Acad. Sci. **651:** 99.
4. Solvason, N., A. Lehuen & J. F. Kearney. 1991. An embryonic source of Ly-1 B cells but not conventional B cells. Int. Immunol. **3:** 543–550.
5. Godin, I. E., J. A. Garcia-Porrero, A. Coutinho, F. Dieterlen-Lièvre & M. A. R. Marcos. 1993. Para-aortic splanchnopleura from early mouse embryos contains B1a cell progenitors. Nature **364:** 67–70.
6. Kantor, A. B., A. M. Stall, S. Adams. L. A. Herzenberg & L. A. Herzenberg. 1992. Differential development of progenitor activity for three B-cell lineages. Proc. Natl. Acad. Sci. USA **89:** 3320–3324.
7. Chong, Y., E. Rabin & H. H. Wortis. 1991. Treatment of murine $CD5^-$ B cells with anti-Ig, but not LPS, induces surface CD5: Two B-cell activation pathways. Int. Immunol. **3:** 467–476.
8. Hayakawa, K., R. R. Hardy, M. Honda, L. A. Herzenberg, A. D. Steinberg & L. A. Herzenberg. 1984. Ly-1 B cell: Functionally distinct lymphocytes that secret IgM autoantibodies. Proc. Natl. Acad. Sci. USA **81:** 2494–2498.
9. Casali, P., S. E. Burastero, M. Nakamura, G. Inghirami & A. L. Notkins. 1984. Human lymphocytes making rheumatoid factor and antibody to ssDNA belong to Leu-1^+ B cells cell-set. Science **236:** 77–80.
10. Hardy, R. R., K. Hayakawa, M. Shimizu, K. Yamasaki & T. Kishimoto. 1984. Rheumatoid factor secretion from human Leu-1^+ B cells. Science **236:** 81–83.
11. Sidman, C. L., L. D. Shultz, R. R. Hardy, K. Hayakawa & L. A. Herzenberg. 1986. Production of immunoglobulin isotypes by Ly-1^+ B cells in viable motheaten mice and normal mice. Science **232:** 1423–1425.
12. Okamoto, M., M. Murakami, A. Shimizu, S. Ozaki, T. Tsubata, S. Kumagai & T. Honjo. 1992. A transgenic model of autoimmune hemolytic anemia J. Exp. Med. **175:** 71–79.
13. Murakami, M., T. Tsubata, M. Okamoto, A. Shimizu, S. Kumagai, H. Imura & T. Honjo. 1992. Antigen-induced apoptotic death of Ly-1 B cells responsible for autoimmune disease in transgenic mice. Nature **357:** 77–80.
14. Murakami, M., T. Tsubata, R. Shinkura, S. Nishitani, M. Okamoto, H. Yoshioka, T. Usui, S. Miyawaki & T. Honjo. 1994. Oral administration of lipopolysaccharides activates B-1 cells in the peritoneal cavity and lamina propria of the gut and induces autoimmune symptoms in an autoantibody transgenic mouse. J. Exp. Med. **180:** 11–122.
15. Hasegawa, K., M. Abe, T. Okada, *et al.* 1989. Are Ly-1 cells responsible for IL-2-hyperresponsiveness of B cells in autoimmune-prone NZB/NZWF1 mice? Int. Immunol. **1:** 99–103.
16. Hitoshi, Y., N. Yamaguchi, S. Mita, *et al.* 1990. Distribution of IL-5 receptor-positive B cells. Expression of IL-5 receptor on Ly-1 $(CD5)^+$ B cells. J. Immunol. **144:** 4218–4225.
17. Fei-Go, N., B. E. Castle, R. Barrett, R. Kastelein, W. Dang, T. R. Mosmann,

K. W. Moore & M. Howard. 1990. Interleukin 10, a novel B cell stimulatory factor: Unresponsiveness of X-linked immunodeficiency B cells. J. Exp. Med. **172:** 1625–1631.

18. Targran, S. R. & F. Shanahan, Eds. 1991. Immunology and Immunopathology of the Liver and Gastrointestinal Tract. Igakushoin, Inc. Tokyo, Japan.
19. Guy-Grand, D. C., C. Griselli & P. Vassalli. 1974. The gut-associated lymphoid system: Nature and properties of large dividing cells. Eur. J. Immunol. **4:** 435–443.
20. Bonneville, M., C. A. Janeway, K. Ito, W. Haser, I. Ishida, N. Nakanishi & S. Tonegawa. 1988. Intestinal intraepithelial lymphocytes are distinct sets of T cells. Nature **336:** 479–481.
21. Goodman, T. & L. Lefrancois. 1988. Expression of $\gamma\delta$T cell receptor on intraepithelial $CD8^+$ intraepithelial lymphocytes. Nature **338:** 855.
22. Holzmann, B. & I. L. Weismann. 1989. Integrin molecules involved in lymphocyte homing to Peyer's patches. Immunol. Rev. **108:** 45–61.
23. Kroese, F. G. M., E. C. Butcher, A. M. Stall, P. A. Lalor, S. Adams & L. A. Herzenberg. 1989. Many of the IgA producing plasma cells in murine gut are derived from self-relenishing precursors in the peritoneal cavity. Int. Immunol. **1:** 75–84.
24. Reap, E. A., E. S. Sobel, P. L. Cohen & R. A. Eizenberg. 1993. Conventional B cells, not B-1 cells, are responsible for producing autoantibodies in lpr mice. J. Exp. Med. **177:** 69–78.
25. Casali, P., S. E. Burastero, J. E. Balow & A. L. Notkins. 1989. High-affinity antibodies to ssDNA are produced by $CD5^-$ B cells in systemic lupus erythematosus patients. J. Immunol. **143:** 3476–3483.
26. Steinberg, B. J., P. A. Smathers & K. Frederiksen. 1982. Ability of *xid* gene to prevent autoimmunity in (NZBXNZW) F1 mice during the course of their natural history, after polyclonal stimulation, or following immunization with DNA. J. Clin. Invest. **70:** 587–597.
27. Ishida, H., T. Muchamuel, S. Sakaguchi, S. Andrade, S. Menton & M. Howard. 1994. Continuous anti-interleukin 10 antibody administration delays onsets of autoimmunity in NZB/WF1 mice. J. Exp. Med. **179:** 305–310.
28. Kanno, K., T. Okada, M. Abe, S. Hirose & T. Shirai. 1992. $CD5^+$ B cells as precursors of $CD5^-$ IgGanti-DNA antibody-producing B cells in autoimmune-prone NZB/WF1 mice. Ann. N.Y. Acad. Sci. **651:** 576–578.
29. Sekigawa, I., T. Okada, K. Noguchi, G. Ueda, S. Hirose, H. Sato & T. Shirai. 1987. Class-specific regulation of anti-DNA antibody synthesis and the age-associated changes in (NZBxNZW) F1 hybrid mice. J. Immunol. **138:** 2890–2895.
30. Data, S. K., H. Patel & D. Berry. 1987. Induction of a cationic shift in IgG anti-DNA autoantibodies. J. Exp. Med. **165:** 1252–1268.
31. Murakami, M., S. Kumagai, M. Sugita, K. Iwai & H. Imura. 1992. In vitro induction of IgG anti-DNA antibody from high density B cells of systemic lupus erythematosus patients by an HLA DR-restricted T cell clone. Clin. Exp. Immunol. **90:** 245–250.
32. Defrance, T., B. Vanbervliet, I. Durand, J. Briolay & J. Banchereau. 1992. Proliferation and differentiation of human $CD5^+$ and $CD5^-$ B cells subsets activated through their antigen receptors or CD40 antigens. Eur. J. Immunol. **22:** 2831–2840.
33. Hirose, S., M. Wakiya, Y. Kawano-Nishi, J. Yi, R. Shinokawa, S. Taki, T. Shimomura, T. Kishimoto, H. Tsurui, H. Nishimuta & T. Shirai. 1993. Somatic diversification and affinity maturation of IgM and IgG anti-DNA antibodies in murine luous. Eur. J. Immunol. **23:** 2813–2820.
34. Tillman, D. M., N-T. Jou, R. Hill & T. N. Marison. 1992. Both IgM and IgG anti DNA antibodies are products of clonally selective B cell stimulation in (NZBxNZW)F1 mice. J. Exp. Med. **176:** 761–779.

B-1 Cellular Origin and V_H Segment Structure of IgG, IgA, and IgM Anti-DNA Autoantibodies in Patients with Systemic Lupus Erythematosus[a]

MARION T. KASAIAN[b] AND PAOLO CASALI[c]

Division of Molecular Immunology
Department of Pathology
Cornell University Medical College
New York, New York 10021

INTRODUCTION

The human B-cell repertoire includes at least three distinct subsets, B-1a ($CD5^+$), B-1b (CD5 "sister"), and B-2 ("conventional") cells, each endowed with discrete phenotypic and functional properties.[1-4] B-2 cells (surface $CD5^-$, CD5 $mRNA^-$) account for the majority of circulating, splenic, and tonsillar B lymphocytes. They arise in the fetal liver, are continuously replenished in adult life by progenitors in the bone marrow,[2-6] and produce specific antibodies elicited by foreign antigen (Ag). B-1a cells (surface $CD5^+$, CD5 $mRNA^+$) emerge from progenitors in the fetal splanchnic district, whereas their B-1b (surface $CD5^-$, CD5 $mRNA^+$) counterparts can originate from the fetal splanchnic district or the bone marrow and persist throughout life by virtue of their self-replenishing nature.[1-4,6] In healthy humans, B-1a and B-1b cells comprise 15 to 30% and 2 to 5%, respectively, of circulating B lymphocytes and constitute the main source of natural autoantibodies.[1,5,7-11] Although they may display relatively high affinity for a specific Ag,[1] most natural autoantibodies are polyreactive; that is, they bind with different affinities to multiple self and foreign Ags that may differ greatly in nature (e.g., proteins, polynucleotides, and polysaccharides).[1,5,7,12]

Because of the self Ag-binding activity of the antibodies they produce, B-1 cells may play an important role in autoimmune phenomena. B-1a cells are greatly increased in number and/or produce autoantibodies in patients with Sjogren's syndrome,[13] primary anti-phospholipid syndrome,[14] and rheumatoid arthritis,[15-17] as well as in autoimmune (NZB × NZW)F1 mice.[18] In rheumatoid arthritis patients, B-1a cells can account for up to 60% of circulating B cells and produce rheumatoid factor autoantibodies that are monoreactive, high affinity, and somatically diversified.[9,10] In contrast, in systemic lupus erythematosus (SLE) patients, B-1a cells are generally not increased in number and do not constitute the main source of

[a] This work was supported by the U.S. Public Health Service Grant AR-40908. This is publication #5 from the Division of Molecular Immunology.

[b] Current address: ImmuLogic Pharmaceutical Corporation, 610 Lincoln Street, Waltham, MA 02154.

[c] Address for correspondence: Paolo Casali, M.D., Division of Molecular Immunology, Department of Pathology, Cornell University Medical College, 1300 York Avenue, New York, New York 10021-4896.

autoantibodies; a high proportion of anti-DNA autoantibodies is produced by lymphocytes that exhibit a surface $CD5^-$ phenotype.[19–21]

We investigated the cellular origin as well as the structure and reactivity of anti-DNA autoantibodies in SLE patients. First, we analyzed B cells selected on the basis of surface CD5 and/or CD5 mRNA expression for the production of these autoantibodies. Second, we generated an IgM, three IgG,[23] and five IgA[24] anti-DNA monoclonal antibodies (mAbs) from three SLE patients, analyzed the mAb reactivity to various self and exogenous Ags, and sequenced the genes encoding the mAb V_H segments. We found that, in these patients, B-1b cells accounted for most or all surface $CD5^-$ anti-DNA autoantibody-producing cell precursors; the remaining anti-DNA autoantibody-producing cell precursors being B-1a cells. We also found that the high-affinity anti-DNA autoantibodies produced by these cells were not only monoreactive, but also, as suggested by others,[22] polyreactive. Finally, not only monoreactive, but also polyreactive anti-DNA mAbs displayed somatic mutations of a nature and with a distribution characteristic of Ag selection.

SEGREGATION OF ANTI-DNA IgM-, IgG-, AND IgA-PRODUCING B-1a, B-1b, AND B-2 CELLS

Human B-1a, B-1b, and B-2 cells express surface CD5 and CD45RA at different densities; B-1a cells are $CD5^+CD45RA^{int}$, B-1b cells are $CD5^-CD45RA^{lo}$, and B-2 cells are $CD5^-CD45RA^{hi}$.[1,8] In order to analyze the segregation of anti-DNA autoantibody-producing cell precursors among these subsets in patients with SLE, we enriched B cells by depletion of T cells and monocytes[1] from the peripheral blood of three women (30, 50, and 67 years old) with clinically active disease and presenting at least four of the American Rheumatism Association (ARA) criteria for SLE. Prednisone (<20 mg/day) was withdrawn from all patients 24 to 48 hr before analysis. The cells were reacted with mAbs to CD45RA, CD5, and CD20 labeled with FITC, PE, and biotin, respectively (Coulter, Hialeah, FL). After further reaction with RED613-streptavidin (GIBCO-BRL Life Technologies, Gaithersburg, MD), they were analyzed by fluorescence flow cytometry using an EPICS Elite flow cytometer (Coulter). In patient A (FIG. 1A), B-1b (**panel c,** lower left rectangle), and B-1a (**panel c,** upper center rectangle) cells were resolved as relatively broad but not discrete shoulders, comprising 9% and 17%, respectively, of total $CD20^+$ (B) lymphocytes. B-2 cells were operationally defined as those $CD5^-$ B cells displaying the highest CD45RA density (**panel c,** lower right rectangle).[1,8] B-1a, B-1b, and B-2 cells were sorted into discrete fractions only minimally overlapping (**panels d, e,** and **f**). B-1a and B-1b cells comprised 20% and 13% of total B lymphocytes, respectively, in patient B, and 3% and 45%, respectively, in patient C (not shown). In each patient, approximately 20% of total B lymphocytes were sorted as B-2 cells.

The sorted B-1a, B-1b, and B-2 lymphocytes were infected with EBV and seeded at 500 cells/microwell in the presence of irradiated feeders.[1,25,26] After three weeks, culture fluids were analyzed for antibody to ssDNA, dsDNA, and tetanus toxoid using specific ELISAs.[1,25,26] In all three patients, the vast majority of the cells committed to the production of IgA, IgG, or IgM to dsDNA or ssDNA segregated with the B-1a and B-1b subsets (FIG. 1B). Few such cells were found in the B-2 fraction, which instead contained 86% of the total cells producing IgG to tetanus toxoid in these patients who had been previously vaccinated but not recently boosted with tetanus toxoid (not shown).

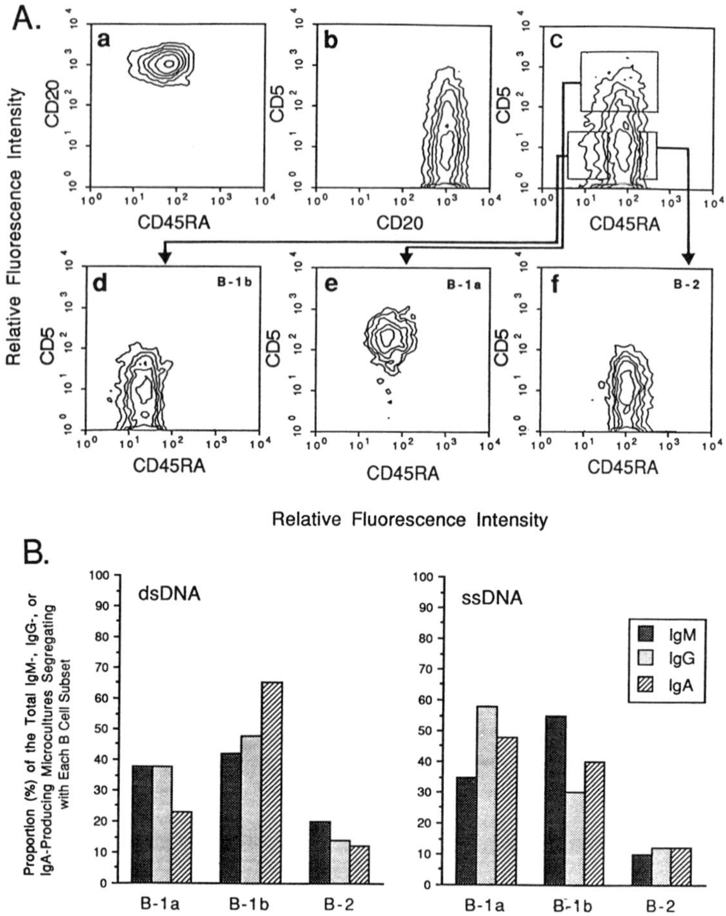

FIGURE 1. **(A)** Analysis of CD5 and CD45RA expression by B lymphocytes of SLE patients and sorting of B-1a, B-1b, and B-2 cells. Enriched peripheral blood B cells from patient A were reacted with PE-labeled mAb to CD20, FITC-labeled mAb to CD45RA, and biotinylated mAb to CD5, followed by RED613-labeled streptavidin, and applied to the fluorescence-activated cell sorter. CD20$^+$ cells were "gated" and analyzed for the expression of surface CD45RA (**panel a**) or CD5 (**panel b**) or both CD45RA and CD5 (**panel c**). B lymphocytes in **panel c** were "gated" as CD5$^-$CD45RA^{lo} (*left rectangle*), CD5$^+$ (*center rectangle*), or CD5$^-$CD45RA^{hi} (*right rectangle*) cells (B-1b, B-1a, and B-2 cells, respectively) and sorted as three discrete fractions. Some of the sorted cells from each fraction were reanalyzed for their surface CD45RA or CD5 fluorescence density. Upon reanalysis, most of the cells in each sorted fraction expressed CD5 and CD45RA densities similar to those expressed by the cells in each respective "gate" before sorting (**panels d, e,** and **f**). **(B)** Segregation of anti-DNA IgM, IgG, and IgA-producing cell precursors with the B-cell subsets in SLE patients. Sorted B-1a, B-1b, and B-2 cells from patients A, B, and C were infected with EBV for 1 hr and then seeded in microcultures at 500/well (96 wells/subset/patient) in the presence of 10^5 irradiated PBMC as feeders. After 3 weeks, the culture fluids were tested for the content of IgM, IgG, and IgA antibodies to ds-DNA or ssDNA. Histograms depict the proportion (%) of total anti-DNA IgM-, IgG-, or IgA-producing microcultures that segregated with each B-cell subset. For instance, the microcultures seeded with B-1a, B-1b, and B-2 cells accounted for 22, 66, and 12%, respectively, of the total microcultures producing IgA to ds-DNA.

GENERATION OF ANTI-DNA MAB-PRODUCING CELL LINES AND MAB ANALYSIS

Some EBV-transformed B lymphocytes producing anti-DNA autoantibodies were selected and cloned by sequential subculturing followed by fusion with the F3B6 cell line.[25,26] Nine anti-DNA mAbs (five IgA, three IgG, and one IgM) were isolated from these lines[23,24] (TABLE 1) and analyzed for binding to various self and foreign Ags using specific ELISAs[1] (FIG. 2). Of the six mAbs selected for binding to ssDNA, two were monoreactive (mAbs 424.F6.24 and 447.8H) and four were polyreactive, binding to ssDNA, dsDNA, and other self and foreign Ags (mAbs 412.66.F1, 410.7.F91, 410.30.F305, and 424.F10.3). Of the three mAbs selected for binding to dsDNA, one bound both dsDNA and ssDNA (mAb 412.67.F1.3), and the other two were polyreactive (mAb 448.9G.F1 and mAb 46). The mAb relative avidities for DNA, measured by competitive inhibition ELISA,[1,26] ranged from 1.6×10^{-7} to 8.0×10^{-10} g/μl for ssDNA, to 2.7×10^{-9} g/μl for dsDNA (TABLE 1).

EXPRESSION OF CD5 MRNA BY ANTI-DNA MAB-PRODUCING CELL LINES

Using a CD5-specific PCR amplification method, we have shown that in healthy subjects, B-1a and B-1b, but not B-2 cells, express high levels of CD5 mRNA.[1] To verify the origin of the anti-DNA IgA mAb-producing B-cell lines, we analyzed CD5 mRNA expression in the hybridomas or their progenitor EBV transformants, using the PCR method we recently devised.[1,10] Briefly, poly(A)$^+$ RNA was purified from EBV transformants, B-cell hybrids, purified human T cells (positive control), or F3B6 cells (negative control), reverse transcribed, and the cDNA used as a template for CD5- or β-actin-specific DNA amplification, using appropriate primers.[1,10] Amplified DNA was fractionated onto agarose gels, blotted onto nylon membranes (GeneScreen, Du Pont Co., NEN Research Products, Boston, MA), and hybridized with internal β-actin-specific or CD5-specific oligonucleotide probes[1,10] labeled with [^{32}P]-γ-ATP (Du Pont Co., NEN Research Products). Identical amounts of mRNA (18 ng) from eight of the nine mAb-producing cell lines, F3B6 cells and the mAb 417.3E.10H-producing cell line of B-2 origin[27] yielded similar amounts of cDNA, as indicated by the comparable levels of amplified β-actin DNA (FIG. 3, **panels A**). The degree of specific DNA amplification was proportional to the dose of input mRNA,[1,10] as shown for β-actin and CD5 DNA amplified from a B-cell clone (mAb 448.9G.F1; FIG. 3, **panels C** and **D**) and by the CD5 DNA amplified from human T cells (FIG. 3, **panels E**). One cell line, mAb 424.F6.24, producing a monoreactive IgA anti-ssDNA, expressed no detectable CD5 mRNA, as did the control mAb 417.3E.10H-producing B-2-derived cell line and the F3B6 fusion partner[10] (FIG. 3, **panels B**). The remaining seven anti-DNA mAb-producing cell lines all expressed high levels of CD5 mRNA, consistent with a B-1a or B-1b cellular origin of anti-DNA mAb-producing cell precursors.

ANTI-DNA MAB V_H, D, AND J_H GENE SEGMENTS AND SOMATIC MUTATIONS

Sequence analysis of the expressed V region gene segments was performed as described.[1,10,23,24] The nucleotide and deduced amino acid sequences of the

TABLE 1. Ag-Binding Features and V, D, and J Gene Composition of Anti-DNA mAbs from SLE Patients

Clone	Patient	Reactivity	Expression of CD5 mRNA	mAb Chains H, L	Relative Avidity (g/μl)[a] ssDNA	Relative Avidity (g/μl)[a] dsDNA	V_H Segment: Closest V_H Gene [Family][b]	V_H Segment: Nucleotide (Amino Acid) Identity[c] %	V_H Segment: Nucleotide Differences CDR R:S	V_H Segment: Nucleotide Differences FR R:S	D Gene[d]	J_H Gene[d]	EMBL/ GenBank Accession No.
mAb 424.F6.24[e]	B	DNA only	Nil	IgA1, κ	7.0×10^{-9}	Nil[f]	VI-3b [V_H1]	93.9 (91.8)	3:3	5:7	DN1	J_H4b	L28049
mAb 447.8H[e]	B	DNA only	Yes	IgA1, κ	7.5×10^{-8}	Nil	V_H2-MC2 [V_H2]	96.6 (93.9)	2:3	4:2	D21-7	J_H4b	L28050
mAb 412.67.F1.3[e]	A	DNA only	Yes	IgA1, κ	8.0×10^{-10}	2.7×10^{-9}	WHG16G [V_H3]	96.6 (92.8)	$\underline{6:0}$[g]	$\underline{2:2}$	DM2	J_H4b	L28051
mAb 412.66.F1[e]	A	Polyreactive	Yes	IgA1, κ	3.0×10^{-8}	Nil	V_H26 [V_H3]	95.9 (92.8)	$\underline{5:0}$[g]	$\underline{2:2}$	DxP'4	J_H4b	L28052
mAb 448.9G.F1[e]	C	Polyreactive	Yes	IgA1, λ	2.2×10^{-9}	4.4×10^{-8}	V71-2 [V_H4]	93.3 (85.8)	$\underline{8:0}$[g]	$\underline{6:6}$	DLR2	J_H4b	L28053
mAb 410.7.F91[h]	C	Polyreactive	Yes	IgG1, κ	5.0×10^{-9}	Nil	$V_H4.21$ [V_H4]	99.3 (99.0)	0:0	1:2	DxP'1	J_H4b	L23518
mAb 410.30.F305[h]	C	Polyreactive	n.d.[i]	IgG3, κ	4.2×10^{-8}	Nil	hv1263 [V_H1]	95.9 (93.8)	$\underline{4:0}$[g]	$\underline{2:6}$	DK4	J_H4b	L23519
mAb 424.F10.3	B	Polyreactive	Yes	IgG3, κ	1.6×10^{-7}	Nil	$V_H4.21$ [V_H4]	94.8 (95.9)	2:3	2:8	DM1 DxP'1	J_H3	L34163
mAb 46	A	Polyreactive	Yes	IgM, λ	1.2×10^{-8}	3.4×10^{-8}	V_H26 [V_H3]	100 (100)	0:0	0:0	D21/10	J_H2	L34164

[a] Relative avidity is expressed as the concentration of fluid-phase Ag (g/μl) inhibiting by 50% the binding of mAb to homologous solid-phase Ag. For each mAb, the binding measured in the absence of soluble Ag represented 100% binding activity.

[b] The sequences of the genomic germline V_H genes have been reported as follows: VI-3b,[45] WHG16G,[46] V_H26(V3-23),[28] V71-2(V4-61),[47] hv1263,[48] $V_H4.21$(V4-34),[29] and V_H2-MC2.[49]

[c] Compared with the germline genomic sequences.

[d] The DN1, DM1, DM2, Dk4, DxP'1, DxP'4, and DLR2 genes have been reported by Ichihara *et al.*[50]; the D21-7 and D21/10 genes have been reported by Buluwela *et al.*[51] The J_H2 and J_H3 genes have been reported by Ravetch *et al.*,[52] and J_H4b by Yamada *et al.*[44]

[e] Nucleotide and deduced amino acid sequences reported by Kasaian *et al.*[24]

[f] Avidity greater than 10^{-4} g/μl.

[g] The liklihood that the excess of putative R mutations in the CDR arose by chance was $p < 0.01$, calculated as described.[24a]

[h] Nucleotide and deduced amino acid sequences reported by Ikematsu *et al.*[23]

[i] Not determined.

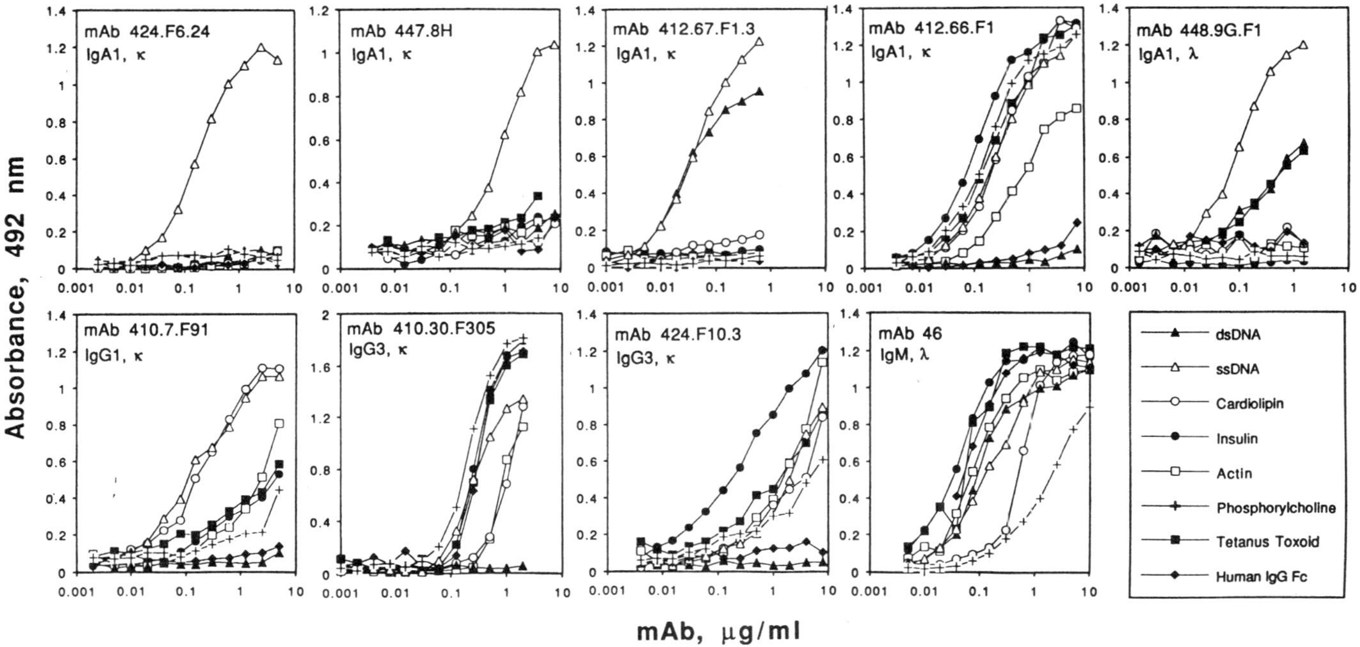

FIGURE 2. Dose-saturable binding (expressed as optical absorbance at 492 nm) to solid-phase Ag of anti-DNA mAbs generated from SLE patients. The following Ags were used: dsDNA (▲), ssDNA (△), cardiolipin (○), insulin (●), actin (□), phosphorylcholine chloride (+), tetanus toxoid (■), and human IgG Fc fragment (◆).

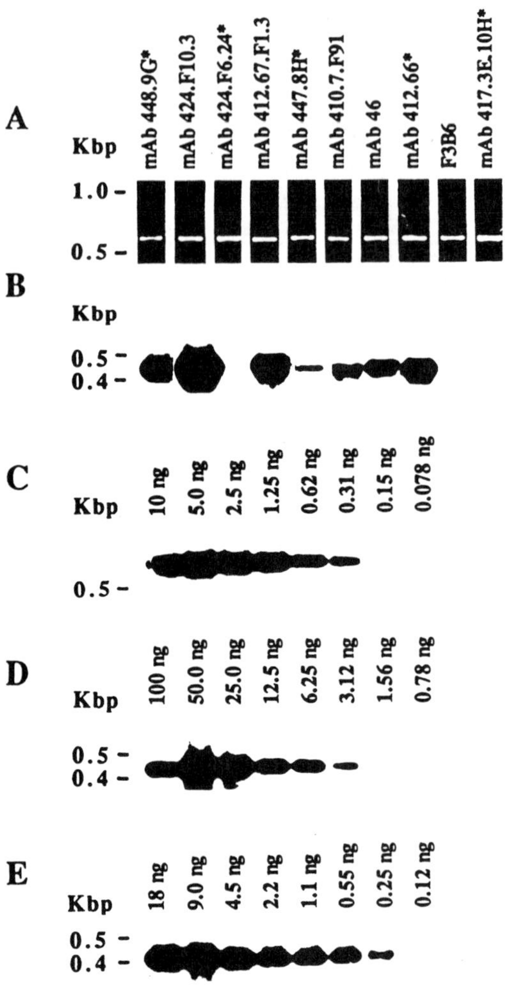

FIGURE 3. Expression of CD5 mRNA by the anti-DNA mAb-producing B-cell lines. The cell lines indicated by asterisks were EBV-transformed lymphoblasts; the remaining cell lines were B-cell hybridomas; F3B6 was the CD5-negative human–mouse heterohybridoma cell fusion partner (negative control); IgG mAb 417.3E.10H-producing cells were EBV-transformed B-2 lymphoblasts[24] (negative control). mRNA (18 ng) from each cell line was reverse transcribed. The cDNA was amplified by PCR using β-actin-specific or CD5-specific primers, and fractionated on a 1.2% agarose gel. **Panels A** depict gels containing amplified β-actin cDNA (~0.6 kbp) stained with ethidium bromide. **Panels B** depict hybridization of the ^{32}P-labeled CD5-specific oligonucleotide probe with fractionated amplified CD5 cDNA (~0.45 kbp). **Panels C** and **D** depict hybridization of ^{32}P-labeled β-actin and ^{32}P-labeled CD5-specific oligonucleotide probes, respectively, to fractionated DNA that had been amplified from cDNA reverse-transcribed from different amounts of mRNA of the mAb 448.9G.F1 mAb-producing cell line. **Panels E** depict hybridization of the ^{32}P-labeled CD5-specific oligonucleotide probe to fractionated, amplified cDNA independently reverse-transcribed from different amounts of purified T cell mRNA.

V_H-D-J_H genes of the polyreactive IgG3 mAb 424.F10.3 and the polyreactive IgM mAb 46 are shown in FIG. 4A–C, and those of the remaining seven mAbs have been reported elsewhere.[23,24] The V_H, D, and J_H genes used by the anti-DNA mAbs and the similarity of the V_H sequences to the closest reported germline genes are summarized in TABLE 1. The nine mAbs used seven distinct V_H genes from four different families: two V_H1, one V_H2, three V_H3 (including two V_H26c[28]), and three V_H4 (including two $V_H4.21$[29]). The sequences of the V_H genes encoding all mAbs but one (mAb 46) displayed a number of differences when compared with those of the closest germline genes; these differences were assumed to be somatic point mutations. In four mAbs, mAb 412.67.F1.3, mAb 412.66.F1, mAb 448.9G.F1, and mAb 410.30.F305, such putative point mutations were in nature and distribution consistent with an Ag selection process.[24,24a] In two mAbs, 412.66.F1 and 412.67.F1.3, the mutations were confirmed by differentially targeted amplification experiments and by cloning and sequencing the corresponding germline genes from autologous PMNs[24] as outlined in FIG. 5. The deduced amino acid squences of the D-J_H segments were divided into CDR3 and FR4 areas (FIG. 4D) according to Kabat *et al.*[30] The CDR3 sequences were highly divergent in composition and length, and the FR4 regions were invariable in length and displayed little diversity.

DISCUSSION

In the present studies, we analyzed the cellular origin, Ag reactivity, and structure of one IgM, three IgG, and five IgA mAbs to DNA generated from three patients with SLE. We provided evidence that, in these patients, (i) B-1a and B-1b cells comprise most of the anti-DNA autoantibody-producing cell precursors; (ii) high-affinity anti-DNA autoantibodies can be not only monoreactive, but also polyreactive; and, finally, (iii) the V_H genes encoding not only monoreactive but also polyreactive autoantibodies bear traces of somatic mutation and Ag selection. Taken together, these findings suggest that B-1 cells originally committed to the production of polyreactive or monoreactive DNA-binding natural autoantibodies can be recruited by Ag and undergo Ig class switch, somatic mutation, and affinity maturation to yield not only monoreactive, but also polyreactive high-affinity autoantibodies.

Previous findings have shown that, in spite of the primary role played by B-1a cells in autoantibody production in rheumatoid arthritis and other autoimmune diseases, these cells are not the main source of anti-DNA autoantibodies in patients with SLE.[19,20] Reap *et al.*[21] reported that in autoimmune MRL/*lpr* mice, the autoantibody-producing cell precursors are bone marrow-derived surface $CD5^-$ cells. These features are inconsistent with those of B-1a cells, but not with those of surface $CD5^-$ B-1b cells which, unlike their B-1a counterparts, can stem from progenitors in the bone marrow.[2,4] Our analysis of the segregation of anti-DNA autoantibody-producing cell precursors in SLE patients demonstrated that B-1b cells can account for the surface $CD5^-$ B cells that have been shown to produce anti-DNA autoantibodies.[19–21] The B-1 origin of the vast majority of the anti-DNA IgA, IgM, and IgG autoantibodies in the three patients was further strengthened by the demonstration of high levels of CD5 mRNA expression, a hallmark of B-1 cells, in all but one of the anti-DNA mAb-producing cell lines examined. These observations, together with our demonstration that B-1a cells account for the vast majority of rheumatoid factor-producing cells in rheumatoid arthritis,[9,10,15] suggest

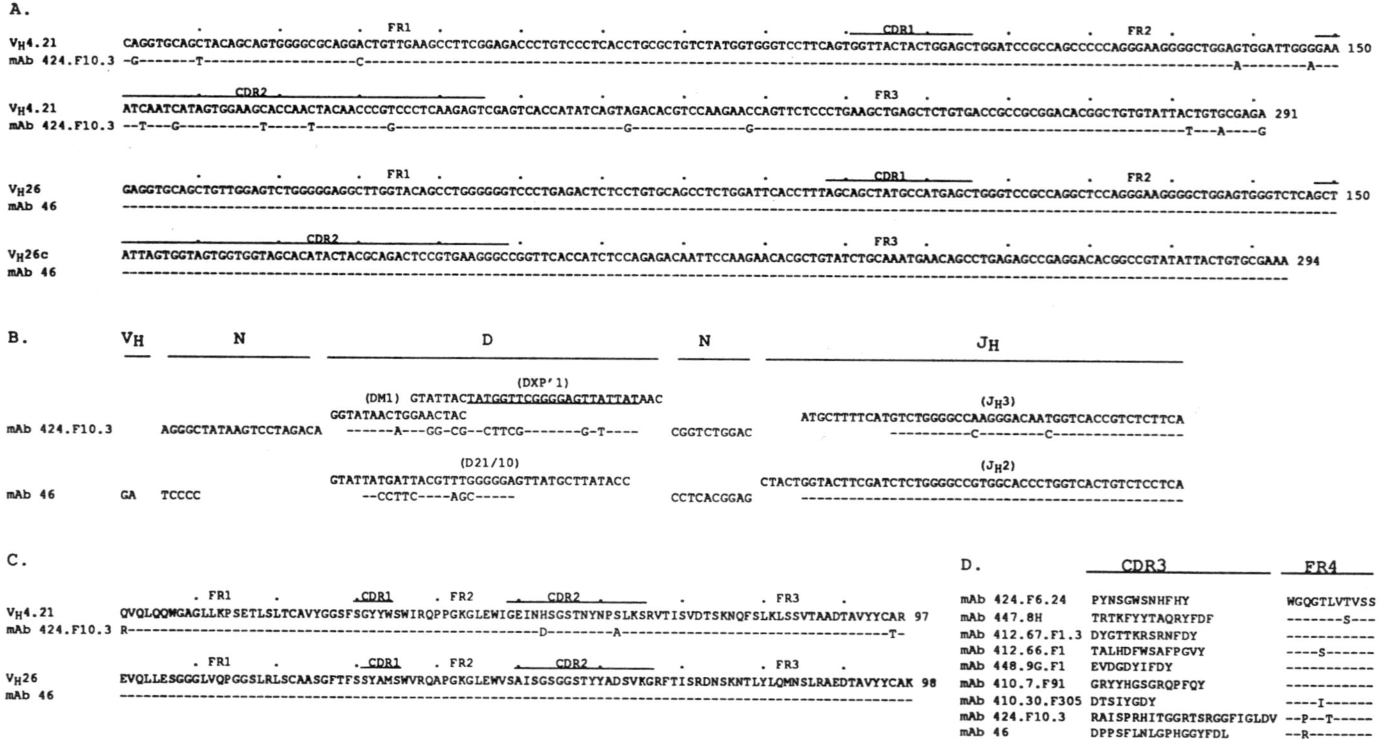

FIGURE 4. Nucleotide (**A** and **B**) and deduced amino acid (**C** and **D**) sequences of the human anti-DNA mAb V_H, D, and J_H genes. In each cluster, the top sequence is given for comparison and represents the closest reported germline gene. Dashes indicate identities. See text for definition of CDR3 and FR4 areas. The mAb 424.F6.24, 447.8H, 412.67.F1.3, 412.66.F1, and 448.9G.F1 V_H-D-J_H gene sequences are as reported in Chang and Casali.[24a] The mAb 410.7.F91 and 410.30.F305 V_H-D-J_H gene sequences are as reported in Ikematsu *et al.*[23]

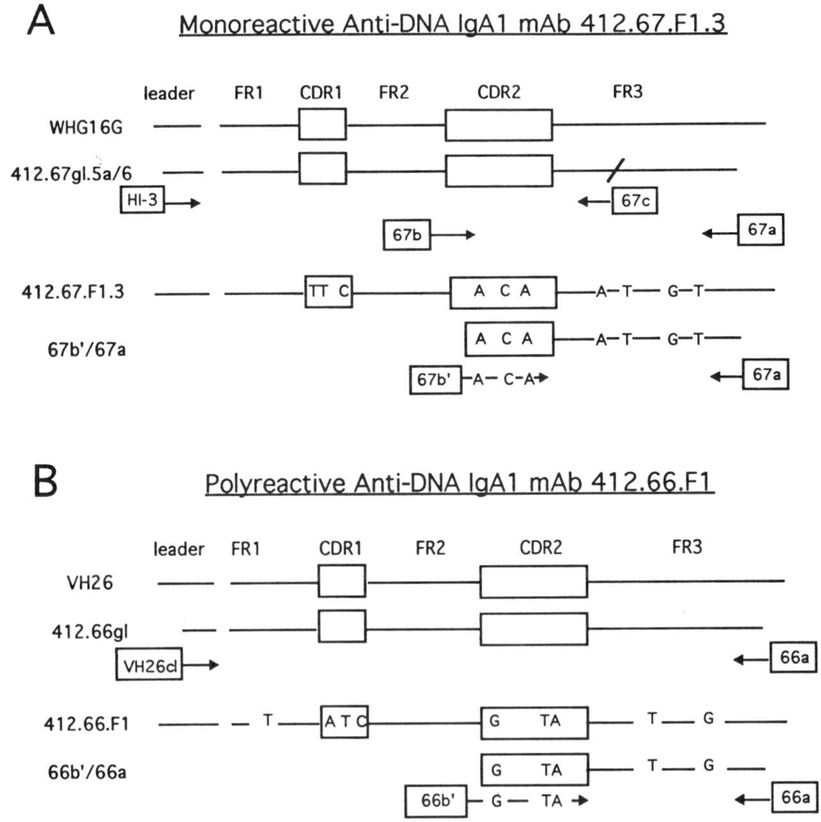

FIGURE 5. Evidence for somatic mutations in monoreactive and polyreactive anti-DNA autoantibodies. **(A)** Monoreactive anti-DNA IgA1 mAb 412.67.F1.3. The sequence of the WHG16G germline gene[26] is schematized at the top of the cluster. 412.67gl.5a/b is the germline V_H gene sequence (identical to WHG16G[24a]) was derived from patient A PMN genomic DNA by sequencing overlapping PCR-amplified portions of the germline gene encompassing: (i) the leader to FR3 stretch, amplified using the sense HI-3 and antisense 67c primers; and (ii) the FR2 to FR3 stretch, amplified using the sense 67b and antisense 67a primers (see Ref. 24). The V_H gene encoding mAb 412.67.F1.3 is schematized according to the sequence reported in Chang and Casali.[24a] the 67b′/67a sequence was derived from the mAb 412.67.F1.3-producing B-cell genomic DNA using the sense 67b′ and antisense 67a primers. This combination of primers did not amplify a product from the genomic DNA of autologous PMN.[24] The combination of sense 67b and antisense 67a primers amplified products from both PMN and hybridoma genomic DNA, but only that amplified from the hybridoma hybridized to an oligonucleotide probe corresponding to the 67b′ sequence.[24] **(B)** Polyreactive anti-DNA IgA1 mAb 412.66.F1. The sequence of the V_H26 germline gene[28] is schematized at the top of the cluster. 412.66gl, the germline V_H gene sequence found to be identical to V_H26,[24] was derived from patient A PMN genomic DNA using the combination of sense V_H26cl and antisense 66a primers. The V_H gene encoding mAb 412.66.F1 is schematized according to the sequence reported by Chang and Casali.[24a] The 66b′/66a sequence was derived from the mAb 412.66.F1-producing B-cell genomic DNA using the sense 66b′ and antisense 66a primers. This combination of primers did not amplify a product from the genomic DNA of autologous PMN.[24] The combination of sense V_H26cl and antisense 66a primers amplified products from both PMN and hybridoma genomic DNA, but only that amplified from the hybridoma hybridized to an oligonucleotide probe corresponding to the 66b′ sequence.[24]

a primary role for B-1 cells in human autoantibody production and demonstrate a common cellular origin for the self-reactive natural antibodies of healthy subjects and the pathogenic autoantibodies of SLE patients.

Among the nine anti-DNA mAbs, three were monoreactive for DNA, and six showed multiple Ag-reactivity with relatively high avidity for DNA. The comparable relative avidities for DNA of the poly- and monoreactive autoantibodies supports a role for both in the pathogenesis of SLE. By molecular analysis of the V_H genes encoding these anti-DNA autoantibodies, we sought to define the role of Ag in the selection of self reactivity. Although any B lymphocyte undergoing somatic hypermutation could by chance give rise to an antibody with high affinity for self Ag, those B cells already committed to the production of self Ag-reactive natural antibodies may have a greater probability of doing so, and of producing pathogenic autoantibodies by clones that escape anergy or deletion.[31,32] Somatic selection processes were traditionally thought to apply only to antibodies produced by "conventional" B-2 cells, but recent evidence that natural antibodies in healthy subjects can be somatically mutated suggests that B-1 cells can participate in affinity maturation.[23,33] In rheumatoid arthritis patients, most B-1a cell-derived high-affinity rheumatoid factor autoantibodies bear the molecular imprints of an Ag-driven selection process, with extensive somatic mutations yielding high R : S mutation ratios in the Ig V segment CDRs.[9,10] Nucleotide substitutions yielding high CDR but not FR R : S mutation ratios were evident not only in a monoreactive (mAb 412.67.F1.3), but also in three polyreactive anti-DNA mAbs (mAbs 412.66.F1, 448.9G.F1, and 410.30.F305).[23,24] These findings substantiate a growing body of evidence[10,23,24,33] that the anti-DNA autoantibodies of SLE undergo affinity maturation,[34–37] and further support our contention that somatic hypermutation and Ag selection can apply to polyreactive antibodies.

The nine mAbs used a broad distribution of V_H germline genes, including two V_H26[28] and two $V_H4.21$.[29] Although both have been reported to encode other SLE-derived anti-DNA autoantibodies,[34,38–40] these germline elements are also over expressed in the normal B-cell repertoire[41–43] and may not display, therefore, any particular association with DNA-binding activity. The J_H4b gene, which was used by seven of the nine anti-DNA mAbs, is also predominantly expressed in both the neonatal and adult B-cell repertoires.[44] Thus, overall, the anti-DNA mAbs used a pool of V_H, D, and J_H genes that overlaps extensively with that used by mAbs reactive to other self and exogenous antigens,[9,12,23] suggesting that the selection of clones expressing certain V_H gene segments is not a factor influencing the production of Ag-selected high-affinity autoantibodies.

The present demonstration that somatically mutated monoreactive and polyreactive anti-DNA autoantibodies are produced by B-1 cells in SLE patients extends our previous observation that these putatively "primordial" B cells display some of the functional features of "conventional" B-2 cells[9,10,23] and suggests that V_H gene segments encoding self-reactive natural antibodies undergo mutation and selection to produce disease-associated autoantibodies. The factors imposing such a selective pressure on B-1 cells and the mechanisms leading to the generation of somatically mutated high-affinity anti-DNA autoantibodies remain to be determined.

REFERENCES

1. Kasaian, M. T., H. Ikematsu & P. Casali. 1992. Identification and analysis of a novel human surface CD5⁻ B lymphocyte subset producing natural antibodies. J. Immunol. **148:** 2690–2702.

2. KANTOR, A. B., A. M. STALL, S. ADAMS, L. A. HERZENBERG & L. A. HERZENBERG. 1992. Differential development of progenitor activity for three B-cell lineages. Proc. Natl. Acad. Sci. USA **89:** 3320–3324.
3. HAUGHTON, G., L. W. ARNOLD, A. C. WHITMORE & S. H. CLARKE. 1993. B-1 cells are made, not born. Immunol. Today **14:** 84–87.
4. HERZENBERG, L. A. & A. B. KANTOR. 1993. B-cell lineages exist in the mouse. Immunol. Today **14:** 79–83.
5. CASALI, P., M. T. KASAIAN & G. HAUGHTON. 1993. B-1 (CD5 B) cells. *In* Autoimmunity: Physiology and Disease. A. Coutinho & M. D. Kazatchkine, Eds.: 57–88. John Wiley & Sons. New York.
6. KANTOR, A. B. & L. A. HERZENBERG. 1993. Origin of murine B cell lineages. Annu. Rev. Immunol. **11:** 501–538.
7. KASAIAN, M. T., H. IKEMATSU & P. CASALI. 1991. $CD5^+$ B lymphocytes. Proc. Soc. Exp. Biol. Med. **197:** 226–241.
8. KASAIAN, M. T. & P. CASALI. 1992. Analysis of the human $CD5^-CD45RA^{low}$ B-cell subset. Ann. N.Y. Acad. Sci. **651:** 59–69.
9. HARINDRANATH, N., I. S. GOLDFARB, H. IKEMATSU, S. E. BURASTERO, R. L. WILDER, A. L. NOTKINS & P. CASALI. 1991. Complete sequence of the genes encoding the V_H and V_L regions of low- and high-affinity monoclonal IgM and IgA1 rheumatoid factors produced by $CD5^+$ B cells from a rheumatoid arthritis patient. Int. Immunol. **3:** 865–875.
10. MANTOVANI, L., R. L. WILDER & P. CASALI. 1993. Human rheumatoid B-1a ($CD5^+$ B) cells make somatically hypermutated high affinity IgM rheumatoid factors. J. Immunol. **151:** 473–488.
11. KASAIAN, M. T. & P. CASALI. 1993. Autoimmunity-prone B-1 (CD5 B) cells, natural antibodies and self recognition. Autoimmunity **15:** 315–329.
12. RIBOLDI, P., M. T. KASAIAN, L. MANTOVANI, H. IKEMATSU & P. CASALI. 1993. Natural antibodies. *In* The Molecular Pathology of Autoimmunity. C. A. Bona, K. Siminovitch, M. Zanetti & A. N. Theophilopoulos, Eds.: 45–64. Harwood Academic Publishers GmbH. Chur.
13. DAUPHINEE, M., Z. TOVAR & N. TALAL. 1988. B cells expressing CD5 are increased in Sjogren's syndrome. Arthritis Rheumatol. **31:** 642–647.
14. VELASQUILLO, M. C., J. ALCOCER-VARCIA, D. ALARCON-SEGOVIA, J. CABIEDES & J. SANCHEZ-GUERRO. 1991. Some patients with primary antiphospholipid syndrome have increased circulating $CD5^+$ B cells that correlate with levels of IgM antiphospholipid. Clin. Exp. Rheum. **9:** 1–5.
15. HARDY, R. R., K. HAYAKAWA, M. SHIMIZU, K. YAMASAKI & T. KISHIMOTO. 1987. Rheumatoid factor secretion from human Leu-1^+ B cells. Science **236:** 81–83.
16. CASALI, P., S. E. BURASTERO, M. NAKAMURA, G. INGHIRAMI & A. L. NOTKINS. 1987. Human lymphocytes making rheumatoid factor and antibody to ssDNA belong to the Leu-1^+ B-cell subset. Science **236:** 77–81.
17. YOUINOU, P., L. MACKENZIE, P. KATSIKIS, G. MERDRIGNAC, D. A. ISENBERG, N. TUAILLON, A. LAMOUR, P. LEGOFF, J. JOUQUAN, A. DROGOU, S. MULLER, B. GENETET, H. M. MOUTSOPOULOS & P. M. LYDYARD. 1990. The relationship between CD5-expressing B lymphocytes and serologic abnormalities in rheumatoid arthritis patients and their relatives. Arthritis Rheumatol. **33:** 339–348.
18. HAYAKAWA, K. & R. R. HARDY. 1988. Normal, autoimmune, and malignant $CD5^+$ B cells: The Ly-1 B lineage? Ann. Rev. Immunol. **6:** 197–218.
19. CASALI, P., S. E. BURASTERO, J. E. BALOW & A. L. NOTKINS. 1989. High affinity antibodies to ss-DNA are produced by $CD5^-$ B cells in SLE patients. J. Immunol. **143:** 3476–3483.
20. SUZUKI, N., R. SAKANE & E. G. ENGLEMAN. 1990. Anti-DNA antibody production by $CD5^+$ and $CD5^-$ B cells of patients with systemic lupus erythematosus. J. Clin. Invest. **85:** 238–247.
21. REAP, E. A., E. S. SOBEL, P. L. COHEN & R. A. EISENBERG. 1993. Conventional B cells, not B-1 cells, are responsible for producing autoantibodies in *lpr* mice. J. Exp. Med. **177:** 69–78.

22. TAN, E. M. 1989. Antinuclear antibodies: diagnostic markers for autoimmune diseases and probes for cell biology. Adv. Immunol. **44:** 93–151.
23. IKEMATSU, H., M. T. KASAIAN, E. W. SCHETTINO & P. CASALI. 1993. Structural analysis of the V_H-D-J_H segments of human polyreactive IgG mAb: Evidence for somatic selection. J. Immunol. **151:** 3604–3616.
24. KASAIAN, M. T., H. IKEMATSU, J. E. BALOW & P. CASALI. 1994. Structure of the V_H and V_L segments of monoreactive and polyreactive IgA autoantibodies to DNA in patients with systemic lupus erythematosus. J. Immunol. **152:** 3137–3151.
24a. CHANG, B. & P. CASALI. 1994. The CDR1 sequences of a major proportion of human germline V_H genes are inherently susceptible to amino acid replacement. Immunol. Today **15:** 367–373.
25. IKEMATSU, H., I. S. GOLDFARB, N. HARINDRANATH, M. T. KASAIAN & P. CASALI. 1992. Generation of human monoclonal antibody-producing cell lines by Epstein-Barr virus (EBV)-transformation of B lymphocytes and a somatic cell hybridization technique. J. Tissue Culture Meth. **14:** 9–12.
26. NAKAMURA, M., S. E. BURASTERO, Y. UEKI, J. W. LARRICK, A. L. NOTKINS & P. CASALI. 1988. Probing the normal and autoimmune B cell repertoire with Epstein-Barr virus: Frequency of B cells producing monoreactive high affinity autoantibodies in patients with Hashimoto's disease and systemic lupus erythematosus. J. Immunol. **141:** 4165–4172.
27. CHAI, S., M. T. KASAIAN, H. IKEMATSU & P. CASALI. 1993. Sequences of V_H-D-J_H genes expressed by human B-1a, B-1b, and B-2 cells: Lack of N segment additions in the V_H segment of natural antibodies produced by B-1b cells. J. Immunol. **150:** 253A.
28. CHEN, P. P., M.-F. LIU, S. SINHA & D. A. CARSON. 1988. A 16/6 idiotype-positive anti-DNA antibody is encoded by a conserved V_H gene with no somatic mutation. Arthritis Rheumatol. **31:** 1429–1431.
29. SANZ, I., P. KELLY, C. WILLIAMS, S. SCHOLL, P. W. TUCKER & J. D. CAPRA. 1989. The smaller human V_H families display remarkably little polymorphism. EMBO J. **8:** 3741–3748.
30. KABAT, E. A., T. T. WU, H. M. PERRY, K. S. GOTTESMAN & C. FOELLER. 1991. Sequences of proteins of immunological interest. Fifth Edit. U.S. Department of Health and Human Services. Washington, DC.
31. NOSSAL, G. V. J. 1992. Cellular and molecular mechanisms of B lymphocyte tolerance. Adv. Immunol. **52:** 283–331.
32. CHEN, C., M. Z. RADIC, J. ERIKSON, S. A. CAMPER, S. LITWIN, R. R. HARDY & M. WEIGERT. 1994. Deletion and editing of B cells that express antibodies to DNA. J. Immunol. **152:** 1970–1982.
33. MARTIN, T., S. F. DUFFY, D. A. CARSON & T. J. KIPPS. 1992. Evidence for somatic selection of natural antibodies. J. Exp. Med. **175:** 983–991.
34. WINKLER, T. H., H. FEHR & J. R. KALDEN. 1992. Analysis of immunoglobulin variable region genes from human IgG anti-DNA hybridomas. Eur. J. Immunol. **22:** 1719–1728.
35. DAVIDSON, A., A. MANHEIMER-LORY, C. ARANOW, R. PETERSON, N. HANNIGAN & B. DIAMOND. 1990. Molecular characterization of a somatically mutated anti-DNA antibody bearing two systemic lupus erythematosus-related idiotypes. J. Clin. Invest. **85:** 1401–1409.
36. DIAMOND, B., J. B. KATZ, E. PAUL, C. ARANOW, D. LUSTGARTEN & M. D. SCHARFF. 1992. The role of somatic mutation in the pathogenic anti-DNA response. Ann. Rev. Immunol. **10:** 731–757.
37. ZOUALI, M. 1992. Development of human antibody variable genes in systemic autoimmunity. Immunol. Rev. **128:** 73–99.
38. VAN ES, J. H., F. H. J. GMELIG MEYLING, W. R. M. VAN DE AKKER, H. AANSTOOT, R. H. W. M. DERKSEN & T. LOGTENBERG. 1991. Somatic mutations in the variable regions of a human IgG anti-double-stranded DNA autoantibody suggest a role for antigen in the induction of systemic lupus erythematosus. J. Exp. Med. **173:** 461–470.
39. DERSIMONIAN, H., R. S. SCHWARTZ, K. J. BARRETT & B. D. STOLLAR. 1987. Relationship of human variable region heavy chain germ-line genes to genes encoding anti-DNA autoantibodies. J. Immunol. **139:** 2496–2501.

40. MANHEIMER-LORY, A., J. B. KATZ, M. PILLINGER, C. GHOSSEIN, A. SMITH & B. DIAMOND. 1991. Molecular characteristics of antibodies bearing an anti-DNA-associated idiotype. J. Exp. Med. **174:** 1639–1652.
41. ADDERSON, E. E., P. G. SHACKELFORD, A. QUINN & W. L. CAROLL. 1991. Restricted Ig H chain V gene usage in the human antibody response to *Haemophilus influenzae* type b capsular polysaccharide. J. Immunol. **147:** 1667–1674.
42. PASCUAL, V. & J. D. CAPRA. 1992. V_H4.21, a human V_H gene segment overrepresented in the autoimmune repertoire. Arthritis Rheumatol. **35:** 11–18.
43. STEWART, A. K., C. HUANG, B. D. STOLLAR & R. S. SCHWARTZ. 1993. High-frequency representation of a single V_H gene in the expressed human B cell repertoire. J. Exp. Med. **177:** 409–418.
44. YAMADA, M., R. WASSERMAN, B. A. RICHARD, S. SHANE, A. J. CATON & G. ROVERA. 1991. Preferential utilization of specific immunoglobulin heavy chain diversity and joining segments in adult human peripheral blood B lymphocytes. J. Exp. Med. **173:** 395–407.
45. SHIN, E. K., F. MATSUDA, H. NAGAOKA, Y. FUKITA, T. IMAI, K. YOKOYAMA, E. SOEDA & T. HONJO. 1991. Physical map of the 3′ region of the human immunoglobulin heavy chain locus: clustering of autoantibody-related variable segments in one haplotype. EMBO J. **10:** 3641–3645.
46. KUPPERS, R., U. FISCHER, K. RAJEWSKI & A. GAUSE. 1992. Immunoglobulin heavy and light chain gene sequences of a human CD5 positive immunocytoma and sequences of four novel V_HIII germline genes. Immunol. Lett. **24:** 57–62.
47. LEE, K. H., F. MATSUDA, T. KINASHI, M. KODAIRA & T. HONJO. 1987. A novel family of variable region genes of the human immunoglobulin heavy chain. J. Mol. Biol. **195:** 761–768.
48. CHEN, P. P., M.-F. LIU, C. A. GLASS, S. SINHA, T. J. KIPPS & D. A. CARSON. 1989. Characterization of two immunoglobulin V_H genes that are homologous to human rheumatoid factors. Arthritis Rheumatol. **32:** 72–76.
49. CAMPBELL, M. J., A. D. ZELENETZ, S. LEVY & R. LEVY. 1992. Use of family-specific leader region primers for PCR amplification of the human heavy chain variable region gene repertoire. Mol. Immunol. **29:** 193–203.
50. ICHIHARA, Y., H. MATSUOKA & Y. KUROSAWA. 1988. Organization of human immunoglobulin heavy chain diversity gene loci. EMBO J. **7:** 4141–4150.
51. BULUWELA, L., D. G. ALBERTSON, P. SHERRINGTON, P. H. RABBITTS, N. SPURR & T. H. RABBITTS. 1988. The use of chromosomal translocations to study human immunoglobulin gene organization: Mapping D_H segments within 35 kb of C_μ gene and identification of a new D_H locus. EMBO J. **7:** 2003–2010.
52. RAVETCH, J. V., U. SIEBENLIST, S. KORSMEYER, T. WALDMANN & P. LEDER. 1981. Structure of the human immunoglobulin μ locus: Characterization of embryonic and rearranged J and D genes. Cell **27:** 583–591.

The Importance of the Diversity Region YYGS Motif in Anti-DNA Antibody Binding to DNA[a]

MAHMOUD MAHMOUDI, DAVID A. BELL, AND EWA CAIRNS

Departments of Medicine and Microbiology and Immunology
University of Western Ontario
London, Ontario, Canada

We investigated the structural basis for DNA reactivity of the normal derived natural human IgMλ monoclonal anti-DNA antibody KIM4.6, which expresses the light chain 4.6.3 idiotype (Id). This antibody, which binds native and denatured DNA, is encoded by unmutated germline genes (V_H chain by V_H 1.9.III, DXP′1, J_H6, and V_L chain by Vλ1 and Jλ3).[1] KIM4.6 V_H chain is cationic (net positive charge 4.73) and tyrosine rich, and its diversity region contains an amino acid motif YYGS. These features can also be found in the V_H chains of IgG monoclonal anti-DNA antibodies derived from human and murine systemic lupus erythematosus (SLE). The YYGS motif is encoded by the human diversity germline gene DXP′1, which is utilized in the D region of 30% of known monoclonal anti-DNA antibodies.[2]

HYPOTHESIS

We hypothesized that the conservation of the YYGS motif in the most variable region of the heavy chains may be of functional significance and that the YYGS motif may be directly involved in binding to DNA.

EXPERIMENTAL APPROACHES AND RESULTS

To test the hypothesis, we generated and molecularly characterized a variant of KIM4.6 hybridoma that secreted IgMλ 4.6.3 Id-positive non-DNA-reactive antibodies. The V_L of the variant antibody was identical to KIM4.6. The V_H chain of the variant used completely different genetic elements ($V_H4.21$, unique D, and J_H5) than KIM4.6 but had a similar cationic charge (4.64). Thus, the loss of the DNA binding in KIM4.6 variant hybridoma could be attributed to the loss of the KIM4.6 heavy chain containing YYGS motif. Further testing was performed to

[a] This work was supported by grants from the Canadian Arthritis Society to Drs. Bell and Cairns.

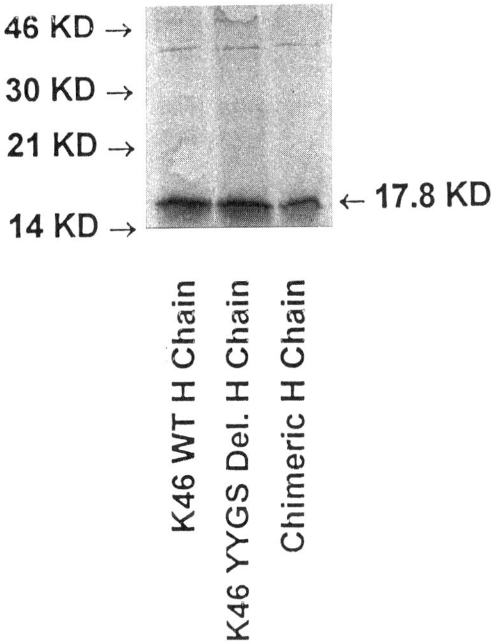

FIGURE 1. SDS-PAGE of different STP-derived heavy chains. Constructs of KIM4.6 WT heavy chain, KIM 4.6 heavy chain without the YYGS encoding sequence, and chimeric heavy chain PCR ligated to T7 and CITE sequence were used as templates for *in vitro* transcription and translation. The STP-derived heavy chains which were labeled with ^{35}S-labeled methionine, separated by SDS-PAGE to confirm the production of the expected size peptide.

directly prove that the YYGS imparts the DNA binding property to KIM4.6. The polymerase chain reaction (PCR) and appropriate primers were used to construct a chimeric KIM4.6 V_H chain gene. In this chimeric construct D and J segments of the wild-type (WT) KIM4.6 were replaced with D and J segments (not encoding for YYGS) from the KIM4.6 variant. We also made another construct by removing the YYGS-encoding sequence from KIM4.6 V_H. These V_H constructs as well as the KIM4.6 V_H variant, KIM4.6 V_H, and KIM4.6 V_L genes were cloned into the phagemid expression vector pComb3. Testing phage surface-expressed Fabs confirmed the importance of the KIM4.6 heavy chain and in particular its D and J region in reactivity to DNA (data not shown). In parallel, some of these constructs (KIM4.6 WT heavy chain, KIM4.6 heavy chain without the YYGS-encoding sequence, and chimeric heavy chain) were PCR-ligated downstream to a pCITE-2-derived T7 promoter and a cap-independent translation enhancer (CITE) sequence. These PCR-ligated templates were then *in vitro* transcribed and translated using the single tube protein system (STP) from Novagen (FIG. 1). The STP-derived heavy chain of KIM4.6 (containing the YYGS motif), bound double-stranded DNA, but the chimeric heavy chain without YYGS did not. The YYGS-deleted KIM4.6 heavy chain had >50% reduced DNA binding when compared with the KIM4.6 WT heavy chain (FIG. 2).

CONCLUSIONS

KIM4.6 monoclonal anti-DNA antibody binds to DNA via its heavy chain. The YYGS motif in the heavy D region is essential but not the sole structural

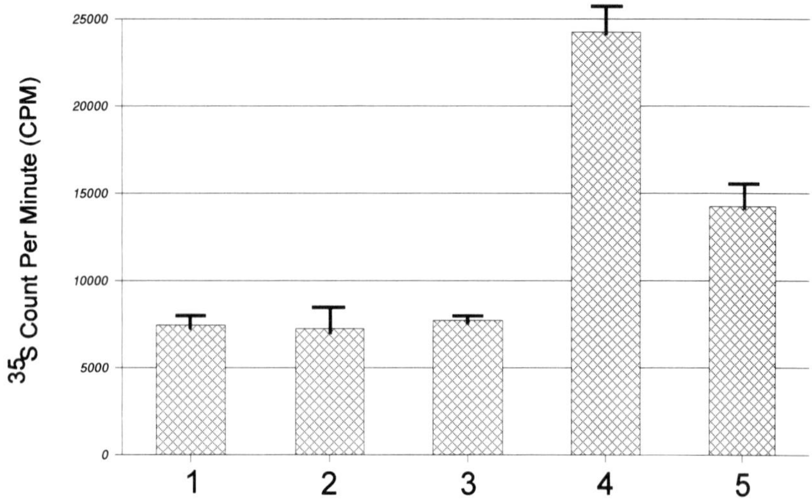

FIGURE 2. Reactivity of STP-derived peptides with double-stranded (ds) DNA. In this assay wells have been coated with dsDNA and then incubated with STP-derived peptides. After extensive washing, the amount of ^{35}S-labeled peptide bound to dsDNA was determined. *Lane 1* represents the negative control which no DNA template was added to the STP system. *Lane 2* shows counts for STP-derived β-galactosidase peptide. *Lane 3* shows the binding of STP-derived chimeric heavy chain. *Lane 4* represents the counts for WT KIM4.6 heavy chain containing YYGS motif. *Lane 5* represents the binding of KIM4.6 heavy chain without YYGS motif.

determinant responsible for DNA reactivity of KIM4.6 antibody. The YYGS motif may also play an important role in DNA binding of other murine and human anti-DNA antibodies.

REFERENCES

1. Cairns, E., P. C. Kwong, V. Misener, P. Ip, D. A. Bell & K. A. Siminovitch. 1989. Analysis of variable region genes encoding a human anti-DNA antibody of normal origin. J. Immunol. **143:** 685.
2. Stewart, A. K., C. Huang, A. A. Long, B. D. Stollar & R. S. Schwartz. 1992. V_H-gene representation in autoantibodies reflects the normal human B-cell repertoire. Immunol. Rev. **128:** 101.

Dual Recognition of Lipid A and DNA by Human Antibodies Encoded by the V_H4-21 Gene

A Possible Link between Infection and Lupus[a]

MYFANWY SPELLERBERG, CAROLINE CHAPMAN, TERRY HAMBLIN, AND FREDA STEVENSON

Molecular Immunology Group
Tenovus Laboratory
Southampton University Hospitals
Southampton SO16 6YD, United Kingdom

The germline repertoire of human V_H genes comprises 100 to 200 V_H gene segments, with 11 to 14 of these assigned to the V_H4 family.[1] The V_H4-21 gene segment, a single member of V_H4, has been found to encode certain autoantibodies, and these antibodies can interact with distinct autoantigens via conventional or unconventional binding sites. In addition, cross-reaction between autoantigen and exogenous antigen indicates first, a biological role for the V_H4-21-encoded antibodies, and second, a possible mechanism for induction of autoimmune disease.

THE V_H4-21 GENE ENCODES ANTI-RED CELL AND ANTI-DNA ANTIBODIES

In 1986 we raised a monoclonal anti-idiotypic antibody (9G4) specific for immunoglobulins encoded by the V_H4-21 gene.[2] This has provided the opportunity to trace utilization of V_H4-21 by a wide range of antibodies and to follow gene expression in a variety of diseases.[3] One unexpected finding was that the V_H4-21 gene is mandatory for autoanti-red cell antibodies against the I/i carbohydrate antigen, which are found in patients with cold agglutinin disease.[4,5] This suggests that interaction with the red cell is via a framework region (FWR) of V_H4-21.[4] In fact, reactivity against red cells is inhibitable with the 9G4 antibody, indicating that the idiotype (Id), which has been located to FWR1,[6] may be close to the red cell binding site.

Idiotypic Ig was raised in sera from patients with infectious mononucleosis (IM) or *Mycoplasma* infection, conditions under which cold agglutinins (CAs) are known to be synthesized and cloned Id-positive B cells from patients with IM secrete anti-i antibodies.[7] Nucleotide sequence analysis of V_H sequences of cloned anti-i/I antibodies from patients with IM, or from patients with cold agglutinin disease arising from a B-cell tumor, revealed utilization of the V_H4-21 gene in

[a] This work was supported by Tenovus, the Wessex Medical Trust, U.K. and by a collaborative grant from NATO.

combination with a variety of light chains and with a wide range of CDR3 sequences,[4,6] again suggesting that a FWR determinant is the key to red cell binding.

A survey of sera from patients revealed that the V_H4-21-associated Id was not raised in a range of autoimmune diseases or in Hodgkin's disease, tuberculosis, or *Klebsiella* infections.[8] It was raised in patients with systemic lupus erythematosus (SLE), however, and was associated with anti-DNA activity.[3] Id-positive Ig was also detectable in the involved renal tissue of some patients with SLE, indicating a possible role in disease pathology.[8] Analysis of anti-DNA-secreting B-cell hybridomas from patients with SLE revealed that 4 out of 10 were Id-positive, and in two out of two cases available for sequence used the V_H4-21 gene.[3] Each of these (one IgM and one IgG) had CDR3 sequences with predominance of basic amino acids, candidates for recognition of DNA.[9] Other cloned anti-DNA antibodies using the V_H4-21 gene segment have been described, and these also have CDR3 sequences containing basic amino acids.[10,11] Recognition of DNA was not inhibited by the anti-Id, indicating that the site of interaction with DNA is distant from the Id, fitting with evidence that the Id is located in FWR1, whereas the DNA recognition sequence lies in CDR3.

CROSS-REACTIVITY BETWEEN DNA AND LIPID A

In an effort to obtain anti-lipid A antibodies for treatment of Gram-negative septicemia, two IgM antibodies were isolated, and both were found to express the V_H4-21 Id.[12] These antibodies also recognized ssDNA and had weak reactivity with red cells. Analysis of the V_H gene sequences revealed utilization of V_H4-21 in germline configuration, with basic amino acids in the CDR3 sequences.[12] This suggested that recognition of a biologically important exogenous antigen (bacterial lipid A) and of an autoantigen (DNA) might be mediated via similar sequences in the CDR3 sequences of the antibodies. This was supported by the finding that recognition of lipid A was not inhibited by the 9G4 anti-Id, mirroring results with DNA, again indicating that the site of interaction is distant from the red cell binding site.

Correlation between binding to DNA and lipid A was sought by profiling a panel of V_H4-21-encoded IgM antibodies against each antigen, and five out of five of the available anti-DNA antibodies also reacted with lipid A (TABLE 1). However, the pattern of reactivity was heterogeneous with two out of five (NE-1 and 216) reacting strongly with both DNA and lipid A, whereas two out of five (RT-79 and RT-79(2)), although similar to NE-1 and 216 in binding to DNA, were weaker against lipid A. Because all four antibodies have a germline sequence and carry a similar overall net positive charge in the CDR3, this would suggest that for interaction with lipid A the position of the arginine residues could be important. Light chains could also have a role, although there is no apparent restriction in V_L gene usage. Finally, one out of five (CAL-4G), also in germline configuration but with a different CDR3, is a strong anti-DNA antibody but bound lipid A only weakly.

DISCRIMINATION BETWEEN RED CELL ANTIGEN, DNA, AND LIPID A

TABLE 1 summarizes the profiles of the five anti-DNA antibodies as compared with the CAs (FS clones). The first point is that the anti-DNA antibodies generally

TABLE 1. Antibody Activities Mediated by V_H4-21-Encoded IgM Molecules

Clone	CDR3	J_H	Cold Agglutinin Titer I	Cold Agglutinin Titer i	Antibody against ssDNA (U/μg)	Antibody against dsDNA (U/μg)	Antibody against Lipid A (U/μg)
Anti-DNA							
NE-1 (Mκb)	**R**HV**R**P**R**VT**R**	J_H6	0	0	381	54	235
216 (MλII)	G**R**MV**R**GAKYN	J_H5		+	209	33	385
RT-79 (MκII)	V**RR**SG**R**VVVPGA**R**N**R**D	J_H3	0	0	348	100	60
RT-79/2 (Mκ)	G**RR**ITGTK**R**H	J_H6	0	0	247	ND	32
CAL-4G (MλIX)	**R**IMYCSGGSCYSAGW**R**	J_H4	0	0	1000	88	2
Cold agglutinins							
FS-1 (MκII)	GQ**R**GSGSDY**R**QGA	J_H3	+++		0	0	0
FS-2 (MκIIIb)	EWGSGAYYPYY	J_H4	++++		0	0	0
FS-3 (MλI)	GT**R**SWYPSD	J_H4		+++	0	0	0
FS-4 (MκIIIa)	**R**LGSGYPN	J_H4	++++		0	0	0
FS-6 (MκIIIb)	ALGDGSTEGLP	J_H4	++++		0	0	0
FS-7 (MκI)	G**R**EQWLV**R**GG	J_H4	++++		0	0	0
FS-9 (MκIIIb)	GPALMITG	J_H4	+++		0	0	0
FS-10 (Mκ)	EDAFS**R**II	J_H4	+++		0	0	0
FS-11 (Mκ)	GDTVVVPS**R**P	J_H4	++		0	0	0

fail to react with red cells, and conversely the CAs do not recognize DNA or lipid A. This discrimination is unlikely to depend on the light chain, since the V_L genes used are quite diverse. Equally, it is unlikely to arise from mutations in the V_H4-21 gene, since FS-7 and all the anti-DNA antibodies use the gene in germline configuration; the other FS lines vary in mutational pattern with no consistent motif that could confer red cell specificity.[4] It is also unlikely that the J_H is involved, since we have a CA (FS-8), not available for the panel, that also uses J_H6.[4] This leaves the CDR3 as a candidate for discrimination between red cell antigen and DNA, as suggested previously from more limited data.[3] The CDR3s (TABLE 1) can be assessed for overall charge by subtracting acidic (D and E) residues from basic (R and K). On this basis, all the anti-DNA antibodies have a net positive charge of +2 or greater, whereas the CAs have −1 (3/9), 0 (3/9), or +1 (3/9). The positive charge is mainly on arginine (R) residues, known to be significant for recognition of DNA by murine anti-DNA antibodies.[13] Interestingly, three of four anti-DNA antibodies were unable to agglutinate red cells, and one of four (216) had only a weak anti-i activity, indicating a negative influence of the positive charge on agglutination. The anti-DNA antibodies recognized ssDNA preferentially as compared to dsDNA, and the titers against ssDNA were similar for all five (TABLE 1), indicating no detectable effect of the different positions of the R residues along the CDR3 sequence. However, avidity for lipid A appears to correlate with distinct CDR3 motifs, with strong binding seen by NE-1 and 216, which have similar lengths (9 and 10 amino acids) and distribution of basic residues (TABLE 1), particularly at the NH_2-terminal end where there is an R-VR common sequence. Some of the common sequence (VR) probably arises from the DXP′1 D segment gene, and the first R is most likely an N addition. Antibodies RT-79 and RT-79(2) each have a double RR near the NH_2-terminus, and three and two further basic residues, respectively. However, CAL-4G differs, with a central sequence derived from the DLR D2-segment gene (YCSGGCYSA) and terminal R residues probably arising from N additions. Lipid A consists of residues probably arising from N additions. Lipid A consists of a β1-6 linked glucosamine disaccharide, with fatty acid side chains attached to the amino groups. Each sugar carries a phosphate group, giving the molecule an overall negative charge and some similarity to the phosphate–sugar backbone of DNA. Clearly, any cross-reactivity betwen lipid A and DNA would be via the sugar–phosphate structure and is likely to involve electrostatic bonds. However, simple charge density may not be the only factor in binding to lipid A, since antibody 216 has a lower overall positive charge (2R and 1K) than RT-79(2) (3R and 1K), but has a greater avidity for lipid A.

DISCUSSION

The V_H4-21 gene is active early in development and is used by a high proportion (~6%) of normal B cells in adult lymphoid tissue,[14] suggesting that it encodes an important component of the functional antibody repertoire. It also encodes autoantibodies, however, and is mandatory for the pathological anti-red cell antibodies against the I/i antigen.[4,5] Additionally, it is involved in a subset of anti-DNA antibodies in patients with SLE. Our findings indicate that the binding to red cells is unconventional, being mediated via a FWR sequence, whereas recognition of DNA is conventional, being via the CDR3 sequence. Therefore a single unmutated gene can encode antibodies that bind through either conventional or unconventional sites; the ability to encode antibodies with two kinds of binding site makes the V_H4-21 gene a super-V_H gene, where specificity can be switched by recombination and without somatic mutation. Proof of this will require manipulation of sequence and expression of recombinant Fab molecules, and this work is in progress.

The fact that anti-DNA activity is associated with basic amino acids in the CDR3 sequence parallels findings in the autoimmune mouse.[13] The cross-reactivity we have observed between DNA and lipid A raises the intriguing possibility that there may be a link between infection and induction of SLE. It could be envisioned that exposure to lipid A during infection with Gram-negative bacteria generates V_H4-21-encoded antibodies that also recognize DNA. In susceptible individuals, immunoglobulin class switching and somatic mutation of the V_H gene could then occur, leading to pathological anti-dsDNA antibodies as identified in patients. A possible pathway to SLE may have been revealed for this V_H gene, but it is known that there are many other V_H and V_L genes involved in encoding anti-DNA antibodies.[15] The crucial question is whether this subset of autoanti-DNA antibodies has a major role in mediating the damaging effects of vasculitis or renal deposition in these patients.

CONCLUSION

The V_H4-21 gene segment, a member of the V_H4 family, encodes the vast majority of anti-red cell antibodies of anti-Ii specificity; it is also used by a subset of anti-DNA antibodies in patients with SLE. Recognition of the two autoantigens appears to be mediated via different sites in the V_H sequence, with red cell binding being via the FWR, and DNA binding via basic acids in the CDR3. The IgM anti-DNA antibodies encoded by the V_H4-21 gene also reacted with bacterial lipid A (5/5), whereas antibodies that were monospecific for red cell antigen (9/9) failed to bind to lipid A. Among the anti-DNA antibodies, avidity for lipd A was high (2/4), moderate (2/4), or low (1/4). Analysis of the deduced amino acid sequences revealed that the anti-DNA antibodies were germline encoded, with CDR3s containing basic amino acids and motifs that correlated with different avidity. These results indicate that there is dual recognition of a common bacterial component, lipid A, and an autoantigen, DNA, which may be mediated by specific sequences in CDR3, thus providing a possible link between infection and SLE.

ACKNOWLEDGMENTS

We wish to thank Professor D. Isenberg, Dr. M. Bieber, and Dr. T. Sasaki for kind gifts of RT-79, 216, and NE-1, respectively.

REFERENCES

1. Sanz, I., P. Kelly, C. Williams, S. Scholl, P. Tucker & J. D. Capra. 1989. The smaller VH gene families display remarkably little polymorphism. EMBO J. **8:** 3741–3748.
2. Stevenson, F. K., M. Wrightham, M. J. Glennie, D. B. Jones, A. R. Cattan, T. Feizi, T. J. Hamblin & G. T. Stevenson. 1986. Antibodies to shared idiotypes as agents for analysis and therapy for human B cell tumors. Blood **68:** 430–436.
3. Stevenson, F. K., C. Longhurst, C. J. Chapman, M. Ehrenstein, M. B. Speller-

BERG, T. J. HAMBLIN, C. T. RAVIRAJAN, D. LATCHMAN & D. ISENBERG. 1993. Utilization of the V_H4-21 gene segment by anti-DNA antibodies from patients with systemic lupus erythematosus. J. Autoimmunol. **16:** 809–825.

4. PASCUAL, V., K. VICTOR, M. SPELLERBERG, T. J. HAMBLIN, F. K. STEVENSON & J. D. CAPRA. 1992. V_H restriction among human cold agglutinins. The V_H4-21 gene segment is required to encode anti-I and anti-i specificities. J. Immunol. **149:** 2337–2344.
5. SILBERSTEIN, L. E., L. C. JEFFRIES, J. GOLDMAN, D. FRIEDMAN, J. S. MOORE, P. C. NOWELL, D. ROELCKE, W. PRUZANSKI, J. ROUDIER & G. J. SILVERMAN. 1991. Variable region gene analysis of pathologic human autoantibodies to the related i and I red blood cell antigens. Blood **78:** 2372–2386.
6. POTTER, K. N., Y. LI, V. PASCUAL, R. WILLIAMS, JR., L. C. BYRES, M. SPELLERBERG, F. K. STEVENSON & J. D. CAPRA. 1993. Molecular characterization of a cross-reactive idiotope on human immunoglobulins utilizing the V_H4-21 gene segment. J. Exp. Med. **178:** 1419–1428.
7. CHAPMAN, C. J., M. B. SPELLERBERG, G. A. SMITH, S. J. CARTER, T. J. HAMBLIN & F. K. STEVENSON. 1993. Autoanti-red cell antibodies synthesized by patients with infectious mononucleosis utilize the V_H4-21 gene segment. J. Immunol. **151:** 1051–1061.
8. ISENBERG, D., M. SPELLERBERG, W. WILLIAMS, M. GRIFFITHS & F. K. STEVENSON. 1993. Identification of the 9G4 idiotope in systemic lupus erythematosus. Br. J. Rheumatol. **32:** 876–882.
9. EILAT, D., D. M. WEBSTER & A. R. REES. 1988. V region sequences of anti-DNA and anti-RNA autoantibodies from NZB/NZW F1 mice. J. Immunol. **141:** 1745–1753.
10. VAN ES, J. H., F. H. J. GMELIG MEYLING, W. R. M. VAN DE AKKER, H. AANSTOOT, R. H. W. M. DERKSON & T. LOGTENBERG. 1991. Somatic mutations in the variable regions of a human IgG anti-double stranded DNA autoantibody suggest a role for antigen in the induction of SLE. J. Exp. Med. **173:** 461–470.
11. HIRABAYASHI, Y., Y. MUNAKATA, O. TAKAI, S. SHIBATA, T. SASAKI & H. SANO. 1993. Human B-cell clones expressing lupus nephritis-associated anti-DNA idiotypes are preferentially expanded without somatic mutation. Scand. J. Immunol. **37:** 533–540.
12. BHAT, N. M., M. M. BIEBER, C. J. CHAPMAN, F. K. STEVENSON & N. N. H. TENG. 1993. Human anti-lipid A monoclonal antibodies bind to human B cells and the i antigen on cord red blood cells. J. Immunol. **151:** 5011–5021.
13. RADIC, M. Z., J. MACKLE, J. ERIKSON, C. MOL, W. F. ANDERSON & M. WEIGERT. 1993. Residues that mediate DNA binding of autoimmune antibodies. J. Immunol. **150:** 4966–4977.
14. STEVENSON, F. K., G. J. SMITH, J. NORTH, T. J. HAMBLIN & M. J. GLENNIE. 1989. Identification of normal B-cell counterparts of neoplastic cells which secrete cold agglutinins of anti-I and anti-i specificity. Br. J. Haematol. **72:** 9–15.
15. DIAMOND, B., J. B. KATZ, E. PAUL, C. ARANOW, D. LUSTGARTEN & M. D. SCHARFF. 1992. The role of somatic mutation in the pathogenic anti-DNA response. Ann. Rev. Immunol. **10:** 731–757.

Immunoglobulin Gene Use by Naturally Occurring Cold Agglutinins

LEIGH C. JEFFERIES, CATHERINE M. CARCHIDI, AND LESLIE E. SILBERSTEIN

Department of Pathology and Laboratory Medicine
University of Pennsylvania Medical Center
Philadelphia, Pennsylvania 19104

Cold agglutinins (CA) are IgM autoantibodies with anti-i/I RBC specificity that may be benign, present in the sera of healthy persons, and are referred to as naturally occurring. Alternatively, anti-i/I CA may be pathogenic and cause immune-mediated hemolytic anemia.[1] In the current study we have begun to investigate the relationship between naturally occurring, nonpathogenic anti-i/I CA and pathogenic anti-i/I CA derived from clonal B-cell expansions. In recent years, several laboratories have demonstrated a highly restricted V_H4 gene use among pathogenic CA from different persons. The heavy chain variable gene, $V_H4.21$ (or a closely related gene) has been found in all pathogenic, monoclonal CA studied to date, suggesting that this gene may play an important role in defining the i/I specificity.[2-4] Our study addressed the question whether the same restricted pattern of V gene usage by pathogenic CA might be found in naturally occurring CA. The simplest hypothesis, that the $V_H4.21$ gene product is the principal determinant of anti-i/I CA activity, would predict that this same gene would encode naturally occurring CA with similar specificity. Alternatively, naturally occurring anti-i/I CA could be encoded by V_H genes other than $V_H4.21$. A schematic diagram illustrating these possibilities is shown in FIGURE 1.

To this end, B cells from two persons were selected from either peripheral blood lymphocytes (PBL) or splenic lymphocytes using a panel of monoclonal anti-idiotypes, which bind to both benign and pathogenic CA.[5] Following EBV transformation,[2] we generated two anti-I secreting clones from two persons: clone 4G (individual CM, PBL) and clone Sp1 (individual SC, spleen). In addition, an anti-i specific clone was isolated from individual SC. Ig V_H and V_L gene use of these clones, determined by nucleotide sequence analysis,[2] is summarized in TABLE 1. Clone 4G expresses a V_H3 gene sequence that is 92% homologous to the germline gene WHG26. Clone Sp1 also expresses a V_H3 gene that is 98% homologous to the fetally rearranged M85/20P1 gene. Another clone, Sp2 (anti-i specificity), from individual SC is 98% homologous to the germline gene $V_H4.21$ gene.

To correlate these findings with circulating anti-i/I CA, we studied anti-i/I CA fractions purified from 15 normal sera to determine the relative expression of $V_H4.21$ versus non-$V_H4.21$ encoded IgM. This was accomplished by using the anti-idiotypic antibody 9G4, which recognizes a heavy chain idiotype highly associated with cold agglutinins encoded by $V_H4.21$ or closely related V_H4 heavy chain genes. We found that these CA fractions contain no or relatively small amounts of 9G4-reactive IgM. Also, for 14 of 14 CA tested, hemagglutination was not inhibited utilizing high concentrations of 9G4, known to inhibit hemagglutination

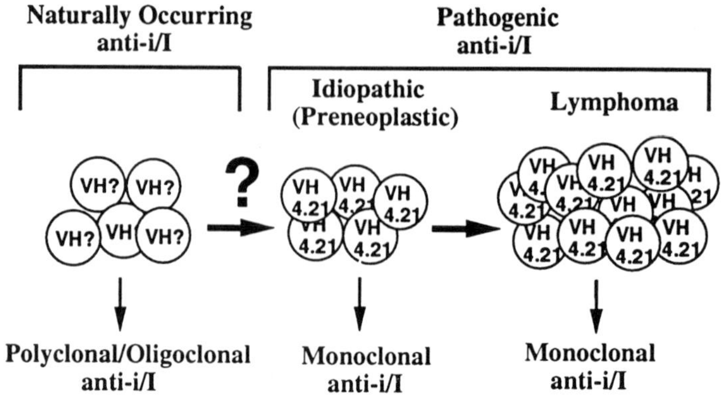

FIGURE 1. The relationship between naturally occurring and pathogenic anti-i/I CA. The origin of pathogenic (e.g., monoclonal) CA derived from clonal B-cell expansions has been previously investigated.[2–4]

by pathogenic, monoclonal anti-i/I CA encoded by the $V_H4.21$ gene (FIG. 2). To evaluate the presence of V_H3 IgM in the CA purified from serum, we used modified protein staph A (SPA), which binds to approximately 90% of V_H3-encoded IgM through Fab-binding sites. Five CA fractions contained large amounts of V_H3-encoded IgM (compared to pooled normal IgM) by virtue of their binding to modified SPA, and absorption of three CA fractions with modified SPA specifically removed anti-i/I binding specificity entirely.[6]

Collectively, the data suggest that to a large extent, naturally occurring anti-i/I CA are encoded by non-$V_H4.21$ genes. Because pathogenic, monoclonal anti-i/I CA exclusively use $V_H4.21$ or a closely related gene, this suggests that natural/nonpathogenic anti-i/I autoantibodies may not necessarily represent the precursors of their monoclonal counterpart (FIG. 1). This also suggests that the mechanisms leading to the development of the two different anti-i/I autoantibody populations may also differ, and that the $V_H4.21$ gene is not uniquely required for anti-i/I specificity.

TABLE 1. Ig V Region Use and Idiotype of Autoantibody-Secreting Clones

Clone	Ig Variable Region Gene V_L	V_H	Idiotype 9G4
4G	$V_\kappa III$	V_H3	Neg
Sp1	$V_\kappa I$	V_H3	Neg
Sp2	V_λ	V_H4	Pos

NOTE: V_H and V_L were determined by nucleotide sequence analysis for clones 4G, Sp1, and Sp2. Expression of idiotype 9G4 was determined by ELISA. Sp2 was bound by 9G4 (O.D. > 1.0). None of the other clonal supernatants were bound by 9G4 (O.D. < 0.02) but were bound by murine anti-IgM as a control to indicate adequate concentrations of clonal IgM present (O.D. > 1.0). Cold agglutinin VOG (a previously described $V_H4.21$ anti-I) was used as a positive control for 9G4 reactivity (O.D. > 1.0). Background values are <0.02.

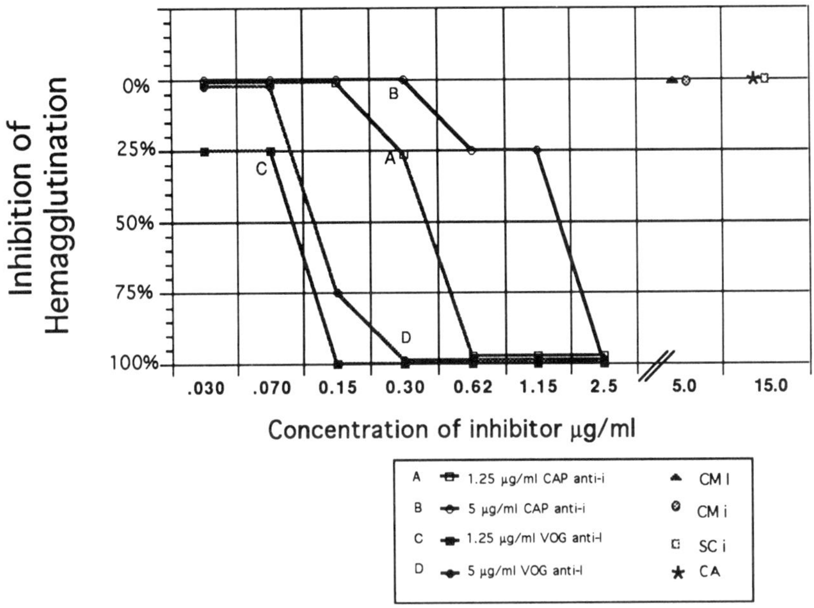

FIGURE 2. Hemagglutination inhibition studies. Serial dilutions of purified VOG and CAP (monoclonal V_H4.21-encoded anti-I and anti-i CA, respectively) were first tested to establish concentrations yielding 4^+ hemagglutination. Aliquots (50 μl) of these antibodies (both at 1.25 and 5.0 μg/ml) were incubated with serial twofold dilutions of anti-idiotypic antibody 9G4 (beginning at 10 μg/ml). After a 30-min incubation at 37°C, ficin-treated RBCs were added, and hemagglutination assessed after a 30-min incubation at 4°C. The 9G4 antibody (>5.0 μg/ml) was also incubated with purified naturally occurring CA (0.20–0.60 μg/ml) for inhibition of hemagglutination. CA were incubated with adult (I) or cord (i) RBCs depending on the predetermined relative specificity. There was no inhibition of HA for any naturally occurring CA tested. CM I and CM i represent anti-I and anti-i enriched CA fractions, respectively, purified from serum of individual CM. SC i represents the anti-i-enriched CA fraction purified from serum of individual SC. * Represents 14 anti-i and anti-I-enriched CA fractions from seven healthy persons.

REFERENCES

1. Pruzanski, W. & K. Shumak. 1977. N. Engl. J. Med. **297:** 538.
2. Silberstein, L. E., L. C. Jefferies, J. Goldman, *et al.* 1991. Blood **78:** 2372.
3. Pascual, V., K. Victor, *et al.* 1991. J. Immunol. **146:** 4385.
4. Pascual, V., K. Victor, M. Spellerberg, T. J. Hamblin, F. K. Stevenson & J. D. Capra. 1992. J. Immunol. **149:** 2337.
5. Jefferies, L., J. Gregg, G. Silverman, C. Carchidi & L. Silberstein. 1992. Clin. Immunol. Immunopathol. **65:** 119–128.
6. Silverman, G. J., M. Sasano & S. B. Wormsley. 1993. Immunomethods **2:** 17–32.

Human Pathogenic Autoantibodies and Their Escape from Deletion[a]

MONCEF ZOUALI

Département d'Immunologie
Institut Pasteur
28, rue Dr Roux
75015 Paris, France

Random combinatorial assembly of Ig variable (V) gene elements leads to generation of a proportion of antibodies with high affinity to self. In normal conditions, the corresponding B cells will be clonally deleted, rendered silent, or switched to a different specificity. In systemic autoimmunity, however, anti-self high-affinity antibodies are overexpressed. For instance, the presence of anti-DNA antibodies is a hallmark of the disease systemic lupus erythematosus (SLE), and a subset of these autoantibodies is involved, at least in part, in the pathogenesis of tissue injury.[1] We have been trying to uncover the molecular choreography that underlies the production of aggressive autoantibodies in SLE. We have derived, from patients with active SLE, a panel of B-cell clones secreting anti-DNA autoantibodies expressing the O-18 idiotype, a marker present in 72% of immune complexes and in 100% of renal eluates (FIG. 1). In contrast to other lupus idiotypes, this marker is present in the circulation of most patients with active lupus nephritis, but not in inactive SLE or healthy subjects.[2] The V_H genes expressed by 12 of the autoantibodies we have studied exhibited a high load of replacement mutations that were clustered in the complementarity-determining regions (CDRs). The junctional (J_H) and diversity (D_H) segments were also mutated. Comparative genomic PCR experiments using DNA extracted from the B-cell clones and germline DNA of the patients of origin showed that the mutations occurred *in vivo* in the corresponding B-cell clonotypes. The high rate and the distribution of the somatic mutations within the V_H genes (FIG. 2) is indicative of somatic selection pressures that were responsible for the production of these autoantibodies.[3–5]

To understand why only a proportion of persons develop SLE, we initiated studies to investigate the recently described mechanism of receptor editing. It is becoming clear that secondary light chain rearrangements are important for expanding the diversity of the antibody repertoire and for escape of autoreactive B cells from clonal deletion and/or anergy.[6,7] We have therefore proceeded to see whether lupus anti-DNA-secreting clonotypes that produce potentially aggressive autoantibodies had incurred secondary rearrangements of their kappa light chain locus.

The human V_κ locus spans over 1600 kb of DNA and contains approximately 76 nonallelic genes arranged in two sections and positioned in opposite directions to one another. Of these, only 40 to 50 are thought to be functional.[8] Combination of this analysis with related studies[9–11] reveals that the 14 productive V_κ-J_κ re-

[a] This work was supported by grants from the ARC and the Institut Pasteur. The author is a senior investigator of the Institut National de la Recherche et de la Santé Médicale (INSERM).

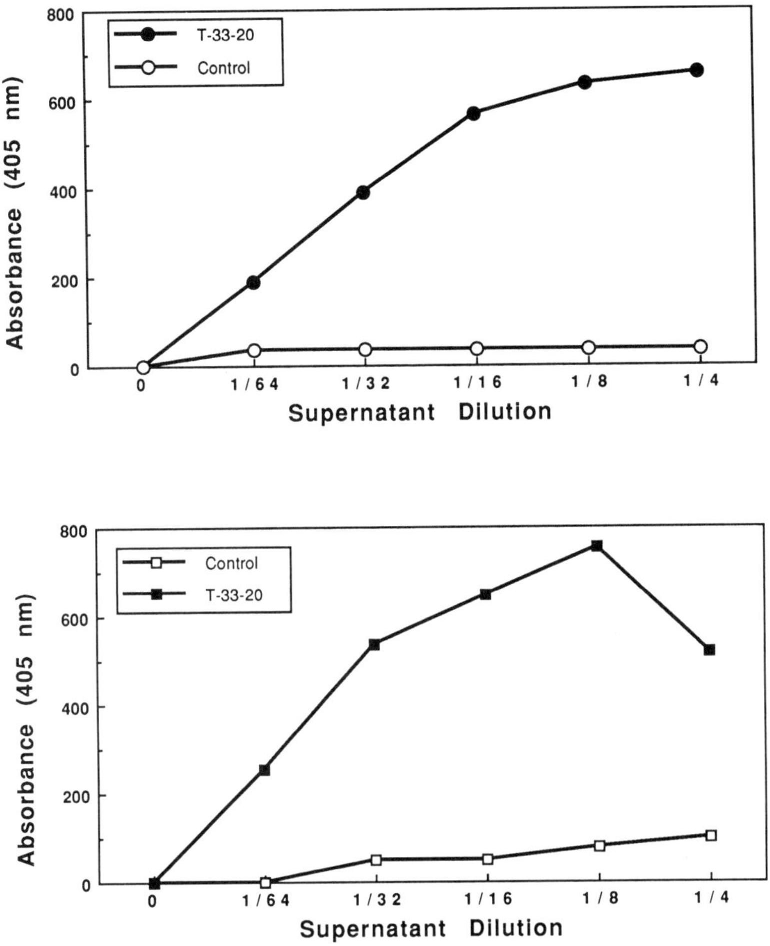

FIGURE 1. DNA binding (*top*) and expression of the O-81 idiotype (*bottom*) in T33-20 lupus antibody. Culture supernatant of the SLE-derived EBV-transformed T33-20 B-cell clone was tested for DNA binding and for O-81 idiotype expression as described previously.[3,4]

arrangements encoding for nephritogenic lupus antibodies arise from at least nine different germline genes. On the basis of the organization of the human V_κ locus,[8] we found that all the 14 rearrangements use genes that are clustered within ~280 kb of the J_κ-proximal portion of the locus. By comparison, a survey of 31 immunization-induced human antibodies revealed no trend toward use of J_κ-proximal V_κ genes. On the contrary, 24 of the rearrangements involved distal V_κ genes that map between 1266 and 1430 kb upstream of J_κ genes.[9,12] It is of further interest that the lupus nephritogenic antibodies frequently use upstream J_κ genes ($J_\kappa 1$ = 29%, $J_\kappa 2$ = 29%). This pattern contrasts with the tendency of immunization-induced antibodies to express downstream J_κ genes. One interpretation of this

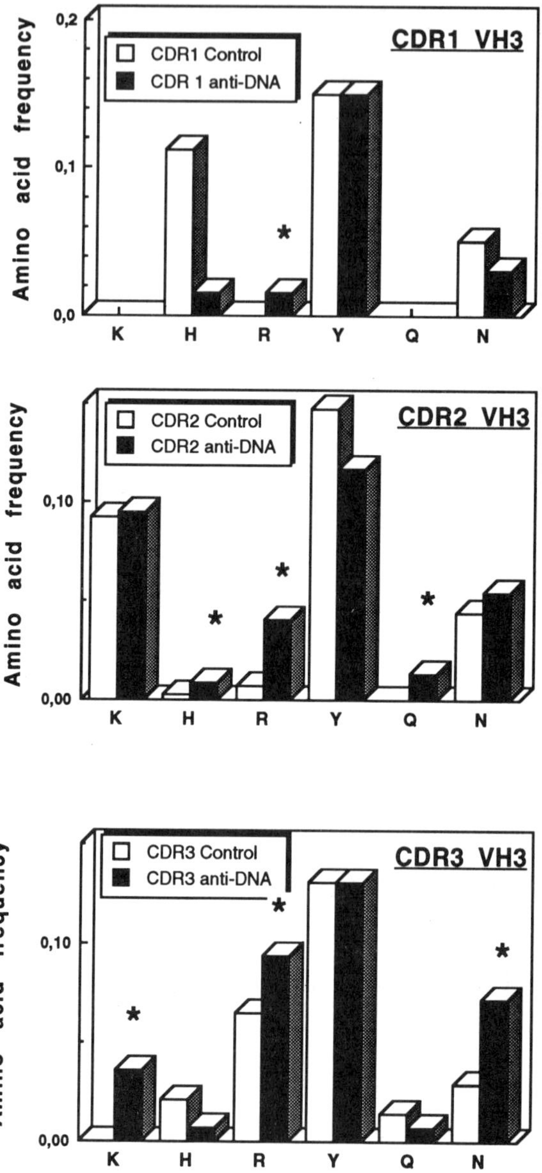

FIGURE 2. Frequency of amino acids that potentially impart DNA binding in two groups of antibodies. The frequencies of six amino acids potentially reactive with DNA were calculated for the CDR1, CDR2, and CDR3 of human V_H3 antibodies. Anti-DNA antibodies are representative of the pathogenic subset of human lupus antibodies,[3–5] and control antibodies are from Marks *et al.*[13] Statistically significant differences between the two groups of antibodies are indicated by stars.

disparity between immunization-induced and lupus autoantibodies is that the latter underwent essentially primary V_κ gene rearrangements. We conclude that, in SLE, B cells are the target of antigenic selection. We also propose that a subset of B cells may be unable to revise their receptors and extinguish their high affinity for self.

REFERENCES

1. EBLING, F. M. & B. H. HAHN. 1989. Pathogenic subsets of antibodies to DNA. Intern. Rev. Immunol. **5:** 79–95.
2. MUROY, T., T. SASAKI, A. HATAKEYAMA, S. SHIBATA, M. SUSUKI, J. SEINO & K. YOSHINAGA. 1990. Clonotypes of anti-DNA antibodies expressing specific idiotypes in immune complexes of patients with active lupus nephritis. J. Immunol. **144:** 3856–3861.
3. ZOUALI, M. 1992. Development of human antibody variable genes in systemic autoimmunity. Immunol. Rev. **128:** 73–99.
4. DEMAISON, C., P. CHASTAGNER, J. THÈZE & M. ZOUALI. 1994. Somatic diversification in the heavy chain variable genes expressed by human autoantibodies bearing a lupus-associated nephritogenic anti-DNA idiotype. Proc. Natl. Acad. Sci. USA **91:** 514–518.
5. CHASTAGNER, P., C. DEMAISON, J. THÈZE & M. ZOUALI. 1994. Clonotypic dominance and variable gene elements of pathogenic anti-DNA antibodies from a single patient with lupus. Scand. J. Immunol. **39:** 165–178.
6. TIEGS, S. L., D. M. RUSSEL & D. NEMAZEE. 1993. Receptor editing in self-reactive bone marrow B cells. J. Exp. Med. **177:** 1009–1020.
7. RADIC, M. Z., J. ERIKSON, S. LITWIN & M. WEIGERT. 1993. B lymphocytes may escape tolerance by revising their antigen receptors. J. Exp. Med. **177:** 1165–1173.
8. HUBER, C., K. F. SCHABLE, E. HUBER, R. KLEIN, A. MEINDL, R. THIEBLE, R. LAMM & H. ZACHAU. 1993. The VK genes of the L regions and the repertoire of VK gene sequences in the human germline. Eur. J. Immunol. **23:** 2868–2875.
9. BENSIMON, C., P. CHASTAGNER & M. ZOUALI. 1994. Human lupus anti-DNA autoantibodies undergo essentially primary V_k gene rearrangements. EMBO J. **13:** 2951–2962.
10. MANHEIMER-LORY, A., J. B. KATZ, M. PILLINGER, C. GHOSSEIN, A. SMITH & B. DIAMOND. 1991. Molecular characteristics of antibodies bearing an anti-DNA-associated idiotype. J. Exp. Med. **174:** 1639–1652.
11. WINKLER, T. H., H. FEHR & J. R. KALDEN. 1992. Analysis of immunoglobulin variable region genes from human IgG anti-DNA hybridomas. Eur. J. Immunol. **22:** 1719–1728.
12. BYE, J. M., C. CARTER, Y. CUI, B. D. GORICK, S. SONGSIVILAI, G. WINTER, N. C. HUGHES-JONES & J. D. MARKS. 1992. Germline variable region gene segment derivation of human monoclonal anti-Rh(D) antibodies. J. Clin. Invest. **90:** 2481–2490.
13. MARKS, J. D., M. TRISTEM, A. KARPAS & G. WINTER. 1991. Oligonucleotide primers for PCR amplification of human Ig variable genes and design of family-specific oligonucleotide probes. Eur. J. Immunol. **21:** 985–991.

V-Gene Repertoire and Hypermutation of Rheumatoid Factors Produced in Rheumatoid Synovial Inflammation and Immunized Healthy Donors[a]

KEITH M. THOMPSON, MARIE BØRRETZEN, INGRID RANDEN, ØYSTEIN FØRRE, AND JACOB B. NATVIG

Institute of Immunology and Rheumatology
The National Hospital
University of Oslo
N-0172 Oslo, Norway

The rheumatoid inflammatory process is characterized by an infiltration of the synovial tissues by dendritic cells, macrophages, and both T and B lymphocytes. There is a proliferation of the synovial lining cells and a striking vascularization consistent with a chronic inflammatory process. The B lymphocytes in the rheumatoid synovium differ from those in the normal circulation. They are activated to produce immunoglobulin, as has been shown by the spontaneous secretion of immunoglobulin *in vitro,*[1,2] and a substantial number of them are engaged in the production of autoantibodies, particularly rheumatoid factors (RFs) of the IgM, IgA, and IgG isotypes. An interesting feature of RF autoantibodies is that they can be induced in healthy persons following infection and immunization,[3-6] and there is evidence that immune complexes are the triggering mechanism for this production.[7] Little is known about the regulation of RF autoantibody production in healthy persons. These antibodies are presumably maintained in normals for physiological reasons, yet should avoid the normal immunological processes resulting in affinity maturation that could result in deleterious effects. T cells specific for Fc are normally deleted,[8] and in the absence of specific T-cell help, RF-producing B cells in healthy persons should be incapable of undergoing somatic hypermutation and affinity maturation.[9,10] If this model is true, RFs in normal physiology should be of low affinity and in germline, or close to germline, configuration.

Over the last few years, we have been examining the V-gene usage and structure of RF autoantibodies isolated as hybridoma clones from the synovial tissues of rheumatoid arthritis (RA) patients.[11-18] We have recently developed this work to compare the structure of RF produced in physiological conditions in order to examine whether there are any abnormalities in the regulation of RF production in RA. To achieve this, we isolated RF-secreting hybridoma clones from the peripheral blood of five healthy donors immunized with mismatched red blood cells (RBC).[19,20] The five donors responded to hyperimmunization with a specific

[a] This work was supported by funding from the Norwegian Women's Health Organization, the Norma and Leon Hess Foundation for Rheumatology Research, and the Norwegian Research Council.

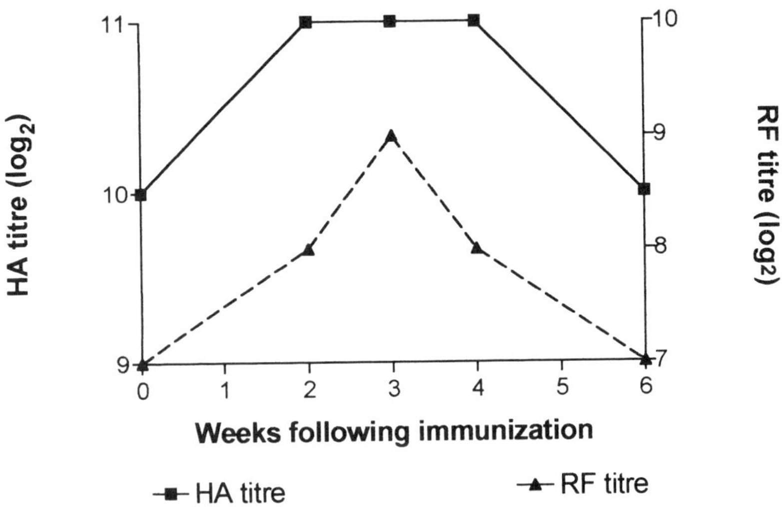

FIGURE 1. Specific anti-RBC and RF levels after immunization—typical kinetic response.

anti-RBC response. In addition, IgM RF antibody levels increased two to fourfold. The rise and fall in specific anti-RBC and RF levels after immunization closely followed each other in each donor (typical kinetic response is shown in FIG. 1). All fusions from the five donors yielded IgM RF-secreting hybridomas, but no IgA RF. In contrast, three fusions from nonimmunized donors yielded no RF-secreting hybridomas. This strongly suggests that the monoclonal RFs are representative of those involved in the specific immune response. These RFs were judged monoreactive for IgG.[19]

V-GENE FAMILIES AND GERMLINE SEGMENTS USED BY RHEUMATOID FACTORS

We investigated the expression of V-gene markers and cross-reactive idiotopes (CRI) by RF, using various murine monoclonal antibodies. These included G6 and G8, which recognize V_H1-associated CRI,[21] and B6 and D12, which recognize V_H3-associated CRI.[22] $V\kappa$ gene family products were determined by C7 (specific for the $V\kappa3$ family) and C6 (specific for the $V\kappa3b$ subfamily). 17.109 was used to identify the product of the $V\kappa3b$ subfamily, Kv325 variable region gene segment.[23] The majority of the RFs from immunized donors expressed at least one of these markers (TABLE 1). Within individual donors, particularly donor MR, were groups of RFs that shared identical patterns of V-gene markers, suggesting a restricted response.

Rheumatoid factors from RA synovial tissues were found to use many diverse germline V-gene segments, both in heavy and light chains. Comparison of the V-gene families used by RFs from RA synovial tissues and from immunized donors shows a different preference in the two conditions (FIG. 2). The most striking difference is in the use of light chain segments of the $V\kappa3$ family, which are used by 87% of RFs from immunized donors, but only 27% of RFs from synovial

TABLE 1. CRI and V Gene Markers Expressed by RFs from Immunized Donors

Donor	Clone No.	V_H	V_L	V_H Ids	V_L Ids
MR	10, 22	1	k3	G6	C6, C7
	6, 18	1	k3	G6	C6, C7, 17.109
	9	1	k	G6, G8	—
	30, 32	1	k3	G6, G8	C6, C7
	Group (a)	1	k3	G6, G8	C6, C7, 17.109
	4	1	k3	G8	C6, C7
	27	3	k	—	—
	5, 41	3	k3	B6, D12	C6, C7
	24	3	k3	D12	C6, C7, 17.109
	20, 39	4	k3	—	C6, C7
	12	nd	k3	—	C6, C7
DI	3	1	k3	—	C6, C7
	2	3	k3	—	C6, C7
	1	3	λ	—	—
	4	nd	k	—	—
FO	1	1	k3	G6, G8	C6, C7
	2	3	λ	—	—
	3	4	λ	—	—
SB	1	4	k3	—	C7
TT	3	1	k3	—	C6, C7
	1	1	k3	G6, G8	C6, C7
	6	3	k3	—	C6, C7
	5, 9	3	k3	B6, D12	C6, C7
	7	nd	k3	—	C6, C7

NOTE: RF clones from five donors (MR, DI, FO, SB, and TT) are shown. Where two or more clones share the same heavy chain markers (V_H Ids) and light chain markers (V_L Ids), they are grouped on one line. Group (a) represents 17 monoclonal RFs.

tissues. This overexpression of $V\kappa3$, together with a distribution of V_H gene segments of the V_H1, V_H3 and V_H4 families, closely resembles the distribution seen among RFs found as M components in lymphoproliferative diseases.[24] In contrast, RFs from the synovial tissues of RA patients predominantly use heavy chain segments of the V_H3 family. These data suggest that RFs found as M components are representative of RFs used in normal physiological responses, but have undergone neoplastic or other transformation. The differences seen in the RFs found in the synovial tissues of RA patients may be related to a particular antigenic stimulus and suggests that a partly different population of B cells are triggered to produce RFs than in normal persons.

SOMATIC HYPERMUTATION IN RHEUMATOID FACTORS FROM RHEUMATOID ARTHRITIS SYNOVIAL TISSUES

By nucleotide sequencing, we have analyzed RFs to examine the role of antigen-driven processes, that is, somatic hypermutation and antigenic selection, in the generation of these autoantibodies. Ten IgM RFs, derived from the synovial tissues

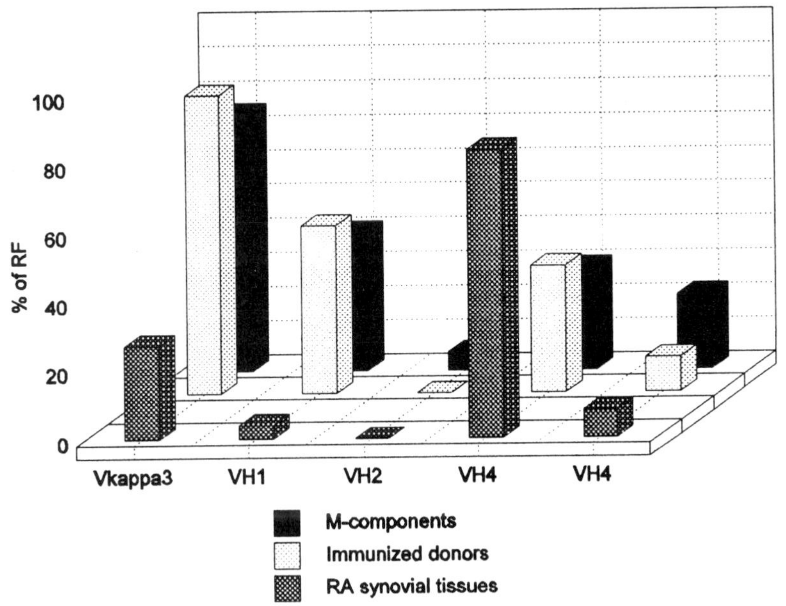

FIGURE 2. Expression of V-gene families by RFs of different origins.

of two RA patients and one juvenile rheumatoid arthritis (JRA) patient, clearly show evidence of having undergone somatic hypermutation, with between 2 and 18 mutations per V_H gene segment. The isolation of three IgG RFs, one from each of these patients, confirmed that class switching occurs in RA and is accompanied by the accumulation of mutations (2, 9, and 11 mutations per V_H segment).[17,18] From one of the RA synovial samples, we obtained two IgM RFs (RF-SJ1 and RF-SJ2) that were clonally related.[12,14,16] The V_H of RF-SJ2 has only two nucleotide differences from its germline counterpart (GL-SJ), while RF-SJ1 has accumulated 18 nucleotide differences. The light chain of RF-SJ1 has also acquired 18 more nucleotide changes than RF-SJ2. Functional affinity studies showed that the extensively mutated antibody had a hundred times higher affinity for human IgG than the RF close to germline.[16] These findings indicate that IgM RFs in RA can undergo affinity maturation as seen in mouse models of autoimmune disease.[25,26]

SOMATIC HYPERMUTATION IN RHEUMATOID FACTORS FROM HEALTHY, IMMUNIZED DONORS

On the basis of their similarity in V gene markers, ten RFs from a single donor (MR) were extensively studied by sequence analysis. All ten RFs use variable heavy chains with closest homology to the DP-10 germline segment[27] and are rearranged to the J_H4 segment (FIG. 3). Nine of these RFs are clonally related (RF-MR2, RF-MR13, RF-MR14, RF-MR16, RF-MR25, RF-MR28, RF-MR30, RF-MR33, and RF-MR37). They have identical CDR3 rearrangements, of equal length and with identical nucleotide junctions between the V_H, D, and J_H segments. The DA1 D segment contributes to each CDR3 with an unidentified adjacent segment that is identical in each clone. The nine clones have nine shared mutations; two of these mutations are also apparent in the remaining clone, RF-MR1.

Nucleotide sequencing of the light chains of five of these RFs confirmed the clonal relationship and showed that all use the Kv325 light chain germline gene rearranged to the $J_\kappa 1$ gene segment with identical V-J junctions (RF-MR2, RF-MR13, RF-MR16, RF-MR28, and RF-MR30). Hypermutation was also evident in the light chains, but to a lesser degree than in the heavy chains. Four of the five clonally related RFs have one shared replacement mutation in the light chain.

RFs from immunized donors thus clearly undergo somatic hypermutation. We have confirmed this by sequencing additional RFs from unrelated donors (unpublished data). This is somewhat surprising, as somatic hypermutation of immunoglobulin variable region genes in B cells needs T-cell help.[28] T cells specific for Fc are normally deleted[8]; thus, specific T-cell help would be absent, and RF-producing B cells in healthy persons should be incapable of undergoing somatic hypermutation. Without T cells specific for IgG Fc, RF-producing B cells may obtain T-cell help when presenting peptides, derived from the antigen part of antibody–antigen complexes, to antigen-specific T cells.[29] Alternatively, RF-producing B cells could present peptides from Ig V regions derived either from their own Ig or from antigen–antibody complexes.[30] The results from this study indicate that such help can support the hypermutation process.

SELECTIVE PRESSURES ACTING ON THE CDRs OF RHEUMATOID FACTORS

In responses to exogenous antigens, the hypermutation process is followed by both positive and negative selection pressures by which B cells with receptors of high affinity for antigen are selected positively (antigenic selection). B cells with

```
          FR1                                      CDR1                     FR2       CDR2                        FR3
           4   13  24  26  27  28  29  30      31  32  33  34  35      37  43      53  54  56  57  59  63      69  70  74  77  79
DP-10:    CTG AAG GCT GGA GGC ACC TTC AGC     AGC TAT GCT ATC AGC     GTG CAA     CCT ATC GGT ACA AAC AAG     ACG ATT GAA AGC GCC
RFMR2:    ... ... ... ... ... ..t ..t ..t     ..t ... A.. ... ...     ... ...     ... ... ... ... ... ...     ... ... ... .A. ...
RFMR13:   ... ... ... ... ... ... ..t ..t     ..t ... A.. ..t ...     ... ...     ... ..t ... ... ... ...     ... ... ... .A. ...
RFMR14:   t.. ... ... ... ... ... ..t ..t     ..t ... A.a ... ...     ... ...     ... ..t ..a ... ... .C.     ... ... ... .A. .T.
RFMR16:   ... ... ... ... ... ... ..t ..t     ..t ... A.. C.t ..t     C.. ...     ... ... .C. ... ... ...     ..a G.. ... .A. ...
RFMR25:   ... ... ... ... ... ... ..t ..t     ..t ... A.a ... ..t     ... ...     ... ..t ... ... ..t ...     ... ... ... .A. ...
RFMR28:   ... ... ... ... ... ... ..t ..t     ..t ... A.. ..t ...     ... ...     ..c ..t ... ... ... ...     ... ... ... ... T..
RFMR30:   ... .G. ... ... ... ... ... ..t     ..t ..c A.. ... ...     ... ...     ... ... ..c ... ... ...     ... ... ... .A. ...
RFMR33:   ... ... ... ..g ... ... ..t ..t     ..t ... A.. ... ...     ... ..g     ... G.t ..c ... ... ...     ... ... ... .A. ...
RFMR37:   ... ... ..c ... ... ... ..t ..t     ..t ... A.. ... ...     ... ...     ... ..t ... ... ... ...     ... ... AC. .A. ...
RFMR1:    ... .C. ... ... .A. ... ... ...     ..t ... A.c ... ...     ... ...     ... ... ... ..t ... ..C     ... ... ... ... ...

                                                     CDR3
          80  82  83  84  88  89  93  94  95        E   G   K   A   G           D   Y   S   N   Y           F   D   Y
DP-10:    TAC GAG CTG AGC TCT GAG GTG TAT TAC nd:GAG GGA AAG GCG GGG    DA1:GAC TAC AGT AAC TAC JH4:TTT GAC TAC
RFMR2:    ... ... ... .A. ... ... ... ..c ..t    ... ... ... ... ...        ... ... ... ..t CC.     ... ... ...
RFMR13:   ... ..a ... ... ... ..a A.a ... ..t    ... ... ... ... ...        ... ... ... ... CC.     ... ... ..t
RFMR14:   ... ... ... ... ... ... A.a ... ..t    ... ... ... ... ...        ... ... ... ... CCg     ..c ... ...
RFMR16:   ... ... ..t .T. ... ... ... ... .Tt    ... ... ... ... ...        ... ... .A. ... CC.     ... ... ...
RFMR25:   ... ... ... ... ... ... A.a ... ..t    ... ... ... ... ...        ... ... ... ... CC.     ... ... ...
RFMR28:   ... ... ... ... ... ... A.a ... ..t    ... ... ... ... ...        ... ... ... ... CC.     ... ... ...
RFMR30:   .G. ... t.. ... ... ... ... ... ..t    ... ... ... ... ...        ... ... ... ... CC.     ... ... ...
RFMR33:   ... ... ... ... ... ... A.a ... ..t    ... ... ... ... ...        ... ..t ... ... CC.     ... ... ...
RFMR37:   ... ... ... ... ... ... A.a ... ..t    ... ... ... ... ...        ... ... ... ... CC.     ... ... ...
RFMR1:    ... ... ... ... C.. ... ..c ... ... nd:GAG GGC CGT AGG ATG GCC GCT AAC CCC                .A. ... ...
```

FIGURE 3. Nucleotide sequences of the V_H of nine clonally related RFs from donor MR and one closely related RF (MR-1). The bases that differ from the germline are shown; periods represent identity to the germline. Mutations resulting in amino acid replacements are shown in upper case.

TABLE 2. Summary of the Number and Location of Replacement (R) and Silent (S) Mutations in the Variable Heavy Regions of RF from Healthy Immunized Donors (HID), Patients with RA (RA RF) and anti-D Antibodies

				Heavy Chain					
				All FR			CDR1+2		
Clone	Mutations from Germline	Mutations from Nearest Progenitor	K_d M	R	S	R:S	R	S	R:S
HID RF									
RF-MR1	9	9	2×10^{-6}	3	1		2	3	
VH Pro I		5		1	2		1	1	
RF-MR30	10	5	5×10^{-6}	2	1		0	2	
VH Pro II		1		0	1		0	0	
RF-MR2	9	3	1×10^{-6}	1	2		0	0	
RF-MR16	16	10	6×10^{-7}	4	2		2	2	
VH Pro III		3		1	1		0	1	
RF-MR37	12	3	2×10^{-6}	2	1		0	0	
RF-MR33	13	4	9×10^{-7}	0	2		1	1	
VH Pro IVa		1		0	0		0	1	
RF-MR25	12	2	2×10^{-6}	0	0		0	2	
RF-MR14	14	4	2×10^{-6}	2	0		1	1	
VH Pro IVb		1		0	0		0	1	
RF-MR13	12	2	nd	0	2		0	0	
RF-MR28	13	3	1×10^{-6}	2	0		0	1	
***All HID RF:**	**12.0**	**56**	$\mathbf{2 \times 10^{-6}}$	**18**	**15**	**1.2**	**7**	**16**	**0.4**
RA RF									
RF-TS1	11	11	8×10^{-8}	6	2		3	0	
Re12	11	11	nd	1	5		2	3	
YES8c	15	15	nd	7	5		3	0	
Ro47	18	18	nd	1	8		7	2	
Ro7	27	27	nd	10	6		7	4	
mAb112/113	16/15	29	$3\text{–}5 \times 10^{-7}$	5	8		11	5	
***All RA RF:**	**16.1**	**111**	$\mathbf{3 \times 10^{-7}}$	**30**	**34**	**0.9**	**33**	**14**	**2.4**
Anti-D									
MS205/210	5/1	5	nd	2	2		1	0	
MS214	1	1	nd	0	1		0	0	
FOM 1	3	3	3×10^{-8}	0	1		2	0	
MS209	6	6	nd	1	3		1	1	
FOM A	6	6	1×10^{-7}	1	2		3	0	
MAD 2	8	8	7×10^{-8}	4	3		1	0	
***All anti-D:**	**4.3**	**29**	$\mathbf{7 \times 10^{-8}}$	**8**	**12**	**0.7**	**8**	**1**	**8.0**

[a] NOTE: For the clonally related group of RFs from MR, mutations are shown away from hypothetical intermediates (progenitors). Among clonally related RFs, mutations common to more than one member are only counted once.

* The mean numbers of mutations and mean K_d are shown, together with the total number of mutations and overall R:S ratios. Data for RA RF are from Refs. 12, 33–35; data for anti-D are from the present study and Ref. 36.

mutated receptors that have low affinity for antigen die by the process of apoptosis.[31] Besides an increase in affinity, positively selected B cells show a characteristic elevation of the ratio of R to S in the CDRs.[25,32] Surprisingly, RFs produced in healthy persons after immunization show a selection against replacement mutations (counterselection) in the CDRs. Without any selective pressures, random mutations in CDR1+2 of the DP-10 gene segment would generate R:S ratios of 3.2. The overall R:S ratio of the CDR1+2 in the V_H region of the RF from immunized donor MR is significantly lower at 0.4 ($p < 0.0027$, Fisher's exact test, two sided) (TABLE 2).

We also analyzed a further group of unrelated RF autoantibodies from four other immunized donors. For seven RFs where we can be reasonably certain of the germline origin, (≤16 mutations away from the closest published germline sequences), the overall R : S ratio in CDR1+2 was only 1.2. It is not clear whether IgM antibodies to exogenous antigens generally show selection for replacement mutations in the CDRs. It is clear, however, that among the IgM anti-Rh(D) antibodies presented in TABLE 2, there is not a strong counterselection against replacement mutations as seen in the RFs from immunized donors. The heavy chains of the seven human monoclonal IgM antibodies to the exogenous Rh(D) antigen have a high R : S ratio in the CDR1+2 (8 : 0), despite a low rate of mutation, but this is not statistically higher than the expected random ratio of 3.6.

RFs from RA patients do not seem to show a similar counterselection of replacement mutations in the CDRs as seen in RFs from immunized donors. TABLE 2 shows that seven IgM RFs, all using the same DP-10 V_H segment as the immunized donor RF but derived from peripheral blood, synovial tissue, and bone marrow of RA patients, have an overall R:S ratio in CDR1+2 of 2.4.[12,33-35] The R : S ratios seen in the heavy chain CDRs of these IgM RFs more closely resemble antibodies to the exogenous Rh(D) antigen.

THE ROLE OF AFFINITY MATURATION IN THE GENERATION OF RHEUMATOID FACTORS

The derivation of this group of clonally related RFs from immunized donor MR allows a detailed examination of the relationship between somatic hypermuta-

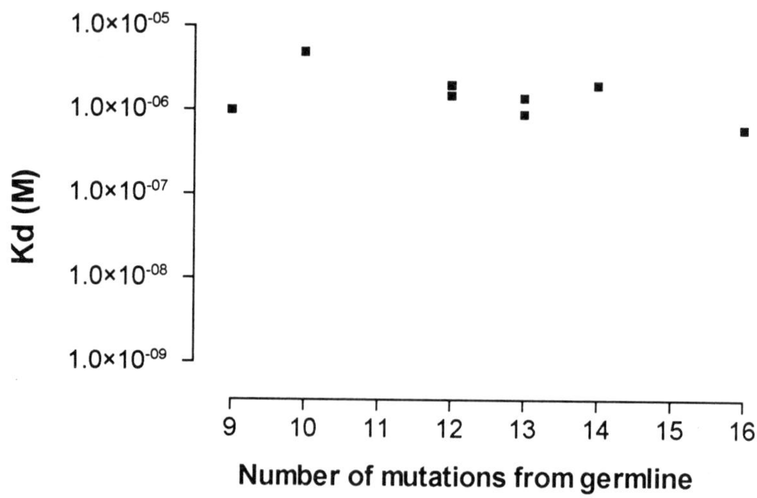

FIGURE 4. The relationship between dissociation constant and number of mutations in the V_H segments of eight clonally related RFs from donor MR. The slope is not significantly different from zero.

tion and increase of affinity. Among this group of RFs, there is very little increase in affinity with the accumulation of mutations (FIG. 4). They have an average K_d of 2×10^{-6} M (range 0.6–5 $\times 10^{-6}$ M). This suggests there is a mechanism operating on RF-producing B cells that constrains the affinity maturation process by tolerizing high-affinity variants. If we compare the affinity of RF from RA patients, that use the same gene segments, the mean affinity is approximately 10-fold higher, that is a K_d of 3×10^{-7} M (range 8×10^{-8} M to 5×10^{-7} M) (TABLE 2). RFs in RA thus do not appear to be subject to such a controlling mechanism.

SUMMARY

We have compared RF from normal, immunized donors and RF derived from the synovial tissues of RA patients. We have found a difference in the preferential use of V_L and V_H gene families. In both conditions, RFs were found to have accumulated somatic mutations. However, there was a striking difference in the patterns of mutation. RFs from normals were characterized by a very low R:S ratio in the CDR1+2, considerably lower than seen among the RA RFs. In addition, there was little increase in affinity with increasing numbers of mutations in a group of clonally related RFs from an immunized normal. This contrasts with RF from RA, where there is evidence of both affinity maturation and class switching. Together these data suggest that in healthy persons there is a controlling mechanism to limit the affinity of RF autoantibodies, and that this is lost in RA. The higher affinity of the RA-derived RF may be of significance in the pathology of the disease.

ACKNOWLEDGMENTS

We would like to thank Ingvild Vatn for skilled technical assistance and Dr. R. Mageed and Prof. R. Jefferis for providing marine monoclonal antibodies used in these studies.

REFERENCES

1. CHATTOPADHYAY, C., H. CHATTOPADHYAY, J. B. NATVIG, T. E. MICHAELSEN & O. J. MELLBYE. 1979. Lack of suppressor cell activity in rheumatoid synovial lymphocytes. Scand. J. Immunol. **10:** 309–316.
2. HOFFMAN, W. L., M. S. GOLDBERG & J. D. SMILEY. 1982. Immunoglobulin G3 subclass production by rheumatoid synovial tissue cultures. J. Clin. Invest. **69:** 136–144.
3. ABRUZZO, J. L. & C. L. CHRISTIAN. 1961. The induction of a rheumatoid factor-like substance in rabbits. J. Exp. Med. **114:** 791.
4. AHO, K. & O. WAGER. 1961. Production of "anti-antibodies" in rabbits. Appearance in rabbit serum of "anti-antibodies" reacting with autogenous and isogenous antibody, following stimulation with protein antigens. Ann. Med. Exp. Biol. Fenn. **39:** 29.
5. LEVINE, P. R. & D. A. AXELROD. 1985. Rheumatoid factor isotypes following immunization. Clin. Exp. Rheumatol. **3:** 147–149.
6. WELCH, M. J., S. FONG, J. VAUGHAN & D. CARSON. 1983. Increased frequency of rheumatoid factor precursor B lymphocytes after immunization of normal adults with tetanus toxoid. Clin. Exp. Immunol. **51:** 299–304.
7. NEMAZEE, D. A. 1985. Immune complexes can trigger specific, T cell-dependent, autoanti-IgG antibody production in mice. J. Exp. Med. **161:** 242–256.

8. BIKOFF, E. K. 1983. Tolerance for self IG at the level of the Ly1$^+$ T cell. J. Exp. Med. **158:** 1868–1880.
9. MAIZELS, N. & A. BOTHWELL. 1985. The T-cell-independent immune response to the hapten NP uses a large repertoire of heavy chain genes. Cell **43:** 715–720.
10. KARVELAS, M. & G. J. NOSSAL. 1992. Memory cell generation ablated by soluble protein antigen by means of effects on T- and B-lymphocyte compartments. Proc. Natl. Acad. Sci. USA **89:** 3150–3154.
11. RANDEN, I., K. M. THOMPSON, J. B. NATVIG, O. FORRE & K. WAALEN. 1989. Human monoclonal rheumatoid factors derived from the polyclonal repertoire of rheumatoid synovial tissue: Production and characterization. Clin. Exp. Immunol. **78:** 13–18.
12. PASCUAL, V., I. RANDEN, K. M. THOMPSON, M. SIOUD, O. FORRE, J. NATVIG & J. D. CAPRA. 1990. The complete nucleotide sequences of the heavy chain variable regions of six monospecific rheumatoid factors derived from Epstein-Barr virus-transformed B cells isolated from the synovial tissue of patients with rheumatoid arthritis. Further evidence that some autoantibodies are unmutated copies of germ line genes. J. Clin. Invest. **86:** 1320–1328.
13. THOMPSON, K. M., I. RANDEN, J. B. NATVIG, R. A. MAGEED, R. JEFFERIS, D. A. CARSON, H. TIGHE & O. FORRE. 1990. Human monoclonal rheumatoid factors derived from the polyclonal repertoire of rheumatoid synovial tissue: Incidence of cross-reactive idiotopes and expression of VH and V kappa subgroups. Eur. J. Immunol. **20:** 863–868.
14. VICTOR, K. D., I. RANDEN, K. M. THOMPSON, O. FORRE, J. B. NATVIG, S. M. FU & J. D. CAPRA. 1991. Rheumatoid factors isolated from patients with autoimmune disorders are derived from germline genes distinct from those encoding the Wa, Po, and Bla cross-reactive idiotypes. J. Clin. Invest. **87:** 1603–1613.
15. PASCUAL, V., K. VICTOR, I. RANDEN, K. M. THOMPSON, M. STEINITZ, O. FORRE, S. M. FU, J. B. NATVIG & J. D. CAPRA. 1992. Nucleotide sequence analysis of rheumatoid factors and polyreactive antibodies derived from patients with rheumatoid arthritis reveals diverse use of VH and VL gene segments and extensive variability in CDR-3. Scand. J. Immunol. **36:** 349–362.
16. RANDEN, I., D. BROWN, K. M. THOMPSON, N. C. HUGHES-JONES, V. PASCUAL, K. VICTOR, J. D. CAPRA, O. FORRE & J. B. NATVIG. 1992. Clonally related IgM rheumatoid factors undergo affinity maturation in the rheumatoid synovial tissue. J. Immunol. **148:** 3296–3301.
17. RANDEN, I., K. M. THOMPSON, S. J. THORPE, O. FORRE & J. B. NATVIG. 1993. Human monoclonal IgG rheumatoid factors from the synovial tissue of patients with rheumatoid arthritis. Scand. J. Immunol. **37:** 668–672.
18. RANDEN, I., V. PASCUAL, K. VICTOR, K. M. THOMPSON, O. FORRE, J. D. CAPRA & J. B. NATVIG. 1993. Synovial IgG rheumatoid factors show evidence of an antigen-driven immune response and a shift in the V-gene repertoire compared to IgM rheumatoid factors. Eur. J. Immunol. **23:** 1220–1225.
19. THOMPSON, K. M., I. RANDEN, M. BORRETZEN, O. FORRE & J. B. NATVIG. 1994. Variable region gene usage of human monoclonal rheumatoid factors derived from healthy donors following immunization. Eur. J. Immunol. **24:** 1771–1778.
20. BORRETZEN, M., I. RANDEN, E. ZDARSKY, O. FORRE, J. B. NATVIG & K. M. THOMPSON. 1994. Control of autoantibody affinity by selection against amino acid replacements in the complementarity determining regions. Proc. Natl. Acad. Sci. USA **91:** 12917–12921.
21. MAGEED, R. A., M. DEARLOVE, D. M. GOODALL & J. JEFFERIS. 1986. Immunogenic and antigenic epitopes of immunoglobulins. XVII—Monoclonal antibodies reactive with common and restricted idiotopes to the heavy chain of human rheumatoid factors. Rheumatol. Int. **6:** 179–183.
22. CROWLEY, J. J., R. A. MAGEED, G. J. SILVERMAN, P. P. CHEN, F. KOZIN, R. A. ERGER, R. JEFFERIS & D. A. CARSON. 1990. The incidence of a new human cross-reactive idiotype linked to subgroup VHIII heavy chains. Mol. Immunol. **27:** 87–94.
23. RADOUX, V., P. P. CHEN, J. A. SORGE & D. A. CARSON. 1986. A conserved human germline V kappa gene directly encodes rheumatoid factor light chains. J. Exp. Med. **164:** 2119–2124.

24. RANDEN, I., K. M. THOMPSON, V. PASCUAL, K. VICTOR, D. BEALE, J. COADWELL, O. FORRE, J. D. CAPRA & J. B. NATVIG. 1992. Rheumatoid factor V genes from patients with rheumatoid arthritis are diverse and show evidence of an antigen driven response. Immunol. Rev. **128:** 49–71.
25. SHLOMCHIK, M. J., A. MARSHAK ROTHSTEIN, C. B. WOLFOWICZ, T. L. ROTHSTEIN & M. G. WEIGERT. 1987. The role of clonal selection and somatic mutation in autoimmunity. Nature **328:** 805–811.
26. SHLOMCHIK, M., M. MASCELLI, H. SHAN, M. Z. RADIC, D. PISETSKY, A. MARSHAK ROTHSTEIN & M. WEIGERT. 1990. Anti-DNA antibodies from autoimmune mice arise by clonal expansion and somatic mutation. J. Exp. Med. **171:** 265–292.
27. TOMLINSON, I. M., G. WALTER, J. D. MARKS, M. B. LLEWELYN & G. WINTER. 1992. The repertoire of human germline V(H) sequences reveals about 50 groups of V(H) segments with different hypervariable loops. J. Mol. Biol. **227:** 776–798.
28. MAIZELS, N. & A. BOTHWELL. 1985. The T-cell-independent immune response to the hapten NP uses a large repertoire of heavy chain genes. Cell **43:** 715–720.
29. ROOSNEK, E. & A. LANZAVECCHIA. 1991. Efficient and selective presentation of antigen-antibody complexes by rheumatoid factor B cells. J. Exp. Med. **173:** 487–489.
30. BOGEN, B. 1993. Processing and presentation of immunoglobulin idiotypes to T cells. Immunologist **1:** 121–125.
31. LIU, Y. J., D. E. JOSHUA, G. T. WILLIAMS, C. A. SMITH, J. GORDON & I. C. MACLENNAN. 1989. Mechanism of antigen-driven selection in germinal centres. Nature **342:** 929–931.
32. TONEGAWA, S. 1983. Somatic generation of antibody diversity. Nature **302:** 575–581.
33. EZAKI, I., H. KANDA, K. SAKAI, N. FUKUI, M. SHINGU, M. NOBUNAGA & T. WATANABE. 1991. Restricted diversity of the variable region nucleotide sequences of the heavy and light chains of a human rheumatoid factor. Arthritis Rheumatol. **34:** 343–350.
34. MANTOVANI, L., R. L. WILDER & P. CASALI. 1993. Human rheumatoid-B-1a ($CD5^+$ B) cells make somatically hypermutated high affinity IgM rheumatoid factors. J. Immunol. **151:** 473–488.
35. YOUNGBLOOD, K., L. FRUCHTER, G. DING, J. LOPEZ, V. BONAGURA & A. DAVIDSON. 1994. Rheumatoid factors from the peripheral blood of two patients with rheumatoid arthritis are genetically heterogeneous and somatically mutated. J. Clin. Invest. **93:** 852–861.
36. BYE, J. M., C. CARTER, Y. CUI, B. D. GORICK, S. SONGSIVILAI, G. WINTER, N. C. HUGHES-JONES & J. D. MARKS. 1992. Germline variable region gene segment derivation of human monoclonal anti-Rh(D) antibodies. Evidence for affinity maturation by somatic hypermutation and repertoire shift. J. Clin. Invest. **90:** 2481–2490.

Analysis of Immunoglobulin Gamma Heavy Chains from Rheumatoid Arthritis Synovium

Evidence of Antigen-Driven Selection

S. LOUIS BRIDGES, JR.,[a,b] BJÖRN E. CLAUSEN,[b,c] JOHN C. LAVELLE,[c] PRISCILLA G. FOWLER,[c] WILLIAM J. KOOPMAN,[a,b,d] AND HARRY W. SCHROEDER, JR.[b,c,d,e,f]

[a]Division of Clinical Immunology and Rheumatology
[c]Division of Developmental and Clinical Immunology
[b]Department of Medicine
[d]Department of Microbiology
[e]Comprehensive Cancer Center
University of Alabama at Birmingham and the Birmingham Veterans Administration Medical Center
Birmingham, Alabama 35294

Rheumatoid arthritis (RA) is a chronic illness of unknown etiology characterized primarily by inflammation and proliferation of the synovial lining of joints. Plasma cells present in the inflammatory infiltrates in synovium may secrete large amounts of immunoglobulin. The center of the antigen-binding site of an antibody is encoded by the CDR3 domains of the heavy and light chains. Transcripts with identity in CDR3 likely arise from the same progenitor B lymphocyte. In heavy chains, CDR3s are formed by V_H-D_H-J_H joins. D_H gene segments contribute significantly to sequence diversity, as there are six potential reading frames (three by inversion and three by deletion).

Previous analyses of the κ light chain and γ heavy chain repertoires in a synovial cDNA library from a patient with long-standing, active RA demonstrated evidence of an antigen-driven response: nonrandom utilization of variable domains, marked somatic mutation, and high R/S ratios in the CDRs.[1,2] Further analysis of the κ light chain repertoire revealed three identical transcripts, which further implicates an antigen-driven response in this patient's synovium.[3]

To search for clonally related V_H gene segments and to analyze D_H reading frames and charge, we have now sequenced CDR3s of 58 randomly isolated $C\gamma^+$ transcripts from this library. The RA synovial V_H family distribution resembles that reported in normal adult tissues: V_H1 24%, V_H2 1%, V_H3 62%, V_H4 10%, V_H5 1%, V_H6 1%. Within each family, however, there appears to be preferential use of particular V_H gene segments frequently expressed during fetal development and in autoantibodies. For example, 7 of 16 (44%) V_H1-containing clones are likely

[f] Address for correspondence: Harry W. Schroeder, Jr., M.D., Ph.D., Division of Developmental and Clinical Immunology, Wallace Tumor Institute 404, University of Alabama at Birmingham, Birmingham, AL 35294.

```
                            VH3 Framework 3
Germline 30P1     GCTGTATCTGCAAATGAACAGCCTGAGAGCCGAGGACACGGCCGTATATTACTGTGCGAAA
         S8P5    *A........................T...........T......T........G.
         S13P1   *A........................T...........T......T........G.
         S20P6   *A........................T...........T......T........G.
Amino acid         LeuTyrLeuGluMetAsnSerLeuArgValGluAspThrAlaValTyrPheCysAlaArg


         N region    Germline DH DXP'1               N       Germline JH4            Framework 4
                    GTATTACTATGGTTCGGGGAGTTATTATAAC           TACTTTGACTAC TGGGGCCAGGGAACCCTGGTCACCGTCTCCTCA
S8P5     ACGCCCAACA  ..G..G..C.C........           CCGTCT              ..T. .................................
S13P1    ..........  ..G..G..C.C........           ......              ..T. .................................
S20P6    ..........  ..G..G..C.C........           ......              ..T. .................................
         ThrProAsn   IleValValArgPheGlyAsp          ArgLeu              Phe TrpGlyGlnGlyThrLeuValThrValSerSer
```

FIGURE 1. Clonally-related V_H gene segments from rheumatoid arthritis synovial tissue. Sequences are shown in comparison to their most homologous germline gene segments. Dots denote nucleotide sequence homology. Amino acid sequences shown are deduced from nucleotide sequences. Asterisks indicate site of termination of transcript.

derived from the gene segment 51P1, which is preferentially expressed in human fetal liver and can encode rheumatoid factors in patients with RA. Three of 67 (4.4%) $C\gamma^+$ clones were found to contain identical sequences (FIG. 1). The presence of three identical V_H sequences corroborates findings in the $V\kappa$ analysis. The clonally expanded heavy chain represents a potential partner for the previously reported clonally expanded κ chain.

Of 39 identifiable D_H gene segments, there was a marked preference for members of the DXP family (64%), followed by the DLR (18%) family. D_H reading frame 1 was used in 69% of the rearrangements. Most of the CDR3s had a neutral charge, with 51 of 59 clones (86%) having mean charge per residue of between −0.20 and 0.30 (Kyte-Doolittle hydrophobicity indices). CDR3 lengths varied between 5 and 26 amino acids, a distribution similar to reports from normal adult peripheral blood lymphocytes (PBLs). J_H utilization resembled that seen in normal PBLs with predominance of the J_H4 and J_H6 gene segments.

In summary, we found additional evidence of clonal B-cell expansion in heavy chain derived from the synovium of a patient with active RA. We speculate that the presence of three identical sequences and the absence of nonidentical clones with shared mutations may reflect the long history of RA in this person. These clones may represent a subset of antibodies from the terminal stage of a highly focused immune response for a particular antigen.

REFERENCES

1. BRIDGES, S. L. JR., S. K. LEE, W. J. KOOPMAN & H. W. SCHROEDER, JR. 1993. Analysis of immunoglobulin gamma heavy chain expression in synovial tissue of a patient with rheumatoid arthritis. Arthritis Rheum. **36:** 631–641.
2. LEE, S. K., S. L. BRIDGES, JR., W. J. KOOPMAN & H. W. SCHROEDER, JR. 1992. The immunoglobulin kappa light chain repertoire expressed in the synovium of a patient with rheumatoid arthritis. Arthritis Rheum. **35:** 905–913.
3. LEE, S. K., S. L. BRIDGES, JR., P. M. KIRKHAM, W. J. KOOPMAN & H. W. SCHROEDER, JR. 1994. Evidence of antigen receptor-influenced oligoclonal B lymphocyte expansion in the synovium of a patient with longstanding rheumatoid arthritis. J. Clin. Invest. **93:** 361–370.

In Situ Hybridization Analysis of Immunoglobulin V_H Gene Family Expression in Rheumatoid Arthritis

CHARLES H. RUNDLE,[a,b] HARRY W. SCHROEDER, JR.,[b,c]
AND WILLIAM J. KOOPMAN[a,b,c,d]

[a]*The Birmingham Veterans Administration Medical Center*
[b]*Department of Microbiology*
[c]*Department of Medicine*
University of Alabama at Birmingham
Birmingham, Alabama 35294

In an effort to elucidate the specificity of autoantibodies in rheumatoid arthritis (RA), *in situ* hybridization was used to compare the frequency of expression of immunoglobulin heavy chain variable (V_H) gene families in normal subjects and RA patients. The sensitivity and specificity of oligonucleotide probes specific for the germline sequences of the second or third framework regions of each of the seven known human V_H gene families[1,2] was established by Southern hybridization to 30 cDNAs representing germline and mutated V_H gene family members. A minimum of 80% of the members of each V_H family were detected. *In situ* hybridization[3,4] to a panel of 17 sequenced B-cell lines representing each V_H family detected immunoglobulin mRNA in each cell line; each probe exhibited excellent specificity (FIG. 1A and B) and similar frequencies for relevant V_H family and isotype probes were observed in these cell lines. Additionally, *in situ* hybridization was found to be more sensitive than assays of immunoglobulin secretion (ELISPOT assay).

Peripheral blood mononuclear cells (PBMCs) from seven normal persons and seven RA patients were stimulated *in vitro* with pokeweed mitogen (PWM) and V_H gene family distribution determined by *in situ* hybridization. Single cell suspensions were prepared from synovia obtained from five additional RA patients and compared by *in situ* hybridization to PWM-stimulated PBMCs from each patient. The frequency of V_H family expression in the PBMCs from each group of subjects was similar and appeared to be the result of stochastic, or undirected, gene utilization. The V_H families with the most numerous germline members were more frequently expressed (V_H1, 16%; V_H3, 42%; V_H4, 20%) than those with fewer members (V_H2, 4%; V_H5, 7%; V_H6, 3%). The V_H7 family, when present, was expressed at a frequency of approximately 3%.

The distribution of V_H gene family expression in RA synovia differed from paired PBMC samples (TABLE 1). In the five patients studied, the V_H7 gene family was found to predominate in three synovia (at least 49% of the total V_H-positive cells), with V_H4 being most frequent (40% of the total V_H-positive cells) in the fourth. In the fifth patient, from whom two synovia were available, no single V_H family predominated, although the V_H1 (FIG. 1C) and V_H3 families exhibited altered ratios of expression in the synovia relative to the blood. Additionally, the

[d] Address for correspondence: Dr. William J. Koopman, Department of Medicine, BDB 420, University of Alabama at Birmingham, Birmingham, AL 35294-0012.

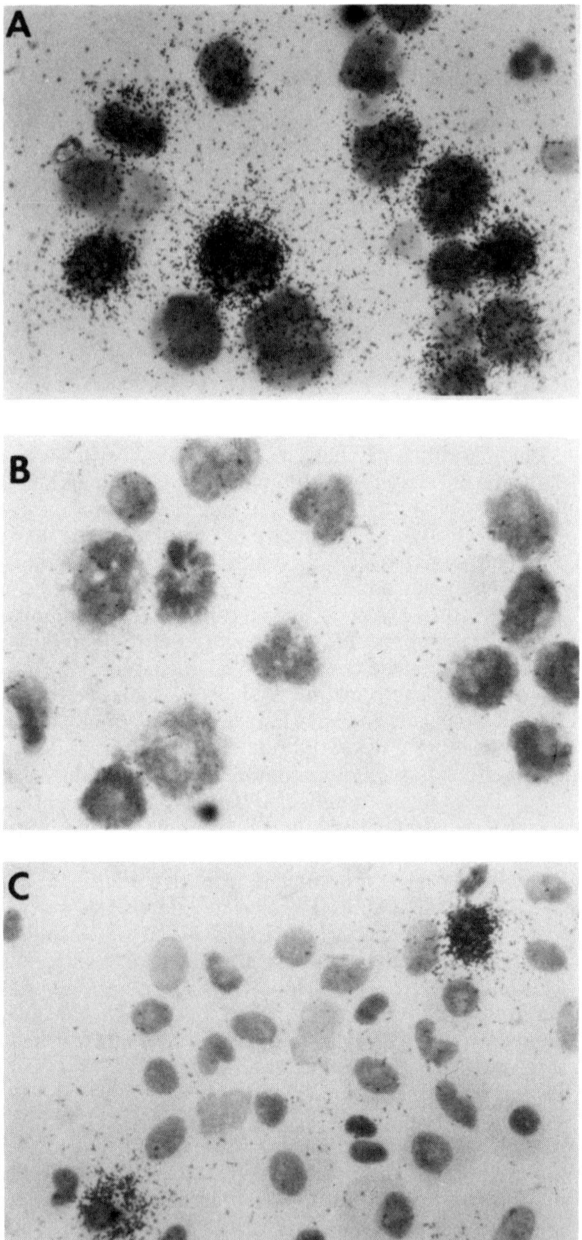

FIGURE 1. *In situ* hybridization of V_H gene family probes. **(A)** V_H7 probe hybridized to V_H7-expressing cell line RF-TS3. **(B)** V_H1 probe hybridized to V_H7-expressing cell line RF-TS3. **(C)** V_H1 probe hybridized to synovial cells from the left knee of RA patient 5 (TABLE 1).

TABLE 1. V_H Gene Family Use in RA Patient Paired Synovia and Peripheral Blood Mononuclear Cells[a]

RA Patient[b]	Isotype				V_H Gene Family							
		% Normalized				% Normalized						
	% Positive	μ	γ	α	% Positive	V_H1	V_H2	V_H3	V_H4	V_H5	V_H6	V_H7
1. Synovium (RK)	**2.1**	8.0	60.9	31.1	**1.5**	2.8	9.0	9.0	5.5	9.0	9.7	55.2
PBMC (PWM+)	**3.3**	30.8	36.6	32.6	**4.1**	16.5	4.1	44.1	23.2	6.3	3.4	2.4
2. Synovium (RW)	**0.9**	10.8	31.2	58.1	**2.0**	0.0	0.0	6.4	14.3	0.0	10.8	68.5
PBMC (PWM+)	**1.1**	22.8	34.2	43.0	**0.8**	12.8	0.0	34.6	19.2	6.4	10.3	16.7
3. Synovium (LE)	**1.5**	17.7	44.2	38.1	**2.0**	7.4	4.0	18.8	12.4	4.5	4.0	49.0
PBMC (PWM+)	**7.9**	7.3	48.2	44.6	**7.8**	17.3	5.7	45.4	18.2	7.6	1.7	4.1
4. Synovium (RK)	**9.7**	5.1	87.5	7.4	**9.6**	17.2	2.4	22.4	40.5	7.0	2.1	8.4
PBMC (PWM+)	**23.7**	8.5	77.7	13.8	**18.7**	15.9	4.9	38.5	27.8	8.6	0.9	3.4
5. Synovium (LK)	**3.6**	20.7	69.2	10.2	**3.9**	31.6	1.3	29.5	23.6	10.4	2.3	1.3
(RK)	**3.8**	15.8	74.0	10.2	**3.0**	34.5	0.3	21.7	28.6	12.8	0.0	2.0
PBMC (PWM+)	**3.8**	34.1	41.9	24.0	**3.6**	10.0	2.5	50.4	21.6	9.1	4.2	2.2

[a] Percentage positive cells by *in situ* hybridization.
[b] Abbreviations: PBMC, peripheral blood mononuclear cells; PWM, 5-day pokeweed mitogen stimulation; K, knee; W, wrist; E, elbow.

V_H family distributions were similar in both synovia in this patient. In each case, the total frequencies of the isotypes and V_H families were comparable, indicating that mutated V_H targets were detected by *in situ* hybridization with V_H family-specific probes specified by germline sequences. The skewed distribution of V_H gene family expression in the synovia is consistent with local immune responses that are antigen driven.

The determination of individual V_H gene or V_H gene family expression may help define antibody specificity,[2] provide insights into the molecular characteristics of immune responses,[1] and their escape from regulation in autoimmune conditions such as RA.[5] *In situ* hybridization appears to be a reliable technique for the determination of expression of the members of multigene families. It is possible to screen samples with large or small numbers of cells, and additional probes may subsequently be used to more accurately identify the contributing V_H gene.

REFERENCES

1. SCHROEDER, H. W., JR. & R. M. PERLMUTTER. 1993. Development of the human antibody repertoire. *In* New Concepts in Immunodeficiency Diseases. Chapter 1. S. Gupta & G. Griscelli, Eds. John Wiley. New York.
2. KIRKHAM, P. M., F. MORTARI, J. A. NEWTON & H. W. SCHROEDER, JR. 1992. Immunoglobulin V_H clan and family identity predicts variable domain structure and may influence antigen binding. EMBO J. **11(2):** 603–609.
3. LONG, A. A., J. MUELLER, J. ANDRE-SCHWARTZ, K. J. BARRETT, R. SCHWARTZ & H. WOLFE. 1992. High specificity *in situ* hybridization. Methods and application. Diagn. Mol. Pathol. **1(1):** 45–57.
4. KRISHAN, T. & R. H. SINGER. 1987. Use of oligonucleotide probes for quantitative *in situ* hybridization to actin mRNA. Anal. Biochem. **166:** 389–398.
5. CHEN, P. P. & D. A. CARSON. 1991. Molecular and genetic studies of human autoantibodies. *In* Molecular Autoimmunity. Chapter 5. N. Talal, Ed. Academic Press. New York.

Immunoglobulin Gene Usage in Diabetes-Associated Human Monoclonal Antibodies Recognizing Glutamate Decarboxylase[a]

WILTRUD RICHTER,[b] KARIN M. JURY,[b]
DANIELA LOEFFLER,[c] THOMAS H. EIERMANN,[c]
AND BERNHARD O. BOEHM[b]

[b]*Department of Internal Medicine I and*
[c]*Department of Transfusion Medicine*
University of Ulm
89070 Ulm, Germany

Islet cell antibodies directed towards the GABA-synthesizing enzyme glutamate decarboxylase (GAD) are a major early marker for islet cell destruction in persons developing type 1 diabetes.[1] Because of the lack of stable monoclonal B-cell lines, these diabetes-associated human autoantibodies have not been accessible for analyses of immunoglobulin (Ig) variable gene usage. We recently obtained seven human monoclonal IgG autoantibodies (MICA 1–7) directed towards GAD by Epstein-Barr virus transformation of B cells derived from peripheral blood of persons with newly diagnosed type 1 diabetes.[2] MICA 1–6 originated from one person and MICA 7 from a second person. The MICA had been used to characterize in detail the autoreactive epitopes in GAD associated with type 1 diabetes. Two pairs of antibodies, MICA 1/3 and MICA 4/6 blocked each other's binding to GAD and showed equal reactivity with a panel of GAD deletion mutants.[3] The aim of this study was to distinguish the MICA on the basis of their Ig gene sequences and to assess the variable gene segment usage observed in GAD-reactive human monoclonal autoantibodies associated with type 1 diabetes.

MATERIALS AND METHODS

The monoclonal cell lines were expanded, and mRNA was isolated from 2×10^6 cells by the guanidine-isothiocyanate method. cDNA synthesis was performed using oligo dT-primers, and Ig-variable regions were amplified by polymerase chain reaction (PCR) using primers of the ImmunoZap cloning system (Stratagene). Sequences of heavy and light chain PCR products were determined by solid-phase DNA sequencing of both strands on an automated DNA sequencer and compared to sequences in the EMBL and Genebank databases. IgG subclasses were determined by ELISA as described.[2]

[a] This work was supported by DFG Ri 707/1-1 and DFG Bo 849/6-1.

RESULTS

Six of the MICA were of subclass IgG1 and one (MICA 5) was of subclass IgG3. Four of the antibodies used lambda (MICA 1, 2, 3, and 6) and three (MICA 4, 5, and 7) used kappa light chains. Sequence comparison demonstrated that all MICAs originated from different B-cell clones. Deduced amino acid V_H sequence is shown in FIGURE 1 in comparison with the closest published germline sequence. Three antibodies used a V_H4 gene, two used V_H1 genes, and one used the V_H5 gene V_H251. The MICA 5 heavy chain could not be amplified with existing primers. Mutations were concentrated in the CDR regions and in FR3. D_H and J_H segments identified in the MICA were heterogeneous. Although all antibodies were directed to GAD and two pairs of them may bind an identical epitope, homology of the heavy chain variable domain (V_H, D_H, and J_H) was less than 80% even in antibodies sharing the same V_H family. V_H similarity to published germline nucleotides ranged from 97% in MICA 3 to 88% in MICA 2 (TABLE 1). N-regions were identified in all heavy chains, and the J_H region was mutated in all MICAs except MICA 4. High mutation rates in J_H regions were found for MICA 2 and 3 (TABLE 1). Light-chain Ig gene usage is summarized in TABLE 1. The three kappa chains represented by the MICA were of three different subtypes. Among the four lambda chains, however, three used V_L2 and $J_L2,3$ genes. MICA 1, 3, and 6, which all used a V_L2 gene in combination with $J_L2,3$ or J_L1, revealed a homology of their lambda variable domain of >91%. Homology of the MICA light chains to published germline V_L sequences ranged from 94.1% to 97.2%. For MICA 1, 3, and 6 lambda chains, the closest partners found in the databases were rearranged light-chain sequences. Two of these sequences were isolates from a person with the autoimmune disease systemic lupus erythematodes (SLE).

TABLE 1. Homology to Published Germline Sequences

MICA	V	Identity	D	J	Identity	C
			Heavy Chain			
MICA 1	V_H4	95.0%	DK1	J_H2	96.0%	IgG1
MICA 2	V_H4	88.0%	DLR2	J_H4b	92.3%	IgG1
MICA 3	V_H1	97.0%	DK4	J_H5b	83.4%	IgG1
MICA 4	V_H5	95.6%	DN1	J_H5b	100 %	IgG1
MICA 6[a]	V_H4	96.5%	DRL2	J_H6b	94.4%	IgG1
MICA 7	V_H1	93.5%	DXP1′	J_H4b	94.3%	IgG1
			Light Chain			
MICA 1	V_L2	98.6%		$J_L2,3$	97.4%	Lambda
MICA 2	V_L1	97.9%		$J_L2,3$	97.4%	Lambda
MICA 3	V_L2	97.3%		J_L1	100 %	Lambda
MICA 4	V_K3	97.9%		J_K4	94.4%	Kappa
MICA 5	V_K2	97.0%		J_K1	91.7%	Kappa
MICA 6	V_L2	98.6%		$J_L2,3$	94.3%	Lambda
MICA 7	V_K1	95.9%		J_K1	94.5%	Kappa

```
VH1

DP 7    QVQLVQSGAEVKKPGASVKVSCKASGYTFT[SYYMH]WVRQAPGQGLEWMG[IINPSGGSTSYAQKFQG]RVTMTRDTSTSTVYMELSSLRSEDTAVYYCAR
MICA 3  ------------------------------[N--IY]--------------[------------EE---]---L----------------------L-----

DP 10   QVQLVQSGAEVKKPGSSVKVSCKASGGTFS[SYAIS]WVRQAPGQGLEWMG[GIIPIFGTANYAQKFQG]RVTITADESTSTAYMELSSLRSEDTAVYYCAR
MICA 7  -----------------------------N[I-SF-]--------------[-----YRP-----N---]------------V--D-----D---M-----

VH 4

DP 71   QVQLQESGPGLVKPSETLSLTCTVSGGSIS[SYYWS]WIRQPPGKGLEWIG[YIYYSGSTNYNPSLKS]RVTISVDTSKNQFSLKLSSVTAADTAVYYCAR
MICA 1  -----------------------------N[-----]--------------[F-SH--I------R--]----L-------I---------G---------
MICA 2  -----------------------------T[-----]-----A----DF--[RVNIA---------T-]--SM----A-KH---R-T------------I

DP 79   QLQLQESGPGLVKPSETLSLTCTVSGGSISS[SYYWG]WIRQPPGKGLEWIG[SIYYSGSTYYNPSLKS]RVTISVDTSKNQFSLKLSSVTAADTAVYYCAR
MICA 6  -------------------------------[RN---]--------------[--H--R----L-----]--------------------------------

VH 5

DP 73   EVQLVQSGAEVKKPGESLKISCKGSGYSFT[SYWIG]WVRQMPGKGLEWMG[IIYPGDSDTRYSPSFQG]QVTISADKSISTAYLQWSSLKASDTAMYYCAR
MICA 4  ----------------------A---RLA-[---A-]--------------[----D----T-------H]---------N---------R-----I-----
```

FIGURE 1. Deduced amino acid sequences of V_H regions of the MICAs compared with published germline counterpart grouped in V_H families. Sequences of MICAs belonging to the same family are shown together with the closest published germline counterparts.[6] CDR1 and CDR2 are boxed.

CONCLUSIONS

All MICA represent different primary B-cell clones, and MICA pairs with similar epitope recognition use V, D, and J regions of different families. In the heavy chains there seemed to be a bias towards the use of the V_H4 family. Expression of monomeric Fab fragments on the surface of lambda phages is currently under way. Surprisingly, the largest gene family, V_H3, was not represented among the MICA. This is in strong contrast to autoantibody repertoires from all other autoimmune diseases described so far and to the normal human Ig repertoire where V_H3 gene usage is very common.[4,5] In the light chains a clear bias towards the use of V_L2 lambda together with $J_L2,3$ lambda genes was found. Two of these light chains varied only slightly compared to lambda isolates from a patient with SLE. Besides the IgG-constant region used by all MICA, mutations clustered in the CDR regions as well as J sequences different from germline argue for antigen-driven immune responses to GAD in type 1 diabetes.

REFERENCES

1. Baekkeskov, S., H. J. Aanstoot, S. Christgau, A. Reetz, M. Solimena, M. Cascalho, F. Folli, H. Richter-Olesen & P. DeCamilli. 1990. Nature **347:** 151.
2. Richter, W., J. Endl, T. H. Eiermann, M. Brandt, R. Kientsch-Engel, C. Thivolet, H. Jungfer & W. A. Scherbaum. 1992. Proc. Natl. Acad. Sci. USA **89:** 8467.
3. Richter, W., Y. Shi & S. Baekkeskov. 1993. Proc. Natl. Acad. Sci. USA **90:** 2832.
4. Sanz, I., P. Casali, J. W. Thomas, A. L. Notkins & J. D. Capra. 1989. J. Immunol. **142:** 4054.
5. Pascual, V. & J. D. Capra. 1990. Adv. Immunol. **49:** 1.
6. Tomlinson, I. M., G. Walter, J. D. Marks, M. B. Llewelyn & G. Winter. 1992. J. Mol. Biol. **227:** 776.

Expression and Regulation of Clonally Ignorant B Cells

JAMES W. THOMAS AND CHRYS HULBERT

Departments of Medicine, Microbiology, and Immunology
Vanderbilt University
Nashville, Tennessee 37232-2681

Administration of insulin for the treatment of diabetes mellitus in humans and immunization of inbred animals results in the production of antibodies that bind the autologous protein. This observation has led to the conclusion that anti-insulin B cells are not clonally eliminated or rendered unresponsive.[1] Such B cells are often referred to as "clonally ignorant," and B cells of this type may provide a pathway to induce autoreactive T cells and to contribute to autoimmune diseases. In this study, we used an anti-insulin mAb from BALB/c mice to examine the genetic origin and structure of clonally ignorant B cells.

Anti-insulin mAb 125 expresses a V_H gene from the small V_H-IX family, and PCR on liver DNA was used to identify the germline V_H-IX progenitor for the expressed anti-insulin. The expressed V_H gene differs in four nucleotides that result in four amino acid replacements when compared to germline.[2] Two replacements are closely associated with CDRH2, glycine to aspartic acid at residue 49 and valine for glutamic acid at residue 56. To directly determine the role of these somatic mutations in insulin binding, heavy chain loss variants of mAb 125 were selected and transfected with a eukaryotic expression vector. This vector contains a human IgG1 constant region and a cloning site for expression of VDJ segments.[3] Oligonucleotide site-directed mutagenesis was used to return the residues at position 49 and 56 to their germline structure. An enzyme-linked immunoassay was used to compare binding of engineered anti-insulin mAb. The results of these experiments are summarized in FIGURE 1. Both murine and human chimeric mAb bind insulin with comparable avidity (I-50, μg/ml of human insulin that inhibits 50% of binding). When individual mutations at positions 49 and 56 are returned to germline, the mAbs bind insulin but with a reduced avidity. When both residues 49 and 56 are returned to germline configuration, the antibody no longer binds insulin. Experiments using human rheumatoid factors to increase valence of the unmutated antibody have not revealed insulin binding.

These data show that somatic mutation is essential to generate high-affinity insulin binding seen in mAb 125. A single mutation at either residue 49 or 56 is sufficient to generate insulin binding from a V_H structure that does not bind insulin. This finding suggests that the progenitor for mAb 125 may be a memory cell generated in response to another antigen. Further, these data indicate that acquisition of affinity maturation through several rounds of cell division and mutation is not essential for this mAb. Work is in progress to determine if use of the memory B-cell pool is a common pathway for generation of anti-insulin and other "clonally ignorant" responses.

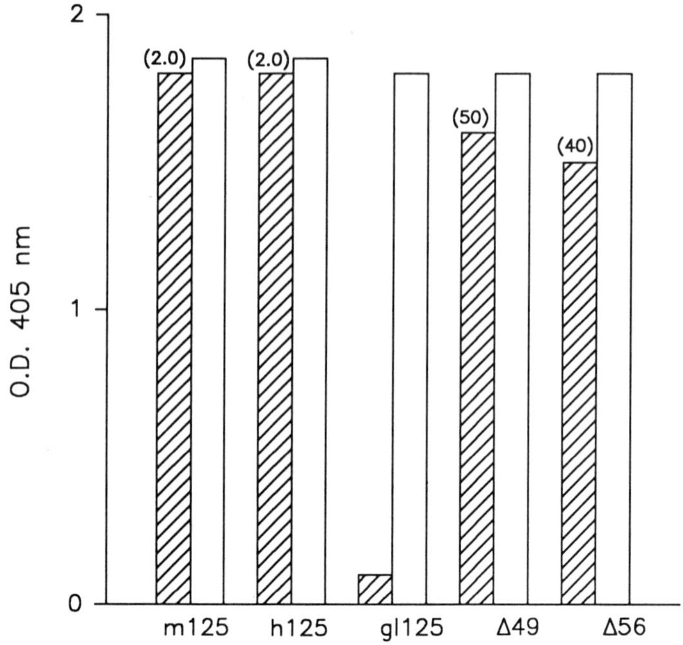

FIGURE 1. Insulin binding (*hatched bars*) and IgG production (*open bars*) are compared for five mAbs engineered from hybridoma 125. Murine (m) and human (h) 125 constructs bind insulin equally well. mAbs with either residue 49 or 56 returned to germline bind insulin but with lower avidity (I-50, μg/ml). When both residues 49 and 56 are returned to unmutated configuration (gl125), the antibody does not bind insulin. Symbols: ▨ anti-insulin (150 μg/ml); □ anti-IgH.

REFERENCES

1. Goodnow, C. G. 1992. Transgenic mice and analysis of B-cell tolerance. Annu. Rev. Immunol. **10:** 489–519.
2. Thomas, J. W. 1992. Anti-insulin and regulatory anti-idiotypic antibodies use the same germ-line V_H-IX gene. Eur. J. Immunol. **22:** 2445–2448.
3. Snyder, J. G., Q. Dinh, S. L. Morrison, E. A. Padlan, M. Mitchell, L.-Y. Yu-Lee & D. M. Marcus. 1994. Structure–function studies on anti-3-fucosyllactosamine (Lex) and galactosylgloboside antibodies. J. Immunol. **153:** 1161–1170.

Analysis of Immunoglobulin V_H Gene Repertoire by an Anchored PCR-Elisa[a]

LAURA Z. RASSENTI,[b] HITOSHI KOHSAKA,[c]
AND THOMAS J. KIPPS[b,d]

[b]*Sam & Rose Stein Institute for Research on Aging*
Division of Hematology/Oncology
Department of Medicine
University of California, San Diego
La Jolla, California 92093

[c]*Division of Immunological Diseases*
Tokyo Medical and Dental University
1-5-45 Yushima, Bunkyo-ku
Tokyo, Japan 113

INTRODUCTION

During B-cell development, immunoglobulin (Ig) genes are formed by recombination and assembly of several gene segments selected from a larger pool of related Ig gene segments.[1,2] To generate an Ig heavy chain, rearrangements involving a heavy chain variable region gene (Ig V_H gene), a diversity (D) segment(s), and a heavy chain joining (J_H) segment must occur. Although there are approximately one hundred in the human haploid genome, the Ig V_H genes are grouped into seven Ig V_H gene subgroups based on similarities in primary structure.[3–7] The rearrangement of these gene segments, combined with somatic mutation and the diversity afforded by the pairing of two chains to form an antibody molecule, provides the humoral immune system a vast and varied repertoire.

Analyses of the Ig V_H genes expressed in the human adult B-cell repertoire reveal that Ig V_H gene subgroups are not used randomly. Ig V_H genes of the Ig V_H3 subgroup are expressed by approximately half of the IgM-expressing B cells of normal adults. The second most widely represented subgroups are Ig V_H4 (17%) and Ig V_H1 (15%), followed by Ig V_H5 (7%). The Ig V_H2, Ig V_H6, and Ig V_H7 families are the least represented, each constituting less than 2% of the expressed IgM repertoire.[8–10]

Both genetic and somatic mechanisms account for the nonstochastic use of Ig V_H gene subgroups. The frequent use of Ig V_H3 genes is due in part to the relatively large number of functional Ig V_H3 genes in the human haploid genome. There also may be nonstochastic use of Ig V_H genes due to advantageous accessibility of particular Ig V_H genes to recombinase, enhanced Ig V_H gene transcriptional activation and/or homology-directed joining, as reviewed.[11] Selection for

[a] This work is funded in part by National Institutes of Health Grants CA-49870 and AG-04100. The authors also acknowledge the excellent technical assistance of Ms. Anthereca S. Edmerson.

[d] Address for correspondence: Thomas J. Kipps, M.D., Ph.D., Department of Medicine, University of California, San Diego, Clinical Sciences Building, Room 131, 9500 Gilman Dr., La Jolla, CA 92093.

or against particular Ig encoded by certain Ig V_H genes also may alter the distribution of Ig V_H gene subgroups represented in the expressed Ig repertoire. In addition, there may exist genetic polymorphism in the relative expression of individual Ig V_H genes or Ig V_H gene subgroups. Finally, transformation of an antibody-producing cell may cause gross alterations in the Ig repertoire due to outgrowth of the neoplastic B-cell clone at the expense of normal Ig-producing cells. For these reasons, repertoire analysis of the expressed Ig V_H genes may offer insight into selective forces acting upon the Ig repertoire and/or potential genetic polymorphism in Ig V_H gene expression. In addition, serial analyses may allow for sensitive detection of evolving B lymphoid dysplasia or of minimal residual B-cell neoplastic disease.

Conventional methods for analyzing the Ig V_H gene repertoire have relied on the immortalization of B cells via transformation with Epstein-Barr virus (EBV), generation of cDNA libraries, or PCR-based cloning techniques. However, these methods are cumbersome and not readily amenable for serial studies of the Ig repertoire of any one individual. What is needed is a rapid and reproducible method for analyses of the Ig repertoire.

We have developed a new technology that uses an anchored RT-PCR and an enzyme-linked immunosorbent assay (ELISA). The anchored PCR technique, developed by Loh,[12] was used recently to examine the distribution of the T-cell receptor V_γ genes in the peripheral blood of patients with rheumatoid arthritis.[13–15] We have adapted this method to analyze the expressed Ig repertoire of polyclonal or monoclonal B-cell populations. The advantages of this technique over conventional techniques are that it is rapid, nonradioactive, and quantitative, because it allows the measurement of the relative concentration of the expressed Ig V_H gene subgroups using standard curves generated using cloned Ig V_H genes representing each of the major Ig V_H gene subgroups.

MATERIALS AND METHODS

Peripheral Blood Mononuclear Cells

Heparinized blood was obtained from normal donors or from patients with B-cell chronic lymphocytic leukemia (CLL).[16] Peripheral blood mononuclear cells (PBMCs) were isolated via ficoll-hypaque density gradient centrifugation. Washed PBMCs were used immediately or viably frozen in fetal calf serum containing 10% dimethylsulfoxide (Sigma, St. Louis, MO) for storage in liquid nitrogen.

RNA Isolation and First-Strand cDNA Synthesis

Total RNA was isolated from 5×10^6 PBMCs using RNAzol B (Cinna/Biotex, Friendswood, TX). First-strand cDNA was synthesized from 1 to 5 μg of total RNA using an oligo-dT primer and Superscript reverse transcriptase (Life Technologies, Inc., Gaithersburg, MD). The resulting cDNA/RNA hybrids were size selected (0.5–3 kb) on agarose gels and phenol extracted. The remaining RNA was hydrolyzed by incubation of the sample at 60°C in 0.3 M NaOH for 30 minutes.

dG-Tailing of First-Strand cDNA and Anchored PCR

To amplify the expressed Ig V_H region regardless of the 5′ sequence, the cDNA was poly dG-tailed with dGTP and terminal deoxytransferase (Boehringer

Mannheim, Indianapolis, IN). To remove free dGTP, the modified cDNA was precipitated twice in ammonium acetate and ethanol. One-fourth of the dG-tailed cDNA was subjected to primary anchored PCR amplification in 100 μl of 10 mM Tris HCl (pH 8.3), 50 mM KCl, 2 mM $MgCl_2$, 0.001% gelatin, 200 μM dNTPs, 1 U of Perfect-Match polymerase enhancer (Stratagene, La Jolla, CA) and 2.5 U of Ampli-Taq™ DNA polymerase (Perkin-Elmer, Cetus Instruments, Norfolk, CT). The primers for analyses of the Ig V_H used by IgM-expressing B cells consisted of a Cμ-specific antisense oligonucleotide (5′-AATTCTCACAGGA-GACGA-3′) and two anchor primers at a 9:1 mixture of an AN primer, 5′-ATTACGGCGGCCGCGGATCC-3′, and ANC primer 5′-ATTACGGCGGCCGC-GGATCCCCCCCCCCCCCC-3′.[12] The Cμ-specific primer was used at a concentration of 1 μM. The amplification consisted of 20 cycles of 95°C for 30 sec, 42°C for 30 sec, and 72°C for 1 min, except for the final cycle. This instead uses an extended 7-min incubation at 72°C. The PCR products (500–750 bp) were size fractionated on 2% NuSieve agarose (Fisher Biotech, Fair Lawn, NJ.). One-third of the purified product was used as the template for a nested PCR. For this, the reaction mixture was the same as the primary anchored PCR, except the primers used were the sense AN primer and a 5′ biotinylated Cμ antisense primer corresponding to a sequence upstream of the initial Cμ primer (5′-TTGGGGCGGATG-CACT-3′). The nested PCR employed 30 cycles at an annealing temperature of 55°C. The DNA products were purified on minicolumns (Magic PCR preps, Promega, Madison, WI) before being distributed onto individual wells of a 96-well preprepared polystyrene ELISA plate (see below) for the ELISA PCR. A schematic outline of this technique is depicted in FIGURE 1.

Preparation of ELISA Plates

Two hundred microliters of a 0.1% solution of glutaraldehyde (Polysciences, Warrington, Pa) in phosphate-buffered saline (PBS) (pH 7.2) were added to individual wells of a polychlorovinyl ELISA microtiter plate (Falcon, Oxnard, CA) at room temperature. After a 15-min incubation, the plates were washed with PBS. Immediately after washing, 200 μl of a 0.1 μg/ml solution of streptavidin (Sigma, St. Louis, MO) in PBS was added to each well. After a 2-hr incubation at 37°C, the plates were washed with PBS. Unbound streptavidin was removed further by adding 0.1% Triton X-100 (Sigma, St. Louis, MO) to each well and then incubating the plates for an additional 30 min. Subsequently, each well was filled with a 1% solution of bovine serum albumin (BSA) (RIA grade, Sigma, St. Louis MO) in PBS, and the plates were allowed to incubate for 2 hours at room temperature before being rinsed once with PBS. These plates can be used immediately for the ELISA-PCR or stored for up to one week at 4°C.

Digoxigenin Labeling of Oligonucleotide Probes and Hybridization

The oligonucleotides corresponding to the sense strand sequences of each of the six Ig V_H gene subgroup leader sequences are listed in TABLE 1. The leader sequences of the Ig V_H1 and Ig V_H7 are similar, precluding use of a leader-specific oligonucleotide to discriminate between these two subgroups. Accordingly, an oligonucleotide specific for a sense strand sequence of Ig V_H1 or Ig V_H7 genes was used to distinguish between these two subgroups (TABLE 1). Each oligonucleotide was labeled with digoxigenin in a 20-μl volume of solution containing 5 μM

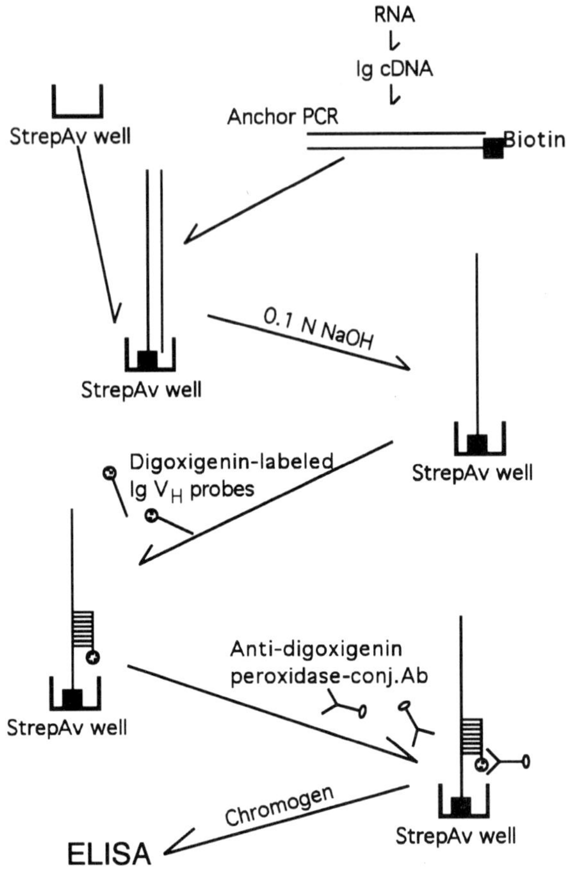

FIGURE 1. Schematic of the PCR ELISA technique. Following anchored PCR, a nested PCR reaction attaches a biotin molecule (*black box*) to the antisense strand of the PCR product. This allows the strand to bind to streptavidin (StrepAv) coated wells of an ELISA plate. The sense strand is removed by alkaline wash, allowing for hybridization of the tethered anti-sense strand with digoxigenin-labeled Ig V_H oligonucleotide probes (represented by the *ball* and *stick figures*). Hybridized and bound oligonucleotides then are detected using peroxidase-conjugated anti-digoxigenin antibodies ("anti-digoxigenin peroxidase-conj. Ab"). The peroxidase-conjugated antibodies are developed with chromogen and the plates are read using an ELISA plate reader. See text for details.

of the oligonucleotide, 200 μM potassium cacodylate, 25 mM Tris HCl (pH 6.6), 250 μg/ml BSA, 5 mM $CaCl_2$, 50 μM digoxigenin-dideoxy UTP, and 50 U of terminal deoxynucleotidyltransferase (Boehringer Mannheim) for 30 min at 37°C.

Capture of the Amplified DNA and Oligonucleotide Hybridization onto the ELISA Plate

The purified PCR product was diluted to 600 μl in 6 × SSC (1 × SSC = 0.15 M NaCl/0.015 M sodium citrate, pH 7.0), 0.1% Tween-20 and 100 μl, and then

TABLE 1. Oligonucleotide Probes Used to Assess Ig V_H Subgroup Distribution

V_H1 LEADER	ATGGACTGGACCTGG
V_H2 LEADER	ATACTTTGTTCCACGCTCCT
V_H3 LEADER	GAGTTTGGGCTGAGCTGG
V_H4 LEADER	CTGGTGGCAGCTCCCAGA
V_H5 LEADER	ATCCTCGCCCTCCTCCTAGC
V_H6 LEADER	TGTCTCCTTCCTCATCTTCC
V_H1 FRW3	AGCACAGCCTACATGGAGCTG
V_H7 FRW3	TCAGCACGGCATATCTGCAGA

distributed into six streptavidin-coated ELISA plate wells. The solution was allowed to react with the plates for 1 hour at room temperature. Subsequently, the plates were washed free of unbound PCR product with PBS. Two hundred microliters of 0.1 N NaOH were added to each well to denature the double-stranded DNA PCR product. After 10 minutes at room temperature, the plates were washed free of the PCR product sense strand with 0.1× SSC wash, followed by two consecutive washes in 6× SSC. Subsequently, 150 μl (0.5 pmole/ml) of the digoxigenin-labeled sense-strand oligonucleotide (in 6× SSC and 0.1% *N*-lauroylsarcosine) was added to each well. The plates were incubated at 65°C for 20 min and then at 42°C for 90 min. The plates were washed three times with 6× SSC, 0.1% *N*-lauroylsarcosine, and then once with blocking buffer (0.5% Genius blocking reagent (Boehringer Mannheim) in 100 mM Tris HCl (pH 7.5), 800 mM NaCl). Each well then received 100 μl of 150 μg/ml peroxidase-conjugated anti-digoxigenin antibody (Boehringer Mannheim) in blocking buffer and allowed to incubate for 30 min at room temperature. The wells were washed once with blocking buffer, twice with 100 mM Tris HCl (pH 7.5), 800 mM NaCl, and then incubated with 100 μl of tetramethylbenzidine (TMB) and peroxidase (Kirkegaard & Perry Laboratories). The reaction was stopped with 1 M *O*-phosphoric acid (H_3PO_4, Fisher). The optical densities of the solutions in the individual wells were measured at 450 nm using a ELISA microplate reader (Molecular Devices, Menlo Park, CA).

Construction of V_H1-V_H6 Leader Vector to Generate Standard Curves

To measure the amount of amplified product, we converted the optical density readings at 450 nm (OD_{450}) into concentration values by using standard curves derived from analyses of the hybridization efficiency of each of the oligonucleotides to the six major Ig V_H leader segments (TABLE 1, FIG. 2). For this, we constructed a 124-bp DNA fragment that contained the leader-specific sequences of each of the six Ig V_H gene subgroups using overlapping oligonucleotides. This 124-bp fragment was ligated into the cloning site of an *Eco*RI-*Hin*dIII cut Bluescript SK plasmid (Stratagene, La Jolla, CA) that previously had been purified free of insert sequences after digestion with the restriction enzymes. This plasmid was amplified via conventional PCR using a T3 sense primer (5′-ATTAACCCTCACTAAAG-3′) and a 5′ biotinylated antisense T7 promoter primer (5′-GATATCACTCAGCA-TAA-3′). The 270-bp band was purified via agarose gel electrophoresis, and the concentration was determined. One hundred microliters of 1:2 serial dilutions ranging from 200 fm to 1.5 fm was covalently captured onto the streptavidin-coated ELISA wells. These wells then were treated as described above. Sets of

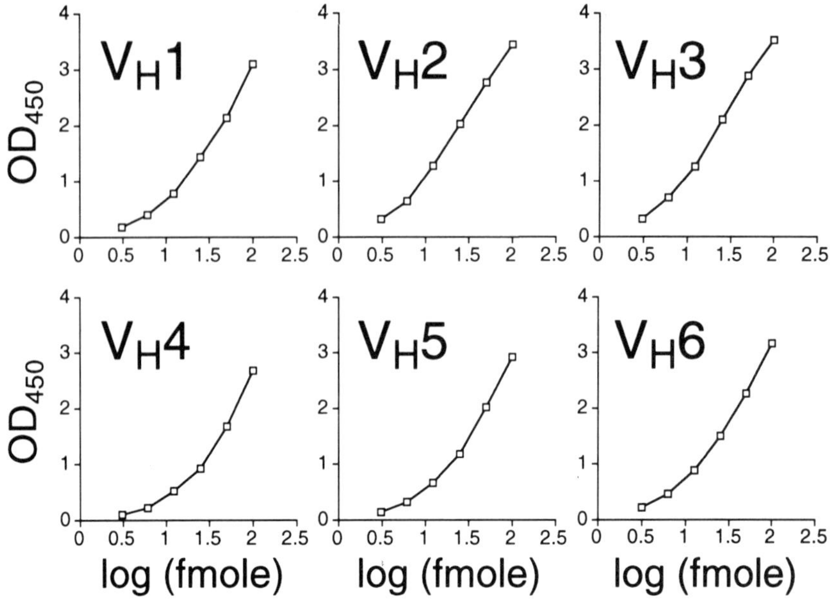

FIGURE 2. Standard curves obtained with the V_H1-6 leader constructs. The Ig V_H subgroup probe examined is indicated in the upper left-hand corner of each graph. The ordinate provides the optical density readings at 450 nm (OD_{450}). The abscissa lists the log concentration of the template in femtomoles. These standard curves are used to convert optical density readings into concentration values. See text for details.

wells coated with each set of dilutions (totaling six) were reacted with digoxigenin-labeled oligonucleotides for each of the Ig V_H leader segments. Six standard curves were generated for the hybridization efficiency of the respective oligonucleotide (FIG. 2).

RESULTS

Fidelity of Ig V_H Gene Subgroup Oligonucleotide Probes

Plasmids containing cloned sequences belonging to the seven Ig V_H families were amplified by conventional PCR using the T3 sense primer and the 5′-biotinylated antisense primer. The PCR product was distributed equally and captured onto streptavidin-coated ELISA plates. After removal of the sense strand by the 0.1 N NaOH wash, digoxigenin-labeled oligonucleotides corresponding to a leader sequence of each of the Ig V_H gene subgroups, or to a sequence in the third framework of the V_H1 or V_H7 subgroup (TABLE 1), were added to sets of wells and allowed to hybridize with the plate-tethered antisense PCR product. Each oligonucleotide leader probe hybridized specifically only to wells containing PCR products of plasmids corresponding to the appropriate Ig V_H gene subgroup. No cross-hybridization was observed between an oligonucleotide probe and a PCR product derived from an irrelevant Ig V_H gene subgroup, except for the Ig V_H1 leader oligonucleotide probe. In addition to hybridizing to the leader sequences of tethered Ig V_H1 genes, the Ig V_H1 leader oligonucleotide probe also annealed with plate-bound Ig V_H7 gene products. Plate-bound Ig V_H genes of the Ig V_H1 subgroup could be distinguished from those of the Ig V_H7 subgroup using the Ig V_H1 and V_H7 third framework oligonucleotide probes.

Use of the V_H1-V_H6 Leader Construct to Generate Standard Curves

Serial dilutions (1 : 2) of the amplified V_H1-V_H6 leader construct ranging from 200 fm to 1.5 fm were distributed onto six sets of ELISA wells. After the removal of the sense strand by alkaline wash, the digoxigenin-labeled oligonucleotide corresponding to any one of the Ig leader sequences (Ig V_H1 through Ig V_H6) was added to each set of wells. This allowed us to generate standard curves for each Ig V_H subgroup probe relating the optical density (OD_{450}) of the wells to the number of femtomoles of gene product initially added to the wells (FIG. 2). Inclusion of such standards in each experiment allows us to relate the observed OD_{450} reading to the estimated concentration of the relevant Ig V_H gene product in the test sample.

Analyses of Polyclonal and Monoclonal B-Cell Populations

We evaluated the distribution of Ig V_H gene subgroups among IgM-expressing blood B cells of normal adults. From the number of mononuclear cells typically isolated from less than 5 ml of blood (5×10^6 cells), we isolated RNA and generated cDNA for the ELISA-PCR as described in the MATERIALS AND METHODS section. The distribution of Ig V_H gene subgroups assessed by this method of three separate donors (A, B, and C, FIG. 3) compares favorably with that noted using conventional techniques to assess the Ig V_H gene repertoire of normal adults.[8–10] Most (range 41–59%) of the Ig V_H genes of each donor's sample were identified as members of the Ig V_H3 subgroup (FIG. 3). The next largest subgroup was Ig V_H4, accounting for 19 to 23% of the expressed Ig V_H genes, followed by the Ig V_H1 and Ig V_H5 gene subgroups, each accounting for 15 to 17% and 7 to 11% of the repertoire, respectively. The V_H2, V_H6, and V_H7 subgroups each accounted for less than 3% of the detected Ig V_H genes.

The variation between different individuals in the relative expression of Ig V_H subgroups exceeded the variation noted between disparate samples of the same individual. For example, donor C consistently had a higher proportion of Ig V_H4 genes detected than donor B, whereas donor B had a higher proportion of Ig V_H1 genes (FIG. 3). Preliminary studies of blood samples from four sets of identical twins, moreover, revealed that the variation between sets of twins exceeds that noted between twin pairs (data not shown). The latter suggest that there may be a genetic basis for individual-to-individual variations in the noted distributions of Ig V_H gene subgroups used by IgM-expressing blood B cells.

We also examined the ability of this technique to assess the Ig V_H gene subgroup expressed in B-cell leukemias. First, representative CLL B-cell populations known to express Ig of each of the seven Ig V_H gene subgroups were analyzed. In each case, the PCR-ELISA technique correctly detected the Ig V_H gene subgroup expressed by the leukemia cell population. Moreover, each analysis revealed a marked skewing in the expressed Ig repertoire, the Ig V_H gene subgroup used by the leukemia cell population accounting for >99% of the detected Ig V_H gene

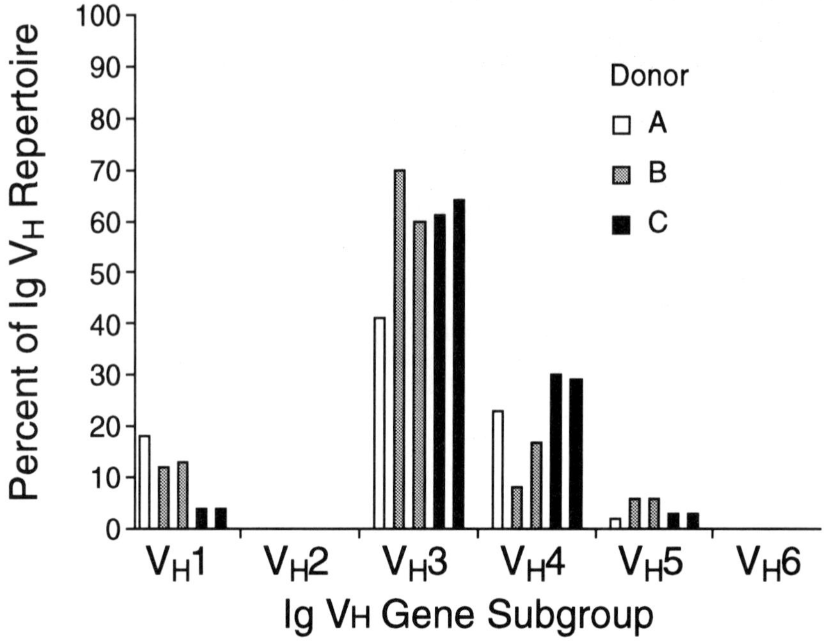

FIGURE 3. Frequency distribution of the Ig V_H gene subgroups used by IgM-expressing blood B cells of three normal donors (**A, B,** and **C**). PCR-ELISA was performed on the one blood specimen of donor A and on two blood samples of donors B and C. Listed on the ordinate is the percent of the detected Ig V_H gene repertoire that corresponds to each of Ig V_H gene subgroups, as indicated on the abscissa below each set of bar histograms. Where there are no bars, the percent of the Ig V_H repertoire corresponding to that Ig V_H subgroup was less than 2%.

subgroups. Finally, reconstitution experiments, mixing CLL B cells with PBMCs from normal adults, indicate that this technique is highly sensitive to changes in the Ig repertoire caused by a minor subpopulation of neoplastic B cells.

This method was used to examine blood samples of 11 randomly selected patients with B-cell CLL-expressing Ig of unknown Ig V_H gene subgroup. Invariably, we detected only one of the seven Ig V_H gene subgroups in the Ig repertoire of each sample. In five of the 11 samples (45%), we detected Ig that exclusively hybridized with oligonucleotides for the V_H1 subgroup. Three samples (or 27% of those tested) exclusively expressed Ig V_H3 genes, two (or 18%) expressed Ig of the V_H4 subgroup, and one (or 9%) expressed only Ig V_H2 genes.

DISCUSSION

We developed a novel technique to analyze the relative concentration of the expressed immunoglobulin V_H genes in monoclonal and polyclonal B-cell populations. This method employs anchored PCR to amplify expressed Ig V_H genes independent of genetic polymorphism or somatic mutation. The detection of Ig V_H genes is simplified by using a nested PCR that enhances the specificity of the reaction and attaches a biotin molecule to the 5′ end of the antisense strand of the PCR product. This allows us to attach the antisense strand of all amplified Ig V_H genes to avidin-coated microtiter plates, facilitating their detection using digoxinin-labeled oligonucleotide probes that, in turn, are detected with peroxidase-labeled anti-digoxinin antibodies in an ELISA. As such, this method does not employ nonlinear detection methods, such as autoradiography, making detection of the relative contribution of the expressed Ig V_H genes more direct and amenable to rapid colorimetric analysis in an ELISA plate reader.

Because this method examines the distribution of Ig V_H genes in Ig cDNA, it is not a direct measure of the relative numbers of B cells expressing a particular Ig V_H gene or Ig V_H gene subgroup. B cells may vary in their relative expression of Ig mRNA.[17] Plasma cells, committed to Ig synthesis, each synthesize much larger quantities of Ig mRNA than does a resting B cell. Moreover, B cells activated in an antigen-driven immune response may be induced to synthesize more Ig mRNA. For these reasons, the relative contribution of Ig V_H genes or Ig V_H gene subgroups may be anticipated to vary depending the state of humoral immune activation or B-cell maturation. Nevertheless, we find that the relative frequency at which Ig V_H gene subgroups are used by IgM-expressing blood B cells to be quite stable for each individual over time. Moreover, the variation in expressed Ig V_H gene subgroups apparently varies more between individuals than it does between identical twin pairs. Perhaps this may reflect the high turnover of B cells and the fact that most IgM-expressing blood B cells co-express IgD. IgD-expressing tonsillar B cells, in contrast to IgD-negative B cells, generally express nonmutated Ig V_H genes and may represent primary B cells.[18,19] Further studies are needed to confirm these initial observations. The PCR-ELISA technique described here should allow for rapid longitudinal analyses of the Ig repertoire used by B cells expressing any one of the major Ig isotypes.

In addition, this technique is sensitive to changes in the Ig repertoire caused by B cell lymphoproliferative disease. This method faithfully detected the correct Ig V_H gene subgroup expressed by each sample of a panel of leukemic B cells expressing each of the seven major Ig V_H gene subgroups. Moreover, each of 11 previously noncharacterized blood samples from patients with B-cell CLL was found to use one Ig V_H gene subgroup, namely that expressed by the leukemia B-cell clone. Finally, reconstitution experiments, mixing leukemia B cells with PBMCs from normal individuals, confirm the sensitivity of this technique for detecting changes in the Ig repertoire resulting from a minor subpopulation of neoplastic B cells. Together, these studies indicate that the PCR ELISA technique should have utility for assessing blood or tissue B cells for clonal imbalances that are associated with B-cell lymphoproliferative disease.

SUMMARY

We developed a novel technique to analyze the relative concentration of the expressed immunoglobulin (Ig) V_H genes using an enzyme-linked immunosorbent assay (ELISA). Expressed Ig cDNA are amplified via anchored PCR and then subjected to a "nested PCR" reaction that attaches biotin to the 5′ end of the antisense strand. This allows us to tether the antisense strand of PCR products onto avidin-coated ELISA plates. Digoxigenin-labeled oligonucleotides specific for the leader sequence sense strand of each major Ig V_H gene subgroup are used to probe the plate-tethered, alkaline-denatured, and single-stranded antisense cDNA. Bound probes then are detected with alkaline-phosphatase-conjugated anti-digoxigenin antibodies. Using this method, we assessed the distribution of Ig V_H genes used by IgM-expressing blood B cells of normal adults. We found the predominant subgroup is V_H3, representing approximately half (range 41–59%) of the expressed IgM repertoire. The next largest subgroups used are V_H4 (19–23%), V_H1 (15–17%), and V_H5 (7–11%). The V_H2, V_H6, and V_H7 subgroups each constitute less than 3% of the expressed IgM repertoire. These results agree with those obtained using traditional and more laborious methods that analyze the distribution of Ig clones in cDNA libraries. In addition, we find that this method compares favorably in sensitivity and specificity to more conventional techniques for assessing the clonality of blood or tissue B-cell populations. This rapid and nonradioactive method should have utility for assessing the Ig repertoires expressed by normal, autoimmune, or neoplastic B-cell populations.

REFERENCES

1. TONEGAWA, S. 1983. Somatic generation of antibody diversity. Nature **302:** 575–581.
2. ALT, F. W., T. K. BLACKWELL & G. D. YANCOPOULOS. 1987. Development of the primary antibody repertoire. Science **238:** 1079–1087.
3. RECHAVI, G., D. RAM, L. GLAZER, R. ZAKUT & D. GIVOL. 1983. Evolutionary aspects of immunoglobulin heavy chain variable region (V_H) gene subgroups. Proc. Natl. Acad. Sci. USA **80:** 855–859.
4. PASCUAL, V. & J. D. CAPRA. 1991. Human immunoglobulin heavy-chain variable region genes: Organization, polymorphism, and expression. Adv. Immunol. **49:** 1–74.
5. KODAIRA, M., T. KINASHI, I. UMEMURA, F. MATSUDA, T. NOMA, Y. ONO & T. HONJO. 1986. Organization and evolution of variable region genes of the human immunoglobulin heavy chain. J. Mol. Biol. **190:** 529–541.
6. TOMLINSON, I. M., G. WALTER, J. D. MARKS, M. B. LLEWELYN & G. WINTER. 1992. The repertoire of human germline V_H sequences reveals about fifty gorups of V_H segments with different hypervariable loops. J. Mol. Biol. **227:** 776–798.
7. VAN DIJK, K. W., F. MORTARI, P. M. KIRKHAM, H. W. J. SCHROEDER & E. C. MILNER. 1993. The human immunoglobulin V_H7 gene family consists of a small, polymorphic group of six to eight gene segments dispersed throughout the V_H locus. Eur. J. Immunol. **23:** 832–839.
8. HUANG, C., A. K. STEWART, R. S. SCHWARTZ & B. D. STOLLAR. 1992. Immunoglobulin heavy chain gene expression in peripheral blood B lymphocytes. J. Clin. Invest. **89:** 1331–1343.
9. COOK, G. P., I. M. TOMLINSON, G. WALTER, H. RIETHMAN, N. P. CARTER, L. BULUWELA, G. WINTER & T. H. RABBITTS. 1994. A map of the human immunoglobulin V_H locus completed by analysis of the telomeric region of chromosome 14q. Natl. Genet. **7:** 162–168.
10. HUANG, C. & B. D. STOLLAR. 1993. A majority of Ig H chain cDNA of normal human adult blood lymphocytes resembles cDNA for fetal Ig and natural autoantibodies. J. Immunol. **151:** 5290–5300.
11. STEWART, A. K. & R. S. SCHWARTZ. 1994. Immunoglobulin V regions and the B cell. Blood **83:** 1717–1730.
12. LOH, E. Y., J. F. ELLIOTT, S. CWIRLA, L. L. LANIER & M. M. DAVIS. 1989. Polymerase chain reaction with single-sided specificity: Analysis of T cell receptor delta chain. Science **243:** 217–220.
13. KOHSAKA, H., P. P. CHEN, A. TANIGUCHI, W. E. OLLIER & D. A. CARSON. 1993. Divergent T cell receptor gamma repertoires in rheumatoid arthritis monozygotic twins. Arthritis Rheumatol. **36:** 213–221.
14. KOHSAKA, H., P. P. CHEN, A. TANIGUCHI, W. E. OLLIER & D. A. CARSON. 1993.

Regulation of the mature human T cell receptor gamma repertoire by biased V-J gene rearrangement. J. Clin. Invest. **91:** 171–178.

15. Kohsaka, H., A. Taniguchi, P. P. Chen, W. E. Ollier & D. A. Carson. 1993. The expressed T cell receptor V gene repertoire of rheumatoid arthritis monozygotic twins: Rapid analysis by anchored polymerase chain reaction and enzyme-linked immunosorbent assay. Eur. J. Immunol. **23:** 1895–1901.
16. Kipps, T. J. 1995. Chronic lymphocytic leukemia and related diseases. *In* Williams Hematology. E. Beutler, M. A. Lichtman, B. S. Coller & T. J. Kipps, Eds.: 1017–1039. 5th edit. McGraw-Hill. New York.
17. Kipps, T. J. 1995. Functions of B lymphocytes and plasma cells in immunoglobulin production. *In* Williams Hematology. E. Beutler, M. A. Lichtman, B. S. Coller & T. J. Kipps, Eds.: 930–938. 5th edit. McGraw-Hill. New York.
18. Kipps, T. J. & S. F. Duffy. 1991. Relationship of the CD5 B cell to human tonsillar lymphocytes that express autoantibody-associated cross reactive idiotypes. J. Clin. Invest. **87:** 2087–2096.
19. Nicholson, I. C., M. J. Brisco & H. Zola. 1995. Memory B lymphocytes in human tonsil do not express surface IgD. J. Immunol. **154:** 1105–1113.

Expression of Immunoglobulin Heavy Chain Variable Genes in B-Chronic Lymphocytic Leukemia

PABLO BERTIN, CONCEPCIÓN RISUEÑO, AND CLAUDIA MELLER

Departamento de Hematología/Oncología
Centro de Investigaciones Médicas
Escuela de Medicina
Pontificia Universidad Católica de Chile
Santiago, Chile

B-Chronic lymphocytic leukemia (B-CLL) is the most prevalent form of leukemia of the Western world and accounts for 30% of all the leukemias. B-CLL represents a malignancy of a subset of B lymphocytes that coexpress CD5 and CD19 and/or CD20 antigens.[1]

Malignant B lymphocytes, like normal B lymphocytes, express surface immunoglobulin (Ig). The variable heavy chain of immunoglobulin (V_H Ig) is encoded by three germline gene segments: variable (V_H), diversity (D), and joining (J_H) segments, which undergo somatic recombination. V_H germline genes are organized in at least seven families based on nucleotide sequence analysis. V_H genes from the same family have 80% or more sequence homology, whereas V_H genes from different families have a homology of less than 70%. The V_H1 family has 20 to 25 members; V_H2 has 5 to 10; V_H3, 25 to 30; V_H4, 6 to 10; V_H5, 2 to 3; V_H6 has 1 member; and V_H7, 6 to 8 members. V_H gene members of the multimember families are interspersed across the entire V_H complex of 1500 to 2000 kb. V_H6, with only one member, is the most D-proximal gene.[2]

Several investigators have reported V_H gene usage in B-CLL. Initially, biased usage of V_H4, V_H5, and V_H6 was reported. Analysis of the larger database demonstrates the most frequent usage of V_H3 proportional to its family size and also demonstrates an overrepresentation of the V_H4, V_H5, and V_H6 families with respect to their family size. Selected genes like 51p1 account for 60% of the V_H1 family usage, while V_H251 of the V_H5 family is rearranged in 20% of B-CLLs.[3] We analyzed the expression of V_H family genes in 31 patients on the basis of a PCR method.

Patients

A total of 31 patients, 17 males and 14 females, with an age range of 28 to 82 (mean: 64) were studied. Only one patient was below the age of 40, a woman with a diagnosis of hairy cell leukemia. All patients had blood drawn, and a Ficoll-hypaque gradient was done to obtain lymphocytes. DNA and RNA were obtained by the guanidintiocyanate method. PCR reactions were carried out with primers for six V_H families and a consensus J_H primer.

```
31 cdr3         GATGT GCGAT ATTGC AGTCG TGGTA ACTGC TACGG GGTGG ACTAC
                        ||| ||||  ||| | |||||  |||| |||
 DLR 2                 gGAT ATTGt AGTgG TGGTA gCTGC TACtc>

15 cdr3         GATGT GCGAT CTTGC AGTCG TGGTA ACTGC TACGG GGTGG ACTAC
                        |||  |||  ||| | |||||  |||| |||
DLR 2                  gGAT aTTGt AGTgG TGGTA gCTGC TACtc>

7 cdr3          TTCGG GGGAG AGGGT TCGGG GAGTT ATTAT AAACC GTACT ACTAC TACTA CGGTA TGGAC
                            | ||| ||||| ||||| ||||| ||
DXP' 1                      AtGGT TCGGG GAGTT ATTAT AA>

19 cdr3         GATCC ACCCC GCCGG TACGA CTAC
                   || |||
DIR                CC ACC>

30 cdr3         CTTTA CTATT ACGAT TTTTG GAGTG GTCCC TTTGG CGGGA ACTGG TTCGA CCCC
                       |||| ||||| ||||| ||||| |
DXP 4                 gTATT ACGAT TTTTG GAGTG G>

25 cdr3         GGCGG GACGG GTGCA GTGTC TGGAG AGGCT GAGTA C
                                | |||||
DK 5                            A GTGTC>

12 cdr3         TATTA CTGTG CGAAA GATAT GTCTC TGCGA GATTA CGATA TTTTG ACTGG TTATT TAGAC TATTA CTACG GTATG GACGT C
                                                     |||| ||||| ||||| ||||| ||| | || |
DXP 1                                                ATTA CGATA TTTTG ACTGG TTA-T TAtA>

17 1 cdr3       GGGGT CGAAC CAAGC CCGCT CTCCC CATCG CAGGT CAAGA TC
                      ||||| ||
DXP'1            <ccc CGAAC CA

24 3 cdr3       GATGT AAACT TTGGG ACCCT TGACT CC
                                      | ||||
D 21/05                               T TGAC

3 6  cdr3       CGCCG TCACA TGTAT TACTA TGTAG ATGAT GGTTA TTATT TTGAC TAC
                             |||| ||||| |||
D6                           GTAT TACTA TGT>

32 cdr3         GATAT GTTCC TGCGA GATTA CGATA TTTTG ACTGG TTATT TAGAC TATTA CTACG GTATG
                                   |||| ||||| ||||| ||||| |||||
D21/05                          g tATTA CGATA TTTTG ACTGG TTATT>
```

FIGURE 1. Alignment of 11 CDR3 DNA sequences and their D gene usage. The length of these CDR3 sequences varies between 24 and 81 nucleotides.

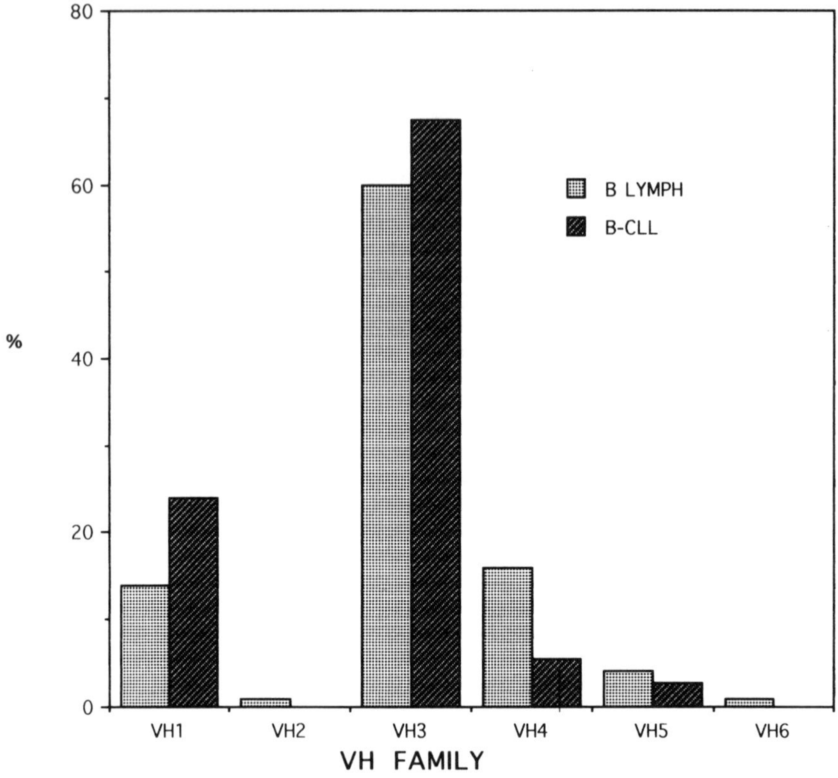

FIGURE 2. Relative frequency of V_H family usage in B-CLL and EBV-transformed PBLs. Our 31 B-CLL cases compared with EBV-transformed monoclonal B-cell lines (B lymphs, 97 lines) that represent the normal repertoire.

PCR Conditions

A reaction volume of 50 μl, with 50 mM KCl; 10 mM Tris-HCl, pH 8.4; 2.5 mM $MgCl_2$; and 200 μM dNTPs was used. Each primer was 0.5 μM and 0.5 μg DNA and 0.5 U of Taq pol were used. The temperatures of the reaction were 94°C for 4 min, 68°C for 1 min 30 sec, 94°C for 1 min (30 cycles), and 72°C for 10 min.

Cloning and Sequencing of PCR-Amplified Material

PCR products were phenol extracted and ethanol precipitated and then ligated into pT7T3 18U *Eco*RI/BAP vector (Pharmacia LKB, NJ). D_H5a-competent cells were transformed (BRL, Gaithersburg, MD). White plates were grown in broth medium and lysed by alkali medium. DNA sequencing was done using Sequenase version 2.0 (USB, Cleveland, OH). Primers for the double-strand sequencing were T3 and T7 (Pharmacia LKB, NJ).

PCR products were obtained as follows: $V_H1 = 9$, $V_H2 = 0$, $V_H3 = 25$, $V_H4 = 2$, $V_H5 = 1$, and $V_H6 = 0$. In 25 patients, only one PCR product was obtained; six cases had two bands. CDR3 sequence analysis of 11 patients so far show a mean length of 15.8 codons, with a range of 8 to 27 (FIG. 1).

V_H gene usage in this group of patients seems to be related to family size (FIG. 2). CDR3 length is longer than the average of fetal cells, which is 11 codons. Our findings suggest that the leukemic cells in B-CLL are more mature than fetal cells.

REFERENCES

1. FOON, K. A., K. R. RAI & R. P. GALE. 1990. Chronic lymphocytic leukemia: New insights into biology and therapy. Ann. Intern. Med. **113(7):** 525–539.
2. BERMAN, J. E., S. J. MELLIS, R. POLLOCK, C. L. SMITH, H. SUH, B. HEINKE, C. KOWAL, U. SURTI, L. CHESS, C. R. CANTOR & F. ALT. 1988. Content and organization of the human Ig V_H locus: Definition of three new V_H families and linkage to the Ig CH locus. EMBO J. **7(3):** 727–738.
3. SCHROEDER, H. W. & G. DIGHIERO. 1994. The pathogenesis of chronic lymphocytic leukemia: Analysis of the antibody repertoire. Immunol. Today **15(6):** 288–294.

In Vivo Isotype Class Switching in CD5$^+$ Chronic Lymphocytic Leukemia B Cells[a]

MARIELLA DONO,[b] SHIORI HASHIMOTO,[b]
MANLIO FERRARINI,[c] AND NICHOLAS CHIORAZZI[b]

[b]*Departments of Medicine*
North Shore University Hospital and
Cornell University Medical College
Manhasset, New York 11030

[c]*Servizio di Immunologia Clinica*
Istituto Nazionale per la Ricerca sul Cancro
Genova, Italy

Antigen specificity in B cells is accomplished through the assembly of distinct V, D, and J gene segments of heavy (H) and light (L) chain genes.[1] These rearrangements, and any accompanying N additions or deletions, are unique to each individual B-cell clone. Thus, these molecular "signature" sequences can be used to detect the association of a particular rearranged V gene with various H-chain classes. In this study we analyzed the circulating B cells of four patients with CD5$^+$ B CLL whose leukemic clones produce IgG antibodies to determine if their CLL-unique V_H sequences were present in B cells producing IgM and IgA.

MATERIALS AND METHODS

RNA Preparation, PCR Analyses, and DNA Sequencing

Total RNA was reverse-transcribed using oligonucleotides specific to the constant (C) region of the human μ, γ, and α H chains. These cDNAs then were amplified twice by polymerase chain reaction (PCR), first using a V_H family-specific and a C_H-specific primer pair followed by a primer specific for the FR2 region of the same V_H plus a J_H consensus primer (FIG. 1). To "fingerprint" the CDR3 length of various Ig V_H genes, the nested J_H primer was end-labeled with [$^{32}\gamma$]phosphorus using T4 polynucleotide kinase.[2,3] After amplification, purified "cold" PCR products were ligated into TA vector, used to transform competent *E. coli,* and sequenced in both directions by the dideoxy chain termination method.

[a] These studies were supported in part by US Public Health Service Grant #AI 10811 from the National Institutes of Health NIAID, by the Joseph Eletto Leukemia Research Fund, the Richard and Nancy Leeds Fund of the Department of Medicine of North Shore University Hospital, and by CNR-ACRO.

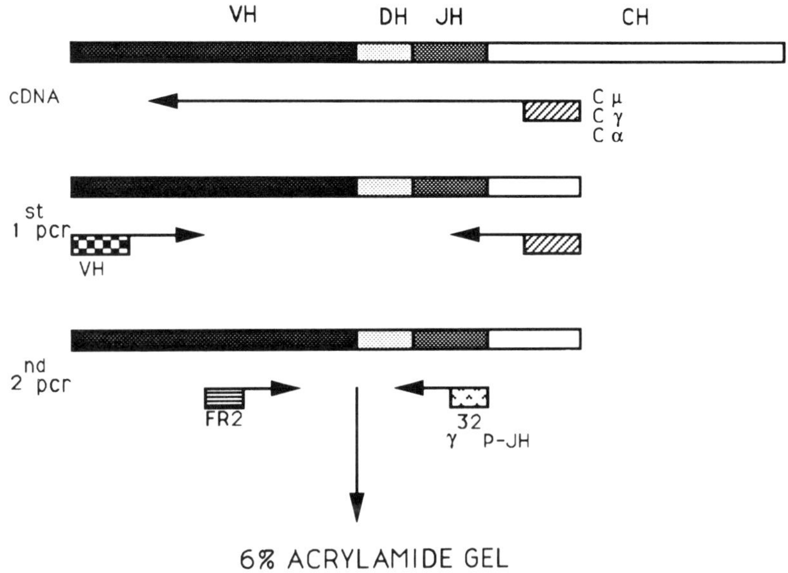

FIGURE 1. The nested PCR procedure used to detect IgM- and IgA-producing B cells that share CDR3 length with the IgG-producing CLL B-cell clone.

RESULTS

Detection of CLL-Related CDR3 Lengths among M and A Isotypes

In a previous study,[4] we characterized the rearranged Ig V_H genes of the IgG-producing B CLL samples used in this work. B cells from the same samples were analyzed for the presence of cells expressing CDR3 H-chain lengths, characteristic of the IgG CLL, among either pre-switch (IgM) or post-switch (IgA) isotypes. FIGURE 1 depicts the molecular approach used. The results of this nested "hot" PCR revealed that, in all samples, CDR3 lengths identical to the CLL cells were associated with μ and/or α H chains. These data strongly suggest a clonal relationship between the IgG-producing CLL clones and the non-IgG producing "normal" B-cell clones.

Sequence Analyses of IgM and IgA cDNA Clones

IgM and IgA cDNA clones generated from cold PCR products of the four patients were sequenced to further establish their clonal relationship with the leukemic B cells (TABLE 1). These analyses indicated that in all four patients the IgM cDNA clones shared identical N insertions and DJ point mutations with the IgG^+ CLL clone. In addition, in three of these cases, IgA cDNA clones displayed the same sequence. Furthermore, other occasional mutations were also observed, some of which led to amino acid replacements in the translated protein (TABLE 1).

TABLE 1. Sequence Analyses of Molecular Clones Derived from IgM- and IgA-Producing Cells of Patients #39 and #57

	CDR2		FR3		CDR3	
	Clone M	Clone A	Clone M	Clone A	Clone M	Clone A
Patient #39						
Position	–	220; 222	293	–		
Point mutation	–	T → C; T → G	A → T	–		
Replacement	–	Y → Q	Q → L	–		
Charge of AA	–	+ → –	– → +	–		
Patient #57						
Position	212; 244	212; 244	–	337	381	381
Point mutation	G → A; A → T	G → A; A → T	–	G → C	T → C	T → C
Replacement	S → N; T → S	S → N; T → S	–	Y → L	–	–
Charge of AA	– → –; – → –	– → –; – → –	–	+ → +	–	–

REFERENCES

1. Alt, F. W., T. K. Blackwell & G. D. Yancopoulos. 1987. Development of the primary antibody repertoire. Science **238:** 1079–1087.
2. Deane, M. & J. D. Norton. 1991. Immunoglobulin gene "fingerprinting": An approach to analysis of B lymphocyte clonality in lymphoproliferative disorders. Br. J. Haematol. **77:** 274–281.
3. Hingorani, R., I. H. Choi, P. Akolkar, B. Gulwani-Akolkar, R. Pergolizzi, J. Silver & P. K. Gregersen. 1993. Clonal predominance of T cell receptors within the $CD8^+$ $CD45RO^+$ subset in normal human subjects. J. Immunol. **151:** 5762–5769.
4. Hashimoto, S., M. Wakai, S. L. Allen, S. M. Lichtman, P. Schulman, V. P. Vinciguerra, M. Ferrarini, J. Silver & N. Chiorazzi. 1995. Somatic diversification and selection of Ig H and L chain variable region genes in IgG^+, $CD5^+$ chronic lymphocytic leukemia B cells. J. Exp. Med. **181:** 1507–1517.

Clonal Heterogeneity in Membrane Expression of CD5 in a Patient with Chronic Lymphocytic Leukemia[a]

SHIORI HASHIMOTO,[b] MARIELLA DONO,[c] VINCENT VINCIGUERRA,[c] JACK SILVER,[c] AND NICHOLAS CHIORAZZI[c]

[b]*The Department of Neurology Tokyo Women's Medical College Tokyo, Japan*

[c]*Departments of Medicine North Shore University Hospital and Cornell University Medical College Manhasset, New York 11030*

Chronic lymphocytic leukemia (CLL) is thought to involve, almost invariably, the expansion of a clone of CD5-expressing human B lymphocytes.[1] These cells have been studied in order to increase understanding about both the normal features of this small subset of human B cells and the abnormal features of the B cells expanded in the disease. Studies in mice suggest that $CD5^+$ B cells develop from progenitors distinct from those for conventional $CD5^-$ B cells,[2] and follow a separate differentiation pathway. Thus, CD5 expression is a frequently used marker of either a genetically[2] or functionally[3] derived B-cell lineage. *In vitro* studies in man have suggested that CD5 expression is not stable, however, and may represent an activation and/or differentiation marker.[4] Studies have not identified alterations in CD5 expression among human B cells *in vivo.*

Recently, our laboratory studied the IgV_H genes of a cohort of CLL patients whose $CD5^+$ B cells produce IgG.[5] In many cases, these B cells used V_H genes that had undergone somatic diversification, either at the time of recombination or subsequently.[6] These somatic events resulted in unique V_H gene sequences that serve as unequivocal markers of the CLL clone. One of these patient's B cells (CLL #030) used the H11 V_H3 gene with characteristic mutations and N region additions. FIGURE 1 illustrates the nucleic acid (A) and amino acid (B) sequences of this B-cell clone.

Although originally this patient's B cells comprised primarily $IgG\kappa^+$ $CD5^+$ cells,[4] over a three-year period the peripheral blood B cells developed two populations of IgG-expressing cells, one co-expressing CD5 and the other not. In order to test the possibility that clonal evolution had resulted in the loss of CD5 among certain of the CLL B cells, the circulating $IgG\kappa$-expressing cells were sorted by FACS into $CD5^+$ and $CD5^-$ fractions. The IgV_H sequences of the two sets of B cells then were analyzed. Surprisingly, both fractions were found to have the

[a] Supported in part by US Public Health Service Grant #AI 10811 from the National Institutes of Health NIAID, and by the Richard and Nancy Leeds Fund of the Department of Medicine of North Shore University Hospital.

A.

```
        |                      CDR 2                               |
H11     CGT ATT AAT AGT GAT GGG AGT AGC ACA ACG TAC GCG GAC TCC GTG AAG GGC
CLL030  A.. ... .GA .C. ... ... G.C ... ... .AC ... ... ... G.. ... ... ...

       |                                                  Framework 3                                                           |
H11    CGA TTC ACC ATC TCC AGA GAC AAC GCC AAG AAC ACG CTG TAT CTG CAA ATG AAC AGT CTG AGA GCC GAG GAC ACG GCT GTG TAT TAC TGT GCA AGA
CLL030 ... ... ... ... ... ... ... ..T ... ... ... ..A ..A ... ... ... ... ... ... ... ... ... ... ... ... ... ..C ... ... ... ... ...

       |             CDR 3                   | |          Framework 4                  |
CLL030 GCT CAC TCG CCT CAC GGC AGC CAC TAT CCC TCT TGG GGC CAG GGA ACC CTG GTC ACC GTC TCC TCA
```

B.

```
            CDR 2            Framework 3                     CDR 3       FR 4
H11    RINSDGSSTTYADSVKG RFTISRDNAKNTLYLQMNSLRAEDTAVYYCAR
CLL030 S.RT..G..N...A... ................................ AHSPHGSHYPS WGQGTLVTVSS
```

FIGURE 1. **(A)** Nucleotide sequences of CLL 030. **(B)** Amino acid sequences of CLL 030.

identical signature CDR2 and CDR3 sequences of the original CD5 B-cell clone (see FIG. 1) that had been determined three years previously.

Next, the $CD5^-$ B-cell subset was analyzed for CD5 mRNA expression by PCR with primer pairs originally defined by Kasaian *et al.*[7] Using this nested PCR approach, both the $CD5^+$ and the $CD5^-$ fractions were found to contain CD5 mRNA. In a separate set of experiments, these $CD5^-$ B cells were shown to express CD5 on their surface membranes when cultured in fetal calf serum containing medium or in the presence of PMA (19% and 58% expression, respectively). Thus it is clear that *in vivo* this clone of B cells has become divergent in its ability to express membrane CD5, although CD5 surface expression can be induced by appropriate T-independent stimuli. These data have implications for the role of membrane CD5 expression as a B-cell subset marker in both experimental and clinical studies.

REFERENCES

1. WANG, C. Y., R. A. GOOD, P. AMIRATI, G. DYMBORT & R. L. EVANS. 1980. Identification of a 69,71 complex expressed on human T cells sharing epitopes with B-type chronic lymphocytic leukemia. J. Exp. Med. **51:** 1539.
2. HAYAKAWA, K., R. R. HARDY & L. A. HERZENBERG. 1985. Progenitors for Ly-1 B cells are distinct from the progenitors of other B cells. J. Exp. Med. **161:** 1554.
3. WORTIS, H. H. 1992. Surface markers, heavy chain sequences and B cell lineages. Int. Rev. Immunol. **8:** 235.
4. ZUPO, S., M. DONO, R. MASSARA, G. TABORELLI, N. CHIORAZZI & M. FERRARINI. 1994. Expression of CD5 and CD38 by human $CD5^-$ B cells: Requirement for special stimuli. Eur. J. Immunol. **24:** 14.
5. WAKAI, M., S. HASHIMOTO, M. DONO, M. OMATA, Z. M. STHOEGER, S. L. ALLEN, S. M. LICHTMAN, P. SCHULMAN, V. P. VINCIGUERRA, B. DIAMOND, M. FERRARINI & N. CHIORAZZI. 1994. IgG^+, $CD5^+$ chronic lymphocytic leukemia B cells. Production of IgG antibodies that exhibit diminished autoreactivity and IgG subclass skewing. Autoimmunity **24:** 1426.
6. HASHIMOTO, S., M. WAKAI, S. L. ALLEN, S. M. LICHTMAN, P. SCHULMAN, V. P. VINCIGUERRA, M. FERRARINI, J. SILVER & N. CHIORAZZI. 1994. Somatic diversification and selection of Ig H chain variable region genes in IgG^+, $CD5^+$ chronic lymphocytic leukemia B cells. J. Exp. Med. **181:** 1507.
7. KASAIAN, M. T., H. IKEMATSU & P. CASALI. 1992. Identification and analysis of a novel human surface $CD5^-$ B lymphocyte subset producing natural antibodies. J. Immunol. **148:** 2690.

Clonal Analysis of IgM$^+$ CD5$^+$ CLL B Cells[a]

EDWARD W. SCHETTINO,[b,c] ELIZABETH W. NEWCOMB,[c]
MILAN POTMESIL,[d] GIORGIO INGHIRAMI,[c]
AND PAOLO CASALI[b,e]

[b]*Division of Molecular Immunology*
Department of Pathology
Cornell University Medical College
New York, New York 10021

[c]*Department of Pathology and*
The Laboratory of Experimental Therapy
[d]*Department of Radiology*
New York University School of Medicine
New York, New York 10016

Chronic lymphocytic leukemia (CLL) is the most frequent form of adult leukemia in Western societies.[1-3] The disease has an annual incidence of 2.7 cases per 100,000 persons, with more than 90% occurring in persons over the age of fifty.[1,4,5] More than 95% of CLLs originate from B cells (B-CLL), while the remainder of cases are of T-cell origin (T-CLL).[4] B-CLL is thought to be characterized by the neoplastic emergence of a single lymphocyte clone that replaces the normal polyclonal B-cell population, yielding a virtually monoclonal accumulation of cells in the peripheral blood. This malignant B-cell clone is, in most cases, surface CD5$^+$, suggesting that B-CLL is a malignancy of B-1a cells.[3,6-10]

The antibodies produced by the CLL B-1a cells are mainly IgM and use a random selection of V_H genes, with an underrepresentation of members of the small V_H families, V_H2, V_H5, and V_H6, and an overrepresentation of those of the large V_H families, V_H1, V_H3, and V_H4.[11] This possibly stochastic V_H gene usage reflects the V_H gene expression observed in the B-cell repertoire of healthy humans. B-CLL J_H gene usage is also reminiscent to that observed in adult B cells; J_H4 is the most frequently used gene followed by J_H6 and J_H5 genes.[12-14] Although the CLL B-cell V_H gene usage resembles that of the normal B-cell repertoire, whether or not such expressed V_H genes are in germline configuration is an unresolved issue. The configuration of CLL B-cell V genes has been addressed by numerous groups,[2,3,6,7,10,12,13,15-18] with a total of 75 V_H sequences being reported.[19,20] Approximately three-quarters of these sequences were unmutated (germline configuration). The remaining sequences were shown to contain nucleotide differences as compared to those of the closest reported germline genes, but these putative mutations were not formally verified by further analyses of the rearranged B-cell and autolo-

[a] This work was supported by the U.S. Public Health Service Grants CA-68541 and AR-40908. This is manuscript #6 from the Division of Molecular Immunology.

[e] Address for correspondence: Paolo Casali, M.D., Division of Molecular Immunology, Department of Pathology, Cornell University Medical College, 1300 York Avenue, New York, NY 10021-4896.

gous germline DNAs,[2,3,6,7,10,12,16–18,21–25] leading to the conclusion that rearranged CLL B-cell V_H genes lack somatic point mutations.[6,7,12,18] This conclusion is consistent with the observations that the V segments of IgM autoantibodies are, in general, unmutated, and with the long-held view that B-1a cells are primordial elements lacking the machinery necessary to hypermutate the expressed Ig V genes.[20,26–28] Our laboratory has recently reported that human rheumatoid B-1a cells can make somatically mutated and antigen-selected high-affinity IgM rheumatoid factors,[19,20] showing that these lymphocytes may display the machinery necessary for somatic hypermutation and can undergo an affinity maturation response.

To investigate the usage and composition of Ig V genes in CLL B cells, we obtained peripheral blood mononuclear cells from four patients fulfilling diagnostic and immunophenotypic criteria for B-CLL. By direct immunofluorescence analysis, we established that more than 99% of B cells, as defined by surface expression of the pan B-cell Ag CD19, were surface CD5 positive, and all B cells were surface IgM positive.[29] To analyze the V_H gene segment utilization by these CLL B cells, poly$(A)^+$ mRNA was reverse-transcribed and first-strand cDNA used as a template in six separate PCR amplification reactions. Each reaction included a sense leader V_H primer, specific for the members of one of the six V_H families, in conjunction with the antisense $C\mu$ oligonucleotide primer. Consistent with the inherent polyclonality of the expressed normal B-cell repertoire, PCR amplification of cDNA transcribed from peripheral blood B cells of a healthy human yielded an amplification product using each of six V_H gene family primers.[29] Consistent with a monoclonal B-CLL composition, only one of the six V_H gene family primers, V_H6, V_H4, and V_H1, yielded an amplification product of approximately 500 base pairs (bp) in three of the four B-CLL samples, 1.19, 1.32, and 1.69, respectively.[29] Amplification of cDNA from the fourth B-CLL sample, 1.42, yielded three distinct products using the V_H1, V_H3, or V_H4 gene family primer.[29] The monoclonality or polyclonality of the CLL B-cell populations was verified by isolating genomic DNA from each of the four CLLs and separately digesting it with *Bam*H I, *Hin*d III, and *Eco*R I for Southern blot hybridization analysis. The hybridization of the filter-immobilized, digested DNA with a radiolabeled 2.2-kb genomic J_H DNA probe[30] showed that three of the CLLs displayed an Ig gene rearrangement pattern consistent with that of a monoclonal B-cell population, and the fourth a polyclonal B-cell population.[29]

The Ig V_H genes PCR amplified from each of the four B-CLL samples were cloned and sequenced, each sequence being derived from four to six independent bacterial isolates. The sequences of the Ig V_H genes in two of the three monoclonal B-CLLs, 1.19 and 1.32, contained 12 and 7 nucleotide differences when compared with those of the V_H6 and $V_H4.15$ germline genes, respectively (TABLE 1). B-CLL 1.69 expressed a Ig V_H1 gene that differed in only one nucleotide from the germline hv1051k gene (TABLE 1). The polyclonal 1.42 B-CLL expressed three distinct Ig V_H genes: Clones A and B expressed a V_H1 and V_H3 gene sequence that differed in 14 (95.2% identity) and 15 (94.9% identity) nucleotides from the germline Humighvlll and H11 genes, respectively (TABLE 1), and clone C expressed a $V_H4.18$ gene in germline configuration (TABLE 1). In each of the six clones, the multiple independent bacterial isolates analyzed yielded identical sequences.

The expressed D-J_H Ig gene region sequences were analyzed by comparison with those of the reported germline D, DIR, and J_H genes.[22,31–37] The best-fit D and/or DIR genes were identified according to the two following criteria: first, priority was given to identity with the V_H or J_H gene sequence when V_H and D, or D and J_H region sequences overlapped; and secondly, if more than one candidate D and/or DIR gene was identified, the germline D gene sequence displaying the

TABLE 1. Structure of B-CLL-Expressed Ig V_H Genes[a]

CLL	CD19/CD5 (%)		Chains H	Chains L	V_H Gene: Closest Germline Gene (family)	V_H Gene: Nucleotide % Identity	CDRs R	CDRs S	FRs R	FRs S	D Gene	J_H Gene
1.19	99.0		μ	κ	V_HVI (V_H6)	96.0	2 (2.4) $p = 0.286$	0	3 (6.1) $p = 0.042$[†]	7	Dxp4*/DA5	J_H6
1.32	99.9		μ	λ	V_H4.15 (V_H4)	97.6	1 (1.1) $p = 0.401$	0	3 (3.4) $p = 0.296$	2	Dir1*/D5′/D_{21-9}	V_H4
1.69	99.9		μ	κ	hv1051k (V_H1)	99.7	0 N/A	0	1 (0.6) $p = 0.244$	0	Dxp1	J_H6
1.42	99.9	A.	μ	κ	Humighvlll (V_H1)	95.2	4 (2.0) $p = 0.089$	0	5 (6.4) $p = 0.165$	2	D_{21-10}/Dk1/Dm1	J_H4
		B.	μ	κ	H11 (V_H3)	94.9	3 (2.5) $p = 0.240$	3	5 (7.9) $p = 0.059$	3	Dxp4/Dxp′1	J_H5
		C.	μ	κ	V_H4.18 (V_H4)	100	0 N/A	0	0 N/A	0	Dxp4	J_H4

[a] R, number of detected and (expected) replacement mutations; S, number of detected silent mutations; p, probability; [†] statistically significant; * inversion of the D gene; N/A, not applicable.

longest stretch of identity (with a minimum match of six nucleotides in a stretch of seven base pairs, or with a minimum match of five nucleotides in a row) to that of the expressed D gene was assigned. D gene utilization is summarized in TABLE 1. Two of the six expressed D segments (1.69 and 1.42 clone C) resulted from the use of a single germline D gene; two (1.19 and 1.42 clone B) resulted from the use of two different germline D genes, possibly by D-D fusion; and two (1.32 and 1.42 clone A) from the use of three different D genes. Three of the B-CLL clones used a J_H4 gene, two J_H6, and one J_H5 (TABLE 1), in full-length or truncated form. All V_H-D-J_H gene sequences contained nontemplated residues (N segment additions) that ranged in size from 4 to 15 nucleotides.[29] All clones but one, 1.32, expressed D genes that were 3′ flanked by N additions, and all clone D genes but that encoding 1.69 were also 5′ flanked by N additions. Segregation of deduced amino acid junctional V_H-D-J_H gene sequences into CDR3 and FR4 as proposed by Kabat *et al.*[38] showed that the FR4 sequences were invariable in length and relatively homogeneous in composition. In contrast, the CDR3 sequences were highly divergent in length, ranging from 14 to 22 amino acids, and composition.

We analyzed a total of six discrete B cell clones in four B-CLL cases (three monoclonal B-cell populations and one polyclonal B-cell population). Two of these clones were in germline (1.42 clone C) or virtual germline (1.69) configuration. The remaining four B-cell clones contained different numbers of nucleotide differences in their Ig V_H gene segments, when compared to those of the closest Ig germline genes. The high degree of conservation of a number of V_H genes, including H11, $V_H4.15$, and V_H6, in the human population suggested that the Ig V_H genes expressed by 1.42 clone B, and the 1.32 and 1.19 clones constituted somatically mutated forms of the germline H11, $V_H4.15$, and V_H6 V_H genes, respectively, or closely related V_H genes. To formally verify the nature of the observed nucleotide differences, we isolated the genomic DNA from patient 1.19 polymorphonuclear cells (PMNs) and selectively amplified it using a V_H6 sense leader primer and a unique anti-sense 23-bp spacer-nonamer primer.[29] The amplified DNA sequence was 100% identical to that of the reported V_H6 germline gene. To formally demonstrate that the nucleotide differences observed in the V_H segment sequences of B-CLL 1.42 clones A and B, compared with those of the germline Humighvlll and H11 genes, respectively, represented somatic point-mutations, we performed PCR amplifications using *ad hoc* designed oligonucleotide primers and genomic DNA from B cells and autologous PMNs. Two unique sets of oligonucleotide primers were synthesized for clone A: (i) the sense CDR2 and anti-sense FR3 primers, specific for the expressed V_H sequence, and differing in three and five nucleotides, respectively, from the germline gene Humighvlll sequence ("mutated" primers), and (ii) the sense CDR2 and anti-sense FR3 primers identical to the corresponding areas on the germline Humighvlll gene ("germline" primers). The "mutated" primer pair yielded an amplification product when used in PCR involving genomic DNA from the B cells but not that from autologous PMNs. Utilization of the "germline" primer combination resulted in the amplification of only PMN genomic DNA. A similar strategy was used for clone B (with one exception, the antisense FR3 primer was identical in both the germline H11 gene and the expressed sequence), yielding similar results. When analyzed, all four of the PCR products displayed the expected sequences.[29] Lastly, the antigen-selected nature of these somatic point mutations were assessed, according to the binomial distribution model and using the replacement to silent (R : S) mutation ratio values inherent to the progenitor germline gene sequence, as proposed by Chang and Casali.[39] The possibility that the excess of R mutations in the CDRs resulted solely from chance was established by the high probability values calculated (TABLE 1), enabling us to conclude that antigen did not exert any positive pressure to select replacement mutations in the B-CLL Ig V_H genes.

These findings show that CLL B cells can be not only monoclonal, as previously reported, but also polyclonal, and express V_H genes in either germline or somatically mutated configuration. In contrast with other reports, the somatically mutated B-CLLs analyzed here did not display any degree of intraclonal diversification, nor did they bear traces of selection by antigen in the expressed Ig V_H genes. Thus, CLL may represent a relatively heterogeneous disease, in which different modalities (mono or polyclonal) and mechanisms of B-cell expansion and/or selection may be at work in different patients. The possible role(s) of antigen and/or genetic accidents in the events leading to such differential B-cell expansion and/or selection need to be determined.

REFERENCES

1. GALE, R. P. & K. A. FOON. 1987. Biology of chronic lymphocytic leukemia. Sem. Hematol. **24:** 209–229.
2. KUPPERS, R., A. GAUSE & K. RAJEWSKY. 1991. B cells of chronic lymphocytic leukemia express V genes in unmutated form. Leuk. Res. **15:** 487–496.
3. SHEN, A., C. HUMPHRIES, P. TUCKER & F. BLATTNER. 1987. Human heavy-chain variable region gene family nonrandomly rearranged in familial chronic lymphocytic leukemia. Proc. Natl. Acad. Sci. USA **84:** 8563–8567.
4. FREEDMAN, A. S. 1990. Immunobiology of chronic lymphocytic leukemia. Hematol./Oncol. Clinics N. Am. **4:** 405–429.
5. TEFFERI, A. & R. L. PHYLIKY. 1992. A clinical update on chronic lymphocytic leukemia. I. Diagnosis and prognosis. Mayo Clin. Proc. **67:** 349–353.
6. RASSENTI, L. Z. & T. J. KIPPS. 1993. Lack of extensive mutations in the V_H5 genes used in common B cell chronic lymphocytic leukemia. J. Exp. Med. **177:** 1039–1046.
7. KIPPS, T. J., E. TOMHAVE, L. F. PRATT, S. DUFFY, P. P. CHEN & D. A. CARSON. 1989. Developmentally restricted immunoglobulin heavy chain variable region gene expressed at high frequency in chronic lymphocytic leukemia. Proc. Natl. Acad. Sci. USA **86:** 5913–5917.
8. LOGTENBERG, T., M. E. M. SCHUTTE, G. INGHIRAMI, J. E. BERMAN, F. H. J. GMELIG-MEYLING, R. A. INSEL, D. M. KNOWLES & F. W. ALT. 1989. Immunoglobulin V_H gene expression in human B cell lines and tumors: Biased V_H gene expression in chronic lymphocytic leukemia. Int. Immunol. **1:** 362–366.
9. MAYER, R., T. LOGTENBERG, J. STRAUCHEN, A. DIMITRIU-BONA, L. MAYER, S. MECHANIC, N. CHIORAZZI, L. BORCHE, G. DIGHIERO, A. MANNHEIMER-LORY, *et al.* 1990. CD5 and immunoglobulin V gene expression in B-cell lymphomas and chronic lymphocytic leukemia. Blood **75:** 1518–1524.
10. DEANE, M. & J. D. NORTON. 1990. Immunoglobulin heavy chain variable region family usage is independent of tumor cell phenotype in human B lineage leukemias. Eur. J. Immunol. **20:** 2209–2217.
11. SCHROEDER, H. W. J. & G. DIGHIERO. 1994. The pathogenesis of chronic lymphocytic leukemia: Analysis of the antibody repertoire. Immunol. Today **15:** 288–294.
12. CAL, J., C. HUMPHRIES, A. RICHARD & P. W. TUCKER. 1992. Extensive and selective mutation of a rearranged V_H5 gene in human B cell chronic lymphocytic leukemia. J. Exp. Med. **176:** 1073–1081.
13. EBELING, S. B., M. E. M. SCHUTTE, K. E. AKKERMANS-KOOLHAAS, A. C. BLOEM, F. H. J. GMELIG-MEYLING & T. LOGTENBERG. 1992. Expression of members of the immunoglobulin V_H3 gene families is not restricted at the level of individual genes in human chronic lymphocytic leukemia. Int. Immunol. **4:** 313–320.
14. SANZ, I. 1991. Multiple mechanisms participate in the generation of diversity of human H chain CDR3 regions. J. Immunol. **147:** 1720–1729.
15. ROUDIER, J., G. J. SILVERMAN, P. P. CHEN, D. A. CARSON & T. J. KIPPS. 1990.

Intraclonal diversity in the V_H genes expressed by $CD5^-$ chronic lymphocytic leukemia-producing pathologic IgM rheumatoid factor. J. Immunol. **144:** 1526–1530.

16. SPATZ, L. A., K. K. WONG, M. WILLIAMS, R. DESAI, J. GOLIER, J. E. BERMAN, F. W. ALT & N. LATOV. 1990. Cloning and sequence analysis of the V_H and V_L regions of an anti-myelin/DNA antibody from a patient with peripheral neuropathy and chronic lymphocytic leukemia. J. Immunol. **144:** 2821–2828.
17. SCHUTTE, M. E. M., S. B. EBELING, K. E. AKKERMANS, F. H. J. GMELIG-MEYLING & T. LOGTENBERG. 1991. Antibody specificity and immunoglobulin V_H gene utilization of human monoclonal $CD5^+$ B cell lines. Eur. J. Immunol. **21:** 1115–1121.
18. MEEKER, T. C., J. C. GRIMALDI, R. O'ROURKE, J. LOEB, G. JULIUSSON & S. EINHORN. 1988. Lack of detectable somatic hypermutation in the V region of the Ig H chain gene of a human chronic B lymphocytic leukemia. J. Immunol. **141:** 3994–3998.
19. HARINDRANATH, N., I. S. GOLDFARB, H. IKEMATSU, S. E. BURASTERO, R. L. WILDER, A. L. NOTKINS & P. CASALI. 1991. Complete sequence of the genes encoding the V_H and V_L regions of low- and high-affinity monoclonal IgM and IgA1 rheumatoid factors produced by $CD5^+$ B cells from a rheumatoid arthritis patient. Int. Immunol. **3:** 865–875.
20. MANTOVANI, L., R. L. WILDER & P. CASALI. 1993. Human rheumatoid B-1a ($CD5^+$ B) cells make somatically hypermutated high-affinity IgM rheumatoid factors. J. Immunol. **151:** 473–488.
21. PRITSCH, O., C. MAGNAC, G. DUMAS, C. EGILE & G. DIGHIERO. 1993. V gene usage by seven hybrids derived from $CD5^+$ B-cell chronic lymphocytic leukemia and displaying autoantibody activity. Blood **82:** 3103–3112.
22. FRIEDMAN, D. F., J. S. MOORE, J. ERIKSON, J. MANZ, P. GOLDMAN, P. C. NOWELL & L. E. SILBERSTEIN. 1992. Variable region gene analysis of an isotype-switched (IgA) variant of chronic lymphocytic leukemia. Blood **9:** 2287–2297.
23. PRATT, L. F., R. SZUBIN, D. A. CARSON & T. J. KIPPS. 1991. Molecular characterization of a supratypic cross-reactive idiotype associated with IgM autoantibodies. J. Immunol. **147:** 2041–2046.
24. BERTIN, P. A. & G. E. MARTI. 1992. Expression of immunoglobulin heavy chain variable gene (V_H) in B-chronic lymphocytic leukemia (B-CLL) and B-prolymphocytic leukemia (B-PLL) cell lines: "restricted" usage of V_H3 family. Ann. N.Y. Acad. Sci. **651:** 464–466.
25. HASHIMOTO, S., M. WAKAI, J. SILVER & N. CHIORAZZI. 1992. Biased usage of variable and constant-region Ig genes by IgG^+, $CD5^+$ human leukemic B cells. Ann. N.Y. Acad. Sci. **651:** 477–479.
26. RIBOLDI, P., M. T. KASAIAN, L. MANTOVANI, H. IKEMATSU & P. CASALI. 1993. Natural antibodies. *In* The Molecular Pathology of Autoimmune Diseases. C. A. Bona, K. Siminovitch, M. Zanetti & A. N. Theophilopoulos, Eds.: 45–64. Harwood Academic Publishers. Langhorne, PA.
27. KASAIAN, M. T., H. IKEMATSU & P. CASALI. 1991. $CD5^+$ B lymphocytes. Proc. Soc. Exp. Biol. Med. **197:** 226–241.
28. SANZ, I., P. CASALI, J. W. THOMAS, A. L. NOTKINS & J. D. CAPRA. 1989. Nucleotide sequences of eight human natural autoantibodies V_H region reveals apparent restricted use of V_H families. J. Immunol. **142:** 4054–4061.
29. SCHETTINO, E. W. & P. CASALI. 1995. Somatic hypermutation but lack of intraclonal diversity and antigen selection in IgM^+ $CD5^+$ CLL B cells. A multiple time point analysis. In preparation.
30. RIBOLDI, P., G. GIAIDANO, E. W. SCHETTINO, T. G. STEGER, D. M. KNOWLES, R. DALLA-FAVERA & P. CASALI. 1994. Two acquired immunodeficiency syndrome-associated Burkitts's lymphomas produce specific anti-i IgM cold agglutinins using somatically mutated $V_H4.21$ segments. Blood **83:** 2952–2961.
31. YAMADA, M., R. WASSERMAN, B. A. REICHARD, S. SHANE, A. J. CATON & G. ROVERA. 1991. Preferential utilization of specific immunoglobulin heavy chain diversity and joining segments in adult human peripheral blood B lymphocytes. J. Exp. Med. **173:** 395–407.
32. RAVETCH, J. V., U. SIEBENLIST, S. KORSMEYER, T. WALDMANN & P. LEDER. 1981.

Structure of the human immunoglobulin μ locus: Characterization of embryonic and rearranged J and D genes. Cell **27:** 583–591.

33. BIRD, J., N. GALILI, M. LINK, D. STITES & J. SKLAR. 1988. Continuing rearrangement but absence of somatic hypermutation in immunoglobulin genes of human B cell precursor leukemia. J. Exp. Med. **168:** 229–245.
34. SCHROEDER, H. W., J. L. HILLSON & R. M. PERLMUTTER. 1987. Early restriction of the human antibody repertoire. Science **238:** 791–793.
35. ICHIHARA, Y., H. MATSUOKS & Y. KUROSAWA. 1988. Organization of human immunoglobulin heavy chain diversity gene loci. EMBO J. **7:** 4141–4150.
36. BULUWELA, L., D. G. ALBERTSON, P. SHERRINGTON, P. H. RABBITTS, N. SPURR & T. H. RABBITTS. 1988. The use of chromosomal translocations to study human immunoglobulin gene organization: Mapping D_H segments with 35 kb of the $C\mu$ gene and identification of a new D_H locus. EMBO J. **7:** 2003–2010.
37. SIEBENLIST, U., J. V. RAVETCH, S. KORSMEYER, T. WALDMANN & P. LEDER. 1981. Human immunoglobulin D segments encoded in tandem multigenic families. Nature **294:** 631–635.
38. KABAT, E. A., T. T. WU, H. M. PERRY, K. S. GOTTESMAN & C. FOELLER. 1991. Sequences of Proteins of Immunological Interest. Fifth Edition. U.S. Department of Health and Human Services. Washington, DC.
39. CHANG, B. & P. CASALI. 1994. The CDR1 sequences of a major proportion of human germline Ig V_H genes are inherently susceptible to amino acid replacement. Immunol. Today **15:** 367–370.

Surface Phenotype and Immunoglobulin Heavy Chain Gene Usage in Chronic B-Cell Leukemias

WATARU IKEMATSU,[a,b,c] HIDEYUKI IKEMATSU,[d]
TERUHISA OTSUKA,[b,c] SEIICHI OKAMURA,[c]
SEIZABURO KASHIWAGI,[d] AND YOSHIYUKI NIHO[c]

[b]*Cancer Center*
[c]*First Department of Internal Medicine and*
[d]*Department of General Medicine*
Faculty of Medicine
Kyushu University
Fukuoka, Japan

Chronic B-cell leukemias are known to be heterogeneous concerning morphology, immunophenotype, and clinical course. Phenotypic heterogeneity suggests that various B-lymphoid leukemic phenotypes might encompass a relatively broad spectrum of B cells. We investigated the surface expression of leukocyte differentiation antigens and the immunoglobulin heavy chain variable region (V_H) gene among Japanese patients with chronic B-cell leukemia with special reference to CD5 antigen expression. The expression of cell surface antigens was examined by two-color flow cytometry analysis using mouse monoclonal antibodies against human leukocyte differentiation antigens as previously described.[1] CD5 was expressed on leukemic cells in approximately 80% (21/26) of cases. Results of leukocyte differentiation antigen expression by CD5 antigen expression status are summarized in TABLE 1. CD22, a restricted B-cell differentiation antigen, was expressed more frequently in CD5-negative ($CD5^-$) cases than CD5-positive ($CD5^+$) cases (80% vs. 33%), and the difference was statistically significant ($p < 0.05$ by chi-square test). Another B-cell-restricted antigen and an activation marker, CD23, is expressed in approximately two-thirds of both $CD5^+$ and $CD5^-$ cases. The myelomonocytic marker CD13 antigen was expressed on the leukemic cells in all 5 $CD5^-$ cases examined, but in none of 21 $CD5^+$ cases. Leukemic cells in $CD5^-$ cases expressed CD11b antigen, an adhesion molecule expressed primarily on monocyte, more frequently than those in $CD5^+$ cases (80% vs. 6%), and the difference was statistically significant ($p < 0.01$ by chi-square test). Other myelomonocytic antigens (CD33 and CD14) were expressed infrequently on the leukemic cells in the present study (0% and 4%, respectively). Little is known about the differential expression of CD45 isoforms, which is the common leukocyte antigen and is an enzyme of phosphatase activity, on chronic B-cell leukemia. Although CD45RA was expressed on the majority of chronic leukemic B cells, CD45RA expression level measured by the fluorescence intensity using flow cytometry was significantly higher in $CD5^-$ cases than in $CD5^+$ cases (93.6 ± 21.5 vs. 38.3 ±

[a] Address for correspondence: Wataru Ikematsu, M.D., Cancer Center, Kyushu University Hospital, 3-1-1 Maidashi, Higashi-ku, Fukuoka, 812 Japan.

TABLE 1. Expression of Differentiation Antigens in Chronic B-Cell Leukemias

	CD5$^+$		CD5$^-$		Total	
Surface Antigen	Number Positive	Number Tested (%)	Number Positive	Number Tested (%)	Number Positive	Number Tested (%)
CD22	7/21[a]	33%	4/5[a]	80%	11/26	42%
CD23	14/21	67%	3/5	60%	17/26	19%
CD13	0/21[b]	0%	5/5[b]	100%	5/26	19%
CD11b	1/18[c]	6%	4/5[c]	80%	5/23	22%
CD33	0/21	0%	0/5	0%	0/26	0%
CD14	1/21	5%	0/5	0%	1/26	4%

[a] $p < 0.05$.
[b] $p < 0.001$.
[c] $p < 0.01$ by chi-square test.

28.4, mean ± S.D.) ($p < 0.05$ by Student's t-test). These data indicate that there are differences in the expression of myelomonocytic antigens, B-lineage-restricted antigens, and the common leukocyte differentiation antigen between CD5$^+$ and CD5$^-$ chronic leukemic B cells. The differences in the expression of these antigens may reflect the biological differences between CD5$^+$ and CD5$^-$ leukemic B cells.

To investigate the structural differences of the Ig variable region in CD5$^+$ and CD5$^-$ chronic B-cell leukemia cases, we cloned and sequenced the expressed V_H gene of 10 CD5$^+$ and 5 CD5$^-$ cases using reverse transcriptase-polymerase chain reaction (PCR) amplification in conjunction with the V_H gene family specific primers and Ig constant region primers as described elsewhere.[2,3] The utilized Ig V_H gene elements and the nucleotide similarities of the expressed V_H genes to the closest published germline genes are listed in TABLE 2. V_H4 family members were found in 6 of 10 (60%) CD5$^+$ cases. A V_H3 member was used by one CD5$^+$ case. No V_H6 member was found to be used in this study. V_H1, V_H2, and V_H3 members were used by two, one, and one CD5$^+$ cases, respectively. A single member of V_H4 family gene, V_H4.18 is used in three cases with almost germline configuration (98.7 to 100% nucleotide similarities with published germline sequence). On the other hand, CD5$^-$ cases used mainly (four of five cases) V_H3 family genes without preferential utilization of the specific member. These results suggest the difference in the V_H gene utilization between CD5$^+$ and CD5$^-$ leukemic B cells.

The sequences of D segments were highly heterogeneous among the cases. Twelve of 15 D segments contained considerable stretches of nucleotides (more than 5 bases) identical to the published germline D genes. We could not identify the putative germline counterpart in three cases, and these were all CD5$^-$ cases. The DXP group gene is reported to be the most frequently used D segment both in fetal liver[4] and adult peripheral B cells.[5] Nevertheless, the DXP group is less frequently expressed in chronic B-cell leukemia in the series of our cases. It is possible that there are differences at the stage of V_H-D-J_H gene rearrangements between the origin of chronic B-cell leukemia and the majority of precursors of the normal fetal and adult B-cell repertoires.

In the previous reports of J_H segment use in normal adult peripheral blood B cells[5] and acute lymphoblastic leukemias,[6] J_H4 was used most frequently, and J_H1 and J_H2 segment were used less frequently (<1%). In the present study, four of five CD5$^-$ cases used the J_H4 segment. In contrast, the J_H4 segment was found in only two of 10 CD5$^+$ cases. J_H1 and J_H2 are used in one and two CD5$^+$ cases, respectively.

TABLE 2. Ig V_H, D, and J_H Genes in Chronic B-Cell Leukemias

Case	H, L Chains	Closest V_H Gene Family	V_H Member	Nucleotide Similarity (%)	Amino Acid Similarity (%)	Closest D	Closest J_H[a]
$CD5^+$ cases							
1	$\gamma1, \kappa$	IV	VH4.18	100	100	DN1	J_H5b
2	μ, κ	IV	VH4.18	99.3	99.0	DM2	J_H1
3	μ, λ	IV	VH4.18	98.7	98.0	DA4	J_H6b
4	μ, λ	IV	VH4.13	99.3	99.0	DK4	J_H2
5	μ, κ	I	51P1	99.3	98.0	D21/10	J_H6b
6	μ, κ	IV	VH4.11	93.1	88.7	DN1	J_H3b
7	$\gamma1, \lambda$	I	VI-4.1B	92.2	85.7	DM1	J_H4b
8	μ, κ	IV	VH4.21	92.1	86.6	DN1	J_H4b
9	μ, λ	II	VII-5	92.1	84.8	DA4	J_H5(a, b)
10	μ, κ	III	VH26c	90.5	82.7	DLR5	J_H2
$CD5^-$ cases							
11	μ, κ	III	V-A77	85.5	75.5	N.I.[b]	J_H4b
12	μ, κ	IV	VH4.21	95.9	92.8	DLR4	J_H4b
13	μ, λ	III	WHG16	95.2	94.9	DLR4	J_H5a
14	μ, λ	III	1.9III	93.7	87.8	N. I.	J_H4b
15	μ, κ	III	VH26c	90.8	83.7	N. I.	J_H4b

[a] J_H subclassification form (a, b) is according to Yamada *et al.*[5]
[b] Not identified.

These results suggest that the preferences of V_H, D, and J_H gene usage may differ between $CD5^+$ and $CD5^-$ leukemic B cells. The differences in not only surface antigen expression but also Ig gene utilization between $CD5^+$ and $CD5^-$ leukemic B cells may indicate a different lineage, maturational stage, or process of malignant transformation.

REFERENCES

1. IKEMATSU, W., H. IKEMATSU, S. OKAMURA, T. OTSUKA, M. HARADA & Y. NIHO. 1994. Surface phenotype and immunoglobulin heavy chain gene usage in chronic B cell leukemias: Expression of myelomonocytic markers in CD5 negative chronic B cell leukemia. Blood **83:** 2602–2610.
2. IKEMATSU, H., N. HARINDRANATH, Y. UEKI, A. L. NOTKINS & P. CASALI. 1993. Clonal analysis of a human antibody response II. Sequences of the V_H genes of human IgM, IgG, and IgA to rabies virus reveal preferential utilization of V_HIII segments and somatic hypermutation. J. Immunol. **150:** 1325–1337.
3. IKEMATSU, H., M. T. KASAIAN, E. W. SCHETTINO & P. CASALI. 1993. Structural analysis of the V_H-D-J_H segments of human polyreactive IgG mAb: Evidence for somatic selection. J. Immunol. **151:** 3604–3616.
4. SANZ, I. 1991. Multiple mechanisms participate in the generation of diversity of human H chain CDR3 regions. J. Immunol. **147:** 1720–1729.
5. YAMADA, M., R. WASSERMAN, B. A. REICHARD, S. SHANE, A. J. CATON & G. ROVERA. 1991. Preferential utilization of specific immunoglobulin heavy chain diversity and joining segments in adult human peripheral blood B lymphocytes. J. Exp. Med. **173:** 395–407.
6. DEANE, M. & J. D. NORTON. 1990. Immunoglobulin heavy chain variable region family usage is independent of tumor cell phenotype in human B lineage leukemias. Eur. J. Immunol. **20:** 2209–2217.

Different Clonal Origin of B-Cell Populations of Chronic Lymphocytic Leukemia and Large-Cell Lymphoma in Richter's Syndrome[a]

ANDRÁS MATOLCSY, PAOLO CASALI,
AND DANIEL M. KNOWLES

Department of Pathology
The New York Hospital–Cornell University Medical College
New York, New York 10021

INTRODUCTION

B-cell chronic lymphocytic leukemia (CLL), the most common adult leukemia in the western hemisphere, displays distinctive clinical, morphologic, and immunophenotypic features.[1] Although CLL generally is a stable disease with an indolent clinical course, morphologic transformation to clinical aggressive diffuse large-cell lymphoma (DLL), referred to as Richter's syndrome,[2] occurs in months to years after the initial diagnosis in about 10% of patients.[3] The most intriguing but still unexplained aspect of Richter's syndrome is the genetic relatedness of the neoplastic B cells to the preceding CLL B-cell populations. To address this issue, we analyzed the relationship between the CLL and DLL B cells by tissue microdissection and PCR amplifications, cloning, and sequencing of the rearranged Ig V_H genes in a patient with Richter's syndrome.

MATERIALS AND METHODS

Patient and Pathologic Samples

An 83-year-old woman with a 2-year history of CLL was admitted in January of 1989 with cervical lymphadenopathy. A diagnosis of DLL was made on the basis of a lymph node biopsy performed at the time. In this lymph node biopsy, small CLL cells were admixed with clusters and sheets of DLL cells (FIG. 1). Both CLL and DLL cell populations were surface CD5 positive.

Microdissection of Richter's Syndrome Large-Cell and Small-Cell Components, and DNA Extraction from Microdissected Tissue

Large and small cells were selectively obtained from glass-slide-mounted frozen tissue sections of Richter's syndrome by microdissection. Cells were microdis-

[a] This study was partially supported by the U.S. Public Health Service Grants EY-06337 to D.M.K. and AR-40809 and CA-68541 to P.C.

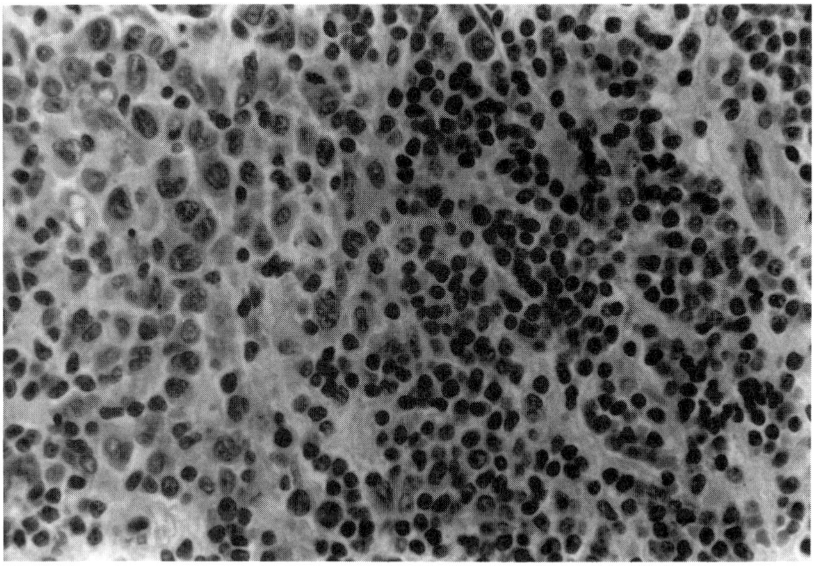

FIGURE 1. Cervical lymph node biopsy specimen showing two discrete cellular lesions consisting of chronic lymphocytic leukemia (*right*) and diffuse large cell lymphoma (*left*) (hematoxylin and eosin, original ×400, reduced by 25%).

sected using a set of Narishige (Image Systems, Columbia, MD) mechanical and hydraulic micromanipulators coupled to an Olympus CK2 inverted microscope (Olympus Optical Co. Ltd., Japan). DLL and CLL cells were identified morphologically in Giemsa-stained sections. Fifteen to 20 large or small cells were placed separately in eppendorf tubes. After addition of 200 μl of 100% ethanol, tubes were centrifuged at 14,000 rpm for 10 min. After removing the supernatant and evaporating ethanol in a Speed Vac SC 100 centrifuge (Savant Instruments Inc., Farmingdale, NY), the cells were resuspended in 100 μl PCR buffer containing 0.5% Tween 20 and 100 μg/ml proteinase-K. After overnight incubation at 55°C, proteinase-K was inactivated by heating at 95°C for 10 minutes. Two, 5, or 10 μl of solution were used for PCR.

PCR Amplification, Cloning, and Sequencing of Ig V_H Genes

DNA was amplified by PCR using Ig V_H gene family-specific sense primers and a consensus antisense J_H primer, as described.[4] PCR products were cloned in PCR 1000 vector using the TA cloning system (Invitrogen Corp., San Diego, CA) following the manufacturer's instructions. DNA sequencing was performed directly from a small-scale plasmid preparation using the Sequenase version 2.0 system (United States Biochemical, Cleveland, OH), according to the manufacturer's instructions. DNA sequences were analyzed using the MacVector version 3.5 software (Eastman Kodak Company, New Haven, CT) and the GeneBank data base. For each B-cell clone, a total of 10 independent isolates were sequenced.

Anaysis of Mutations

We calculated the number of expected replacement (R) mutations in the complementary determining regions (CDR) and framework regions (FR) of the Ig V_H gene sequences using the formula R_{CDR} or $R_{FR} = n \times (CDR\ R_f \text{ or } FR\ R_f) \times (CDR_f \text{ or } FR_f)$, where n is the total number of observed mutations; R_f is the replacement frequency inherent to each Ig V_H gene,[5,6] and CDR_f and FR_f are the relative size of the CDRs and FRs. A binominal probability model was used to evaluate whether the excess of R mutations in CDRs and the scarcity of silent (S) mutations in the FRs were due to chance only.[5,6] $p = \{n!/[k!(n - k)!]\} \times q^k \times (1 - q)^{n-k}$, where q = probability a R mutation will localize to CDR or FR ($q = CDR_{rel}$ or $FR_{rel} \times CDR\ R_f$ or $FR\ R_f$), and k = number of observed R mutations in the CDRs or FRs.

RESULTS

CLL and DLL B Cells Rearrange Different Ig V_H, D, and J_H Genes

Two different PCRs, involving leader-J_H and FR1-J_H oligonucleotide primer pairs, were performed to amplify the rearranged genomic Ig V_H DNA segments of microdissected CLL and DLL cells. The PCR products were cloned and sequenced. The DNA sequences derived from each sample were aligned with those of reported germline V_H, D, and J_H genes. The V_H2 and V_H6 family-specific sense primers amplified, each in conjunction with the antisense consensus J_H primer, DNA from the CLL cells. The sequence of the rearranged V_H2 gene (clone A) showed 99.3% identity to that of the $V_{II\text{-}5}$ germline gene (FIG. 2A, TABLE 1).[7] The rearranged V_H6 gene (clone B) was identical to the sequence of the germline V_H6 gene (FIG. 2B, TABLE 1).[9] DNA from the DLL cells was amplified using the V_H1 family-specific FR and J_H primers but not by the V_H2 and V_H6 FR1 or any other V_H gene family-specific primers, each in conjunction with the J_H primer. The Ig V_H gene sequence of this neoplastic DLL clone was 97.3% identical to that of the germline DP-14 gene (FIG. 2C, TABLE 1).[10] The rearranged D gene and J_H gene sequences were also different in the CLL and DLL cells. Clone A rearranged a DLR3 gene with a J_H4 gene, clone B rearranged an unidentified D gene with a J_H4 gene; clone C rearranged a D21/9 gene with a J_H6 gene. The difference between the V_H-D-J_H gene sequences rearranged by CLL cells (clones A and B) and autologous DLL cells (clone C) indicated that the DLL cells emerged as the expansion of a clone distinct from that of the CLL cells (FIG. 2 and TABLE 1).

Lack of Intraclonal Diversity in CLL and DLL B-Cell Clones

For each rearranged Ig gene clone (clone A, B, and C) from 10 to 12 independent bacterial colonies were sequenced. In no clone did independent isolates show nucleotide differences in the V_H-D-J_H sequences, suggesting that clonal diversification is not operative in the CLL or DLL cells.

Somatic Mutations in the CLL and DLL Clones Are Not Selected by Antigen

In the absence of negative or positive selective pressure on a gene product, nucleotide changes yielding amino acid replacement (R mutations) and nucleotide

FIGURE 2. Nucleotide and amino acid sequences of the rearranged Ig V_H genes of the microdissected CLL and DLL B cells of a patient with Richter's syndrome. The sequences are those of the leader-J_H and FR1-J_H PCR amplification products. In each cluster, the top amino acid and nucleic acid sequence is given for comparison and represent that of the closest reported germline V_H gene. Dots indicate identities. The replacement amino acid residue is given below the replacement nucleotide change. Solid lines above each cluster depict the CDR sequences. The numbers above the deduced amino acid sequences refer to the Ig amino acid residues according to Kabat *et al.*[8]

changes not yielding amino acid replacement (S, or silent mutations) are randomly distributed throughout the coding sequence. If a DNA segment displays a number of R mutations higher than that expected by chance, it is likely that a positive pressure was exerted on the gene product to select for these mutations, such as is observed in the V segment CDRs of affinity-mature antibodies. Conversely, if a DNA segment displays a number of R mutations

TABLE 1. Ig V_H Gene Analysis of Microdissected CLL and DLL Cells in Richter's Syndrome[a]

	Clone Detected in			Ig V_H Gene											
							CDR1 and CDR2			FR1, FR2 and FR3					
Clone	CLL Cells	DLL Cells	Intraclonal Diversity	Closest Germline Gene	V_H Family	Nucleotide Identity (%)	R	*p*	S	R	*p*	S	D_H Gene	J_H Gene	Ig V_H Gene Rearrangement
A	Yes	No	No	V_{II}-5	V_H2	99.3	0 (0.00)	—	0	2 (2.25)	0.36	0	LR3	J_H4	Stop at codon 100B
B	Yes	No	No	V_H6	V_H6	100	0 (0.00)	—	0	0 (0.00)	0.00*	0	ND	J_H4	In frame
C	No	Yes	No	DP-15	V_H1	99.2	0 (0.00)	—	1	1 (1.64)	0.33	0	21/9	J_H6	In frame

[a] R, number of detected and (expected) R mutations; S, number of detected S mutations; *p*, probability; *, statistically significant; ND, not detected.

lower than that expected by chance, it is likely that a positive pressure was exerted on the gene product to select against R mutations so that the protein structure is preserved, such as is observed in the V segment FRs of functional antibodies. The number of expected R mutations and the probability that the R mutations in the Ig V segment CDRs or FRs arose by chance was calculated in the two CLLs (clone A and B) and in the DLL (clone C) clones (TABLE 1). In most clones, the rearranged Ig V_H gene FRs displayed a scarcity or absence of R mutations that is consistent with the preservation of the optimal structure of the secreted Ig molecules. The Ig V_H gene segments of the two CLL (clones A and B) and the DLL (clone C) cell clonotypes displayed a number of putative R mutations that was not different from that expected by chance alone, suggesting that no positive pressure for selection of R mutations was applied to these Ig V_H-D-J_H gene products.

DISCUSSION

Elements of the B-cell lineage display the unique feature of rearranging their Ig V-(D)-J genes. Sequence analysis of these genes has been used to construct a genealogic tree of the process of genetic diversification and clonal selection in antibody responses to self and exogenous antigens; it has also been used to dissect the clonal origin and relatedness of different malignant B cells. The present analysis of the productively rearranged Ig V_H-D-J_H gene sequences of microdissected CLL and DLL B cells from a patient with Richter's syndrome revealed that (i) DLL and CLL cells emerged as independent discrete elements and (ii) that productively rearranged V_H-D-J_H genes by DLL B cells accumulated somatic point-mutations but did not display features associated with intraclonal diversification and/or selection by antigen.

Earlier studies have demonstrated that CLL and the supervening DLL cells can be clonally related.[11,12] The present findings, in conjunction with those of another recent report,[13] suggest an alternative modality of development of Richter's syndrome where DLL B cells emerge as an expansion of a discrete B-cell clone different from preexisting CLL clones. This pattern of evolution can be explained by at least two different mechanisms. The two neoplastic CLL and DLL clones might have derived from a common (early pro-B-cell) progenitor or different progenitors. In the first case, according to the concept of multistep carcinogenesis,[14] a common progenitor, possibly at a developmental stage preceding that of pro-B cell, would take the first step toward carcinogenesis. The cellular progeny would subsequently evolve as discrete B-cell clones by rearranging different V_H, D, and J_H genes and expressing different V_H-D-J_H sequences to yield discrete CLL and DLL clonal populations. This hypothesis is supported by the findings of Nakamine *et al.*,[15] showing a case of Richter's syndrome in which the CLL cells expressed Ig λ and the DLL cells expressed Ig κ light chains, but both malignant clones shared the identical trisomy 12 karyotypic abnormality. A similar mechanism has also been revealed in lymphoblast crisis of chronic myelogenous leukemia where the lymphoblasts and the malignant myeloid cells express the same clonal karyotypic markers.[16] The second modality of Richter's syndrome emergence is independent transformation events in discrete B-cell clones yielding discrete CLL and DLL clonal populations. This possibility is supported by the finding that primary and secondary cancers may display discordant genetic aberrations.[17]

Besides serving as clonal markers, the rearranged Ig V gene sequences can provide important information regarding B-cell clonal evolution and role of antigen in clonal selection. Antigen contact is thought to occur primarily through amino acid residues of the Ig V segment CDRs, and clustering of R mutations in CDRs is an indicator of antigenic selection.[5,6,18] The A and B CLL clone Ig V gene sequences showed 99.3% and 100% identity to the germline VII-5 and DP-15 genes, respectively. Because of a stop signal at codon 100B, however, the Ig V_H gene rearrangement of clone A is nonproductive. Since the productively rearranged CLL clone B displayed no somatic mutations, it appears that Richter's syndrome CLL clones are not different from those CLL clones that do not transform (reviewed in Refs. 19 and 20). The Richter's syndrome DLL clone C Ig V_H gene rearrangement displayed 99.2% identity with that of the corresponding putative germline gene. The nature and distribution of these mutations, however, were inconsistent with antigen selection. In conclusion, our present findings suggest that morphologic transformation and clonal expansion of neoplastic B cells in Richter's syndrome can be associated with a process of somatic hypermutation, but such a process may occur independently of selection by antigen. Finally, in these Richter's syndrome CLL/DLL B cells, somatic hypermutation did not underlie a detectable degree of intraclonal diversification.

SUMMARY

Richter's syndrome is defined as the morphologic transformation of chronic lymphocytic leukemia (CLL) into diffuse large-cell lymphoma (DLL). To determine the clonal nature of the two malignancies, we microdissected the CLL and DLL cells from a lymph node of Richter's syndrome and analyzed the sequences of the rearranged Ig V_H-D-J_H genes of the two lymphomas. Using the Ig V_H-D-J_H sequence as a marker of clonality, we delineated the clonal relationship of the CLL and DLL cells. The microdissected CLL and DLL cells productively rearranged different V_H, D, and J_H genes, suggesting that these DLL B cells emerge as discrete elements independent of the CLL B-cell population. The productively rearranged Ig V gene sequence of the CLL clone was 100% identical to the V_H6 germline gene, but the rearranged Ig V_H gene of the DLL clone was somatically point-mutated based on comparison of its sequence with those of reported germline genes. In the DLL clone, the random distribution and nature of the somatic point-mutations suggests a lack of antigen selection; the identity of the somatic point-mutations in multiple independent isolates of the same B-cell clone suggests a lack of intraclonal diversity. Thus, Richter's syndrome DLL B cells are monoclonal and can emerge as discrete elements independent of the preexisting CLL cells; antigen selection and clonal diversification are not necessarily associated with the events leading to this aggressive neoplastic transformation.

REFERENCES

1. FOUCAR, K. 1992. B cell chronic lymphocytic and prolymphocytic leukemia. *In* Neoplastic Hematopathology. D. M. Knowles, Ed.: 1181–1208. Williams & Wilkins. Baltimore, MD.
2. RICHTER, M. N. 1928. Generalized reticular cell sarcoma of lymph nodes associated with lymphatic leukemia. Am. J. Pathol. **4:** 285–292.

3. FOUCAR, K. & R. E. RYDELL. 1980. Richter's syndrome in chronic lymphocytic leukemia. Cancer **46:** 118–134.
4. DEANE, M. & J. D. NORTON. 1990. Immunoglobulin heavy chain variable region family usage is independent of tumor cell phenotype in human B lineage leukemias. Eur. J. Immunol. **20:** 2209–2217.
5. CHANG, B. & P. CASALI. 1994. A major proportion of the CDR1 sequences of human V_H genes display CDR1 inherently susceptible to amino acid replacement. Immunol. Today **15:** 3670–3680.
6. RIBOLDI, P., G. L. GAIDANO, E. W. SCHETTINO, T. G. STEGEN, D. M. KNOWLES, R. DALLA-FAVERA & P. CASALI. 1994. Two AIDS-associated Burkitt's lymphomas produce anti-i IgM cold agglutinins utilizing somatically mutated V_H4.21 segments. Blood **83:** 2952–2961.
7. SHIN, E. K., F. MATSUDA, H. NAGAOKA, Y. FUKITA, T. IMAI, K. YOKOYAMA, E. SOEDA & T. HONJO. 1991. Physical map of the 3′ region of the human immunoglobulin heavy chain locus: Clustering of autoantibody-related variable segments in one haplotype. EMBO J. **10:** 3641–3645.
8. KABAT, E., T. T. WU, M. REID-MILLER, H. M. PERRY, K. S. GOTTESMANN & C. FOELLER. 1991. Sequences of Proteins of Immunological Interest. Fifth Edition. U.S. Government Printing Office. Washington, DC.
9. BULUWELA, L. & T. H. RABBITTS. 1988. A V_H gene is located with 95 kb of the human immunoglobulin heavy chain constant region genes. Eur. J. Immunol. **18:** 1843–1845.
10. TOMLINSON, I. M., G. WALTER, J. D. MARKS, M. B. LLEWELYN & G. WINTER. 1992. The repertoire of human germline V_H sequences reveals about fifty groups of V_H segments with different hypervariable loops. J. Mol. Biol. **227:** 776–798.
11. CHEREPAKHIN, V., S. M. BAIRD, G. W. MEISENHOLDER & T. J. KIPPS. 1993. Common clonal origin of chronic lymphocytic leukemia and high-grade lymphoma of Richter's syndrome. Blood **82:** 3141–3147.
12. MATOLCSY, A., G. INGHIRAMI & D. M. KNOWLES. 1994. Molecular genetic demonstration of the diverse evolution of Richter's syndrome (chronic lymphocytic leukemia and supervening large cell lymphoma). Blood **83:** 1363–1372.
13. MATOLCSY, A., P. CASALI, F. BAI & D. M. KNOWLES. 1995. Transformation of chronic lymphocytic leukemia B cells to high-grade non-Hodgkin's lymphoma in Richter's syndrome is not associated with somatic diversification of the expressed Ig V_H gene. Submitted for publication.
14. WEINBERG, R. A. 1989. Oncogenes, antioncogenes, and the molecular basis of multistep carcinogenesis. Cancer Res. **49:** 3713–3721.
15. NAKAMINE, H., A. S. MASIH, W. G. SANGER, R. S. WICKERT, D. W. MITCHELL, J. O. ARMITAGE & D. D. WEISENBURGER. 1992. Richter's syndrome with different immunoglobulin light chain types. Molecular and cytogenetic features indicate a common clonal origin. Am. J. Clin. Pathol. **97:** 656–663.
16. VOGLER, L. B., W. M. CRIST, P. C. VINSON, A. SARRIF, M. G. BRATTAIN & M. S. COLEMAN. 1979. Philadelphia chromosome-positive pre-B-cell leukemia presenting as blast chrisis of chronic myelogenous leukemia. Blood **54:** 1164–1170.
17. CHUNG, K. Y., T. MUKHOPADHYAY, J. KIM, A. CASSON, J. Y. RO, H. GOEPFERT, W. K. HONG & J. A. ROTH. 1993. Discordant p53 gene mutations in primary head and neck cancers and corresponding second primary cancers of the upper aerodigestive tract. Cancer Res. **53:** 1676–1683.
18. SHLOMCHIK, M. J., A. MARSHAK-ROTHSTEIN, C. B. WOLFOWICZ, T. L. ROTHSTEIN & M. G. WEIGERT. 1987. The role of clonal selection and somatic mutation in autoimmunity. Nature **328:** 805–811.
19. KIPPS, T. J. & D. A. CARSON. 1993. Autoantibodies in chronic lymphocytic leukemia and related systemic autoimmune diseases. Blood **81:** 2475–2487.
20. SCHROEDER, H. W. & G. DIGHIERO. 1994. The pathogenesis of chronic lymphocytic leukemia: Analysis of the antibody repertoire. Immunol. Today **15:** 288–294.

V_H Gene Repertoire of Mantle Cell Lymphomas

W. VARADE,[a,b] E. MARIN,[b] M. MILANO,[b] R. INSEL,[b]
S. SWERDLOW,[c] AND M. WILLIAMS[d]

[b]*Department of Pediatrics*
University of Rochester
Rochester, New York 14642

[c]*Department of Pathology*
University of Pittsburgh
Pittsburgh, Pennsylvania 15213

[d]*Department of Internal Medicine*
University of Virginia
Charlottesville, Virginia 22908

Mantle cell lymphoma (MCL, formerly centrocytic lymphoma) is a non-Hodgkins lymphoma of small cleaved $CD5^+$ B cells.[1] Approximately 70% of these tumors demonstrate a reciprocal translocation of the bcl-1 locus on chromosome 11 to the IgH locus on chromosome 14, t(11;14) (q13;q32). The bcl-1 locus lies just upstream of the putative bcl-1 oncogene, PRAD1, which encodes the cyclin D1 protein. Unlike other B-cell malignancies (chronic lymphocytic leukemia [CLL], acute lymphoblastic leukemia [ALL], follicular lymphoma [FL]), V_H gene usage, the nature of the V_HDJ_H rearrangements, and the presence or absence of somatic mutation (SM) in the V_H genes of MCL have not been characterized.

CLL and ALL use a restricted set of V_H genes, with bias for those V_H genes expressed early in fetal development. CLL, like MCL, expresses the CD5 surface marker. V_H genes in most CLLs appear to be unmutated or only slightly mutated, though the V_H251 gene may show a high degree of SM.[2] In contrast, V_H gene expression in FL, a $CD5^-$ lymphoma, reflects the complexity of the V_H gene families and resembles the repertoire of adults with a high frequency of SM. To address similar issues in MCL, we have begun to amplify by PCR and determine the nucleotide (nt) sequence of V_HDJ_H rearrangements from DNA of affected lymph nodes of patients with MCL diagnosed by histologic criteria and previously characterized for the PRAD1/bcl-1 t(11;14)[3,4] (also, unpublished data).

Five V_H3- and one V_H4-expressing rearrangements were detected in the six MCLs studied (TABLE 1). All had an open reading frame. The V_H genes of three MCLs showed 100% homology to the germline V3-21, V3-11, and V3-9 genes. 89–81b had three differences compared to V4-34, all leading to CDR amino acid (aa) replacements. 89–85 had two silent mutations in FR and one replacement mutation in both CDR2 and FR3 compared to V3-23. 92–14 had five replacement mutations in CDR and four replacement and one silent mutation in FR3 compared to p3/V3-30. The overall R/S SM ratio was 13 to 3. Four of the V_H genes showed

[a] Address for correspondence: William S. Varade, M.D., Division of Pediatric Nephrology, University of Rochester Medical Center, 601 Elmwood Ave., Box 777, Rochester, NY 14642.

TABLE 1. Cyclin D1/bcl-1 Rearrangement Status, V_H Gene Usage, Relation of V_H Gene to Fetal V_H Repertoire and Somatic Mutations of V_H Genes Expressed by Mantle Cell Lymphomas

MCL	Cyclin D1/bcl-1 Translocation	Germline V_H Gene (% Homology)	Fetal V_H	Somatic Mutations (Replacement/Silent)
89–79	+ (PRAD1)	V3-21 (100)	—	0
89–81b	—	V4-34 (98.9)	—	3/0
89–82	+ (PRAD1)	V3-11 (100)	+ (FL18-3)	0
89–85	+ (bcl-1)	V3-23 (98.5)	+ (30P1)	2/2
89–91	+ (bcl-1)	V3-9 (100)	+ (FL13-29)	0
92–14	+ (bcl-1)	p3 (V3-30) (96.7)	+ (56P1)	8/1

a high degree of homology to V_H genes expressed in the fetal repertoire. The five MCLs carrying a t(11;14) all expressed unique members of the V_H3 family.

The average length of CDR3 was 15 aa (range: 10–20 aa) (TABLE 2). Dlr2 was expressed in three rearrangements, once in reverse complement form. J_H4b was expressed in three rearrangements, J_H6 in two and J_H3b in one. Trimming of coding sequence at the sites of rearrangements was seen at 4 out of 6 3′V_H, 5′D, and 5′J_H segments and at all 3′D segments. The J_H segments of 89–81b and 92–14 contained SM. P nt were seen in three rearrangements at the V_H-D junction. N nt insertions were present at all V_H-D joins and 4 out of 6 D-J_H joins.

Thus, V_H3 genes appeared to be preferentially rearranged in MCL. Other than one V_H4 gene, the smaller V_H families were not represented, in contrast to the bias seen in CLL and ALL. SM was seen in 50% of the V_H genes sequenced. Replacement mutations occurred more frequently than translationally silent changes and were clustered in the CDR and FR3, suggesting Ag stimulation and SM before malignant transformation in at least some of these rearrangements. This suggests that MCL may arise from B cells either before Ag stimulation and SM or after chronic Ag stimulation, as has been suggested for CLL. The presence of the cyclin D1/bcl-1 (t11;14) was associated with the expression of a V_H3 gene but did not correlate with the presence of SM or the expression of fetal type V_H genes. The Dlr2 and J_H4 gene segments were preferentially rearranged in these MCL with significant modifications at the sites of rearrangement in CDR3. The nature of the V_H gene repertoire and presence of SM suggests that the MCLs may arise from precursors at a stage of B-cell development/activation intermediate between CLL and FL.

TABLE 2. Characteristics of CDR3 of Mantle Cell Lymphomas

MCL	CDR3 Length (Codons 95-102)	D	J_H	nt Cut Back				P nt	N nt	
				3′V	5′D	3′D	5′J		V_H-D	D-J_H
89–79	51 nt (17aa)	21-10	3b	0	2	9	0	—	2	7
89–81b	54 nt (18aa)	lr2	6c	2	1	4	0	—	2	0
89–82	30 nt (10aa)	lr2	4b	0	12	1	6	3′V_H-2 bp	1	0
89–85	60 nt (20aa)	xp4	6b	1	0	5	6	5′D-2bp	3	1
89–91	36 nt (12aa)	k1	4b	3	6	10	3	—	5	12
92–14	39 nt (13aa)	lr2(R/C)	4b	0	0	15	2	3′V_H-1 bp	2	6

REFERENCES

1. SWERDLOW, S. H. & M. E. WILLIAMS. 1993. Pathol. Ann. **28**(Part 2): 171–197.
2. CAI, J., C. HUMPHRIES, A. RICHARDSON & P. W. TUCKER. 1992. J. Exp. Med. **176:** 1073–1081.
3. WILLIAMS, M. E., T. C. MEEKER & S. H. SWERDLOW. 1991. Blood **78:** 493–498.
4. WILLIAMS, M. E., S. H. SWERDLOW, C. L. ROSENBERG & A. ARNOLD. 1993. Leukemia **7:** 241–245.

V_H Gene Use by CD5$^+$ AIDS-Associated B-Cell Lymphoproliferations

VALERIE L. NG,[a,b,c] MARK H. HURT,[b]
BRIAN G. HERNDIER,[d] AND MICHAEL S. McGRATH[b,c]

[b]*Department of Laboratory Medicine*
[c]*Department of Medicine, and*
[d]*Department of Pathology*
University of California, San Francisco
and
San Francisco General Hospital
San Francisco, California 94110

The immunoglobulins (Igs) produced by human CD5$^+$ B-cell malignancies (i.e., chronic lymphocytic leukemia, CLL) often use germline or minimally modified V_H genes.[1] During a prospective study in which we extensively immunophenotyped B-cell lymphomas occurring in persons infected with the human immunodeficiency virus, type 1 (HIV-1), and with the acquired immunodeficiency syndrome (AIDS), we had observed a high proportion of large-cell lymphomas that expressed cell surface CD5 (i.e., of eight AIDS-associated large-cell lymphomas analyzed, four stained brightly and three had weak or intermediate staining with anti-CD5). It was of interest to determine if CD5$^+$ B-cell malignancies arising in HIV-1-infected persons similarly used germline or minimally modified V_H genes.

V_H genes of three different CD5$^+$ B-cell malignancies occurring in HIV-1-infected persons were analyzed using PCR or RT-PCR with 5′ V_H family-specific and 3′ consensus J_H primers. PCR of DNA extracted from a CLL occurring in an HIV-1-infected patient demonstrated use of a V_H3 gene that was 99.7% homologous to a previously identified V_H3 germline gene, 3–11 (i.e., 22-2B[2]). The single nucleotide difference observed could have been attributable to an individual-specific allelic variant of the V_H gene or to a PCR artifact. A monoclonal AIDS-associated large-cell lymphoma containing only CD5$^+$/CD20$^+$ B cells was then analyzed. RT-PCR revealed use of a V_H3 gene 93.5% homologous to a previously identified germline V_H3 gene, DP-42.[3] RT-PCR of RNA extracted from a polyclonal CD5$^+$ large-cell B-cell lymphoma yielded products with only the V_H1/J_H primer pair; 15 unique V_H1 sequences were identified from 18 subclones, and all 15 V_H1 sequences had the highest homology to only three different previously identified V_H1 germline genes (i.e., 21-2,[2] 1-2,[4] 1-3b[4]). Although one clone was 99.7% related to a previously identified V_H1 germline gene, 21-2,[2] nine of the 15 unique clones were <95% homologous to the three previously identified V_H1 germline genes; the nucleotide differences in the tumor Ig V_H genes as compared to their most homologous previously identified germline V_H gene were scattered throughout

[a] Address for correspondence: Valerie L. Ng, Ph.D., M.D., 2M9 Clinical Laboratories, San Francisco General Hospital, 1001 Potrero Avenue, San Francisco, CA 94110.

and did not occur in a pattern typical of programmed modification (i.e., somatic hypermutation).

To compare with "normal" V_H1 gene use in HIV-1-infected persons, RT-PCR was performed on RNA extracted from a reactive lymph node obtained from an AIDS patient; this lymph node contained approximately 10% $CD5^+/CD20^+$ B cells but lacked any evidence of tumor. Although V_H amplifications were observed with all V_H family-specific primers, only the PCR products obtained with the V_H1/J_H primer pair were further analyzed. Fifteen subclones were obtained from V_H1/J_H-primed products; one clone was 100% homologous to a previously identified V_H1 germline gene, 1-2,[4] whereas 10 clones were <95% homologous to five previously identified V_H1 germline genes (i.e., 21-2,[2] hvlf10t,[5] 1-2,[4] DP-10,[3] and DP-21[3]); for the majority of these Ig V_H genes, the nucleotide changes as compared with their most homologous previously identified germline V_H genes did not occur in a pattern characteristic of programmed modification.

In summary, our observation of virtual germline Ig V_H gene use by the malignant B cells in a single CLL that arose in an HIV-1-infected person was consistent with that observed for the majority of Igs produced by malignant B cells of CLL arising in non-HIV-infected persons. In contrast, the majority of V_H genes used by non-CLL, AIDS-associated, malignant $CD5^+$ B cells were significantly modified from that of germline and occurred in a pattern inconsistent with that typical of programmed modification.

REFERENCES

1. Kipps, T. J. & D. A. Carson. 1993. Autoantibodies in chronic lymphocytic leukemias and related systemic autoimmune diseases. Blood **81:** 2475.
2. Berman, J. E., S. J. Mellis, R. Pollock, C. L. Smith, H. Suh, B. Heinke, C. Kowal, U. Surti, L. Chess, C. R. Cantor & F. W. Alt. 1988. Content and organization of the human Ig V_H locus: Definition of three new V_H families and linkage to the Ig C_H locus. EMBO J. **7:** 727.
3. Tomlinson, I., G. Walter, J. D. Marks, M. B. Llewelyn & G. Winter. 1992. The repertoire of human germline V_H sequences reveals about fifty groups of V_H segments with different hypervariable loops. J. Mol. Biol. **227:** 776.
4. Shin, E. K., F. Matsuda, H. Nagoaka, Y. Fukita, T. Imai, K. Yokoyama, E. Soeda & T. Honjo. 1991. Physical map of the 3′ region of the human immunoglobulin heavy chain locus: Clustering of autoantibody-related variable segments in one haplotype. EMBO J. **10:** 3641.
5. Soto-Gil, R. W., T. Olee, B. K. Klink, T. P. Kenny, D. L. Robbins, D. A. Carson & P. P. Chen. 1992. A systematic approach to defining the germline gene counterparts of a mutated autoantibody from a patient with rheumatoid arthritis. Arthritis Rheum. **35:** 356.

Cellular Origin, Antigen Reactivity, and V_H Segment Structure of IgM mAbs from AIDS Lymphomas[a]

PIERSANDRO RIBOLDI,[b] GIANLUCA GAIDANO,[c]
EDWARD W. SCHETTINO,[b,d] DANIEL M. KNOWLES,[d]
RICCARDO DALLA-FAVERA,[c] AND PAOLO CASALI[b,d,e]

[b]*Department of Pathology and*
Kaplan Comprehensive Cancer Center
New York University School of Medicine
New York, New York 10016

[c]*Department of Pathology*
College of Physicians and Surgeons
Columbia University
New York, New York 10032

[d]*Department of Pathology*
Cornell University Medical College
New York, New York 10021

INTRODUCTION

Stimulation of B lymphocytes by antigen (Ag) leads to cell activation, proliferation, somatic hypermutation of the expressed Ig V genes, and, eventually, clonal selection of those B cells bearing a mutated surface receptor displaying the highest affinity for Ag. Evidence for persistent clonal B-cell expansion and proliferation by foreign and self Ags has been provided in chronic infectious diseases and autoimmune disorders.[1-4] In addition, it has been suggested that Ag stimulation also plays a role in the clonal B-cell selection associated with development of B-cell tumors, such as follicular lymphomas.[5-7] Similarly, the generalized B-cell expansion and hyperactivity observed in HIV-infected patients may be at least in part triggered by Ag and might prepare the ground for the development of B-cell lymphomas, particularly those that arise in the absence of other putative lymphomagnetic cofactors, such as EBV.

[a] This work was supported by U.S. Public Health Service Grants CA-68541 to P.C., CA-37895 to R.D-F., and the Ministry of Health, I.S.S., AIDS Projects 1993, Grant 8206-07 (Italy) to P.R. G.G. was supported by a fellowship for AIDS research from the Ministry of Health, I.S.S. (Italy). This is publication #7 from The Division of Molecular Immunology.

P.R. current address: Dept. of Internal Medicine, Infectious Diseases, and Immunopathology, University of Milan, 20122 Milan, Italy.

[e] Address for correspondence: Dr. Paolo Casali, Division of Molecular Immunology, Department of Pathology, Cornell University Medical College, 1300 York Avenue, New York, NY 10021.

Because the assessment of a potential role for Ag in the clonal B-cell selection associated with B-cell malignant proliferation entails the definition of the specificity and of V_H segment structure of the tumor-derived antibody, we established *in vitro* two cell lines from two patients with AIDS-associated Burkitt's lymphoma (BL) and analyzed the Ag specificity and the structure of the V_H segment of the produced IgM mAbs.[8] Here, we extend those findings to the analysis of a third AIDS-BL cell line and to the analysis of the B-cell origin of these three AIDS-BL clones. Two AIDS-derived BL cell lines produced anti-i IgM cold agglutinins and utilized somatically mutated V_H4-21 genes.[8] The third AIDS-BL cell line made an IgM antibody of undetermined specificity and used a putatively mutated V_HIII gene. Two IgM-producing BLs originated from B-1 cells and one from B-2 cells.

AIDS-BL CELL LINES

The features of the HBL-1, HBL-2, and HBL-3 cell lines and respective mAbs are summarized in TABLE 1. The cell lines were established from bioptic specimens of three AIDS patients with BL, diagnosed by histologic and immunophenotypic criteria. The monoclonality of the AIDS BL cell lines and the absolute relatedness to the respective tumors was formally established by the analysis of the Ig H and L chain gene rearrangements and that of the *c-myc* oncogene translocations.[9] HBL-1 cells were positive for EBV and negative for HIV and HTLV-1 DNA sequences. HBL-2 and HBL-3 were negative for genetic sequences of the three viruses.

CD5 mRNA EXPRESSION

To determine the lymphocyte origin of the cell lines, we analyzed its CD5 mRNA expression, using a PCR amplification method we recently devised *ad hoc*.[4] As we previously showed, CD5 mRNA is characteristically expressed by cells of the B-1 population, be they surface $CD5^+$ (B-1a cells) or surface $CD5^-$ (B-1b cells).[10] HBL-1 and HBL-2 cells expressed detectable amounts of CD5 mRNA suggesting a B1 cell origin, while HBL-3 cells did not (FIG. 1B). The specificity of the PCR method was detailed in our original report[10] and was further supported by the proportionality between amount of input mRNA and amount of CD5 cDNA amplified (FIG. 1A and C).

ANTIGEN-SPECIFICITY OF THE IgM mAbs

The Ag-specificity of the HBL-1, HBL-2, and HBL-3 IgM mAbs was determined using specific ELISA[4,10,11] and self and exogenous Ags among the most commonly recognized by natural antibodies and autoantibodies, including IgG Fc fragment, human thyroglobulin, ssDNA, phosphorylcholine, insulin, human recombinant TNF-α, TNF-β, and IL-1β, i/I blood group determinants, myelin-associated glycoprotein, β-galactosidase and LPS from *E. coli,* tetanus toxoid, HIV-1, recombinant glycoproteins of CMV and parvovirus B19, *Haemophilus influenzae* type b capsular polysaccharide, and *Pneumococcus* polysaccharides. The HBL-1 IgM mAb did not bind any of the Ags tested. The HBL-2 and HBL-3

TABLE 1. Features of the AIDS-BL Cell Lines and Related mAbs

	Tumor						mAb			V_H Segment							
							Chains				Nucleotide (Amino Acid) Identity[a]	Nucleotide Differences in:					
												CDR		FR			
Cell Line	Source	Histology	EBV DNA	HIV DNA	HTLV-I DNA	*c-myc*	H	L	Ag Specificity	V_H Gene	%	R	S	R	S	D Gene[b]	J_H Gene
HBL-1	Peripheral blood	SNCC (L3)	Yes	No	No	t(8;14)	μ	κ	ND[c]	Hum5igvh[d] (V_HIII)	94.6 (91.0)	5 (2.9)[e]	2	4 (9.1)[e]	5	DLR3	J_H3
HBL-2	Pleural fluid	SNCC	No	No	No	t(8;14)	μ	λ	i 4[f]	V_H4-21[d]	93.5 (88.6)	5 (3.2)[e]	1	7 (10.1)[e]	6	20P3	J_H6
HBL-3	Liver mass	SNCC	No	No	No	5(8;22)	μ	λ	i 0.6[f]	V_H4-21[d]	84.2 (80.4)	11 (7.8)[e]	5	8 (26.2)[e]	22	DXP′1 (Inverted)	J_H5

[a] Compared with the genomic germline sequences.
[b] DLR3 and DXP′1 genes have been reported by Ichihara *et al.*[13]; gene 20P3 has been reported by Schroeder *et al.*[14]
[c] Not determined.
[d] The complete sequence of the genomic germline Hum5igvh gene is not published and is available from GenBank under accession number M99401. The sequence of the genomic germline V_H4-21 gene has been reported by Sanz *et al.*[12]
[e] Expected numbers of R mutations calculated as reported in Chang and Casali.[20]
[f] Smallest amount (ng) of IgM mAb agglutinating 10^7 papain-treated cord RBCs in a 100-μl reaction volume.

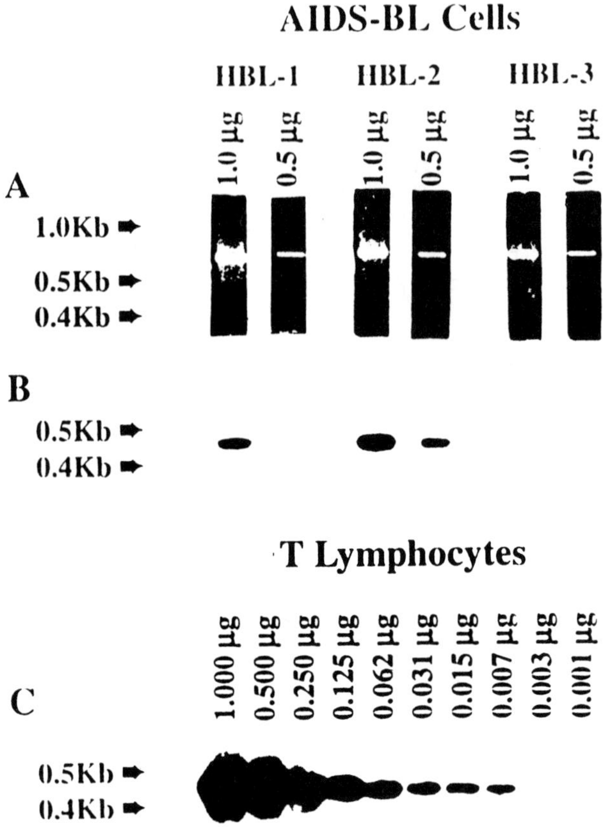

FIGURE 1. Expression of CD5 mRNA by AIDS-BL cells. mRNA from AIDS-BL cells (1.0 and 0.5 μg) and T lymphocytes (1.0 to 0.001 μg) were reverse transcribed. cDNAs were amplified by PCR using β-actin or CD5-specific primers. **(A)** Ethidium bromide-stained gel of amplified β-actin DNA (~0.6 kb). Panels **B** and **C**, hybridization of the ^{32}P-labeled "internal" CD5-specific oligonucleotide probe with the fractionated CD5 DNA amplified from cDNA independently reverse-transcribed from different amounts of purified mRNA from AIDS-BL cells and T lymphocytes, respectively.

IgM mAbs strongly and uniquely bound to the i determinant on human erythrocytes, as shown by their agglutination of cord (smallest agglutinating dose: 4 and 0.6 ng/10^7 RBC, respectively), but not adult (no agglutination by HBL-2; HBL-3 mAb smallest agglutinating dose, 38.8 ng/10^7 RBC) papain-treated human RBCs.

IgM mAb V_H Genes

To determine the structure of the V_H segments of the HBL-1, HBL-2, and HBL-3 IgM mAbs, we cloned and sequenced the respective cell line $C\mu$ cDNA.[8]

FIGURE 2 depicts the nucleotide (A) and deduced amino acid (B) sequences of the HBL-1, HBL-2, and HBL-3 IgM mAb V_H genes and those of the closest reported germline V_H genes. The nucleotide differences are summarized in TABLE 1. The HBL-1 mAb V_H gene sequence was 94.6% identical to that of Hum5igvh gene, a member of the V_HIII family. Nine of the 16 nucleotide differences displayed by the HBL-1 gene sequence resulted in putative amino acid replacements yielding replacement to silent (R : S) mutation ratios of 2.5 in the CDRs and 0.8 in the framework regions (FRs). The HBL-2 and HBL-3 mAb V_H gene sequences were 93.5% and 84.2% identical, respectively, to that of V_H4.21 gene, a member of the V_HIV family.[12] Twelve of the 19 nucleotide differences displayed by the HBL-2 V_H gene sequence resulted in putative amino acid replacements yielding R : S mutation ratios of 5.0 in the CDRs and 1.1 in the FRs. Nineteen of the 46 nucleotide differences displayed by the HBL-3 V_H sequence resulted in putative amino acid replacements yielding R : S mutation ratios of 2.2 in the CDRs and 0.3 in the FRs.

The structure of the CDR3s of the HBL-1, HBL-2, and HBL-3 IgM mAbs is depicted in FIGURE 2 (panel C, nucleotide sequence compared to the closest germline D and J_H genes; panel D deduced amino acid sequences). The HBL-1, HBL-2, and HBL-3 mAb D gene sequences displayed some identity with those of the DLR3, the expressed fetal 20P3 and the DXP'1 D genes,[13,14] respectively, and were juxtaposed with the untruncated and mutated form of the J_H3 gene (HBL-1), and truncated and mutated forms of J_H6 and J_H5 genes (HBL-2 and HBL-3, respectively).

SOMATIC MUTATIONS IN THE HBL-3 IgM mAb V_H GENE

The high number of nucleotide differences displayed by the HBL-3 V_H gene sequence when compared with that of the V_H4-21 germline gene and the presence of mutations in the HBL-3 mAb J_H5 segment, in conjunction with the conservation of the V_H4-21 gene in humans, strongly suggested that the HBL-3 mAb V_H segment consisted of a somatically mutated form of the V_H4-21 gene. Thus, we amplified by PCR DNA from the HBL-3 cell line or autologous fibroblasts using the sense V_H4-21 FR1 primer in conjunction with the antisense V_H4-21 FR3 primer, both encompassing sequences shared by the V_H4-21 segment and the expressed HBL-3 V_H gene (FIG. 3). The two combined primers amplified DNA from both fibroblasts and HBL-3 cells, yielding products of appropriate molecular size.[8] The same antisense V_H4-21 FR3 primer was also used to amplify fibroblast DNA in conjunction with the sense HBL-3 leader primer, encompassing a stretch of the leader sequence of HBL-3 V_H gene and differing in only one nucleotide from that of the corresponding area of the V58 gene, the V_HIV family member displaying the highest degree of identity with the V_H4-21 gene. The amplified product of appropriate molecular size (~400 bp) was obtained.[8] The three DNA amplification products were analyzed for their ability to hybridize the [γ-^{32}P]-labeled HBL-3 CDR1 oligonucleotide, encompassing a stretch of the HBL-3 V_H gene FR1-CDR1 sequence that displayed seven putative mutations when compared with corresponding germline V_H4-21 gene sequence. The oligonucleotide strongly hybridizes with the DNA amplification product from the HBL-3 cell line but not with the ~260 or ~400 bp DNA amplified from the autologous fibroblasts.[8]

A

```
                 5'  .  Leader .                 .
5igvh    -24 GCTATTTTAAAAGGTGTCCAATGT   -1
HBL-1    -27 TGTT-----------------------

                                                      FR1                                           CDR1
5igvh    GAGGTGCAGCTGGTGGAGTCTGGGGGAGGCTTGGTACAGCCAGGGCCGTCCCTGAGACTCTCCTGTACAGCTTCTGGATTCACCTTTGGGTATTATCCTA 100
HBL-1    ---------------------------------------------GG----GA---------------------C--------------TG-A---A---

                         FR2                                                     CDR2
5igvh    TGAGCTGGGTCCGCCAGGCTCCAGGGAAGGGGCTGGAGTGGGTAGGTTTCATTAGAAGCAAAGCTTATGGTGGGACAACAGAATACGCCGCGTCTGTGAA 200
HBL-1    --A----------------------------------------------------------C---------A---------G-----------------

                                                            FR3                                          3'
5igvh    AGGCAGATTCACCATCTCAAGAGATGATTCCAAAAGCATCGCCTATCTGCAAATGAACAGCCTGAAAACCGAGGACACAGCCGTGTATTACTGTACTAGA 300
HBL-1    ------------------G----------------------------------------------------------------C----------T-----

                 5' . Leader  .                 .
HBL-3    -24 ACAACTCCCAGATGCGTCCTGTCC   -1
HBL-2    -24 G--G----------G---G-----

                                                      FR1                                           CDR1
VH4.21   CAGGTGCAGCTACAGCAGTGGGGCGCAGGACTGTTGAAGCCTTCGGAGACCCTGTCCCTCACCTGCGCTGTCTATGGTGGGTCCTTCAGTGGTTACTACT 100
HBL-3    ------G-C-----A--A-----------------AC--A--------------A-A--------------A-----A-----G---C---A----CT--
HBL-2    --------A---G-C----------A-------------------------------------------A-A---------A-G----A--A-----C--

                         FR2                                                     CDR2
VH4.21   GGAGCTGGATCCGCCAGCCCCCAGGGAAGGGGCTGGAGTGGATTGGGGAAATCAATCATAGTGGAAGCACCAACTACAACCCGTCCCTCAAGAGTCGAGT 200
HBL-3    ---CA------------A--------------T------------------G-G--T---A-----AG-G------T-CG--------------------
HBL-2    -------------------------------------------------C----------------A------------T--------------------

                                                            FR3                                  3'
VH4.21   CACCATATCAGTAGACACGTCCAAGAACCAGTTCTCCCTGAAGCTGAGCTCTGTGACCGCCGCGGACACGGCTGTGTATTACTGTGCGAGA 291
HBL-3    T-G---T-----T--T-----------T--A----------GTT---CT--------------------T-----A-----T--------G
HBL-2    -G--------A------------------------------------AG--------------------------T---------------
```

(In the figure, "VH4.21" is printed as $V_H4.21$.)

B

```
                 .  FR1    .        .CDR1     . FR2      .   CDR2 .          .         . FR3    .         .
5igvh    EVQLVESGGGLVQPGPSLRLSCTASGFTFGYYPMSWVRQAPGKGLEWVGFIRSKAYGGTTEYAASVKGRFTISRDDSKSIAYLQMNSLKTEDTAVYYCTR 100
HBL-1    ---------------G-E------------E-T-N-------------------P--E------------------------------------A---S-

                 .  FR1    .        .CDR1     . FR2      .   CDR2 .          .         . FR3    .
VH4.21   QVQLQQWGAGLLKPSETLSLTCAVYGGSFSGYYWSWIRQPPGKGLEWIGEINHSGSTNYNPSLKSRVTISVDTSKNQFSLKLSSVTAADTAVYYCAR 97
HBL-3    --D---------QT----T----------RD-L-T----T----------SDYN-KS--T-------S------------S-T--------------
HBL-2    ----E---E--------------I---T-ND-S----------------D-----N-----------A--I-----------K--------------
```

C

```
               5'  .         .         .                                    .         .         .         .     3'
DLR3           AGCATATTGTGGTGGTGATTGCTATTCC                 ATGCTTTTGATGTCTGGGGCCAAGGGACAATGGTCACCGTCTCTTCA  JH3
HBL-1       GGCCCT------AACAC-T--C-C-                       --A--------A-------------------------------------  HBL-1

                   .                                .         .         .         .         .         .         .
20P3           ACGTGGGAGCTACT                ATTACTACTACTACTACGGTATGGACGTCTGGGGGCAAGGGACCACGGTCACCGTCTCCTCA  JH6
HBL-2           G-CCTT--G---G                                     ------------CCA----C-----------------G---  HBL-2

                   .         .         .                      .         .         .         .         .         .
DXP'1(INV)     GTTATAATAACTCCCCGAACCATAGTAATAC               ACAACTGGTTCGACTCCTGGGGCCAAGGAACCCTGGTCACCGTCTCCTCA  JH5
HBL-3                  --AGAGG-A--TTC-GA-----ACAGTGGAAGACTT           -----G--------G----A-T-----------------  HBL-3
```

D

```
         _____CDR3_____  FR4
JH3                      WGQGTMVTVSS  JH3
HBL-1    GPHINTWLYTFDI   -----------  HBL-1
JH6                      -----T-----  JH6
HBL-2    GLRYVDV         -AK--------  HBL-2
JH5                      -----L-----  JH5
HBL-3    KEANSDNNSGRLDS----N------    HBL-3
```

FIGURE 2. Nucleotide and deduced amino acid sequences of the V_H (**A** and **B**), D, and J_H (**C** and **D**) genes utilized by the HBL-1, HBL-2, and HBL-3 IgM mAbs. The top sequence is given for comparison and represents the published germline V_H, D, and J_H genes displaying the highest degree of identity to the expressed V_H-D-J_H genes. Dashes indicate identity. Solid lines on the top of each cluster depict CDRs.

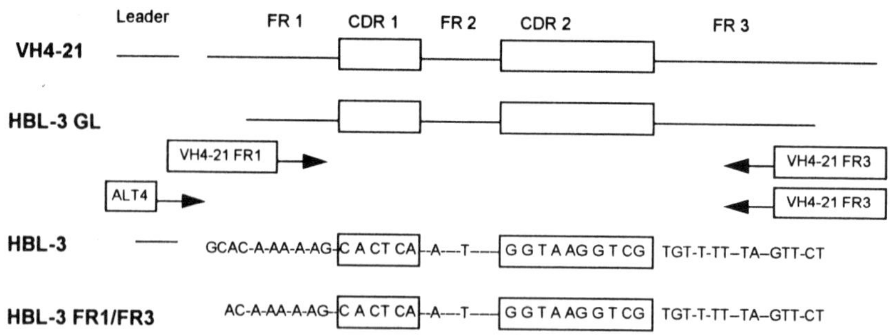

FIGURE 3. Evidence that the expressed HBL3 V_H segment is somatically mutated. Continuous lines depict sequence identities. Boxes depict CDRs or primer sequences. Using the V_H4-21 FR1 and V_H4-21 FR3 oligonucleotide primers appropriate products were amplified by priming genomic DNA from fibroblasts (HBL-3 GL) and HBL-3 cells (HBL-3 FR1/FR3). Analysis of these products showed that HBL-3 GL sequence was identical to that of the V_H4-21 gene, and the HBL-3 FR1/FR3 sequence was identical to that of the expressed HBL-3 V_H segment.

To identify the autologous germline V_H gene that putatively gave rise to the expressed HBL-3 V_H gene, the product amplified from fibroblast DNA using the sense V_H4-21 FR1 and the antisense V_H4-21 FR3 primers was cloned and sequenced. The sequences of eight independent clones were all identical to each other and to that of the V_H4-21 germline gene throughout the overlapping area (FIG. 3; HBL-3GL sequence). DNA amplified from the HBL-3 cell line using the sense V_H4-21 FR1 and the antisense V_H4-21 FR3 primers was also cloned and sequenced. The sequences of three independent clones were identical to each other and to that of the expressed HBL-3 V_H gene throughout the overlapping area (FIG. 3; HBL-3 FR1/FR3 sequence). These experiments proved that the expressed HBL-3 V_H segment was somatically mutated and suggested that V_H4-21 is the germline gene that gave rise to it.

SUMMARY AND CONCLUSIONS

In the present studies we analyzed the Ag specificity, V_H gene structure, and cellular origin of three IgM mAb-producing cell lines established *in vitro* from bioptic specimens of three AIDS patients with BL. We found that (i) both HBL-2 and HBL-3 IgM mAbs were cold agglutinins highly specific for the i blood group determinant, a self Ag the expression of which is dominant in the early stages of life, and both mAbs used somatically point-mutated V_H4-21 segments; (ii) HBL-1 IgM mAb, the Ag-specificity of which has not been determined, used a putatively mutated member of the V_HIII family; and (iii) both HBL-1 and HBL-2, but not HBL-3, cells expressed CD5 mRNA, suggesting a B-1 cell origin.

The utilization of V_H4-21 by the HBL-2 and HBL-3 cold agglutinins is consistent with the usage of this gene segment by all the reported pathogenic except the naturally occurring cold agglutinins.[15–18] This restricted V_H gene usage may reflect an inherent affinity of the germline V_H4-21 gene product for the i/I carbohydrate structure, and, perhaps, an overrepresentation of V_H4-21 in the human early and late B-cell repertoire.[15,19] Consistent with both an early and late developmental expression of the V_H4-21 gene is the B-1 and B-2 cellular origin of the two V_H4-21$^+$ cold agglutinins reported here. Thus, the two cold agglutinin autoantibodies possibly emerged at different stages of the natural history of the B-cell repertoires of these patients and might display a different temporal relationship, as discussed below, to the time of emergence of the respective tumoral cells.

The somatically mutated status of the HBL-2 and HBL-3 mAb V_H segments was suggested by the monomorphism of the human V_H4-21 gene, the extension of the nucleotide differences to the, in general, highly conserved J_H segment; and it was formally proved in HBL-3 mAb. Positive selection by Ag of the R mutations in the HBL-2 and HBL-3 mAb V_H segments was suggested by the differential R : S mutation ratios in the CDRs and FRs (HBL-2 mAb, 5.0 and 1.1, respectively; HBL-3 mAb, 2.2 and 0.3, respectively) but not substantiated by appropriate statistical analysis according to the binomial distribution model.[20] In fact, changes in the structure of the V_H segment CDRs may be irrelevant to a self Ag selection of the HBL-1 and HBL-3 mAbs, as it has been recently postulated that the structural correlate for i/I Ag binding is provided by an unmutated sequence of the V_H4-21 FR1 (Capra *et al.*, unpublished findings). Consistent with the suggested primary role of structurally conserved V_H4-21 in binding the i Ag, both HBL-2 and HBL-3 mAb V_H segments displayed significant lower numbers of R mutations in the FRs than theoretically expected (p = 0.03 and p = 0.00000002, respectively).

In these AIDS-associated BLs, it is unclear whether the initiation of the anti-i/I autoantibody response constituted a crucial event in the neoplastic transformation. The B-1 cell origin of the HBL-1 and HBL-2 mAbs is consistent with an early (fetal or neonatal, when B-1 cells are predominant in the B-cell repertoire) emergence of the respective B-cell clonotype, perhaps preceding the occurrence of the genetic accident, *c-myc* translocation; the B-2 origin of the HBL-3 mAb would be equally consistent with a similar and relatively early emergence of the respective B-cell clonotype or with a later (adult) origin, when B-2 cells are predominant. In the latter case, *c-myc* translocation might have preceded the rearrangement and expression of the Ig genes in this B-cell clone. Knowledge of the sequential order of activating, proliferating, and transforming events is crucial for a better understanding of the molecular pathogenesis of AIDS BL.

REFERENCES

1. Shlomchick, M. J., A. Marshak-Rothstein, C. B. Wolfowicz, T. L. Rothstein & M. G. Weigert. 1987. The role of clonal selection and somatic mutation in autoimmunity. Nature **328:** 805–811.
2. Shlomchick, M. J., M. Mascelli, H. Shan, M. Z. Radic, D. Pisetsky, A. Marshak-Rothstein & M. G. Weigert. 1990. Anti-DNA antibodies from autoimmune mice arise by clonal expansion and somatic mutation. J. Exp. Med. **171:** 265–292.
3. van Es, J. H., H. Aanstoot, F. H. J. Gmelig-Meyling, R. H. W. M. Derksen & T. Logtenberg. 1992. A human systemic lupus erythematosus-related anti-cardiolipin/single-strand DNA autoantibody is encoded by somatically mutated variant of the developmental restricted 51P1 V_H gene. J. Immunol. **149:** 2234–2240.
4. Mantovani, L., R. L. Wilder & P. Casali. 1993. Human rheumatoid B-1a ($CD5^+$ B) cells make somatically hypermutated high affinity IgM rheumatoid factors. J. Immunol. **151:** 473–488.

5. Friedman, D. F., E. A. Cho, J. Goldman, C. E. Carmack, E. C. Besa, R. R. Hardy & L. E. Silberstein. 1991. The role of clonal selection in the pathogenesis of an autoreactive human B cell lymphoma. J. Exp. Med. **174:** 525–537.
6. Bahler, D. W. & R. Levy. 1992. Clonal evolution of a follicular lymphoma: Evidence for antigen selection. Proc. Natl. Acad. Sci. USA **89:** 6770–6774.
7. Zelenetz, A. D., T. T. Chen & R. Levy. 1992. Clonal expansion in follicular lymphoma occurs subsequent to antigenic selection. J. Exp. Med. **176:** 1137–1148.
8. Riboldi, P., G. Gaidano, E. W. Schettino, T. G. Steger, D. M. Knowles, R. Dalla-Favera & P. Casali. 1994. Two acquired immunodeficiency syndrome-associated Burkitt's lymphomas produce specific anti-i IgM cold agglutinins using somatically mutated V_H4-21 segments. Blood **83:** 2952–2961.
9. Gaidano, G., N. Z. Parsa, V. Tassi, P. Della-Latta, R. S. K. Chaganti, D. M. Knowles & R. Dalla-Favera. 1993. In vitro establishment of AIDS-related lymphoma cell lines: Phenotypic characterization, oncogene and tumor-suppressor gene lesions, and heterogeneity in Epstein-Barr virus infection. Leukemia **7:** 1621–1629.
10. Kasaian, M. T., H. Ikematsu & P. Casali. 1992. Identification and analysis of a novel $CD5^-$ B lymphocyte subset producing natural antibodies. J. Immunol. **148:** 2690–2702.
11. Nakamura, M., S. E. Burastero, A. L. Notkins & P. Casali. 1988. Human monoclonal rheumatoid factor-like antibodies from CD5 $(Leu\text{-})^+$ B cells are polyreactive. J. Immunol. **140:** 4180–4186.
12. Sanz, I., P. Kelly, C. Williams, S. Scholl, P. Tucker & J. D. Capra. 1989. The smaller human V_H gene families display remarkably little polymorphism. EMBO J. **8:** 3741–3748.
13. Ichihara, Y., H. Matsuoka & Y. Kurosawa. 1988. Organization of human immunoglobulin heavy chain diversity gene loci. EMBO J. **7:** 4141–4150.
14. Schroeder, H. W., Jr., J. L. Hilson & R. M. Perlmutter. 1987. Early restriction of the human antibody repertoire. Science **238:** 791–793.
15. Pascual, V., K. Victor, M. Spellerberg, T. J. Hamblin, F. K. Stevenson & J. D. Capra. 1992. V_H restriction among human cold agglutinins. The V_H4-21 gene segment is required to encode anti-I and anti-i specificities. J. Immunol. **149:** 2337–2344.
16. Silberstein, L. E., L. C. Jefferies, J. Goldman, D. Friedman, J. S. Moore, P. C. Nowell, D. Roelcke, W. Pruzanski, J. Roudier & G. J. Silverman. 1991. Variable region gene analysis of pathologic human autoantibodies to the related i and I red blood cell antigens. Blood **78:** 2372–2386.
17. Leoni, J., J. Ghiso, F. Goni & B. Frangione. 1991. The primary structure of the Fab fragment of protein KAU, a monoclonal immunoglobulin M cold agglutinin. J. Biol. Chem. **266:** 2836–2842.
18. Jefferies, L. C., C. M. Carchidi & L. E. Silberstein. 1993. Naturally occurring anti-i/I cold agglutinins may be encoded by different V_H3 genes as well as the V_H4-21 gene segment. J. Clin. Invest. **92:** 2821–2833.
19. Pascual V. & J. C. Capra. 1992. V_H4-21, a human V_H gene segment overrepresented in the autoimmune repertoire. Arthritis Rheum. **35:** 11–18.
20. Chang, B. & P. Casali. 1994. The CDR1 sequences of a major proportion of human germline V_H genes are inherently susceptible to amino acid replacement. Immunol. Today **15:** 367–373.

Analysis of Multiple Myeloma Third Complementarity-Determining Regions Reveals Characteristics of Prenatal B Cells

L. A. KUNKEL, R. VESCIO, J. CAO, C. H. HONG, A. KIM,
G. J. SCHILLER, A. K. LICHTENSTEIN,
AND J. R. BERENSON

UCLA School of Medicine
Division of Hematology/Oncology
and
D.V.A. W.L.A.
Los Angeles, California 90024

Characterization of the immunoglobulin (Ig) heavy chain third complementarity-determining region (CDR3) in B-cell malignancies can suggest the stage in development that the transforming event occurs. Analysis of the CDR3 in fetal, neonatal, and adult B cells has demonstrated differences in the composition of gene rearrangements during lymphocyte development. Fetal/neonatal B cells, in comparison to adults, generally show (1) a lack of or very short nongermline nucleotides (N-regions) in the V_H-D and D-J_H junctions, (2) few palindromic events (P elements), (3) Ig recombination by fusing short stretches of sequence homology, and (4) a bias in composition of certain V_H and D gene families.

Multiple myeloma (MM) is a malignancy characterized by the accumulation of plasma cells that secrete large amounts of homogeneous Ig. There is little data on the composition of Ig genes expressed by the malignant clones or the process of gene recombination in MM. To investigate from which subset the neoplastic cells in MM are derived, we analyzed the CDR3 sequences in 22 patients (TABLE 1). The VDJ genes were cloned and sequenced through use of the polymerase chain reaction on RNA using 5′ V_H family specific primers and 3′ C_H primers.

The CDR3s averaged 30 base pairs (12–51 base pairs). There is a paucity of N regions in D-J_H joining (in 64% of cases) and even when present (36%) the average was short (n = 1.6 base pairs). V_H-D joint N-regions were also short (n = 3.3 base pairs) and were present except in two cases. P elements were rare (1 out of 22 cases). Although the length is similar to adult CDR3, the composition of nucleotides in the MM sequence is different. The majority of nucleotides contributing to the CDR3 region of MM patients appears to be derived from fusion of multiple D genes (TABLE 2). Only five in 22 (23%) MM patients expressed D segments that could be derived from one single germline gene. Therefore, in the majority of MM clones, CDR3 diversity cannot be explained by classical canonical recombination, and alternative mechanisms must be used. These events include fusion of two germline D_H genes in a straight orientation (11 of 17), fusion with inversion of at least one of the participating genes (6 of 17), and fusion with DIR genes (6 of 17). These fusions occur in areas of 2–4 base pairs of homology. One clone exhibited fusion of three D_H genes (WAD-5), and another clone had four genes fused (WAD-6). In some clones, more than one mechanism was involved.

TABLE 1. Sequences of the CDR3s of 22 Multiple Myeloma Patients[a]

Patient	N	D	N	J_H
WAD-1[a]	GACTTG	GGGTGGGAGGCTAGACTC		TGG
WAD-2	GAGAGCCTGCCTA	ATTACGATCTTTTGAATGGCTATCATAAT	TGG	TTCGACCC TGG
WAD-3	CT	TATTTGGTGGTGCCTGGTACTCGATTGTGCCTTAT	CAG	GACTAC TGG
WAD-4	AGACGTT	TCACCAGTGACTGGAAC	CTCG	ACGCTTTTGATGTC TGG
WAD-5	TC	AGGATACTGTGGGAGTCCCAGTTGTTACAGTTATTAT		TACTACGGTATGGACGTC TGG
WAD-8		GGGCAGTATTATCACGATAGTGCTGGTGATTACTACTC	ACCCGCT	GAGTTCTTCCAACAC TGG
WAD-9	GG	GGATATAGGAGTGGCTGATTA[b]		GAATATTTCCAGCAC TGG
WAD-10	G	AGAATTTTTATCTTGATGGCACTCAATCGAGAAACTTCTTCTTCGATTTC		TGG
WAD-11	T	CTTGGAGGAAG		TGGTTCGACCCC TGG
WAD-12	GATGG	GAATTACTTTGATAGTGTTGGTTATTACTACGTGGGGA	TT	TGACTAC TGG
WAD-13	CTTAAA	TACTTGACGGTAGTGGTTATGCTGCTCCTGTC		TGG
WAD-14	G	GAAATAATGCC		TACTACTACTACGGCATGGACGTC TGG
WAD-15	GTGA	GATTTAGTAGCGGTTTTTAC	TTTAAA	TTGACTAC TGG
WAD-17	GGAG	GAGGTAGCGGTAATTACAAGGGGAAAATT		GACTCT TGG
WAD-18	GC	CACAACTTTTTATACTGGGTATTCGGCGATGGGCCTCC		TGG
WAD-19	GGCA	GGGGTATAGCGCGAGGTGGTAC	TCTGCGC	CTCGACCCC TGG
WAD-20	TT	TTCCCATATAGAAGCAGCTGGTAC	AGC	TGACTAC TGG
WAD-21	GATGT	TAGCGGCAGCT		TTGACTTCT TGG
WAD-22		TTCTATACCGGTGGG		GACTTC TGG
WAD-23	CAACTG	GAGGAT	G	ATGCTTTGATATC TGG
WAD-24	CGTGA	GAGTCCTGGAGTCCCATTGAT		GCTTTTGATGTC TGG
WAD-25	GACCC	CCGACAGTCTACGTGGTGCTTATGATTTC		TGG

[a] Sequences have been submitted to Gene Bank, accession number pending.
[b] X indicates possible P nucleotide.

TABLE 2. Summary of V_H Gene, D Gene, J_H Gene Segments, and C_H Chain Usage

Patient	V Family	D Family[a]	J Family	C Chain
WAD-1	V_H1	DIR(r)DIR(r)	J4	G
WAD-2	V_H1	DXP4/DXP(r)	J5	G
WAD-3	V_H2	DIR2/D21.5	J4	G
WAD-4	V_H2	D21.5/DM1	J3	A
WAD-5	V_H3	DL4/DHQ52/D21-9	J6	G
WAD-8	V_H3	DXP'1/D21.9	J1	G
WAD-9	V_H3	DLR2/DLR5	J1	G
WAD-10	V_H3	D23.7/DLR3/DA1/DK1	J4	A
WAD-11	V_H3	DIR(r)	J5	G
WAD-12	V_H3	D21.9/DHQ52	J4	G
WAD-13	V_H3	D21.9/DA1(r)	J4	G
WAD-14	V_H3	D21.9(r)	J6	G
WAD-15	V_H4	DM2/DN1	J4	A
WAD-17	V_H4	DM2/(r)/DXP'1	J4	G
WAD-18	V_H4	DL5/DIR(r)	J4	G
WAD-19	V_H4	DN1	J5	A
WAD-20	V_H4	DHQ52/DN1	J4	G
WAD-21	V_H4	DN1	J4	G
WAD-22	V_H5	DLR5/DIR(r)	J4	A
WAD-23	V_H5	DK1	J3	A
WAD-24	V_H5	DIR/DIR	J3	G
WAD-25	V_H6	DIR/D21.9	J4	G

[a] Reverse sequences are indicated by (r).

D genes with irregular spacer signals are used with an approximately twofold greater frequency than in normal adult clones. All major D families were expressed. The DXP was the family most commonly used (30%). DHQ52 generally expressed in fetal development and rarely in adults was noted in 14% of MM sequences. V_H segment usage showed a predominance of V_H3 (8 of 22) and V_H4 (6 of 22). The combined usage of V_H5 and V_H6 was 18% consistent with the data of transient overexpression in fetal development. The J_H gene usage was less restricted than normally seen in adult populations (e.g., J_H3 was used in three cases and J_H1 in two cases). J_H4 was the most commonly used J_H segment (12 out of 22 patients); furthermore, some degree of base pair deletion of the 5′ end was evident in all J_H4 patients. There was a mechanism of D-J fusion, mediated by homologous sequences similar to mechanisms used in D-D fusion. The sequences of five of these patients demonstrated extreme truncation of the 5′ ends with deletion of 14 germline nucleotides. At the remaining 5′ end, there is a nine base pair segment (TCTGGGGCC) homologous with a region in the reverse sequence of the DIR gene.

The results suggest that the malignant clone in MM may be derived from a B-cell progenitor that has rearranged the Ig gene in a manner seen in fetal or neonatal development such as lack of and short N-regions, few P events, and a bias in V, D, J. Furthermore, the majority of cases are not generated by the classical canonical sequences, but use unconventional mechanisms such as D-D fusion, inverted sequences, and DIR genes. Our data also suggest a mechanism that accounts for the paucity of N regions: homologous recombination is a major contributing factor to the generation of D-J fusions in these clones, and these fusions characteristically lack N region insertion. These CDR3 sequences can be exploited as clonal markers for this B-cell neoplasm and will be useful for detecting residual disease. In addition, with the observation that these clones rearrange the CDR3 by nonclassical mechanisms, they may prove to be useful in studies of the recombinase mechanisms.

REFERENCES

1. Sanz, I. 1991. Multiple mechanisms participate in the generation of diversity of human H chain CDR3 regions. J. Immunol. **147:** 1720.
2. Yamada, M., R. Wasserman, B. A. Reichard, *et al.* 1991. Preferential utilization of specific immunoglobulin heavy chain diversity and joining segments in adult human peripheral blood B lymphocytes. J. Exp. Med. **173:** 395.
3. Shroeder, H. W., Jr. & J. Y. Wang. 1990. Preferential utilization of conserved immunoglobulin heavy chain variable gene segments during human fetal life. Proc. Natl. Acad. Sci. USA **87:** 6146.

Hodgkin's Disease: Clonal Ig Gene Rearrangements in Hodgkin and Reed-Sternberg Cells Picked from Histological Sections

RALF KÜPPERS,[a] KLAUS RAJEWSKY,[a] MIN ZHAO,[b]
GÜNTHER SIMONS,[b] RALF LAUMANN,[a]
ROBERT FISCHER,[b] AND MARTIN-LEO HANSMANN[b]

[a]*Institute for Genetics*
Weyertal 121
and
[b]*Department of Pathology*
J. Stelzmannstr. 9
University of Cologne
50931 Cologne, Germany

Hodgkin's disease (HD) is characterized by a small number of putative malignant mononuclear Hodgkin and multinucleated Reed-Sternberg cells (HRS cells; usually less than 1% of the tumor) amidst a major population of lymphocytes and histiocytes. Although Hodgkin's disease has been known for a long time, the clonality and the lineage derivation of the putative malignant HRS cells is still a matter of controversy.[1] To investigate HRS cells for monoclonality as well as a possible B-lineage origin, we used a method that we had recently established.[2] Single HRS cells were micromanipulated from frozen sections of three cases of HD representing three of the four histopathologically defined subtypes of HD: lymphocyte predominant (LP), nodular sclerosis (NS), and mixed cellularity (MC). V_H and V_κ gene rearrangements from individual cells were amplified by PCR (R. Küppers, K. Rajewsky, M. Zhao, G. Simons, R. Laumann, R. Fischer, M.-L. Hansmann; manuscript in preparation).

For 30 single HRS cells of the NS case, 12 PCR products were obtained. Sequencing of eight of these products revealed an identical in-frame $V_HD_HJ_H$ rearrangement for seven cells and one nonfunctional rearrangement (TABLE 1).

For the case of MC HD, 37 PCR products were obtained from 25 HRS cells. Among 22 PCR products that were sequenced, two nonfunctional V_H and two nonfunctional V_κ rearrangements were obtained repeatedly (TABLE 1), with five cells harboring two to four of these rearrangements. A V_H1 rearrangement obtained from four cells harbors a deletion of 47 base pairs within the V_H gene.

For the case of LP HD, 28 HRS cells were analyzed. A nonfunctional V_H5 family gene rearrangement with a deletion of codons 28 and 29 was obtained from six cells (TABLE 1). For three cells (two of which had also given rise to a V_H5 PCR product), a potentially functional V_H4 rearrangement with an insertion of six base pairs between codons 49 and 50 was obtained. Nucleotide differences of the V_H and J_H sequences of both rearrangements compared to the most homologous germline genes, and the detection of intraclonal diversity indicates that both rearrangements harbor somatic point mutations.

TABLE 1. Summary of Single-Cell Analyses of Three Cases of Hodgkin's Lymphoma for Immunoglobulin V Region Rearrangements

Case	Clonal Rearrangements	In-Frame	Remark
Nodular sclerosis	7 × VH3	+	
Mixed cellularity	4 × VH1	−	47-bp deletion
	7 × VH3	−	
	5 × VK1	−	
	2 × VK4	−	
Lymphocyte predominant	6 × VH5	−	6-bp deletion
	3 × VH4	+	6-bp insertion

DISCUSSION

In all three cases a clonal origin of the HRS cells and their derivation from the B-cell lineage could be demonstrated. The detection of somatic mutations and intraclonal diversity in the case of LP HD identifies a B cell as the progenitor of the tumor in this patient, and more specifically a germinal center B cell because somatic hypermutation is selectively taking place in B cells proliferating in the germinal center microenvironment.[2] Because only nonfunctional V_H and V_k gene rearrangements were obtained from the HRS cells of the case of MC HD, the disease in this patient may have resulted from the malignant transformation of a B-cell progenitor that is distinct from that of the classical, cytoplasmic μ chain expressing pre-B cells. The situation is less clear in the case of NS HD, in which only a potentially functional heavy chain joint was identified. Inasmuch as we may have missed a corresponding light chain gene rearrangement, it cannot be decided whether the transformed progenitor cell was a pre-B or a B cell in this patient. Interestingly, in three of the five V_H rearrangements amplified from the three cases of HD, deletions or insertions were detected within the V_H genes. Further analyses will reveal whether unusual deletions or insertions are frequently occurring in HRS cells. While the present results demonstrate the HRS cells can be derived from B-lineage cells at various stages of development, they do not exclude the possibility of other lineage origins in other cases of HD.

REFERENCES

1. Drexler, H. G., D. B. Jones, V. Diehl & J. Minowada. 1989. Hematol. Oncol. **7:** 95–113.
2. Küppers, R., M. Zhao, M.-L. Hansmann & K. Rajewsky. 1993. EMBO J. **12:** 4955–4967.

Production of Antigen-Specific Human Antibodies from Mice Engineered with Human Heavy and Light Chain YACs[a]

AYA JAKOBOVITS, LARRY L. GREEN,
MARGARET C. HARDY,
CATHERINE E. MAYNARD-CURRIE, HIROHISA TSUDA,
DONNA M. LOUIE, MICHAEL J. MENDEZ,
HADI ABDERRAHIM, MASATO NOGUCHI,
DOUGLAS H. SMITH, YONGJUN ZENG,
NATHANIEL E. DAVID, HITOSHI SASAI, DAN GARZA,
DANIEL G. BRENNER, JOANNA F. HALES,
RYAN P. McGUINNESS, DANIEL J. CAPON,
AND SUE KLAPHOLZ

Cell Genesys, Inc.
322 Lakeside Drive
Foster City, California 94404

INTRODUCTION

One of the major obstacles in the development of therapeutic applications in humans for mouse and mouse-derivatized monoclonal antibodies is their intrinsic immunogenicity that reduces dramatically the efficacy and safety of the administered antibodies. The full therapeutic benefit of monoclonal antibodies can be properly assessed only with the availability of fully human monoclonal antibodies that could be administered repeatedly. One of the most appealing approaches towards this goal is the creation of a mouse engineered with unrearranged human immunoglobulin (Ig) chains, capable of producing a large repertoire of human antibodies, in the absence of mouse antibodies. Such a mouse could be used to generate high-affinity, and high-specificity, fully human monoclonal antibodies directed against a broad spectrum of antigens, including human proteins.

Here we describe a novel strategy to create a mouse strain in which the majority of the antibodies produced are fully human, by inactivating the mouse Ig genes and introducing into the mouse germline large fragments of the human heavy and light chain loci (FIG. 1).

Using the technology we have recently developed,[1] large fragments of the human heavy and kappa light chain loci cloned on yeast artificial chromosomes (YACs) have been introduced into the mouse germline via fusion of yeast spheroplasts with mouse embryonic stem (ES) cells.[2] These fragments, which contain the structural genes for variable, joining, diversity, and constant domains as well

[a] The work was supported in part by Xenotech Inc., the joint venture between Cell Genesys Inc. and JT Immunotech.

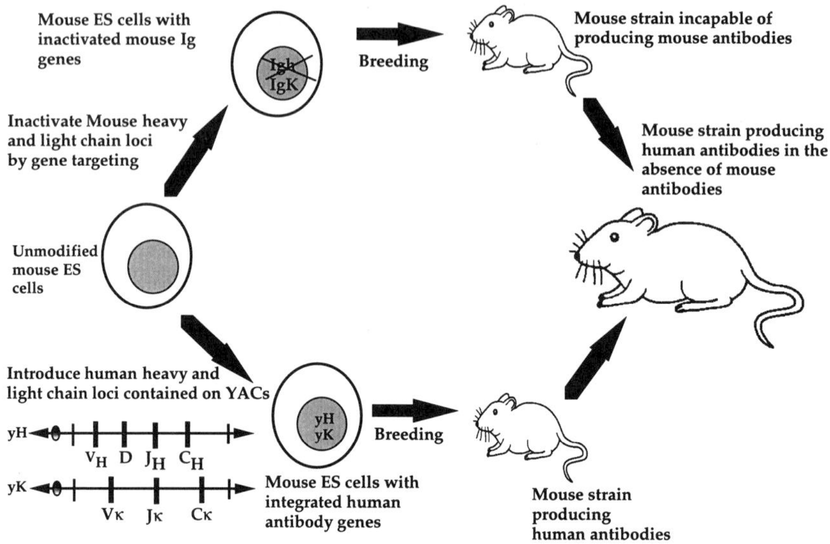

FIGURE 1. An outline of the strategy used to generate Xenomouse strain, producing fully human antibodies but not endogenous mouse antibodies.

as critical regulatory elements, proved sufficient for the production of a broad adult-like repertoire of human Ig and capable of mounting significant levels of human antibodies in mice, reflecting a utilization of a repertoire in many ways similar to adult humans.[2] Moreover, the breeding of such mice with animals lacking functional mouse antibody genes[3] led to a mouse strain in which fully human antibodies are produced at levels approaching those observed in normal serum.[2] Such mice may be exploited to develop human monoclonal antibodies for the treatment of human disease.

INTRODUCTION OF HUMAN HEAVY AND KAPPA CHAIN YACs INTO ES CELLS AND MICE

YACs containing variable and constant sequences from the human heavy and kappa chain loci (FIG. 2), identified from the Washington University human-YAC library, were characterized and determined to be in intact germline configuration.[4] The human heavy chain YAC harbors a 220-kb DNA insert containing the μ and δ constant regions, all six functional joining genes (J_H1-6), the major diversity (D) cluster, the intronic enhancer, and five variable (V) genes from four human V_H families: V_{VI}, V_I (2 genes: $V_{I\text{-}2}$, $V_{I\text{-}3}$), $V_{IV\text{-}4}$ and $V_{II\text{-}5}$. The 170-kb DNA insert in the human kappa YAC contains the kappa-deleting element (Kde), the intronic and 3′ enhancers, the C_κ constant region, all five functional joining genes (J_κ1-5), and the three V_κ variable genes in the B cluster (B1, B2, B3). Selection for the incorporation of these YACs in ES cells deficient for hypoxanthine phosphoribosyl transferase (HPRT) was conferred by a human HPRT expression cassette,[5] targeted into the right vector arm of each YAC.[4]

HPRT-positive ($HPRT^+$) ES cell colonies were evaluated by Southern blot analysis for the presence and integrity of the human immunoglobulin YACs (TABLE 1A). Seven $HPRT^+$ ES clones obtained by fusion with yH1-containing yeast spheroplasts were analyzed for human heavy chain-specific sequences using probes spanning the 220-kb insert. Five clones analyzed were shown to contain all of the *Hin*dIII fragments detected by probes for the $C\mu$, $C\delta$, J_H, D, and V_H genes found in the parental yH1 YAC. Deletions within the D or V_{II} regions were detected in the other two clones (TABLE 1A). A single YAC integration was detected in all but two clones, in which multiple copies were detected (TABLE 1A). Eleven $HPRT^+$ ES clones obtained from fusion with yK1-containing yeast spheroplasts were similarly analyzed, using probes spanning the kappa chain insert, including probes for the Kde and the C_κ, J_κ, and three V_κ genes (B3, B2, and B1) (TABLE 1B). Ten $HPRT^+$ ES clones analyzed contained all of the *Hin*dIII fragments detected by these probes in the parental yK1 YAC, and all but four of the clones were shown to contain a single integrated YAC (TABLE 1B). These results, together with pulsed-field gel analysis of these clones,[4] strongly suggested that the single YAC integrated in the yH1- and yK1-ES cell clones is structurally intact.

Yeast genomic sequences were detected in four yH1 and in five yK1 ES cell clones, whereas no yeast sequences were detected in the remaining clones (TABLES 1A and B).

Five yH1- and five yK1-containing ES cell clones were microinjected into C57BL/6J blastocysts, and mice containing the respective YACs in their germline were generated. Both YAC and yeast genomic DNA (when present) were transmitted in unaltered form.

Our results indicate that both the yH1 and yK1 YACs were integrated into ES cells at a high frequency (1 $HPRT^+$ colony/10^6 fused cells), primarily as an intact, single-copy, and apparently unrearranged human insert. When present, the

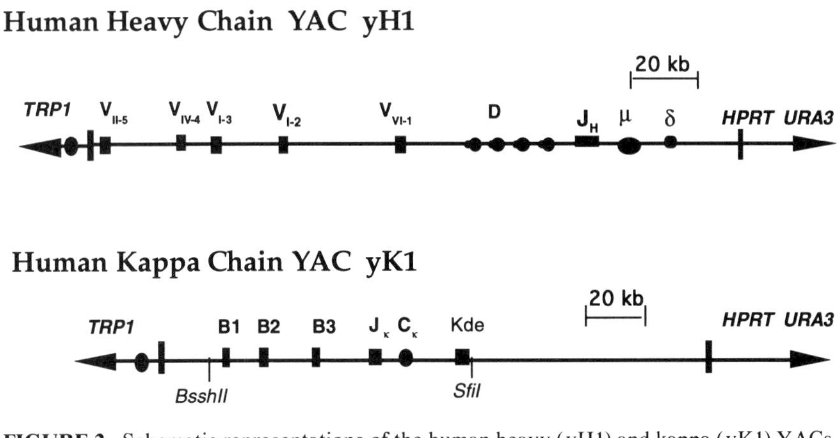

FIGURE 2. Schematic representations of the human heavy (yH1) and kappa (yK1) YACs, retrofitted with a HPRT minigene. The locations of specific Ig regions are indicated along with YAC vector elements: ◀, telomere; ●, centromere; ■, *Eco*RI cloning site and yeast-selectable markers, TRP1 and URA3 (YAC arms are not shown to scale).

TABLE 1A. Structural Integrity of Human Heavy Chain-Derived YAC in ES Clones, Generated from Fusion with yH1-Containing Yeast Spheroplasts[a]

yH1 ES Cell Clone	Cδ 7.8 kb	Cμ 11 kb	J_H 10 kb	D 9.7, 8.5, 7 kb	V_{VI} ~28 kb	V_I >12, 1.7 kb	V_{IV} 8 kb	V_{II} 4.8 kb	Yeast Genomic Sequences	YAC Copy Number
2B	+	+	+	+	+	+	+	+	–	1
2C	+	+	+	+	+	+	+	+	+	2
125A	+	+	+	+	+	+	+	+	+	1
125E	+	+	+	+	+	+	+	+	–	1
3A	+	+	+	+	+	+	+	+	+	1
3B	+	+	+	9.7 kb 8.5 kb	+	+	+	+	+	~8
5C	+	+	+	+	+	+	+	–	–	1

[a] HPRT$^+$-ES clones were analyzed for the presence of human heavy chain-specific sequences and yeast genomic sequences. Southern blot analysis, probes used, and determination of YAC copy number were as described in Green *et al.*[2] The presence of the expected fragment sizes for the specific probe (as shown) and of varied levels of yeast genomic sequences are indicated as (+). Altered-sized fragments are indicated. In the case of yH1-ES clone 3B, the 7-kb doublet was deleted. ND, not determined.

TABLE 1B. Structural Integrity of Human Kappa Light Chain Derived YAC in ES Clones, Generated from Fusion with yK1-Containing Yeast Spheroplasts[a]

yK1 ES Cell Clone	Kde 2.5 kb	C_κ 5.8 kb	J_κ 5 kb	B3 4.8 kb	B2 5.1 kb	B1 2.8 kb	Yeast Genomic Sequences	YAC Copy Number
4.4	+	+	5.5 kb	4.7 kb	4.9 kb	+	+	1
5.2	+	+	+	+	+	+	−	1
8.2A	+	+	+	+	+	+	−	1
9.2E	+	+	+	+	+	+	+	2
9.2C	+	+	+	+	+	+	−	1
10.2B	+	+	+	+	+	+	+	1
9.2H	+	+	+	+	+	+	+	~3
9.2F	+	+	+	+	ND	+	+	1
10.2C	+	+	+	+	ND	+	+	1
10.2A	+	+	+	+	+	+	+	2
10.2CA	+	+	+	+	+	+	+	2

[a] HPRT$^+$-ES clones were analyzed for the presence of human kappa chain-specific sequences and yeast genomic sequences. Southern blot analysis, probes used, and determination of YAC copy number were as described in Green *et al.*[2] The presence of the expected fragment sizes for the specific probe (as shown) and of varied levels of yeast genomic sequences are indicated as (+). Altered-sized fragments are indicated. ND, not determined.

cointegrated yeast genomic DNA apparently did not interfere with proper mouse development. These results demonstrate the utility and reproducibility of ES cell–yeast spheroplast fusion technology for the functional introduction of large and complex loci into the mouse germline.

PRODUCTION OF ANTIGEN-SPECIFIC HUMAN ANTIBODIES BY yH1-, yK1-BEARING MICE

Expression of the inherited human heavy and kappa chain YACs by mouse B cells was demonstrated by flow cytometry and ELISA. In all yH1-containing mouse strains analyzed, 1.5–2.5% of the B220$^+$ cells expressed human μ chain (hμ). Approximately half of this population had no detectable surface mouse μ, indicating that the production of human heavy chains can at least partially exclude the expression of the mouse μ chains. The majority of the B220$^+$, hμ^+ population also expressed surface human δ chain (hδ). Similarly, in all yK1-bearing mouse strains analyzed, 5 to 9% of the B220$^+$ splenocytes expressed surface human κ chain (hκ). The majority (80%) of the hκ^+ population did not express mouse kappa (mκ) or lambda (mλ), indicating that the human kappa chain was able to exclude expression of the mouse light chain genes.

Human heavy and kappa chain proteins were detected by ELISA in sera derived from the respective yH1- and yK1-containing mice. Human μ protein was detected in all yH1-bearing mice at concentrations of 0.2–3.8 μg/ml. Similarly, human κ protein, at levels of 8 to 30 μg/ml, was detected in all mice containing intact and unrearranged yK1. Thus, significant immunoglobulin chain expression was observed in mice derived from independent YAC-containing ES cell clones, which are likely to represent different YAC integration sites.

Heavy and kappa chain-expressing mouse strains were bred to generate human antibody-producing (HuAb) mice containing one copy each of yH1 and yK1. Flow cytometry analysis of B splenocytes, derived from five different HuAb strains, revealed $B220^+$ populations containing 0.8 to 3.9% $h\mu^+$ or 1.4 to 4.8% $h\kappa^+$ cells, as well as a population of $B220^+$ cells (0.06–0.27%) that simultaneously expressed both human μ and κ chains. The presence of fully human antibodies ($h\mu/h\kappa$), at levels of 0.3 to 0.7 mg/ml, in sera from HuAb mice was demonstrated by ELISA, based on the capture of the $h\mu$ chain and the detection of the $h\kappa$ chain.

To determine whether HuAb mice can mount a specific human antibody response, mice were immunized with tetanus toxin C fragment (tet C) or with a human protein, IgE antibody. Upon immunization, tet C-specific or IgE-specific $h\mu$ and $h\kappa$ antibodies were readily detected in the serum of immunized mice. The human origin of these antigen-specific antibodies was confirmed by an ELISA in which each of the antigens was used to capture an anti-$h\mu$ or anti-$h\kappa$ employed to detect the bound species (FIG. 3). Furthermore, upon fusion of splenocytes derived from these immunized mice with P3X63-Ag8.653 myeloma cells, mouse hybridomas producing tet C or IgE-specific fully human antibodies were generated.[2] These results indicate that HuAb mice, although containing only a small fraction of B cells coexpressing $h\mu$ and $h\kappa$, can mount an antigen-specific response leading to the generation of fully human monoclonal antibodies.

HUMAN IG YACS DIRECT THE PRODUCTION OF A DIVERSE, ADULT-LIKE REPERTOIRE OF HUMAN ANTIBODY SEQUENCES IN MICE

The diversity of the human antibody repertoire, directed by the human heavy and light chain YACs, was determined by analysis of immunoglobulin transcripts from mouse spleen. Colony hybridization analysis using V and J region-specific probes (TABLES 2A and B) and sequencing of human heavy and kappa chain cDNA clones revealed broad usage of V, D, and J segments contained in the YACs.[2] All J_H and J_κ segments were represented, with a frequency similar to that detected in adult human B cells.[6,7] In 18 heavy chain cDNAs, 10 different D segments were represented (N1, 21/9, XP4, M1, XP1, K1, N4, LR2, A4, A1).[2] Three of the four V_H families contained in yH1 (V_{VI}, V_{IV}, V_I) were represented. All three of the V_κ variable segments (B1, B2, and B3) were represented, with B3 the most frequently used. Frequent N segment addition (83%) was observed in the heavy chain cDNAs, with a majority having N segments at both the 5′ and 3′ D segment junctions. The average length of N regions was 6.1 nucleotides. The majority of the in-frame heavy chain clones contained a complementary determining region (CDR3) sequence of 10 or more amino acids, with some as long as 18 residues (average 12). CDR3 sequences, for the in-frame κ transcripts were 9 to 10 amino acids in length.

The pattern of V_H, D, and J_H use demonstrated is reminiscent of adult human B cells[6,7] in several respects. First, there is an absence of position-biased usage of V and D genes, in particular, the V_{VI} and DQ52 segments, which are preferentially employed during human fetal development.[8,9] In addition, the frequency of usage of specific J_H genes, the extent and length of N segments, and the length of the heavy and kappa CDR3 region, all appear to recapitulate the diversity seen in adult humans. These results suggest that the human immunoglobulin sequences contained on these YACs include all the elements necessary to produce a normal adult human heavy chain repertoire.

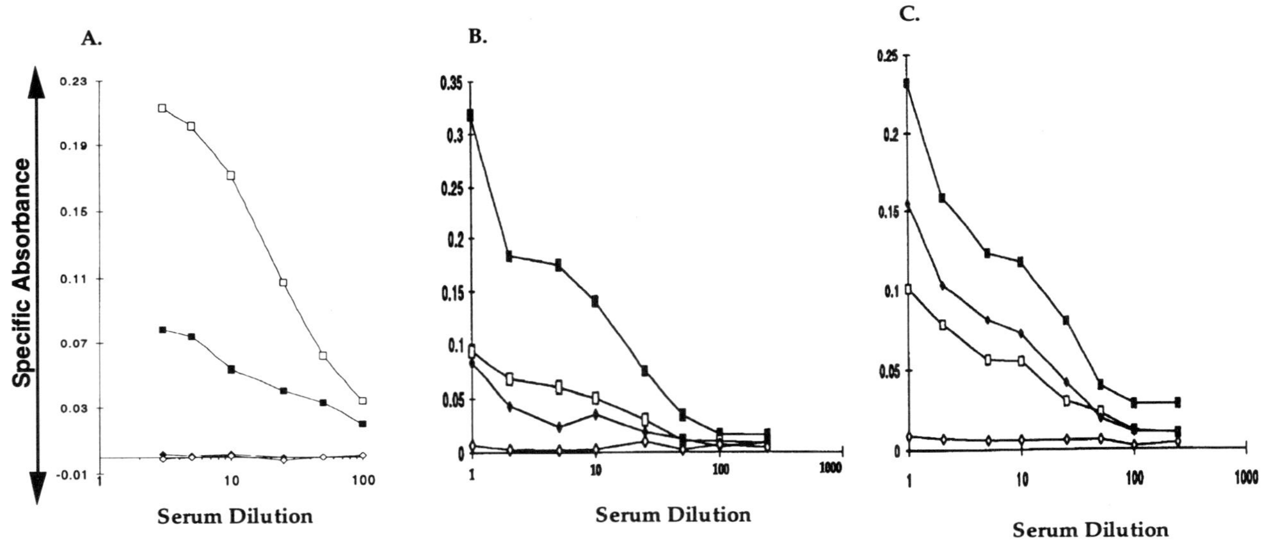

FIGURE 3. Production of tet C-specific human polyclonal **(A)** or monoclonal **(B, C)** antibodies by HuAb mice. **(A)** Sera obtained from HuAb strain nonimmune (◇, ◆) or hyperimmune (□, ■) to tet C were analyzed by ELISA for tet C-specific hμ (□) and hκ (■) antibodies as described in Green *et al.*[2] Supernatants derived from three individual hybridomas (■, □, ◆, control ◇), generated by fusion of hyperimmune HuAb strain splenocytes with mouse myeloma cells,[2] were analyzed for the presence of fully (hμ/hκ) human monoclonal antibodies **(B)**, and for specificity of the generated human antibodies to tet C **(C)**, detected by ELISA, using tetanus toxin to capture and anti-hκ to detect.[2] The hybridoma secreting a tetanus toxin-specific monoclonal antibody containing only human heavy chain (◇) was used as a control in **B** and **C**.

TABLE 2A. Repertoire Analysis of Human Heavy Chain Transcripts Expressed in Transgenic Mice[a]

	V_H	V_{IV}	V_I	V_{VI}	Total	Human PBL[b]
J_1	0	0	1	0	1 (1%)	1%
J_2	0	0	1	0	1 (1%)	0%
J_3	0	5	7	3	15 (18%)	9.1%
J_4	0	8	16	21	45 (54%)	52.5%
J_5	0	0	3	1	4 (5%)	15.2%
J_6	0	3	7	7	17 (21%)	22.2%
Total	0 (0%)	16 (19%)	35 (42%)	32 (39%)	83 (100%)	100%

[a] Human μ-specific mRNAs were amplified by PCR, cloned and analyzed by hybridization to V_H- and J_H- region specific probes, respectively, as described in Green *et al.*[2] Results of colony hybridizations showing V_H and J_H (as compared to that detected in human peripheral B cells described in Yamada *et al.*[6]) gene utilization. Results are given only for colonies hybridizing to a V_H, a J_H, and a $C\mu$ probe. A similar J, V utilization pattern was obtained by nucleotide sequences of unique human heavy chain cDNA clones.[2]

[b] Yamada *et al.*[6]

THE HUMAN IG YACS RESTORE B-CELL DEVELOPMENT AND ANTIBODY PRODUCTION IN IG-INACTIVATED MICE

The results described above demonstrate that the YAC-encoded human immunoglobulin genes are productively rearranged and expressed in the presence of functional mouse antibody genes. Nevertheless, the preferential expression of the mouse Ig genes indicated the need for their inactivation to increase the production of fully human antibodies in mice.

The ability of yH1 to restore proper B-cell development and antibody production was examined in mice in which both heavy chain alleles were functionally inactivated by gene-targeted J_H region ($\Delta J_H/\Delta J_H$).[3] In $\Delta J_H/\Delta J_H$ mice, bearing one copy of yH1 (yH1;$\Delta J_H/\Delta J_H$), reconstitution of mature B cells was detected in

TABLE 2B. Repertoire Analysis of Human Kappa Chain Transcripts Expressed in Transgenic Mice[a]

	B1	B2	B3	Total
J_1	4	13	108	125 (47%)
J_2	10	5	53	68 (26%)
J_3	0	2	29	31 (12%)
J_4	1	9	25	35 (13%)
J_5	0	0	4	4 (1%)
Total	15 (6%)	29 (11%)	219 (53%)	263 (100%)

[a] Human κ-specific mRNAs were amplified by PCR, cloned, and analyzed by hybridization to V_κ- and J_κ- region specific probes, respectively, as described in Green *et al.*[2] Results of colony hybridizations showing V_κ and J_κ gene utilization. Results are given only for colonies hybridizing to V_κ and J_κ probes. A similar J, V utilization pattern was obtained by nucleotide sequences of unique human kappa chain cDNA clones.[2]

bone marrow, spleen, and blood at levels corresponding to 58%, 55%, and 30% of the $B220^+$ population in wild-type mice, respectively. Virtually all mature B cells detected expressed hμ on their surface, with no detectable mouse μ, with the majority of the $B220^+/h\mu^+$ cells also expressing hδ, indicating proper expression and regulation of the human constant region in these mice.

The orderly differentiation of bone marrow–derived B cells in yH1;$\Delta J_H/\Delta J_H$ mice was evaluated by three-color flow cytometry using antibodies to the cell surface markers CD43, HSA, and BP-1, which define early B-cell subpopulations.[10] Such analysis revealed that proper B-cell development was substantially restored by yH1. The maturation of pro-B cell to pre-B cell, however, a point at which V-DJ rearrangement is initiated, was found to be less efficient than in normal mice.[2] These findings suggested that lower frequency of rearrangement, which may result from the limited number of V genes in yH1, can lead to the detected incomplete B-cell development.

ELISA analysis of serum from yH1;$\Delta J_H/\Delta J_H$ mice demonstrated the production of hμ antibodies at a concentration of 350 μg/ml, a level 510-fold greater than that produced by yH1-transgenic mice, and only several-fold below normal human IgM serum levels.

These results demonstrate that the human heavy chain YAC can reconstitute normal B-cell development in ΔJ_H-inactivated mice and induce the proper expression and assembly of human hμ chains with mouse light chains to form functional antibodies.

We further examined the ability of the heavy chain YAC, in conjunction with the human kappa chain YAC, to restore normal B-cell development and antibody production in the absence of functional mouse heavy and kappa light antibody genes. HuAb strains were bred with mice that are homozygous for inactivated heavy ($\Delta J_H/\Delta J_H$) and kappa ($\Delta C_\kappa/\Delta C_\kappa$, derived by deletion of the mouse Cκ region) chain alleles, in which the expression of mouse heavy and kappa chain expression is completely blocked. The generated mouse strain, Xenomouse, contained the human heavy and kappa YACs and lacked functional mouse heavy and kappa chains. Analysis of peripheral blood lymphocytes and splenocytes from Xenomouse indicated the presence of mature $B220^+$ cells, 45% of which are coexpressing exclusively hμ and hκ chains, and 55% expressing the hμ and mλ chains. No coexpression of mλ and hκ was detected, indicating that each of the light chains can completely exclude the expression of the other. The majority (75%) of the antibodies detected in Xenomouse serum were fully human antibodies (350 μg/ml) with a minority of hμ/mλ antibodies. The level of B cells expressing only fully human antibodies, and the level of serum hμ/hκ antibodies in Xenomouse are several hundred-fold higher than those detected in HuAb strains, confirming that inactivation of the mouse Ig genes greater increased the level of fully human antibodies and suggesting Xenomouse as a valuable source for the production of human monoclonal antibodies.

SUMMARY

Our paper describes the introduction of large fragments of both the human heavy and light chain Ig genes into the mouse germline to create a mouse strain capable of producing a broad repertoire of antigen-specific, fully human antibodies.

The human immunoglobulin gene sequences were functional in the context of the mouse machinery for antibody recombination and expression, either in the presence or absence of functional endogenous genes. This was demonstrated by their ability to undergo diverse rearrangement, to be expressed at significant levels, and to exclude expression of mouse immunoglobulins irrespective of their copy number or site of integration. The decrease in susceptibility to influence by adjacent genomic sequences may reflect the greater size, variable gene content, or structural integrity of the human Ig YACs and/or the presence of unidentified but important regulatory elements needed for optimal expression of the human immunoglobulin genes and their correct regulation.

Our results show that mouse B cells coexpressing human heavy and kappa chains, upon immunization, can produce antigen-specific, fully human antibodies. Furthermore, the human heavy and kappa chain YACs induced differentiation and maturation of the growth-arrested B-cell lineage in mice with inactivated endogenous Ig genes, leading to the production of a diverse repertoire of fully human antibodies at levels approaching those in normal serum. These results suggest the potential value of these mice as a source of fully human antibodies for human therapy. Furthermore, it is expected that such mice would lack immunological tolerance to and thus readily yield antibodies to human proteins, which may constitute an important class of targets for monoclonal antibody therapy.

Our findings suggest that the introduction of even larger portions of the human heavy and light chain loci, which should be achievable with the ES cell–yeast spheroplast fusion technology described, will result in strains of mice ultimately capable of recapitulating the full antibody repertoire characteristic of the human humoral response to infection and immunization. The present and future mouse strains may prove to be valuable tools for studying the molecular mechanisms and regulatory sequences influencing the programmed assembly and expression of human antibodies in the normal immune response, as well as the abnormal response characteristic of autoimmune disease and other disorders.

The strategy we have described for the introduction of large segments of the human genome into mice in conjunction with the inactivation of the corresponding mouse loci may also have broad applicability to the investigation of other complex or uncharacterized loci.

ACKNOWLEDGMENTS

We thank Eva Ramirez and Veronica Lubeck for figure preparations.

REFERENCES

1. Jakobovits, A., A. L. Moore, L. L. Green, G. J. Vergara, C. E. Maynard-Currie, H. A. Austin & S. Klapholz. 1993. Nature **362:** 255–258.
2. Green, L. L., M. H. Hardy, C. E. Maynard-Currie, H. Tsuda, D. M. Louie, M. J. Mendez, H. Abderrahim, M. Noguchi, D. H. Smith, Y. Zeng, N. E. David, H. Sasai, D. Garza, D. G. Brenner, J. F. Hales, R. P. McGuinness, D. J. Capon, S. Klapholz & A. Jakobovits. 1994. Nature Genetics **17:** 13–21.
3. Jakobovits, A., G. J. Vergara, J. L. Kennedy, J. F. Hales, R. P. McGuinness, D. E. Casentini-Borocz, D. G. Brenner & G. R. Otten. 1993. Proc. Natl. Acad. Sci. USA **90:** 2551–2555.
4. Mendez, M. J., H. Abderrahim, M. Noguchi, N. E. David, M. H. Hardy, L. L. Green, H. Tsuda, S. Yoast, C. E. Maynard-Currie, D. Garza, R. Gemmil, A. Jakobovits & S. Klapholz. 1995. Genomics **26:** 294–307.

5. REID, L., R. G. GREGG, O. SMITHIES & B. KOLLER. 1990. Proc. Natl. Acad. Sci. USA **87:** 4299–4303.
6. YAMADA, M., R. WASSERMAN, B. A. REICHARD, S. SHANE, A. J. CATON & G. ROVERA. 1991. J. Exp. Med. **173:** 395–407.
7. MARKS, J. D., M. TRISTEM, A. KARPAS & G. WINTER. 1991. Eur. J. Immunol. **21:** 985–991.
8. SCHROEDER, H. W., JR. & J. Y. WANG. 1990. Proc. Natl. Acad. Sci. USA **87:** 6146–6150.
9. CUISINIER, A.-M., V. GUIGUO; L. BOUBLI, M. FOUGEREAU & C. TONNELLE. 1989. Scand. J. Immunol. **30:** 493–497.
10. HARDY, R. R., C. E. CONDIE, S. A. SHINTON, J. D. KEMP & K. HAYAKAWA. 1991. Resolution and characterization of pro-B and pre-pro-B cell stages in normal bone marrow. J. Exp. Med. **173:** 1213–1225.

Class Switching in Human Immunoglobulin Transgenic Mice

FIONA A. HARDING AND NILS LONBERG

GenPharm International
297 N. Bernardo Avenue
Mountain View, California 94043

INTRODUCTION

To generate mice that express a repertoire of human sequence antibodies, we disrupted the endogenous mouse heavy and κ light chain loci by homologous recombination in embryonic stem cells[1,2] and bred the resulting B-cell-deficient animals with mice harboring germline configuration human immunoglobulin transgenes.[3] These transgenes are artificial miniloci that represent compressed, or edited, versions of the natural immunoglobulin loci. Despite the complexity of the natural Ig loci—exemplified by the large number of different types of transcriptional, recombinational, and coding elements contained within them—we and others have shown that subsets of the intact immunoglobulin loci can undergo V(D)J joining. In addition, these relatively short sequences can also functionally complement mutations of the endogenous mouse loci by rescuing B-cell development.[3–9] It thus appears to be possible to construct relatively small artificial miniloci without discarding important functional sequences.

CONSTRUCTION OF HEAVY CHAIN MINILOCUS TRANSGENES

The natural human heavy chain locus includes ~50 functional V gene segments, ~30 D gene segments, 6 J gene segments, and 9 constant region gene segments.[10,11] Associated with each of these protein-coding segments are *cis*-acting transcriptional regulatory elements and/or recombination signal sequences. In germline chromosomal DNA these elements are spread out over a distance of 1.3 Mb at the telomere of chromosome 14. We have constructed and microinjected several different heavy chain transgenes from subsets of these sequences. Two of the transgenes, HC1 and HC2, are depicted in FIGURE 1. They include one and four functional variable (V) segments, respectively; 10 and 15 diversity (D) segments, respectively; all six joining (J_H) segments; and both the μ and $\gamma 1$ constant region segments.[4,5] All of these coding segments are human. The miniloci include human *cis*-acting regulatory sequences—such as the J_H-μ intronic enhancer and the μ and $\gamma 1$ switch sequences—that are closely linked to the coding segments. They also include an additional enhancer element derived from the 3′ end of the rat IgH locus.[12] Both constructs are propagated as single inserts that can be isolated from plasmid vector sequences by digestion with *Not*I. The smaller construct, HC1, is a 62-kb insert, whereas the larger construct, HC2, is an 80-kb insert.

The HC1 transgene contains the single functional V gene segment V_H5-51. This V gene segment is one of two members of the V_H5 family that have an open reading frame, and it is the only V_H5 gene that is known to contribute to the human repertoire.[13] In addition to this gene, the larger transgene construct, HC2,

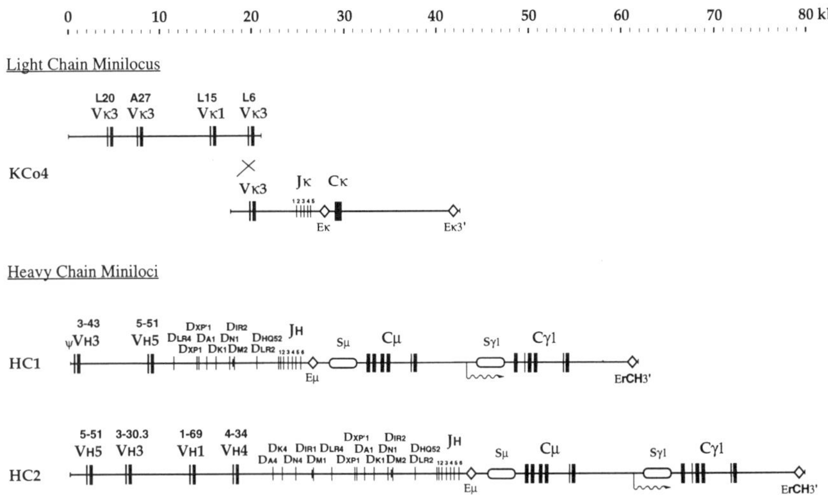

FIGURE 1. Human immunoglobulin minilocus transgenes. The constructs are described in detail elsewhere.[3,5] The light chain minilocus, KCo4, is created by co-injection of two overlapping DNA fragments, each of which is propagated within a different plasmid. One of the DNA fragments includes four different functional κ variable segments. The second fragment includes one of these same variable sequences together with all five human Jκ segments and the human κ constant region segment. This fragment also includes both the κ intronic enhancer and the 3′ κ enhancer. All of the sequences in these two inserts are of human origin. The HC1 heavy chain minilocus contains a single functional variable gene segment, 10 functional D segments, all six J_H segments, the μ and $\gamma 1$ constant region segments, as well as the μ and $\gamma 1$ switch regions. It also contains the human heavy chain J-μ intronic enhancer and the rat heavy chain 3′ enhancer. The HC2 plasmid insert is similar to the HC1 insert; however, it includes three additional variable gene segments and five additional D segments, as well as a further 0.7 kb of human $\gamma 1$ upstream sequence not included in HC1. All of the sequences in HC1 and HC2 are human with the exception of the rat 3′ enhancer. The entire 63-kb HC1 and 80-kb HC2 transgenes are isolated from vector sequences by digestion with *Not* I. The start site of the human $\gamma 1$ preswitch sterile transcript is indicated by the wavy arrow below HC1. V, variable segment; D, diversity segment; J, joining segment; C, constant region gene; S, switch region; E, enhancer. Transgenic mice: DNA fragments were purified by agarose gel electrophoresis and microinjected into the pronuclei of half-day (C57BL/6J × CBA/J)F2 embryos. The resulting mice and their progeny were bred with mice containing targeted mutations in the endogenous mouse heavy and κ light chain loci to obtain mice containing human heavy and κ light chain loci in the absence of functional mouse heavy and κ light chain loci.

includes three other functional V genes: V_H1-69, V_H3-30.3, and V_H4-34. V_H1-69 is one of eight members of the V_H1 family for which VDJ rearrangements have been reported. V_H30.3 is a member of the largest V_H family, V_H3, which comprises 22 known functional genes.[14] V_H4-34 is a member of the V_H4 family and encodes the 9G4 idiotype.[15] Although these four genes comprise only 8% of the entire library of functional V gene segments, they represent frequently rearranged genes, and together they account for approximately one-quarter of the reported rearranged gene sequences in the literature.[16]

CONSTRUCTION OF κ LIGHT CHAIN MINILOCUS TRANSGENES

The human κ locus consists of ~30 functional V gene segments, divided between two clusters separated by 800 kb, five J gene segments, and a single constant region gene segment.[17] The entire locus spans ~1.9 Mb and is located on chromosome 2.[18] Like the rodent heavy chain locus, the κ locus contains a J-C intronic enhancer as well as a 3′ enhancer.[19] Both of these enhancers have been identified in the human κ locus[20] and have been included in our light chain transgene, KCo4[3] (FIG. 1).

The KCo4 κ-light chain transgene is derived from two individually cloned DNA fragments. The fragments together comprise four functional Vκ segments, five J segments, the Cκ exon, and both the intronic and downstream enhancer elements. All of the sequences are of human origin. Because the two fragments share a common 3-kb sequence, they can potentially integrate into the mouse genome as a contiguous 43-kb transgene following homologous recombination between the overlapping sequences. DNA sequence analysis of light chain transcripts from two independent lines reveals that all four of the transgene-encoded V segments are incorporated into the light chain repertoire. There is no expression bias in favor of the J proximal V segment, which is the only V that is physically linked to Cκ before microinjection.[3] Expression of all transgene-encoded V regions and a lack of bias for the J proximal V region gene suggest that the coinjection strategy resulted in homologous recombination and cointegration of the two fragments at one site in the genome.

COMPLEMENTATION OF MOUSE IGH AND IGκ MUTATIONS BY HUMAN TRANSGENES

We bred the transgenic animals with two different strains of mice containing independent mutations in the endogenous heavy and κ light chain loci. The mutant heavy chain strain (J_HD) contains a deletion of the entire J_H region.[1] Homozygous J_HD mice do not have any B cells in the periphery. In the bone marrow, there is a blockade of B-cell development at the pro-B-cell stage. The mutant κ light chain strain (JKD) includes a deletion of both the Jκ and the Cκ gene segments.[2] Homozygous JKD mice do not express any detectable κ chain and have fewer peripheral B cells than normal mice. The human sequence minilocus transgenes, HC1, HC2, and KCo4 partially restore the B-cell compartment in mice carrying disrupted endogenous heavy and κ light chain loci. We find human IgM-expressing cells in the spleen, lymph nodes, peritoneal cavity, and bone marrow of the KCo4/HC2 mice that are homozygous for both the J_HD and JKD mutations (double-transgenic/double-deletion mice). Although the peritoneal cavity contains the normal number of B cells, the absolute number of transgenic B cells in the bone marrow and spleen is typically 10 to 50% of normal.[3,5] The reduction could be the result of a retardation in transgene-dependent B-cell development. This would have less impact on compartments consisting largely of self-renewing B cells, such as the peritoneum, than on the bone marrow and spleen, which require a constant supply of newly formed cells. Whatever is responsible for the reduced size of the conventional B-cell population, once the transgenic cells are formed they appear to be functional.

To test the ability of the transgenic B cells to participate in an immune response, we immunized the double-transgenic/double-deletion mice with human protein antigens and monitored serum levels of antigen specific immunoglobulins. The initial antibody response against human IgE emulsified in complete Freund's adjuvant was a human IgMκ α-IgE response. Upon boosting, the initial IgM response was followed three weeks later by the emergence of antigen-specific human IgGκ.[3] This lag before the appearance of IgG antibodies is consistent with an association between class switching and a secondary response to antigen. Repeated exposure to human antigens leads to relatively high serum titers of specific human IgG. In serum from a transgenic mouse immunized with recombinant human CD4, human IgG reactivity to antigen is detectable at a dilution of one part in ten thousand.

SOMATIC MUTATION AND CLASS SWITCHING

Class switch recombination of introduced heavy chain transgenes has been previously reported.[21,22] Nevertheless, because the *trans*-switch recombination events described in those experiments involved transgene-encoded μ and endogenous γ switch sequences, they do not completely define the *cis*-acting sequences required for switching. Although we also observe *trans*-switching in our system, our introduced heavy chain minilocus transgenes include both μ and γ1 switch regions and are thus capable of undergoing intra-transgene class switch recombination. Furthermore, transgene-autonomous class switching in the transgenic mice appears to be taking place in the same population of B cells that are involved in the secondary response to antigen, suggesting that the human transgene class switch could be regulated in mouse cells. This is demonstrated by the observation that somatic mutation of the heavy chain transgenes is largely confined to the products of class switching. We determined the V region nucleotide sequence of 47 heavy chain cDNA clones from a single HC1 transgenic animal; 24 of the clones were μ and 23 were γ.[5] The density of nongermline-encoded nucleotides in the μ population is low (0.1%) and is indistinguishable from a Poisson distribution (FIG. 2, *top panel*). However, the overall density of nongermline-encoded nucleotides within the γ population is an order of magnitude higher (1.2%) and does not fit a Poisson distribution (FIG. 2, *bottom panel*). Instead, it appears that the γ clones are derived from at least two distinct populations: a minority (approximately 20%) of the clones resemble the μ population and contain few nongermline-encoded nucleotides, while the majority (approximately 80%) are derived from cells that have undergone multiple rounds of somatic mutation. The distribution of these somatic mutations across the transgene V regions is not uniform. The highest density of nongermline-encoded nucleotides occurs within the CDR1 region, and specific "hot spots" accumulate the most mutations. Curiously, these "hot spots" correspond well to codons that are frequently mutated in normal human cells (FIG. 3). The four codons that are found to be most frequently mutated in human B cells are also the four most frequently altered codons in the transgenic B cells. It thus appears that the pattern of somatic mutation is an intrinsic property of the V region sequence and that it is unlikely to be dependent on a template provided by other V, or V-like, sequences elsewhere in the genome. We also observe some suppression of replacement changes within framework regions, and an apparent selection for replacement changes in CDR regions. For a set of 68 γ sequences encoding V_H5-51 or V_H3-30.3,[5] the observed replacement-to-silent (R/S) ratios for framework and CDR regions are 2.7 and 6.4, respectively. The expected random R/S ratios for these same sequences are 4.0 and 4.0. Thus, the confinement of somatic mutation to γ sequences, the similarity of the pattern of somatic muta-

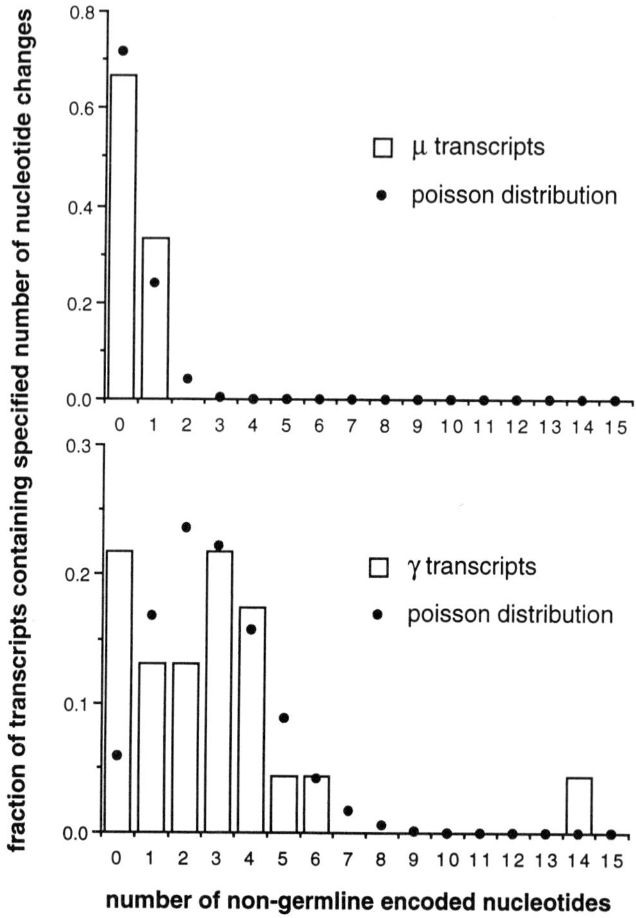

FIGURE 2. Density of observed nucleotide changes per cDNA clone in μ and γ transcripts from a single transgenic mouse. The nucleotide sequence was determined of μ and γ cDNA clones isolated from the same RNA sample taken from the spleen of a single HC1 transgenic mouse.[5] *Top panel*: data from 24 μ sequences shown as open bars; Poisson distribution for this average density shown as filled circles. *Bottom panel*: data from 23 γ sequences shown as open bars; Poisson distribution for this average density shown as filled circles.

tions in human and transgenic B cells, and the distortion of R/S ratios in CDR and framework regions all indicate that the transgenic cells are participating in a secondary response and that class switching of the human sequences is part of this response.

In normal mice there is abundant evidence that class switching is regulated by sequences associated with each of the individual non-μ switch regions. Individual antigen-activated B cells enter distinct maturation programs determined by the levels and types of cytokines and other switch factors they are exposed to, and marked by the isotype of the class switch they undergo. For example, LPS and interleukin 4 (IL-4) can specifically induce switching to IgG1 and IgE in mouse B cells. Sterile transcripts, which initiate upstream of the switch regions, and can be induced by cytokines that also specifically induce switch recombination,[23] may be controlled by the same *cis*-acting sequences that control switching. Sequences that are closely linked to these sterile transcripts are clearly important; for example, targeted deletion of the first exon of the mouse γ1 and γ2b sterile transcript, together with upstream flanking sequences, eliminates class switching to γ1 and γ2b, respectively.[24,25] These results establish that upstream *cis*-acting sequences define the functionality of the individual switch regions and are necessary for class switching. Our observation—that class switching within the HC1 transgene is

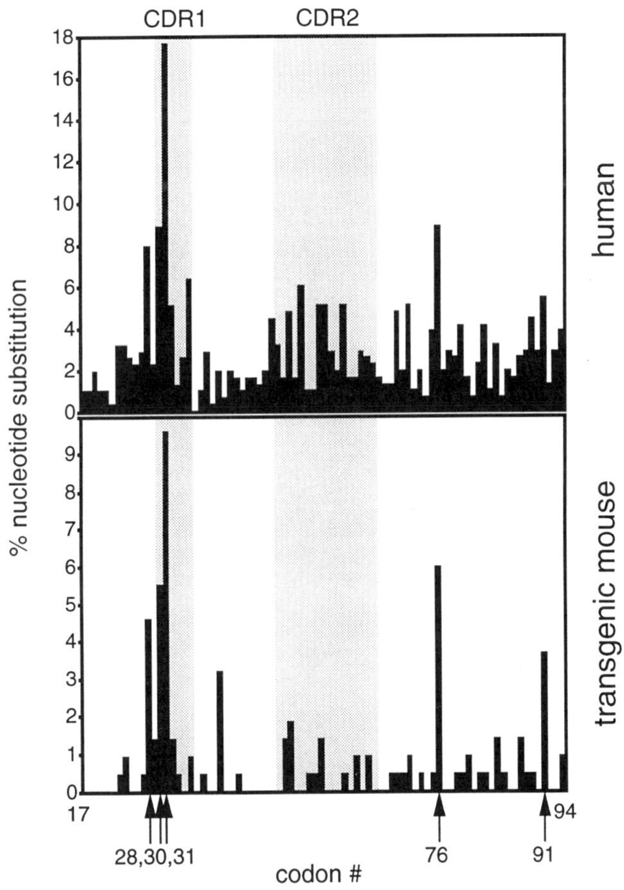

FIGURE 3. Distribution of somatic mutations within V_H251 transcripts expressed in humans and transgenic mice. Somatic mutation "hot spots" within the V_H251 gene are similar in authentic human B cells and B cells from transgenic mice. *Top*: distribution of nongermline-encoded nucleotides with 106 authentic human V_H251 sequences from the published literature.[31-34] *Bottom*: distribution for 73 V_H251 sequences (49 γ and 24 μ cDNA clones) from HC1 transgenic mice.[5]

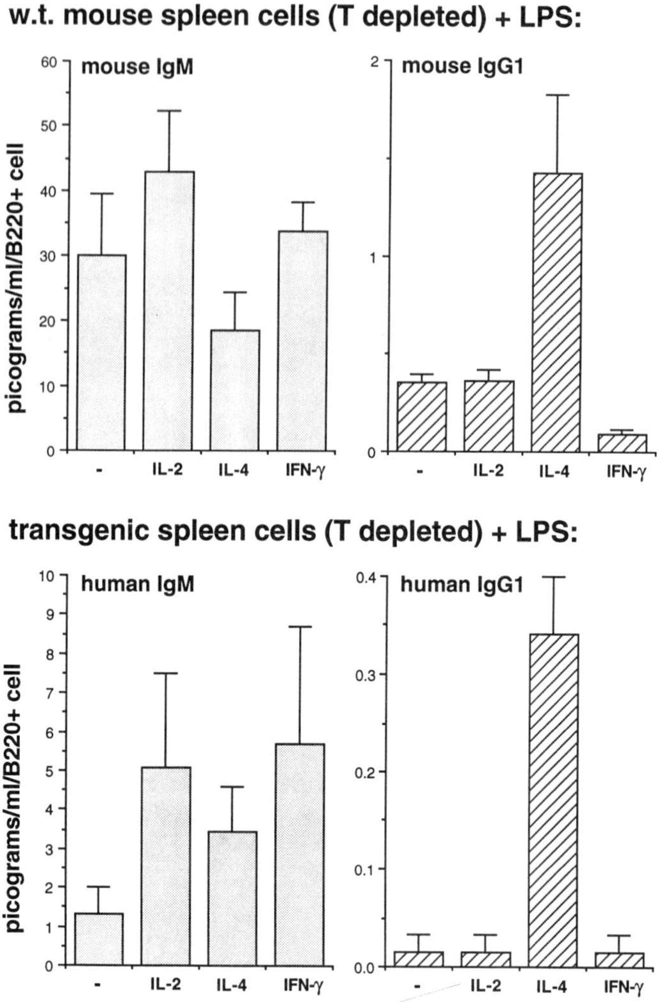

FIGURE 4. Cultured transgenic B cells undergo regulated class switching to the transgene-encoded human γ1 isotype. Splenic B cells from wild-type control, and double deletion/double transgenic mice (lines 2550/4436) were tested for immunoglobulin secretion in response to the thymus-independent B-cell mitogen, LPS, alone and in conjunction with various cytokines. Splenocytes were enriched for B cells by complement-mediated lysis of T cells. B-enriched cells were plated in 24-well plates at 2×10^6 cells per well in 1.5 ml of 10% FCS in RPMI-1640. LPS was added to all wells at 10 μg/ml. IL-2 was added at 50 units/ml, and IL-4 was added at 10 ng/ml. Cultures were incubated at 37°C, 5% CO_2 for 10 days, then supernatants were assayed for human and mouse IgM and IgG by ELISA. All reagents for ELISA were polyclonal antiserum (Jackson Immunologicals; West Grove, PA), except the capture anti-human IgM, which was a monoclonal antibody (The Binding Site, Birmingham, U.K.).

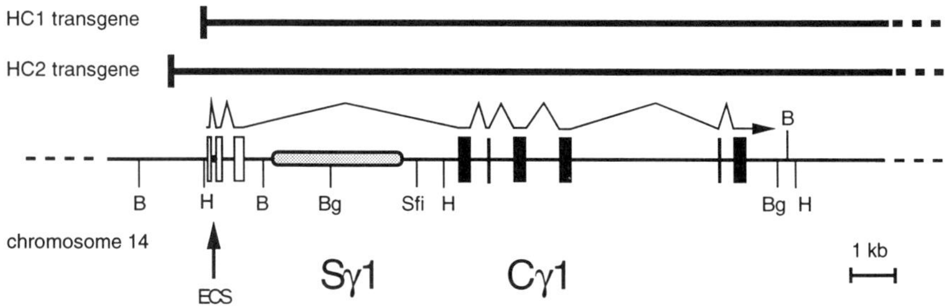

FIGURE 5. Human γ1 isotype switch control region. The germline chromosomal organization of the human γ1 gene and switch region, including exons encoding the preswitch sterile transcript. The chromosomal sequences included in the HC1 and HC2 transgenes are indicated above the map. *Open boxes,* sterile transcript exons; *filled boxes* (Cγ1), γ1 coding exons; shaded ellipse (Sγ1), switch region; ECS, evolutionarily conserved sequence.[27-29] Restriction sites: B, *Bam*HI; H, *Hin*dIII; Bg, *Bgl*II; Sfi, *Sfi*I.

largely confined to cells involved in secondary responses and does not occur randomly across the entire B-cell population—suggests that the minimal sequences contained within the transgene are sufficient. This result is supported by *in vitro* experiments with cultured transgenic B cells from double-transgenic/double-deletion mice. Transgenic B cells can be specifically induced to secrete human γ1 by LPS and IL-4 under the same conditions that induce mouse γ1 in wild-type mouse B cells (FIG. 4). Because the γ sequences included in the HC1 construct begin only 116 nucleotides upstream of the start site of the γ1 sterile transcript (the HC2 transgene includes an additional 0.7 kb of upstream sequence), the switch regulatory region appears to be compact (FIG. 5). Further experiments are required to determine if the observed regulation is correct in detail. However, structural comparisons of mouse and human γ genes, as well as differences in IgG subclass responses, suggest that the individual subclasses evolved independently in the two lineages.[26] It is therefore surprising that a human γ1 switch region is regulated at all in mouse cells *in vivo*. A possible explanation for this can be obtained from an inspection of the nucleotide sequences upstream of the human and mouse switch regions (FIG. 6). The upstream sequence has been previously reported to include a region of nucleotide sequence that is conserved across all of the mouse and human γ genes.[27-29] This so-called "evolutionary conserved sequence" (ECS) is coincident with a fragment of DNA from the mouse γ1 gene that has been reported in the literature to confer IL-4-inducible transcription on a reporter gene and is presumed to include the IL-4-inducible enhancer that regulates mouse γ1 switching.[30] The sequence homology matrices presented in FIGURE 6 show that, within this conserved region, the mouse γ1 gene is more closely related to the human γ genes than it is to the other mouse γ genes. Thus, in terms of regulation of class switching, the human γ genes appear to be the functional homologues of the mouse γ1 gene.

SUMMARY

We have introduced human germline-configuration heavy and kappa light chain minilocus transgenes into mice that have been engineered so that their endogenous

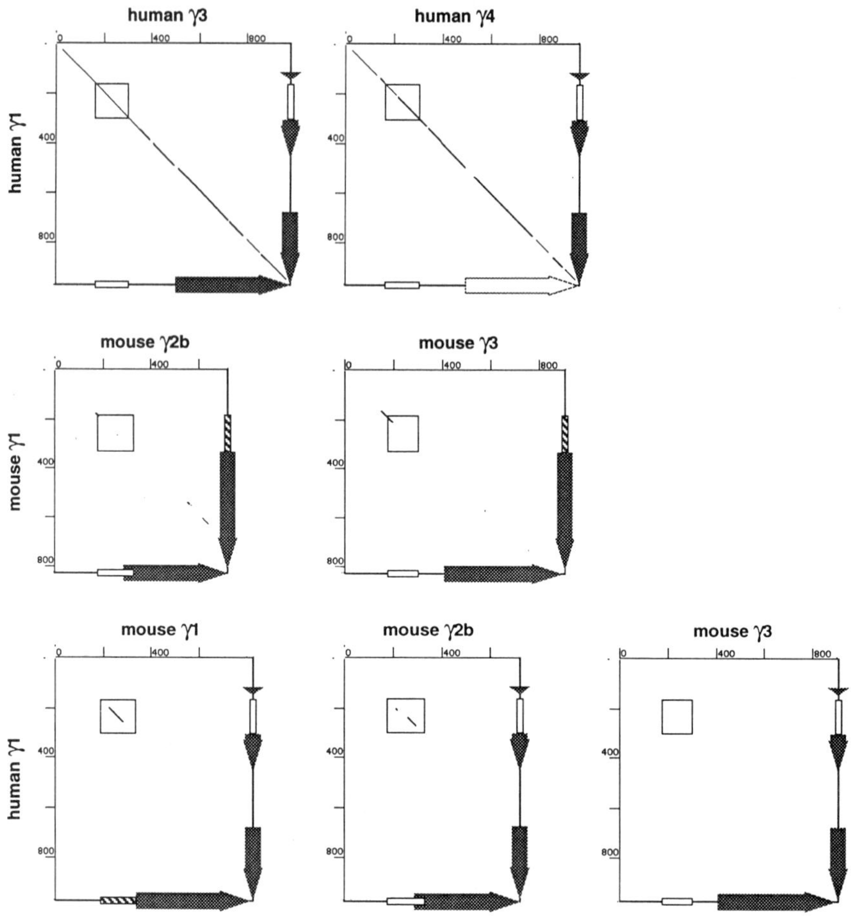

FIGURE 6. Sequence homology between human and mouse γ switch control regions. The nucleotide sequences of the sterile transcript exons and immediate upstream flanking regions of the human γ1, γ3, and γ4 genes—and the mouse γ1, γ2b, and γ3 genes—are compared using a nucleotide identity matrix algorithm. Each point represents a 51-bp window of one sequence with 65% or greater nucleotide identity to the corresponding 51 bp of the other sequence. By this very stringent criterion it can be seen that the individual human γ upstream regions are more closely related to each other than the individual mouse sequences are to each other. And, the human genes include sequences that are homologous to sequences within the mouse genes. The shaded arrows represent sterile transcript exons that have been reported in the literature (human γ1 & γ3,[27] mouse γ1,[35–37] mouse γ2b,[38] mouse γ3.[28,39] The exon for human γ4 is represented by an unshaded arrow because its position is derived from that of the related human γ3 gene.[29] The striped box upstream of the mouse γ1 gene represents the 150-bp region shown by Xu and Stavnezer[30] to be sufficient for IL-4-inducible transcription. The corresponding sequence for each of the other γ genes is represented by an open box. The rectangle within each of the matrices indicates the sequence comparison of these corresponding regions. This region includes those upstream sequences that share the greatest degree of sequence identity between the mouse and human genes.

heavy and kappa light chain loci are inactive. The two human transgenes are inserted by pronuclear microinjection, while the two endogenous mouse genes are disrupted by homologous recombination in embryonic stem cells. The resulting animals contain four unlinked genetic modifications and must rely on the introduced transgenes for the development of the B-cell lineage and for the generation of an antibody repertoire. The heavy chain transgene includes both the human μ and the human $\gamma 1$ constant region gene segments, as well as upstream switch region sequences. Although mouse B cells and human B cells exhibit species-specific differences in the induction of γ isotype expression, the transgenic mouse B cells appear to undergo regulated switching to human $\gamma 1$ both *in vivo* and *in vitro*. This observation defines a subset of the heavy chain constant region that is sufficient for class switching, and implies that the human $\gamma 1$ switch region includes a core of sequence that is functionally homologous to those *cis*-acting regulatory elements that direct mouse γ switching.

ACKNOWLEDGMENT

We thank Condie Carmack for comments on the manuscript.

REFERENCES

1. Chen, J., M. Trounstine, F. W. Alt, C. Kurahara, F. Young, J. F. Loring & D. Huszar. 1993. Int. Immunol. **5:** 647–656.
2. Chen, J., M. Trounstine, C. Kurahara, F. Young, C.-C. Kuo, Y. Xu, J. F. Loring, F. W. Alt & D. Huszar. 1993. EMBO J. **12:** 821–830.
3. Lonberg, N., L. D. Taylor, F. A. Harding, M. Trounstine, K. M. Higgins, S. R. Schramm, C.-C. Kuo, R. Mashayekh, K. Wymore, J. G. McCabe, D. Munoz-O'Regan, S. L. O'Donnell, E. S. G. Lapachet, T. Bengoechea, D. M. Fishwild, C. E. Carmack, R. M. Kay & D. Huszar. 1994. Nature **368:** 856–859.
4. Taylor, L. D., C. E. Carmack, S. R. Schramm, R. Mashayekh, K. M. Higgins, C.-C. Kuo, C. S. Woodhouse, R. M. Kay & N. Lonberg. 1992. Nucleic Acids Res. **20:** 6287–6295.
5. Taylor, L. D., C. E. Carmack, D. Huszar, K. M. Higgins, R. Mashayekh, G. Sequar, S. R. Schramm, C-C. Kuo, S. L. O'Donnell, R. M. Kay, C. S. Woodhouse & N. Lonberg. 1994. Int. Immunol. **6:** 579–591.
6. Green, L. L., M. C. Hardy, C. E. Maynard-Currie, H. Tsuda, D. M. Louie, M. J. Mendez, H. Abderrahim, M. Noguchi, D. H. Smith, Y. Zeng, N. E. David, H. Sasai, D. Garza, D. G. Brenner, J. F. Hales, R. P. McGuinness, D. J. Capon, S. Klapholz & A. Jakobovitz. 1994. Nature Genetics **7:** 13–21.
7. Chen, J., R. Lansford, V. Stewart, F. Young & F. W. Alt. 1993. Proc. Natl. Acad. Sci. USA **90:** 4528–4532.
8. Brüggemann, M., H. M. Caskey, C. Teale, H. Waldmann, G. T. Williams, M. A. Surani & M. S. Neuberger. 1989. Proc. Natl. Acad. Sci. USA **86:** 6709–6713.
9. Wagner, S. D., G. T. Williams, T. Larson, M. S. Neuberger, D. Kitamura, K. Rajewsky, J. Xian & M. Brüggemann. 1994. Nucleic Acids Res. **22:** 1389–1393.
10. Cook, G. P., I. M. Tomlinson, G. Walter, H. Reithman, N. P. Carter, L. Buluwela, G. Winter & T. H. Rabbitts. 1994. Nature Genetics **7:** 162–168.
11. Hofker, M. H., M. A. Walter & D. W. Cox. 1989. Proc. Natl. Acad. Sci. USA **86:** 5567–5571.
12. Pettersson, S., G. P. Cook, M. Brüggemann, G. T. Williams & M. S. Neuberger. 1990. Nature **344:** 165–168.
13. Humphries, C. G., A. Shen, W. A. Kuziel, J. D. Capra, F. R. Blattner & P. W. Tucker. 1988. Nature **331:** 446–449.

14. Tomlinson, I., G. Walter, J. D. Marks, M. B. Llewelyn & G. Winter. 1992. J. Mol. Biol. **227:** 776–798.
15. Pascual, V., K. Victor, D. Lelsz, M. B. Spellerberg, T. J. Hamblin, K. M. Thompson, I. Randen, J. Natvig, J. D. Capra & F. K. Stevenson. 1991. J. Immunol. **146:** 4385–4391.
16. Griffiths, A. D., M. Malmqvist, J. D. Marks, J. M. Bye, M. J. Embleton, J. McCafferty, M. Bair, K. P. Holliger, B. D. Gorick, N. C. Hughes-Jones, H. R. Hoogenboom & G. Winter. 1994. EMBO J. **13:** 3245–3260.
17. Weichhold, G. M., R. Ohnheiser & H. G. Zachau. 1993. Genomics **16:** 503–511.
18. Klein, R., R. Jaenichen & H. G. Zachau. 1993. Eur. J. Immunol. **23:** 3248–3271.
19. Meyer, K. B. & M. S. Neuberger. 1989. EMBO J. **8:** 1959–1964.
20. Judde, J-G. & E. E. Max. 1992. Mol. Cell. Biol. **12:** 5206–5216.
21. Gerstein, R. M., W. N. Frankel, C. L. Hsieh, J. M. Durdik, S. Rath, J. M. Coffin, A. Nisonoff & E. Selsing. 1990. Cell **63:** 537–548.
22. Durdik, J., R. M. Gerstein, S. Rath, P. F. Robbins, A. Nisonoff & E. Selsing. 1989. Proc. Natl. Acad. Sci. USA **86:** 2346–2350.
23. Stavnezer, J., G. Radcliffe, Y.-C. Lin, J. Nietupski, L. Berggren, R. Sitia & E. Severinson. 1988. Proc. Natl. Acad. Sci. USA **85:** 7704–7708.
24. Jung, S., K. Rajewsky & A. Radbruch. 1993. Science **259:** 984–987.
25. Zhang, J., A. Bottaro, S. Li, V. Stewart & F. W. Alt. 1993. EMBO J. **12:** 3529–3537.
26. Callard, R. E. & M. W. Turner. 1990. Immunol. Today **11:** 200–202.
27. Sideras, P., T.-R. Mizuta, H. Kanamori, N. Suzuki, M. Okamoto, K. Kuze, H. Ohno, S. Doi, S. Fukuhara, M. S. Hassan, L. Hammarstrom, E. Smith, A. Shimizu & T. Honjo. 1989. Int. Immunol. **1:** 631–642.
28. Rothman, P., S. Lutzker, B. Gorham, V. Stewart, R. Coffman & F. W. Alt. 1990. Int. Immunol. **2:** 621–627.
29. Mills, F. C., J. S. Brooker & R. D. Camerini-Otero. 1990. Nucleic Acids Res. **18:** 7305–7316.
30. Xu, M. & J. Stavnezer. 1992. EMBO J. **11:** 145–155.
31. Cai, J., C. Humphries, A. Richardson & P. W. Tucker. 1992. J. Exp. Med. **176:** 1073–1081.
32. Ebeling, S. B., M. E. M. Schutte & T. Logtenberg. 1993. J. Immunol. **151:** 6891–6899.
33. Rassenti, L. Z. & T. J. Kipps. 1993. J. Exp. Med. **177:** 1039–1046.
34. van der Stoep, N., J. van der Linden & T. Logtenberg. 1993. J. Exp. Med. **177:** 99–107.
35. Xu, M. & J. Stavnezer. 1990. Dev. Immunol. **1:** 11–17.
36. Mowatt, M. R. & W. A. Dunnick. 1986. J. Immunol. **136:** 2674–2683.
37. Goodman, D. J., C. Gaff & S. Gerondakis. 1993. Int. Immunol. **5:** 199–208.
38. Lutzker, S. & F. W. Alt. 1988. Mol. Cell. Biol. **8:** 1849–1852.
39. Gerondakis, S., C. Gaff, D. J. Goodman & R. J. Grumont. 1991. Immunogenetics **36:** 392–400.

Isolation of Human Anti-Red Blood Cell Antibodies by Repertoire Cloning[a]

DON L. SIEGEL[b]

Blood Bank/Transfusion Medicine Section
Department of Pathology and Laboratory Medicine
University of Pennsylvania School of Medicine
Philadelphia, Pennsylvania 19104

The production of human anti–red blood cell (RBC) antibodies *in vitro* from immunized persons would facilitate the genetic analysis of the human auto- and alloimmune response to RBC antigens, provide useful serological reagents, and aid in the design of model systems to study immune modulation. Despite the development of murine hybridoma technology nearly 20 years ago,[1] the production of human RBC antibodies *in vitro* has been hampered by the absence of reliably effective procedures for immortalizing human B lymphocytes.[2,3]

Most recently, molecular approaches have been developed that bypass the need for cell transformation.[4–6] Collectively, these techniques, referred to as "repertoire cloning," seek to immortalize immunoglobulin (Ig) genes rather than the cells from which they were derived. cDNA is prepared from the mRNA of the rearranged Ig genes from donor B cells, and heavy and light chain gene segments are amplified by the polymerase chain reaction (FIG. 1). The heavy and light chain DNA is cloned into one of a number of available plasmid expression vectors,[7–9] a procedure which, after transformation into *E. coli,* results in the accumulation of Ig Fab fragment in the bacteria's periplasmic space. After co-infection with M13 helper phage, phagemid particles are generated that contain the plasmid DNA and incorporate ~one to three copies of Ig Fab fragment as part of their coat protein. Libraries of such phage display particles have been panned with purified, immobilized antigen to adsorb phage bearing specificities of interest. Adsorbed phage particles are eluted and allowed to reinfect *E. coli* cultures for further propagation and enrichment. In this manner, human antibodies have been isolated against a large variety of soluble, purified autoantigens[10–13] and alloantigens[14–18] with potential therapeutic and/or diagnostic utility.[19,20]

Our goal has been to apply this technology to the isolation of RBC-specific auto- and alloantibodies.[21,22] Because the majority of RBC antigens of interest are defined only serologically and their molecular identities are either unknown (as in the case recognized by warm-reactive autoantibodies[3,23]) or are known but not purifiable in a nondenatured state that retains their antigenic properties (as in the

[a] Supported by grants from the National Institutes of Health (No. K08-HL02621), National Blood Foundation, Lupus Foundation of the Delaware Valley, and the March of Dimes Birth Defects Foundation.

[b] Address for correspondence: Don L. Siegel, Ph.D., M.D., Department of Pathology & Laboratory Medicine, 6-55 Founders Pavilion, Hospital of the University of Pennsylvania, 3400 Spruce Street, Philadelphia, PA 19104.

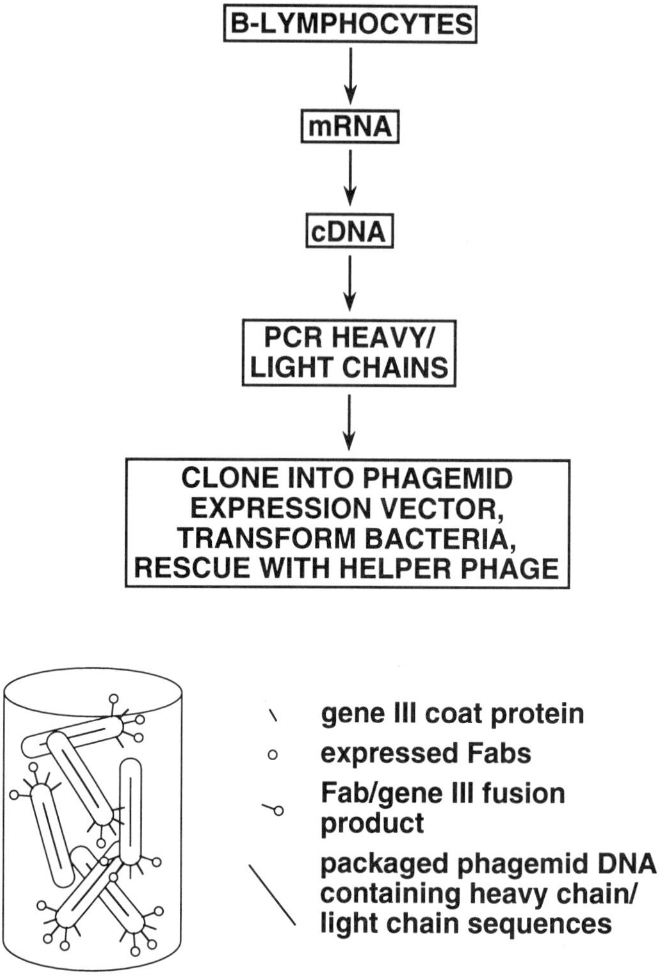

FIGURE 1. Expressing human immune repertoires on filamentous phage.

case of the Rhesus [Rh] antigens[24]), purified RBC auto- and alloantigens cannot be prepared, attached to a solid phase, and used to adsorb immunoreactive phage particles. We therefore set out to test the feasibility of using intact RBCs of defined phenotype to selectively adsorb out phage bearing specificities of interest from polyclonal phage mixtures (FIG. 2).[22] We prepared two sets of phage particles—one set expressing a human anti-Rh(D) alloantibody (derived from an unstable human heterohybridoma cell line) and the other set expressing a human anti-tetanus toxoid antibody (derived from a phage library panned for anti-tetanus toxoid reactivity[25]). We then constructed an artificial library comprising one anti-Rh(D)-expressing phage particle per 10^4 anti-tetanus toxoid-expressing phage particles and asked whether conditions could be found in which anti-Rh(D)-bearing phage particles

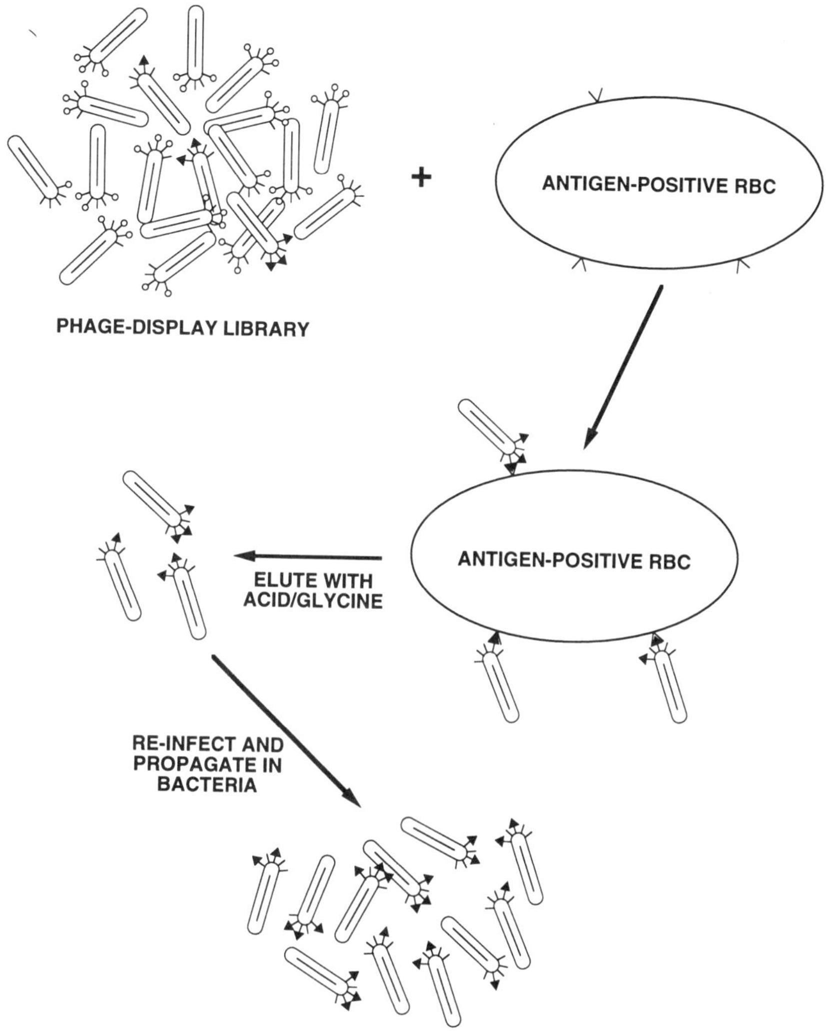

FIGURE 2. Panning phage display libraries by cell-surface enrichment using intact cells.

would be selectively enriched and propagated by panning on intact Rh(D)-positive RBCs. The following experiments summarize the biochemical, flow cytometric, and electron microscopic analyses used to develop the approach.

PRODUCTION AND FUNCTIONAL CHARACTERIZATION OF PHAGE DISPLAYED ANTIBODIES

The phagemid vector used in our studies was pComb3 (kindly provided by Dr. Carlos Barbas, Scripps Research Institute, La Jolla, CA), which is engineered to facilitate the surface expression of Fab molecules via fusion of the F_d chain with the carboxy terminus of M13 gene III coat protein.[8] As provided, the vector contained the heavy and light chain DNA encoding a human Igκ anti-tetanus toxoid antibody previously isolated from an immunized donor by repertoire cloning.[25] Phage particles expressing human anti-Rh(D) specificity were prepared from the mRNA of an unstable IgG1λ anti-Rh(D)-secreting heterohybridoma (H1121GS; kindly provided by Dr. Tom Setcavage, Ortho Diagnostics, Raritan, NJ) as described.[22] Production of recombinant antibody was assessed by performing nitrocellulose colony lifts and immunoblots of lysed bacterial colonies (FIG. 3A). Only bacteria transformed with pComb3/anti-Rh(D) stained with anti-λ antibody, whereas only those transformed with the anti-tetanus toxoid construct stained with anti-κ antibody. Dot blots of phage produced by their respective bacterial clones reacted similarly (FIG. 3B).

To demonstrate that the anti-Rh(D)-expressing phagemid particles retained their binding specificity, a flow cytometric assay was employed using Rh(D)-negative RBCs and anti-tetanus toxoid-bearing phage particles as negative controls (FIG. 4A). Phage particles bearing anti-Rh(D) bound to Rh(D)-positive RBCs, yet bound to Rh(D)-negative RBCs at the same background level as anti-tetanus toxoid-expressing phage particles bound to either Rh(D)-negative or -positive cells. That anti-Rh(D) phage particles were indeed binding to Rh(D) antigens present on the Rh(D)-positive cells was verified by the ability to block anti-Rh(D) phage binding when RBCs were preincubated with the parent IgGλ anti-Rh(D) heterohybridoma supernatant (FIG. 4B). Furthermore, the specific interaction of anti-Rh(D) phage with Rh(D)-positive RBCs was visualized by transmission electron micros-

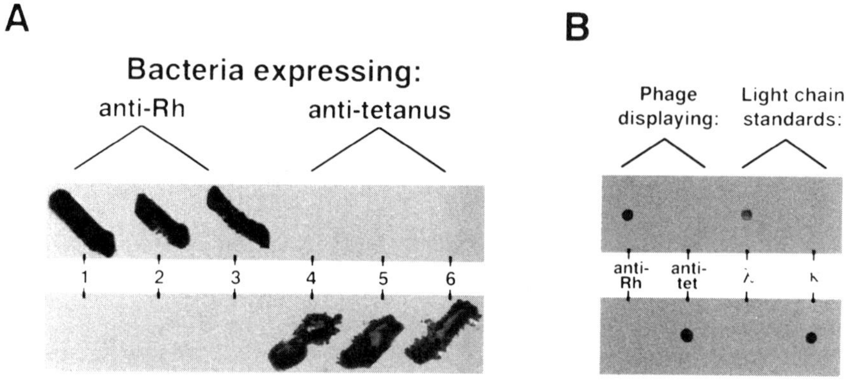

FIGURE 3. **(A)** Bacterial expression of human anti-Rh(D) and anti-tetanus Fab/gene III fusion proteins. Three bacterial clones transformed with anti-Rh(D)/pComb3 ("1," "2," "3") or anti-tetanus/pComb3 ("4," "5," "6") plasmid DNA were grown on nitrocellulose and developed with HRP-conjugated goat anti-human λ light chain antibody (to detect recombinant anti-Rh(D), *upper panel*) or HRP-conjugated goat anti-human κ light chain antibody (to detect recombinant anti-tetanus, *lower panel*). **(B)** Dot blots of M13 phagemids expressing human anti-Rh(D) or anti-tetanus Fab fragments on their surfaces. In the positions indicated, duplicate 1-μl samples of light chain standards (100 ng) or phagemids (10^9 pfu) were spotted onto strips of nitrocellulose. Blots were blocked and incubated with enzyme conjugated-goat anti-human λ or κ antibody to verify phage display of anti-Rh(D) and anti-tetanus, respectively.

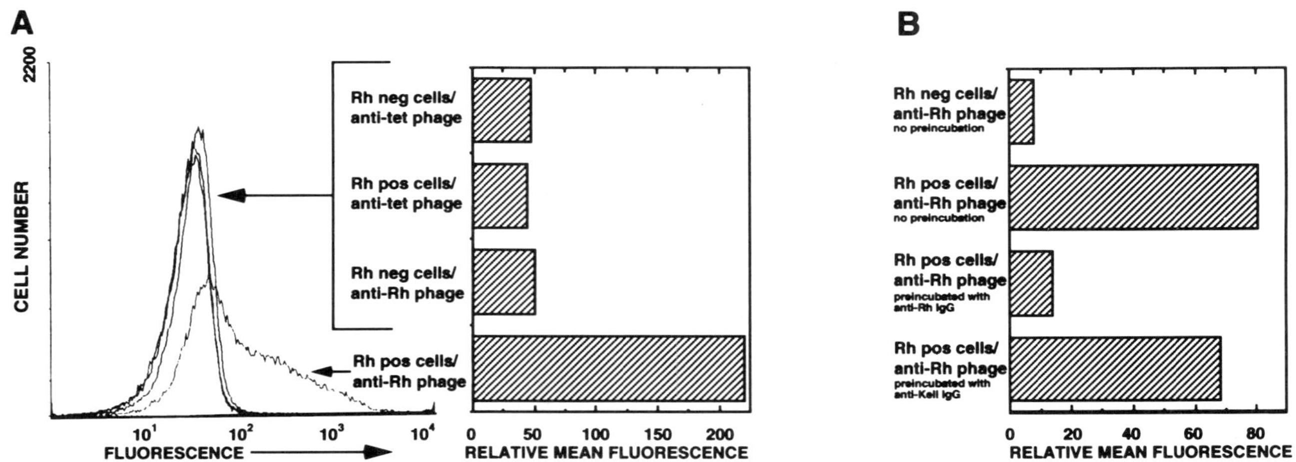

FIGURE 4. **(A)** Binding of antibody-expressing phage particles to Rh(D)-negative or -positive red cells as measured by flow cytometry. Phage particles were incubated with intact red cells, washed, and detected with biotinylated sheep anti-M13 antibody followed by FITC-conjugated streptavidin. **(B)** Inhibition of binding of anti Rh(D)-expressing phage particles to Rh(D)-positive Kell-positive red cells that were preincubated with anti-Rh(D) IgG heterohybridoma supernatant. Preincubation with human monoclonal anti-Kell (an unrelated red cell antigen present on these cells) does not inhibit anti-Rh(D) phage binding.

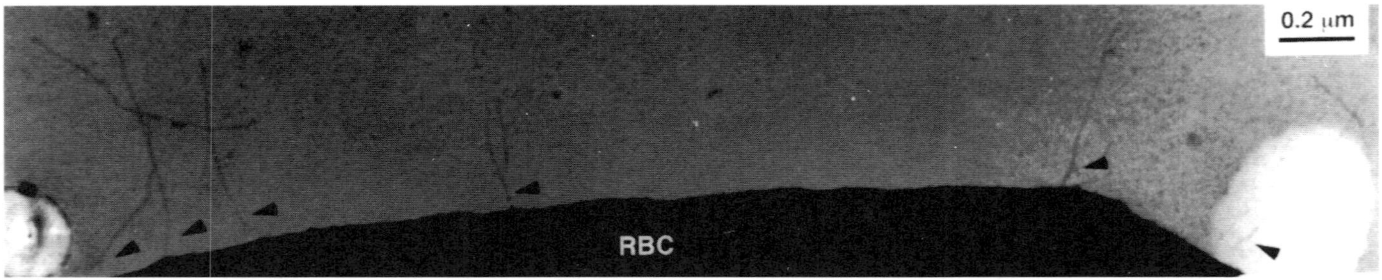

FIGURE 5. Transmission electron microscopic images of anti-Rh(D)-expressing phage bound to the surface of an Rh(D)-positive red cell. Binding occurs at the tips of phage (*arrowheads*) where the Fab/gene III fusion complex is located (×50,000). (Bar = 0.2 μm.)

copy to verify that the ~0.5-μm filamentous structures were binding via their tips where the gene III/Fab fusion protein resides (FIG. 5).[22]

PANNING PHAGE LIBRARIES FOR RBC-REACTIVE PHAGE

Having verified the expression of anti-Rh(D) and anti-tetanus toxoid Fab molecules on the surfaces of their respective phage particles, an artificial library was constructed containing one anti-Rh(D)-expressing phage per 10^4 that express anti-tetanus toxoid. A protocol was developed that facilitated the efficient enrichment for anti-Rh(D)-specific phage by performing rounds of panning, elution, and propagation under an extensive array of incubation, washing, and elution conditions.[22] A quantitative assessment of our optimized panning protocol demonstrated that

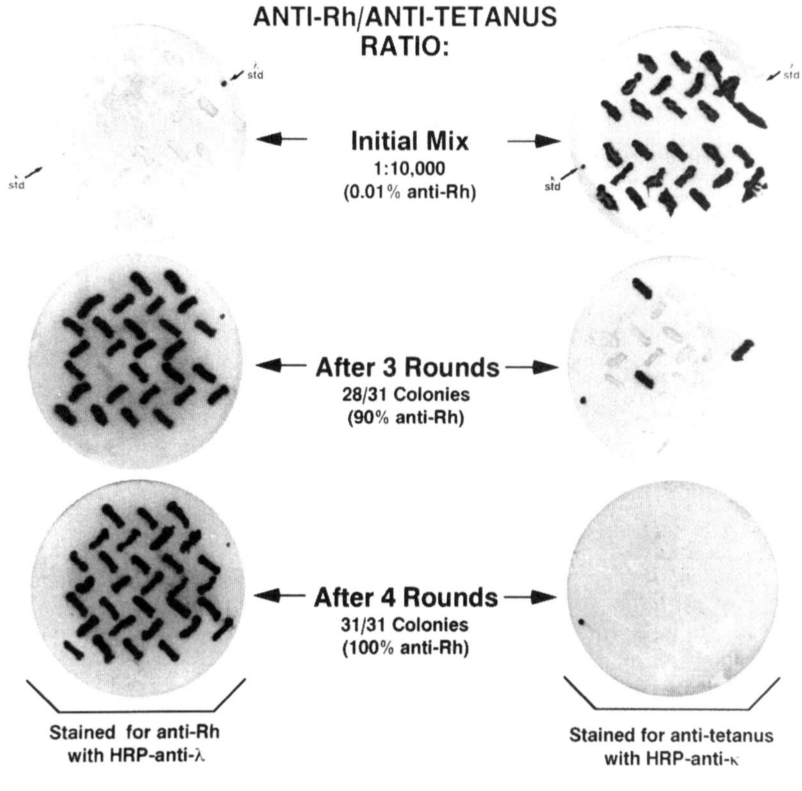

FIGURE 6. Quantification of panning efficiency. Thirty-one randomly chosen colonies from bacterial cultures reinfected with phage derived from the initial anti-Rh(D)/anti-tetanus mix (*top row*) and from the third and fourth rounds of panning (*middle and bottom rows*) were streaked in duplicate and stained for anti-Rh(D) production (*left column*) or anti-tetanus production (*right column*). One-microliter spots of λ and κ human light chain standards were applied as controls to each filter in the positions indicated.

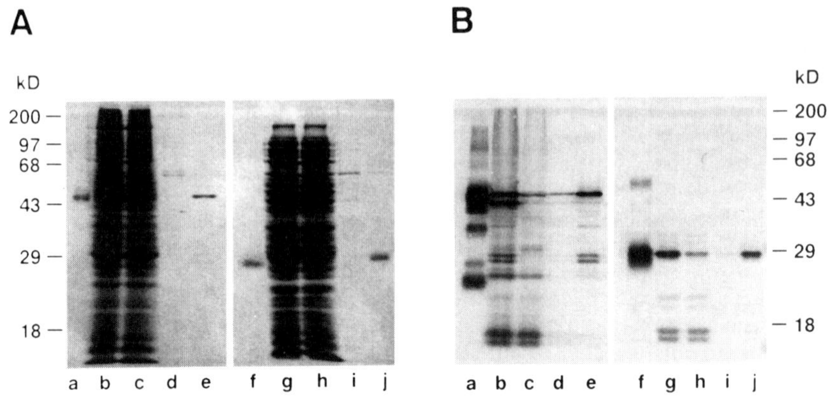

FIGURE 7. SDS-polyacrylamide gel electrophoresis of recombinant anti-Rh(D) Fab fragment purification by nickel-chelation affinity chromatography. **(A)** Coomassie blue-stained gel; **(B)** immunoblot of "A" stained with HRP-conjugated goat anti-human λ light chain antibody. *Lanes a–e,* unreduced samples; *lanes f–j,* samples reduced with 40 mM dithiothreitol. Samples are human IgG Fab fragment standard (*a, f*), column load (*b, g*), column flow through (*c, h*), 10 mM imidazole column eluate (*d, i*), and 250 mM imidazole column eluate containing purified anti-Rh(D) Fab molecules (*e, j*). Molecular size markers (kDa) as indicated. Amounts of protein samples applied to each lane in "A" were normalized to the original bacterial extract volume to permit comparison among lanes *b–e* and *g–j* and to facilitate visualization of the Fab purification process. Amounts of protein samples in "B" were 1/50 of those in "A" to compensate for the increased sensitivity of immunoblotting over protein staining.

after three rounds of panning, 28 of 31 bacterial colonies (90%) were positive for anti-Rh(D), and after a fourth round, 31 of 31 colonies (100%) were producing anti-Rh(D) antibody (FIG. 6).

PRODUCTION OF SOLUBLE, RECOMBINANT ANTI-RH(D) FAB MOLECULES

To carry our model system one step further, bacterial colonies producing anti-Rh(D)/geneIII fusion products were used to express soluble anti-Rh(D) Fabs by genetically unlinking the F_d chain from the gene III protein.[26] This allowed us to develop efficient methods for the isolation, purification, and characterization of recombinant anti-RBC Fab molecules and evaluate their potential utility as analytical reagents and serological RBC-typing reagents. A number of approaches were used to isolate the anti-Rh(D) Fab molecules from the bacterial periplasmic space. Nitrogen cavitation was found to be far superior to cell disruption by sonication, freeze-thawing, and osmotic shock in terms of speed, reproducibility, yield of recombinant protein (~1 mg/liter of bacterial culture), and retention of antibody/antigen binding.[22] Fab protein was then affinity purified from the crude periplasmic extracts by adsorption on nickel-derivitized agarose (via a hexahistidine tail on the carboxy terminus of the heavy chain F_d fragment) and elution with imidazole at neutral pH. SDS-PAGE and immunoblots revealed a product of M_r ~ 50 kDa (under nonreducing conditions) and $M_r \sim 26$ kDa (under reducing conditions), similar to authentic chemically produced, commercially obtained, pooled human IgG Fab fragment (FIG. 7). The purified molecules were shown to retain their binding specificity for Rh(D)-positive RBCs by flow cytometry[22] and to discriminate between Rh(D)-negative and -positive RBCs by using conventional blood-banking serological agglutination assays (FIG. 8).

FUTURE DIRECTIONS

Having demonstrated the feasibility of panning phage display libraries using intact cells and having developed the appropriate biochemical and immunologic

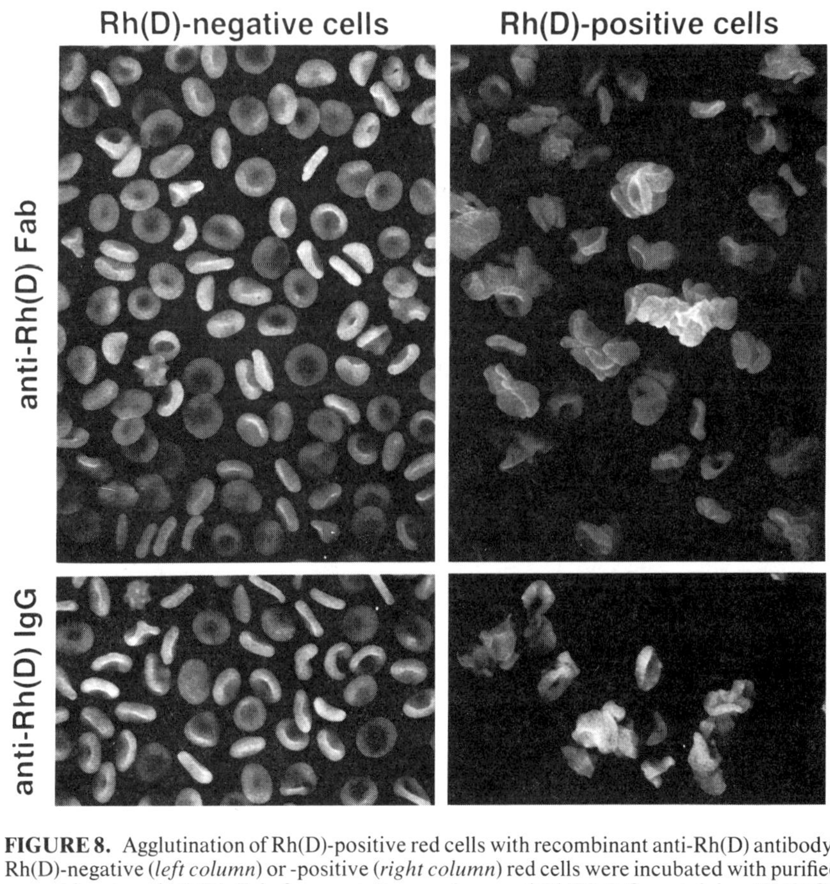

FIGURE 8. Agglutination of Rh(D)-positive red cells with recombinant anti-Rh(D) antibody. Rh(D)-negative (*left column*) or -positive (*right column*) red cells were incubated with purified recombinant anti-Rh(D) Fab fragment (*top row*) or anti-Rh(D) IgG-containing H1121GS hybridoma supernatant (*bottom row*). Red cells were washed, incubated with goat anti-human $F(ab')_2$ antibody, and prepared for scanning electron microscopy (original magnification ×1500; reduced by 10%).

assays to monitor cell-surface enrichment, we created a series of phage display libraries from patients undergoing hemolysis as a result of their acquiring anti-RBC auto- and/or alloantibodies. Our approach consists of performing several rounds of "negative panning" on antigen-negative RBCs followed by a single incubation with antigen-positive RBCs before each amplification step. The initial set of negative incubations serves to deplete the libraries of undesired specificities and "pan-reactive" phage before positive selection. We are currently screening bacterial clones obtained after several rounds of such selections for antibody specificities that match those present in high-titer in the patients' sera.

Using this approach, we will use the human red cell as a model system for expanding our current understanding of auto- and alloimmune responses. For example, the major obstacle in studying the pathogenesis of warm autoimmune hemolytic anemia, an IgG-mediated form of autoimmune hemolysis,[27] has been the inability to obtain sufficient quantities of autoantibodies that are homogeneous in structure and specificity.[3] Consequently, our understanding of the biochemistry of the relevant autoantigen(s) and the nature of the corresponding pathogenic immune response has been limited by serological observations and the analyses of minute quantities of polyclonal immunoglobulin molecules eluted off of the surface of patient red cells.[23] Autoreactive phage particles derived from phage display libraries made from patients with warm autoimmune hemolytic anemia would produce both the gene segments as well as the expressed autoantibodies that can be used to analyze the clonality and genetic origin of the pathologic autoantibodies and the relevant autoantigenic structures.[3,23] Similarly, phage display libraries produced from patients undergoing hemolytic transfusion reactions could serve as model systems to study the human immune response to foreign antigens as well as reveal the immune repertoires directed against specific red cell antigens of clinical importance. For example, analyses of the immune response to the Rh(D) antigen could lead to an understanding of the immunologic composition of Rh(D)-immune globulin and aid in the design of studies to dissect the mechanisms by which Rh(D)-immune globulin exerts its immunomodulatory effect. From a practical standpoint, recombinant anti-RBC alloantibodies may eventually provide an economical, endless source of blood bank typing reagents that would resolve current ethical concerns regarding the hyperimmunization of human donors with foreign red cells.[28] It is anticipated that repertoire cloning by cell-surface enrichment will also prove useful for the analyses of the human immune response to nonerythroid, membrane-bound antigens that are structurally uncharacterized, unpurifiable in a nondenatured state, or are components of macromolecular complexes that lose their antigenicity when disassembled.

REFERENCES

1. Köhler, G. & C. Milstein. 1975. Continuous cultures of fused cells secreting antibody of predefined specificity. Nature **256:** 495–497.
2. Williamson, R. A., M. A. A. Persson & D. R. Burton. 1991. Expression of a human monoclonal anti-(rhesus D) Fab fragment in *Escherichia coli* with the use of bacteriophage λ vectors. Biochem. J. **277:** 561–563.
3. Siegel, D. L. & L. E. Silberstein. 1992. Red cell antigen specificities and autoimmune hemolytic anemias. *In* Protein Blood Group Antigens of the Human Red Cell: Structure, Function, and Clinical Significance. P. C. Agre & J.-P. Cartron, Eds.: 193–208. The Johns Hopkins University Press. Baltimore, MD.
4. Winter, G. & C. Milstein. 1991. Man-made antibodies. Nature **349:** 293–299.
5. Lerner, R. A., C. F. Barbas, A. S. Kang & D. R. Burton. 1991. On the use of

combinatorial antibody libraries to clone the "fossil record" of an individual's immune response. Proc. Natl. Acad. Sci. USA **88:** 9705–9706.
6. HOOGENBOOM, H. R., J. D. MARKS, A. D. GRIFFITHS & G. WINTER. 1992. Building antibodies from their genes. Immunol. Rev. **130:** 41–67.
7. KANG, A. S., C. F. BARBAS, K. D. JANDA, S. J. BENKOVIC & R. A. LERNER. 1991. Linkage of recognition and replication functions by assembling combinatorial antibody Fab libraries along phage surfaces. Proc. Natl. Acad. Sci. USA **88:** 4363–4366.
8. BARBAS, C. F., A. S. KANG, R. A. LERNER & S. J. BENKOVIC. 1991. Assembly of combinatorial antibody libraries on phage surfaces: The gene III site. Proc. Natl. Acad. Sci. USA **88:** 7978–7982.
9. HOOGENBOOM, H. R., A. D. GRIFFITHS, K. S. JOHNSON, D. J. CHISWELL, P. HUDSON & G. WINTER. 1991. Multi-subunit proteins on the surface of filamentous phage: Methodologies for displaying antibody (Fab) heavy and light chains. Nucleic Acids Res. **19:** 4133–4137.
10. HEXHAM, J. M., J. FURMANIAK, C. PEGG, D. R. BURTON & B. R. SMITH. 1992. Cloning of a human autoimmune response: Preparation and sequencing of a human antithyroglobulin autoantibody using a combinatorial approach. Autoimmunity **12:** 135–141.
11. GRIFFITHS, A. D., M. MALMQVIST, J. D. MARKS, J. M. BYE, M. J. EMBLETON, J. MCCAFFERTY, M. BAIER, K. P. HOLLIGER & B. D. GORICK. 1993. Human anti-self antibodies with high specificity from phage display libraries. EMBO J. **12:** 725–734.
12. CHAZENBALK, G. D., S. PORTOLANO, D. RUSSO, J. S. HUTCHISON, B. RAPOPORT & S. MCLACHLAN. 1993. Human organ-specific autoimmune disease: Molecular cloning and expression of an autoantibody gene repertoire for a major autoantigen reveals an antigenic immunodominant region and restricted immunoglobulin gene usage in the target organ. J. Clin. Invest. **92:** 62–74.
13. CHA, S., P. S. LEUNG, M. E. GERSHWIN, M. P. FLETCHER, A. A. ANSARI & R. L. COPPEL. 1993. Combinatorial autoantibodies to dihydrolipoamide acetyltransferase, the major autoantigen of primary biliary cirrhosis. Proc. Natl. Acad. Sci. USA **90:** 2527–2531.
14. CATON, A. J. & H. KOPROWSKI. 1990. Influenza virus hemagglutinin-specific antibodies isolated from a combinatorial expression library are closely related to the immune response of the donor. Proc. Natl. Acad. Sci. USA **87:** 6450–6454.
15. BURTON, D. R., C. F. BARBAS, M. M. A. PERSSON, S. KOENIG, R. M. CHANOCK & R. A. LERNER. 1991. A large array of human monoclonal antibodies to type 1 human immunodeficiency virus from combinatorial libraries of asymptomatic seropositive individuals. Proc. Natl. Acad. Sci. USA **88:** 10134–10137.
16. ZEBEDEE, S. L., C. F. BARBAS, Y.-L. HOM, R. H. CAOTHIEN, R. GRAFF, J. DEGRAW, J. PYATI, R. LAPOLLA, D. R. BURTON, R. A. LERNER & G. B. THRONTON. 1992. Human combinatorial antibody libraries to hepatitis B surface antigen. Proc. Natl. Acad. Sci. USA **89:** 3175–3179.
17. BARBAS, C. F., J. E. J. CROWE, D. CABABA, T. M. JONES, S. L. ZEBEDEE, B. R. MURPHY, R. M. CHANOCK & D. R. BURTON. 1992. Human monoclonal Fab fragments derived from a combinatorial library bind to respiratory syncytial virus F glycoprotein and neutralize infectivity. Proc. Natl. Acad. Sci. USA **89:** 10164–10168.
18. WILLIAMSON, R. A., R. BURIONI, P. P. SANNA, L. J. PARTRIDGE, C. F. BARBAS & D. BURTON. 1993. Human monoclonal antibodies against a plethora of viral pathogens from single combinatorial libraries. Proc. Natl. Acad. Sci. USA **90:** 4141–4145.
19. CROWE, J. E. J., B. R. MURPHY, R. M. CHANOCK, R. A. WILLIAMSON, C. F. I. BARBAS & D. R. BURTON. 1994. Recombinant human respiratory syncytial virus (RSV) monoclonal antibody Fab is effective therapeutically when introduced directly into the lungs of RSV-infected mice. Proc. Natl. Acad. Sci. USA **91:** 1386–1390.
20. BARBAS, C. F. I., D. HU, N. DUNLOP, L. SAWYER, D. CABABA, R. M. HENDRY, P. L. NARA & D. R. BURTON. 1994. In vitro evolution of a neutralizing human antibody to human immunodeficiency virus type 1 to enhance affinity and broaden strain cross-reactivity. Proc. Natl. Acad. Sci. USA **91:** 3809–3813.
21. SIEGEL, D. L. & L. E. SILBERSTEIN. 1992. Production of human red cell antibodies in bacteria by repertoire cloning. Transfusion **32:** 45S.

22. SIEGEL, D. L. & L. E. SILBERSTEIN. 1994. Expression and characterization of recombinant anti-Rh(D) antibodies on filamentous phage: A model system for isolating human red blood cell antibodies by repertoire cloning. Blood **83:** 2334–2344.
23. SIEGEL, D. L. & L. E. SILBERSTEIN. 1994. Structural analyses of red cell autoantibodies. *In* Immunobiology of Transfusion Medicine. G. Garratty, Eds.: 387–399. Dekker. New York.
24. AGRE, P. & J.-P. CARTRON. 1991. Molecular biology of the Rh antigens. Blood **78:** 551–563.
25. PERSSON, M. A. A., R. H. CAOTHIEN & D. R. BURTON. 1991. Generation of diverse high-affinity human monoclonal antibodies by repertoire cloning. Proc. Natl. Acad. Sci. USA **88:** 2432–2436.
26. BARBAS, C. F. & R. A. LERNER. 1991. Combinatorial immunoglobulin libraries on the surface of phage (Phabs): Rapid selection of antigen-specific Fabs. Methods: A Companion to Methods Enzymol. **2:** 119–124.
27. SIEGEL, D. L. & L. E. SILBERSTEIN. 1991. Acquired hemolytic anemia. *In* Current Diagnosis 8. R. B. Conn, Ed.: 559–563. W. B. Saunders. Philadelphia.
28. CHERNOFF, A. I., H. G. KELIN & L. A. SHERMAN. 1989. Research opportunities in transfusion medicine: Report from an American Association of Blood Banks Think Tank. Transfusion **29:** 711–742.

Genetic and Functional Characterization of Human Autoantibodies Using Combinatorial Phage Display Libraries

MARIA ZEIDEL,[a,b] ELENA REY,[a] JOE TAMI,[b]
MICHAEL FISCHBACH,[a] AND IÑAKI SANZ[a,c]

Department of Medicine
[a]Division of Rheumatology
[b]Division of Hematology
The University of Texas Health Science Center
San Antonio, Texas 78284

The study of human autoantibody responses has been hampered by the limitations that conventional hybridoma technology imposes on the generation of human monoclonal antibodies. This is even more significant for some autoantibodies for which the frequency of specific B-cell precursors is very low. For instance, the frequency of B-cell precursors for the SLE-specific anti-Smith antibodies (anti-Sm) has been shown to be less than 1 : 30,000 splenocytes in the autoimmune MRL\lpr\lpr mice and undetectable in normal mice even after LPS stimulation. This situation is in marked contrast with the higher precursor frequency established for most autoantigens including DNA (1 : 5,000).[1] As a consequence, only five human IgM anti-Sm antibodies and no IgG anti-Sm antibodies have been thus far generated. In order to circumvent these limitations, we have applied newly developed *in vitro* technology, namely combinatorial phage display libraries, to the generation of human monoclonal autoantibodies. In this paper, we demonstrate that this approach is capable of generating functional human autoantibodies from peripheral blood lymphocytes of patients with systemic lupus erythematosus (SLE) and myasthenia gravis (MG).

MATERIAL AND METHODS

Peripheral blood mononuclear cells were isolated by Ficoll-Hypaque gradient from 50 ml of peripheral blood obtained from one anti-Sm + SLE patient and one MG patient that were untreated at the time the venopuncture was performed. Total RNA was extracted by the single-step protocol, and first-strand cDNA was synthesized with random hexamers and RNase H^--reverse transcriptase (BRL-Gifco) in a final volume of 30 μl as described. Two microliters of the cDNA were used to amplify the seven V_H families and six $V\kappa$ families by the polymerase chain reaction (PCR). The variable region primers were selected to anneal to the first framework region of all V_H genes and all $V\kappa$ genes under annealing temperatures

[c] Address for correspondence: Iñaki Sanz, M.D., Department of Medicine, Division of Rheumatology, The University of Texas Health Science Center at San Antonio, 7703 Floyd Curl Drive, San Antonio, TX 78284-7874.

of 10 to 15°C below the calculated melting temperature of the amplimers.[2] For the constant regions, a $C\gamma 1$ primer corresponding to the hinge region and a $C\kappa$ primer extending to the last residue of the $C\kappa$ exon were used to amplify rearrangements of IgG1/κ immunoglobulin RNA. After digestion with the appropriate restriction enzymes, the PCR fragments were cloned into the pComb3 vector (kindly provided by Drs. D. Burton and C. Barbas, Scripps Research Foundation) as described.[3] Combinatorial libraries were panned against affinity-purified B/B′ polypeptides obtained from ImmunoVision (Arizona) for the SLE library and against purified acetyl choline receptor (AChR) from the electric fish *Torpedo californica* for the MG library. After five consecutive rounds of panning, soluble Fab fragments were generated as described[3] and tested by ELISA against the above antigens. Clones giving OD410 readings of at least 1.0 at 20 minutes with values at least 10-fold above background values were selected for further study. Four such clones from each library were tested in competitive ELISA assays against the corresponding antigen and control antigens including Ro, La, BSA, ovalbumin, lysozyme, and thyroglobulin; and a relative affinity constant was determined as the molar concentration giving 50% of binding inhibition.

RESULTS

Anti-Sm Library

Four anti-Sm clones (Fati-1, CS-51, CS-59, and CS-61) were analyzed as described. The relative affinity constant of these clones ranged from 5×10^6 M^{-1} to 5×10^7 M^{-1}. No binding inhibition was obtained with any of the control antigens except for Ro, which was able to inhibit the binding of clones Fati-1 and CS-61 to the Sm antigen by 30% at molar concentrations of 10^7 M^{-1}. We then tested the anti-Sm clones in Western blot assays against purified B/B′ polypeptides, Ro, and La proteins and showed that the recombinant clones were able to recognize specifically the B/B′ polypeptides that constitute the main target of anti-Sm antibodies[4] (FIG. 1). The nucleotide sequence of anti-Sm clones revealed that all four antibodies were encoded by V_H genes belonging to the V_H4 family with identity scores ranging from 93% to 95% as compared to the best germline V_H genes showing the highest homology (Fati-1/VH4-33, CS-51/VH4-11, CS-59 Humighca, and CS-61/VH4-21). The replacement to silent mutation rate was higher than 3.0 in the hypervariable regions (CDRs) and lower than 3.0 in the framework regions (FRs) in a pattern consistent with antigen-driven somatic mutation. CDR3 regions were very heterogeneous and were contributed by several D genes including the newly described DIR4 and DIR5 genes[5] (Fati-1: inverted DIR5 + DXP4 + J_H4,, CS-51: DK1 + inverted D3 + J_H5,, CS-59: inverted DIR4 + J_H5,, CS-61: DXP4 + J_H4).

Anti-AChR Library

Four clones from the MG library (BCH-20, BCH-26, BCH-27, and BCH-39) were characterized as described. Relative affinity constants ranged from 5×10^7 M^{-1} to 5×10^8 M^{-1}. No inhibition of binding to the AChR protein was obtained with any of the control antigens. These clones were tested for their ability to specifically stain a human rabdomyosarcoma cell line (TE671) that expresses high

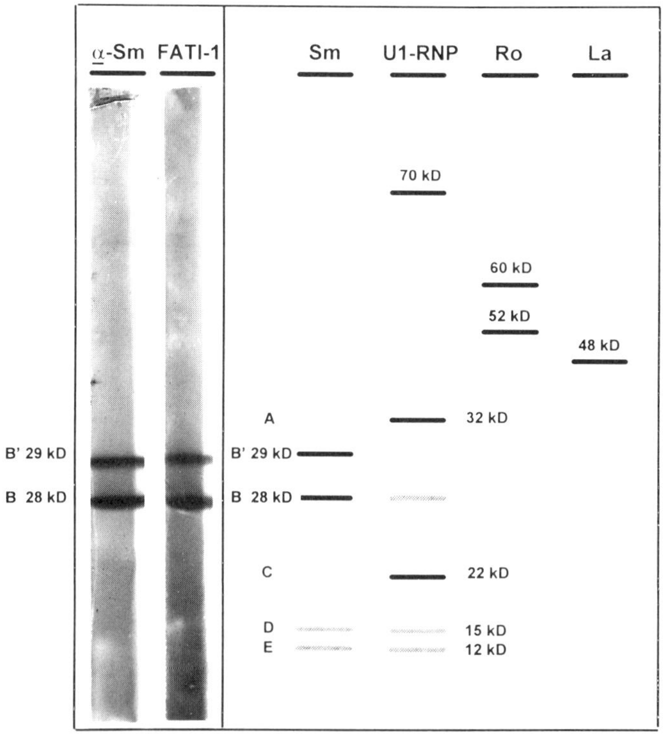

FIGURE 1. Representative example of the antigenic profile identified by recombinant anti-Sm Fabs in Western blot experiments. Purified Sm-B/B′ polypeptides 4, Ro, and La proteins were loaded in a single well and separated by SDS-PAGE, electrotransferred to Immobilon PVDF membranes, and probed with either a positive anti-Sm antiserum (α-Sm) or with recombinant Fab (FATI-1). The secondary antibody used for detection was a FITC-labeled goat anti-human IgG(κ + λ) (Pierce). Recombinant anti-Sm Fabs specifically recognized the B/B′ polypeptides that constitute the main antigenic target of anti-Sm antibodies (28/29-kDa bands) in a pattern identical to the one obtained with the positive control. The right half of the figure represents the expected pattern of immunoblot reactivity obtained with anti-Sm, anti-U1-RNP, anti-Ro, and anti-La antibodies, respectively. Negative control anti-AChR Fabs produced and purified in the same way as FATI-1 gave a negative result in the same experiment (not shown).

levels of the AChR.[6] All four clones were able to stain the muscle cells (FIG. 2) but not other cell lines including fibroblasts and multiple myeloma cell lines. Furthermore, the anti-Sm Fabs failed to produce any staining of TE671 cells. DNA sequencing of the anti-AChR clones revealed that the V_H genes used by these antibodies belong to the V_H1 (BCH-20 and BCH-39) and V_H3 families (BCH26 and 27) with homologies ranging from 86 to 93% as compared with the closest GL V_H genes (BCH20/VH1-L33, BCH26/VH3-23, BCH27/VH3-21, BCH39/VH1-8). In this case, the replacement to silent mutation ratio was higher than 3.0 in the CDRs only for clones 20 and 26, while showing values above 3.0 in the FRs of clones 27 and 39. The incidence of replacement mutations in the latter

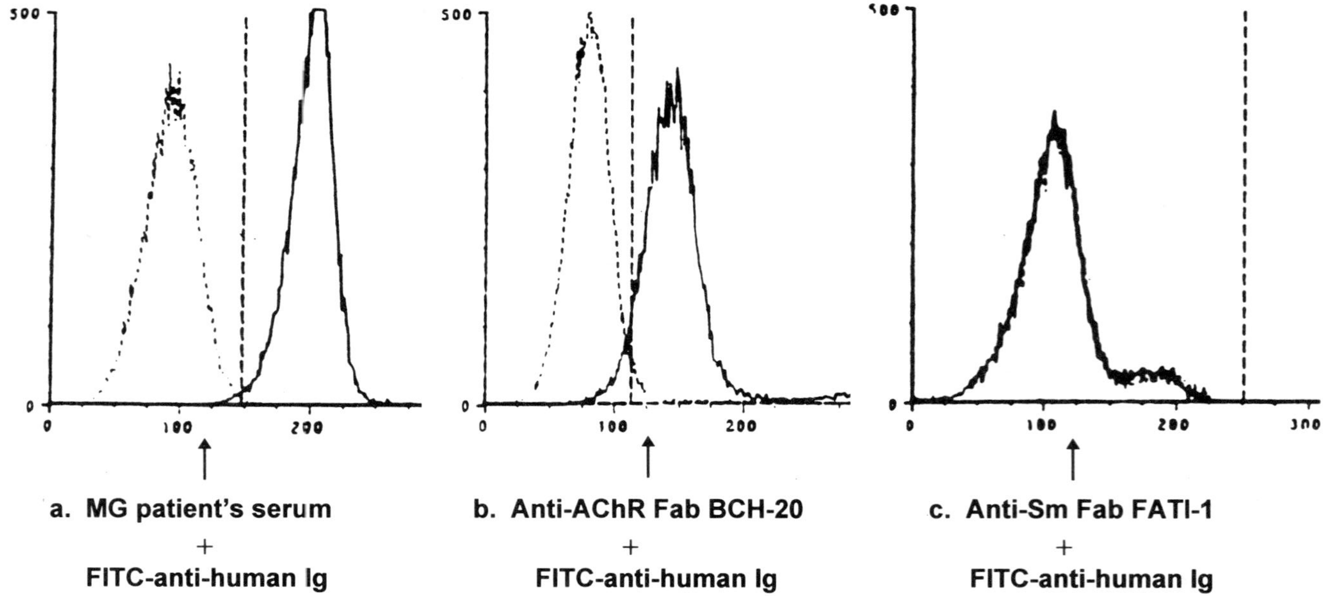

FIGURE 2. Flow cytometry analysis of human rabdomyosarcoma cell line TE671. The FACS profiles of the staining obtained with either anti-Sm Fab FATI-1 **(a)**, anti-AChR Fab BCH-20 **(b)** or positive serum from MG patients **(c)** are shown. Although virtually 100% of TE671 cells were stained with the anti-AChR antibodies (both recombinant and from patients' sera), no staining was obtained with anti-Sm Fabs. The secondary antibody used for detection was a FITC-labeled goat anti-human IgG(κ + λ).

two clones was particularly high in the FR3 region. The CDR3 regions were contributed by multiple germline D genes as for the anti-Sm clones (BCH20: DFL16 + J_H4,, BCH26: DIR4 + D4 + J_H4,, BCH27: DIR5 + J_H3,, BCH39: DXP'1 + J_H6).

DISCUSSION

Herein we report the characterization of human autoantibodies with combinatorial phage display libraries from both SLE and MG patients. Perhaps the most relevant aspect of our work is the demonstration that this technology is indeed capable of rescuing from just the peripheral blood lymphocytes (PBLs) of autoimmune patients the antibody genes responsible for the generation of autoantibodies. Indeed, the anti-Sm clones described in this report represent the first human IgG anti-Sm antibodies ever generated. Strikingly, all four anti-Sm clones used V_H genes from the V_H4 family including the V_H4-21 gene that is overrepresented in the autoimmune repertoire.[7] Depsite the limited size of our sample, we believe this restriction to be relevant inasmuch as most human antibody responses including the AChR response display greater V_H representation. This genetic restriction is more relevant in light of the marked restriction of the anti-Sm response to only 25% of SLE patients as well to 25% of the MRL/lpr autoimmune mice. Recent experiments with IgH congenic MRL/lpr mice have indeed showed that the V_H haplotype plays a critical role in this response[8] and suggests that V_H polymorphism may be at least partially responsible.

The anti-AChR studies suggest that the pComb system is able to isolate functional autoantibodies that can now be tested for pathogenicity both *in vitro* and *in vivo* by studying their ability to induce disease in susceptible animals. As opposed to the anti-Sm response, anti-AChR were heterogeneous in terms of V_H family utilization. Both types of antibodies showed significant diversity in the CDR3 regions, probably accounting for differences in the fine specificity of these molecules.

REFERENCES

1. Cohen, P. L., E. W. Shore, R. Rapoport, S. Caster, R. A. Eisenberg & D. S. Pisetsky. 1985. Cell Immunol. **96:** 448.
2. Tomlinson, I. A., G. Walter, J. D. Marks, M. B. Llewelyn & G. Winter. 1992. The repertoire of human germline V_H sequences reveals about fifty groups of V_H segments with different hypervariable loops. J. Mol. Biol. **227:** 776.
3. Barbas III, C. F., T. A. Collet, W. Amberg, P. Roben, J. M. Binley, D. Hoekstra, D. Cababa, T. M. Jones, R. A. Williamson, G. R. Pilkington, N. L. Haigwood, E. Cabezas, A. C. Satterthwait, I. Sanz & D. Burton. 1993. Molecular profile of an antibody response to HIV-1 as probe by combinatorial libraries. J. Mol. Biol. **230:** 812.
4. Steitz, J. A., D. L. Black, K. A. Gerke, A. Parker, D. Kramer, D. Frendewey & W. Keller. 1983. Functions of the abundant U snRNPs. *In* Structure and function of major and minor small nuclear ribonucleoprotein particles. M. L. Birnsteil, Ed.: 115. Springer. Heidelberg, FRG.
5. Sanz, I., S-S. Wang, G. Meneses & M. Fischbach. 1994. Molecular characterization of human Ig heavy chain DIR genes. J. Immunol. **152:** 3958.
6. Luther, M. A., R. Schoepfer, P. Whiting, B. Casey, Y. Blatt, M. S. Montal,

M. MONTAL & J. LINDSTROM. 1989. A muscle receptor is expressed in the human cerebellar medulloblastoma cell line TE671. J. Neurosci. **9:** 1082–1096.
7. PASCUAL, V. & J. D. CAPRA. 1992. V_H4-21, a human V_H gene segment overrepresented in the autoimmune repertoire. Arthritis Rheum. **35:** 11.
8. HALPERN, M. D., S. Y. CRAVEN, P. L. COHEN & R. A. EISENBERG. 1993. Regulation of anti-Sm autoantibodies by the immunoglobulin heavy chain locus. J. Immunol. **151:** 7268.

Repertoire Cloning of Human Lupus Autoantibodies[a]

G. J. SILVERMAN,[b] S. BARBAS,[c] P. ROBEN,[b]
AND D. R. BURTON[c]

[b]*Department of Medicine*
University of California, San Diego
La Jolla, California 92093

[c]*The Scripps Research Institute*
La Jolla, California 92037

Our understanding of human antibody responses has greatly lagged behind that of the murine immune system, in large part because of the poor cloning efficiency of human B lymphocytes. A newly developed and largely untried technology, repertoire cloning, offers an alternative with great potential advantages. In this approach monoclonal antibodies are selected from combinatorial expression libraries created using antibody gene mRNA from lymphoid sources, thereby bypassing the need to create B-cell lines. In the current report, we describe certain refinements of repertoire-cloning methods that facilitate the study of pathologic autoantibody responses in human systemic lupus erythematosus (SLE).

MATERIALS AND METHODS

Using previously published methods,[1-3] we created combinatorial IgG libraries in the phage-display vector, pComb3, using peripheral blood lymphocytes (PBLs) from SLE and healthy donors. Among these donors was a set of monozygotic twins discordant for SLE. The libraries contained $>10^7$ transformants for both H (heavy) and L (light) chains. Libraries were biopanned against microtiter wells precoated with either purified human placental DNA or double-stranded (ds) DNA that was S1 nuclease–treated to remove single-stranded (ss) ends. Phage and soluble Fabs were treated with DNAse, RNAse, and hydroxyapatite to remove any bacterial DNA that might interfere with the detection of DNA-binding capacity. Following affinity purification of the IgG Fab, binding fine specificity was assessed using ELISA plates precoated with purified human placental DNA, ssDNA dsDNA, or a panel of control antigens. Indirect immunofluorescence *Crithidia lucilae* assays were performed using FITC-conjugated, affinity-purified antibodies to L chains. Binding affinity was determined by direct binding and competitive inhibition assays. Idiotype expression was assessed by direct binding assays and quantitated by inhibition of IgG dsDNA-binding capacity. The V_H region-specific anti-idiotypic antibodies, anti-18/2 (that identifies the 16/6 cross-reactive idiotype [CRI]) and the B6 antibody were the kind gifts of Drs. B. D.

[a] This work was supported in part by National Institutes of Health Grant AI34001. GJS is a recipient of a Biomedical Sciences Award from the Arthritis Foundation and is a Scholar of the Lupus Foundation of America, Inc.

Stollar and R. Mageed, respectively. The VλIIIa antibody that identifies a subset of VλIII L chains was the kind gift of Dr. A. Solomon.

RESULTS AND DISCUSSION

Six different IgG anti-DNA antibodies were obtained from one SLE library and two additional IgG autoantibodies were obtained from the second SLE library. However, antibodies with high-affinity dsDNA-binding activity could not be obtained from libraries from healthy persons that included an identical twin discordant for lupus who has a high serum titer of IgG anti-ssDNA antibodies. Of six autoantibodies tested, all have an apparent K_D of 10^{-7} to 10^{-8} M in a competitive ELISA, and three are also reactive in the *Crithidia* assay that identifies high-affinity anti-dsDNA antibodies diagnostic for SLE.

Nucleic acid sequence analyses demonstrated that all V_H regions are highly homologous to genes previously reported from anti-DNA lupus cell lines. The recombinant antibodies use six different V_H3s, one V_H5, and one V_H1 gene rearrangement. The V_H3 antibodies derive from four different V_H germline gene segments. These V_H regions are 91.6 to >99% identical to the closest known germline genes. In serologic studies, there was complete concordance between the idiotypes expressed on recombinant anti-DNA antibodies and those expressed on the anti-dsDNA antibodies in the circulation of the donor. In particular, monoclonal autoantibodies from one of the SLE patients expressed either the B6 or the 16/6 V_H-associated CRI. These anti-CRIs could each inhibit about 11% of the IgG anti-dsDNA activity in the donor's serum. From the other SLE patient, one autoantibody expressed both the B6 V_H-associated CRI and the λIIIa V_L-associated CRI. The B6 anti-idiotype could inhibit 20%, and the λIIIa anti-idiotype could inhibit >40% IgG anti-dsDNA-binding activity in the donor's serum. The 16/6 CRI was neither on a recombinant antibody nor on the serum autoantibodies from this person.

In summary, the binding specificities, affinities, and idiotypic characteristics of these anti-DNA autoantibodies suggest that these rescued antibodies are representative of the *in vivo* disease-associated response. These preliminary data establish the feasibility of using repertoire cloning to isolate IgG anti-dsDNA antibodies from the PBL or SLE patients. Future studies will extend these observations and apply this approach to analyses of responses to other targets of autoimmune attack.

REFERENCES

1. Barbas, C. F., A. S. Kang, R. A. Lerner & S. J. Benkovic. 1991. Assembly of combinatorial antibody libraries on phage surfaces: The gene III site. Proc. Natl. Acad. Sci. USA **88:** 7978–7982.
2. Sasano, M., D. R. Burton & G. J. Silverman. 1993. Molecular selection of human antibodies with an unconventional bacterial B-cell antigen. J. Immunol. **151:** 5822–5839.
3. Barbas, C. F., III, T. A. Collet, W. Amberg, P. Roben, J. M. Binley, D. Hoekstra, D. Cababa, T. M. Jones, R. A. Williamson, G. R. Pilkington, N. L. Haigwood, A. C. Satterthwait, I. Sanz & D. R. Burton. 1993. Molecular profile of an antibody response to HIV-1 as probed by combinatorial libraries. J. Mol. Biol. **230:** 812–823.

Substrate-Driven Formation of a Proteolytic Antibody Light Chain[a]

QING-SHENG GAO,[b] MEI SUN,[b] SONIA TYUTYULKOVA,[b]
DAVID WEBSTER,[c] ANTHONY REES,[c]
ALFONSO TRAMONTANO,[d] RICHARD J. MASSEY,[d]
AND SUDHIR PAUL[b,e]

[b]*Department of Anesthesiology and Eppley Cancer Research Institute University of Nebraska Medical Center Omaha, Nebraska 68198-6830*

[c]*Department of Biochemistry University of Bath Bath BA2 7AY, United Kingdom*

[d]*IGEN, Inc. Rockville, Maryland 20852*

Efficient catalysis by autoantibodies to vasoactive intestinal polypeptide (VIP)[1] and DNA[2] by human autoantibodies has been reported, but the genesis of these activities and identity of antigens inciting formation of the catalysts is not clear. In analogy with the evolution of enzymes, the extraordinary rate of sequence diversification occurring in antibody variable (V) region genes over the course of the immune response could result in the formation of catalytic sites. The key prediction of this hypothesis is that immunization with natural antigens, as opposed to commonly used transition state mimics, should lead to development of catalytic activity. Here, we describe efficient catalysis by the recombinant light chain subunit of an anti-VIP antibody.

PCR-amplified light chain cDNA was inserted into a bacterial expression vector and the recombinant protein purified by means of its affinity for metals (FIG. 1). The purified light chain preparations contained a single protein migrating with an apparent pI between 5.2 and 5.8 on isoelectric focusing gels. The molecular mass of the light chain was 27 kDa, as assessed by SDS-polyacrylamide gel electrophoresis, and the protein was immunoblotted by anti-light chain and anti-*c-myc* antibodies. (The recombinant protein contains a 10-residue *c-myc* tag.) The complete nucleotide sequence of the L chain is available in GenBank (accession No. L34775). A linear increase in [tyr^{10}-^{125}I]VIP hydrolysis was observed (5.3 to 49.4% available peptide) at increasing recombinant light chain concentrations (1.5 to 15 nM). The reaction was saturable at increasing substrate concentration, and the initial rates were consistent with Michaelis-Menten kinetics (TABLE 1). The K_m value for the light chain (0.2 μM) was 1900-fold lower than that for trypsin-

[a] This work was supported by National Institutes of Health Grants HL44126 and A131268 and a contract with IGEN, Inc.

[e] Address for correspondence: Department of Anesthesiology and Eppley Cancer Research Institute, University of Nebraska Medical Center, 600 South 42nd St., Box 986830, Omaha, NE 68198-6830.

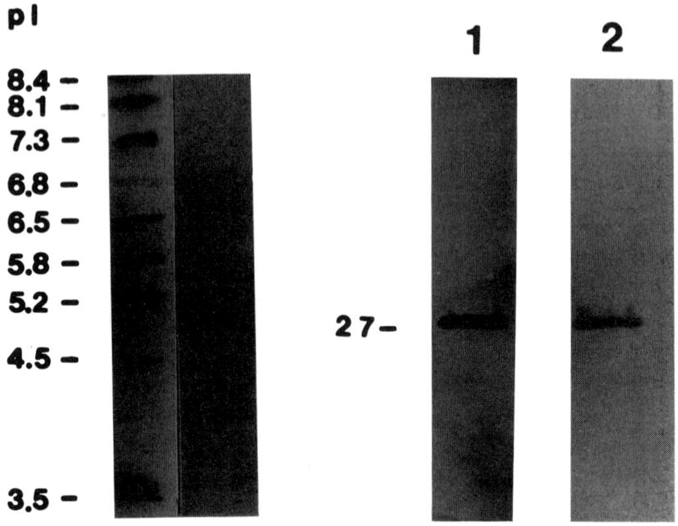

FIGURE 1. IEF (*left*) and SDS-PAGE (*right*) of purified recombinant anti-VIP light chain. IEF gel, silver-stained pI protein markers (*lane 1*) and light chain (*lane 2*); SDS-PAGE, silver-stained (*lane 1*) and anti-*c-myc* antibody stained (*lane 2*) light chain. A hybridoma cell line (clone c23.5) secreting a monoclonal antibody was made from a mouse hyperimmunized with a VIP-keyhole limpet hemocyanin conjugate.[5] cDNA preparation was by reverse transcriptase-polymerase chain reaction from RNA isolated from 3×10^7 c23.5 hybridoma cells using forward (GAGTCATTCTGCGGCCGCCTCATTCCTGTTGAAGCTCTTGAC) and back primers (GTCCTCGCAACTGCGGCCAGCCGGCCATGGCCGAYGTNGTNATGACNCAGAC) and inserted into pCANTAB5his$_6$ via *Sfi*I and *Not*I restriction sites.[6] Following transformation of *E. coli* (HB2151), the recombinant L-chain was induced with 1 mM IPTG and purified by two cycles of chromatography on chelating-fast-flow Sepharose (Pharmacia) charged with nickel sulfate.[7]

catalyzed VIP hydrolysis determined under identical conditions. The kinetic efficiency (k_{cat}/K_m) of light-chain-catalyzed VIP hydrolysis is only 53-fold lower than that of trypsin, a highly evolved protease. Treatment of 30 μM nonradioactive VIP with 2 μM light chain for 6 hr resulted in 32% peptide hydrolysis, estimated using an HPLC assay[3] corresponding to 4.8 catalyst turnovers. This value is close to the turnover determined by radiometric analyses. The K_m value for the

TABLE 1. Kinetic Parameters for VIP Hydrolysis by Recombinant Light Chain

Catalyst	Substrate	K_m (M)	k_{cat} (min^{-1})	k_{cat}/K_m ($min^{-1}M^{-1}$)
L chain	VIP	0.2×10^{-6}	1.1×10^{-2}	5.5×10^{4}
Trypsin	VIP	3.8×10^{-4}	1.1×10^{3}	2.9×10^{6}

NOTE: (tyr^{10}-^{125}I)VIP (0.15 nM) breakdown rates were estimated by partitioning into 10% trichloroacetic acid-soluble and insoluble fractions[1] after incubation with 5 nM light chain or trypsin and increasing concentrations of unlabeled VIP (1–2000 nM) Michaelis-Menten equation (Enzfitter, Biosoft Elsevier).

recombinant and light chain is lower by 8.5-fold than its counterpart purified from the parent antibody under denaturing conditions.[4]

The distinctive property of the light chain not shared by proteolytic enzymes is an ability to recognize the ground state of VIP with high affinity. The lower K_m of the light chain for VIP compared to the nonhomologous substrate is consistent with the clonal selection theory; that is, the light chain has been selected during the immune response to VIP because it recognizes the inciting antigen with high affinity. The turnover of the L chain is substantially lower than that of trypsin. Thus, it remains unclear whether the immune system can actively select for catalysis. An independent evolution of the hydrolytic and antigen-binding functions remains possible. Even so, it is clear that the hydrolytic function is preserved in the affinity-matured light chain. The immune system should prove, therefore, a rich source of catalysts that combine a hydrolytic function with strong substrate binding affinity.

ACKNOWLEDGMENTS

We are grateful to John McCafferty and Rodger Smith for providing the expression vector and helpful suggestions in cDNA cloning.

REFERENCES

1. Paul, S., D. J. Volle, C. M. Beach, D. R. Johnson, M. J. Powell & R. J. Massey. 1989. Catalytic hydrolysis of vasoactive intestinal peptide by human autoantibody. Science **244:** 1158–1162.
2. Shuster, A. M., G. V. Gololobov, O. A. Kvashuk, A. E. Bogomolova, I. V. Smirnov & A. G. Gabibov. 1992. DNA hydrolyzing autoantibodies. Science **256:** 665–667.
3. Paul, S., M. Sun, B. Mody, S. H. Eklund, C. M. Beach, R. J. Massey & F. Hamel. 1991. Cleavage of vasoactive intestinal peptide at multiple sites by autoantibodies. J. Biol. Chem. **266:** 16128–16134.
4. Sun, M., L. Li, Q-S. Gao & S. Paul. 1995. Localization of protease activity in antibody subunits. Ann. N.Y. Acad. Sci. This volume.
5. Paul, S., M. Sun, R. Mody, H. K. Tewary, S. Mehrotra, T. Gianferrara, M. Meldal & A. Tramontano. 1992. Peptidolytic monoclonal antibody elicited by a neuropeptide. J. Biol. Chem. **267:** 13142–13145.
6. Hoogenboom, H. R., A. D. Griffiths, K. S. Johnson, D. J. Chiswell, P. Hudson & G. Winter. 1989. Multi-subunit proteins on the surface of filamentous phage: Methodologies for displaying antibody (Fab) heavy and light chains. Nucl. Acids Res. **19:** 4133–4137.
7. McCafferty, J., K. J. Fitzgerald, J. Earnshaw, D. J. Chiswell, J. Link, R. Smith & J. Kenton. 1994. Isolation of murine antibody fragments which bind a transition state analogue by phage display. Appl. Biochem. Biotechnol. In press.

Catalytic Activity of Anti-Thyroglobulin Antibodies[a]

LAN LI,[b] SRINIVAS KAVERI,[c] SONIA TYUTYULKOVA,[b]
MICHEL D. KAZATCHKINE,[c] AND SUDHIR PAUL[b,d]

[b]*Department of Anesthesiology and Eppley Cancer Research Institute University of Nebraska Medical Center 600 South 42nd Street, Box 986830 Omaha, Nebraska 68198-6830*

[c]*INSERM U28 Hôpital Broussais 75014, Paris, France*

Autoantibodies to thyroglobulin (Tg), a 660-kDa precursor of thyroid hormones, are found in more than 90% of patients with Hashimoto's thyroiditis.[1] The thyroid-damaging potential of anti-Tg antibodies has been shown by induction of histological lesions in mice implanted with anti-Tg antibody-secreting hybridomas[2] and adoptive transfer of thyroiditis by antiserum alone[3] or in combination with lymph node cells.[3] However, anti-Tg antibodies do not fix complement or participate in antibody-dependent cellular cytotoxicity,[4,5] and the mechanism underlying their tissue toxicity is not known. Catalytic activities can develop by natural means in antibody-combining sites.[6,7] Here, we report efficient catalysis by autoantibodies from a patient with Hashimoto's thyroiditis.

Thyroglobulin (Tg) was radiolabeled with ^{125}I using chloramine-T.[8] SDS electrophoresis of ^{125}I-Tg followed by autoradiography showed a single radioactive band at 330 kDa, consistent with observations that SDS dissociates native Tg into two identical 330-kDa subunits. Affinity-purified Tg-specific autoantibodies[9] were isolated from a patient with Hashimoto's thyroiditis (code DEM) and incubated with ^{125}I-Tg. This resulted in disappearance of the 330-kDa Tg monomer and formation of a major 15-kDa product and minor 125-kDa, 60-kDa, and 25-kDa products, indicating that the antibodies must cleave several peptide bonds in Tg. The reaction rate increased linearly with increasing Tg antibody concentrations, estimated from the rate of disappearance of the 330-kDa ^{125}I-Tg band. Background ^{125}I-Tg hydrolysis in buffer without antibodies was not detected. In control experiments, the hydrolytic activities were removed by immunoadsorption with immobilized anti-human IgG. IgG depleted of Tg-specific antibodies (IgG in the unbound fraction from the Tg-Sepharose column) did not hydrolyze ^{125}I-Tg. The composite kinetics of Tg-antibody activity computed from initial rate data obtained at increasing concentrations of Tg were consistent with Michaelis-Menten kinetics

[a] SP, LL, and ST were supported by National Institutes of Health Grants AI31268 and HL44126 and a contract with IGEN, Inc., and SK and MDK by INSERM and CNRS, France, the Central Swiss Red Cross Laboratory, Switzerland, and Hyland Division, Baxter Healthcare Corp., USA.

[d] Corresponding author.

TABLE 1. Kinetic Parameters for Thyroglobulin (Tg) Hydrolysis by Catalytic Autoantibodies

Catalyst	Substrate	K_m (nM)	k_{cat} (10^{-2} min^{-1})	k_{cat}/K_m (min^{-1} M^{-1})
Tg-Ab	Tg	3.9×10^{-8}	3.9×10^{-3}	1.0×10^{5}

NOTE: Hydrolysis of Tg was estimated from the intensities of the ^{125}I-labeled Tg 330-kDa band observed by SDS-PAGE and autoradiography following incubation of ^{125}I-labeled Tg (10,000 CPM) mixed with varying concentrations of unlabeled Tg (1–300 nM) for 5 hr with Tg-specific antibodies (150 nM).

(TABLE 1). The deduced K_m value (39 nM) is remarkably low and consistent with high-affinity Tg recognition by the antibodies. The reaction rate (k_{cat}) is slow, but by virtue of efficient substrate recognition, the antibody kinetic efficiency (k_{cat}/K_m) is in the range observed for conventional proteases.

This kinetic efficiency of the antibodies is derived exclusively from potent Tg binding, since the turnover rate is slow. Thus, the specificity of the antibodies is manifested primarily at the binding step. A similar antigen-specific catalytic site capable of slow turnover has been inferred from studies on a recombinant antibody light chain raised against a polypeptide antigen by experimental immunization.[10]

Because the catalytic autoantibodies recognize Tg with high affinity, they may selectively deplete Tg. Taken together with previous observations of increased VIP-hydrolyzing antibodies in asthma patients and DNA-hydrolyzing autoantibodies in lupus patients, the demonstration of catalysis by Tg-specific autoantibodies from a patient with Hashimoto's thyroiditis points to the possibility of a pathophysiological role for catalytic antibodies in autoimmune disease.

ACKNOWLEDGMENTS

We thank Al Tramontano and Richard Massey for discussion and suggestions.

REFERENCES

1. BIGAZZI, P. E. & N. R. ROSE. 1985. Autoimmune thyroid disease. *In* The Autoimmune Diseases. N. R. Rose & I. R. Mackay, Eds.: 161–199. Academic Press. Orlando, FL.
2. YOKOCHI, T., Y. INOUE, M. FUKADA, M. KAWAI, K. YOSHIKAWA, Y. SUZUKI & N. KATO. 1991. Histological and functional changes in the thyroid glands of mice implanted with hybridomas secreting monoclonal autoantibody against mouse thyroglobulin. Autoimmunity **10:** 125–131.
3. CLAGETT, J. A., C. B. WILSON & W. O. WEIGLE. 1974. Interstitial immune complex thyroiditis in mice. The role of autoantibody to thyroglobulin. J. Exp. Med. **140:** 1439–1456.
4. ADLER, T. R., G. N. BEALL, J. G. CURD, D. C. HEINER & U. K. SHABHARWAL. 1984. Studies of complement activation and IgG subclass restriction of anti-thyroglobulin. Clin. Exp. Immunol. **56:** 383–389.
5. BOGNER, U., H. SCHLEUSENER & J. R. WALL. 1984. Antibody-dependent cell mediated cytotoxicity against human thyroid cells in Hashimoto's thyroiditis but not Graves' disease. J. Clin. Endocrinol. Metab. **59:** 734–738.
6. PAUL, S., D. J. VOLLE, C. M. BEACH, D. R. JOHNSON, M. J. POWELL & R. J. MASSEY. 1989. Catalytic hydrolysis of vasoactive intestinal peptide by human autoantibody. Science **244:** 1158–1162.

7. SHUSTER, A. M., G. V. GOLOLOBOV, O. A. KVASHUK, A. E. BOGOMOLOVA, I. V. SMIRNOV & A. G. GABIBOV. 1992. Science **256:** 665–667.
8. SELMI, S. & B. ROUSSET. 1988. Identification of two subpopulations of thyroid lysosomes: Relation to the thyroglobulin proteolytic pathway. Biochem. J. **253:** 523–532.
9. DIETRICH, G. & M. D. KAZATCHKINE. 1990. Normal immunoglobulin G (IgG) for therapeutic use (intravenous Ig) contains cross-reactive idiotype of human anti-thyroglobulin autoantibodies. J. Clin. Invest. **85:** 620–625.
10. GAO, Q. S., M. SUN, S. TYUTYULKOVA, D. WEBSTER, A. REES, A. TRAMONTANO, R. J. MASSEY & S. PAUL. Substrate-driven formation of a proteolytic antibody light chain. This volume.

Localization of Protease Activity in Antibody Subunits[a]

MEI SUN, LAN LI, QING-SHENG GAO,
AND SUDHIR PAUL[b]

Department of Anesthesiology
University of Nebraska Medical Center
600 South 42nd Street, Box 986830
Omaha, Nebraska 68198-6830

Human autoantibodies have been demonstrated to catalyze the hydrolysis of vasoactive intestinal peptide (VIP).[1] The two subunits, light (L) and heavy (H) chains, could provide different contributions in recognition and chemical transformation of the substrate by catalytic antibodies. Here, a monoclonal antibody raised against VIP[2] was reduced and alkylated,[3] and its subunits were purified on two Superose-12 gel filtration columns (Pharmacia) connected in series.[4] We chose to perform the chromatography under denaturing conditions (6 M guanidine chloride) in order to eliminate any possibility of copurification of adventitious enzymes. The gel filtration permitted baseline separation of heavy and light chains. Both proteins were electrophoretically homogeneous as assessed by SDS-polyacrylamide gel electrophoresis and immunoblotting with antibodies to heavy and light chains.[4] Renaturation was achieved by removal of guanidinium chloride by dialysis. The renatured L-chain hydrolyzed tyr^{10}-^{25}I-labeled VIP. The concentration at which the L-chain refolding was conducted governed the magnitude of its hydrolytic activity, with higher specific activity observed at lower L-chain concentrations (TABLE 1), suggesting a negative influence of aggregation during refolding. Inclusion of VIP (25-fold molar excess) permitted refolding of the L-chain at higher concentrations (2 μM) with retention of hydrolytic activity. L-chains refolded in acetonitrile or glycerol (25% vol/vol) solutions did not display detectable hydrolytic activity. The refolded H-chain was devoid of VIP-hydrolyzing activity. A control L-chain purified from an irrelevant monoclonal antibody (myeloma IgG_{2a}, κ, UPC10; Sigma) in exactly the same way as the anti-VIP light chain was without activity. The hydrolytic activity coeluted with the L-chain monomer peak (25–30 kDa) obtained by rechromatography of the protein (FIG. 1), and the specific hydrolytic activities of two fractions spanning the peak were similar (12,830 and 14,412 CPM/μg protein). The reaction kinetics were investigated using tyr^{10}-^{125}I-labeled VIP mixed with unlabeled VIP at varying concentrations as substrate.[1] The initial rates for L-chain-catalyzed VIP hydrolysis were consistent with the Michaelis-Menten equation. Over the course of 18 hours, the reaction products accumulated to concentrations 14-fold greater than the light chain concentration. Thus, the L-chain is capable of substrate turnover, indicating that it is a true catalyst. The K_m value of the L-chain was 200-fold lower than that of trypsin assayed under identical conditions, indicating higher affinity recognition of VIP by the former catalyst.

[a] This work was supported by National Institutes of Health Grants HL44126 and AI31268.
[b] Corresponding author.

This study demonstrates catalysis by the L-chain subunit of an antibody formed in response to the ground state of VIP. The low K_m value for the L-chain is consistent with the tendency of the immune system to generate high-affinity antibodies over the course of the immune response. The VIP-hydrolyzing kinetic efficiency of this L-chain (6.8×10^3 min^{-1} M^{-1}) is only 426-fold lower than trypsin, a highly evolved proteolytic enzyme. These observations support the hypothesis that catalytic L-chains are formed and selected as a routine component of the immune response against a polypeptide antigen. Lymphocytes synthesize L-chains in excess over H-chains,[5] and free L-chains uncomplexed with H-chains are found in the blood of healthy humans[6] and patients with some B-lymphocyte tumors.[7] An antigen-specific catalytic function of L-chains *in vivo* is possible in view of the results reported in the present study.

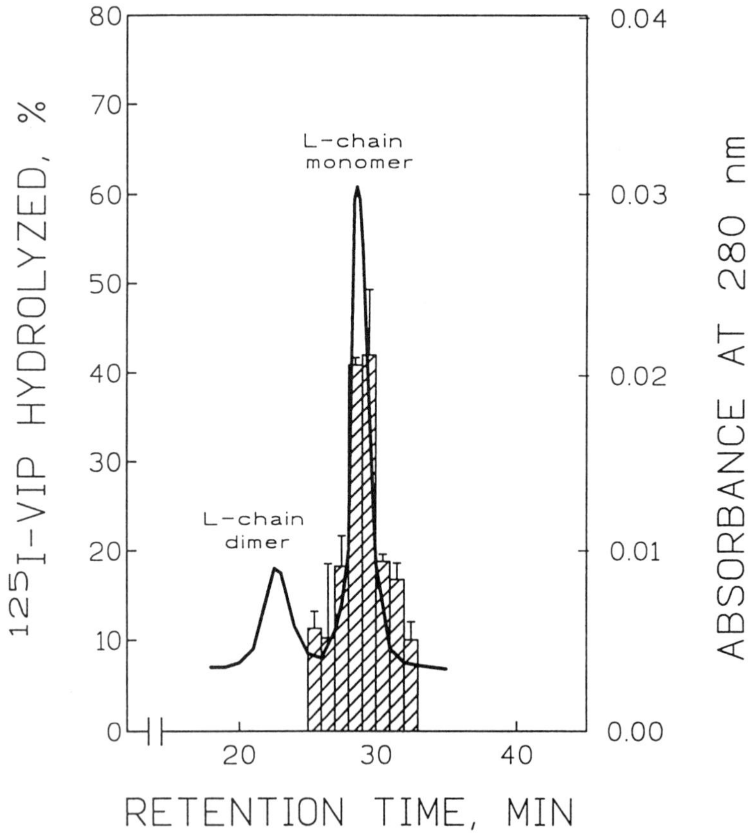

FIGURE 1. Coelution of VIP-hydrolyzing and renatured L-chain from a gel filtration column. Renatured L-chain (100 μg) was rechromatographed on a Superose-12 gel filtration column in 50 mM Tris-HCl, 0.5 M NaCl, 0.025% Tween-20, 0.02% NaN_3, pH 7.3 (A280, –). Column fractions corresponding to the L-chain monomer peak were dialyzed against assay diluent, and tyr^{10}-^{125}I-labeled VIP hydrolytic activity was measured (▨).

TABLE 1. [tyr^{10}-^{125}I]VIP Hydrolysis by Antibody Light Chain

Antibody Subunit[a]	Protein Concentration during Refolding	% VIP Hydrolyzed/10 nM Protein[c]
L-chain	1.6 μM	4.4 ± 0.8
	0.4 μM	20.0 ± 1.5
	0.16 μM	55.2 ± 0.1
	2 μM + 50 μM VIP[b]	48.0 ± 0.8
H-chain	0.16 μM	−2.5 ± 0.2
Nonimmune L-chain	0.4 μM	0.3 ± 0.6

[a] The L- and H-chains of an anti-VIP antibody (clone c23.5) were purified by gel filtration in 6 M guanidine chloride and refolded at varying concentrations by dilution with 50 mM Tris-HCl, pH 7.3, 0.025% Tween-20, 0.02% sodium azide, and dialysis against this buffer.

[b] Refolding was in 50 μM of VIP, followed by two rounds of gel filtration to remove the peptide.

[c] Total available ^{125}I-labeled VIP, ~50 pM. The data are means of three replicates ± SD.

REFERENCES

1. PAUL, S., D. J. VOLLE, C. M. BEACH, D. R. JOHNSON, M. J. POWELL & R. J. MASSEY. 1989. Catalytic hydrolysis of vasoactive intestinal peptide by human autoantibody. Science **244:** 1158–1162.
2. PAUL, S., M. SUN, R. MODY, H. K. TEWARY, S. MEHROTRA, T. GIANFERRARA, M. MELDAL & A. TRAMONTANO. 1992. Peptidolytic monoclonal antibody elicited by a neuropeptide. J. Biol. Chem. **267:** 13142–13145.
3. FLEISCHMAN, J. B., R. H. PAIN & R. R. PORTER. 1962. Reduction of γ-globulins. Arch. Biochem. Biophys. Supp. **1:** 174.
4. SUN, M., L. LI, Q-S. GAO & S. PAUL. 1994. Antigen recognition by an antibody light chain. J. Biol. Chem. **269:** 734–738.
5. HOPPER, J. E. & E. PAPAGIANNES. 1986. Evidence by radioimmunoassay that mitogen activated human blood mononuclear cells secrete significant amounts of light chain Ig unassociated with heavy chain. Cell. Immunol. **101:** 122.
6. SØLLING, K. 1977. Normal values for free light chains in serum in different age groups. Scand. J. Clin. Lab. Invest. **37:** 21.
7. NELSON, M., R. D. BROWN, J. GIBSON & D. E. JOSHUA. 1992. Measurement of free kappa and lambda chains in serum and the significance of their ratio in patients with multiple myeloma. Br. J. Haematol. **81:** 223.

Use of a Phage Expression System to Produce Fabs from Human Autoanti-DNA Antibodies Encoded by the V_H4-21 Gene Segment

C. J. CHAPMAN, C. I. MOCKRIDGE,
M. B. SPELLERBERG, D. A. ISENBERG, T. J. HAMBLIN,
AND F. K. STEVENSON

Molecular Immunology Group
Tenovus Laboratory
Southampton University Hospitals
Tremona Road
Southampton SO16 6YD, United Kingdom

The V_H4-21 gene has been reported to encode anti-red cell autoantibodies of I/i specificity[1] as well as a number of IgM and IgG anti-DNA antibodies from patients with systemic lupus erythematosus (SLE).[2] Analysis of the nucleotide sequences of a large panel of monoclonal antibodies that utilize this gene suggests that discrimination between the different autoantigens appears to arise, at least in part, from the CDR3 region of the heavy chain. To test this hypothesis, we have amplified the V_H and V_L regions from three autoantibodies that use the V_H4-21 gene, two of which bind to DNA. We have cloned and expressed these genes as Fab fragments in the pHEN phage expression system, initially to investigate the role of the heavy and light chains on the ability to bind to DNA.

METHODS

V_H and V_L genes were amplified from three autoantibodies: D5, an IgG anti-DNA antibody; RT79, an IgM anti-DNA antibody already described[2]; and FIF17B, an IgM anti-red cell, non-DNA-binding antibody that uses a germline V_H4-21 V_H gene and V_kIIIb V_L gene. These genes were amplified using primers specific for the relevant V_H and J_H, and V_K and J_K genes and ligated into the pHEN expression system[3] in various combinations. Heterodimeric Fab fragments were produced in HB2151 *E. coli*, isolated from the bacterial periplasm, affinity purified on Affigel coupled to sheep anti-human Fab′γ, and treated with DNAse1 before assessment of their concentration and binding activity by ELISAs.

RESULTS

Recombinant Fab fragments were produced by combining the V_H genes from the anti-DNA antibodies with the homologous or with a heterologous non-DNA binding light chain 17B. Similarly, the non-DNA binding antibody was assembled

TABLE 1. Recombinant Fab Fragments Expressed in pHEN

Fab	V_H Gene	CDR3	J_H	V_L Gene	J_k
$D5_HD5_L$	V_H4.21 (M)	GAPNFGNYY**K**A**RR**G	WFDP J_H5	V_kIIIb (M)	J_k5
$D5_H17B_L$	V_H4.21 (M)	GAPNFGNYY**K**A**RR**G	WFDP J_H5	V_kIIIb (GL)	J_k1
$RT79_HRT79_L$	V_H4.21 (GL)	V**RR**SG**R**VVVPAAP**R**N**R**D	AFDI J_H3	V_k2 (M)	J_k5
$RT79_H17B_L$	V_H4.21 (GL)	V**RR**SG**R**VVVPAAP**R**N**R**D	AFDI J_H3	V_kIIIb (GL)	J_k1
FIF_H17B_L	V_H4.21 (GL)	GDIVATMG	GFDI J_H3	V_kIIIb (GL)	J_k1
FIF_HD5_L	V_H4.21 (GL)	GDIVATMG	GFDI J_H3	V_KIIIb (M)	J_k5
FIF_HRT79_L	V_H4.21 (GL)	GDIVATMG	GFDI J_H3	V_k2 (M)	J_k1

NOTE: The different combinations of heavy and light chain genes assembled are shown. Basic amino acid residues are highlighted. Key: M, mutated gene used; GL, germline gene used.

TABLE 2. Amounts of Fab Produced by Recombinant Constructs and Anti-DNA Binding Specificity as Compared to the Parental Molecule

Fab/Ig	Fab/Ig Conc[a]	Anti-ssDNA Binding
D5 IgG	0.8	+++++
$D5_H D5_L$ Fab	8.2	++++
$D5_H 17B_L$ Fab	0.4	ND
RT79 IgM	1.2	+++++
$RT79_H RT79_L$ Fab	1.2	+++
$RT79_H 17B_L$ Fab	0.3	++
FIF IgM	10	−
$FIF_H 17B_L$ Fab	1.9	−
$FIF_H D5_L$ Fab	1.2	ND
$FIF_H 79_L$ Fab	0.8	ND

NOTE: Amounts are expressed as μg/50 ml culture (periplasmic fraction) for Fabs and μg/ml culture supernatant for IgG. Fab fragments were detected using a capture ELISA. Sheep anti-human Fdγ or ssDNA was coated onto an ELISA plate, and Fabs were detected using goat anti-human kappa-HRP. Anti-ssDNA activity is expressed as relative E492 in the ELISA. Key: − = <0.01; + = 0.01–0.1; ++ = 0.1–0.25; +++ = 0.25–0.5; ++++ = 0.5–1.0; +++++ = >1.0; ND, not tested.

as a Fab fragment, and its heavy chain expressed in combination with the light chains from the DNA-binding antibodies, see TABLE 1.

The Fab fragments were assayed for the presence of both heavy and light chains by capture ELISA, and in all cases correctly assembled Fabs were detected.

Assembly of the heavy and light chains, in the same combination as used by the original antibody, resulted in similar DNA-binding specificities as those of the parent molecules.

DISCUSSION

The V_H4-21 gene is known to encode the heavy chain variable region of a number of anti-DNA antibodies as well as anti–red cell antibodies. Analysis of the V_H and V_L sequences used by these antibodies suggests that discrimination between the two autoantigens appears to be conferred by the heavy chain, and in particular by the CDR3 region.

Fab constructs containing the V_H and V_L sequences encoding human anti-DNA antibodies, or controls, have been produced in the pHEN expression system. These mimic the DNA-binding activities of the parent antibodies. Initial experiments exchanging the light chains between these molecules indicate that part of the anti-DNA binding properties of these V_H4-21-expressing Fab fragments is provided by the heavy-chain gene. The contributions from the amino acid changes that are present in the V_L sequences of D5 and RT79 are also being investigated. Following this, mutagenesis experiments will be carried out to delineate the individual amino acids responsible for their binding properties. Identification of the amino acids involved in recognition of DNA could have implications for therapeutic intervention.

REFERENCES

1. CHAPMAN, C. J., M. B. SPELLERBERG, G. A. SMITH, S. J. CARTER, T. J. HAMBLIN & F. K. STEVENSON. 1993. Autoanti-red cell antibodies synthesized by patients with infectious mononucleosis utilize the V_H4-21 gene segment. J. Immunol. **151:** 1051–1061.
2. STEVENSON, F. K., C. LONGHURST, C. J. CHAPMAN, M. EHRENSTEIN, M. B. SPELLERBERG, T. J. HAMBLIN, C. T. RAVIRAJAN, D. LATCHMAN & D. ISENBERG. 1993. Utilisation of the V_H4-21 gene segment by anti-DNA antibodies from patients with systemic lupus erythematosus. J. Autoimmunol. **16:** 809–825.
3. HOOGENBOOM, H. R., A. D. GRIFFITHS, K. S. JOHNSON, D. J. CHISWELL, P. HUDSON & G. WINTER. 1991. Multi-subunit proteins on the surface of filamentous phage: Methodologies for displaying antibody (Fab) heavy and light chains. Nuc. Acids Res. **19:** 4133–4137.

The Development of a Bacteriophage Expression Vector for Cloning Immunoglobulin Variable Region Genes from B-Cell Genomic DNA[a]

GREGORY A. DENOMME,[b,c] ZHILI SONG,[b]
AND JOHN G. KELTON[b,c]

[b]*Department of Pathology*
McMaster University, HSC 3H35
1200 Main Street West
Hamilton, Ontario Canada, L8N 3Z5

[c]*The Canadian Red Cross BTS*
Hamilton Centre
400 King Street East
Hamilton, Ontario, Canada, L8N 1C6

Unique immunoglobulin variable (Ig V) region gene sequences for antibodies are reflected by specific reactivity for antigen. Techniques such as *in vitro* mutagenesis are used to characterize the specific amino acid residue responsible for specificity and have been used to develop higher affinity antibodies from bacteriophage libraries. To some extent, mutagenesis capitulates affinity maturation of memory B cells.[1] Therefore, the isolation of memory B cells clonally related to the active expressed repertoire for a given antibody specificity may also provide clues regarding the key residues contributing to antibody specificity. Also, these sequences may be clinically useful for passive immunization.[2] Memory B cells produce little or no soluble immunoglobulin and are considered to be in a resting state in the absence of stimulating antigen. The quantity of mRNA transcripts produced by resting cells is considerably less than that of plasma cells, and mRNA transcripts are thus poorly represented in combinatorial libraries. Therefore, developing a more complete bacteriophage library necessitates cloning both the active and memory B-cell pool.[3] This can be accomplished by Ig V region PCR amplification of only variable regions (V_HDJ_H and V_LJ_L), using genomic B-cell DNA. A derivative of pComb3, termed pC3KG, has been developed to express these amplified genes. It was constructed by cloning the first constant domain of IgG (cGH1) and the constant domain of kappa (cK) into pComb3 with the restriction sites used for Ig V gene cloning moved to the 5′ region of the constant domains. The cGH1 domain includes an extra *Nhe* I site at the 3′ end so that soluble Fab can be produced.

cGH1 and cK domains were PCR amplified using a standard PCR reaction buffer (Perkin-Elmer Cetus, Norwalk, CT) from reverse-transcribed mRNA with

[a] This work was supported by a grant from Miles Canada/Canadian Red Cross Society Research and Development Fund. G. A. Denomme is supported by a CRCS Blood Services Career Development Fellowship Award.

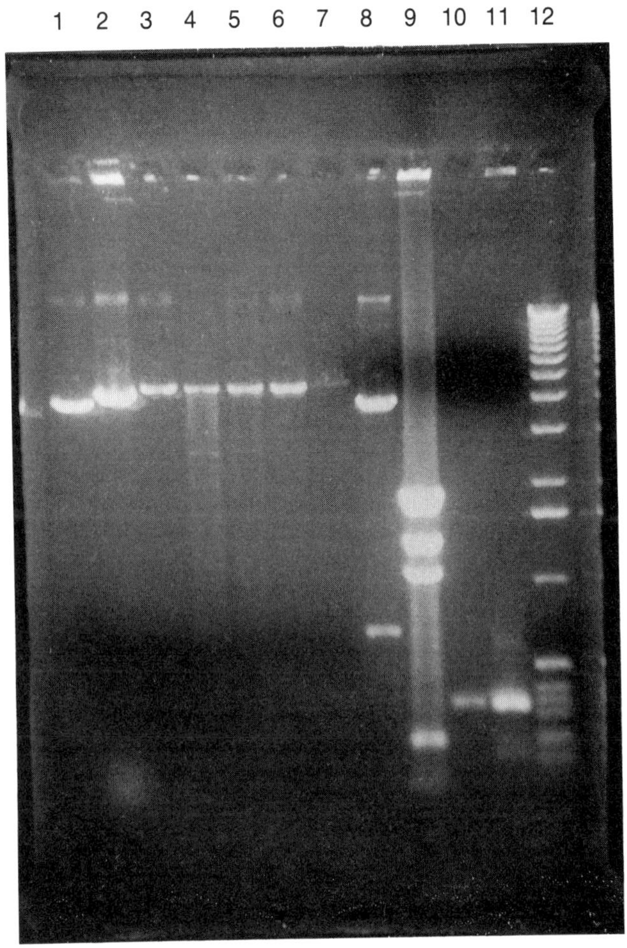

FIGURE 1. Restriction enzyme mapping of pC3KG phagemid vector. From left to right: pComb3, lane 1–*Xba* I; pC3K, lane 2–*Xba* I; pC3KG, lane 3–*Xba* I, lane 4–*Xho* I, lane 5–*Sac* I, lane 6–*Spe* I, lane 7–*Bst*E II (cGH1 domain), lane 8–*Rsa* I (cK domain), lane 9–*Nhe* I (cGH1 orientation); PCR products, lane 10–cGH1, lane 11–cK; lane 12–1-kb marker (Gibco/BRL).

an antisense primer in the hinge region of IgG1 or the 3′ end of the kappa region. The primers for cGH1 amplification were 5′-GCGCCGACTAGTACCAAGG-GCCCATCGGTC (codons 116–121, restriction enzyme site underlined) and 5′-GCGCCGTCTAGA GCTAGCGTGAGTTTTGTCACAAGA (237–229), and for cK, 5′-GCGCCGTCTAGAGCTGCACCATCTGTCTTCACT (110–116) and 5′-GCGCCGACTAGTACTAACACTCTCCCCTGTTGAA-3′ (214–209). PCR-amplified fragments were phenol extracted, digested, and gel isolated using Gene-clean (BIO101, Inc., La Jolla, CA). cK was ligated into *Xba* I-digested and dephosphorylated pComb3, and recombinant phagemids were verified by digestion using

Rsa I (two of four cut sites in cK). The cK-containing phagemid was then used to clone cGH1. The presence of cGH1 was verified using *Bst*E II restriction enzyme digestion (one cut site in cGH1). Phagemids with the correct cGH1 orientation were identified using *Nhe* I digestion (presence of a 690 bp fragment). Ig V region amplified fragments can be cloned into pC3KG using the identical restriction enzymes designed for pComb3.

REFERENCES

1. Hawkins, R. E., S. J. Russel & G. Winter. 1992. Selection of phage antibodies by binding affinity: Mimicking affinity maturation. J. Mol. Biol. **226:** 889–896.
2. Lerner, R. A., C. F. Barbas III, A. S. Kang & D. R. Burton. 1991. On the use of combinatorial antibody libraries to clone the "fossil record" of an individual's immune response. Proc. Natl. Acad. Sci. USA **88:** 9705–9706.
3. Gherardi, E. & C. Milstein. 1992. Original and artificial antibodies. Nature **357:** 201–202.

Index of Contributors